Digital Signal Processing In High-Speed Optical Fiber Communication Principle and Application

Jianjun Yu · Nan Chi

Digital Signal Processing In High-Speed Optical Fiber Communication Principle and Application

清華大學出版社
TSINGHUA UNIVERSITY PRESS

Springer

Jianjun Yu
Fudan University
Shanghai, China

Nan Chi
Fudan University
Shanghai, China

ISBN 978-981-15-3097-5 ISBN 978-981-15-3098-2 (eBook)
https://doi.org/10.1007/978-981-15-3098-2

Jointly published with Tsinghua University Press
The print edition is not for sale in China Mainland. Customers from China Mainland please order the print book from: Tsinghua University Press.

This Springer imprint is published by the registered company Springer Nature Singapore Pte Ltd.
The registered company address is: 152 Beach Road, #21-01/04 Gateway East, Singapore 189721, Singapore

Introduction

We are in the age of information and we communicate with anyone at anytime and anywhere about anything. All information is delivered in communication systems and carried by a certain medium at different frequencies. The carrier frequency can range from a few MHz to several hundred THz, as shown in Fig. 1. An optical communication system refers to a system that uses high-frequency electromagnetic waves in the infrared, visible, or ultraviolet regions of the electromagnetic spectrum to communicate. With the invention of low-loss optical fiber, optical communication systems have been widely used since 1980 and rapidly changed the structure of communication networks around the world.

Conventional optical communication systems generally use intensity modulation and direct detection (IM/DD), in which the transmission electrical signal is used to drive an intensity modulator to generate on/off keying (OOK) optical signal at the transmitter, and the envelope detection is performed at the receiver to recover the transmitted signal. Although this strategy has the advantages of simple structure and low cost, the capacity of the transmission system is limited due to that the OOK intensity modulation format has low spectral efficiency. Therefore, the optical communication system with coherent detection has become a research hotspot

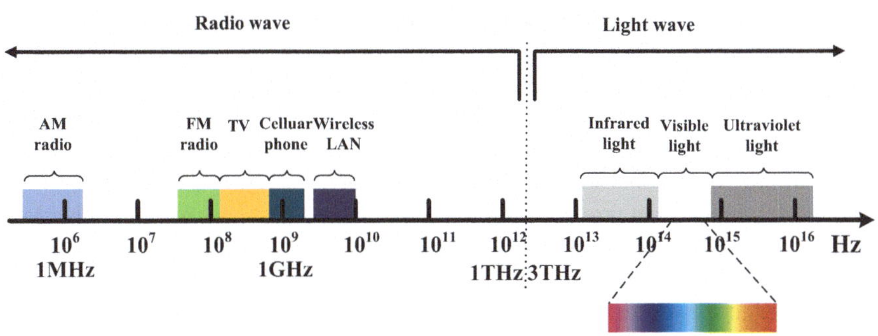

Fig. 1 Spectrum resources

nowadays due to capacity demand. In a coherent optical communication system, information is transmitted at the transmitter by modulating the frequency or phase of the optical carrier, and at the receiver, the transmitted signal is detected by using a homodyne or heterodyne technique with digital signal processing (DSP). Hence, DSP is crucial for modern optical fiber communication. Moreover, even if we use intensity modulation in the transmitter, DSP is still useful if the modulated signal is not OOK signals. For example, if the optical signals at the transmitter are pulse amplitude modulation (PAM), discrete multi-tone (DMT), or careless amplitude phase (CAP), DSP is necessary at the receiver to recover the data. This book aims to introduce modern optical communication systems with DSP to readers. We hope that readers can have an overall picture of the concept, implementation, and technology of DSP for direct detection or coherent optical communication system.

The Development and Current Status of Optical Communication

Optical communication has a long history. Broadly speaking, any communication method that uses light as a medium for information transmission can be defined as optical communication. Based on this definition, China is the first country in the world to conduct optical communications in the history that dates back to ancient times when the Chinese already knew how to use fireworks to transmit information. With the evolution of civilization, communication has also taken place by means of signal lights and semaphores. In 1880, Bell even invented the optical telephone. At this time, however, the development of optical communication has come to an end, and analog electrical communication technology has dominated in the field of communication.

It was not until the 1950s that people again investigate lightwave as a carrier for communication, though at this time there was no suitable coherent lightwave source and transmission medium. The invention of the laser in 1960 solved the problem of the lightwave source. However, the loss of the fiber is still very high, for example, the loss of the best fiber was greater than 1000 dB/km, which limits the practical use of fiber substantially. In 1966, British-Chinese scientist Charles Kuen Kao and G. A. Hockham of STC labs jointly proposed that fiber can be used as a communication transmission medium if the loss of fiber can be smaller than 20 dB/km by removing impurities in glass fiber. This research has an epoch-making significance and brought about a revolution in the field of communications, and Dr. Charles Kuen Kao won the 2009 Nobel Prize in Physics for this achievement. In 1970, Corning announced that they had produced the world's first low-loss fiber, reducing fiber loss to 17 dB/km. After that, through the intense efforts of many researchers, the transmission loss of the optical fiber has dropped significantly. So far, the loss of the optical fiber has reached the theoretical limit, as low as 0.15 dB/km.

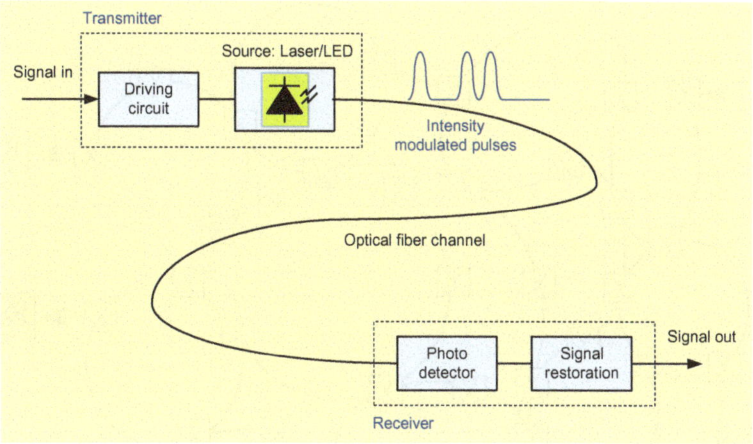

Fig. 2 The optical communication system model proposed by Charles Kuen Kao and G. A. Hockham

Dr. Charles Kuen Kao and G. A. Hockham first proposed a point-to-point optical communication system model, as shown in Fig. 2. At the transmitter, a laser or LED is used as a light source, and an input signal is used to intensity-modulate the lightwave of the laser or the LED, and then the modulated optical signal is transmitted through a fiber of several kilometers, detected by a photodetector at the receiver end, and then converted into an electrical signal. Finally, the electrical signal is recovered and measured. Dr. Charles Kuen Kao predicted that the maximal bit rate of the optical communication system could reach 1 Gb/s. Compared with the maximum communication capacity of the communication system of 100 Mbps/s in the 1970s, the predicted optical communication system showed significant capacity expansion.

Researchers have seen the considerable development prospects of optical communication systems and have devoted themselves to the research of optical communication technologies. New technologies and new devices have emerged one after another, making the communication capacity of optical communication systems increased by hundreds and thousands times within half a century, which is still growing. The structure of today's optical communication systems is very complex. Figure 3 shows a schematic diagram of a wide-area optical network with a coverage of more than one thousand kilometers and a bit rate of transmission capacity of ~Tb/s. In the optical network, an optical amplifier is used to amplify the optical energy, and an optical add/drop multiplexer (ROADM) and an optical cross-connector that can be connected to the upper and lower channels are reconfigured.

The optical communication system has evolved from the originally proposed model to the complex and practical system. It has undergone several technological innovations, and each technological innovation has led to an increase in system capacity. Figure 4 shows the changes in the transmission capacity of optical communication experiments and commercial systems from 1980 to 2010. It can be

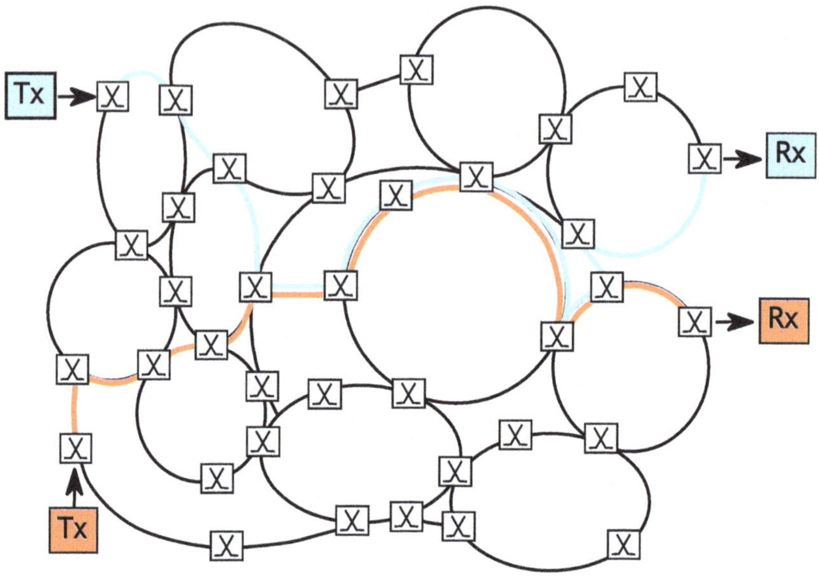

Fig. 3 The modern optical fiber communication system

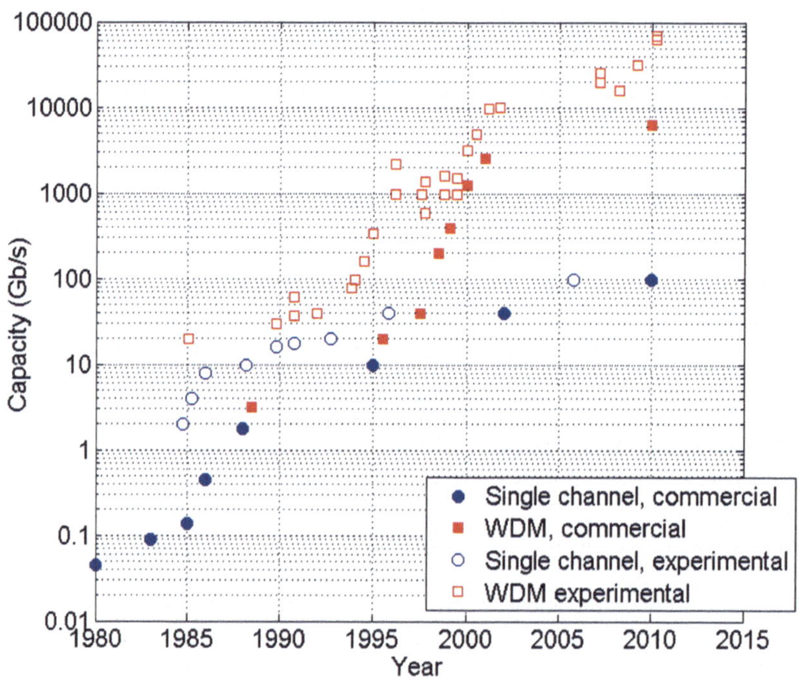

Fig. 4 Optical communication system capacity from 1980 to 2010

seen that the transmission capacity of optical communication systems has grown tremendously within half a century.

The first generation of practical optical communication systems was put into commercial use in 1978. The system works at 0.8 μm and uses multimode fiber with a bit rate of 20 ~ 100 Mb/s. With the advent of single-mode fibers and the development of lasers and photodetectors, the second-generation optical communication system was put into commercial operation in 1987. The system operates at 1.3 μm and uses single-mode fiber with a bit rate of up to 1.7 Gb/s. The third-generation optical communication system works at 1.55 μm. Because the minimum loss of the fiber appears at around 1.55 μm, the bit rate of the system can reach more than 10 Gb/s. These three generations above of optical communication systems use intensity modulation/direct detection (IM/DD), and their system capacity encountered bottlenecks in the late 1980s. Researchers began to explore new directions to further improve system performance that have a major impact on the development of optical communications. The researchers' ideas are mainly classified into two categories. One is to develop new devices to increase the transmission distance. Optical amplifiers were born during this period, especially the Erbium-doped fiber amplifier (EDFA) is one of the most important inventions in the history of optical communication. It is still the most widely used amplifier in optical communication systems. Another idea is to modifiy the system structure and explore new technologies like wavelength division multiplexing (WDM) technology and coherent optical communication technology. The fourth-generation optical communication system is characterized by the use of optical amplifiers and wavelength division multiplexing. At this time, although coherent optical communication systems have emerged, due to the complexity of the system and the level of the devices, direct detection of optical communication systems remain as the most popular schemes in the 1990s. After a period of silence, the coherent optical communication with DSP technology has become a research hotspot in the twenty-first century due to the development of the device level and the need for system expansion, and further improved the transmission rate of the system.

We focus on the development of optical communication systems from the 1990s to the present, as summarized in Table 1. In the late 1980s, WDM technology was introduced, but due to the additional insertion loss of the WDM multiplexer, the transmission rate of the wavelength division multiplexing system was limited. In the early 1990s, the rapid commercialization of EDFA solved the insertion loss problem. EDFA can provide a large power gain, and its amplification bandwidth is high, which can simultaneously amplify multiple signals, which is very suitable for WDM systems. As a result, the development of WDM technology has embarked on a fast lane, and the data rate of optical communication systems has rapidly multiplied.

Table 1 Development of commercial optical communication systems from 1990 to the present

1995 History	2000 History	2005 Deploying	2010 Planned	2015 Needed	2020 Needed
2.5-10 Gb/s channel rate	10 Gb/s channel rate	40 Gb/s channel rate	100 Gb/s channel rate	400 Gb/s channel rate	1 Tb/s channel rate
8,16,40 WDM channels	100 WDM channels	80 WDM channels	100 WDM channels	50 WDM channels	50 WDM channels
20-160 Gb/s capacity	1 Tb/s capacity	3.2 Tb/s capacity	10 Tb/s capacity	15-20 Tb/s capacity	25-50 Tb/s capacity
SE = .025-.05	SE = 0.2	SE = 0.8	SE = 2	SE = 3-4	SE = 5-10

In the 1990s, the single-channel rate of optical communication systems was increased from 2.5 to 10 Gb/s, mainly by improving the stability of the laser output wavelength and optical filtering techniques such as flat-top filters. During this period, optical communication systems was dominated by direct detection. The simple OOK intensity modulation format and direct detection technology were used to transmit multiple channels of data on one fiber through WDM technology. The total communication rate of the direct detection system reached 160 Gb/s. In 2000, by increasing the degree of wavelength division multiplexing, the number of transmission channels on a single fiber was increased to 100, achieving a total communication rate of 1 Tb/s. Then, by adopting a spectrally more efficient modulation format (such as dual binary, DPSK, DQPSK, and PDM-QPSK which can be directly detected or coherently detected), the single-channel communication rate is further increased from 10 Gb/s to 40 Gb/s. In order to increase the system capacity, higher order modulation formats (such as PM-8QAM, PM-16QAM, and PM-64QAM) can be applied for coherent detection. Currently, commercial WDM systems have achieved a single-carrier 400 Gb/s bit rate with a total system capacity of 20 Tb/s or more.

Coherent optical communication with DSP technology improves spectral efficiency by adopting high-order modulation formats such as phase shift keying modulation (PSK), QAM, polarization multiplexing modulation, and orthogonal frequency division multiplexing (OFDM), thereby improving system single-channel transmission rate and communication capacity. By adopting new technologies, the spectrum efficiency of optical communication systems could be increased. Throughout the 1990s to the present, the spectrum efficiency has increased by almost 10 times every ten years. The spectrum efficiency in the past 20 years is shown in Fig. 5. According to this trend, the spectrum efficiency in 2020 will reach 100 bit/s/Hz. It seems difficult to imagine how to achieve such high spectral efficiency.

In recent years, researchers have seen a series of breakthroughs in high-speed coherent optical communication systems. In 2007, Coreoptics first realized the transmission of QPSK signals of 100 Gb/s polarization multiplexing over 1600 km. In 2008, NEC Labs and AT&T Labs combined PDM-RZ-8PSK modulation and coherent detection to achieve 161 × 114 Gb/s DWDM transmission via 662 km ultra-low-loss fiber, creating a record that a C-band optical bandwidth (4.025 THz) is with a capacity of 17 Tb/s. In 2010, NTT completed the single-channel

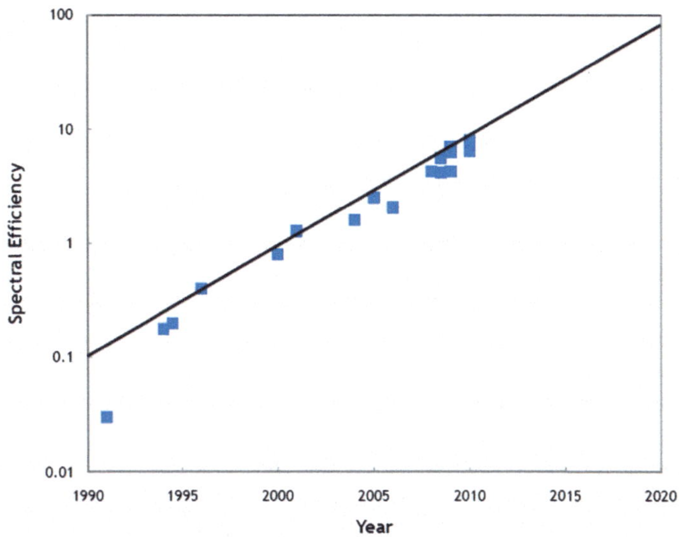

Fig. 5 Changes in spectral efficiency from 1990 to 2010

transmission of 160 Gb/s PDM-16QAM signals on optical fibers with a length of 3123.9 km, which is the longest transmission distance achieved by the current 16QAM modulation format. In 2012, NTT Labs in Japan adopted PDM-64QAM to achieve a 240 km transmission of 102.3 Tb/s C+L band optical signals. Recently people to use multicore and multimode, over 5 Pbit/s capacity has been demonstrated, which is the fifty times of the capacity record in 2013.

At present, coherent optical communication with DSP has become the mainstream scheme in the field of optical communication. Coherent detection was adopted in almost all experiments on high-speed transmission. 100 G and 400 G per channel coherent detection systems have been widely deployed all over the world. The single-channel transmission rate of optical communication systems with 1 Tb/s will be commercialized soon. It is foreseeable that coherent optical communication technology has a bright future, and the future optical communication field will be the world of coherent optical communication systems.

Signal Degradation in Optical Communication Systems

When an optical signal is transmitted in an optical communication system, noise existing in the optoelectronic device, as well as loss, dispersion, and nonlinearity of the optical fiber may cause damage to the signal, and thus the signal may inevitably deteriorate. The compensation for these factors that cause signal degradation largely determines the communication capacity of the system. In this section, we will briefly describe the signal degradation and its solutions in optical communication systems.

The main cause of signal degradation in an optical communication system is the interference caused by the noise of the optoelectronic device, and the signal attenuation and signal distortion caused by the optical fiber transmission. The noise of optoelectronic devices comes mainly from optical amplifiers. Signal attenuation, also known as fiber loss, is one of the important characteristics of fiber optics, which largely determines the maximum distance that a signal in an optical communication system can transmit without an optical amplifier and optical repeater. The signal distortions caused by optical fibers come in two basic classes: linear distortion and nonlinear distortion. Linear distortion refers to the broadening of the optical pulse caused by the dispersion of the fiber, and the nonlinear distortion refers to the signal distortion caused by the nonlinear effects of the optical fiber. The dispersion and nonlinear effects of the fiber, when the system transmission distance is long enough, will have a significant impact on signal transmission and limit the transmission capacity of the system.

Since Dr. Charles Kuen Kao made the first fiber that can be used for communication in 1966, in the next decade or so, the loss of multimode fiber has been reduced to 0.6 ~ 0.7 dB/km around 1.3 μm. The loss of single-mode fiber is reduced to 0.2 dB/km around 1.55 μm. Figure 6a shows the change of loss spectrum of multimode fiber from 1972 to 1982, and Fig. 6b shows the loss spectrum of single-mode fiber using different methods in 1982 (MCVD Modified Chemical Vapor Deposition method, OVD Outside Vapor Deposition method, VAD vapor-axial epitaxy method). Although the loss of fiber at that time was already low, for long-distance transmission systems, fiber loss was still a key factor limiting system capacity growth.

In the 1990s, with the invention of optical amplifiers and its rapid commercialization, the problem of fiber loss was completely solved. Long-distance transmission can be achieved by compensating for fiber losses with optical amplifiers. Erbium-doped fiber amplifier (EDFA) is one of the most important optical amplifiers. It has been widely used in optical communication systems. Its structure is shown in

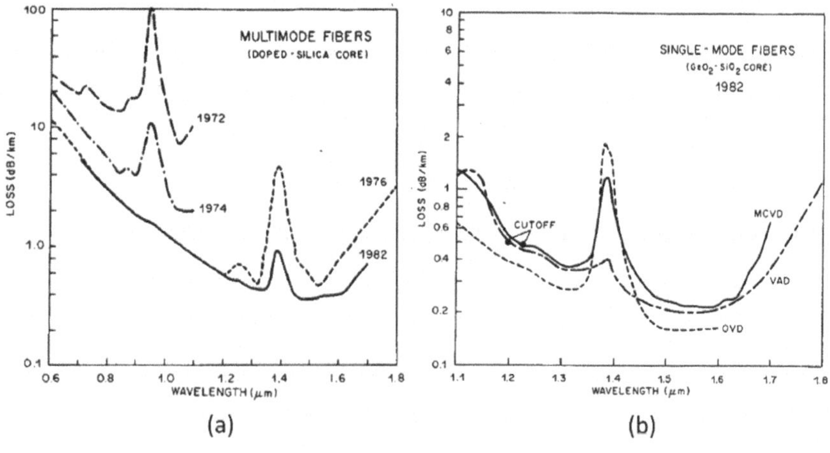

Fig. 6 **a** Loss spectrum of multimode fiber, **b** Loss spectrum of single-mode fiber

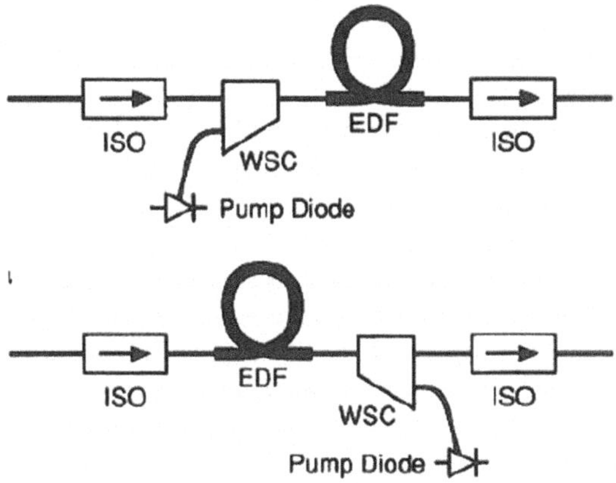

Fig. 7 Basic structure of an Erbium-doped fiber amplifier

Fig. 7. EDFA uses Erbium-doped single-mode fiber as the gain medium, and the population inversion occurs under the excitation of pump light, and the stimulated radiation amplification is realized by signal light induction. EDFA has excellent performances such as high gain, high power, and wide bandwidth. It is very suitable for WDM systems and promotes the commercialization of WDM technology. It has brought a huge revolution in the field of optical communication, which has doubled the communication capacity in ten years. It is worth mentioning that when using an optical amplifier (EDFA) in a WDM system, the non-uniform gain of the WDM link within the EDFA amplification bandwidth will result in a difference in channel quality. In this case, the equalization technique is very necessary.

The invention of optical amplifiers solves the problem of fiber loss, but another important parameter of fiber, fiber dispersion, still limits communication capacity and communication distance. Fiber dispersion is the broadening of the light pulse transmitted in the fiber as the transmission distance increases. When the distance of transmission is long enough, adjacent optical pulses may overlap due to broadening, resulting in a false decision of the receiver, and thus fiber dispersion limits the information-carrying capacity of the optical fiber.

Fiber dispersion mainly is chromatic dispersion (including material dispersion and waveguide dispersion), modal dispersion, and polarization dispersion. Chromatic dispersion refers to the broadening of the optical pulse caused by the difference in group delays in which different frequency (wavelength) components of the source spectrum propagate in the fiber. The material dispersion is caused by the nonlinear change of the refractive index of the material with the change of wavelength. And the waveguide dispersion is due to the nonlinear variation of the propagation constant of the guided wave mode with the change of wavelength. The modal dispersion refers to the dispersion caused by the different propagation constants of different modes at the same wavelength. In a single-mode fiber, there are two polarization modes in which the degenerate polarization directions are orthogonal. When the fiber has birefringence, the propagation speeds of the two degenerate modes are not the same, causing random spreading of optical pulses, so it is called polarization dispersion. Strictly speaking, polarization dispersion is also a kind of modal dispersion. In general, in multimode fibers, there are modal dispersion and chromatic dispersion, but mainly modal dispersion. In single-mode fibers, there are material dispersion and waveguide dispersion, which are generally dominated by material dispersion.

Most fiber-optic communication systems use single-mode fiber as the transmission medium. Therefore, we mainly introduce the dispersion compensation methods for single-mode fiber. The system's dispersion requirements have been the main driving force behind the development of single-mode fiber, as shown in Fig. 8. In order to solve the problem of dispersion of optical fibers, researchers have continuously improved the structure and parameters of optical fibers and manufactured a variety of new optical fibers.

We know that the zero-dispersion point of ordinary single-mode fiber is around 1.3 μm, but the optical communication system operating at 1.3 μm is limited by the fiber loss around 1.3 μm (typically 0.5 dB/km); the minimum loss of single-mode fiber is near 1.55 μm, but the fiber dispersion at 1.55 μm is very high (typically 17 ps/(nm km)). Based on the analysis and calculation of the dispersion formation mechanism, the researchers improved the structure and parameters of the common single-mode fiber and shifted the zero-dispersion wavelength to the vicinity of 1.55 μm to achieve a dispersion-shifted fiber (DSF) with zero dispersion and low attenuation. By adopting DSF, the communication distance and data rate of the optical communication system can be simultaneously increased.

With the commercialization of WDM technology, dispersion-shifted fibers induce serious nonlinear problems. The nonlinear effects of optical fibers can be divided into two categories according to their physical characteristics. The first category covers nonlinear inelastic scattering processes, namely stimulated Raman

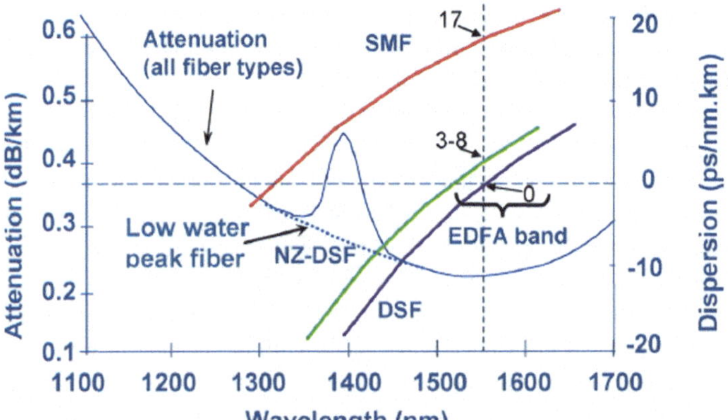

Fig. 8 Dispersion characteristics and attenuation characteristics of various single-mode fibers

scattering (SRS) and stimulated Brillouin scattering (SBS). The second type is the change of refractive index related to the light intensity in an optical fiber, including self-phase modulation (SPM), cross-phase modulation (XPM), and four-wave mixing (FWM). The current nonlinear effect that has the greatest impact on optical communication systems is four-wave mixing, which we will focus on.

The WDM system requires both high input power and low dispersion, resulting in the four-wave mixing effect (FWM) that produces a new spectrum. The new spectrum can interfere with the operation of the WDM system, resulting in system performance degradation. Four-wave mixing is a third-order nonlinear effect in an optical fiber, which is similar to intermodulation distortion in electrical systems. In a multi-channel system, the mixing of three optical frequencies produces a fourth optical frequency $fg = fi + fj - fk$. The lower the dispersion is, the higher the new frequency energy generated by the four-wave mixing effect, the greater the interference to the communication. When WDM technology is applied to the DSF, a severe four-wave mixing effect is produced. For a system with N channels, the number of new frequencies generated by four-wave mixing is $M = 1/2\ (N^3-N^2)$, that is, for two-channel and three-channel systems, 2 and 9, respectively, will be generated. The new frequency is shown in Fig. 9. Figure 10 shows the optical

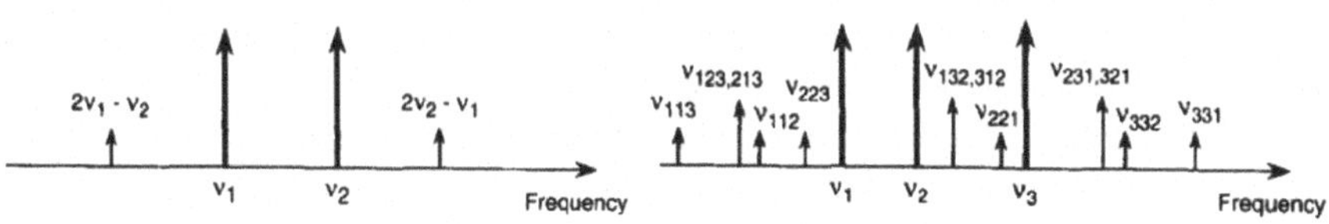

Fig. 9 Schematic diagram of four-wave mixing effect for two-channel and three-channel WDM systems

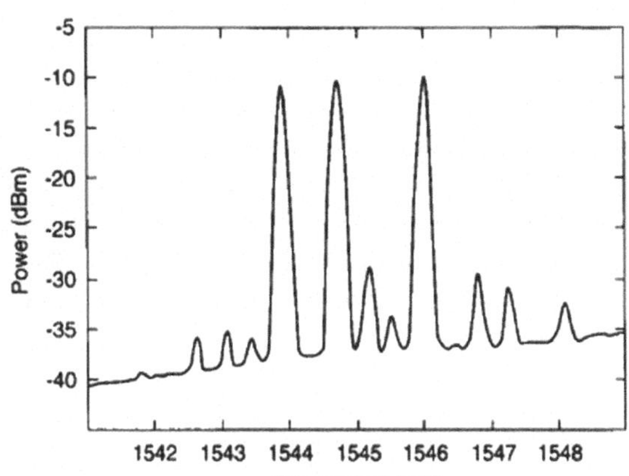

Fig. 10 Output optical power spectrum of a 3 mW/channel, three-channel WDM system after 25 km DSF transmission

power spectrum measured at the output after a 25-km DSF transmission for a three-channel WDM system with an input power of 3 mW per channel.

Due to the severe four-wave mixing effect when applying WDM technology on DSF, DSF is not suitable for WDM systems and is therefore not widely used. In order to suppress the four-wave mixing effect which is generated when WDM technology is applied within the EDFA bandwidth, a certain degree of dispersion is necessary, but at the same time, the dispersion must be small enough to reduce the dispersion loss, so researchers designed a new type of fiber, the non-zero dispersion-shifted fiber (NZDSF). In the NZDSF design, the zero-dispersion point is moved from 1.55 μm to the spectrum beyond the amplification bandwidth of the EDFA. As shown in Fig. 8, the typical dispersion value at 1.55 μm is $3 \sim 8$ ps/(ns km). The application of NZDSF is very suitable for WDM systems, while the standard single-mode fiber has a high dispersion limit at 1.55 μm and a nonlinear effect due to four-wave mixing.

Since the nonlinear effect is inversely proportional to the effective area of the fiber, increasing the effective area of the fiber can reduce the nonlinear effect. Most NZDSFs have an effective area of 50 μm^2, which is much smaller than standard single-mode fibers. In order to increase the effective area, the researchers designed and manufactured a large effective area of NZDSF, called Large Effective Area Fiber (LEAF), with an effective area of 72 μm^2. Therefore, when transmitting the same optical power, LEAF can reduce the power density and nonlinear effects in the fiber and is more suitable for use in WDM systems.

A large number of dispersion-shifted fibers have been laid around the world for single-wavelength transmission systems. When upgrading to WDM systems, four-wave mixing has become a serious problem requiring dispersion management. Dispersion management is to suppress the FWM by properly arranging the fibers with different dispersion characteristics to obtain a locally higher but overall lower dispersion. The lower average dispersion broadens the light pulse, and the higher local dispersion can destroy the phase matching relationship of the carrier frequency at which the FWM intermodulation products are formed. One of the methods of dispersion management is to use passive dispersion compensation, that is, inserting an optical fiber with negative dispersion characteristics in the fiber link to cancel the cumulative dispersion of the transmission fiber. The inserted negative dispersion characteristic fiber is called a dispersion compensation fiber (DCF). After this method, the total dispersion of the system is zero, but the absolute dispersion at each frequency point in the fiber is not necessarily zero. The non-zero dispersion destroys the phase matching between the wavelength channels, thus destroying the condition of FWM generation.

At present, optical communication systems have evolved from analog communication to digital communication. Digital signal processing (DSP) such as dispersion compensation, carrier recovery, and polarization demultiplexing at the digital end is relatively mature. However, the nonlinear effect of optical fiber still limits signal rate and transmission distance. Researchers have done a lot of research on the compensation of nonlinear effects and have achieved a series of results. Volterra filtering and digital back propagation (DBP) have been proposed for nonlinear compensation and have formed a theoretical system. In 2011, the researchers used Volterra filters to mitigate nonlinear effects. The simulation realized a 14 GBd/s DP-16QAM signal transmission of 500 km and a 28 GBd/s DP-QPSK signal transmission of 1200 km. In 2012, Bell Labs used a digital phase conjugate to compensate for the nonlinear effects of the fiber. The experiment demonstrated the transmission of 40-Gb/s CO-OFDM-16QAM signals over 10,400 km. The NEC lab America used the simplified DBP algorithm to effectively compensate the nonlinear effect of the fiber. T. Kobayashi uses RF pilot to compensate for the nonlinear effect, achieving 1,200 km transmission of 538 Gb/s/ch PDM-64QAM SC-FDM signal. The compensation algorithm for nonlinear effects is still in the experimental stage. Due to the complexity of the algorithm and the limitations of the device, it is still not practical. We hope that in the near future, breakthroughs will be made in this area so that the system capacity can be increased.

Optical Communication System

The basic components of an optical communication system are shown in Fig. 11. The optical communication system is mainly composed of three basic units: an optical transmitter, a fiber channel, and an optical receiver. This block diagram is only a simplified diagram of the principle. The actual optical system also includes some optical interconnect and optical signal processing devices, such as fiber jumpers, optical couplers, optical beam splitters, optical amplifiers, regenerative repeaters, and so on.

The function of the optical transmitter is to convert the electrical signal into an optical signal and inject the optical signal into the optical fiber for transmission, generally consisting of a light source, a modulator, and a channel coupler. Light-emitting diodes (LEDs) and semiconductor laser diodes (LDs) are used as light sources because their emission wavelengths are adapted to the fiber channel. The communication channel is an optical fiber, and its basic characteristic parameters are loss and dispersion. In order to achieve high-speed long-distance transmission, the fiber is required to have low-loss and low-dispersion characteristics. The function of the optical receiver is to convert the optical signal output from the optical fiber into an electrical signal, generally consisting of a channel coupler, a photodetector, and a demodulator. A semiconductor photodiode is used as a photodetector because its response characteristic is adapted to the optical communication wavelength. The design of the demodulator depends on the modulation format

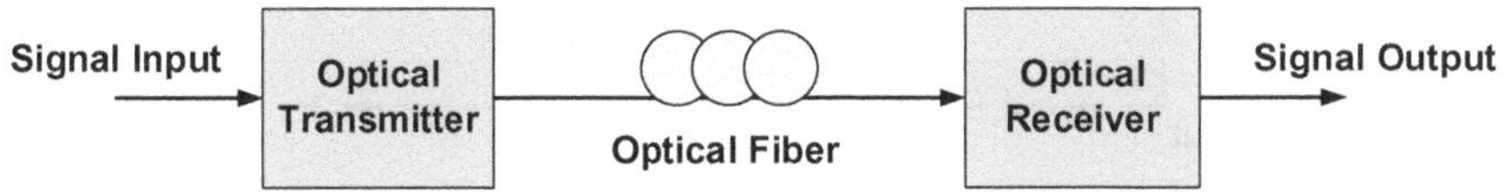

Fig. 11 Block diagram of the optical communication system

used by the system. According to the demodulation method, the optical communication system can be classified into a direct detection optical communication system and a coherent detection optical communication system.

Direct Detection Optical Communication Systems Without DSP

Most of the commercial optical communication systems are still direct detection systems without DSP at the receiver side. In this section, we will briefly introduce the composition and principle of direct detection optical communication systems. The block diagram of the direct detection optical communication system is shown in Fig. 12. The transmitter of the direct detection optical communication system uses intensity modulation, and the receiver is a direct detection receiver. The input electrical signal is intensity-modulated by the drive circuit and converted into an optical signal for transmission on the optical fiber. The light source generally adopts a light-emitting diode (LED) or a semiconductor laser diode (LD), while the light source is selected according to different performance requirements of the system. After the optical signal is transmitted on the optical fiber, it is input to the optical receiver, and the optical signal is converted into an electrical signal by the photodetector. After the electrical signal is amplified, the signal is shaped by the signal recover, and finally the electrical signal is output. The photodetector can use a photodiode (PIN) or an avalanche photodiode (APD), where PIN is more commonly used.

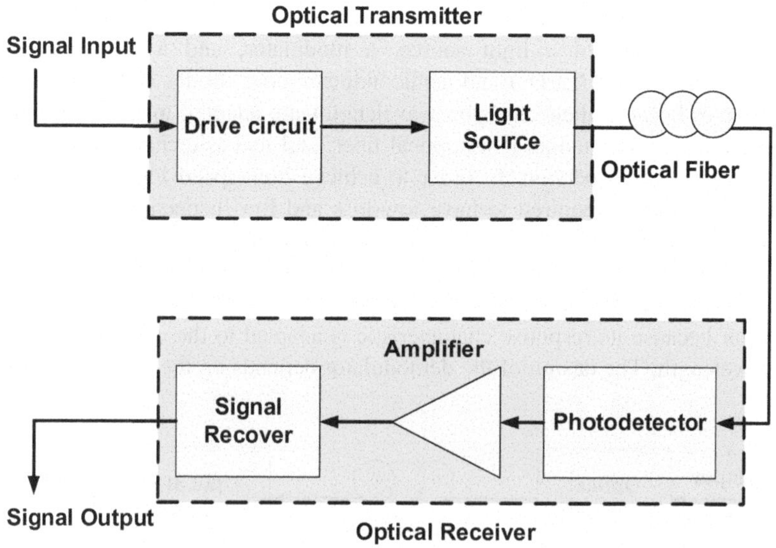

Fig. 12 Block diagram of direct detection optical communication system

The direct detection optical communication system without DSP has been widely used due to its simple structure and low cost. Most of the commercial optical communication systems with a bit rate of 40 Gb/s or lower are direct detection systems, nonetheless the direct detection receiver has low sensitivity and low-frequency band utilization, which cannot give full play to the superiority of fiber-optic communication. Today, with the rapid increase of information, direct detection of optical communication systems without DSP has gradually failed to meet people's communication requirements, and the era of optical communication systems with DSP has arrived.

Direct Detection Optical Communication Systems With DSP

For short-reach data center interconnect or passive optical networks (PON) in the near future, the bit rate can go up to 100 Gb/s. In these networks, the cost, size, and power consumption are very important. The coherent detection system can easily get this bit rate, but it is more expensive, has larger size, and needs more power. Therefore, coherent detection is difficult in employing these cost-sensitive networks. IM/DD is still a better choice. But if we use low spectral efficiency OOK modulation, the baud rate is very high. For 100 Gb/s optical signal, the baud rate is over 100 Gbaud if we consider forward error correction (FEC) overhead. The required bandwidth of the optical and electrical components is up to 50 GHz. The components with this bandwidth are very expensive. Therefore, modulation formats with high spectral efficiency such as PAM, DMT, and CAP are introduced. Usually, to generate these signals, DSP is needed at both transmitter and receiver sides. However, if the bandwidth of components is large enough, PAM does not need DSP. But PAM with DSP usually can get better performance because PAM is sensitive to nonlinear effects in optical and electrical components. The distortion caused by these nonlinear effects can be compensated by DSP.

Coherent Detection Optical Communication System With DSP

The coherent optical communication system was developed around the world in the 1980s, but its system structure is more complicated at higher costs. Due to the limitation of the devices (digital to analog converter or analog to digital converter) at the time and the advent of EDFA, WDM techniques brought direct detection optical communication system's capacity to a new record, so the coherent optical communication has been silent for a long time. In the twenty-first century, due to the development of the devices, coherent optical communication has come back. The coherent optical communication system has high sensitivity, long relay distance, good selectivity, large communication capacity, and many other advantages such as that various modulation methods could be adopted, thus it has a good application prospect.

Figure 13 shows the block diagram of a coherent optical communication system. The main difference from the direct detection system is the modulation and detection. The transmitter of the coherent optical communication system generally adopts external modulation. We can modulate the optical carrier by frequency, phase, or amplitude. The specific advanced modulation formats will be introduced in detail in Chap. 2. The direct detection system generally employs IM/DD. The receiver of the coherent system uses coherent detection, and the main difference from direct detection is the addition of the local oscillator source. The local oscillator and the received signal light are mixed by an optical mixer, and the signal is down-converted from the optical carrier frequency to the microwave carrier frequency, and then the center frequency of the signal is detected by the photodetector, which is the difference of the signal light and the local oscillator light frequency. Then, the intermediate frequency signal is amplified by the intermediate frequency amplifier and demodulated to obtain the baseband signal. If the signal light is equal to the frequency of the local oscillator, this detection is called "homodyne detection"; if the signal light and the local oscillator light are not equal, it is called "heterodyne detection". Coherent detection technology is the core technology of the coherent optical communication system. Next, we will focus on the principle of coherent detection.

The general process of coherent detection is that the transmitted optical signal at the receiver and the local oscillator generated by the local oscillator are interfered by the optical mixer, and then the photocurrent is output via the photodetector, and finally the photocurrent is processed by DSP and the baseband signal is generated. Since the IF signal can carry the amplitude, frequency, or phase information, the coherent detection is suitable for all modulation formats.

Coherent detection can be divided into heterodyne detection and homodyne detection according to the local oscillator frequency and the signal light frequency are different or the same. The former after optical/electrical (O/E) converter obtains an intermediate frequency (IF) signal, and the second demodulation is required to convert it to be a baseband signal. The latter is directly converted into a baseband signal after the O/E converter, but it requires the local oscillator frequency to match the signal frequency.

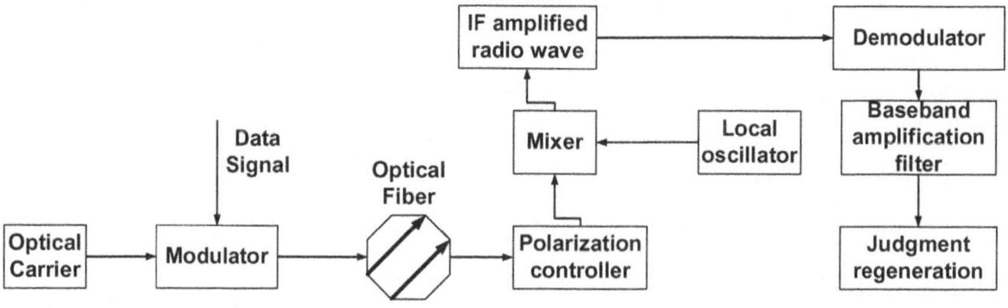

Fig. 13 Block diagram of a coherent optical communication system

Homodyne Detection

For homodyne detection where the local oscillator frequency is equal to the signal optical frequency, the IF frequency f_{IF} is zero. The corresponding photocurrent expression after O/E is

$$I(t) = R(P_s + P_{LO}) + 2R\sqrt{P_s P_{LO}}\cos(\varphi_s - \varphi_{LO}) \quad (1)$$

In the above formula, R is the responsivity of the photodetector, P_s and P_{LO} are the power of the signal lightwave and the local oscillator lightwave, respectively, and φ_s and φ_{LO} are the phases of the signal lightwave and the local oscillator lightwave, respectively. The first term of (1) is an almost constant DC term that is easily filtered out. The last term of (1) contains the transmitted information and provides it to the decision circuit application. Considering the phase lock of the local oscillator on the phase of the signal, it can be seen that the expression of the photocurrent obtained by the homodyne detection is

$$I_p(t) = 2R\sqrt{P_s P_{LO}} \quad (2)$$

We know that the expression of the signal current in the case of direct detection is

$$I_{dd} = RP_s(t) \quad (3)$$

By comparing (2) and (3), one significant advantage of homodyne detection can be obtained: the magnitude of the output photocurrent is proportional to the product of the signal optical power and the local oscillator power after coherent mixing in homodyne detection. And the local oscillator power is much larger than the signal light power, which greatly improves the sensitivity of the receiver. In addition, it is clear from Eqs. (2) and (3) that with the homodyne detection, the average electrical signal power is increased by 4 P_{LO}/P_s times. Since 4 P_{LO}/P_s can be artificially set to be much larger than 1, this increase can be several orders of magnitude. Although the shot noise is also increased, the SNR is still improved in the case of homodyne detection.

From (1), we can also observe another advantage of homodyne detection: since the last term contains the phase of the signal light, it is possible to transmit information by modulating the phase of the optical carrier. Direct detection does not allow phase or frequency modulation due to the phase information after directed detection in a photodiode.

The disadvantage of homodyne detection also comes from a high sensitivity to phase. We can see from (1) that its last term contains the phase φ_{LO} of the local oscillator, which obviously should be controlled. In the ideal circumstance, both φ_s and φ_{LO} should be a constant value except for intentionally modulating φ_s. But in fact, both φ_s and φ_{LO} are randomly floating with time. Although the optical PLL can make the difference $\varphi_s - \varphi_{LO}$ of φ_s and φ_{LO} nearly constant, it is not simple to implement such a phase-locked loop, and it also makes the design of the homodyne detection optical receiver quite complicated.

Another disadvantage of homodyne detection is that since the homodyne detection matches the transmitter frequency to the frequency of the receiver's local oscillator, this imposes stringent requirements on the two sources. But these problems can be solved by using advanced DSP or heterodyne detection.

Heterodyne Detection

For heterodyne detection where the local oscillator frequency and the signal optical frequency are not equal, the intermediate frequency f_{IF} falls within the microwave region. The expression of the photocurrent in this situation is

$$I(t) = R(P_s + P_{LO}) + 2R\sqrt{P_s P_{LO}}\cos(\omega_{IF}t + \varphi_s - \varphi_{LO}) \tag{4}$$

In the above formula, ω_{IF} is the IF angular frequency. Similarly, the first term in Eq. (4) is easily filtered out as an almost constant DC term. Thus, the heterodyne signal is given by the alternating term in equation (4), namely

$$I_{ac}(t) = 2R\sqrt{P_s P_{LO}}\cos(\omega_{IF}t + \varphi_s - \varphi_{LO}) \tag{5}$$

For heterodyne detection, information can be transmitted by modulating the amplitude, phase or frequency of the optical carrier. In addition, similar to homodyne detection, the local oscillator amplifies the received signal, which can significantly improve the sensitivity of the receiver and improve the signal-to-noise ratio SNR. Heterodyne detection improves the signal-to-noise ratio SNR by a factor of 2 (or 3 dB) compared to homodyne detection. This reduction is called a 3 dB heterodyne detection penalty. The 3 dB heterodyne detection penalty can be seen from the signal power proportional to the square of the current.

However, the advantage of 3 dB heterodyne detection cost is that the receiver design is greatly simplified because the optical phase-locked loop is no longer required. Although the random variations of φ_s and φ_{LO} still need to be controlled by a narrow linewidth semiconductor laser diode, in the discussion of the next section, we can see that the asynchronous demodulation technique can relatively alleviate the linewidth requirement. In summary, in practical applications, heterodyne detection is more advantageous than homodyne detection. Therefore, coherent detection optical communication generally adopts heterodyne detection.

The electrical signal obtained by heterodyne detection is a microwave intermediate frequency signal, which needs to be demodulated to the baseband. The demodulation of the optical signal can employ a synchronous or asynchronous scheme.

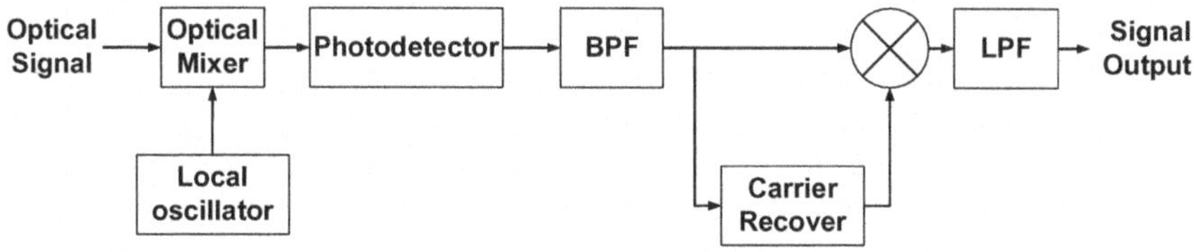

Fig. 14 Block diagram of the heterodyne asynchronous receiver

Figure 14 shows a block diagram of the heterodyne asynchronous receiver. Heterodyne asynchronous demodulation does not require recovery of the IF microwave carrier, so the design of the receiver is much simpler. In the outer heterodyne asynchronous demodulation, the output signal after the bandpass filter is restored to the baseband signal by using an envelope detector and a lowpass filter which is immediately following it. The signal received by the decision circuit is exactly I_f, obtained by Eq. (6). Then I_d can be given by

$$I_d = |I_f| = \left[(I_p \cos Ø + i_c)^2 + (I_p \sin Ø + i_s)^2\right]^{1/2} \tag{6}$$

The main difference between the heterodyne asynchronous receiver and the heterodyne synchronous receiver is that the heterodyne asynchronous receiver can affect the signal by the in-phase and inverting quadrature components of the receiver noise. Therefore, the signal-to-noise ratio (SNR) of asynchronous demodulation is reduced. And because of the SNR reduction, the sensitivity of the asynchronous receiver also decreases. However, the amount of decrease in sensitivity is quite small, approximately 0.5 dB. And in the case of asynchronous demodulation, the linewidth requirements for the transmitter and local oscillator are not that strict. Therefore, in the coherent optical communication system, the heterodyne asynchronous receiver is a practical solution and plays an important role.

Advantages of Coherent Detection

The advantages of coherent detection can be summarized in the following four aspects:

(1) It can significantly improve receiver sensitivity and increase non-relay transmission distance. As we mentioned earlier, the magnitude of the output photocurrent after coherent detection is proportional to the magnitude of the product of the signal light power and the local oscillator power. Since the local oscillator power is much larger than the signal light power, the sensitivity of the receiver is greatly improved. The high sensitivity of coherent detection makes the coherent receiver more suitable for the detection of weak light signals, while direct detection is only suitable for the detection of stronger signals. Under the

same conditions, the coherent receiver improves the sensitivity by about 20 dB compared with the conventional receiver and can achieve high performance close to the shot-noise limit. Therefore, the high sensitivity of the coherent receiver also increases the non-relay transmission distance of the optical signal.

(2) It leads to good selectivity and large communication capacity. Another major advantage of coherent detection is the ability to increase the selectivity of the receiver. In the coherent heterodyne detection, the mixed light of the signal lightwave and the local oscillator is detected, so only the noise in the IF band can enter the system, and other noises are filtered by the narrow bandwidth microwave IF amplifier. It can be seen that the heterodyne detection has good filtering performance. In addition, due to the excellent wavelength selectivity of coherent detection, the coherent receiver can substantially reduce the frequency separation of the frequency division multiplexing system, that is, dense wavelength division multiplexing (DWDM), which replaces the large frequency spacing of the conventional optical multiplexing technology. Small frequency spacing achieves the potential advantage of higher transmission capacity.

(3) A variety of modulation methods are available. In direct detection systems, only the intensity modulation can be used to modulate the lightwaves. In the coherent detection system, in addition to the amplitude modulation of the lightwave, frequency shift keying (FSK) or phase shift keying (PSK) modulation formats can be performed, that is, a plurality of modulation formats can be adopted, which is advantageous for flexible engineering applications, although the system complexity is increased in this scenario. Relative to the change of optical power in the IM/DD system, coherent detection can detect all the information carried by the amplitude, frequency, phase, and polarization state of the optical signal, so coherent detection is a holographic detection technique, which is not available in the IM/DD system.

(4) It can improve bandwidth utilization. Optical coherent detection technology converts optical frequency signals of the order of frequencies up to 10^{14} Hz into electrical intermediate frequency (RF) signals of the order of frequency 10^{9} Hz, while electrical IF filters have unmatched frequency selection performance of optical filters. The wavelength (frequency) selectivity of the optical receiver is greatly improved. In principle, the system can adopt a dense wavelength division multiplexing technology with a channel spacing as small as $1\sim10$ GHz, and the utilization of the optical fiber bandwidth can be further improved by the multi-channel multiplexing technology.

The Trend of Optical Communication System

If you consider signal transmission bandwidth requirements to grow by 40–50%/year, and computing and storage capacity to grow at a rate of 60% per year, these factors will directly lead to an increase in Internet bandwidth at a rate of 60%.

According to Moore's Law, the demand for single-channel 1 Tb/s will appear in 2020. It can be seen that large-capacity transmission is undoubtedly the most important target of the transmission network, and only the optical communication system has the potential to carry such large-capacity information. At present, 100 Gb/s optical transmission systems are being largely deployed, and single-carrier 400 G transmission is becoming a commercial product. 800 Gb/s, 1 Tb/s, and even 10 Tb/s per channel transmission is an optional single-channel rate after 400 G.

The invention of fiber amplifiers has promoted the use of WDM technology in optical transmission systems, and with the widespread adoption of WDM equipment, the single-channel rate has also doubled, enabling transmission capacity to increase hundreds of thousands. However, after the single-channel rate has exceeded 10 G, the simple NRZ pattern and direct detection technology have been difficult to meet the long-distance transmission requirements. Therefore, in the future commercial 800 G or 1T systems, new modulation and demodulation technologies and detection technologies are needed. Coherent optical communication systems will become inevitable. In order to further increase the single-channel rate, such as 1T or higher, considering the availability of high-speed optoelectronic devices, especially the ADC bandwidth limitation, the current research direction mainly includes high-order modulation format, Nyquist WDM technology and orthogonal frequency division multiplexing (OFDM) technology.

The use of high-order modulation formats improves spectral efficiency and enables higher bit rate transmissions with limited bandwidth. However, high-order modulation formats such as 64QAM are more sensitive to nonlinear distortion caused by optoelectronic device bandwidth limitations, and the accuracy requirements for laser linewidth and high resolution of the ADC are also increased.

Optical OFDM technology uses multi-carrier dense modulation, and there is no crosstalk between each carrier, so it can achieve higher spectral efficiency. At present, multi-subcarriers still suffer from stability, performance, and cost problems. It is necessary to study new methods to generate high-performance, high-reliability multi-subcarriers. At the same time, when detecting OFDM modulated signals, multiple subcarriers need to be detected at the same time (generally three need to be detected), which also has certain requirements on the ADC bandwidth of the receiver.

To meet the ADC bandwidth requirements, the researchers proposed Nyquist WDM technology. Nyquist WDM technology uses strong filtering technology, and the bandwidth occupied by each subcarrier is equal to the baud rate of each subcarrier signal. Through strong filtering techniques, the crosstalk between subcarriers is already small, so each subcarrier receiver needs to detect the subcarrier at the receiving end, thus reducing the bandwidth requirement of the ADC. However, because of the strong filtering, it also brings a large optical signal-to-noise ratio (OSNR) cost.

In order to reduce the baud rate of the optical signals, people investigate to use multicore and multimode fiber to deliver high-speed optical signals. New fiber with more cores or modes is invented. Few-mode multicore fibers (FM-MCFs) have been reported with up to 19-core and 6-mode FM-MCF (114 spatial channels). The transmission capacity can go up to 10 Pb/s with an average data rate per spatial channel of 89.1 Tb/s and average spectral efficiency of 1099.9 b/s/Hz. The key components such as multimode or multicore optical amplifier, multiplexer, demultiplexer are becoming mature. But fiber splicing and stability are not well solved. Since 2007, coherent detection communication systems with DSP have become a hot research direction due to their transmission capacity. With the development of broadband optical and electrical device and the emergence of new DSP technologies, coherent optical communication systems with DSP is still the sole solution for large-capacity system. For short reach such as data center interconnect or passive optical networks (PON), multilevel modulation formats in the transmitter to reduce baud rate and direct detection with DSP are the main research areas. It is anticipated that single channel with bit rate up to 100 G will be widely deployed very soon.

Structure of the Book

This book includes 19 chapters.

Chapter 1, advanced modulation. This chapter mainly introduces the principle and methods of generating various modulation signals.

Chapter 2, the basic algorithm of single-carrier coherent optical transmission system. The basic algorithms of single-carrier coherent optical transmission system are introduced, including I/Q imbalance compensation, orthogonal normalization, dispersion compensation, clock recovery, channel equalization, carrier recovery, and phase compensation algorithm. For channel equalization, blind equalization algorithm based on the statistical properties of signal is introduced, from classic CMA algorithm to CMA algorithm in polarization multiplexing signal. However, CMA algorithm has constant amplitude, so it cannot make a good recovery to the high-order signal. This book further introduces CMMA algorithm which aims at high-order modulation of inconstant, this algorithm utilizes angle and the modulus value. Further more, there are too many reference levels of CMMA, so we introduce an improved cascade multimode algorithm MCMMA algorithm. MCMMA changes the diversified angle and modulus values to the orthogonal real and imaginary components, which reduces the reference level and has more obvious performance in high-order signal optimization. In addition, ICA algorithm (Independent component analysis) is introduced. At the end of this chapter, the basic digital signal processing algorithm introduced in this chapter is arranged and compared, and the relevant constellation diagrams are displayed, showing more intuitively the functions of each algorithm.

Chapter 3, quasilinear coherent optical transmission system and digital signal processing. This chapter focuses on the study of quasilinear coherent optical detection system based on advanced digital signal processing technology, including the front-end linear pre-equalization algorithm. The bandwidth limitation of the device and the nonlinear damage of the optical fiber link are two important factors that restrict the transmission of high-speed optical signal. The former limits the bandwidth and baud rate generated by the signal, while the latter limits the transmission distance of high-speed optical signal. This chapter first introduces the basic digital signal processing of coherent optical communication, and then introduces a new front-end digital time-domain pre-equalization scheme based on this. Finally, simulation and experiment are conducted to verify the compensation performance of time-domain data pre-equalization for coherent optical communication.

Chapter 4, the exploration of the high spectral efficiency super-Nyquist wavelength division multiplexing system. Aiming at high spectral efficiency and high-speed optical transmission, a new multimode blind equalization algorithm for super-Nyquist spectrum compression signal is firstly introduced. This algorithm can effectively suppress noise and crosstalk between channels, so as to realize super-Nyquist WDM signal transmission. At the same time, the experimental results show that the algorithm has better anti-noise, anti-filter, and anti-channel interference performances than the ordinary equalization algorithm. In this chapter, the experiment of super-Nyquist signal digital generation and 4-carrier 400 G optical transmission based on front-end filtering and ultra-high-speed PDM-QPSK coherent transmission system up to 110 GBaud will be introduced. By using this system, the world's first 400 G optical transmission system based on a single-carrier super-Nyquist, and the world's highest baud rate of 110 GBaud was created. The carrier interval was compressed to 100 GHz, the transmission capacity reached 8.8 Tb/s, and the signal spectrum efficiency was greater than 4 bit/s/Hz. Finally, the performance of super-Nyquist filter signal in multiple reconfigurable optical division and interpolation multiplexers (ROADMS) is experimentally studied. We also demonstrated the high anti-filtering property of the 9QAM super-Nyquist signal in the experiment. The content of Chap. 4 provides effective theoretical and technical support for realizing high-frequency efficient optical transmission.

Chapter 5, the production and processing of all-light Nyquist signals. All-optical signal processing technologies are explored, mainly for the generation, transmission, and signal detection of all-optical Nyquist signals. First, the principle of all-optical Nyquist signal pulse generation, signal modulation, and multiplexing mechanism will be introduced. This chapter also provides the theoretical basis for realizing frequency locked, linear phase, and equal amplitude optical comb based on a single MZM. Then, the first full light Nyquist signal generation and coherent detection system is introduced. The full light Nyquist QPSK signals up to 62.5 GBaud, 75 GBaud, and 125 GBaud are generated and coercively detected by the full light comb spectrum. Finally, the research work on polarization-multiplexed all-light Nyquist signal long-range and high-order modulation schemes is introduced, including the production, transmission, and full-band coherent detection of 37.5 and 62.5 Gbaud all-light Nyquist PDM-QPSK and 16QAM signals. Aiming at all-optical Nyquist signals, we can not only realize the generation and transmission of high-frequency efficient optical signals, but also improve the signal generation and processing rate by breaking through the bandwidth limitation of electronic devices.

Chapter 6, research on nonlinear compensation algorithm of fiber channel. In this chapter, the nonlinear compensation algorithm is introduced, and the performance of nonlinear compensation is verified from theory, simulation, and experiment. Then, a new DBP nonlinear compensation algorithm based on improved log-step distribution is introduced, and its compensation effect in Nyquist wavelength division multiplexing system (N-WDM) is studied and analyzed. Finally, an improved digital nonlinear compensation algorithm based on logarithmic step size is introduced.

Chapter 7, research of probability shaping technology. The probabilistic shaping technique introduced in this chapter is an optimization one for advanced modulation formats, which has the advantages of high transmission capacity and low system complexity. Based on the basic principles of probabilistic shaping, this chapter explores the optimization of RoF system from the perspectives of algorithm, simulation, and experiment.

Chapter 8, ultra-high baud rate optical signal transmission technology. This chapter introduces and analyzes the experiment of ultra-high baud rate optical signal transmission. This chapter mainly introduces three different experiments from the perspective of experimental setup and analysis of experimental results. The first is an experiment of 110 Gbaud polarization-multiplexed QPSK signal transmission of 3000 km. The second experiment is 128 GBaud polarization-multiplexed QPSK signal transmission of 10,000 km. The third experiment is 128 GBaud polarization-multiplexed 16QAM signal long-distance transmission experiment. In this chapter, the feasibility and prospect of ultra-high baud optical transmission system are proved by experiments.

Chapter 9, advanced modulation code optical signal transmission technology. This chapter also verifies from the experimental perspective. First, it verifies the generation experiment of 34 GBaud PM-256QAM signal, and then it verifies the transmission experiment of single-carrier 400 G PM-256QAM signal. Experimental results show the advantages and prospects of advanced modulation code optical signal transmission technology.

Chapter 10, carrierless amplitude phase modulation (CAP) techniques. This chapter mainly introduces the short distance high-speed optical transmission. Firstly, the modulation and demodulation principle and realization method of new efficient modulation format CAP are introduced, including the generation and reception of single-band CAP-mQAM and the generation and reception of multi-band and multi-order CAP signal. Then, through experimental research, a WDM-CAP-PON multi-user access network based on multi-order multi-band CAP modulation was introduced, and its high-speed access performance was verified, and the high-speed multi-order multi-band CAP signal was first demonstrated for the WDM-CAP-PON experimental system. Then, the experiment of wireless access network transmission based on high-order modulation CAP-64QAM is introduced. Then, a high-speed CAP-64QAM system based on DML and digital equalization is introduced. CAP-64QAM signal is equalized with the improved DD-LMS. Finally, CAP long-distance transmission of 100 G signal with dispersion compensation is introduced. The digital signal processing of CAP, SSB signal generation, experimental equipment, and results are introduced in detail. In addition, the performance of DDMZM pre-dispersion, dispersion compensation fiber, SSB, and IQ modulator are compared.

Chapter 11, PAM4 signal modulation and detection techniques based on digital signal processing. This chapter mainly introduces PAM4 signal modulation and corresponding digital signal processing algorithm and PAM4 high-speed transmission system experiment. First, PAM4 principle, DD-LMS algorithm, and pre-dispersion compensation principle are introduced. In addition, a simple lookup table (LUT) algorithm is introduced, which can pre-distort the signal at the transmitter end and resist all kinds of nonlinear damage in the direct detection system. Then, the four-channel IM/DD 112-Gbit/s PAM-4 system is introduced, and a series of algorithms such as dispersion pre-compensation (pre-CDC), pre-equalization, lookup table (LUT), and DD-LMS are applied to realize the 400 Gbit/s high-capacity high-speed system. Finally, the polarization-multiplexed 400 G PAM4 signal coherent system is introduced, this system use time division multiplexing (TDM) and polarization multiplexing (PDM), and the use of pre-emphasis technology to overcome the photoelectric device bandwidth limit, to achieve 120 GBaud PDM-PAM4 signal transmission, its data rate is up to 480-Gb/s, and can still reach 400-Gb/s net bit rate considering 20% FEC overhead. These results show that it has a broad application prospect in short distance access network and metropolitan area network.

Chapter 12, optical OFDM principle. In this chapter, the basic principle of optical OFDM (orthogonal frequency division multiplexing) system technology is introduced, and the basic structure and principle of optical OFDM system based on direct detection and optical OFDM system based on coherent detection are, respectively, introduced.

Chapter 13, basic digital signal processing technology of direct detection OFDM. This chapter mainly studies the basic digital signal processing technology of direct detection OFDM system. First is the introduction of the system principle. Secondly, the research on the elimination of DDO-OFDM sub carrier interbeat effect based on Half-cycled technology is introduced, and the experimental equipment and results are, respectively, presented. Then, the research of direct detection of high-order QAM-OFDM signal transmission is introduced, the experimental equipment and results are also introduced, and the experimental results are analyzed and summarized. Finally, the research of DDO-OFDM signal transmission based on DFT-spread is introduced. The optimization of training sequence in the system is introduced, and then the effects of pre-emphasis and DFT-spread in DDO-OFDM are compared.

Chapter 14, intensity modulation and direct detection of the high-speed optical fiber access system. In this chapter, the bandwidth shortage, strong nonlinear effect, and fiber dispersion in the direct detection system of intensity modulation are investigated systematically, and the corresponding solutions are proposed from the aspects of system structure, modulation coding technology, and digital signal processing algorithm. First, Nyquist modulation and super-Nyquist modulation techniques with high spectral efficiency are introduced, and then Volterra series and equal-like coding are introduced. Then, the modulation technology of DFT-S OFDM with low power peak average ratio and high spectral efficiency is introduced. Based on this technology, a series of advanced digital signal processing algorithms, including pre-equalization, nonlinear compensation, DD-LMS, and other technologies, are combined to realize the transmission of single-mode fiber signals of over 100 Gb/ S over 320 km.

Chapter 15, high-speed optical fiber access system based on I/Q modulation direct detection. This chapter introduces an optical access system architecture based on I/Q modulator that generates independent double-sideband signals in detail and carries out direct detection at the receiver end, respectively. By adopting a new Image elimination algorithm, the generation and detection of single-carrier 64 Gb/s CAP4 signals and multi-carrier 300 Gb/s DFT-S OFDM signals based on two independent sideband can be successfully realized. The 240 Gb/s 16QAM DFT-S OFDM signal based on two independent sideband can be transmitted over 160 km in single-mode fiber without dispersion compensation.

Chapter 16, forward error correction code. This chapter introduces several common coding techniques in FEC algorithms. The application of LDPC cascade TCM encoding modulation technique in OFDM-ROF system and Turbo equalization technique to reduce the dispersion effect of optical transmission system is introduced. Forward error correction (FEC) technology finds and corrects the error caused by reasons of the dispersion and nonlinearity by adding a small amount of redundant information in the signal optical transmission process. In addition, the influence of dispersion and nonlinearity on transmission system performance is reduced. The FEC technology lowers the OSNR tolerance at the receiver end through the sacrifice of the signal transmission rate, thereby it gains coding gain, reduces the error rate, and improves the reliability of communication system.

In Chap. 17, the principle and key technology of high spectral efficiency four-dimensional modulation. This chapter introduces the basic principle and key technology of high spectral efficiency 4-dimensional modulation. Firstly, the distribution and performance analysis of 2-D and 3-D constellations are introduced, and then the principle and realization of 4-D multi-order modulation are studied. Then it expands from 4 dimensions to multiple dimensions and introduces the design consideration of multi-dimensional modulation constellations. Finally, the performance of the typical multi-dimensional multi-order constellation diagram is analyzed.

Chapter 18, machine learning algorithm in optical communication system. This chapter mainly introduces the machine learning algorithm applied in optical communication system. Firstly, support vector machines are introduced. Based on the principle of structural risk minimization, it minimizes the actual risk of the learning machine by properly selecting the subset of functions and the discriminant functions in the subset, thus ensures the small error classifier obtained from the limited training samples, and the test error of the independent test set is still small. Then, BP neural network, which is one of the most widely used neural network models, is introduced. It is a multi-layer feedforward network trained by error back propagation algorithm. BP network can learn and store a large number of input–output pattern mapping relationships without revealing the mathematical equations describing such mapping relationships in advance. Then, the clustering algorithm is introduced. This paper mainly introduces the principle, process, presentation, and analysis of k-means clustering algorithm. Based on the principle, the application of clustering algorithm in anti-nonlinear is introduced with examples, and the principle and experimental results are analyzed and demonstrated. This chapter focuses on the current research hotspots of artificial intelligence and machine learning, introduces and shows the application of machine learning in the field of optical communication. At the same time, it also explains the role and value of machine learning algorithm in the field of optical communication.

Chapter 19 mainly introduces the principle of KK algorithm and the latest research progress.

Contents

1 Single-Carrier Advanced Modulation Formats 1
1.1 Summary of Modulation Formats 1
1.2 Optical Modulator 6
1.2.1 Phase Modulator 6
1.2.2 Mach–Zehnder Modulator (MZM) 8
1.2.3 IQ Modulator 9
1.2.4 Electro-absorption Modulator (EAM) 11
1.3 Single-Carrier High-Order Modulation 11
1.3.1 Implementation of QPSK 13
1.3.2 Implementation of 8PSK 13
1.3.3 Implementation of 8QAM 16
1.3.4 Implementation of 16QAM 17
1.3.5 High-Order QAM 18
1.3.6 Comparison of Multi-dimensional Multi-order Modulation Formats 22
1.4 Software-Defined Optical Transceiver 23
1.4.1 Software-Defined Multi-modulation Formats Transceiver (SPOT) 24
1.4.2 Software-Defined Polarization Conversion Transceiver 25
1.4.3 Software-Defined Transceiver of Adaptive Multiplexing PON 27
1.5 Conclusion 29
References 29

2 Basic Digital Signal Processing for Single-Carrier Signals 31
2.1 Introduction 31
2.2 I/Q Imbalance Compensation and Orthogonal Normalization . . . 34
2.2.1 Gram–Schmidt Orthogonalization Process (GSOP) 38
2.2.2 Löwdin Orthogonalization 39

2.3 Fiber Chromatic Dispersion Compensation 40
2.3.1 Dispersion Compensation Overview 41
2.3.2 Dispersion Compensation Basic Structure 43
2.3.3 Frequency-Domain Equalization Algorithm 45
2.3.4 Time-Domain Equalization Algorithm 49
2.4 Clock Recovery Algorithm 53
2.5 Channel Dynamic Equalization Algorithm and Polarization Demultiplexing 54
2.5.1 The Basic Principle of Bland Equalization 54
2.5.2 Classical CMA Algorithm 57
2.5.3 CMA Algorithm for Polarization Multiplexing Signals 60
2.5.4 CMMA Algorithm for High-Order Non-constant Mode Modulation 65
2.5.5 Modified Cascade Multi-modulus Algorithm (MCMMA) 69
2.5.6 Independent Component Analysis (ICA) 72
2.6 Carrier Recovery Algorithm 80
2.6.1 Frequency Offset Compensation Algorithm 81
2.6.2 V-V Based Frequency Offset Estimation Algorithm . . . 82
2.6.3 Frequency Offset Estimation Algorithm Based on FFT 84
2.6.4 Frequency Offset Estimation Algorithm Based on Quartic Power Operation 87
2.7 Phase Compensation Algorithm 90
2.7.1 V-V Based Phase Compensation Algorithm 90
2.7.2 Feedforward Based Phase Rotation Algorithm 92
2.7.3 A Recovery Algorithm Based on the Combination of Modified V-V Phase Bias Algorithm and Maximum Likelihood 94
2.7.4 Algorithm Demonstration and Summary 100
References 101

3 Quasi-linear Coherent Optical Transmission System and Digital Signal Processing 105
3.1 Introduction 105
3.2 Theoretical Model and Damage Mechanism of Coherent Optical Transmission System 106
3.3 Research on Time-Domain Digital Pre-equalization Technology with Band Limit Signal 108
3.3.1 Digital Time-Domain Pre-equilibrium Principle Based on Receiver-Side Adaptive Equalizer 109
3.3.2 Linear Digital Pre-equalization Implementation Method for Coherent Optical Communication System 112

3.3.3 Analysis of Time-Domain Digital Pre-equalization Simulation Results for Coherent Optical Communication 113
3.3.4 Experimental Results 118
3.4 Conclusion 121
References 123

4 Super-Nyquist Wavelength Division Multiplexing System 125
4.1 Introduction 125
4.2 Multi-modulus Blind Equalization Algorithms for Super-Nyquist Signals 127
4.2.1 Principle of Multi-modulus Blind Equalization Algorithm 127
4.2.2 Analysis of Simulation Results of Multi-modulus Blind Equalization Algorithm 131
4.2.3 Experimental Comparative Study of Multi-modulus Blind Equalization Algorithm 134
4.3 Four-Carrier Digital Super-Nyquist Signal Generation and 400 G Transmission 138
4.3.1 The Principle of Digital Super-Nyquist Signal Generation 138
4.3.2 Signal Generation and Transmission Experiment for 4 × 512 Gb/s 4-Carrier Super-Nyquist Channels 140
4.4 400 G Transmission Experiment of Single-Carrier 110-GBaud Super-Nyquist Filtered Signal 144
4.4.1 20 × 440 Gb/s Super-Nyquist WDM Long Distance Transmission Experiment 144
4.4.2 Experiment of Ten Channels Super-Nyquist WDM Signal Transmission over ROADMs Link 147
4.4.3 Transmission Experiment of Single-Carrier 400 G Signal Based on 128.8 GBaud PDM-QPSK in the Terrestrial Fiber Link 150
4.5 Conclusions 155
References 155

5 All-Optical Nyquist Signal 157
5.1 Introduction 157
5.2 The Generation and Signal Multiplexing Principle of All-Optical Nyquist Pulse 159
5.2.1 The Basic Theory of High-Quality Sinc-type Nyquist Pulse Generation 159

5.2.2 Principle of All-Optical Nyquist Signal Multiplexing and Coherent Detection . . . 160
5.2.3 Principle of All-Optical Nyquist Pulse Generation Based on MZM and Bandpass Filter . . . 162
5.3 Generation and Coherent Detection Experiment of Single-Polarized 125 GBaud All-Optical Nyquist QPSK Signal . . . 164
5.4 Polarization-Multiplexed All-Optical Nyquist Signal Long-Distance Transmission Experiment . . . 169
5.5 Summary . . . 174
References . . . 175

6 Nonlinear Compensation in Optical Fiber . . . 177
6.1 Introduction . . . 177
6.2 Principle of DBP Fiber Nonlinear Compensation Algorithm . . . 178
6.2.1 Nonlinear Compensation Experiment Based on 4 × 160.8 Gb/s Wavelength Division Multiplexing PDM-QPSK Signal . . . 180
6.2.2 Improved Digital Nonlinear Compensation Algorithm Based on the Logarithmic Step Size . . . 185
6.3 Experimental Setup and Results . . . 194
6.4 Summary . . . 198
References . . . 198

7 Probabilistic Shaping . . . 201
7.1 Introduction . . . 201
7.2 Principle of Probabilistic Shaping . . . 202
7.3 Simulation of Probabilistic Shaping . . . 206
7.3.1 Simulation Setup . . . 206
7.3.2 Simulation Result and Discussion . . . 207
7.4 Principle of PAS Scheme . . . 210
7.5 Comparison of PS and Hybrid-QAM . . . 211
7.6 Experiment of PS Transmission . . . 213
7.6.1 Experimental Setup . . . 213
7.6.2 Comparison of PS-64QAM and Hybrid-32/64QAM . . . 215
7.7 Conclusion . . . 220
References . . . 220

8 High Baud Signal Transmission . . . 223
8.1 Introduction . . . 223
8.2 110-GBaud Polarization-Multiplexed QPSK Signal for 3000 Km Transmission . . . 226
8.2.1 Experimental Setup . . . 226
8.2.2 Experiment Results and Analysis . . . 228
8.2.3 Performance of the Multi-link System . . . 229

8.3 128 GBaud Polarization Multiplexing QPSK Signal Transmission Over 10,000 Km 235
8.3.1 Experimental Setup 235
8.3.2 Experimental Result and Analysis 236
8.4 Long-Distance Transmission of 128-GBaud Polarization-Multiplexed 16QAM Signal 239
8.4.1 Experimental Setup 239
8.4.2 Experimental Results and Analysis 240
8.5 Conclusion 241
References 242

9 Advanced Modulation Code Optical Signal Transmission Technology 245
9.1 Introduction 245
9.2 Trade-Offs Between Coding Overhead and Bandwidth Limitations 246
9.3 Single-Carrier 400 G PM-256QAM Signal Transmission Experiment 248
9.4 Conclusion 251
References 252

10 Carrierless Amplitude and Phase Modulation 253
10.1 Introduction 253
10.2 Modulation and Demodulation Principles of CAP 254
10.2.1 Modulation and Demodulation Principles of Single-Band CAP 254
10.2.2 Modulation and Demodulation Principles of Multi-band CAP 256
10.3 Multi-band Multi-level CAP for WDM-PON Access Network 258
10.4 Application of CAP-64QAM in Wireless Access Network 262
10.5 DML-Based 60 Gb/s CAP-64QAM Transmission Experiment 265
10.6 100 G CAP Long-Distance Transmission with Dispersion Compensation 269
10.6.1 Digital Signal Processing of CAP 270
10.6.2 Generation of SSB Signal 271
10.6.3 Experimental Setup and Results 272
10.6.4 Comparison Between Pre-CD and DCF with DDMZM 273
10.6.5 Comparison Between Pre-CD and SSB with DDMZM 274

10.6.6 Comparison Between DDMZM and IQ Modulator 275
10.6.7 Conclusion 276
10.7 Summary 277
References 278

11 PAM4 Signal Modulation and Digital Signal Processing-Based Detection Technology 281
11.1 Introduction 281
11.2 Principle and Related Algorithms of PAM4 Modulation 282
11.2.1 Principle of PAM4 Modulation 282
11.2.2 DD-LMS Algorithm 283
11.2.3 Principle of Dispersion Pre-compensation 284
11.2.4 Lookup Table Algorithm 285
11.3 PAM4 High-Speed Transmission System 286
11.3.1 4-Lane IM/DD 112.5 Gbit/s PAM4 Transmission System 286
11.3.2 50 Gbit/s and 64 Gbit/s PAM4 PON Downlink Transmission System 289
11.3.3 4-Lane IM/DD 112 Gbit/s PAM4 Dispersion Pre-compensation Transmission System 291
11.3.4 400 Gbit/s PDM-PAM4 Signal Generation and Coherent Detection 294
11.4 Summary 299
References 300

12 Optical OFDM 303
12.1 Introduction 303
12.2 Basic Structure of Direct Detection Optical OFDM System 306
12.3 Structure and Basic Principles of Coherent Detection Optical OFDM Systems 307
12.4 Summary 331
References 331

13 Direct Detection OFDM 335
13.1 Introduction 335
13.1.1 System Principle 339
13.2 Study on Elimination of SSMI in DDO-OFDM Based on Half-Cycled Technique 343
13.2.1 Experimental Setup and Results 343
13.2.2 Summary 347
13.3 Research on High Order QAM-OFDM Signal Modulation Transmission by Direct-Detect 347
13.3.1 Experimental Setup 347
13.3.2 Experimental Results and Analysis 349
13.3.3 Summary 355

13.4 Short Distance Transmission of Large Capacity DDO-OFDM Based on DFT-Spread . . . 355
13.4.1 Optimization of Training Sequences in Large Capacity DDO-OFDM Systems Based on DFT-Spread . . . 355
13.4.2 Comparison of Pre-equalization and DFT-Spread Techniques in High-Capacity DDO-OFDM . . . 362
13.4.3 Summary . . . 369
References . . . 370

14 Intensity Modulation Direct Detection High-Speed Fiber Access System . . . 373
14.1 Introduction . . . 373
14.2 High Spectral Efficiency Modulation Technique . . . 375
14.2.1 Nyquist Modulation Technique . . . 376
14.2.2 Faster Than Nyquist Modulation Technique . . . 383
14.3 Nonlinear Compensation Technique . . . 388
14.3.1 Nonlinear Compensation Technology Based on Volterra Series . . . 389
14.3.2 Nonlinear Compensation Technique Based on Quasi-balanced Coding and Detection . . . 391
14.4 Single Sideband Modulation System . . . 396
14.5 High-Speed Wavelength Division Multiplexing System . . . 403
14.6 Conclusions . . . 405
References . . . 406

15 High-Speed Fiber Access System Based on Direct Detection of I/Q Modulation . . . 407
15.1 Introduction . . . 407
15.2 Independent Sideband Modulation Direct Detection System Based on I/Q Modulator . . . 408
15.3 Image Elimination Algorithm Based on Training Sequence . . . 411
15.3.1 Principle of Image Elimination Algorithm Based on Training Sequence . . . 411
15.3.2 Experimental Setup . . . 413
15.3.3 Experimental Result . . . 415
15.4 Image Elimination Algorithm Based on Adaptive Blind Equalization . . . 421
15.4.1 Principle of Image Elimination Algorithm Based on Adaptive Blind Equalization . . . 421
15.4.2 Experimental Setup and Results . . . 423
15.5 Summary . . . 425
References . . . 426

16 Forward Error Correction 427
16.1 Introduction 427
16.2 Block Code 428
16.2.1 Linear Block Code 429
16.2.2 Cyclic Code 431
16.2.3 BCH Code 434
16.2.4 Rs Code 437
16.2.5 Parity-Check Code 438
16.3 Turbo Code 440
16.3.1 Decoding of Turbo Code 440
16.3.2 Iteration Decoding of Turbo Code 442
16.3.3 MAP Decoding 443
16.3.4 MAP Equalization Technology 444
16.3.5 Turbo Iterative Equalization of OFDM Signal 445
16.3.6 MIMO-CMA Equalization Algorithm 447
16.4 LDPC Code 454
16.4.1 Principle of LDPC Code 454
16.4.2 Principle of LDPC-TCM OFDM Optical Millimeter-Wave Signal Transmission System at 60 GHz 456
16.4.3 Experimental Results and Discussion 459
16.4.4 Experimental Conclusion 460
16.5 Concatenated Coding 461
16.6 Summary 463
References 463

17 High Spectral Efficiency Optical Four-Dimensional Modulation . . . 467
17.1 Introduction 467
17.2 Constellation Point Distribution and Performance Analysis of Two-Dimensional and Three-Dimensional Constant Modulus Modulation 469
17.3 Principle and Implementation of Four-Dimensional Multi-level Modulation 472
17.3.1 Basic Principles of Four-Dimensional Multi-order Modulation 472
17.3.2 Implementation of Four-Dimensional Multi-order Modulation 474
17.4 Design Basis for Multi-dimensional Multi-order Modulation Constellation 482
17.4.1 "Cluster" Problem 483
17.4.2 "Ball" Problem 485

17.5 Typical Multi-dimensional Multi-order Constellation Performance Analysis 485
17.5.1 N = 2 486
17.5.2 N = 4 488
17.6 Experimental Demonstration Based on Geometric Shaped Signal 494
17.6.1 A Constellation-Designed Method for 32QAM with Geometric and Probabilistic Shaping (G/PS 32QAM) 494
17.6.2 Experimental Demonstration Based on PS-Star-16QAM 497
17.6.3 CAP-PS-9QAM Gray Mapping Scheme in IM/DD System 499
17.7 Summary and Outlook 500
References 501

18 Machine Learning Algorithm in the Optical Communication System 503
18.1 Introduction 503
18.2 Support Vector Machines 505
18.2.1 Margin and Support Vector 506
18.2.2 Dual Problem 507
18.2.3 Kernel Function 508
18.2.4 SVM-Based Modulation Format Recognition 509
18.3 BP Neural Network 509
18.3.1 BP Neuron 510
18.3.2 BP Network 512
18.3.3 OSNR Estimator Based on BP Neural Network 513
18.4 Clustering Algorithm 514
18.4.1 Principle of K-Means Clustering Algorithm 514
18.4.2 Algorithm Flow 515
18.4.3 Algorithm Display and Analysis 516
18.5 Application of Clustering Algorithm in Nonlinearity 518
18.5.1 Application Principle 518
18.5.2 Result Analysis 520
References 521

19 Kramers-Kronig Receiver in Direct Detection Systems 523
19.1 Introduction 523
19.2 Principle of KK Receiver 525
19.3 Simulation Setup and Results 527
19.3.1 Simulation Setup 527
19.3.2 Simulation Results and Discussions 528

19.4 Experimental Setup and Results 531
19.4.1 Experimental Setup 531
19.4.2 Experimental Results and Discussions 533
19.5 Conclusions 539
References 540

Chapter 1
Single-Carrier Advanced Modulation Formats

To build a flexible, low-cost, high-capacity optical routing-based wavelength division multiplexing fiber network, choosing the right modulation format is crucial. Currently, the commercially available 40 Gb/s optical communication systems are all based on the binary modulation format with the wavelength division multiplexing channel interval of 50 GHz and maximum spectral efficiency of 0.8 bit/s/Hz. Therefore, in the field of optical communication, all the modulation formats more advanced than the light intensity modulation (OOK) have won the pre-selection of this modulation format competition. In order to further increase the system capacity, reduce the cost, and realize the long-distance transmission of high-speed and large-capacity optical networks, the advanced optical modulation format has become a hot spot in the research of optical communication technology in the world. Nowadays, the research on single-carrier advanced optical modulation format mainly focuses on multi-dimensional multi-order modulation format. The research organizations at home and abroad optical communication have conducted a lot of experiments and have entered a relatively mature research stage. In this chapter, we first introduce the single-carrier optical modulation format [1], then elaborate on the principle of optical modulators that implement optical modulation, analyze the implementation of single-carrier advanced modulation formats such as QPSK, 16QAM, and high-order QAM [2–13], finally present the software-defined transceiver (SDT) [14–20] in single-carrier optical communication.

1.1 Summary of Modulation Formats

In a single-mode fiber, there are four physical quantities that can be used to transmit information in the optical domain: intensity, phase, frequency, and polarization. Correspondingly, we classify the modulation formats into four catalogs: intensity, phase, frequency, and polarization data modulation format, as shown in Fig. 1.1. It should be noted that this classification method does not require a constant amplitude

J. Yu and N. Chi, *Digital Signal Processing In High-Speed Optical Fiber Communication Principle and Application*,
https://doi.org/10.1007/978-981-15-3098-2_1

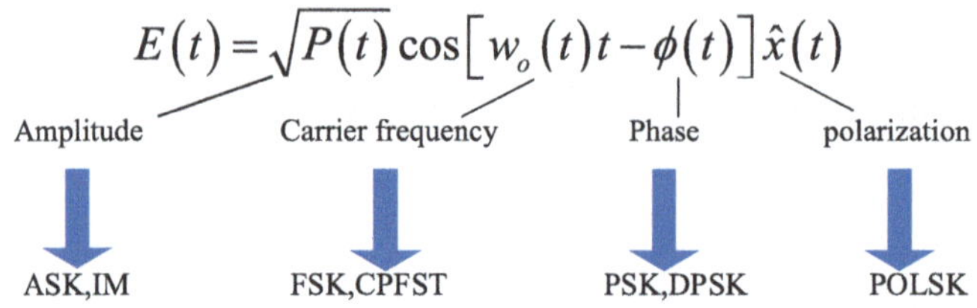

Fig. 1.1 Relationship between optical modulation method and electric field parameters of optical signals

value for phase modulation or a constant phase for intensity modulation. This classification method is determined by the physical quantity of the data information. As a typical example, DPSK is a phase modulation format, regardless of whether the DPSK signal is transmitted in constant envelope transmission or in zero return. On the other hand, CSRZ is an intensity modulation format regardless of whether the phase of the optical field is inverted.

The intensity and phase modulation formats are now widely used in high-speed optical communication systems, but the application of polarization shift keying (Pol-SK) is relatively rare. This is mainly because the polarization variations in the fiber are random, and Pol-SK requires flexible polarization management at the receiver side. For intensity modulation or phase modulation and direct detection of the receiver, polarization management is required only when the polarization mode dispersion is severe. If the receiver sensitivity of Pol-SK can be significantly improved, the increased complexity at the receiver side with Pol-SK data modulation is acceptable. In addition, polarization can often be used to improve the spectral efficiency of signals in research experiments. For example, different signals of the same wavelength can be transmitted by two orthogonal polarization states (polarization-multiplexing) and use alternating polarization states between adjacent WDM channels to reduce the nonlinear effects between channels and polarization-interleaving. Due to the application of coherent detection technology and digital signal processing technology, the complexity introduced by the wavelength-dependent random polarization change in the transmission fiber is transferred to the offline digital signal processing, where the complexity of the receiver is reduced. As a result, the polarization multiplexing has been widely used in high-speed coherent optical communication, and the current 100 Gb/s optical transmitter chip is based on QPSK-based polarization multiplexing. In addition, the derived hybrid modulation technique is an organic combination of the above two or several modulation methods. Taking OOK and DPSK as examples, Fig. 1.2 shows the classification of the most important single-carrier optical modulation formats available today.

Through multi-level signal processing, M symbols can be mapped to log 2 (M) data bits and transmitted at the symbol rate of *R*/log 2 (M) (*R* is bit rate), reducing the symbol rate required for transmission. In general, there is no relationship between the data symbols assigned to a symbol and the symbol to be transmitted and

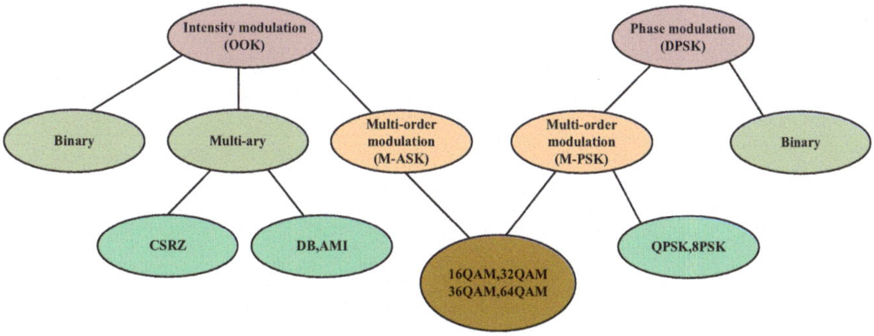

Fig. 1.2 Advanced optical modulation formats

the next symbol to be transmitted, also known as memoryless modulation. Multi-level signal processing improves spectral efficiency, but at the expense of reduced noise immunity. Using multi-level signal processing, the data rate of a single channel can exceed the limitation of high-speed optoelectronic devices. In other words, at a fixed data rate, the symbol rate required for multi-level signal processing is relatively low. Dispersion signal distortion (such as chromatic dispersion or polarization mode dispersion) and the application of digital electrical signal processing are very advantageous.

Multi-order intensity modulation, multi-order phase modulation, and multi-order hybrid modulation are discussed in high-speed optical transmission. Multi-order intensity modulation (M-ASK) has not shown superiority in fiber-optic transmission applications, mainly because M-ASK receiver has a large amount of back-to-back sensitivity loss relative to OOK. For example, 4-ASK introduces signal noise in the square-law detection due to inconsistent signal amplitude levels, so 4-ASK can have a loss of approximately 8 dB. DQPSK is one of the most promising optical modulation formats. DQPSK maps bit sequences into symbol sets such as $\{0, \pi/2, -\pi/2, \pi\}$, as shown in Table 1.1, by loading information into each other. The signal is transmitted in the orthogonal phase. Quadrature amplitude modulation (QAM) is a multi-order hybrid modulation. Unlike M-ASK and M-PSK, QAM simultaneously loads information on the amplitude and phase of the optical field. Therefore, the QAM modulation format can provide a higher modulation rate and higher spectral efficiency. At the same time, the distance between adjacent constellation points in QAM modulation is also the largest in the same modulation mode, which makes the QAM receiver have the advantage of high sensitivity. However, nothing is absolutely perfect. The advantages of QAM modulation also make the transmitter and receiver more complex, and more requirements for digital signal processing are needed. So far, a large part of the international high-speed and large-capacity transmission experiments utilize QAM modulation technology for signal transmission.

If a bit in the modulation format can be represented by more than two symbols and the allocation between the redundant symbol and the transmitted bit is independent of the data signal, we call this modulation format a pseudo-multi-level data modulation

Table 1.1 Encoding mappings for different modulation formats

Data Seq.	0	0	1	0	1	1	1	0	0	1	0	1
DQPSK	0		+π/2		π		+π/2		−π/2		−π/2	
CSRZ	0	0	+1	0	+1	−1	+1	0	0	−1	0	−1
DB	0	0	+1	0	−1	−1	−1	0	0	−1	0	+1
AMI	0	0	+1	0	−1	+1	−1	0	0	+1	0	−1

format. If the allocation of the residual bits is determined by the transmitted data information, in general, we call it the data modulation of the associated code, which is the most important form of the partial response data modulation format.

The most common pseudo-multi-order modulation format is carrier-suppressed zero-return CSRZ, because it is the easiest one to generate. It encodes information on the 0 and 1 according to the intensity, whereas the phase change is π/bit, which is independent of the data information. The most important part of response data modulation format is the optical duobinary modulation format. For CSRZ, the information is passed through the intensity level {0, 1}, while the π phase shift occurs only when 1 bit is separated by an odd number of 0 bits. This association between the auxiliary phase change and the information encoding is the characteristic of the partial response modulation format. However, it should be emphasized that because the square-law direct detection is insensitive to the optical phase in the receiver side, phase information is generally not used for information detection. In order to overcome the nonlinearity of optical fibers, some research organizations at home and abroad are working on more advanced line coding schemes (including pseudo-multiple order and correlation coding).

Figure 1.3 shows the four most important symbol diagrams commonly found in optical communications, including nearly all the modulation schemes in Fig. 1.2. Figure 1.3a shows the symbol of OOK. If the waveform is idealized, and the sample is the best time sample, the symbol map will capture the amplitude and phase information of all data symbols and then maps them to the complex plane of the light domain. If bit transitions are meaningful in the study of modulation formats, then the curves that do not conform to the rules in the symbol set can also be displayed in the symbol map, which provides a way for the study of bit transition. The dotted line in Fig. 1.3a shows this bit transition, and it is a modulation scheme of chirp OOK. However, the symbol graph does not display the dynamic process of transition between symbols, such as the rise and fall times of the pulse or the duration of the light pulse (in the symbol diagram of Fig. 1.3a, we cannot judge the dotted line. Is C-NRZ or CRZ). Figure 1.3b shows the symbol diagram of CSRZ, DB (dual binary),

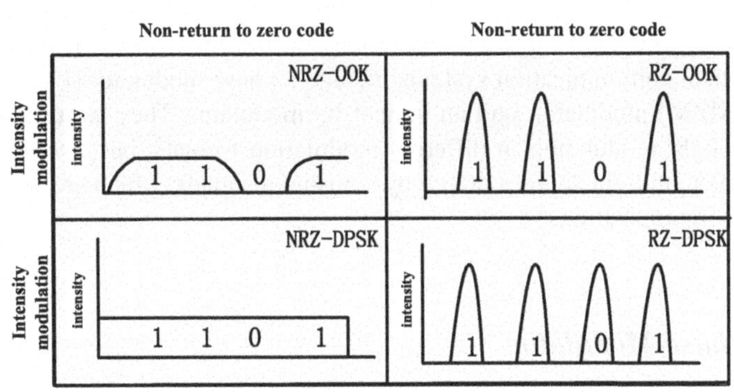

Fig. 1.3 Symbol diagrams of different modulation formats in a complex optical field

and AMI (alternate mark flip code). Figure 1.3c, d is symbolic diagrams of DPSK and DQPSK, respectively. For ease of comparison, the average optical power of the two-phase modulations are equal, as are the two intensity modulation formats (a) and (b) in Fig. 1.3. The symbol diagrams in Fig. 1.3 are all two-dimensional, including only light intensity and optical phase information. Multi-dimensional symbol constellations are likely to add polarization information of light to the symbol map.

When the information is loaded onto the intensity, phase, or polarization of the light pulse, a non-return-to-zero or return-to-zero code is obtained, the latter indicates that the light pulse will return to zero in a bit time slot. In contrast, non-return-to-zero code allows the intensity to remain constant after successive data bits.

1.2 Optical Modulator

In an optical communication system, an optical modulator is a conversion device that converts an electrical signal carrying information into an optical signal and modulates the electrical signal onto an optical carrier generated by the light source to realize optical signal transmission. At present, the most widely used light modulator is based on the electro-optical effect of the material, where the refractive index of the electro-optic material changes with the applied external voltage to achieve modulation of the phase, frequency, or amplitude of the laser. Such a modulator has advantages such as high modulation rate and wide frequency bandwidth. These features are suitable for high-speed communication systems. The most commonly used electro-optical crystal is lithium niobate ($LiNbO_3$) crystal.

There are two types of optical modulation transmitters: direct modulation transmitter and external modulation transmitter. Direct modulation transmitters have a highly device-specific ripple effect, and the ripple effect of the laser broadens the optical spectrum, which is very detrimental to the isolation of dense wavelength division multiplexing channels and can exacerbate signal distortion due to fiber dispersion effects. There are few applications in high-speed coherent optical communication today. An externally modulated transmitter typically consists of one or more basic external light modulators. At present, external light modulators commonly used in optical communication systems include a phase modulator (PM), a Mach–Zehnder (MZM) modulator, and an optical IQ modulator. They are used not only to achieve light modulation in different modulation formats, but also to generate multi-wavelength light sources driven by sinusoidal signals, which are widely used in optical communication.

1.2.1 Phase Modulator

The working mechanism of the phase modulator is based on the change of the propagation speed and phase of the light wave generated by the refractive index of the lithium niobate ($LiNbO_3$) crystal with the change of the applied external electric field. The electro-optic effect experiment found that the refractive index of the crystal material has a complicated relationship with the applied electric field. It can be approximately assumed $\Delta n \sim (R|E|+\gamma|E|^2)$, where the first term represents a linear relationship with E, which is known as the Pockel effect, and the second term stands for a square relationship with E, which is called the Kerr effect. The Pockel effect is mainly applied in electro-optic phase modulators.

The structure of the optical phase modulator integrated with lithium niobate crystal is shown in Fig. 1.4. It can be seen that the phase modulator is a strip waveguide formed by diffusion technique on a lithium niobate substrate, with electrodes applied above and below the waveguide. When a modulation voltage is applied to the electrode, the refractive index of the waveguide changes due to the electro-optical effect, and the phase changes with the modulation voltage after the light wave passes through the electrode, therefore, the phase modulation function is performed.

According to the electro-optical effect, the modulation factor of the phase modulator $\varphi_{\mathrm{PM}}(t)$ is a function of the λ wavelength of the input light, the length L of the interaction of the light and the electrode, and the change Δn_{eff} in the effective refractive index. When only the first-order Pockel effect is concerned, the phase change can be treated as a linear function of the input drive voltage $u(t)$, as shown in formula (1.1).

$$\varphi_{\mathrm{PM}}(t) = \frac{2\pi}{\lambda} \cdot k \Delta n_{eff} L \cdot u(t) \tag{1.1}$$

At the same time, the half-wave voltage V_π of the main characteristic parameter of a modulator is defined, that is, the modulation voltage required when the phase of the modulator changes π inversion. It can be expressed as

$$V_\pi = \frac{\lambda}{2k \Delta n_{eff} L} \tag{1.2}$$

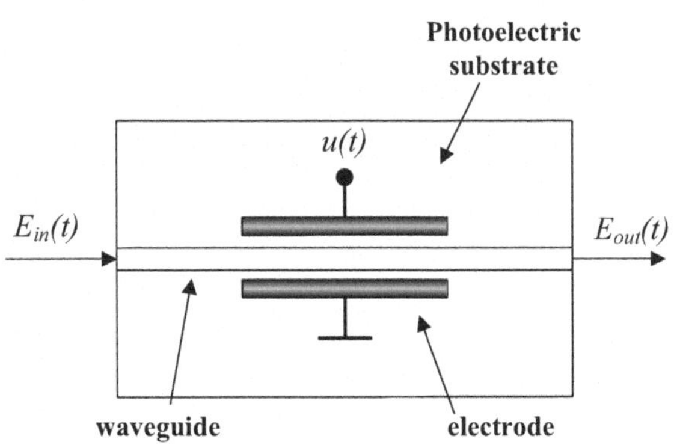

Fig. 1.4 Structure of Lithium niobate crystal integrated optical phase modulator

The relationship between the modulation factor and the input drive voltage and half-wave voltage can be obtained from Eqs. (1.1) and (1.2):

$$\varphi_{PM}(t) = \frac{u(t)}{V_\pi}\pi \tag{1.3}$$

1.2.2 Mach–Zehnder Modulator (MZM)

MZM is an optical modulator based on the principle of interference. Its structure is shown in Fig. 1.5.

A Mach–Zehnder modulator can be produced by combining two phase modulators in parallel on a $LiNbO_3$ crystal substrate. A pair of parallel strip waveguides were fabricated on the lithium niobate substrate, and a 3 dB Y-branch waveguide was connected at both ends, and the two sides of the waveguide and the intermediate surface electrode were connected. The beam entering from the input is split into two beams of equal power through a 3 dB Y-branch waveguide and coupled into two parallel straight waveguides with identical structural parameters. Under the action of the upper and lower arm drive signals, the two-phase modulations are, respectively, performed, and then the two phase-modulated waves are mutually interfered by the second 3 dB Y-branch waveguide to be converted into intensity modulation. In the Mach–Zehnder modulator, the upper and lower arms are driven not only by the RF signal, but also by the DC bias voltage, which interacts to determine the operating state of the Mach–Zehnder modulator.

First, we can get the transfer function of the Mach–Zehnder modulator by analyzing the principle of the phase modulator, which can be expressed as

$$E_{\text{out}} = E_{\text{in}} \cdot \frac{1}{2} \cdot \left(e^{j\varphi_1(t)} + e^{j\varphi_2(t)}\right) \tag{1.4}$$

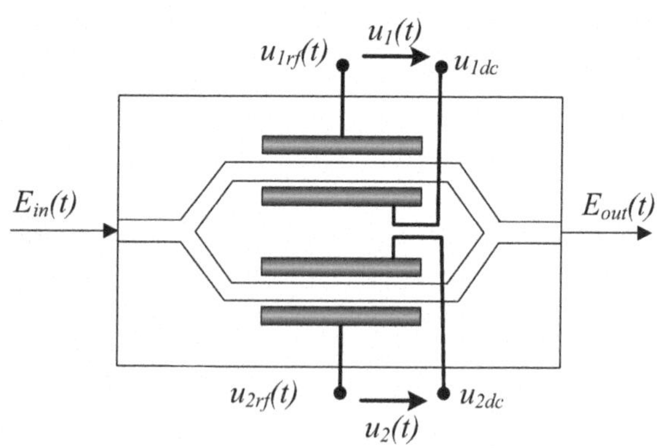

Fig. 1.5 Structure of Mach–Zehnder modulator

where $\varphi_1(t)$ and $\varphi_2(t)$, respectively, indicate the phase shift of the upper and lower arms of the MZM. According to the phase modulator principle, the phase shift of the upper and lower arms can be expressed as

$$\varphi_1(t) = \frac{u_1(t)}{V_{\pi 1}}\pi,\ \varphi_2(t) = \frac{u_2(t)}{V_{\pi 2}}\pi \tag{1.5}$$

$V_{\pi 1}$ and $V_{\pi 2}$ are drive voltage indicating the phase shift of the upper and lower arms, respectively, which are called half-wave voltages. $u_1(t)$ and $u_2(t)$ indicate the external voltage of the upper and lower arms, respectively, including the RF drive voltage $u_{\mathrm{rf}}(t)$ and DC bias voltage $u_{\mathrm{dc}}(t)$.

When MZM works in push-push mode, the phase shifts of the upper and lower arms are exactly the same. The upper and lower arms only increase the phase shift and are still phase modulated for the signal.

When MZM works in push-pull mode, the phase shift between the arms is reversed, that is, $\varphi_1(t) = -\varphi_2(t)$ and $u_1(t) = -u_2(t) = 1/2u(t)$. At this moment, an intensity modulated optical signal will appear at the output. The input and output optical signals are expressed as

$$\begin{aligned} E_{\mathrm{out}}(t) =& \frac{1}{2} \cdot E_{\mathrm{in}}(t) \cdot \left(e^{j\varphi_1(t)} + e^{j\varphi_2(t)}\right) \\ =& \frac{1}{2} \cdot E_{\mathrm{in}}(t) \cdot \left[\cos(\varphi_1(t)) + \cos(\varphi_2(t)) + j \cdot (\sin(\varphi_1(t)) + \sin(\varphi_2(t)))\right] \\ =& E_{\mathrm{in}}(t) \cdot \left(\frac{\Delta\varphi_{\mathrm{MZM}}(t)}{2}\right) = E_{\mathrm{in}(t)} \cdot \cos\left(\frac{u(t)}{2V_\pi}\pi\right) \end{aligned} \tag{1.6}$$

When intensity modulation is performed, the modulator's DC bias is at the quadrature position, which is $-V_\pi/2$, and the drive voltage varies over $0 - V_\pi$. When the DC offset is at the lowest point of the power transfer function, the drive voltage varies in $2V_\pi$, and has a π phase shift when the drive voltage passes the lowest point.

1.2.3 IQ Modulator

The optical IQ modulator consists of two MZMs and a 90° phase shifter, and its integrated devices have been commercialized and widely used in high-speed coherent optical communication. The principle is shown in Fig. 1.6: After the input optical signal is introduced at the input terminal, it is equally divided into two signals—in-phase shunt and quadrature shunt—and transmitted along different paths. Then, one of the two signals passes through the phase modulator, so that the two optical signals have a relative phase difference of 90°. The MZM of the in-phase shunt and the quadrature shunt operates in a push-pull mode, and the DC offset is at the lowest point of the power transfer function. Finally, the two signals interfere with the output coupler at the output and merge into one useful signal.

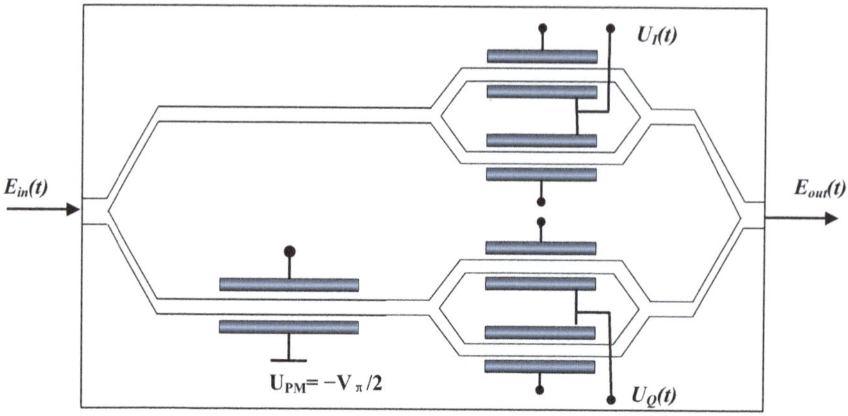

Fig. 1.6 Optical IQ modulator

In Fig. 1.6, the phase difference produced by MZM in the in-phase shunt and quadrature shunt can be expressed as follows (Fig. 1.7):

$$\varphi_I(t) = \frac{u_I(t)}{V_{\pi 1}}\pi, \varphi_Q(t) = \frac{u_Q(t)}{V_{\pi 2}}\pi \tag{1.7}$$

Regardless of the influence of the insertion loss, and the driving voltage of the PM is UPM = −Vπ/2, the transfer function of the optical IQ modulator can be expressed as

$$E_{\text{out}}(t) = \frac{1}{2} \cdot E_{\text{in}}(t) \cdot \left(\cos\left(\frac{\varphi_I(t)}{2}\right) + j\sin\left(\frac{\varphi_Q(t)}{2}\right)\right) \tag{1.8}$$

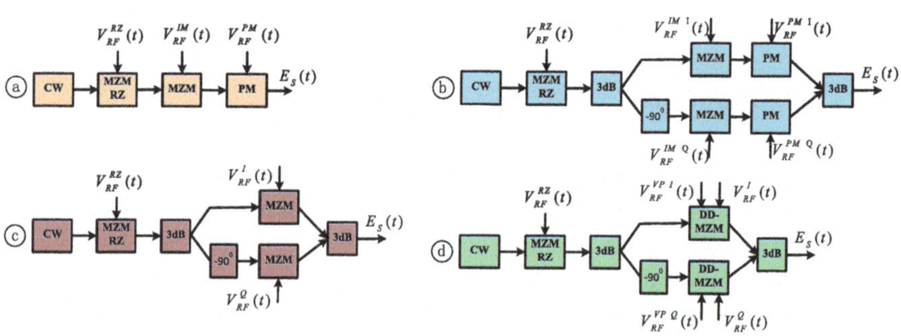

Fig. 1.7 Different multi-order modulation transmitter

1.2.4 Electro-absorption Modulator (EAM)

EAM is a modulator of a pin semiconductor structure. With an external voltage, the bandgap can be modulated to change the absorption characteristics of the device. A distinguishing feature of EAM is its low drive voltage. Now, it has a modulation rate of up to 40 Gb/s, and in laboratory studies, modulation rates of up to 80 Gb/s. However, as working with direct modulation lasers, it also produces residual chirp effect, which is rarely used in high-speed coherent optical communications studied today. The absorption characteristics of EAM are determined by the wavelength, and the dynamic extinction ratio (defined by the ratio of the maximum modulated optical power to the minimum modulated optical power), which usually does not exceed 10 dB. EAM has a limited power response. The insertion loss introduced between the EAM fiber and the fiber is approximately 10 dB. In EAM, the laser diode is integrated with the chip, which reduces the loss of the fiber-to-chip input interface and reduces the size of the transmitter. This modulation format is called EMLs (optical absorption laser modulator), which has an output power of the order of 0 dBm and a modulation rate of 10 Gb/s. Another way to reduce the insertion loss of EAMs is to integrate the laser diode with the semiconductor optical amplifier, which can even produce fiber gain. The optical transfer function of EAM is expressed as follows:

$$E_{\text{out}}(t) = E_{\text{in}}(t) \cdot \sqrt{d(t)} \cdot \exp\left(\frac{ja}{2} \ln[d(t)]\right) \tag{1.9}$$

where $E_{\text{in}}(t)$ denotes the input optical signal, and the power transfer function $d(t) = (1 - m) + m \cdot data(m)$, m is the modulation index of the modulator, and $data(m)$ is the electrical modulation signal. In order to make the power transfer function $d(t)$ greater than 0, in general, the value of the input data can only vary between 0 and 1. Therefore, we can get the power of the output signal as $P_{\text{out}} = |E_{\text{out}}(t)|^2$, where P_{out} can be expressed as follows:

$$P_{\text{out}} = P_{\text{in}}(t) \cdot d(t) = P_{in}(t) \cdot ((1 - m) + m \cdot data(t)) \tag{1.10}$$

1.3 Single-Carrier High-Order Modulation

With the rapid development of Internet services and multimedia applications, network traffic is rapidly growing at an exponential rate, which requires the network to have high bit rate data transmission capability and high throughput cross capability. In order to meet the ever-increasing bandwidth requirements of next-generation Dense Wavelength Division Multiplexing (DWDM) systems, high spectral efficiency transmission has never been more important. Multi-dimensional multi-order modulation formats based on polarization multiplexing and coherent detection are considered to be the best transmission technology to meet the rising bandwidth demand, due to their advantages in terms of higher spectral efficiency, loss resistance, and dispersion effects.

In 2007, Gnauck used PDM-RZ-DQPSK modulation technology, direct detection technology, and hybrid EDFA/Raman fiber amplification technology to achieve transmission of 160 × 160 Gb/s (25.6 Tb/s) DWDM signals over 240 km standard single-mode fiber in the C+L band with 50-GHz channel spacing. s) Transmission of DWDM signals over 240 km standard single-mode fiber. Such a large-capacity transmission has aroused people's enthusiasm for research on multi-dimensional multi-order modulation formats. However, if the order of the modulation format is too high, the distance between the constellation points will be very close, and the decision of the receiver becomes very difficult. Consequently, the improved spectrum efficiency cannot result in an enhanced system capacity or higher baud rate. Multi-level intensity modulation (M-ASK) has not shown superiority in fiber transmission applications. Therefore, the research on single-carrier multi-dimensional multi-order modulation format mainly focuses on modulation formats such as QPSK, 8PSK, 8QAM, and 16QAM to facilitate applications in long-distance and large-capacity communication systems.

The single-carrier optical multi-order modulation transmitter mainly has the following basic forms (as shown in Fig. 1.8): a. series mode; b. IQ modulation mode composed of MZM and PM; c. IQ modulation mode of single-arm driving MZM; d. IQ modulation method composed of two-arm MZM. In order to fulfill the optical modulation of multi-dimensional multi-order signals, the mapping of the original electrical signals to the optical carrier parameters must be accomplished. To achieve this objective, it is mainly based on the serial configuration of MZM and PM. **Based on the system topology, there are mainly two schemes**. The serial scheme is based on the cascading idea, in which the modulation of the above-mentioned optical signal is realized by modulating the first-order level of the signal of the input light field; in parallel, multiple parallel connections are made. The MZM modulator implements mapping of multiple binary digital signals to the above optical signals. In addition, we can also realize the modulation of electrical signals to optical signals by means of serial and parallel mixing according to the integration needs. It should be pointed out that different modulation transmitters result in different performances. In practical systems, we need to compare different modulation transmitters to find the optimum solution.

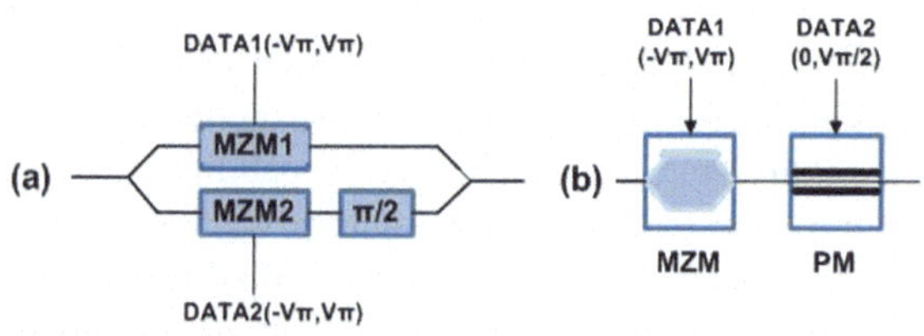

Fig. 1.8 Structures of QPSK modulation: **a** parallel **b** cascade

As an efficient modulation method, QPSK, 8PSK, 8QAM, and 16QAM are considered good candidates for 100 Gbps transmission systems. If you want to achieve an even larger capacity transmission, you need to use a higher order modulation format such as 32QAM, 64QAM, 256QAM, and so on. In this section, we mainly investigate the generation methods of the above several modulation formats in detail.

1.3.1 Implementation of QPSK

QPSK is a quaternary phase shift modulation format, which is the most widely used multiary modulation format in high-speed optical communication. It transmits four phase shifts (the amplitude remains the same) at a symbol rate which is equal to half the total bit rate. With the two MZMs, the required QPSK signal can be easily generated.

The QPSK modulation structure is shown in Fig. 1.9. There are two forms: parallel and cascade [2]. In the parallel scheme (shown in Fig. 1.9a), the two MZMs in the double-balanced structure are offset at 0, and the amplitude of the drive signal fluctuation is 2Vπ, thus generating two BPSK signals (two BPSK signals are, respectively, $E_1 = E_{\text{in}}e^{j\pi c_k}$ and $E_2 = E_{\text{in}}e^{j\pi(d_k+0.5)}$, c_k and d_k are two input data bit sequences), the QPSK signal is generated by the interference of the two orthogonal BPSK signals, and the output is $E_{\text{out}} = E_{\text{in}}\left(e^{j\pi c_k} + e^{j\pi(d_k+0.5)}\right)$; and in the serial scheme, the QPSK signal is at 0.5 π phase modulation. The rotation of the device is realized. Now the output signal of QPSK is $E_{\text{out}} = E_{\text{in}}e^{j\pi(c_k+0.5d_k)}$, where d_k and d_k are two input data bit sequences.

Although the topology of these two implementations is relatively simple, it can be realized by using optoelectronic devices that have been commercialized. The former is more frequently used in practical applications, mainly because the phase modulator converts the jitter of the driving current into an optical signal. The jitter of the phase degrades the performance of the QZK modulator in the form of a series MZM in a bandwidth limited system.

1.3.2 Implementation of 8PSK

PDM-8PSK is a very attractive multi-dimensional multi-order modulation format, mainly because (1) by adding a first-order (0, π/4) phase modulator (PM) after a QPSK modulator, it produces the 8PSK signal we need; (2), since the amplitude value of 8PSK remains constant, compared to other 64-ary modulation formats, such as 8QAM, 8PSK has better fiber nonlinear tolerance; (3), the developed DSP algorithm for QPSK is also applicable to 8PSK.

Figure 1.10 is a block diagram of the 8PSK transmitter structure. The 8PSK transmitter in the serial structure shown in Fig. 1.10a is composed of two MZMs (MZM1 and MZM2) and two phase modulators (PM1 and PM2) [2]. The first three

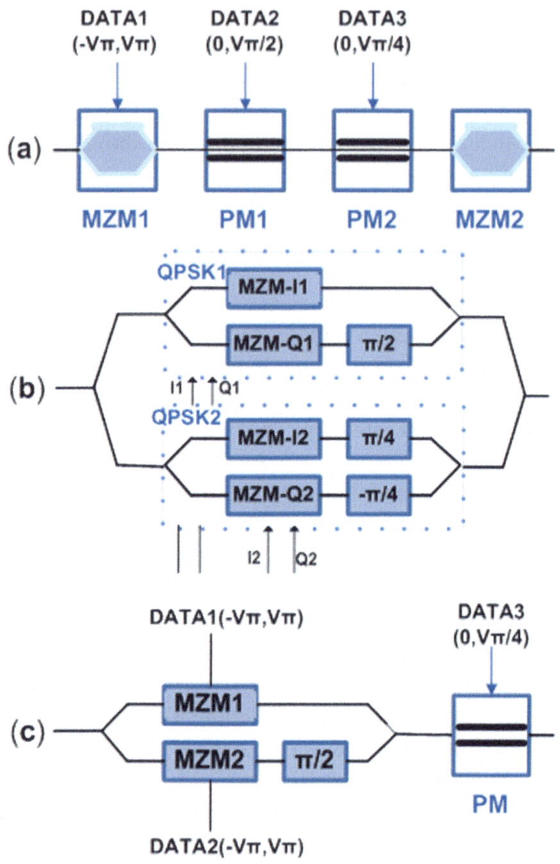

Fig. 1.9 Block diagram of 8PSK transmitter

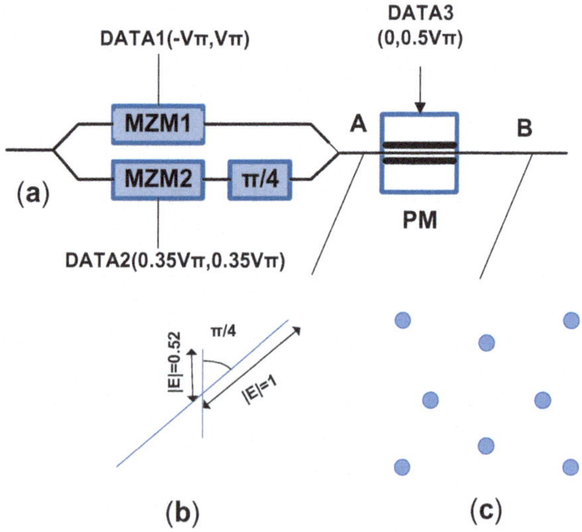

Fig. 1.10 Structure of 8QAM modulation

modulators MZM1, PM1, and PM2 provide phase modulation of 0/π, 0/0.5 π, and 0/0.25 π, respectively, and the outputs after three modulators are, respectively, $E_{\text{out}} = E_{\text{in}}e^{j\pi c_k}$, $E_{\text{out}} = E_{\text{in}}e^{j\pi(c_k+0.5d_k)}$ and $E_{\text{out}} = E_{\text{in}}e^{j\pi(c_k+0.5d_k+0.25e_k)}$. The middle three-way input data bit sequence, the whole structure is equivalent to adding a first-order (0, π/4) phase modulator (PM) behind a QPSK modulator, and finally achieve 8PSK modulation. MZM2 is responsible for shaping the RZ-8PSK signal with a duty cycle of 50%. If only NRZ-8PSK needs to be generated, MZM2 can be omitted. To generate a PDM-8PSK signal, simply add a level 1 polarization-multiplexing device after Fig. 1.10a.

However, this strategy of adding a primary (0, π/4) phase modulator to the 8PSK signal after the QPSK modulator is not perfect. The reason is that in this conventional mode of generating 8PSK signals, phase shift changes will cause nonlinear effects, and to some extent, unnecessary signal chirp will be introduced. Figure 1.10b is a new method of generating an 8PSK signal that produces an 8PSK signal by using a new modulator (QPMZM) with four parallel MZM structures [3]. In Fig. 1.10b, each MZM in the QPMZM structure has a bias point of 0, and the NRZ data sequence drive signal has a fluctuation amplitude of 2Vπ, so each arm generates a BPSK signal, and the bias voltages of these arms are different. Specifically, nπ/4 ($n = -1, 0, 1, 2$). As shown in the figure, the optical phase offsets of MZM-I1, MZM-Q1, MZM-I2, and MZM-Q2 are 0, π/2, π/4, and −π/4, respectively. Therefore, when the input three-way data bit sequence is encoded to generate I1, I2, Q1, Q2 (the encoding rule is: I1 = c_k, Q1 = d_k, when I1 XOR Q1 = 1, I2 = d_k, Q2 = e_k; When I1 XOR Q1 = 0, I2 = e_k, Q2 = e_k), two QPSK signals can be generated by combining [MZM-I1, MZM-Q1] and [MZM-Q2, MZM-I2], respectively, two QPSK signals. The optical phase offset of the signal becomes π/4, and then the two QPSK signals are superimposed to generate an 8PSK signal. The 8PSK signal generated by this method does not generate excessive frequency chirp, and therefore, the transmission between symbols is an ideal linear orbit, and the transmitter is highly integrated. Figure 1.10c is a combination of the two methods of Fig. 1.10. It is a serial-to-parallel structure mode. First, the QPSK signal is generated by the parallel mode, and then the π/4 phase modulation is performed. Under the action of the rotation of the device, an 8PSK signal is generated.

From the brief analysis of the above implementation, we can know that using the modulation device that is now commercially available, we can achieve 8PSK modulation by cascading. This method of synthesizing signals is easy to implement, but the device integration is not high, and the performance is not high. It is not fully optimized; the synthesis of signals by parallel methods can reduce or even eliminate the nonlinear effects caused by frequency chirp, optimize the integration and performance, and impose more stringent requirements on the cost and process of optical devices. It is worth noting that although the nonlinear effect due to frequency chirp can be reduced using the parallel method, the nonlinear effect caused by the lanthanum strontium hydride device itself cannot be eliminated (the power transfer function of the MZM modulation device has sinusoidal characteristics). Currently, we can employ predistortion (transmitter side) or use coherent reception and matching related algorithms (receiver side) to reduce the nonlinear effects caused by the device itself. In the context of existing processes and technologies, it is not practical to synthesize signals by the parallel method of Fig. 1.10b. The signals synthesized by the serial method of Fig. 1.10a also need to be improved in terms of performance and integration. Although the method of generating Fig. 1.10c inherits the advantages of the previous two methods to a certain extent, the improvement in performance is greater than the system complexity brought about by the system. Therefore, in the current actual experiments, most of them use the generation methodology shown in Fig. 1.10a.

1.3.3 Implementation of 8QAM

The generation of the 8QAM signal is not as straightforward as the 8PSK because the amplitude is also modulated while the phase of the light field is modulated. The required 8QAM signal can be easily generated using the power domain arbitrary waveform generator (AWG), but it is difficult to achieve in full-light situations. Figure 1.11 is an 8QAM modulation scheme based on serial structure under all-optical conditions [2]. The 8QAM modulator consists of a $\pi/4$ biased double-balanced MZM modulator and a (0, $\pi/2$) phase modulator. The $\pi/4$ biased double-balanced MZM modulator is similar to the QPSK modulator: the two balanced MZMs in the structure are biased at 0; the difference is that the MZM2 drive signal fluctuates by only 0.7 $V\pi$. The constellation diagram generated by superimposing the two signals is shown in Fig. 1.11b. Then, under the rotation of the 0.5 π phase modulator, the 8QAM constellation shown in Fig. 1.11c is generated to produce the 8QAM signal. It is worth noting that the MZM2 fluctuation amplitude in the figure is set to $V\pi$, and a 5.7 dB optical attenuator is cascaded thereafter to achieve the same modulation effect.

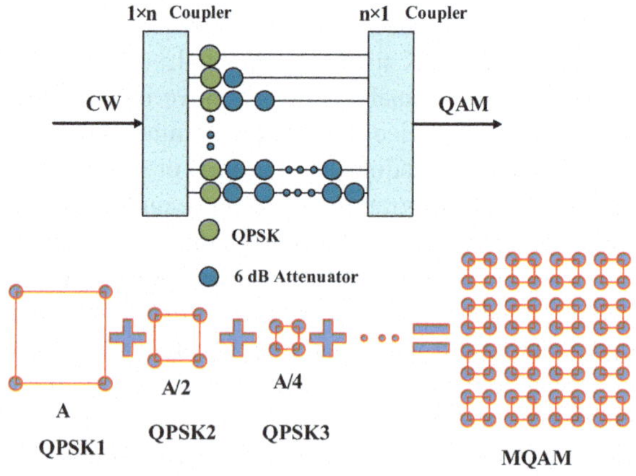

Fig. 1.11 Structure of high-order QAM

Another way to generate 8QAM is to use digital analog converter (DAC) to generate 8QAM. Currently, DAC bandwidth can get up to 40 GHz and sample rate over 120 GSa/s. It means that we can generate over $120 \times 3 \times 2 = 720$ Gb/s polarization-multiplexing optical signals.

1.3.4 Implementation of 16QAM

Currently, there are three main strategies to implement 16QAM modulation. One is to use an arbitrary waveform generator (AWG), where the input binary electrical signal is sent to the AWG to generate a fourth-order intensity signal, and the I- and Q-lights with phase differences of 90° are multi-order intensity modulated by an IQ modulator, respectively. Thus, the IQ light aliasing produces a constellation point of 0 to $\pi/2$ interval, and the AWG outputs an electrical signal to drive the cascaded phase modulator, so that the phase of the optical signal output by the IQ modulator is rotated, thereby causing four constellation points to be distributed. In the constellation area, the 16QAM optical signal is finally generated [5].

The other strategy is to utilize QPMZM, which has a structure similar to Fig. 1.10b. The BPSK signal is generated by MZM on each arm of the modulator, and the BPSK signals on the two arms are combined to form a QPSK (QPSK1, QPSK2) signal, which is attenuated. The two QPSK signals have different amplitudes, and then the two QPSK signals are coupled to generate 16QAM signal. The general model is shown in Fig. 1.12, where when $n = 2$, the 16QAM optical signal is modulated [4].

In these two schemes, new optical devices such as AWG and QPMZM are required, which imposes stringent requirements on the technological level of optical devices. The 16QAM signal can also be generated by a new generation method using commercially available optical devices, as shown in Fig. 1.13 [2].

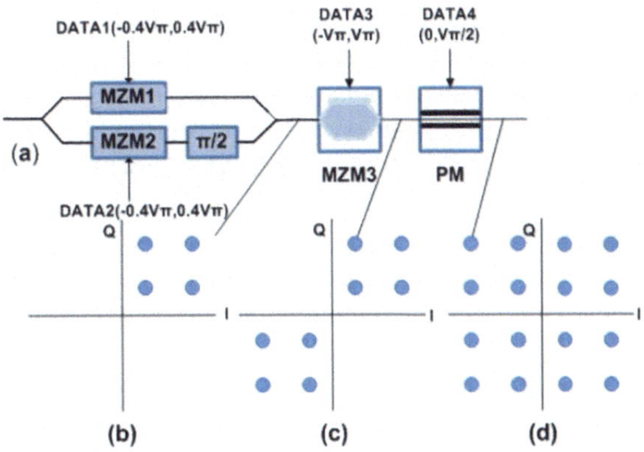

Fig. 1.12 Structure of parallel 16QAM

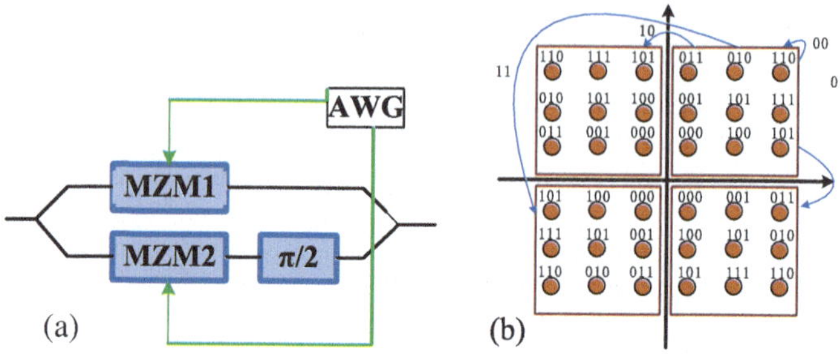

Fig. 1.13 **a** Schematic diagram of 36QAM **b** 36QAM bit mapping based on orthogonal difference coding

The offset of the double-balanced MZM in the scheme of Fig. 1.13 is π/2, where MZM1 and MZM2 are both biased at 0.6 π, and the driving fluctuation peak-to-peak value is 0.8 π. The double-balanced MZM structure produces a right-angle 4QAM constellation point as shown in Fig. 1.13b. After passing the 4QAM signal through the phase modulator MZM3 set to (0, π), a special 8QAM constellation point shown in Fig. 1.13c is generated, and finally, under the rotation of the (0, π/2) phase modulator, 16QAM constellation points are generated. From the 16QAM generation process, we can see that, similar to the 8QAM scheme shown in Fig. 1.11, the scheme first uses the double-balanced MZM to generate the constellation points in a certain quadrant, and then realizes the rotation transformation of the constellation points through the phase modulator, finally making the constellation point fill the entire space.

1.3.5 High-Order QAM

In order to rationally utilize limited bandwidth resources and further improve the spectral efficiency of optical signal transmission, higher order QAM modulation has received much attention in recent years. At present, QAM modulation formats with higher spectral efficiency that have been modulated in the laboratory are mainly 32QAM, 36QAM, 64QAM, 128QAM, 256QAM, 512QAM, and 1024QAM.

36QAM [6, 7]

Due to the QPMZM method described in Sect. 1.3.4, only QAM signals with order of 4^m can be generated. Therefore, the current 32QAM is implemented by the AWG. A commercial AWG can generate a 6th-order intensity signal. Therefore, for 32QAM, the in-phase component and the quadrature component cannot be the highest order simultaneously. Otherwise, a 36QAM signal is generated.

By commercial AWG, a 36QAM electrical signal can be easily generated. In addition, 36QAM can achieve higher spectral efficiency than 32QAM through a suitable coding scheme. Therefore, with the development of technologies such as pre-equalization and post-equalization, 36QAM has also gained certain attention in recent years. Figure 1.14a is the schematic diagram for the generation of optical 36QAM: through the AWG, the 6th order intensity signal is generated simultaneously in the I and Q paths, and the 6th order analog signal is output; the AWG output and its output delay inversion signal, respectively, drive the IQ modulation. The I and Q paths of the device generate a light 36QAM signal. Although, theoretically, 36QAM can achieve higher spectral efficiency than 32QAM, it requires very complex multi-dimensional signal encoding/decoding technology. In the reported transmission experiment about 36QAM, in order to simplify signal processing, the receiving side maps the 36QAM symbols to 5 bits (same as 32QAM). Figure 1.14b is a schematic diagram of the bit map of 36QAM based on orthogonal differential coding [6]. Generally speaking, the first two bits of the mapped bits are determined by the phase position of the measured symbols, and the last three bits are determined by each quadrant. The position of the nine constellation points is determined.

At the transmitter end, the linear band-limited effect caused by the AWG and IQ modulators is compensated in the frequency domain based on the measured transfer function and can be compensated by static pre-equalization techniques

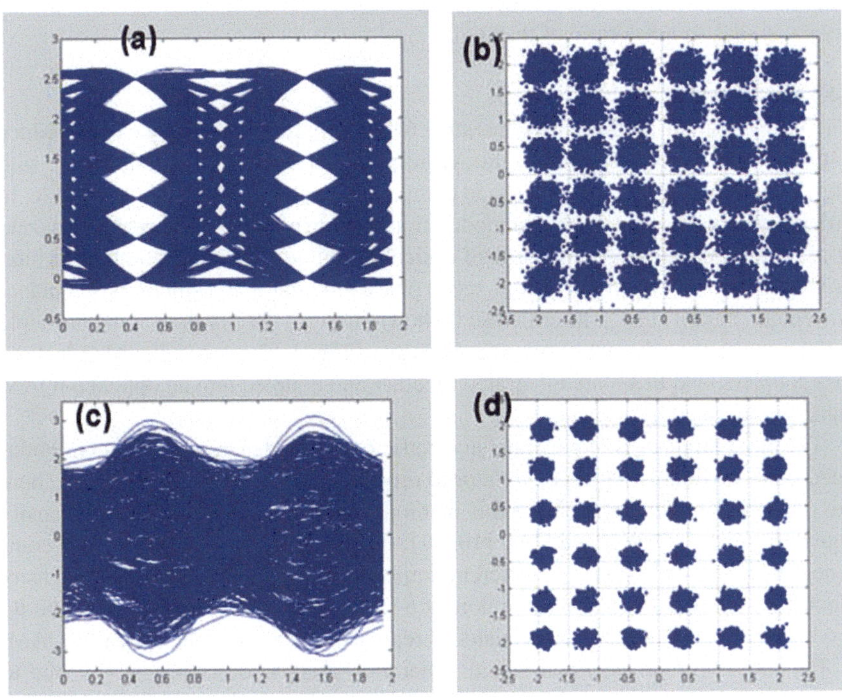

Fig. 1.14 **a** Waveform of electric field after post-equalization **b** optical constellation after post-equalization **c** waveform of electric field after pre- and post-equalization **d** optical constellation after pre- and post-equalization

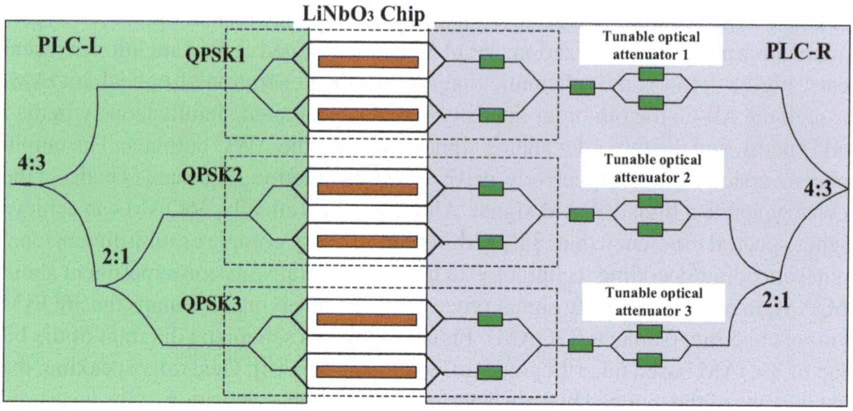

Fig. 1.15 Structure of 64QAM optical modulator

without the need for a feedback loop. As shown in Fig. 1.15 [7], the effect of using both pre-equalization and post-equalization is better than using only post-equalization. This is mainly because, for the filter band-limited effect of the transmitter, the post-equalization enhances the noise component, thereby reducing the optical signal-to-noise ratio of the signal.

64QAM [8–11]

There are two main methods for generating 64QAM optical signals. One is to produce 64QAM signals with all-optical light. At this point, an integrated optical module with high-speed photoelectric response and complex optical structure is indispensable. In order to obtain the integrated light modulation module of 64QAM, Hiroshi Yamazaki et al. developed a 64QAM integrated optical modulator [8] using the hybrid integration technology of $LiNbO_3$ high-speed phase modulator and silicon-based planar waveguide circuit (PLC). Its structure is shown in Fig. 1.16. As shown, the principle is the same as the schematic structure diagram of Fig. 1.12: six parallel MZMs, low-loss couplers, and PLCs are integrated together and coupled into an optical 64QAM signal by three QPSK signals.

The other strategy is to generate an electric 64QAM signal, which is then modulated onto the light by an IQ modulator to produce an optical 64QAM signal. There are two main implementations of this method: (a) producing an 8th-order intensity signal through the AWG and DAC [10, 11]; (b) using an electrical coupler to combine three electrical signals of different amplitudes to obtain an 8th-order intensity electrical signal. The in-phase component and the quadrature component of the IQ modulation are, respectively, modulated to realize modulation of the optical 64QAM.

For the 8th-order electrical signal generation method described in (b), due to the reflection of the connector, the cable, and the electrical coupler, it is difficult to obtain high quality and high speed through a single discrete component in a normal laboratory environment. **Order electrical signal**. However, this difficulty can be overcome by using the latest popular RoF technology to produce high quality

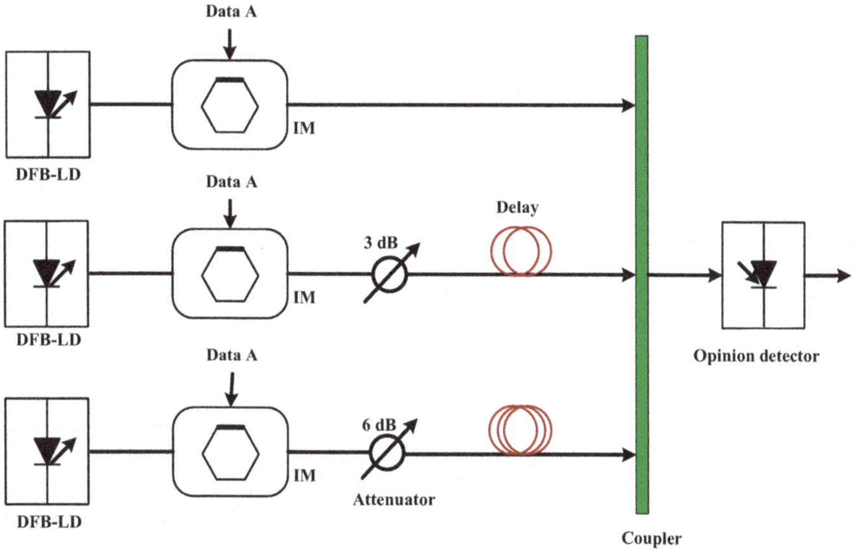

Fig. 1.16 Schematic of generation of E-O-E 8 order electrical signal

multi-level optical signals. Figure 1.17 is the schematic diagram of the EOE scheme [9]. After three different light sources are modulated by different IMs, two of them are attenuated by 3 dB and 6 dB, respectively, and then three optical signals with different amplitudes are generated and are coupled by optical couplers to generate the 8th order optical signal; then the 8th order optical signal is passed through a PD detector to obtain an 8th order electrical signal. In the structure, the optical delay can be used to perform synchronous control of three signals. It is worth mentioning that in order to drive the IQ modulator to generate the optical 64QAM signal, the 8th order electrical signal needs to meet certain amplitude requirements. Therefore, the generated 8th order electrical signal can be amplified by an electric amplifier and then injected into the IQ modulator. The drive produces a light 64QAM signal. With this configuration, a stable 8th-order electrical signal can be generated, and its rate is limited only by the bandwidth of the optoelectronic device such as IM, PD, and the like.

Three schemes for implementing optical 64QAM modulation are introduced. The first one is an all-optical modulation scheme, which is almost independent of the rate limit of the electrical device, and can realize high-speed, high-stability optical 64QAM modulation, but its integration technology is complex and expensive. The process is complicated and the cost is high; while the AWG can easily realize the modulation of the optical 64QAM, and by adjusting the order of the output signal of the AWG, modulation of various optical modulation formats such as 32QAM and 128QAM can be realized, but this structure is easily limited by DAC rate and accuracy. Although the E-O-E method can generate high-speed and stable 8th-order

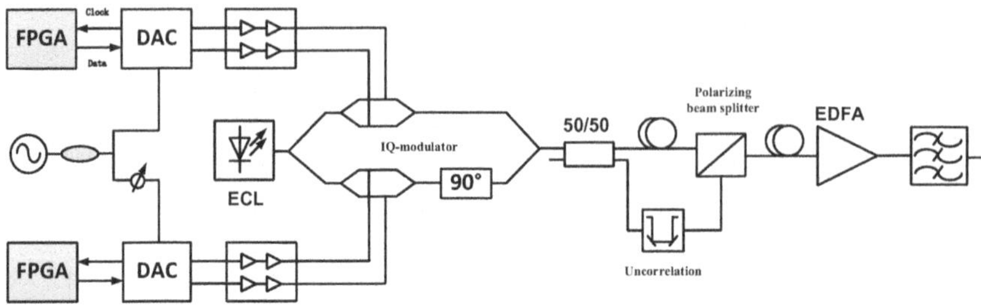

Fig. 1.17 Schematic of software-defined multi-modulation formats transceiver

electrical signals, due to the photoelectric conversion in the structure, the power consumption is large, and the signals of three different amplitudes need to be synchronized, where synchronization control is difficult. Therefore, what kind of structure is chosen to implement optical 64QAM modulation requires trade-offs between cost, speed, stability, complexity, and power consumption. But so far, the 64QAM optical transmission experiment reported mainly uses AWG to achieve modulation.

1024QAM and Higher [12, 13]

At present, 1024QAM and higher order QAM can be generated by commercial AWG or DAC [12, 13]. The implementation of high-order QAM modulation imposes stringent requirements on device levels and processes. Therefore, in the current laboratory environment, most transmission experiments rely on AWGs and DACs.

1.3.6 Comparison of Multi-dimensional Multi-order Modulation Formats

Based on the brief analysis of the above-mentioned multi-dimensional multi-order modulation format implementation, we can know that by using the commercially available modulation device, we can realize the modulation of QPSK, 8PSK, 8QAM, 16QAM by cascading. The integration of this method of synthesizing signals is not high, and the performance is not fully optimized, but it is easier to implement. By synthesizing the signal by the parallel method, the nonlinear effect caused by the frequency chirp can be reduced or even eliminated, the integration degree and performance are optimized, and also the cost and process of optical devices are imposed more stringent requirements, referring to Table 1.2 for the specific comparisons. It is worth noting that although the nonlinear effect due to frequency chirp can be reduced by the parallel method, the nonlinear effect caused by the lanthanum strontium hydride device itself cannot be eliminated (the power transfer function of the MZM modulation device has sinusoidal characteristics). Currently, we can use predistortion (transmitter side) or use coherent reception and matching related algorithms (at receiver side) to reduce the nonlinear effects caused by the device itself.

Table 1.2 Comparison of multi-dimensional multi-order modulation formats

Modulation format	Method	System complexity	Total number of modulators required (MZM, PM)	Integration	Degree of device requirements	If it been implemented in device engineering?
QPSK	Cascade	Simple	2	Low	Low	Yes
	Parallel	Simple	2	Higher	Low	Yes
8PSK	Cascade	Simple	3	Low	Low	Yes
	Parallel	Complex	4	High	High	No
8QAM	Cascade	Complex	3	Low	Low	Yes
16QAM	AWG cascade	Complex	3	Low	High	No
	QPMZM	Complex	4	High	High	No
	Coupling cascade	Complex	4	Low	Low	Yes

For higher order QAM signals, such as 36QAM, 64QAM, and 1024QAM, in order to realize the generation of optical signals, the baseband electrical signals are mainly generated by the AWG and the DAC, and then modulated by the IQ modulator to the light. However, with the concept of "green" being implanted into the hearts of people, more and more attention is being paid to the size and power consumption of devices. The development trend of the integrated high-order QAM modulator is going mainstream in the device development.

1.4 Software-Defined Optical Transceiver

For the next generation of optical transmission systems, increasing bit rate and spectral efficiency is the most fundamental requirement. Therefore, most researches on current optical transmission are focused on improving bit rate and spectral efficiency on diverse services. However, in addition to the high requirements for speed and frequency efficiency, future optical networks also highlight the need for flexibility, and this has received increasing attention. Universally configurable transmitters and receivers supported by software-defined technologies have important implications for the optimal utilization of optical transmission systems and networks, enabling better resource allocation. With software-defined transceivers, optical transmission systems are able to adjust to optimal configuration states in terms of applications, channel requirements, and quality of service.

The rapid development of software-defined optical transceivers (SDOT) is mainly driven by two drivers. On the one hand, recent advances in optoelectronic signal processing have facilitated the advances in transmission technology, maximizing the available transmission fiber bandwidth. On the other hand, the rapid development of real-time DSPs has enabled it to eliminate the obstacles of offline processing. Many new transmission records have contributed significantly to the maturity of SDOT technology [14]. Recently, software-defined optical multi-format receivers have been reported [15], which can convert eight modulation formats within 5 ns without data loss. In addition to SDOT in single-carrier transmission systems, real-time software-defined OFDM transceivers have also been reported [16]. This transceiver uses 64-point Fourier transform (IFFT) and sixteen channel subcarriers of 16QAM data. Signal modulation technology is capable of reaching real-time line rates of 101.5 Gbit/s. The following mainly describes a software-defined transceiver (SDT) in single-carrier optical communication.

1.4.1 Software-Defined Multi-modulation Formats Transceiver (SPOT)

As for software-defined transceivers in single-carrier communication, adapting to multi-format modulation, polarization selection, and forward error correction is a key technology, which is also an urgent problem to be addressed.

Figure 1.18 is a schematic diagram of a typical multi-modulation format transceiver. SPOT generates 8 modulation formats in real time, namely BPSK, QPSK, 4PAM, 6PAM, 8PSK, 16QAM, 32QAM, and 64QAM, and supports baud rates up to 28GBaud. The optical field of the external cavity laser is modulated by an IQ modulator that is electrically driven by the output signals of two 6-bit resolution digital-to-analog converters (DACs) at 28 GHz. At the same time, two FPGA boards that generate pseudo-random sequences are used. Each DAC has a 1:128 divider to provide a 218.75 MHz clock signal to the FPGA. The I and Q channels are independent of each other. These data streams drive the FPGA lookup table generation (LUT) 6-bit binary input, while the 6-bit binary output representing a character drives the DAC at a 28 GBaud baud rate. In order to obtain equidistant constellation points in the optical domain, the nonlinear transmission effects of the Mach–Zehnder Modulator (MZM) should be considered when storing the characters in the LUT, and the irregularly distributed electrical output can be converted to equidistant **distributed optical domain signals**. Switching between different modulation formats is achieved

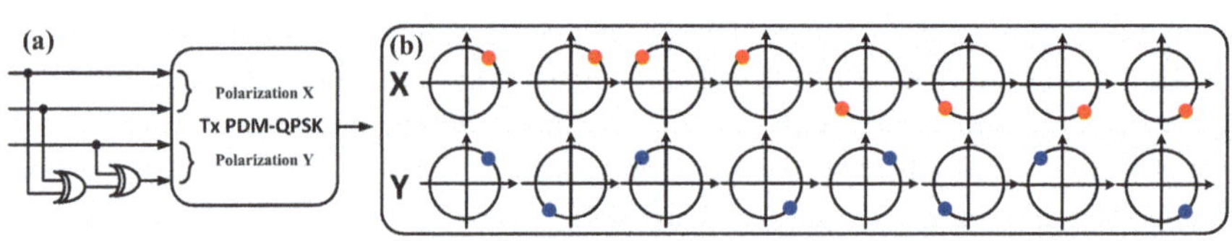

Fig. 1.18 **a** PS-QPSK utilizing hardware configuration of PDM-QPSK **b** correlated output state of PS-QPSK signal in the X-and Y-polarization directions

by overwriting the contents of the LUT. The transform modulation format requires only a single clock cycle (baud rate 28 GBaud, period 5 ns). The 7% FEC guarantees that the signal can always be recovered with BER $\leq 2 \times 10^{-3}$.

1.4.2 Software-Defined Polarization Conversion Transceiver

Transceivers based on software-defined polarization conversion (PS) also have typical applications in long-haul transmission systems [17]. Polarization-converted QPSK format has been widely concerned, and PS-QPSK has better resistance to optical noise than polarization-multiplexed (PDM)-QPSK. In addition, experiments have shown that in the 100 Gbit/s WDM transmission mode, the nonlinear fault tolerance rate is also improved. Despite these advantages, the spectral efficiency of PS-QPSK is not as good as PDM-QPSK, which limits its application in 100 Gb/s optical networks. However, PS-QPSK has a very compelling advantage in that it can be supplemented with the traditional modulation format used in adaptive bit rate systems. In such systems, the data rate of the optical path is based on the capacity requirements and the physical loss that needs to be compensated dynamically. The PS-QPSK can operate at the same baud rate in the same transmitter hardware configuration, and only requires a minor adjustment of the DSP at the receiver side.

The PS-QPSK format is encoded in an 8-state 4-D constellation with 3 bits per character, since this allows two bits to be encoded together on the same polarization (there are four possible states) and the other bit reflects the polarization of the light. For the PS-QPSK signal, not all four QPSK characters can be used if X is independent of the other four characters Y, as shown in Fig. 1.19b. Therefore, a PS-QPSK character can take only 8 states, while a PDM-QPSK character can take 16 states. The special advantage of this representation is that, in the same hardware configuration as PDM-QPSK, a simple scheme for generating PS-QPSK is given [18], referring to Fig. 1.19a.

PS-QPSK has different polarization separation and equalization algorithms compared with PDM-QPSK because its QPSK symbols in two polarization directions are not completely independent. The classical algorithm for polarization demultiplexing in the PDM-QPSK system is the constant modulus algorithm (CMA), which has been widely used in adaptive blind equalizers. However, CMA cannot be used

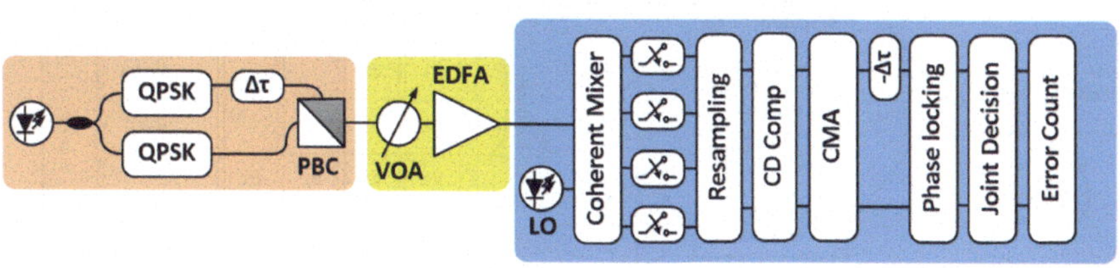

Fig. 1.19 Schematic of software-defined PS-QPSK transceiver

for polarization demultiplexing in PS-QPSK systems. Recently, an improved constant modulus algorithm has been proposed to allow polarization demultiplexing and PS-QPSK equalization [19]. Another solution is delaying the associated data stream driven IQ modulator of the PS and Q polarization of the PS-QPSK transmitter by several symbols, which introduces independence between the polarization branches, and the PS-QPSK signal is like PDM-QPSK. The signal can be easily demultiplexed by the autocorrelation receiver using a conventional CMA. After polarization demultiplexing, the symbol delay is digitally removed to recover the true PS-QPSK signal, as shown in Fig. 1.20. Finally, the Viterbi-based algorithm is used for carrier phase recovery, and the signal is detected by a decision mode based on the minimum Euclidean distance of the received signal point.

Figure 1.20 shows the SDOT schematic based on PS-QPSK. The light generated from the laser source is split into two branches and fed into two QPSK modulators operating at 28 GBaud. According to Fig. 1.19a, the four input data streams that drive the two IQ modulators are the associated binary sequences provided by the FPGA board. A tunable optical delay line is introduced on a branch prior to polarization combining, as previously described.

The setting of SDOT based on PS-QPSK in WDM transmission scenario is shown in Fig. 1.21. The transmitter consists of 80 distributed feedback (DFB) lasers that are divided into 50 GHz intervals and separated into two independently modulated,

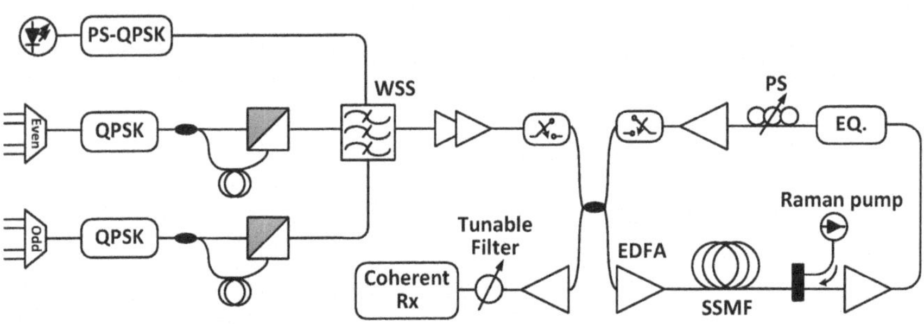

Fig. 1.20 WDM configuration of software-defined PS-QPSK transceiver

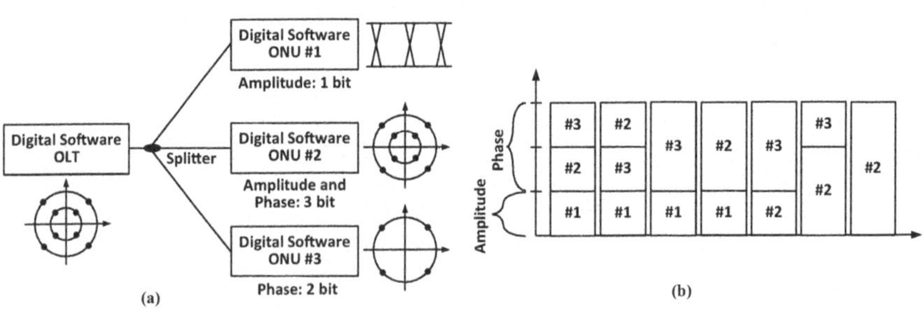

Fig. 1.21 **a** Constellation sharing method **b** application model of digital software PON

spectrally interleaved parity comb filters and a tunable co-laser. The light generated from the parity comb filter is sent to two QPSK modulators operating at 28 GBaud. The modulator is driven by a PRBS sequence at a rate of 28 Gb/s. Polarization multiplexing finally generates PDM-QPSK data at a rate of 112 Gb/s by dividing the QPSK data into two substreams and adding a delay of 300 symbols to them, and then combining them into an optical polarization combiner. The 28 GBaud PS-QPSK signal and the two parity PDM-QPSK signals are fed into the three input ports of a 50 GHz grid wavelength selective switch (WSS). Depending on the configuration of the WSS, the output can be a complete PDM-QPSK signal or a PDM-QPSK signal with a PS-QPSK test signal inserted in the middle of the C-band. As shown in Fig. 1.21, the resulting multiplexed signal is fed into a two-stage erbium-doped fiber amplifier (EDFA) and then sent to a recirculating loop that includes four SSMFs with a 100 km span. All dispersion is digitally compensated by the receiver, while fiber loss is compensated by a hybrid Raman-Erdotron fiber amplifier. Power equalization is achieved by inserting a dynamic gain equalizer (DGE) at the end of the loop. This software-defined polarization conversion scheme can be seen as a powerful complement to software-defined multi-modulation format schemes.

1.4.3 Software-Defined Transceiver of Adaptive Multiplexing PON

In addition to multiple modulation formats and polarization conversion, multiplexing is another potential direction for implementing SDOT, especially in passive optical networks based on digital software. A digital software-based PON that combines software-defined technology with passive optical access will be a strong candidate for future access solutions. Digital signal processing sections are included in optical line terminals (OLTs) and optical network units (ONUs) in such PONs. Therefore, different modulation formats, multiplexing techniques, and compensation methods can be applied and switched. This provides efficiency of use and broad flexibility for a wide range of service requirements with limited system bandwidth. In addition, because the small hardware changes will increase the total capacity of the system, it is very cost-effective in the long run.

Figure 1.22 shows the principle of constellation sharing, a novel multiplexing method based on digital software PON. Constellation sharing allows the OLT to send different data to multiple ONUs simultaneously. Therefore, this function is similar to OFDM and single-carrier subcarrier multiplexing (SCM). In this method, the capacity of the ONU is divided by a number of bits of multi-level modulated signals transmitted by the OLT. The allocation of this bit is determined by the discrimination rules used by each ONU. Figure 1.22a shows an example of a digital software PON system application constellation sharing method. The QAM signal is sent from the OLT to all ONUs via a passive optical splitter. ONU #1, 2, and 3 distinguish signals by the following methods: only by amplitude; by amplitude and phase; only by

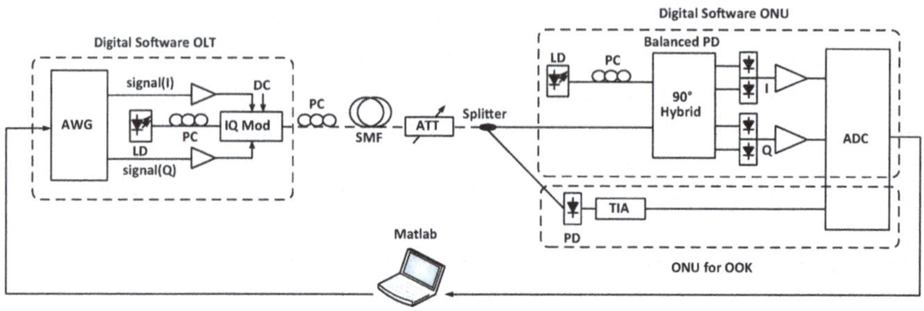

Fig. 1.22 Schematic of software-defined transceiver used for PON

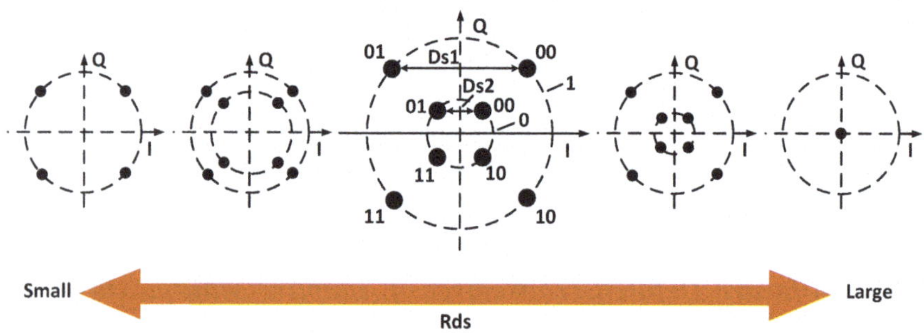

Fig. 1.23 Constellations of different modulation formats of digital software-defined PON

phase. When each ONU receives all the data it can discern, the frame is recovered and sorted bit by bit, depending on the destination address of the frame header to determine whether to accept or discard. With this method, different modes of bit allocation will be possible, as shown in Fig. 1.22b.

The schematic diagram of SDOT based on constellation sharing is shown in Fig. 1.23, assuming that the condition here is a digital software-based OLT connected to a digital software-based ONU and an ONU that can only accept OOK signals. The digital software OLT includes an IQ modulator. The digital software ONU includes a coherent receiver with a 90-degree mixer and two pairs of balanced detectors. The DSP signal processing is performed offline by MATLAB. To simulate an OOK-ONU, a photodiode (PD) is used.

The modulation format used is shown in Fig. 1.23. Of the 3 bits transmitted, 1 bit is modulated in the amplitude direction, and the remaining 2 bits are phase direction modulation. Ds1 and Ds2 are the distances between the outer circle and the inner circle, respectively. Here, a new parameter Rds is introduced to adjust the ratio of the pitch of the points, that is, Ds1/Ds2. In a digital software-based passive optical network, Rds can be adaptively changed, as shown in Fig. 1.23. As the Rds increases, the distance between the inner and outer rings increases, and the constellation map eventually becomes the open key control (OOK). Therefore, a large Rds value improves the receiving sensitivity in the amplitude direction. In contrast, as Rds becomes smaller, the distance between the inner and outer rings becomes smaller, and the constellation map eventually becomes quadrature phase shift keying (QPSK). Therefore, a small Rds value improves the receiving sensitivity in the phase direction. Hence, there is a trade-off between the amplitude and phase reception sensitivity. The ideal point of this trade-off can be found by adjusting the Rds value. Generally speaking, in a PON system, the usage and transmission environment of the user from the OLT to each ONU are different. Consequently, the required bandwidth is time-varying, and the necessary power budget associated with each ONU location is different. In digital software-based PONs, different Rds change modulation formats depending on actual needs, and can mitigate the effects due to different transmission environments. This highly flexible constellation-based sharing SPOT offers many advantages for PON systems that need to handle time-varying bandwidth requirements and different transmission environments.

1.5 Conclusion

In the field of optical communication, all modulation formats more advanced than OOK can be used as advanced modulation formats. This chapter introduces the advanced modulation format of the single-carrier at the transmitter end in high-speed coherent optical communication. The single-carrier multi-dimensional multi-order modulation format reduces the baud rate of signal transmission by transmitting multiple bits of data information in each symbol to achieve high spectral efficiency transmission. In this chapter, we mainly analyze and compare the transmitter implementations of QPSK, 8PSK, 8QAM, and 16QAM. When choosing the right optical multi-tone transmitter, you need to weigh the impact of factors such as cost, performance, and integration. For the multi-modulation format optical modulation transmitter, the research community has now reached a consensus that the successfully developed integrated chips basically adopt the structure shown in Fig. 1.9a. The maximum rate of a single carrier has reached 800 Gb/s, and the commercial rate has reached 600 Gb/s. The modulation format used is 32QAM and 64QAM mixed modulation format.

References

1. J. Peter, Winzer and Rene´-Jean Essiambre, Advanced optical modulation formats. Proc. IEEE **94**(5), 952–985 (2006)
2. J. Yu et al., Ultra-high-capacity DWDM transmission system for 100G and Beyond. IEEE Commun. Mag. **48**(3), S56–S64 (2010)
3. T. Sakamoto et al., *Electro-Optic Synthesis of 8PSK by Quad-Parallel Mach-Zehnder Modulator*, OFC'09, Paper no. OTuG4
4. T. Sakamoto et al., *High-bit-rate Optical QAM*, OFC'09, Paper no. OWG5

5. A.H. Gnauck et al., 10 × 112 Gb/s PDM 16-QAM Transmission Over 630 km of Fiber with 6.2-b/s/Hz Spectral Efficiency, Proc. OFC 2009, Paper no. PDPB8
6. X. Zhou, J. Yu, M.-F. Huang, Y. Shao, T. Wang, L. Nelson, P. Magill, M. Birk, P.I. Borel, D.W. Peckham, R. Lingle, B. Zhu, 64-Tb/s, 8 b/s/Hz, PDM-36QAM transmission over 320 km using both pre- and post-transmission digital signal processing. J. Lightwave Technol. **29**(4), 571–577
7. J. Yu, X. Zhou, 16 × 107-Gb/s 12.5-GHz-Spaced PDM-36QAM transmission over 400 km of standard single-mode fiber. IEEE Photon. Technol. Lett. **22**(17), 1312–1314
8. H. Yamazaki, T. Yamada, T. Goh, Y. Sakamaki, A. Kaneko, *64QAM Modulator with a Hybrid Configuration of Silica PLCs and LiNbO$_3$ Phase Modulators for 100-Gb/s Applications*, ECOC'09, Paper no. 2.2.1
9. J. Yu, X. Zhou, S. Gupta, Y.-K. Huang, M.-F. Huang, A novel scheme to generate 112.8-Gb/s PM-RZ-64QAM optical signal. IEEE Photon. Technol. Lett. **22**(2), 115–117 (2010)
10. J. Yu, Z. Dong, H.-C. Chien, Y. Shao, N. Chi, 7-Tb/s (7 × 1.284 Tb/s/ch) signal transmission over 320 km Using PDM-64QAM modulation. IEEE Photon. Technol. Lett. **24**(4), 264–266
11. A.H. Gnauck, P.J. Winzer, A. Konczykowska, F. Jorge, J.-Y. Dupuy, M. Riet, G. Charlet, B. Zhu, D.W. Peckham, Generation and transmission of 21.4-Gbaud PDM 64-QAM using a novel high-power DAC driving a single I/Q modulator. J. Lightwave Technol. **30**(4), 532–536
12. M.-F. Huang, D. Qian, J.P. Ezra, *SO.S3-Gb/s PDM-I024QAM-OFDM Transmission Using Pilot-Based Phase Noise Mitigation*, OECC 2011, pp. 752–753
13. Y. Koizumi, K. Toyoda, M. Yoshida, M. Nakazawa, 1024 QAM (60 Gbit/s) single-carrier coherent optical transmission over 150 km. Optics Exp. **20**(11), 12508–12514
14. W. Freude, R. Schmogrow, B. Nebendahl, et al., Software-defined optical transmission, in *Proceedings of 13th International Conference on Transparent Optical Networks (ICTON)*
15. R. Schmogrow, D. Hillerkuss, M. Dreschmann et al., Real-Time software-defined multiformat transmitter generating 64QAM at 28 GBd. Photon. Technol. Lett. IEEE **22**(21), 1601–1603 (2010)
16. R. Schmogrow, M. Winter, B. Nebendahl, et al., 101.5 Gbit/s real-time OFDM transmitter with 16QAM modulated subcarriers, in *Proceedings of Optical Fiber Communication Conference and Exposition (OFC/NFOEC)*, March 6–10, 2011, Los Angeles (CA), USA, Paper OWE5
17. J. Renaudier, O. Bertan-Pardo, H. Mardoyan, et al., Experimental comparison of 28Gbaud polarization switched-and polarization division multiplexed-QPSK in WDM long-haul transmission system, in *Proceedings of 37th European Conference and Exhibition on Optical Communication (ECOC)*
18. M. Karlsson et al., Which is the most power-efficient modulation format in optical links? Opt. Express **17**, 10814–10819 (2009)
19. P. Johannisson et al., Modified constant modulus algorithm for polarization-switched QPSK. Opt. Exp **19**, 7734 (2011)
20. N. Iiyama, S. Kim, T. Shimada, et al., Co-existent downstream scheme between OOK and QAM signals in an optical access network using software-defined technology, in *Proceedings of Optical Fiber Communication Conference (OFC) 2012*, Paper JTh2A.53

Chapter 2
Basic Digital Signal Processing for Single-Carrier Signals

2.1 Introduction

In the past ten years, with the application of communication technologies such as video conferencing and the popularity of mobile Internet, the amount of information has exploded, and higher transmission performance is needed, including high spectral efficiency, high speed, and long-distance optical transmission. Higher requirements have been proposed for the physical layer that is the basis of the entire communication system. Commercial optical communication systems generally use intensity modulation/direct detection (IM/DD), that is, the transmitter modulates the intensity of optical carrier, and the receiver performs envelope detection on the optical carrier. Although this structure has the advantages of simplicity, easy integration, etc., only a few modulation formats such as ASK/PSK can be used, the modulation rate is limited, and upgrading is difficult. In this context, coherent optical communication technology, which was highly hoped more than 20 years ago, has once again been promoted to the desktop. Coherent optical communication systems that explore long-distance and high-spectral efficiency transmission of high-speed optical signals have become an important topic in the field. Digital coherent optical communication is the development direction of high-speed optical communication in the future. By combining coherent reception and high-speed digital signal processing, it can effectively balance various damages such as chromatic dispersion, polarization mode dispersion, and nonlinearity in optical links, which greatly improves transmission performance and increases transmission rate and distance. The use of sophisticated digital signal processing technology can also deal with laser linewidth, frequency offset, and various non-ideal devices in the electrical domain, thus avoiding the use of a series of complex devices such as phase-locked loops [1, 2]; Coherent optical communication systems significantly improve spectral efficiency by encoding in various dimensions, and with the development of high-speed, large-bandwidth

J. Yu and N. Chi, *Digital Signal Processing In High-Speed Optical Fiber Communication Principle and Application*,
https://doi.org/10.1007/978-981-15-3098-2_2

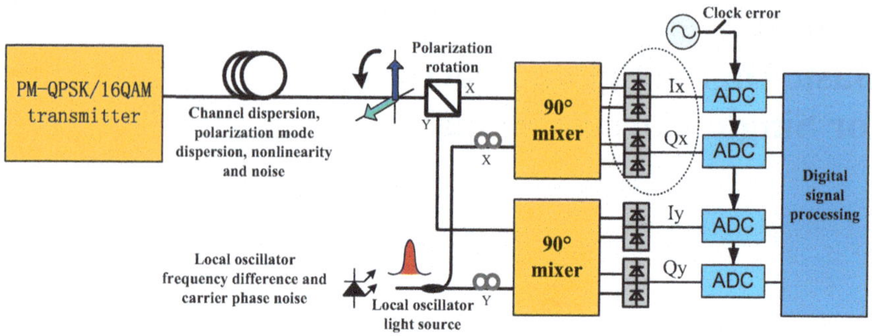

Fig. 2.1 Basic structure of single-carrier coherent receiver

digital-to-analog/analog-to-digital (ADC/DAC) devices, highly integrated and flexible digital coherent light detection technology can effectively achieve high-speed, long-distance, and large-capacity optical communication [3, 4].

The basic structure of the single-carrier coherent receiver is shown in Fig. 2.1. The optical signal $E_R(t)$ transmitted through the optical fiber is separated by the polarization beam splitter (PBS) at the front end of the coherent light receiving beam to obtain two orthogonal polarized lights $E_{R,X}(t)$ and $E_{R,Y}(t)$. Then, two 90° mixers are used to beat the two polarized lights and the receiver local oscillator E_{LO} to achieve optical coherent reception.

The internal structure of the 90 degree hybrid is shown in Fig. 2.2. It consists of four 3 dB couplers and a 90° phase retarder. The 90 degree hybrid is ideally used to output four channels of light in the X-polarization direction. The electric field components of the signal are expressed as

$$\begin{pmatrix} E_1 \\ E_2 \\ E_3 \\ E_4 \end{pmatrix} = \frac{1}{\sqrt{2}} \begin{pmatrix} 1 & 1 \\ 1 & -1 \\ 1 & j \\ 1 & -j \end{pmatrix} \begin{pmatrix} E_{R,x} \\ \frac{1}{\sqrt{2}} E_{LO} \end{pmatrix} = \frac{1}{\sqrt{2}} \begin{pmatrix} E_{R,x} + \frac{1}{\sqrt{2}} E_{LO} \\ E_{R,x} - \frac{1}{\sqrt{2}} E_{LO} \\ E_{R,x} + \frac{j}{\sqrt{2}} E_{LO} \\ E_{R,x} - \frac{j}{\sqrt{2}} E_{LO} \end{pmatrix} \tag{2.1}$$

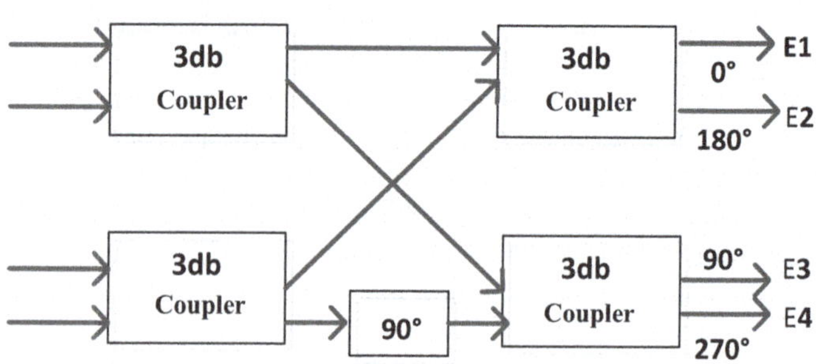

Fig. 2.2 90 degree hybrid structure diagram

Then, the four signals output by the corresponding 90 degree hybrid in each polarization direction are subjected to balanced light detection to obtain in-phase and quadrature components of the polarization direction light detecting current. Taking the X-polarization state as an example, by balancing the receiver photoelectric conversion, the in-phase and quadrature components of the photocurrent after DC-blocking are expressed as follows, respectively:

$$I_X = R\sqrt{2 \cdot P_{LO}} A_x(t) \cos\{2\pi(f_0 - f_{LO})t + \theta_{r,x}(t) - \varphi_{LO}(t)\} \tag{2.2}$$

$$Q_X = R\sqrt{2 \cdot P_{LO}} A_x(t) \sin\{2\pi(f_0 - f_{LO})t + \theta_{r,x}(t) - \varphi_{LO}(t)\} \tag{2.3}$$

where R is the responsivity of the photodiode, P_{LO} and $A_x(t)$, respectively, correspond to the transmit power of the local oscillator signal and the magnitude of the electric field of the received optical signal in the X-polarization direction, f_{LO} and f_0 correspond to the local oscillator source frequency and the originating laser frequency. $\varphi_{LO}(t)$ and $\theta_{r,x}(t)$, respectively, corresponding to the phase of the local oscillator and the phase of the signal on the X-polarization state with the modulation phase of the original signal and the various phase impairments in the fiber transmission. It can be known from Eqs. 2.2 and 2.3 that the balanced reception by coherent detection can completely preserve the amplitude and phase information carried in the electrical signal after photoelectric conversion. But due to the introduction of the local oscillator light source, the frequency and phase of the electrical signal receive the disturbance of the local oscillator frequency f_{LO} and phase $\varphi_{LO}(t)$. Therefore, the corresponding module is designed and estimated in the DSP processing unit at the back end to estimate and compensate for the above disturbance. Finally, the two orthogonal continuous electrical signals of each polarized light are sampled and quantized by the ADC to obtain four discrete digital sequences, which are sent to the DSP unit for processing. From the above process of coherent reception, in addition to the frequency and phase shift introduced by the local oscillator source, other types of system damage will inevitably exist, as indicated in the Fig. 2.3.

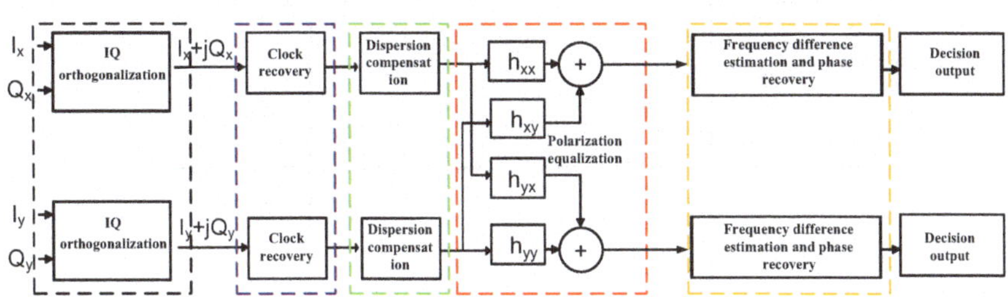

Fig. 2.3 Basic process of digital signal processing

2.2 I/Q Imbalance Compensation and Orthogonal Normalization

Firstly, in the process of optical coherent reception through 90 degree hybrid and photodiode balanced detection, the I/Q two-way bias point is set incorrectly, the internal 3 dB coupler splitter ratio of the mixer is asymmetric or photodiode response, and the rate mismatch that is mentioned above are the reasons causing the amplitude and phase mismatch between the I and Q signals, also known as IQ imbalance.

IQ imbalance on the constellation points of the received signal is shown in Fig. 2.4 under the ideal scenario where the local oscillator frequency is exactly the same as the originating laser frequency where α and θ represent the amplitude and phase mismatch factors, respectively. It is obvious that the existence of α and θ will cause the data in the received Q path to be mixed with the I way information and the amount of mixed information changes with the mismatch factor. Therefore, in the DSP processing flow shown in Fig. 2.3, the received signal is first orthogonalized and normalized to compensate for the above imbalance phenomenon.

Second, the two orthogonal continuous electrical signals of each polarized light must be sampled and quantized by the ADC before being input to the DSP processing unit. But in general, the sampling clock of the ADC is provided by a local clock. Since the local clock is independent of the originating clock and the clock oscillator itself may have undesirable characteristics, there is a significant difference in frequency and phase between the transceiver and the clock. Therefore, the DSP processing flow then introduces a clock recovery module to eliminate clock misalignment due to ADC sampling clock mismatch.

According to the transmission function of the fiber, the static damage of the channel caused by the dispersion of the fiber and the polarization mode dispersion effect will affect the received signal under the premise of neglecting the nonlinear effect of the fiber. In addition, the polarization rotation introduced by the PBS of the front end of the optical coherent receiver will also cause the received signal to

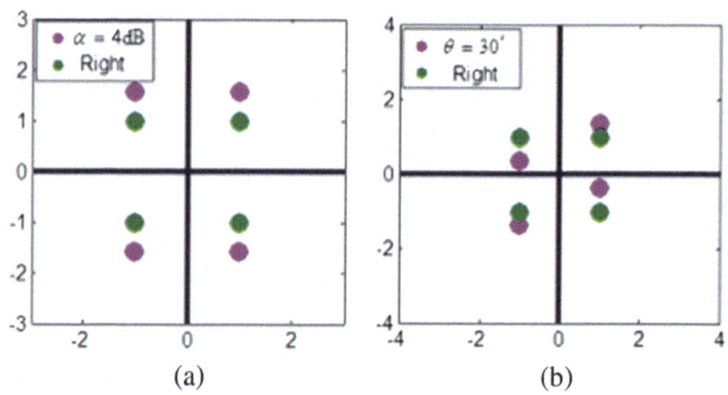

Fig. 2.4 Signal constellation with orthogonal unbalanced damage. **a** Amplitude mismatch **b** Phase mismatch

be affected by polarization crosstalk, that is, the received one-way polarized light may carry the original information on the two polarized lights at the transmitting end. Therefore, dispersion compensation and polarization equalization modules are introduced in the DSP processing flow to eliminate the above effects.

Finally, due to the influence of the light-emitting device manufacturing process, the frequency between the local oscillator source and the originating laser is difficult to maintain completely consistent, and the linewidth of the laser also introduces a corresponding phase shift. A large amount of phase noise is added. The phase component introduced by the frequency offset is a cumulative value. As the sequence number of the symbol increases continuously, the constellation of the received signal appears to be rotated and diverged as shown in Fig. 2.5; and the phase deviation component introduced by the local oscillator source causes the rotated constellation of received signal at a fixed angle as shown in Fig. 2.6. It can be seen that the frequency difference introduced by the local oscillator and the damage of the phase–phase noise of the carrier phase noise cannot be ignored, and it also has a significant impact on the transmission signals of other modulation formats. Therefore, the carrier recovery module in the DSP processing flow first compensates the frequency offset by using the carrier frequency offset estimation algorithm, and then further corrects the phase offset caused by the line width and the residual frequency offset of the local oscillator by way of the carrier phase recovery algorithm, thereby realizes the correctly judged and demodulated received signal.

The subsequent sections of this chapter will introduce each of the principles and implementations of the DSP algorithms used in each module in the complete DSP processing flow, as shown in Fig. 2.3.

Ideally, in coherent detection, the I and Q paths of the received signal are completely orthogonal, but due to the incorrect setting of the I and Q bias points, the

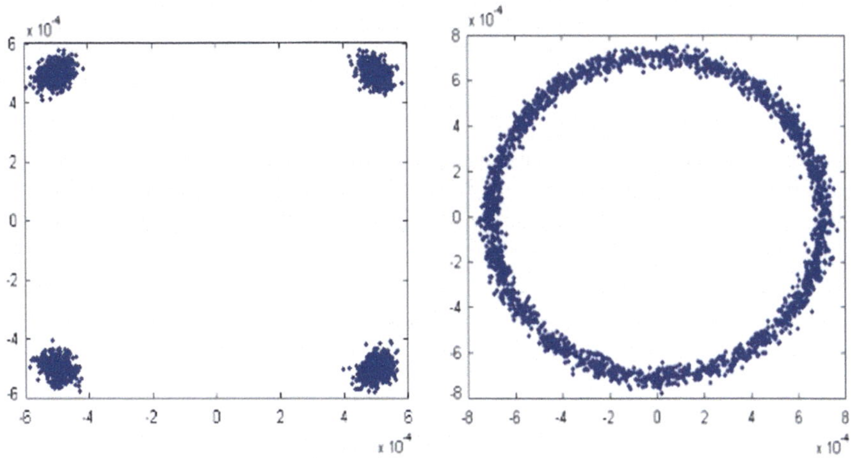

Fig. 2.5 The influence of the frequency offset of the local oscillator on the constellation point of the received signal

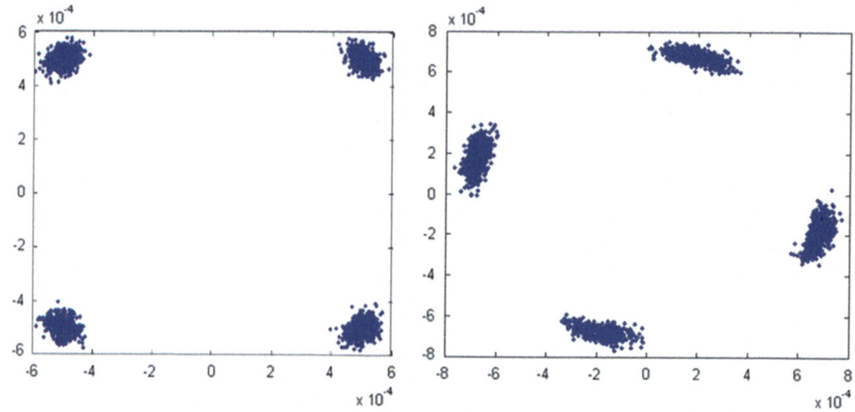

Fig. 2.6 The influence of the local oscillator on the received signal constellation point

asymmetric 3 dB coupler splitting ratio, the photodiode responds rate and polarization controller offset mismatch, factors such as above will produce imbalance of amplitude and phase of the received signal, which will destroy the orthogonality of the two paths of I and Q, deteriorate the performance of the system, and affect the normal operation of the subsequent DSP modules. At the same time, in practice phase diversity receivers are susceptible to imperfect optical 90-degree hybrid, resulting in DC offset and amplitude and phase errors in the received signal. These cause quadrature imbalance, and the received signal needs to be orthogonalized by a corresponding algorithm to compensate for the above imbalance phenomenon. At the same time, normalization is used to compensate for the change in responsiveness of the photodetector.

Ideally, the 90 degree hybrid balanced received signal after reception as

$$\begin{bmatrix} r_I \propto \mathrm{Re}\{E_s E_{LO}^*\} \\ r_Q \propto \mathrm{Im}\{E_s E_{LO}^*\} \end{bmatrix} \tag{2.4}$$

where E_s is the electric field component of the input optical signal; E_{LO}^* is the conjugate of the electric field component of the local oscillator; r_I and r_Q are the I and Q components of the photodetection current, respectively; Re and Im are the real and imaginary parts, respectively.

Since the quadrature imbalance effect introduces the amplitude and phase mismatch between the I and Q signals, the signal output after the 90 degree hybrid is balanced and received as

$$\begin{bmatrix} r_I \propto \mathrm{Re}\{E_s E_{LO}^*\} \\ r_Q \propto \alpha \mathrm{Im}\{E_s E_{LO}^* \cdot \exp(j\theta)\} \end{bmatrix} \tag{2.5}$$

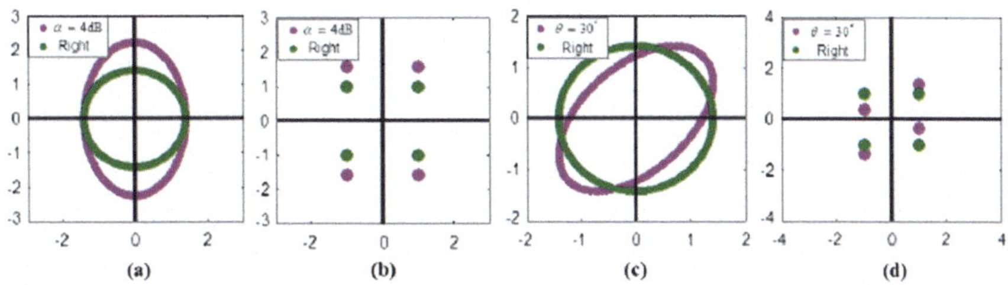

Fig. 2.7 Signal constellation with orthogonal imbalanced damage, amplitude mismatch factor is 4 dB: **a** The signal has frequency offset damage **b** The signal has no other damage, and the phase mismatch factor is 30°. **c** Signal with frequency offset damage **d** Signal without other damages

Among them, α is the amplitude mismatch factor between I and Q; θ is the phase mismatch factor between I and Q. From the constellation diagram, the two types of damage observed, respectively, are shown in Fig. 2.7.

Regardless of the channel damage, the frequency and phase deviation of the laser at the transceiver end, taking the electric field component of the X-polarization state into the above equation, we can obtain

$$\begin{bmatrix} X_I \propto \mathrm{Re}\{E_{s\_x}E_{LO}^*\} = \cos(\pi e_{v,I}) \\ X_Q \propto \alpha \mathrm{Im}\{E_{s\_x}E_{LO}^* \cdot \exp(j\theta)\} = \alpha[\cos(\theta)\cos(\pi e_{v,Q}) + \sin(\theta)\cos(\pi e_{v,I})] \end{bmatrix} \tag{2.6}$$

Among them, $e_{v,I}$ and $e_{v,Q}$ are the original bit sequences. It can be seen from the above equation that due to the existence of the phase mismatch factor, the data in the received Q path is mixed with the information of I path, and the amount of mixed information changes with θ, which will damage the system performance and the subsequent DSP module.

In the literature [5], the relationship between the system BER and the variation of different phase and amplitude mismatch factors after the introduction of the orthogonal unbalance effect is studied by simulation. The simulation data is 112 Gbps PM-DQPSK signal, OSNR = 16.5 dB, CD = 800 ps/nm, PMD = 10 ps, sampling clock deviation SCO = 1000 ppm, transceiver frequency offset = 3 GHz, transceiver line width = 1 MHz, receiving end clock recovery, adaptive equalization, carrier recovery processing, data decision, and statistical system BER. The resulting graph is shown in Fig. 2.8.

As can be seen in the above figure, when the quadrature unbalanced phase mismatch factor $\theta \geq 30°$ and the amplitude mismatch factor $\alpha \geq 3\,\mathrm{dB}$, the system BER begins to deteriorate significantly, and the system needs the corresponding DSP algorithm to compensate for the signal distortion caused by the quadrature imbalance.

In practice, the Gram–Schmidt orthogonalization process (GSOP) and the Löwdin orthogonalization process are usually used to compensate for the quadrature imbalance. Here is an introduction to the two orthogonalization methods.

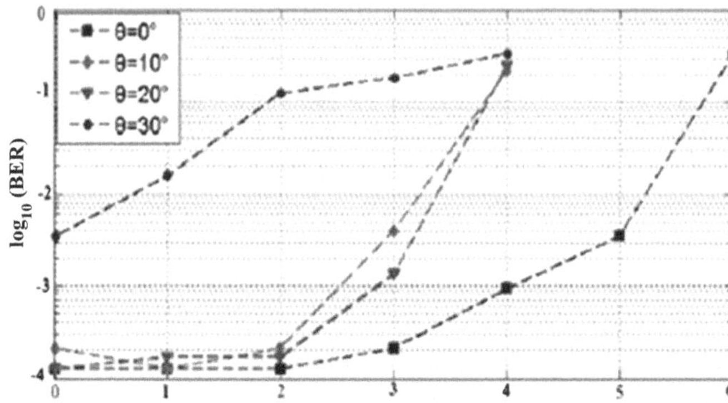

Fig. 2.8 System error rate in the case of introducing different amplitude and phase mismatch factors

2.2.1 Gram–Schmidt Orthogonalization Process (GSOP)

The GSOP algorithm can convert the sample values that do not satisfy the orthogonality into a series of orthogonal sample values, thereby restoring the orthogonality of the I and Q sample sequences. The GSOP algorithm is currently the most commonly used and effective method of compensating for quadrature imbalance. The specific implementation of the algorithm is as follows:

Suppose $r_I(t)$ and $r_Q(t)$ represent the non-orthogonal IQ components of the received signal, while $I_{\text{out}}(t)$ and $Q_{\text{out}}(t)$ represent the two new IQ components that are orthogonal after processing by the GSOP algorithm, then

$$
\begin{aligned}
I_{\text{out}}(t) &= \frac{r_I(t)}{\sqrt{P_I}} \\
Q'(t) &= r_Q(t) - \frac{\rho r_I(t)}{P_I} \\
Q_{\text{out}}(t) &= \frac{Q'(t)}{\sqrt{P_Q}}
\end{aligned}
\tag{2.7}
$$

Correlation coefficient $\rho = E\{r_I(t) \cdot r_Q(t)\}$, $P_I = E\{r_I^2(t)\}$, $P_Q = E\{Q'^2(t)\}$

The GSOP algorithm is an efficient algorithm for correcting orthogonal imbalances. It can fix the I and Q imbalance effects. The comparison of the signal constellation before and after using this algorithm is shown in Fig. 2.9.

The theoretical compensation range of the GSOP algorithm for the phase mismatch problem is $(-90°, +90°)$, and the theoretical compensation range corresponding to the amplitude mismatch problem is $(-\infty, +\infty)$. In practice, due to the influence of noise, dispersion, and other factors, the compensation range of phase mismatch of GSOP algorithm is about $\pm 60°$.

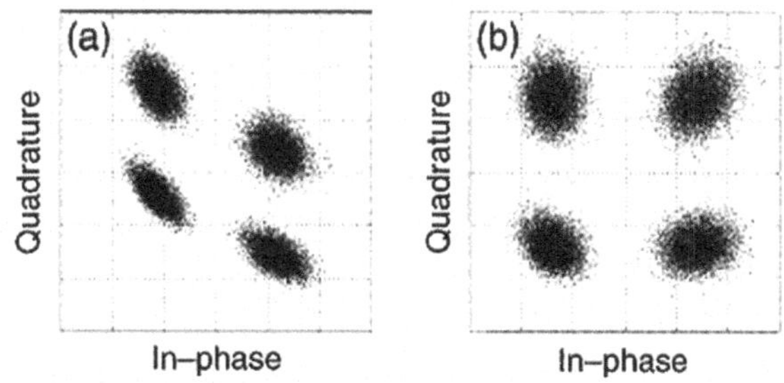

Fig. 2.9 Received signal constellation when the phase mismatch factor is 30°. **a** does not use the GSOP algorithm, **b** uses the GSOP algorithm to compensate for the quadrature imbalance

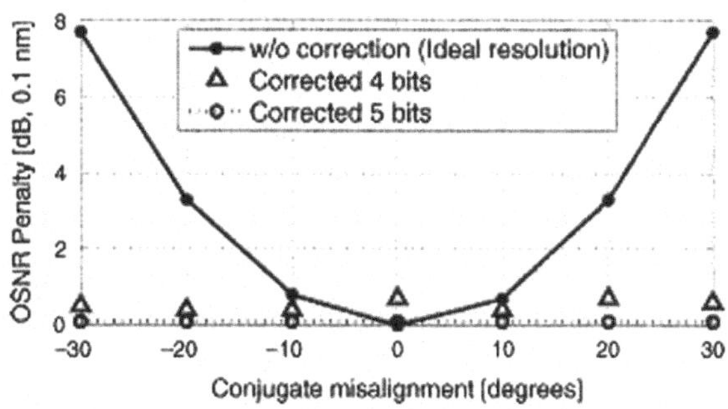

Fig. 2.10 System OSNR loss with different phase mismatch coefficients

The performance of the system using the GSOP algorithm is compared in the literature [ii]. Figure 2.10 shows the relationship between the OSNR loss and the phase mismatch coefficient of the GSOP algorithm.

It can be seen from the figure that when the GSOP algorithm with a resolution of 4 or 5 bits is implemented at the receiving end, the system performance damage caused by the quadrature imbalance can be restored intact.

2.2.2 Löwdin Orthogonalization

Although the GSOP algorithm is an effective method for compensating for quadrature imbalance for current coherent transmission systems, this algorithm increases the influence of quantization noise on the received signal. In contrast, when a symmetric orthogonalization method is employed, the quantization noise is uniformly distributed on the IQ component. The Löwdin orthogonalization process uses a symmetric orthogonalization method. Here is an introduction to the Löwdin orthogonalization process.

It is assumed here that the IQ components of the transmitted signals are s_I and s_Q, which are independent zero mean. Let the received signals be r_I and r_Q, and the mixer will bring the orthogonal unbalance effect. The phase mismatch coefficient is θ, then there are $r_I = s_I \cos(\theta) + s_Q \sin(\theta)$ and $r_Q = s_Q \cos(\theta) + s_I \sin(\theta)$. Thus, we can get the correlation matrix R as

$$R = \begin{bmatrix} \langle r_I^2 \rangle \langle r_I r_Q \rangle \\ \langle r_Q r_I \rangle \langle r_Q^2 \rangle \end{bmatrix} = \begin{bmatrix} 1 & \sin(2\theta) \\ \sin(2\theta) & 1 \end{bmatrix} \quad (2.8)$$

The purpose of the Löwdin orthogonalization method is to generate a series of vectors with a minimum mean square error. If the symmetric matrix is R, then the optimal transformation matrix is $L = R^{-1/2}$, which is to satisfy this orthogonal condition $L^2R = I$. Here, L can be expressed as

$$L = \begin{bmatrix} \frac{\cos(\theta)}{\cos(2\theta)} & -\frac{\tan(2\theta)}{2\cos(\theta)} \\ -\frac{\tan(2\theta)}{2\cos(\theta)} & \frac{\cos(\theta)}{\cos(2\theta)} \end{bmatrix} \quad (2.9)$$

With the increase of the spectral density brought by the high-order modulation format, the Löwdin orthogonalization method can be used to replace the traditional GSOP method to further improve the system performance.

2.3 Fiber Chromatic Dispersion Compensation

The chromatic dispersion of optical fibers belongs to the static linear damage of optical fibers. On the one hand, due to the characteristics of optical fiber materials, when the optical signals transmit signals in optical fibers, the propagation group speeds of different frequency components are inconsistent. This effect is analogous to wireless communication. The multipath effect causes the time domain of the signal to broaden, thereby degrading the signal quality. In the actual system, the dispersion will cause the eye diagram to be blurred, the clock component to disappear, and the inter-symbol interference to cause signal distortion. On the other hand, since dispersion compensation is a kind of static linear damage [6, 7], for a certain rate of optical signal, regardless of the modulation format, as long as the fiber transmission length is constant, the dispersion is constant. Therefore, the dispersion compensation is usually placed in the second step of the digital signal processing equalization.

For single-mode fiber-optic transmission systems, chromatic dispersion is primarily determined by the material dispersion and waveguide dispersion of the fiber. For the early direct modulation direct detection optical communication system, the dispersion compensation module such as dispersion compensation fiber and fiber grating is usually used for optical dispersion compensation. This optical dispersion compensation typically compensates for dispersion in the fiber using a negative dispersion coefficient medium. Developed into a digital coherent optical communication system, digital signal processing can completely replace the module function of optical dispersion compensation. By deriving the time-domain or frequency-domain transfer function of the fiber in reverse, it is easy to perform dispersion compensation in the time domain or frequency domain.

Since dispersion has a significant influence on high-speed fiber-optic communication systems, dispersion compensation technology has become a research hotspot in the field of optical communication. At present, there are mainly three types of schemes for suppressing dispersion in optical fibers: one is dispersion compensation in the optical domain, for example, dispersion compensation fiber (DCF) [5] or photonic crystal fiber (PCF) [8] is used in the transmission link; The electric field is used for dispersion compensation, mainly for various signal processing algorithms [9–11]; the third is to use a new signal modulation format to compress sidebands, such as joint single sideband modulation (SSB) and optical OFDM formats [12, 13]. This chapter focuses on the electric domain dispersion compensation algorithm.

2.3.1 Dispersion Compensation Overview

Since the transmitted signal in the optical fiber contains different frequency or mode components, the signal pulse will be broadened due to different group speeds after being transmitted through the optical fiber, causing signal distortion. This physical phenomenon is called dispersion. Mechanically, fiber dispersion is divided into material dispersion, waveguide dispersion, and modal dispersion. The first two dispersions are mainly present in single-mode fibers, and the latter dispersion is mainly present in multimode fibers. The ubiquity of fiber dispersion causes distortion of the transmitted signal pulses, thereby limiting the transmission capacity and transmission bandwidth of the fiber. For single-carrier coherent optical communication systems, optical domain compensation based on dispersion-compensated fiber (DCF) or digital signal processing (DSP) can compensate for fiber dispersion, but the latter is less expensive. The configuration adjustment is more flexible [14], so the fiber dispersion compensation module is an indispensable part of the digital signal processing flow.

Firstly, based on the mechanism of fiber dispersion effect, the dispersion compensation algorithm is analyzed in detail. From the point of view of the physical properties of dispersion, when light acts as an electromagnetic wave and interacts with the bound electrons of the dielectric, the response of the medium is usually related to the frequency ω of the light wave. Therefore, mathematically, the transmission constant β can be expanded to become a Taylor series at a fixed angular frequency ω_0.

$$\begin{aligned}\beta(\omega) =& n(\omega)\frac{\omega}{c} = \beta_0 + \beta_1(\omega - \omega_0) + \frac{1}{2}\beta_2(\omega - \omega_0)^2 \\ &+ \ldots + \beta_m = \left(\frac{\mathrm{d}^m\beta}{\mathrm{d}\omega^m}\right)_{\omega=\omega_0} \quad (m = 0, 1, 2\ldots)\end{aligned} \tag{2.1}$$

The parameters β_1, β_2 are related to the refractive index n, and their relationship is as follows:

$$\beta_1 = \frac{n_g}{c} = \frac{1}{v_g} = \frac{1}{c}\left(n + \omega\frac{\mathrm{d}n}{\mathrm{d}\omega}\right) \tag{2.2}$$

$$\beta_2 = \frac{1}{c}\left(2\frac{\mathrm{d}n}{\mathrm{d}\omega} + \omega\frac{\mathrm{d}^2 n}{\mathrm{d}\omega^2}\right) \tag{2.3}$$

where n_g is the group refractive index, v_g is the group velocity, and the optical pulse envelope moves at the group velocity. The parameter β_2 represents the group velocity dispersion, which is related to pulse broadening. This phenomenon is called group velocity dispersion (GVD), and β2 is the GVD parameter.

In fiber-optic communication systems, the bandwidth of the modulated signal is much smaller than the optical carrier frequency. In addition, since the refractive index change of the fiber is relatively small within the bandwidth of the modulated signal, the quadratic coefficient in the Taylor expansion is sufficient to reflect the broadening effect of the optical pulse [15].

The degree of pulse broadening caused by fiber dispersion is generally quantified using a dispersion coefficient D in units of ps/nm/km. The chromatic dispersion coefficient D gives the degree of pulse broadening T (in ps) of the optical signal pulse with a unit spectral width ξλ transmitted through a 1 km optical fiber. It can be expressed by the following formula:

$$\Delta T = D \times \xi\lambda \times L \tag{2.4}$$

At the same time, the dispersion coefficient D and the GVD parameter β_2 satisfy the following relationship:

$$D = \frac{\mathrm{d}\beta_1}{\mathrm{d}\lambda} = -\frac{2\pi c}{\lambda^2}\beta_2 \approx -\frac{\lambda}{c}\frac{\mathrm{d}^2 n}{\mathrm{d}\lambda^2} \tag{2.5}$$

The nonlinear Schrödinger equation for optical pulse single-mode fiber transmission is given in [16], as follows:

$$\mathrm{j}\frac{\partial A}{\partial z} = -\mathrm{j}\frac{\alpha}{2}A + \frac{\beta_2}{2}\frac{\partial^2 A}{\partial T^2} - \gamma|A|^2 A \tag{2.6}$$

where A is the slowly varying amplitude of the pulse envelope, which is the time in the reference frame as the pulse moves at the group velocity v_g.

$$\tau = \frac{T}{T_0} = \frac{t - z/v_g}{T_0} \tag{2.7}$$

At the same time, the normalized amplitude U is introduced using the following definition:

$$A(z, t) = \sqrt{P_0}\mathrm{e}^{-\alpha z/2} U(z, \tau) \tag{2.8}$$

The exponential factor in the equation reflects the fiber loss. Using the Eqs. (2.6)–(2.8) and ignoring the last nonlinear term in Eq. (2.6), the normalized amplitude

U(z,τ) of the optical pulse is obtained:

$$j\frac{\partial U}{\partial z} = \frac{\beta_2}{2T_0^2}\frac{\partial^2 U}{\partial \tau^2} \tag{2.9}$$

Substituting the formula (2.5) into the above formula, taking $T_0 = 1$ to obtain

$$\frac{\partial U(z,\tau)}{\partial z} = \frac{-j}{2}\left(\frac{-D\lambda^2}{2\pi c}\right)\frac{\partial^2 U(z,\tau)}{\partial \tau^2} = j\frac{D\lambda^2}{4\pi c}\frac{\partial^2 U(z,\tau)}{\partial \tau^2} \tag{2.10}$$

The above theoretical derivation leads to the partial differential equation of the influence of fiber dispersion on the signal envelope U(z,τ), which is the basis of all fiber dispersion equalization algorithms, where z represents the transmission distance, τ represents the normalized time parameter, and D represents the fiber. The dispersion coefficient, λ represents the wavelength of the light wave, and c represents the speed of light.

A direct solution to the partial differential equations of Eq. (2.10) is to perform a Fourier transform (FFT) to obtain the frequency-domain transfer equation G(z, w):

$$G(z,w) = \exp(-j\frac{D\lambda^2}{4\pi c}\omega^2) \tag{2.11}$$

where ω represents any frequency component.

It is easy to know from Eq. (2.11) that the all-pass filter 1/G(z,w) can be approximated by the digital filter of the infinite impulse response filter (IIR) recursive structure [3] or the FIR non-recursive structure [15], to realize directly compensation of dispersion in the frequency domain. Although the number of filter taps required for dispersion compensation using the IIR filter is much smaller than that of the FIR filter, the inherent recursive feedback structure of the IIR filter makes it almost impossible to achieve in high-speed parallel signal processing, and the phase response is also difficult to design an IIR filter that fully complies with this condition based on the Eq. (2.11), so the FIR type filter is generally used to compensate for the dispersion of the fiber.

2.3.2 *Dispersion Compensation Basic Structure*

In order to better compensate for signal impairments during signal transmission, the signal processing module is usually divided into two parts, the first part of which compensates for polarization-independent losses, such as dispersion, and then the polarization-dependent losses in the second part. Polarization rotation and polarization mode dispersion are compensated. Only the polarization-independent dispersion compensation section is discussed in this section (Fig. 2.11).

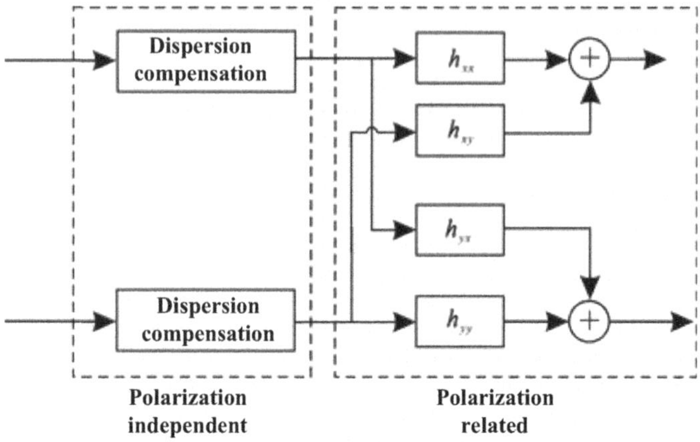

Fig. 2.11 Block diagram of the fiber linear loss compensation algorithm

The reason for adopting this hierarchical structure is that it is considered that the two sub-modules in the module can be operated at different rates after the layered processing, which undoubtedly brings great convenience to the implementation of digital signal processing, such as milliseconds. The lower dispersion is approximately constant and the polarization mode dispersion changes with time. Therefore, the tap coefficients of the filter need not be updated frequently during the dispersion compensation process, which greatly simplifies the computational complexity of the system.

In addition, this algorithm structure is only suitable for the linear damage of the compensation system. For a series of nonlinear damage such as nonlinear phase noise, additional nonlinear filters need to be added for compensation. This structure can well compensate for the chromatic dispersion and polarization mode dispersion loss of the system, and in order to reduce the complexity of the algorithm, both the time-domain equalization algorithm and the frequency-domain equalization algorithm can be used.

It can be known from the above section that the corresponding FIR filter can be designed by the Eq. (2.11) for frequency-domain dispersion compensation. In addition to the direct equalization of the dispersion in the frequency domain as described above, the further Fourier transform is performed on the Eq. (2.11), and the time-domain impulse response is easily obtained, as follows:

$$g(z,t) = \sqrt{\frac{c}{\mathrm{j}D\lambda^2 z}}\exp\left(\mathrm{j}\frac{\pi c}{D\lambda^2 z}t^2\right) \tag{2.12}$$

Therefore, in the time domain, a filter matching the impulse response can also be designed to perform time-domain equalization on the fiber dispersion. The first part of the dispersion compensation is to use the time-domain or frequency-domain equalization algorithm, Fig. 2.11 can be further subdivided into two types of structure

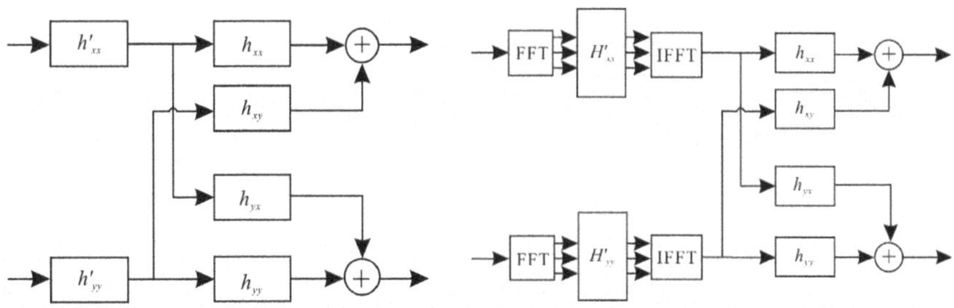

(a) Time domain equalization structure (b) Time-frequency hybrid equalization structure

Fig. 2.12 Double layer structure

as shown in Fig. 2.12a, b. Figure 2.12a shows a pure time-domain equalization structure and Fig. 2.12b shows a mixed structure of frequency–time-domain fusion. The computational complexity and performance comparison of the two structures will be detailed later.

2.3.3 *Frequency-Domain Equalization Algorithm*

The theory shows that the frequency-domain equalization is less complex when the dispersion to be compensated is relatively large. By inverting the sign of the dispersion coefficient D in the dispersion frequency-domain transfer function given by Eq. (2.11), the frequency-domain transfer function of the dispersion frequency-domain compensation filter is obtained:

$$G(z, w) = \exp\left(\mathrm{j}\frac{D\lambda^2}{4\pi c}\omega^2\right) \tag{2.13}$$

As shown in Fig. 2.13, the received time-domain signal symbols are first truncated into several sub-blocks (assuming the symbol length of each sub-block is set to L_f), and the FFT of length L_f is used for each sub-block. The signal is transformed into the frequency domain, and then directly multiplied in the frequency domain with the frequency-domain transfer function as shown in Eq. (2.13) to obtain a dispersion-compensated frequency-domain signal, and then the resulting frequency-domain signal is transformed back into the time domain by an IFFT operation.

The algorithm is described in detail below in accordance with the method of frequency-domain equalization. Assume that the frequency-domain equalizer order is Mf and the FFT point number L_f ($L_f > M_f$). The samples output in the frequency domain are

$$Y_f(1, \ldots L_f) = X_f(1, \ldots L_f)^T \cdot W_f(1, \ldots L_f) \tag{2.14}$$

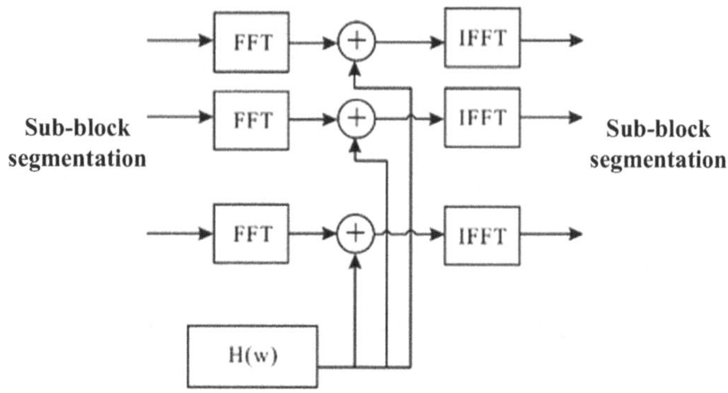

Fig. 2.13 Block diagram of the frequency-domain dispersion compensation algorithm

In the above formula, X_f, Y_f, and W_f are frequency-domain input samples, frequency-domain output samples, and frequency-domain equalizer tap coefficients, respectively, and each vector length is L_f where X_f is the frequency-domain value of two consecutive time-domain sample blocks, one is the sample to be calculated (new), the other is the previous sample (old). After the calculation is completed, the portion of the output $y(k)$ corresponding to the old sample is removed. Here, when $L_f = 2 * M_f$, the size of the old sample block is the same as the size of the new sample block (Fig. 2.14).

M_f: The required equalizer order M_f:

$$M_f = \left\lceil \frac{D\lambda^2 z}{cT^2} \right\rceil + 1 \tag{2.15}$$

where z is the transmission distance, D is the dispersion coefficient of the fiber, λ is the wavelength of the light wave, c is the speed of light, and T is the sample period of the ADC sample.

The detailed process of the algorithm is as follows:

1. Calculate the coefficients of the fixed frequency-domain equalizer offline, storage in group.

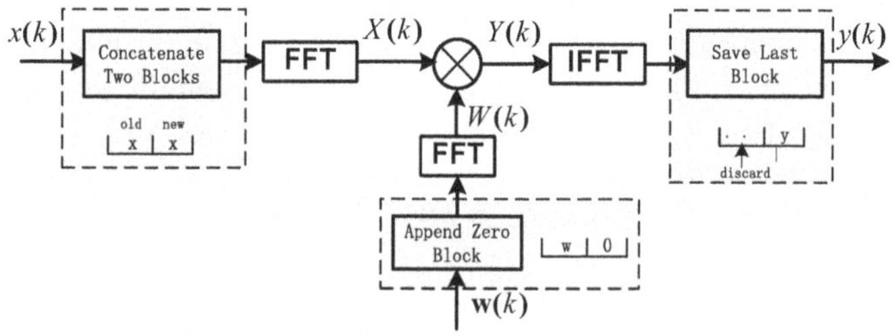

Fig. 2.14 Schematic diagram of frequency-domain filtering method for fixed coefficient FIR filter

2. Perform an FFT operation on the input samples to obtain the frequency-domain value. If the number of FFT points is not 2 M_f, set the number of FFT points to L_f (such as 4096), and input the samples of consecutive L_f. Each sample includes the real part and the imaginary part.

$$X_f(k) = \sum_{n=1}^{L_f} (x_r(n) + jx_i(n)) \cdot exp\left(-2j\pi \cdot (k-1) \cdot \frac{(n-1)}{L_f}\right), \quad 1 \le k \le L_f \tag{2.16}$$

3. Calculate the frequency-domain value of the output:

The formula is the same as (2.14), i.e., $Y_f(1,\ldots L_f) = X_f(1,\ldots L_f)^T \cdot W_f(1,\ldots L_f)$

Among them, $W_f(1,\ldots L_f)$ is the frequency-domain coefficient after the FFT transform of the time-domain coefficient $w(1,\ldots Mt, 0, 0, \ldots 0)$, and the number of the following 0 is also $(L_f - M_t)$. The frequency-domain coefficients are stored offline after calculation. If the number of FFT points is not 2M$_f$, set the number of FFT points to L_f.

4. Calculate the IFFT to get the time-domain value of the output value:

$$x_i(n) = \sum_{k=1}^{2M_f} Y_f(k) \cdot exp\left(2j\pi \cdot (k-1) \cdot \frac{(n-1)}{2M_f}\right) \quad 1 \le \text{n} \le 2M_f \tag{2.17}$$

The single-carrier based frequency-domain equalization method reduces the computational complexity and improves the transmission quality by dividing the signal portion into several sub-blocks. In the case where the equalizer tap coefficient is large, the frequency-domain equalization is obviously due to the time-domain equalization method.

However, in the long-distance optical communication system, the signal sub-block should be large enough to compensate for the dispersion effect generated during the transmission process, and the guard interval is added between the sub-blocks to reduce transmission efficiency. Therefore, the literature [4] proposed the handover. The overlay domain equalization algorithm (O-FDE) generates higher transmission rates by forming an overlap between sub-blocks to improve transmission efficiency.

Figure 2.15 shows the system schematic of the overlapping frequency-domain equalization algorithm (O-FDE). Since the O-FDE algorithm does not use the guard interval, the transmission efficiency is not reduced. The specific process is analyzed as follows.

First, the received signal is divided into Nc symbols and FFT transformed into a frequency-domain signal. Note that there is overlap in this grouping mode, that is, a part of symbols are allocated in two groups. Then through a frequency-domain equalizer (FDE) with a fixed tap coefficient, the weight of the frequency-domain transfer function is determined by the following formula:

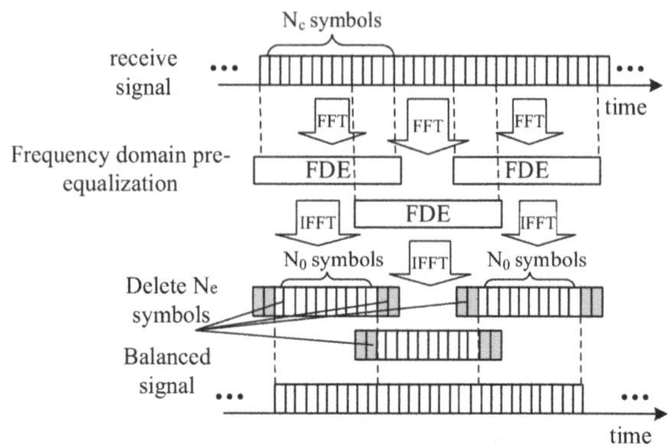

Fig. 2.15 Schematic diagram of the O-FDE algorithm

$$g(f) = \exp\left(-j\frac{\pi cLDf^2}{(f_c - f)^2}\right) \approx \exp\left(-j\frac{\pi cLDf^2}{f_c^2}\right) \tag{2.18}$$

where c is the speed of light, L is the transmission distance, D is the fiber dispersion coefficient, f is the baseband signal frequency, and fc is the center frequency of the optical signal. Due to $f_c \gg f$, the above explanation was made.

Finally, each group of frequency-domain signals is, respectively, subjected to Nc point IFFT transformation. At this time, because the two groups of signals overlap, the Ne symbols of each group are removed and then recombined to obtain the frequency-domain equalization signal. The symbols in the gray part of the figure represent interference between symbol blocks (IBI) and therefore need to be removed during the reassembly process. So far, the main flow of the O-FDE algorithm has been completed.

In addition, the value of Ne needs to be noted. Too small values will cause interference between symbol blocks (IBI). Too large value of Ne will reduce the efficiency of the operation. The range of Ne is given in the literature:

$$N_e > \frac{cLDf_{\max}F_s}{f_c^2} \tag{2.19}$$

where $f_{\max}$ is the maximum frequency occupied by the signal spectrum and Fs is the sampling frequency at the time of analog-to-digital conversion. In [4], the single-carrier coherent optical signal with a rate of 4320 km is transmitted without using the optical dispersion compensation. The experimental results show that when Ne satisfies the Eq. (2.19), the Q factor reaches 11.6 dB.

In addition, in the single-carrier coherent detection system, there is also a class of frequency-domain equalization algorithms based on training sequence assistance [6, 17]. This kind of method considers dispersion as a kind of linear damage, and can use the method of adding training at the transmitting end, using various mature channel estimation algorithms to estimate the tap coefficients of the frequency-domain equalizer, thereby performing dispersion compensation.

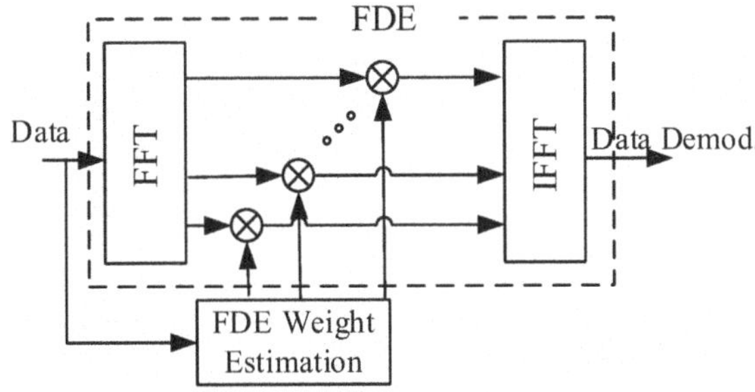

Fig. 2.16 Structure of the FDE algorithm based on the training sequence

The following frequency-domain equalization structure is adopted in [6]. The processing at the receiving end is similar to the previous method, except that the weight determining method of the frequency-domain equalizer is different from before. It adds pilots to the signal block at the transmitting end, determines the tap coefficients of the frequency-domain equalizer by the minimum mean error algorithm (MMSE) in the pilot time, and then uses the fixed coefficient equalizer for frequency-domain dispersion compensation.

In this experiment, by transmitting a single-carrier optical signal of 25 Gbit/s, it is proved that the frequency-domain equalization method has high tolerance for dispersion equalization (Fig. 2.16).

In addition, the literature [17] further shows that the frequency-domain equalizer tap coefficients can also be obtained by means of zero-forcing algorithm (ZF) or matched filter (M_f), and on this basis, the 2*2 MIMO case is analyzed. The dispersion compensation effect of various algorithms is not elaborated here.

2.3.4 Time-Domain Equalization Algorithm

For the time-domain equalization algorithm, the impulse response of the dispersion time-domain compensation filter is obtained by inverting the sign of the dispersion coefficient D in the dispersion time-domain impulse response of Eq. (2.12), as follows:

$$g(z,t) = \sqrt{\frac{c}{\mathrm{j}D\lambda^2 z}}\exp(-\mathrm{j}\varphi(t)), \quad \varphi(t) = \frac{\pi c}{D\lambda^2 z}t^2 \tag{2.20}$$

From the impulse response given by the above equation, the corresponding system response time is infinite and non-causal. Considering that such characteristics may cause aliasing of some sampling frequencies, the time-domain equalization algorithm further cuts the impulse response time of the system. It is a finite length to overcome the phenomenon of frequency aliasing. It is assumed that the receiver samples every T seconds for the received symbol sequence and frequency aliasing may occur when the sampling frequency exceeds the Nyquist frequency $\omega n = \pi/T$. The impulse response

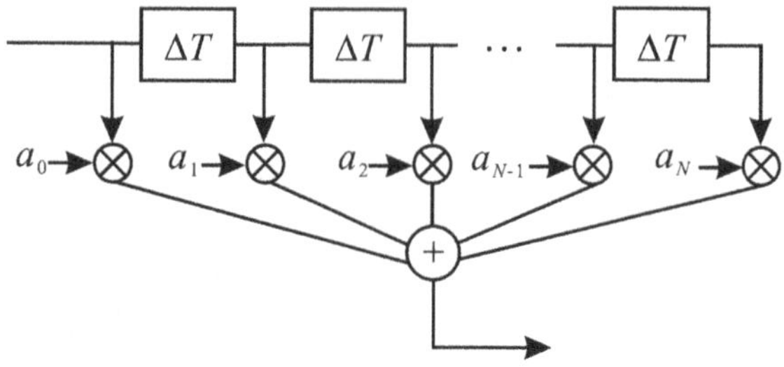

Fig. 2.17 Time-domain dispersion compensation filter structure

is described in the form of a rotation vector whose angular frequency is given by Eq. (2.21):

$$\omega = \frac{\partial \varphi(t)}{\partial t} = \frac{2\pi c}{D\lambda^2 z}t \tag{2.21}$$

When the amplitude of ω exceeds the Nyquist frequency ωn, aliasing occurs, and it is known that time-domain truncation is required in order to avoid frequency aliasing:

$$-\frac{|D|\lambda^2}{2cT} \ll t \ll \frac{|D|\lambda^2}{2cT} \tag{2.22}$$

After the above processing, the impulse response time is truncated to a finite time length, and then the non-recursive structure tap delay FIR filter can be used to implement the time-domain dispersion compensation filter. The filter structure is shown in Fig. 2.17.

The total number of taps of the filter is N, and the weight of the tap is given by

$$a_k = \sqrt{\frac{\mathrm{j}cT^2}{D\lambda^2 z}}\exp(-\mathrm{j}\frac{\pi cT^2}{D\lambda^2 z}k^2)\left\lfloor \frac{N}{2} \right\rfloor \ll k \ll \left\lfloor \frac{N}{2} \right\rfloor N = 2 \times \left\lfloor \frac{|D|\lambda^2 z}{2cT^2} \right\rfloor + 1 \tag{2.23}$$

where $\lfloor\cdot\rfloor$ denotes the rounding operation, it can be seen that the time-domain equalization method not only gives a simple closed interval solution to the filter tap weight, but also gives the upper bound of the number of taps required by the algorithm.

Literature [18] further verified the feasibility of the above time-domain equalization filter. The simulation results show that when the number of filter taps adopts the maximum truncation window length, the weight coefficient of the filter tap is optimal, and the dispersion compensation effect is also the best. However, in the actual coherent receiving system, if the time-domain equalization is used to compensate the fiber dispersion, the maximum number of filter taps given by the Eq. (2.22) can be further cut off under the condition that the system performance is hardly deteriorated. Although the filter tap weight coefficient is suboptimal in this case compared to the maximum number of taps, it provides the possibility to further optimize the system implementation complexity.

This method can be seen as a discrete sampling of the response of the dispersion pulse, known as the time-domain discrete sampling method (TSM). It adds an all-pass dispersion compensation filter to the signal processing process, but this method does not consider the influence of noise. A time-domain equalization method based on the Wiener filter is proposed in [19]. The method can suppress noise while dispersing compensation, and is suitable for multi-rate systems.

The figure shows the basic model of the optical communication system, where $x(n)$ and $\hat{x}(n)$ represent the input information sequence and the output estimation sequence, respectively, and L represents the upsampling and downsampling multiples of the input and output. In order to satisfy the Nyquist sampling law, $F(e^{j\omega T})$ represents the spectral response function of the raised root cosine function (RRC), $G(e^{j\omega T})$ represents the transfer function of the dispersive channel, and $H(e^{j\omega T})$ is the filter response function to be sought. The whole system considers the influence of additive white Gaussian noise.

It is known from the previous analysis that the dispersion signal transfer function $G(e^{j\omega T})$ is

$$G(e^{j\omega T}) = e^{-jk(\omega T)^2}, \omega T \in [-\pi, \pi] \quad k = \frac{D\lambda^2 z}{4\pi c T^2} \tag{2.24}$$

where z is the transmission distance, T is the sampling period, D is the dispersion coefficient of the fiber, λ is the wavelength of the light wave, and c is the speed of light.

As can be seen from Fig. 2.18, the Wiener filter and the subsequent L-time downsampling are designed to better recover the original sequence. The cost function J is defined here as the mean square error function of the estimated sequence $\hat{x}(n)$ and the actual sequence $x(n)$.

$$J = E\left[\|e(n)\|^2\right] = E\left[\left\|\hat{x}(n) - x(n)\right\|^2\right] \tag{2.25}$$

For the sake of analysis, the sampling multiple $L = 2$ is taken here. It can be seen from the figure that $x_0(n)$ represents the signal before the downsampling, i.e., $\hat{x}(n) = x_0(2n)$, then

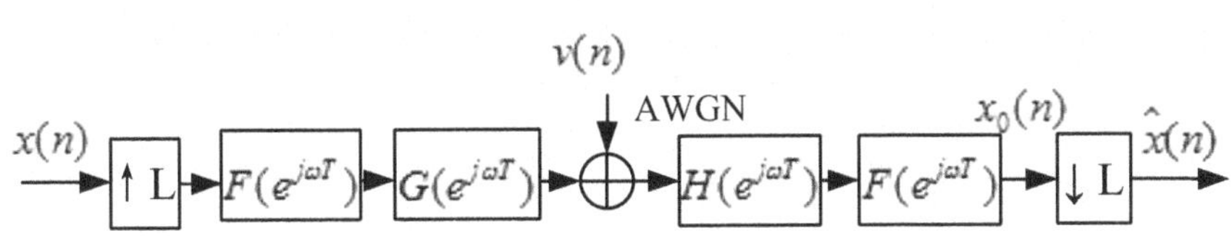

Fig. 2.18 Basic structure of an optical communication system

$$J = E[\|x_0(2n) - x(n)\|^2] \tag{2.26}$$

According to the system structure schematic diagram, it is easy to get $x_0(n) = \sum_k x(k)a(n-2k) + \sum_r v(r)b(n-r)$, where $a(n)$ and $b(n)$, respectively, represent the entire channel impulse response of the signal and noise, and this formula is substituted into (2.26):

$$J = E\left[\left\|\sum_k x(k)a(2n-2k) + \sum_r v(r)b(2n-r) - x(n)\right\|\right]^2 \tag{2.27}$$

Simplify the Eq. (2.27), let $\frac{\partial J}{\partial h^*(m)} = 0 \quad m = -\frac{N-1}{2}, \ldots 0 \ldots, \frac{N-1}{2}$, so that J gets the minimum value. After a series of operations, you can get

$$(R_q + \eta R_f)h = q^* \tag{2.28}$$

$h = \left[h\left(-\frac{N-1}{2}\right), \ldots, h(0), \ldots, h\left(\frac{N-1}{2}\right)\right]^T$, $q = \left[q\left(-\frac{N-1}{2}\right), \ldots, q(0), \ldots, q\left(\frac{N-1}{2}\right)\right]^T$, $q(n) = f(n) \otimes g(n) \otimes f(n)$, $\eta = \frac{P_v}{P_x}$, R_f, and R_q are correlation matrices of $f(n)$ and $q(n)$. Finally, h can be obtained. The time-domain representation of the Wiener filter is

$$h_{\text{Wiener}}(\eta) = (R_q + \eta R_f)^{-1} q^* \tag{2.29}$$

If ss is difficult to obtain η, you can omit ηR_f and simplify it to:

$$h_{\text{Wiener}}(\varepsilon) = (R_q + \varepsilon I_{N\times N})^{-1} q^* \tag{2.30}$$

In the literature [], the 10.7 GBaud 16QAM coherent optical signal is transmitted by simulation, where D = 17 ps/nm/km, λ = 1553 nm, and z = 4000 km. Compared with the previous time-domain sampling method (TSM), the simulation finds that the in the filter, the number of taps N = 111, the error performance of the Wiener filter is better, and the filter length required to reach the error threshold is smaller.

Regarding the time-domain equalization dispersion compensation algorithm, a least square error based dispersion compensation algorithm (LSM) is proposed in recent literature [20], which is based on the frequency response of the actual dispersion compensation filter and the ideal dispersion compensation filter. The squared error is minimized, and the FIR filter is designed. The simulation analysis shows the superiority of the least squares compensation algorithm (LSM) compared with other traditional time-domain sampling equalization (TSM) and frequency-domain sampling equalization (FSM).

2.4 Clock Recovery Algorithm

After photodetection, the electrical signal is digitized by a sampling and digital-to-analog converter ADC. However, in practical systems, since the local sampling clock is not synchronized with the transmitter signal clock, the sampling point of the ADC is not the optimal sampling point of the signal in most cases. On the other hand, due to the instability of the local clock source itself, it is also possible to cause sampling errors of the system. This sampling error includes both phase error and sampling frequency error. The clock error of the sampled signal, on the one hand, introduces inter-sampling interference due to the imperfection of the sampling point; on the other hand, the jitter of the sampling clock also causes fluctuations and fluctuations in signal performance. Therefore, in order to achieve optimal digital signal recovery, a clock recovery module is needed in the actual system to eliminate the influence of clock error. Considering that the dispersion will cause the clock component to disappear, usually, the clock recovery module is placed after the dispersion compensation or interacts with the dispersion compensation module to form a unified equalization feedback module.

In fact, clock recovery algorithms are a common requirement of communication systems and have been widely used in early wired communication and wireless communication systems [3, 6, 7, 15]. In optical communication systems, some classical clock recovery algorithms are also widely used, which proves the universality of these algorithms in coherent optical communication systems [3, 6, 7, 15]. At present, in the coherent optical communication system, several clock recovery algorithms mainly used include feedforward and feedback, and the typical algorithms are as follows:

(1) The digital square filtering recovery algorithm proposed by Martin et al. [21]. This algorithm uses a feedforward clock synchronization structure to redefine the timing error phase by extracting the asynchronous sampling sequence of the signal. Clock, find the best interpolation point to achieve signal clock recovery. This method requires 4 times the resampling of the signal, and the computational complexity is high.
(2) The feedback time-domain clock recovery algorithm proposed by Gardner [22]. This algorithm uses a feedback clock synchronization structure to estimate the phase of the digital clock source by recalculating the feedback error by calculating the timing error. The estimation of the timing error can track the frequency jitter of the signal, and the algorithm can achieve dynamic clock recovery. On the other hand, the Gardner clock recovery algorithm requires only twice the signal sampling and low algorithm complexity, and is widely used in digital signal processing modules of coherent optical communication systems.

Other commonly used clock recovery algorithms include the Godard algorithm [23] and the Muller algorithm [24].

2.5 Channel Dynamic Equalization Algorithm and Polarization Demultiplexing

Since the light field has two independent polarization states, to improve the transmission rate and spectrum efficiency of the optical fiber communication system, optical light with different polarization states can carry different signal bits, so polarization multiplexing (PDM) [5] is realized. The use of polarization multiplexing technology can double the communication rate and spectral efficiency of the system.

On the other hand, with the development of amplifiers, dispersion compensation technology, and nonlinearity management technology, the transmission rate of optical fiber communication system grows rapidly in recent years. Polarization mode dispersion (PMD) becomes the prime factor of restriction system bit distance product [8]. In the transmission process of optical signal, if PMD exists, two optical pulses will be separated with different delay, resulting in pulse distortion, namely inter-symbol interference (ISI) [9]. The damage caused by PMD was first demonstrated and reported by Poole et al. [10]. PMD has become a major obstacle to the development of ultra-high speed and super capacity optical communication systems.

In the real system, the birefringence of optical signals in optical fiber transmission shows the phenomenon of multiple cascade random coupling, which changes constantly according to the environmental temperature, vibration, stress, pressure, and other factors.Therefore, in the actual fiber channel, PMD changes constantly, it is an urgent need to simultaneously equalize the channel dynamically and complete polarization demultiplexing. Since the polarization multiplexing system is actually a 2×2 multiple input and multiple output (MIMO) system, the signal processing can be realized by the traditional channel equalization algorithm, such as constant modulus algorithm (CMA) and direct decision least mean square (DD-LMS), and 2×2 MIMO multiplexing algorithm.

In the following, we will first introduce the basic principle of channel blind equalization, discuss how to realize polarization multiplexing at the receiver end. We will also introduce the implementation of popular CMA to realize polarization demultiplexing of constant envelope modulated signal and compensate for PMD at the same time. Aimed at high-order unconstant modulus QAM modulation, we will introduce the cascading multimode algorithm (CMMA) and the modified cascaded multi-modulus algorithm.

2.5.1 The Basic Principle of Bland Equalization

Channel equalization means that the equalizer at the receiver end produces the opposite characteristics of the channel, which is used to offset the inter-code interference caused by the propagation characteristics of the channel and improve the performance of the communication system. A feasible equalization method is to adopt a

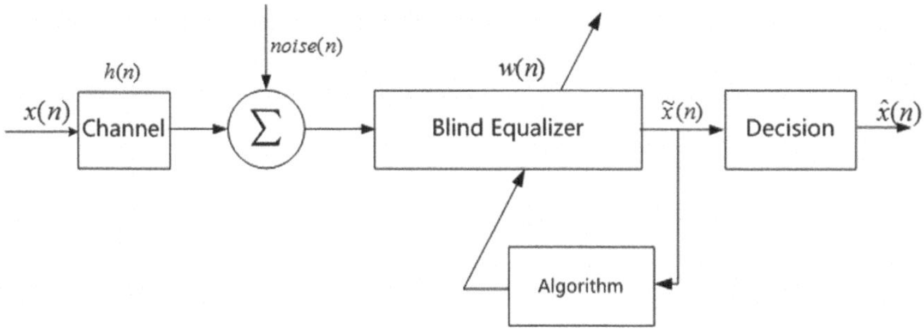

Fig. 2.19 The schematic diagram of blind equalization

data-aided method, which sends a known training sequence before transmitting effective information, and then estimates the characteristics of the channel according to the received signal at the receiving end. However, the blind equalization algorithm does not need to rely on the training sequence, but only uses the received signals to carry out adaptive equalization on the channel, which can improve the effective information transmission rate [11]. Next, we will introduce the basic principle of blind equalization.

The basic principle of blind equalization is shown in Fig. 2.19.

Where $x(n)$ is the originally transmitted signal, it is assumed to be independent and uniformly distributed, stable and irrelevant [11]. $h(n)$ is the unit impulse response describing the channel, which is often unknown. noise(n) is the additive white Gaussian noise (AWGN). $y(n)$ is the originally transmitted signal which deteriorates after transmission. Then, $y(n)$ is sent into the blind equalizer with unit impulse response for processing, the output after processing is $\tilde{x}(n)$, and $\hat{x}(n)$ is the signal output by the decision.

According to the above model, the input signal of the equalizer $y(n)$ can be expressed as

$$y(n) = x(n) * h(n) + noise(n) \tag{2.31}$$

where $*$ reveals for convolution.

When the noise is ignored, the input signal of the equalizer can be expressed as

$$y(n) = x(n) * h(n) \tag{2.32}$$

In order to recover the original transmitted signal from $y(n)$, we need to conduct deconvolution. When the transmitted signal $x(n)$ is a known signal, namely, training sequence, we can calculate $h(n)$ through the observation of $y(n)$, this is a data-assisted approach. However, unlike the data-assist approach, blind equalization does not require training sequences, so it can effectively improve the data transmission rate and reduce communication delay. However, only one parameter is known in

Eq. (2.32), which makes the solution much more difficult. This method of estimating without the aid of data is called blind equalization or blind deconvolution.

The blind equalizer in Fig. 2.19 is an adaptive linear filter whose output can be expressed as

$$\tilde{x}(n) = w(n) * y(n) \tag{2.33}$$

What we want to see is that the equalized signal is exactly the same as the original signal. The relation between the unit impulse response of the optimal blind equalizer and the unit impulse response of the channel can be obtained by substituting Eq. (2.32) into Eq. (2.33):

$$\delta(n) = h(n) * w(n) \tag{2.34}$$

where $\delta(n)$ is the unit pulse sequence.

In practice, $w(n)$ is commonly realized by a finite impulse response (FIR) filter. The length of the FIR filter is L, and the tap coefficient $\mathbf{W}(n)$ is expressed as

$$\mathbf{W}(n) = [w_0(n), w_1(n), \ldots, w_{L-1}(n)]^T \tag{2.35}$$

The input signal $\mathbf{Y}(n)$ is expressed as

$$\mathbf{Y}(n) = [y(n), y(n-1), \ldots, y(n-L+1)]^T \tag{2.36}$$

The output signal $\tilde{x}(n)$ is expressed as

$$\tilde{x}(n) = w(n) * y(n) = \sum_{i=0}^{L-1} w_i(n)y(n-i) = \mathbf{Y}^T(n)\mathbf{W}(n) = \mathbf{W}^T(n)\mathbf{Y}(n) \tag{2.37}$$

where T represents the transpose.

The optimal blind equalizer is infinite, while the length of FIR filter is finite, which will inevitably lead to error, that is, $\tilde{x}(n)$ is only the estimated value of $x(n)$, and the error signal $e(n)$ is

$$e(n) = \tilde{x}(n) - x(n) \tag{2.38}$$

The training process of the blind equalizer is to find a set of optimal tap coefficients $\mathbf{W}(n)$ and to make $\tilde{x}(n)$ and $x(n)$ approximate. When the error signal $e(n)$ approaches 0, the FIR filter tap coefficient is optimal.

2.5.2 Classical CMA Algorithm

According to the basic knowledge of digital communication theory, phase shift keying M-PSK modulation is constant amplitude modulation, and the average transmitting power of the modulated signal is constant. If the equalizer estimates that the signal tends to the original $x(n)$, $\tilde{x}(n)$ should also be constant amplitude. Therefore, we introduce the cost function $J(\mathbf{W_n})$ of CMA and use the mean square error to measure the error:

$$J(\mathbf{W_n}) = E[(|\tilde{x}(n)|^2 - R_2)^2] \tag{2.39}$$

where R_2 is a positive constant, which is the desired radius of convergence. Our goal is to find the optimal tap coefficient $\mathbf{W}(n)$, namely

$$\mathbf{W}(n)_{\text{opt}} = \underset{\mathbf{W}n}{\arg\min}(J(\mathbf{W}n)) = \underset{\mathbf{W}n}{\arg\min}(E[(|\tilde{x}(n)|^2 - R_2)^2]) \tag{2.40}$$

For Eq. (2.40), the steepest descent method is used to obtain the extreme value, and we get

$$\mathbf{W}(n+1) = \mathbf{W}(n) - \mu \frac{\partial J[\mathbf{W}(n)]}{\partial \mathbf{W}(n)} \tag{2.41}$$

where

$$\frac{\partial J[\mathbf{W}(n)]}{\partial \mathbf{W}(n)} = 2E\left[(|\tilde{x}(n)|^2 - R_2)\frac{\partial |\tilde{x}(n)|^2}{\partial \mathbf{W}(n)}\right] \tag{2.42}$$

Since $\tilde{x}(n) = w(n) * y(n) = \mathbf{Y}^T(n)\mathbf{W}(n)$,

$$\frac{\partial |\tilde{x}(n)|^2}{\partial \mathbf{W}(n)} = 2\mathbf{Y}^*(n)\tilde{x}(n) \tag{2.43}$$

Substitute Eq. (2.43) into Eq. (2.42), we get

$$\frac{\partial J[\mathbf{W}(n)]}{\partial \mathbf{W}(n)} = 4E\left[(|\tilde{x}(n)|^2 - R_2)\mathbf{Y}^*(n)\tilde{x}(n)\right] \tag{2.44}$$

Using stochastic gradient instead of expected value, the iterative calculation formula of CMA can be obtained:

$$\mathbf{W}(n+1) = \mathbf{W}(n) + \mu\tilde{x}(n)[R_2 - |\tilde{x}(n)|^2]\mathbf{Y}^*(n) \tag{2.45}$$

where μ is the step size of iteration, and a small positive number is usually selected. It determines the speed and accuracy of convergence.

When the ideal equilibrium is reached, the partial derivative:

$$\frac{\partial J[\mathbf{W}(n)]}{\partial \mathbf{W}(n)} = 0 \tag{2.46}$$

After processed by the equalizer, the distortion-free transmission condition for $y(n)$ is satisfied:

$$\tilde{x}(n) = \mathrm{A}x(n - n_{\text{delay}}) \tag{2.47}$$

where A is a non-zero constant which represents the multiple of linear amplification on the amplitude, and n_{delay} represents the time delay. When $x(n)$ is modulated with a constant amplitude, the irrelevant A can be set to 1, and the delay is equivalent to the addition of a fixed phase shift, i.e.,

$$\tilde{x}(n) = x(n)e^{j\theta_{\mathrm{d}}} \tag{2.48}$$

where θ_{d} stands for fixed phase shift.

From Eqs. (2.44) and (2.46), we get

$$E[|\tilde{x}(n)|^2\mathbf{Y}^*(n)\tilde{x}(n)] = R_2E[\mathbf{Y}^*(n)\tilde{x}(n)] \tag{2.49}$$

Expanding the vector form into the form of components, we get

$$E[|\tilde{x}(n)|^2y^*(n-i)\tilde{x}(n)] = R_2E[y^*(n-i)\tilde{x}(n)],\ i = 0, 1, \ldots, L-1 \tag{2.50}$$

Namely, in Eq. (2.49), the corresponding elements on the left and right sides of the equal sign are equal.

From Eq. (2.32), we obtain

$$y(n-i) = \sum_m x(n-i-m)h(m) \tag{2.51}$$

For M-PSK modulation, the constellation points of data symbols in $x(n)$ are distributed symmetrically on the complex plane, with

$$E(x(n)) = 0 \tag{2.52}$$

$$E\left(|x(n)|^2\right) = \sigma_x^2 \tag{2.53}$$

Since the original signal $x(n)$ is distributed independently, there is

$$E(x(m)x(n)) = \delta_{mn}\sigma_x^2 \tag{2.54}$$

where

$$\delta_{mn} = \begin{cases} 1, & m = n \\ 0, & m \neq n \end{cases} \tag{2.55}$$

is the Kronecker function.

Therefore, if and only if $m = i$ can the item contribute to the left and right sides of Eq. (2.50), namely

$$E[|\tilde{x}(n)|^2 y^*(n)\tilde{x}(n)] = \mathrm{k}E\left[|\tilde{x}(n)|^4\right] \tag{2.56}$$

$$E[y^*(n)\tilde{x}(n)] = \mathrm{k}E\left[|\tilde{x}(n)|^2\right] \tag{2.57}$$

where k is the determined amplitude gain introduced by the channel.

By substituting Eqs. (2.56) and (2.57) into Eq. (2.50), we obtain

$$R_2 = \frac{E\left[|\tilde{x}(n)|^4\right]}{E\left[|\tilde{x}(n)|^2\right]} \tag{2.58}$$

So far, we have completed the theoretical derivation of CMA constant modulus algorithm.

Example CMA equalization algorithm for single-polarization QPSK

Assuming $x(n)$ is the signal after QPSK modulation. When it passes through the channel with the unit impulse response of $h(n)$, it changes to $y(n)$ with the addition of additive white Gaussian noise noise(n). Now we use three CMA algorithms with different step lengths to realize the blind equalization of $y(n)$.

Figure 2.20 is the QPSK constellation before CMA equalization, Fig. 2.21 is the QPSK constellation after CMA equalization, and Fig. 2.22 is the error curve of CMA

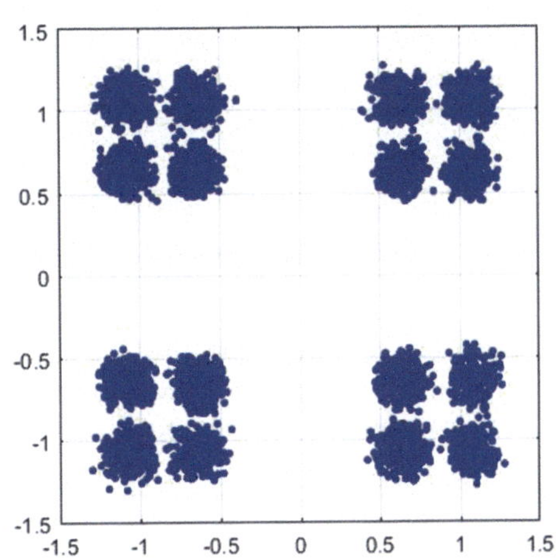

Fig. 2.20 QPSK constellations before CMA equalization

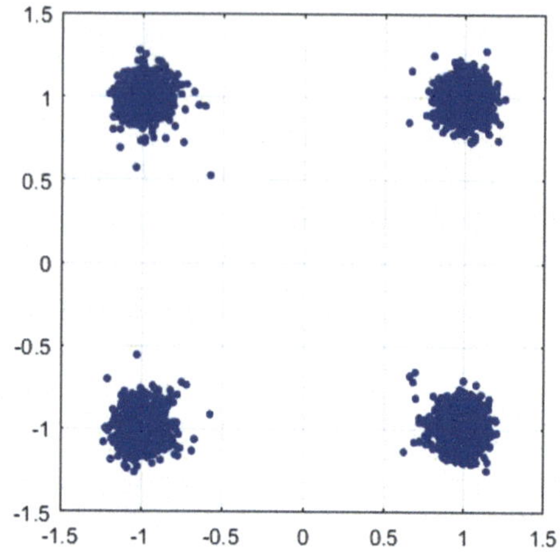

Fig. 2.21 QPSK constellations after CMA equalization

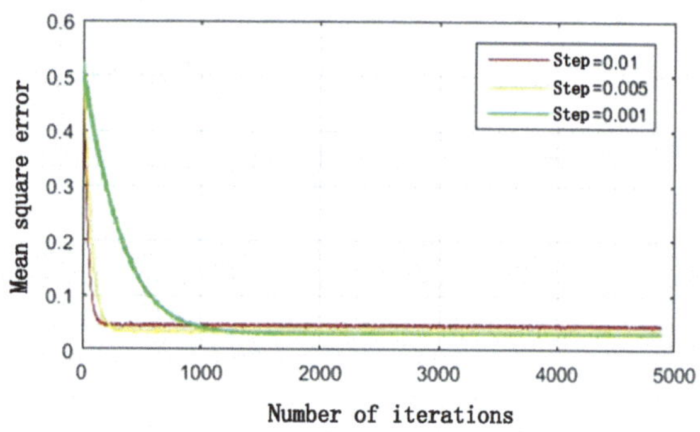

Fig. 2.22 The error curve of CMA equalization

equalization. It can be seen from the constellation diagram that the constellation points are more concentrated after CMA equalization, which is very helpful for our decision. From the error curve of CMA, the convergence of the algorithm can be accelerated if a large step length is adopted, but a large residual error will be brought. If you want to reduce the error of CMA equalization, you should use the small step length, but this will make the convergence slow.

2.5.3 CMA Algorithm for Polarization Multiplexing Signals

In the last section, we introduced the basic principle of CMA and applied it to the blind channel equalization of traditional QPSK modulation. In this section, we will introduce its application in the polarization demultiplexing of optical communication systems.

2.5.3.1 Basic Principle of Polarization Demultiplexing

In optical communication system, polarization demultiplexing is often used to improve the system capacity. The basic idea of polarization demultiplexing is to use two independent and orthogonal polarization states of light as two non-interference channels to transmit data, thus doubling the information transmission rate. Therefore, the receiver needs polarization demultiplexing, that is, separating two original signals from the received signal. CMA is almost the most popular algorithm in constant modulus modulation polarization demultiplexing. Next, we will discuss the basic principles of polarization demultiplexing.

The Jones matrix of ideal optical fiber transmission [12] is

$$\mathbf{T} = \begin{bmatrix} \sqrt{\alpha}e^{i\delta} & -\sqrt{1-\alpha} \\ \sqrt{1-\alpha} & \sqrt{\alpha}e^{-i\delta} \end{bmatrix} \tag{2.59}$$

where α and δ, respectively, represent the spectral ratio and the phase difference between the two polarization states.

Since the fiber is ideal, the T matrix satisfies the unitary condition. The polarization multiplexed optical signal can be expressed as

$$E_{in}(t)^T = \left[E_{in,x}(t), E_{in,y}(t)\right]^T \tag{2.60}$$

where superscript T represents the transpose of the matrix. $E_{in}(t)$ can be understood as the signal to be sent at the transmitter end. After the signal is transmitted through optical fiber, the optical signal at the receiver end can be expressed as

$$\begin{bmatrix} E_x(t) \\ E_y(t) \end{bmatrix} = \mathbf{T}\begin{bmatrix} E_{in,x}(t) \\ E_{in,y}(t) \end{bmatrix} \tag{2.61}$$

As can be seen from Eq. (2.61), two polarization components that should be independent have been mixed together, so they need to be separated by digital signal processing algorithm. It is assumed that the transmitter adopts m-PSK modulation, and its transmitted signal is of constant envelope. Since its envelope does not contain any information, each polarization component can be normalized to

$$\left|E_{in,x}(t)\right|^2 = \left|E_{in,y}(t)\right|^2 = 1 \tag{2.62}$$

We hope that by linear superposition of $E_x(t)$ and $E_y(t)$, we can get the separated signal E_X and E_Y. Let's write E_X as

$$\begin{aligned} E_X(t) &= rE_x(t) + kE_y(t) \\ &= (r\sqrt{\alpha}e^{j\delta} + k\sqrt{1-\alpha})E_{in,x}(t) \\ &= (-r\sqrt{1-\alpha} + k\sqrt{\alpha}e^{-j\delta})E_{in,y}(t) \end{aligned} \tag{2.63}$$

where r and k are complex numbers. To simplify the calculation, we define

$$X \equiv r\sqrt{\alpha}e^{j\delta} + k\sqrt{1-\alpha} \quad (2.64)$$

$$Y \equiv -r\sqrt{1-\alpha} + k\sqrt{\alpha}e^{-j\delta} \quad (2.65)$$

Since the original signal is a constant amplitude modulated signal, if the signal processing algorithm can recover two signals of different polarization states without distortion, a necessary condition is that the signal power after separation should also be constant. The power of the signal E_X is

$$|E_X(t)|^2 = \left|XE_{in,x}(t)\right|^2 + \left|YE_{in,y}(t)\right|^2 + 2\left|XYE_{in,x}(t)E_{in,y}(t)\right|\cos\theta(t) \quad (2.66)$$

$$\theta(t) = \arg\left[\frac{YE_{in,y}(t)}{XE_{in,x}(t)}\right] \quad (2.67)$$

Since the equation in (2.66) is time-varying, there must be $\left|XYE_{in,x}(t)E_{in,y}(t)\right| \equiv 0$, and it's easy to know $X = 0$ or $Y = 0$.

For the case of $Y = 0$, there is

$$\frac{r_x}{k_x} = \frac{\sqrt{\alpha}}{\sqrt{1-\alpha}}e^{-j\delta} \quad (2.68)$$

Thus, we can derive

$$E_X = \frac{k_x}{\sqrt{1-\alpha}}E_{in,x} \quad (2.69)$$

If the complete ideal separation will satisfy $|E_X(t)|^2 = 1$, there is

$$r_x = \sqrt{\alpha}e^{-j\delta+j\varphi_x} \quad (2.70)$$

$$k_x = \sqrt{1-\alpha}e^{j\varphi_x} \quad (2.71)$$

where φ_x is a real number. The original signal can then be recovered:

$$E_X = E_{in,x}e^{j\varphi_x} \quad (2.72)$$

$$E_Y = -k_x^* E_x + r_x^* E_y = E_{in,y}e^{-j\varphi_x} \quad (2.73)$$

In the case of $X = 0$, we can make a similar derivation and finally get

$$E_X = E_{in,y}e^{i\varphi_y} \quad (2.74)$$

$$E_Y = -E_{in,x}e^{-i\varphi_y} \tag{2.75}$$

2.5.3.2 Polarization Demultiplexing CMA Algorithm

In actual communication systems, CMA algorithm is commonly used for polarization demultiplexing [13]. The block diagram of the algorithm is shown in Fig. 2.23.

The matrix of the algorithm is described as

$$\begin{bmatrix} E_X \\ E_Y \end{bmatrix} = \mathbf{p}\begin{bmatrix} E_x \\ E_y \end{bmatrix} \tag{2.76}$$

where the matrix **p** can be written as

$$\mathbf{p} = \begin{bmatrix} p_{xx} \; p_{xy} \\ p_{yx} \; p_{yy} \end{bmatrix} \tag{2.77}$$

And, the matrix **p** must satisfy the unitary condition, namely

$$p_{xy} = -p_{yx}^{*} \tag{2.78}$$

$$p_{yy} = p_{xx}^{*} \tag{2.79}$$

$$|p_{xx}|^2 + |p_{xy}|^2 = 1 \tag{2.80}$$

Therefore, in Fig. 2.23, there are only two independent variables.

CMA goes through a series of iterations, and $|E_X(n)|^2 \to 1$. Once $|E_X(n)|^2 \to 1$, it means that for the case of $Y = 0$, there is $p_{xx}(n) \to r_x, p_{xy}(n) \to k_x, p_{yx}(n) \to -k_x^{*}$, $p_{yy} \to r_x^{*}$. At this point, the X port in Fig. 2.23 outputs the X-polarization component, and the Y port outputs the Y-polarization component. In the case of $X = 0$, the

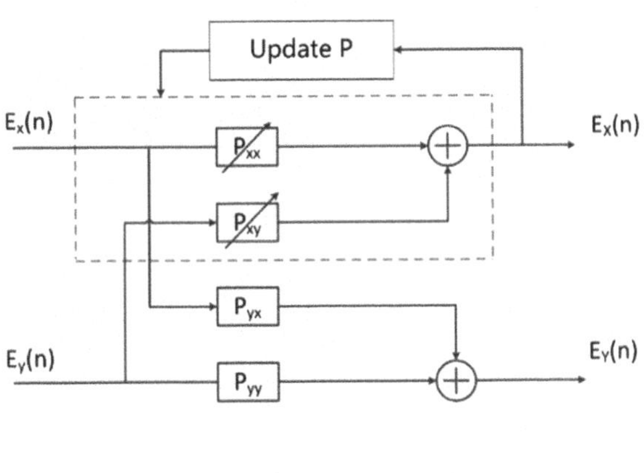

Fig. 2.23 Block diagram of CMA algorithm satisfying the unitary condition

X port will output the Y-polarization component, and the Y port will output the X-polarization component.

Similar to the CMA algorithm of single-polarization signal in the previous section, the iterative updating equation [11] of the CMA algorithm is used, and the elements of the matrix can be iteratively updated as [12].

$$p_{xx}(n+1) = p_{xx}(n) + \mu(1 - |E_X(n)|^2)E_X(n)E_x^*(n) \tag{2.81}$$

$$p_{xy}(n+1) = p_{xy}(n) + \mu(1 - |E_X(n)|^2)E_X(n)E_y^*(n) \tag{2.82}$$

In fact, the matrix will no longer satisfy the unitary condition because the actual fiber is not ideal. Therefore, two updating equations need to be added [12].

$$p_{yy}(n+1) = p_{yy}(n) + \mu(1 - |E_Y(n)|^2)E_Y(n)E_y^*(n) \tag{2.83}$$

$$p_{yx}(n+1) = p_{yx}(n) + \mu(1 - |E_Y(n)|^2)E_Y(n)E_x^*(n) \tag{2.84}$$

The principle block diagram of CMA algorithm at this time is shown in Fig. 2.24. The tap coefficients of the four filters are independent of each other.

Furthermore, if the polarization mode dispersion of the fiber also needs to be equalized, the filter will change from single tap to multiple tap in the polarization demultiplexing CMA algorithm, so that the interaction between the symbols before and after the received signal can be considered to realize the compensation of polarization mode dispersion.

Define the input signal vector as

$$E_k(n)^T = [E_k(n), E_k(n), \ldots, E_k(n-m)]^T \tag{2.85}$$

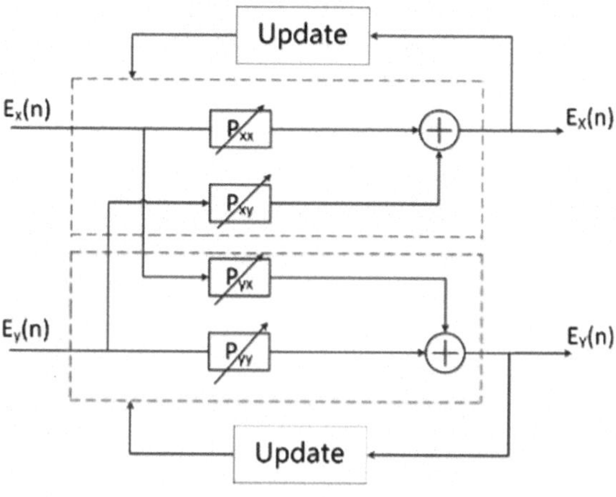

Fig. 2.24 Block diagram of CMA algorithm not satisfying the unitary condition

The filter tap coefficient vector is defined as

$$p_{ij}(n)^T = \left[p_{ij,0}(n), p_{ij,1}(n), \ldots p_{ij,m}(n)\right]^T \tag{2.86}$$

where i, j, k represents arbitrary x-polarization or y-polarization, and m represents the tap number of the filter.

In this case, the tap coefficient of CMA algorithm should adopt the following updating Eq. [12]:

$$\mathbf{p}_{xx}(n+1) = \mathbf{p}_{xx}(n) + \mu(1 - |E_X(n)|^2)E_X(n)\mathbf{E}_x^*(n) \tag{2.87}$$

$$\mathbf{p}_{xy}(n+1) = \mathbf{p}_{xy}(n) + \mu(1 - |E_X(n)|^2)E_X(n)\mathbf{E}_y^*(n) \tag{2.88}$$

$$\mathbf{p}_{yx}(n+1) = \mathbf{p}_{yx}(n) + \mu(1 - |E_Y(n)|^2)E_Y(n)\mathbf{E}_x^*(n) \tag{2.89}$$

$$\mathbf{p}_{yy}(n+1) = \mathbf{p}_{yy}(n) + \mu(1 - |E_Y(n)|^2)E_Y(n)\mathbf{E}_y^*(n) \tag{2.90}$$

Using Eqs. (2.87)–(2.90), we can not only compensate the polarization mode dispersion, but also achieve polarization demultiplexing.

Example: Using CMA blind equalization algorithm to demultiplex PDM-QPSK signals

Let the transmission PDM-QPSK signal be $x(n)$.The receiver will receive the signal $y(n)$ by means of coherent detection and recover the received signal by means of digital signal processing. It is assumed that the received signal has no error in timing and its dispersion has been perfectly compensated by the dispersion compensation algorithm. Now the CMA blind equalization algorithm is used to demultiplex PDM-QPSK signal (Fig. 2.25).

2.5.4 CMMA Algorithm for High-Order Non-constant Mode Modulation

As described in the previous section, the CMA algorithm is very effective in polarization multiplexing—constant mode modulation depolarization. The recovery ability of CMA-based blind equalization algorithm to signals is close to the performance of DD-LMS algorithm, so it is regarded as an independent equalization algorithm [14].

However, for equalizations of PDM-8QAM, PDM-16QAM, and other non-constant mode modulation, the classical CMA algorithm seems to be somewhat inadequate [14, 15]. Therefore, the error signal in CMA algorithm will not go to zero, even for the ideal 8QAM or 16QAM signal, that is, there is still extra noise after equalization. The emergence of cascade multimode algorithm solves this problem [3]. In

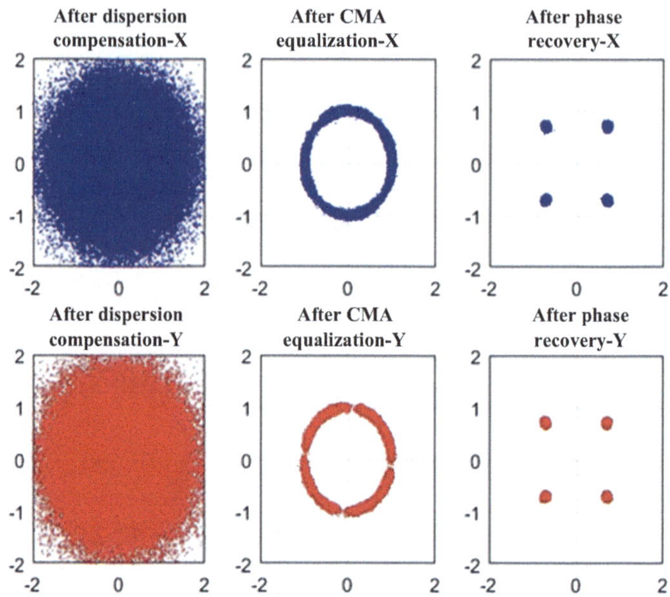

Fig. 2.25 Equalizing PDM-QPSK signals with CMA

this algorithm, multiple reference rings are introduced by cascade, and the final bit error rate approaching 0 is obtained, as shown in Fig. 2.26. CMMA algorithm introduces two reference circles of radius $R_{ref1} = 0.5(R_1 + R_2)$ and $R_{ref2} = 0.5(R_1 - R_2)$ in the form of cascade. Among them, R_1 and R_2 are the circumferential radius of the points of the ideal 8QAM constellation.

In the CMMA algorithm, the whole equalization is divided into two steps. In the first step, the circle with radius R_{ref1} was used as a reference, and the intermediate residual error ε_1 was obtained after iteration. It can be seen that the residual error does not approach zero. Therefore, in the second step, the circle with radius R_{ref2} was used as a reference, and then the residual error ε_1 was used for the iterative operation. Finally, the residual error obtained tended to zero, achieving a perfect equilibrium.

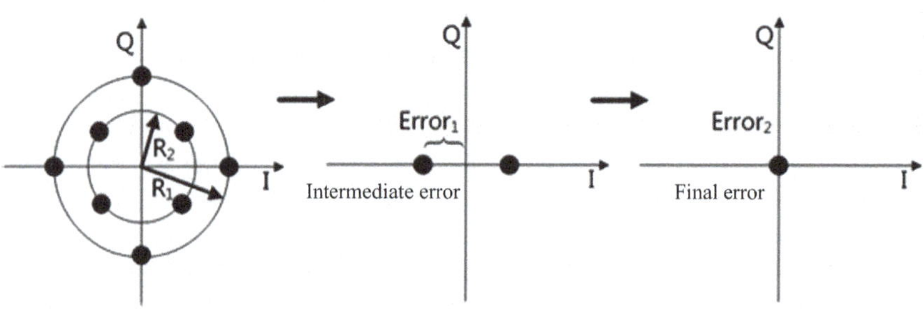

Fig. 2.26 Schematic diagram of CMMA principle to equalize 8QAM signal

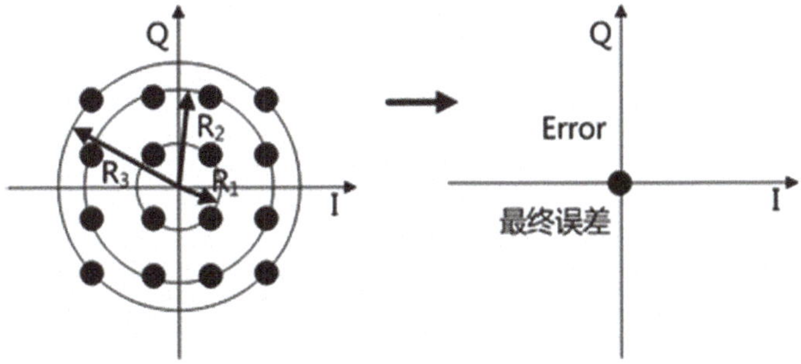

Fig. 2.27 Schematic diagram of CMMA principle to equalize 16QAM signal

When CMMA is used to equalize 16QAM signal, the calculation principle of the error signal is shown in Fig. 2.27. Similarly, the CMMA algorithm introduces three reference circles with radii of $R_{ref1} = 0.5(R_1+R_2)$, $R_{ref2} = 0.5(R_3-R_1)$, and $R_{ref3} = 0.5(R_3 - R_2)$ in a cascade. Where, R_1, R_2, and R_3 are the radius of the circumference of the three-circle constellation point of the ideal 16QAM constellation point.

In contrast, the classical CMA algorithm only uses a reference circle when calculating the error signal [11]. Therefore, it is not surprising that it cannot perfectly balance such multimode signals as PDM-8QAM, and PDM-16QAM.

The algorithm flow of CMMA is shown in Fig. 2.28. Similar to CMA equalizer, CMMA equalizer can be regarded as an adaptive equalizer with a butterfly structure composed of four filters.

In the CMMA algorithm, the update equation of the filter tap coefficient is as follows:

$$h_{xx}(k) \rightarrow h_{xx}(k) + \mu\varepsilon_x(i)e_x(i)\hat{x}(i-k) \tag{2.91}$$

$$h_{xy}(k) \rightarrow h_{xy}(k) + \mu\varepsilon_x(i)e_x(i)\hat{y}(i-k) \tag{2.92}$$

$$h_{yx}(k) \rightarrow h_{yx}(k) + \mu\varepsilon_y(i)e_y(i)\hat{x}(i-k) \tag{2.93}$$

Fig. 2.28 CMMA algorithm flow chart

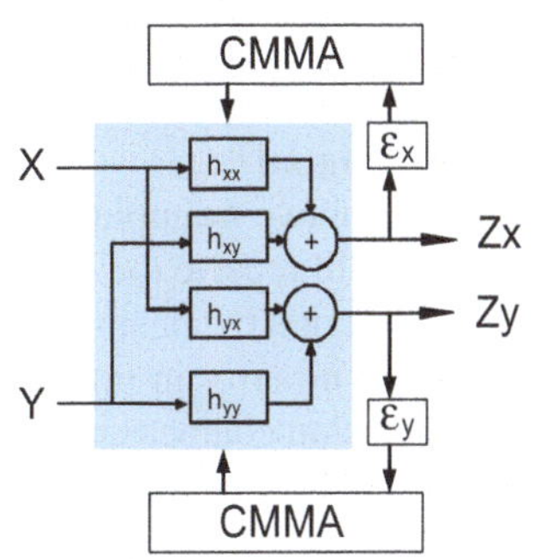

$$h_{yy}(k) \to h_{yy}(k) + \mu \varepsilon_y(i) e_y(i) \hat{y}(i-k) \tag{2.94}$$

For 8QAM signal, there is

$$\varepsilon_{x,y} = \left| \left| Z_{x,y} \right| - R_{ref1} \right| - R_{ref2} \tag{2.95}$$

where *sign* is the sign function and $Z_{x,y}$ is the output of any polarization state after equalization.

For the 16QAM signal, there is

$$\varepsilon_{x,y} = \left| \left| \left| Z_{x,y} \right| - R_{ref1} \right| - R_{ref2} \right| - R_{ref3} \tag{2.97}$$

$$\begin{aligned} e_{x,y}(i) &= sign(\left| \left| \left| Z_{x,y}(i) \right| - R_{ref1} \right| - R_{ref2} \right| - R_{ref3}) \\ &\cdot sign(\left| \left| Z_{x,y}(i) \right| - R_{ref1} \right| - R_{ref2}) \\ &\cdot sign(\left| Z_{x,y}(i) \right| - R_{ref1}) \cdot sign(Z_{x,y}(i)) \end{aligned} \tag{2.98}$$

Compared with CMA, CMMA algorithm can improve SNR performance significantly in 8QAM and 16QAM modulation formats, but it also reduces the robustness of the filter convergence process. This is because the CMMA algorithm relies on the correct judgment of the transmitting signal radius. Since the interval between different rings of QAM signal is smaller than the minimum symbol interval, there will be a lot of errors in the judgment of ring radius when there is a lot of noise or serious signal distortion. One solution is to use the CMA algorithm for pre-convergence at the beginning. After the pre-convergence is completed, the system is processed by multimode algorithm. Since in practical application, it generally needs two CMMA algorithms to achieve a better balance effect, the use of one CMA in the initial stage for pre-convergence and then conduct CMMA process will not cause the increase of implementation complexity. For higher order QAM modulation, such as 32QAM and 64QAM, the complexity of polarization demultiplexing using multimode algorithms is very high. We can reduce the complexity by selecting only two or three inner loops for error feedback calculations. Since the radius difference between QAM inner rings is usually larger than that between outer rings, it can also increase the robustness of convergence.

Example: Using CMMA blind equalization algorithm to demultiplex PDM-16QAM signal

As in the above example, set the transmission PDM-16QAM signal to be $x(n)$. The receiver will receive the signal $y(n)$ by means of coherent detection and recover the received signal by means of digital signal processing. It is assumed that the received signal has no error in timing and its dispersion has been perfectly compensated by the dispersion compensation algorithm. Since 16QAM signal is not a constant mode signal, CMMA blind equalization algorithm is used to demultiplex PDM-16QAM signal.

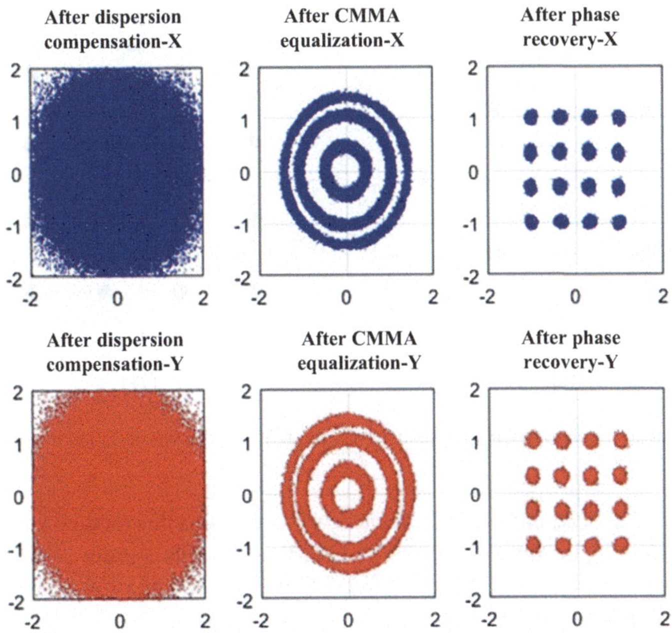

Fig. 2.29 Using CMMA algorithm to demultiplex PDM-16QAM signal

As shown in Fig. 2.29, after CMMA equalization, the original chaotic 16QAM constellation diagram becomes three circles. Then, after frequency offset estimation and phase carrier recovery, the signal finally converges to the position of 16QAM constellation point.

2.5.5 *Modified Cascade Multi-modulus Algorithm (MCMMA)*

Linear equalizer based on classic CMA algorithm in the back end has been widely used in the time-domain equalization, but the effect of constant modulus algorithm for multiple order modulation signal is not very significant, this is because in a multistage modulation system bit-rate signal amplitude is not constant, and balanced constellation points will occur after rotation, additional phase deviation rectification is needed.

Reference [4] proposed a new multimode algorithm for quadrate modulation signals based on statistical rules of constellation points, as shown in Fig. 2.30. This method utilizes the statistical law of constellation point distribution to modify the traditional multimode algorithm, and transforms the single error function depending on the radius of convergence into two orthogonal error functions, thus improving

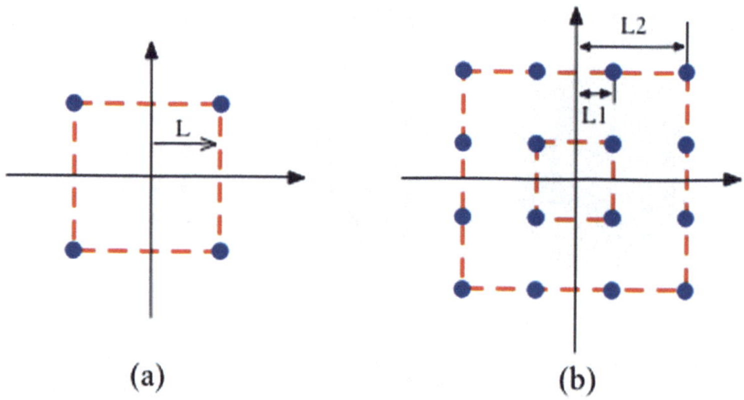

Fig. 2.30 The new equalization scheme: **a** MMA and **b** MCMMA constellation diagrams

the performance of equalizer. This chapter further deepens this method and proposes an equalization algorithm for high-order QAM modulation system called modified cascade multi-modulus algorithm (MCMMA) [6], as shown in Fig. 2.30b. Its error function is expressed as follows:

$$\begin{aligned} \varepsilon_I &= ||||y_I(i)| - Am_1| - Am_2| \ldots - Am_{n-1}| - Am_n \\ \varepsilon_Q &= \Big|\Big|\Big|\big|y_Q(i)\big| - Am_1\Big| - Am_2\Big| \ldots - Am_{n-1}\Big| - Am_n \end{aligned} \tag{2.99}$$

where

$$\begin{aligned} y_I(i) &= \text{real}(y(i)) \\ y_Q(i) &= \text{imag}(y(i)) \\ Am_1 &= (L_1 + L_2)/2 \\ Am_2 &= (L_3 - L_1)/2 \\ &\ldots \\ Am_{\text{n}-1} &= (L_n - L_{n-2})/2 \\ Am_n &= (L_n - L_{n-1})/2 \end{aligned} \tag{2.100}$$

In the above formula, $L_1, L_2, \ldots, L_n$ is the orthogonal coordinate module value in the coded signal constellation diagram. The equalizer coefficient update and output correction of the scheme are as follows:

$$\begin{aligned} y_I(i) &= \underset{11}{\vec{H}}(i) \otimes \underset{I}{\vec{X}}(i) + \underset{12}{\vec{H}}(i) \otimes \underset{Q}{\vec{X}}(i) \\ y_Q(i) &= \underset{21}{\vec{H}}(i) \otimes \underset{I}{\vec{X}}(i) + \underset{22}{\vec{H}}(i) \otimes \underset{Q}{\vec{X}}(i) \end{aligned} \tag{2.101}$$

$$
\begin{aligned}
M_I(i) =& sign(|\ldots|y_I(i) - A_1| - A_2 \ldots| - A_{n-1}) \\
& \ldots sign(|y_I(i)| - A_1) sign(y_I(i)) \\
M_Q(i) =& sign(|\ldots|y_Q(i) - A_1| - A_2 \ldots| - A_{n-1}) \\
& \ldots sign(|y_Q(i)| - A_1) sign(y_Q(i))
\end{aligned}
\tag{2.102}
$$

$$
\begin{aligned}
\vec{H}_{11}(i+1) &= \vec{H}_{11}(i) + \mu \varepsilon_I M_I(i) \vec{X}_I^*(i) \\
\vec{H}_{12}(i+1) &= \vec{H}_{12}(i) + \mu \varepsilon_I M_I(i) \vec{X}_Q^*(i) \\
\vec{H}_{21}(i+1) &= \vec{H}_{21}(i) + \mu \varepsilon_Q M_Q(i) \vec{X}_I^*(i) \\
\vec{H}_{22}(i+1) &= \vec{H}_{22}(i) + \mu \varepsilon_Q M_Q(i) \vec{X}_Q^*(i)
\end{aligned}
\tag{2.103}
$$

In the above formula, sign is the symbol for sign evaluation, and "⊗" is the convolution symbol.

Since the single error function is extended to two orthogonal error functions, MCMMA algorithm has many advantages. First of all, the I and Q transfer function of QAM signal can be independently updated, thus shows better crosstalk resistance. Secondly, since the radius of convergence is changed to the orthogonal coordinate module value in the signal constellation diagram, the phase information of each signal element is ignored by the error function.

Therefore, the phase rotation module after CMMA scheme can be removed to reduce the computational complexity of system equilibrium. Moreover, compared with CMMA, MCMMA scheme can reduce the reference modulus of the algorithm. On the one hand, the computational complexity is further simplified; on the other hand, when the modulation order of the system increases, the space between the reference modules becomes smaller. Under the same signal-to-noise ratio condition, the output of the error function is larger and not convergent easily. The following table shows the number of reference modulus values required for two schemes with different modulation orders. It can be found that with the improvement of QAM signal modulation order, the number of reference modulus required by MCMMA scheme decreases more obviously, so it can be predicted that MCMMA scheme is more suitable for high-order QAM modulation system.

It should be noted that, in order to improve the equalization effect, pre-convergence module is still introduced into the equalizer based on MCMMA. The pre-convergence module is a special MCMMA scheme with a modulus of 1, namely the MMA scheme in Ref. [4] (Table 2.1).

Table 2.1 Number of reference module values required by two equalization schemes for high-order QAM modulation signals

	QPSK	16QAM	32QAM	64QAM
CMMA	1	3	5	9
MCMMA	1	2	3	4

2.5.6 Independent Component Analysis (ICA)

2.5.6.1 Summary of ICA

Polarization multiplexing technology can double the speed of coherent optical communication system and also double the spectrum efficiency of the system. As a cost, the receiver needs to demultiplex and separate the data from the two polarization states. In general, polarization demultiplexing is realized by digital signal processing. Previously, some feasible methods have been proposed for polarization demultiplexing [6], such as channel estimation [17] or sending a training sequence for the equalizer [18]. Later, using CMA to polarization demultiplex was more popular [11].

Although CMA is a blind equalization algorithm that does not need training sequence, and the communication efficiency is higher than that of the equalization method of data auxiliary class. However, it also has some disadvantages. For example, two polarized signals may converge to the same channel. This is caused by the selection of cost function [19]. Moreover, CMA is not designed for non-constant amplitude modulation schemes such as higher order QAM signals, so it seems that a better polarization demultiplexing performance will be achieved if a better cost function is chosen.

In the previous section, we introduced a cascaded multimode equalization algorithm CMMA to realize blind equalization of high-order QAM modulation for polarization demultiplexing. In this section, we will introduce another class of algorithms for polarization demultiplexing—independent component analysis.

As described in the previous section, after the signal is transmitted through the optical fiber, the two independently polarized signals are mixed together. Therefore, the mixed signal needs to be separated at the receiver end. Similar to "blind equalization", the separation of signals that can only be separated from the observed signals without prior knowledge of the transmitted signal and the channel transfer function is called "blind separation".

ICA is a special kind of algorithm to separate signals. It can separate signals from linear combinations of several signals [13, 25]. At present, many ICA based blind signal separation algorithms have been proposed [26]. The representative theoretical bases of these algorithms are maximization of information [27], stable neural network [28], nonlinear principal component analysis [29], maximum likelihood estimation [4, 30], maximum negative entropy algorithm [1], etc.

Although ICA algorithm is well known in signal processing field, it is not widely used in optical communication system. In optical communication systems, ICA is commonly used to solve polarization demultiplexing for negligible polarization mode dispersion. Reference [2] firstly recommends the use of ICA algorithm for polarization demultiplexing in optical communication systems [2], and proves that ICA algorithm will not have a singular value problem and its performance is similar to CMA. References [31] and [32] have studied an ICA algorithm called art-bounded Blind Source Separation, while Ref. [32] has a simplified algorithm based on the Kurtosis optimization, which uses ICA to solve the polarization multiplexing of

PDM-16QAM signals. In Ref. [33], ICA is used to solve any constellation signal with multiple amplitudes, and its convergence rate is higher than that of CMA.

Next, we will first describe the data preprocessing process of signal blind separation, and briefly introduce the basic principle of ICA represented by the classical FastICA algorithm, and then discuss the application of ICA algorithm in optical communication system polarization demultiplexing.

2.5.6.2 Signal Preprocessing

Data preprocessing is needed before signal blind separation. The most useful preprocessing method is to remove the mean value and Whitening or Sphering [26].

Since most of the blind separation algorithms assume that the source signal is zero mean, the mean of the signal needs to be removed before separation. Let's say the mean of the random variable x is 0, so we can substitute $x_0 = x - E(x)$ for x. In practice, since the sample mean is an unbiased estimate of the population mean, the sample mean is used to replace the mean of the random variable x. Suppose the random vector $\mathbf{x}(t) = [x_1(t), x_2(t), \ldots x_n(t)]^T, t = 1, 2, \ldots, N$ is N samples of the random variable $\mathbf{x}$, then the algorithm of de-averaging is

$$x_{0-\text{mean},i}(t) = x_i(t) - \frac{1}{N}\sum_{x=1}^{N} x_i(t), \quad i = 1, 2, \cdots N \tag{2.104}$$

The whitening of a random vector $\mathbf{x}$ is a linear transformation of $\mathbf{x}$:

$$\widetilde{\mathbf{x}} = \mathbf{T}\mathbf{x} \tag{2.105}$$

So that the correlation matrix $\mathbf{R}_{\mathbf{x}} = E\left[\widetilde{\mathbf{x}}\widetilde{\mathbf{x}}^{\mathbf{H}}\right] = \mathbf{I}$ of the transformed random vector $\widetilde{\mathbf{x}}$ is the identity matrix. So each of the components of $\widetilde{\mathbf{x}}$ satisfy $E\left[\widetilde{x_i}\widetilde{x_j}\right] = \delta_{ij}$.The whitening process removes the correlation between the components in the mixed signal, that is, makes their second-order statistics independent. The transformation Matrix $\mathbf{T}$ in formula (2.105) is also known as the Whitening Matrix.

In general, Whitening cannot restore the independence between the original signal components from mixed signals, but only the independence of their second-order statistics. Although Whitening treatment is not enough to completely separate the signals of each component, in general, it is better to separate the signals after albino treatment.

Whitening processing can usually be realized by decorating the eigenvalues of the correlation matrix of mixed signals.

Let the eigenvalue decomposition (EVD) of vector $\mathbf{x}$ of mixed signal be R_x, and according to the property of the correlation matrix, the eigenvalue decomposition (EVD) of r exists, which can be expressed as [26]

$$R_x = Q\sum^{2} Q^T \tag{2.106}$$

where the matrix $\sum^2$ is a diagonal matrix whose diagonal elements are exactly the eigenvalues of R_x. Moreover, the columns of the matrix R_x are orthonormal eigenvectors corresponding to these eigenvalues.Therefore, the Whitening matrix T can be expressed as

$$T = \sum^{-1} Q^T \tag{2.107}$$

Let $\widetilde{\mathbf{x}} = \mathbf{Tx}$, there is

$$R_{\widetilde{x}} = \left[\widetilde{x}\widetilde{x}^T\right] = TE\left[xx^T\right]T^T = TR_xT^T \tag{2.108}$$

Substitute (2.106) and (2.107) into (2.108), and we have

$$R_{\widetilde{x}} = \left(\sum^{-1} Q^T\right)\left(Q\sum Q\right)^T\left(\sum^{-1} Q\right)^T = I \tag{2.109}$$

Therefore, the linear transformation $\widetilde{\mathbf{x}} = \mathbf{Tx}$ removes the correlation between the components in the mixed signal.

As with de-averaging, the correlation matrix of observed mixed signals can only be obtained from some samples. Let $\mathbf{x}(1), \ldots, \mathbf{x}(n)$ to be the sample of random vector of mixed signals, and the correlation matrix can be estimated as

$$\widehat{\mathbf{R}}_{\mathbf{x}} = \frac{1}{N-1}\sum_{i=1}^{N} \mathrm{x}(i)\mathrm{x}(i)^T \tag{2.110}$$

$\widehat{\mathbf{R}}_{\mathbf{x}}$ satisfies Hermitian symmetry and is a nonnegative definite matrix, its eigenvalue is a nonnegative real number, and $\widehat{\mathbf{R}}_{\mathbf{x}}$ also has EVD. After the whitening transformation, the correlation matrix of mixed signal vector samples is the unit matrix.

2.5.6.3 FastICA Algorithm Based on Negentropy

According to the central limit theorem (CLT), the distribution of the sum of independent random variables would be close to a normal distribution. Therefore, the starting point of ICA is to make the distribution of the separated signal "least like" a normal distribution. But how do you measure whether a random variable looks like a normal distribution? For ICA estimation, the most classical non-Gaussian Measure is kurtosis. However, the direct use of kurtosis as a non-Gaussian measure is too sensitive to the value of samples and is not a stable algorithm. Another non-Gaussian measure is negentropy, which, according to information theory, is the maximum entropy of all random variables with the same variance. This means that the normal distribution is the most random distribution. Negative entropy is defined as

$$J(\mathbf{x}) = H(\mathbf{x}_{\text{gauss}}) - H(\mathbf{x}) \tag{2.111}$$

If and only if the random variable $\mathbf{x}$ obeys normal distribution, the equal sign is satisfied. So it's worth noting that even though it's called negative entropy, it tends to be nonnegative. For any reversible linear transformation of $\mathbf{x}$, negative entropy doesn't change. Due to the complexity of the calculation of negative entropy, it can be approximately calculated as [26].

$$J(x) \propto [E\{G(x)\} - E\{G(\upsilon)\}]^2 \tag{2.112}$$

where G is a non-quadratic function, and υ is a random variable that obeys the standard normal distribution. For the random variable x, we assume that it has a zero mean and unit variance.

In general, the following functions have been proved to be effective as G:

$$G_1(x) = \frac{1}{a} \lg \cosh ax \tag{2.113}$$

$$G_2(x) = -e^{-x^2/2} \tag{2.114}$$

where a is a constant and satisfies.

In general, a is 1.

The fixed point ICA algorithm adopts the iterative calculation method to calculate the maximum value of the Eq. (2.8), which is specifically described as

S1. Whiten the data to $\mathbf{x}$

S2. Select the initial value of matrix W that separates the data

S3. Select non-quadratic function G and update iteratively $W \leftarrow E\{xg(W^T x)\} - E\{g'(W^T x)\}W$

S4. Repeat S3 until convergence

where g is the derivative of G and g' is the derivative of g.

Example: FastICA

Two independent signals are randomly mixed in the form of linear combinations. Now FastICA algorithm is used to separate these two signals.

As shown in Fig. 2.31, the two signals before mixing are sinusoidal wave and rectangular pulse sequence. The mixed signal contains both signals (because the mixing process is a linear combination). We select $g(x) = \tanh(x)$ and perform S1–S4 of the FastICA algorithm to separate the signal. It can be seen that after the iterative calculation, the mixed signal is well de-mixed.

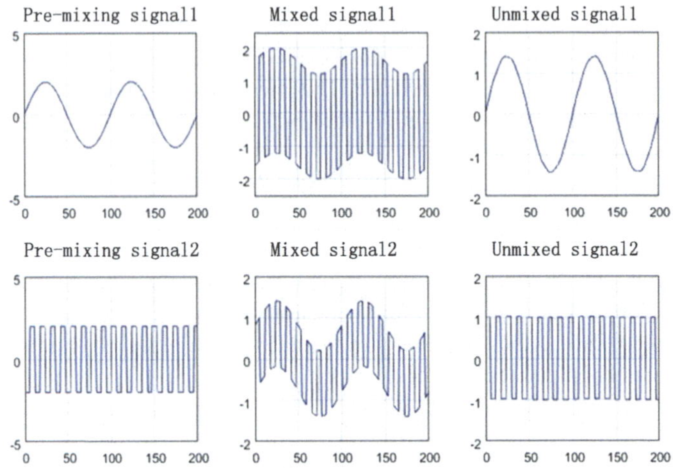

Fig. 2.31 Separation of two signals using FastICA algorithm

2.5.6.4 ICA Algorithm for Complex Numbers Based on Maximum Likelihood Estimation

Next, we introduce the application of ICA algorithm in polarization demultiplexing. Since a prerequisite of ICA algorithm is which signals are statistically independent before mixing, the singular value problem existing in CMA (if two polarization signals converge to the same channel, the output signal is not independent) will not occur [33]. In Ref. [33], an ICA algorithm based on the complex form of maximum likelihood estimation is proposed. The convergence rate of this algorithm is very fast and there is no singular value problem.

First, polarization mode dispersion and polarization-dependent losses (PDL) are ignored and polarization demultiplexing is carried out directly, that is, two polarization state signals are separated from the received signal. Similar to CMA, single-tapped ICA can also extend the form of multi-tapped, so that ICA can also cause depolarization in the presence of PMD and PDL.

We first introduce the single tap ICA algorithm.

The model of the ideal optical fiber is redrawn here as Fig. 2.32. At time k, the independent and identically distributed complex data symbol $\mathbf{a_k} = \left[a_k^{(X)}, a_k^{(Y)}\right]^T$ of two polarization directions becomes $\mathbf{s_k}$ due to the influence of complex additive white Gaussian noise and random phase rotation. The sampled output signal is $\mathrm{x_k} = \mathrm{A_k s_k}$.

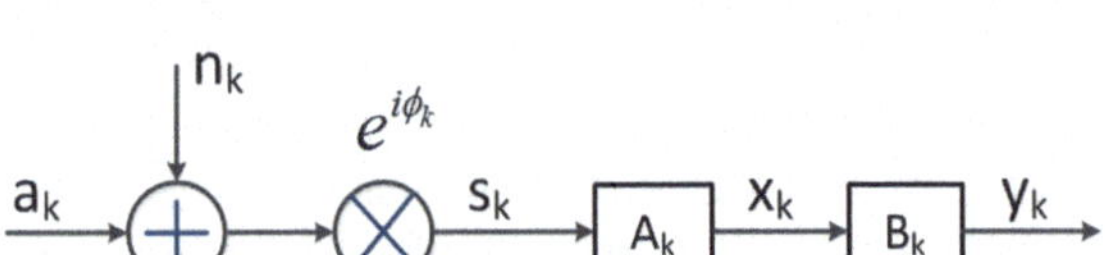

Fig. 2.32 Schematic diagram of fiber channel model

The complex matrix A_k satisfies the unitary condition. In addition, it can be approximately considered to remain unchanged during observation time as compared to the large-scale time when polarization changes [33]. The purpose of the polarization demultiplexing algorithm is to find the demultiplexing matrix $\mathbf{B}_k$ quickly so that the output symbol $\mathbf{y_k} = \mathbf{B_k x_k}$ is a better estimate of $\mathbf{s_k}$.

Let the distribution of time k be known, and this distribution is independent of k. At this time, the distribution of received symbol $x_k = A_k s_k$ can be written as [33]

$$p_{\mathbf{X}}(\mathbf{X}_k|A_k) = \left|\det \mathbf{A}_k^{-1}\right|^2 p_{\mathbf{S}}\left(\mathbf{A}_k^{-1}\mathbf{x}_k\right) \tag{2.115}$$

Substitute the estimated value $\mathbf{B}_k$ of $\mathbf{A}_k^{-1}$ into Eq. (2.115), and get

$$p_{\mathbf{X}}(\mathbf{x}_k|\mathbf{B}_k) = |\det \mathbf{B}_k|^2 p_s(\mathbf{B}_k\mathbf{x}_k) \tag{2.116}$$

The ultimate goal of ICA algorithm based on maximum likelihood estimation is to find a matrix $\mathbf{B}_k$ that maximizes the likelihood function shown in Eq. (2.116). Take the logarithm of Eq. (2.116) we get

$$\Lambda(\mathbf{B}_k) = \log p_{\mathbf{X}}(\mathbf{x}_k|\mathbf{B}_k) = \log|\det \mathbf{B}_k|^2 + \log p_{\mathbf{S}}(\mathbf{y}_k) \tag{2.117}$$

Note that if the matrix $\mathbf{B}_k$ is close to the singular matrix, the determinant in Eq. (2.117) will be close to zero, and the value of the log-likelihood function will be very small, so the singular value problem in CMA can be avoided.

The random steepest descent method is adopted to calculate the matrix $\mathbf{B}_k$ iteratively:

$$\mathbf{B}_{k+1} = \mathbf{B}_k + \mu\mathbf{G}(\mathbf{B}_k),\ \mathbf{G}(\mathbf{B}) = \frac{\partial\Lambda}{\partial\mathbf{B}^*} \tag{2.118}$$

where μ is the step size.

And if we simplify this a little bit, we get

$$\mathbf{B}_{\mathrm{k}+1} = \mathbf{B}_{\mathrm{k}} + \mu\mathbf{G}(\mathbf{B}_{\mathrm{k}})\mathbf{B}_{\mathrm{k}}^H\mathbf{B}_{\mathrm{k}} \tag{2.119}$$

It can be proved that $\mathbf{G}$ in Eq. (2.119) can be simplified to [33]

$$\widetilde{\mathbf{G}}(\mathbf{B}) = \left[I - yy^H + f(y)y^H - yf(y)^H\right]B^{-H} \tag{2.120}$$

where

$$f(y) = \frac{1}{\mathrm{p_S}(y)}\frac{\partial \mathrm{ps}(y)}{\partial \mathbf{s}^*} \tag{2.121}$$

Detailed modeling and analysis of probability density function $p_S(s)$ of M-point constellation map can be referred to [34]. Let the symbols transmitted from the M-element constellation diagram be c_l, subject to additive white Gaussian noise and random deviation, s becomes

$$s = (c_l + n)e^{i\phi} \tag{2.122}$$

where n obeys the complex Gaussian distribution, the mean is 0, the variance is $2\sigma^2$, and ϕ obeys the uniform distribution on $0 \sim 2\pi$.

The conditional probability describing the transfer function can be expressed as

$$p_{S|C}(s|c_l) = \frac{1}{2\pi\sigma^2}\exp\left(-\frac{|c_l|^2 + |s|^2}{2\sigma^2}\right)I_0\left(\frac{|c_l s|}{\sigma^2}\right) \tag{2.123}$$

where I_0 is the 0th order modified Bessel function.

For complex numbers, the joint probability density function:

$$p_S(\mathbf{s}) \approx p_S(s^{(\mathrm{X})})p_S(s^{(\mathrm{Y})}) \tag{2.124}$$

Assuming that the prior probability of all the original symbols is equal, and using the full probability formula, there is

$$p_S(s) = \frac{1}{M}\sum_{l=1}^{M} p_{S|C}(s|c_l) \tag{2.125}$$

Using the conclusion of Eq. (2.124) and the hypothesis of Eq. (2.125), we can obtain $\mathbf{f} = \left[f(s^{(X)}), f(s^{(Y)})\right]^T$, namely

$$f(s) = \frac{1}{2\sigma^2}\frac{\sum_{l=1}^{M} p_{S|C}(s|c_l)\left[\frac{I_1\left(|c_l s|/\sigma^2\right)}{I_0\left(|c_l s|/\sigma^2\right)}|c_l|e^{i\angle s} - s\right]}{\sum_{l=1}^{M} p_{S|C}(s|c_l)} \tag{2.126}$$

Finally, the matrix $\mathbf{B}_k$ can be calculated iteratively by using Eqs. (2.119), (2.120), and (2.126).

Although formula (2.126) can be suitable for any modulation format, the formula is quite complex, and the calculation of its exponential term and Bessel function may become a bottleneck in the real-time communication system. For M-PSK modulation, since it belongs to constant mode modulation, the probability density function in formula (2.126) will disappear. Moreover, in practical usable systems, the signal-to-noise ratio is sufficient for I_1/I_0 to approach 1, so it is reasonable to replace this ratio with a constant value, which can be calculated by the magnitude of the sign. In conclusion, Eq. (2.126) can be simplified as follows:

$$f_{M-PSK}(s) \approx \frac{1}{2\sigma^2}\left(D|c_1|e^{i\angle s} - s\right) \tag{2.127}$$

where

$$D = \frac{I_1\left(|c_1|^2/\sigma^2\right)}{I_0\left(|c_1|^2/\sigma^2\right)} \tag{2.128}$$

Reference [33] also studies the initial value of the matrix $\mathbf{B}_k$, because the selection of the initial value determines the convergence speed of the algorithm.

Example: complex single tap ICA algorithm based on maximum likelihood estimation

Two independent QPSK modulated signals are randomly mixed in the form of linear combination. The ICA algorithm based on maximum likelihood estimation is used to separate the two signals.

It can be seen from Fig. 2.33 that the ICA algorithm based on likelihood estimation for complex single tap can better separate randomly mixed QPSK signals. In the single tap algorithm, $\mathbf{B}_k$ is a 2×2 matrix. After demulsification, the signal has become four constellation points of QPSK. After frequency deviation estimation and phase carrier recovery, the signal can be completely restored to QPSK signal.

Single-tapped ICA can conduct polarization demultiplexing in the absence of PMD. However, in real systems, PMD is often not negligible. At this point, you can extend single-tapped ICA to multiple-tapped ICA. In Ref. [7], the block diagram of the multi-tapped ICA algorithm is similar to the block diagram of CMA algorithm. FIR filter is also used to filter the signal and the tap coefficient $\mathbf{B}_k$ is updated as

$$\mathbf{B}_{k+1} = \mathrm{N}_k \mathbf{B}_k \tag{2.129}$$

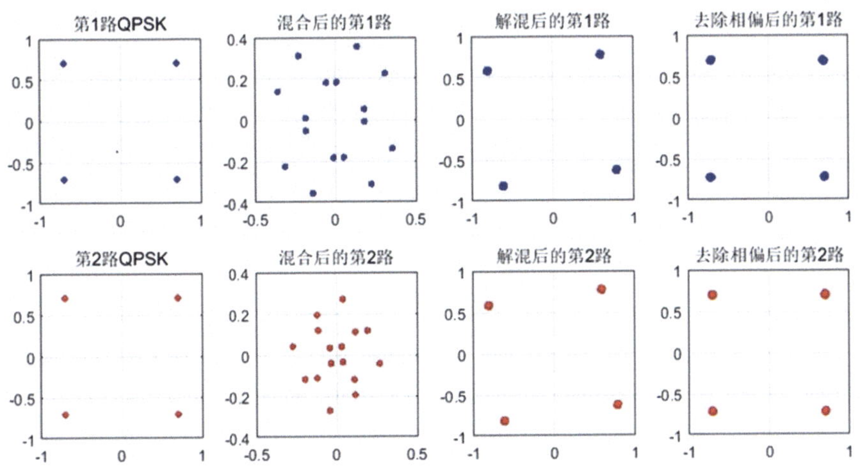

Fig. 2.33 Using single-tapped ICA to separate two QPSK signals

where

$$N_{1,1} = 1 + \mu(1 - |y_1|^2) \tag{2.130}$$

$$N_{1,2} = \frac{\mu|a|}{2\sigma^2}(e^{i\varphi_1}y_2^* - y_1e^{-i\varphi_2}) - \mu y_1 y_2^* \tag{2.131}$$

$$N_{2,1} = \frac{\mu|a|}{2\sigma^2}(y_1^* e^{i\varphi_2} - e^{-i\varphi_1}y_2) - \mu y_1^* y_2 \tag{2.132}$$

$$N_{2,2} = 1 + \mu(1 - |y_2|^2) \tag{2.133}$$

Whereas y_1 and y_2 are the output after equilibrium, φ_1 and φ_2 are the angle of amplitude of y_1 and y_2, respectively.

In addition, in Ref. [35], the matrix is directly dimensionalized and a simplified multi-tapped ICA algorithm is proposed. It takes L samples of two polarized signals and arranges them in a column to form a vector, where $\mathbf{B}_k$ becomes a square matrix of $2L \times 2L$.According to the updating formula in Ref. [34] to update $\mathbf{B}_k$, the polarization demultiplexing is realized.

2.6 Carrier Recovery Algorithm

Carrier recovery algorithm usually consists of two parts: frequency offset estimation and phase recovery. In the practical communication system, the frequency of laser sources is not locked. Moreover, there may be a certain frequency shift between LO and signal light, ranging from several megahertz to hundreds of megahertz or even GHz; on the other hand, the signal light and the local light could have the frequency shift effect due to the change of environmental temperature and other conditions. Such frequency deviation will introduce a large phase rotation to the signal light so that the phase information of the signal itself is submerged. In addition, due to the existence of laser linewidth, phase noise is introduced, which changes randomly at a certain rate, resulting in the tailing, extension, and aliasing of constellation points. Both frequency and phase offset will lead to the degradation of signal quality.

In the optical coherent receiver, the LO and the received signal are sent into hybrid for homodyne and heterodyne detection to get the baseband signal. Ideally, the LO and the optical carrier must operate at the same frequency. However, limited by the current production process of optical devices, there is a certain frequency deviation in each laser; at the same time, the optical communication system is usually long-distance transmission, frequency drift inevitably occurs between the LO and optical signal carrier, which is called frequency offset. In the phase modulation system, the frequency offset is usually converted into the phase offset, which could make the constellation of the received signal rotates in different degrees. Therefore, it is very important to remove frequency offset so that the transmitted signal can be recovered.

For QAM signal analysis, it can be expressed as

$$S(t) = A(t)\cos(\omega_c t + \phi) - B(t)\sin(\omega_c t + \phi) \tag{2.134}$$

Multiply the following two orthogonal carriers at the receiving end:

$$C_c(t) = \cos(\omega_c t + \hat{\phi}) \tag{2.135}$$

$$C_N(t) = -\sin(\omega_c t + \hat{\phi}) \tag{2.136}$$

Then low-pass filtering is carried out to generate in-phase component and orthogonal component as follows:

$$Y_I = \frac{1}{2}A(t)\cos(\phi - \hat{\phi}) - \frac{1}{2}B(t)\sin(\phi - \hat{\phi}) \tag{2.137}$$

$$Y_Q = \frac{1}{2}B(t)\cos(\phi - \hat{\phi}) - \frac{1}{2}A(t)\sin(\phi - \hat{\phi}) \tag{2.138}$$

If there is phase offset, i.e., $\Delta\phi = \phi - \hat{\phi} \neq 0$, then $\cos(\Delta\phi) < 1$ and $\sin(\Delta\phi) > 0$. The phase offset will reduce the power of the received signal component by $\cos^2(\Delta\phi)$, which can lead to bit errors. Moreover, such phase offset will introduce interference between the in-phase and quadrature components. Since the average power of a (T) and B (T) is approximately equal, a small error could cause a significant degradation of demodulation performance.

Take 16QAM as an example, we can observe the influence of frequency offset and phase offset, which is shown in the Fig. 2.34.

2.6.1 Frequency Offset Compensation Algorithm

The decision assist and decision feedback mechanism based on phase-locking loop (PLL) are widely used in the carrier phase recovery of RF communication system. Since the phase change of RF carrier is slow, digital PLL can keep up with the phase change. In the high-speed optical communication system, the phase change speed of the optical carrier is much faster than that of the radio frequency carrier. Only the feedforward based phase recovery can be realized by parallel processing and pipeline structure in a practical system. Relatively speaking, the speed of the frequency shift is slow. DPLL can be used to estimate the frequency shift of transmitter and local oscillator laser in high-speed optical communication system.

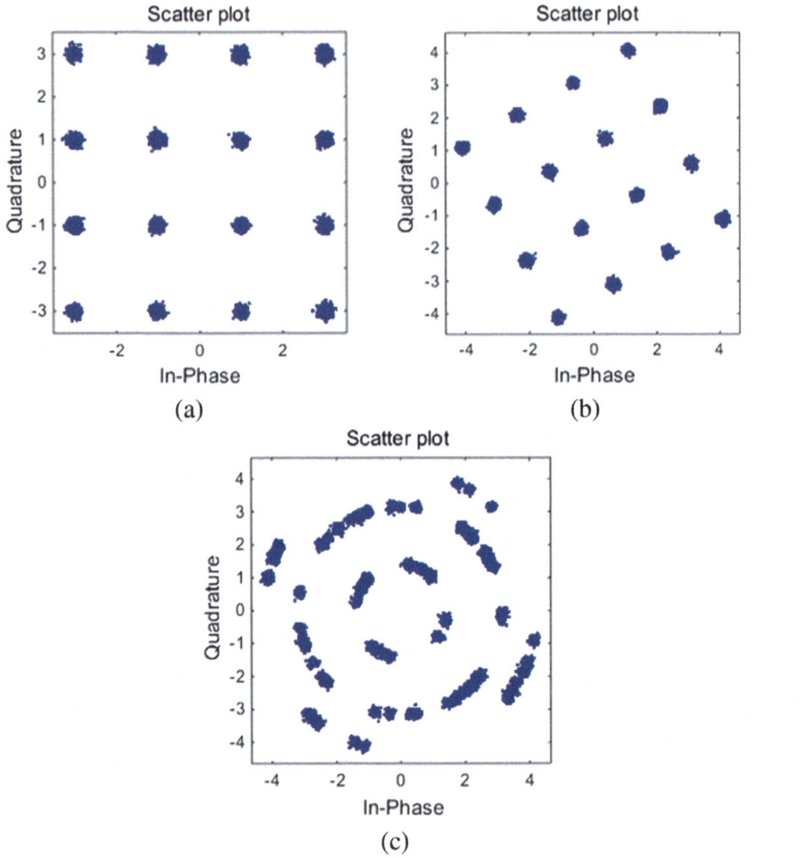

Fig. 2.34 **a** w/o frequency offset and phase offset **b** only phase offset **c** w/ frequency offset and phase offset

2.6.2 V-V Based Frequency Offset Estimation Algorithm

Feedforward carrier recovery based on the classical Viterbi–Viterbi algorithm [22] is widely used in the phase shift keying (PSK) modulation system. In general, the algorithm first calculates the M-power of the signal, where M = 2 N, N is the modulation order of the phase modulation signal. This process removes the data modulation phase. Then, by calculating the phase rotation rate, the frequency offset of the optical signal carrier and the local oscillator can be obtained. After the frequency offset is removed, M-power operation is performed on the signal again to remove the data modulation phase. Because the change rate of phase noise is much slower than other additive noises, the phase noise can be estimated by averaging multiple adjacent symbols. The following is a simple proof of the algorithm.

Here, we only consider the influence of frequency offset and laser phase noise. For the phase modulation signal, the nth received symbol S_n can be expressed as

$$S_n = \exp(j(a_n + 2\pi \Delta v T_s + \theta_n)) \quad (2.139)$$

a_n is the phase of the original signal, Δv is the frequency offset between LO and laser source at the transmitter, T_s is the sampling time interval, and θ_n is the laser phase noises. Since the amplitude of the phase-modulated signal is constant and we are not interested in the amplitude of the signal, the amplitude is normalized in expressions.

For this algorithm, the first step is to estimate the frequency offset of the signal, because the frequency offset must be estimated before the phase estimation. As the frequency offset will cause the phase difference between adjacent samples, the frequency offset can be calculated only by estimating the phase difference between consecutive samples.

The block diagrams of frequency offset estimation in a QPSK transmission system are shown in Fig. 2.35. The precondition of the algorithm is that the difference between the modulation phases of QPSK signal is four times a constant phase value. Firstly, the phase difference between the two adjacent symbols is quadrupled so that the modulation phase can be removed. Then, the noise phase of the first M symbols is averaged to obtain the frequency offset estimation.

The algorithm flow is presented as follows:

First, the received signal S_n is multiplied by the complex conjugation of S_{n-1}. The obtained complex d_n is the phase difference of two symbols

$$d_n = S_n \times S_{n-1}^* = \exp\{j[a_n - a_{n-1} + \Delta\varphi + (\theta_n - \theta_{n-1})\}] \quad (2.140)$$

Semiconductor lasers are commonly used as transmitters and local oscillator lasers, and their linewidth ranges from 100 kHz to 10 MHz. The change of the phase noise of the optical carrier is much slower than that of the modulation phase. Therefore, the phase noise difference between the two adjacent symbols can be ignored, i.e., $\theta_n - \theta_{n-1} \approx 0$.

Subsequently, the coding information contained in the signal phase needs to be removed. For the M-PSK signal, M-power operation is needed for the complex signal. Here, we consider QPSK signals. In this way, we only need to perform the fourth power operation on the complex, d_n, obtained above:

$$d_n^4 = \exp[j(4b_n + 4 \times \Delta\varphi)] = \exp(j \times 4b_n) \times \exp(j \times 4 \times \Delta\varphi) \quad (2.141)$$

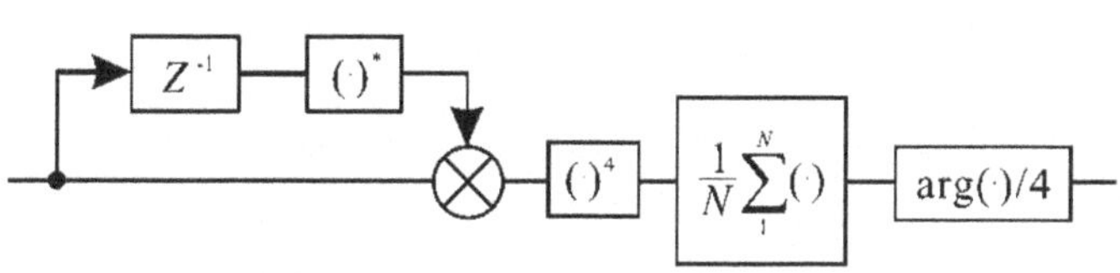

Fig. 2.35 Block diagrams of frequency offset estimation system

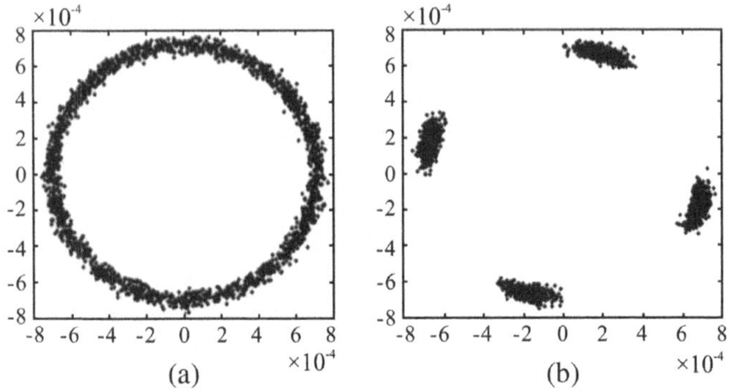

Fig. 2.36 QPSK constellations after frequency offset estimation **a** before FOE **b** after FOE

Here, $b_n = \varphi_n - \varphi_{n-1}$. For QPSK signal, $b_n \in \{0, \pm\pi/2, \pm\pi\}$ So, $\exp(j\times 4\theta_n) = 1$.

$$d_n^4 = \exp(j \times 4 \times \Delta\varphi) \tag{2.142}$$

We accumulate the adjacent dozens of data, and do average operation to eliminate the burst error. The N samples used for accumulation are symmetrically distributed around the nth symbol, so the final result can be regarded as an unbiased estimation.

Finally, we need to find the phase of the average, because we are only interested in its phase. In the previous 4-th power operation, the obtained phase needs to be divided by 4 to obtain phase difference estimation. Each symbol needs to remove the cumulative phase offset, which is caused by frequency offset, so that the data without frequency offset, S_n', can be gotten. Finally, we should further estimate phase to get the correct data. The phase estimation algorithm will be described in the next chapter.

The result of frequency offset estimation algorithm is shown in Fig. 2.36.

2.6.3 Frequency Offset Estimation Algorithm Based on FFT

The above-mentioned frequency offset recovery algorithm based on V-V is focused on constant mode signals. FFT based frequency offset algorithm is a frequency-domain based frequency offset recovery algorithm, which estimates the frequency offset by the frequency component corresponding to the peak value of the signal spectrum after Fourier transform [23]. The following is a simple proof of the algorithm.

Assuming that the signal is M-PSK modulated, the *n*th received symbol is expressed as follows:

$$S_n = I_n + jQ_n = \exp[j(\theta_{d,n} + \theta_{l,n} + 2\pi\Delta f n T_s)] + N_n \tag{2.143}$$

Here, N_n is system noise, T_s is symbol cycle, $\theta_{d,n}$ is the phase of the moduated signal, $\theta_{l,n}$ is the phase noise of LED. Δf is the frequency offset. For QPSK signal, $\theta_{d,n} = \{-3\pi/4, \pi/4, \pi/4, 3\pi/4\}$.

The radiation angle of the received signal is as follows:

$$\varphi_{d,n} = \theta_{d,n} + 2\pi\Delta fnT_s - 2\pi m = \langle\varphi_{d,n}\rangle + \theta^{'}_{d,n} \tag{2.144}$$

Constant m makes $\varphi_{d,n}$ change in section $[-\pi, +\pi]$, $\langle\varphi_{d,n}\rangle$ is the average of phase. After FFT,

$$FFT\{\varphi_{d,n}\} = FFT\{\langle\varphi_{d,n}\rangle\} + FFT\{\theta'_{d,n}\} \tag{2.145}$$

From Eq. (2.145), it can be obtained that

$$\langle\varphi_{d,n}\rangle = \langle\theta_{d,n}\rangle + \langle 2\pi\Delta fnT_s\rangle - \langle 2\pi m\rangle = 2\pi\Delta fnT_s - 2\pi\langle m\rangle \tag{2.146}$$

In order to simplify the calculation of <M>, the constant I is introduced. In this case, it can be ensured that m and are $2\pi\Delta fnT_s$ independent of each other.

Table 2.2 discusses four cases of i. For example, there were two possibilities for the value of $\theta_{d,n}$, when $i = 4k + 1$. The first is that $m = k + 1$ and $\theta_{d,n} = 3/4\pi$, the probability of occurrence is 1/ 4; the second is that $m = k$, the probability of occurrence is 1/ 4, where $\langle m\rangle = 3/4 \times k + 1/4 \times (k + 1) = (4k + 1)/4 = i/4$.

Combining Eq. (2.146) with Table 2.2, Eq. (2.146) can be transformed into

$$\langle\varphi_{d,n}\rangle = 2\pi\Delta fnT_s - \frac{i}{2}\pi \tag{2.147}$$

Based on Eqs. (2.144) and (2.147), $\theta'_{d,n} = \varphi_{d,n} - \langle\varphi_{d,n}\rangle = \theta_{d,n} - \frac{i-4m}{2}\pi$, where $\theta_{d,n}$ is a random value in $\{-3\pi/4, \pi/4, \pi/4, 3\pi/4\}$.

Based on Eq. (2.147), we can draw the curves of $\langle\varphi_{d,n}\rangle$/ arg[exp($j2\pi\Delta fnT_s$)] versus T, when $\Delta f > 0$ (Fig. 2.37). $\langle\varphi_{d,n}\rangle$ is marked as a solid line, which is the phase information of QPSK signal itself. arg[exp($j2\pi\Delta fnT_s$)] is marked as a dotted line, which is the additional phase induced by Δf [24].

Table 2.2 Values of ⟨m⟩ in four cases

i	$\theta_{d,n}$	m	Possibility	m
4 k	$\pm\pi/4, \pm 3\pi/4,$	k	1	i/4
4 k + 1	$\neq 3\pi/4$ $= 3\pi/4$	k k + 1	3/4 1/4	i/4
4 k + 2	$= -\pi/4, -3\pi/4$ $= \pi/4, 3\pi/4$	k k + 1	1/2 1/2	i/4
4 k + 3	$\neq -3\pi/4$ $= -3\pi/4$	k + 1 k	3/4 1/4	i/4

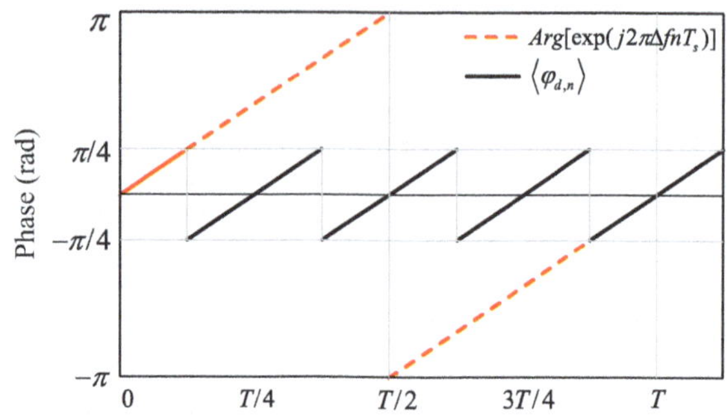

Fig. 2.37 The $\langle\varphi_{d,n}\rangle$ and $\arg[\exp(j2\pi\,\Delta fnT_s)]$ as functions of T for QPSK signal

It can be seen that the slopes of $\langle\varphi_{d,n}\rangle$ and $\arg[\exp(j2\pi\,\Delta fnT_s)]$ are the same, but the period of $\arg[\exp(j2\pi\,\Delta fnT_s)]$ is that of the $\langle\varphi_{d,n}\rangle$ 4 times. Therefore, the peak value appears at the first harmonic frequency, $4|\Delta f|$, after FFT. Of course, there will be a small peak at the higher harmonic frequency, but the peak at $4|\Delta f|$ is the largest. In this way, by observing the spectrum of the received signal phase, we can divide the frequency corresponding to the peak value by 4 to get the frequency offset value of the signal. Figure 2.38 shows the diagrams of the algorithm, in which (a) is the basic diagram of FFT based frequency offset algorithm, and (b) is the optimized frequency offset compensation algorithm. Since the frequency offset compensation

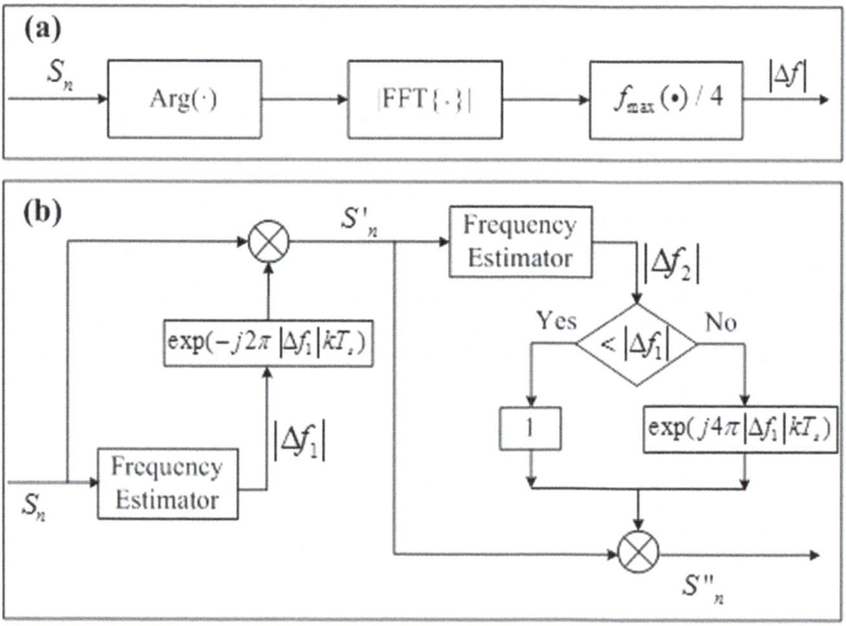

Fig. 2.38 Diagrams of frequency offset algorithm based on FFT: **a** simplification **b** optimization

algorithm based on FFT cannot know the direction of frequency offset, it needs to make a decision after the first frequency offset compensation [36].

2.6.4 Frequency Offset Estimation Algorithm Based on Quartic Power Operation

When the range of frequency difference between the signal light and the local oscillator light is (−3.0 GHz, +3.0 GHz), the quartic method is used. The schematic diagram of the quartic frequency offset estimation algorithm is shown in the following figure.

If the phase of the received signal is $\theta_k = \theta_S(k) + \Delta\omega k T_i + \theta_n + \theta_{ASE}$. Here, $\theta_S(k)$ is the original signal phase, θ_n is the phase caused by laser linewidth, θ_{ASE} is phase noise. As Fig. 2.39 shows, by taking the conjugate of the signal V_{k-1} and multiplied with V_k, we can get

$$\begin{aligned} V_k V_{k-1}^* &= \exp\left[j(\theta_k - \theta_{k-1})\right] \\ &= \exp\{j\Delta\theta_s\} \cdot \exp\{j\Delta\omega T_i\} \cdot \exp\{j\Delta\theta_{ASE}\} \end{aligned} \tag{2.148}$$

Because θ_n is a slowly varying signal, the difference between the front and back samples is zero. In an ideal case, the value of $\Delta\theta_s$ is $\{0, \frac{\pi}{2}, \pi, -\frac{\pi}{2}\}$, $(V_k V_{k-1}^*)^4$ can remove $\Delta\theta_s$. The frequency offset is also a slow change compared with the symbol rate. After averaging the calculated results of continuous N_f samples, the influence of Gaussian noise can be eliminated. Then, the main value of the radiation angle is taken, and the phase caused by the frequency offset, $\Delta\varphi = k\Delta\omega T_i$, is obtained. By subtracting $\Delta\varphi = k\Delta\omega T_i$ from the original phase of the k-th symbol, the signal phase after eliminating the influence of frequency offset can be obtained. Similarly, the next block of NF data also uses the same process described above to estimate the frequency offset value. Since the phase damage caused by frequency offset is accumulated with the number of symbols, the system needs to store the accumulated phase up to the current symbol. In order to prevent the infinite growth of the accumulation phase, the accumulation phase needs to be calculated with mode 2π, which will not affect the result of frequency offset compensation.

Algorithm decomposition:

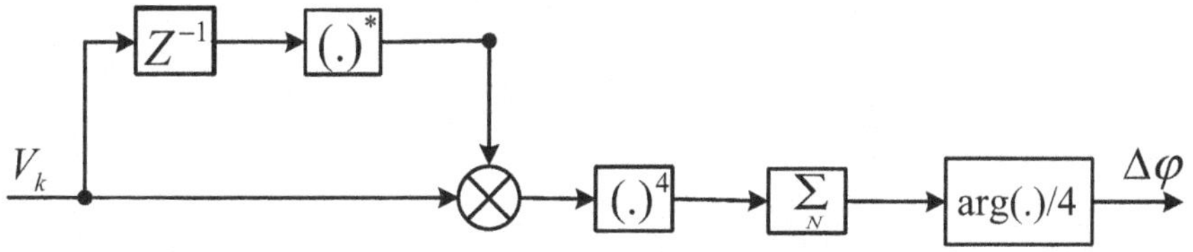

Fig. 2.39 Schematic diagram of quartic frequency offset estimation algorithm

(1) The value interval of the CMA output is used for subsequent operations (including frequency offset estimation, phase offset estimation, decision, decoding, etc.), that is, to change the equalization output from 2 times sampling to 1 times sampling.

(2) Conjugatation and multiplication: Decompose Eq. (2.148) into real and imaginary part:

$$\begin{aligned}
&\text{X.pol} = V_{f_x_r} + jV_{f_x_i} \\
&V_{k,x}V^{*}_{k-1,x} = (X'_{I,k} + jX'_{Q,k}) \cdot (X'_{I,k-1} - jX'_{Q,k-1}) \\
&\text{Real part}: V_{f_x_r} = X'_{I,k} \cdot X'_{I,k-1} + X'_{Q,k} \cdot X'_{Q,k-1} \\
&\text{Imaginary part}: V_{f_x_i} = X'_{Q,k} \cdot X'_{I,k-1} - X'_{I,k} \cdot X'_{Q,k-1}
\end{aligned} \tag{2.149}$$

Similarly,

$$\begin{aligned}
&\text{Y. pol} = V_{f_y_r} + jV_{f_y_i} \\
&\text{Real part}: V_{f_y_r} = Y'_{I,k} \cdot Y'_{I,k-1} + Y'_{Q,k} \cdot Y'_{Q,k-1} \\
&\text{Imaginary part}: V_{f_y_i} = Y'_{Q,k} \cdot Y'_{I,k-1} - Y'_{I,k} \cdot Y'_{Q,k-1}
\end{aligned} \tag{2.150}$$

(3) The fourth power operation (It can be divided into two square operations.): X. pol can be expressed as follows:

$$\begin{aligned}
V_{f_2_x} &= V_{f_2_x_r} + jV_{f_2_x_i} \\
&= (V_{f_x_r} + jV_{f_x_i})^2 \\
&= V^2_{f_x_r} - V^2_{f_x_i} + 2jV_{f_x_r} \cdot V_{f_x_i}
\end{aligned} \tag{2.151}$$

$$\begin{aligned}
V_{f_4_x} &= V_{f_4_x_r} + jV_{f_4_x_i} \\
&= (V_{f_2_x_r} + jV_{f_2_x_i})^2 \\
&= V^2_{f_2_x_r} - V^2_{f_2_x_i} + 2jV_{f_2_x_r} \cdot V_{f_2_x_i}
\end{aligned} \tag{2.152}$$

X. pol:

$$\begin{aligned}
V_{f_2_y} &= V_{f_2_y_r} + jV_{f_2_y_i} \\
&= \left(V_{f_y_r} + jV_{f_y_i}\right)^2 \\
&= V^2_{f_y_r} - V^2_{f_y_i} + 2jV_{f_y_r} \cdot V_{f_y_i}
\end{aligned} \tag{2.153}$$

$$\begin{aligned}
V_{f_4_y} &= V_{f_4_y_r} + jV_{f_4_y_i} \\
&= \left(V_{f_2_y_r} + jV_{f_2_y_i}\right)^2 \\
&= V^2_{f_2_y_r} - V^2_{f_2_y_i} + 2jV_{f_2_y_r} \cdot V_{f_2_y_i}
\end{aligned} \tag{2.154}$$

(4) **Average**: Take N_f samples for arithmetic average after the operations in (2.152) and (2.154). It can be seen from (2.149) that, ideally, after conjugation and quartic, the phase of each result is $\Delta\omega T$. That is, the value of each result is the same. However, in the presence of phase noise, the actual value is some values near the ideal value, and the average result of these values is $\Delta\omega T$.

After the average operation, the following four values are obtained (it is not necessary to divide by NF, because the real part and the imaginary part are the sums of NF. NF will be removed when taking the phase angle in the next step:

$$\sum_{Nf} V_{f\_4\_x\_r}, \sum_{Nf} V_{f\_4\_x\_i}, \sum_{Nf} V_{f\_4\_y\_r}, \sum_{Nf} V_{f\_4\_y\_i} \tag{2.155}$$

(5) Taking the amplitude angle: First, calculate the angle according to the following formula, and judge the quadrant based on the sign of $\sum_{Nf} V_{f\_4\_x\_r}, \sum_{Nf} V_{f\_4\_x\_i}, \sum_{Nf} V_{f\_4\_y\_r}, \sum_{Nf} V_{f\_4\_y\_i} \sum_{Nf} V_{f\_4\_y\_i}$. Taking X. pol as an example:

(a) if the real part $\sum_{Nf} V_{f\_4\_x\_r} < 0$, and the imaginary part $\sum_{Nf} V_{f\_4\_x\_i} < 0$, the argument is $\Delta\omega T_{x,4} = \Delta\omega T_{x,1,4} - \pi$;
(b) if the real part $\sum_{Nf} V_{f\_4\_x\_r} < 0$, and the imaginary part $\sum_{Nf} V_{f\_4\_x\_i} > 0$, the argument is $\Delta\omega T_{x,4} = \Delta\omega T_{x,1,4} + \pi$;
(c) if the real part $\sum_{Nf} V_{f\_4\_x\_r} > 0$, the argument is $\Delta\omega T_{x,4} = \Delta\omega T_{x,1,4}$.
(d) Otherwise, if the real part is zero, when the imaginary part is larger than zero, $\Delta\omega T_{x,4} = \frac{\pi}{2}$; when the imaginary part is smaller than zero, $\Delta\omega T_{x,4} = -\frac{\pi}{2}$. That is, the value of frequency offset estimation is $[-2\pi, +2\pi]$):

$$\Delta\omega T_{x,1,4} = arctg\left(\frac{\sum_{Nf} V_{f\_4\_x\_i}}{\sum_{Nf} V_{f\_4\_x\_r}}\right), \quad \Delta\omega T_{y,1,4} = arctg\left(\frac{\sum_{Nf} V_{f\_4\_y\_i}}{\sum_{Nf} V_{f\_4\_y\_r}}\right) \tag{2.156}$$

Then divide the corrected angle by 4 to get the frequency offset value:

$$\Delta\omega T_x = \frac{1}{4} \cdot \Delta\omega T_{x,4}, \quad \Delta\omega T_y = \frac{1}{4} \cdot \Delta\omega T_{y,4} \tag{2.157}$$

Frequency offset correction and sample value output:

Multiply the values of the two polarization states after equalization by $e^{-j\Delta\omega kT}$, where k is the sample number. Because of the uncertainty of the initial phase of the first sample, the absolute phase values of all samples are not determined, but the relative phase values are determined.

$$\begin{aligned} V_{k\_x} &= V_{fout\_k\_x\_r} + jV_{fout\_k\_x\_i} \\ &= (X'_{I,k} + jX'_{Q,k}) \cdot (\cos(\Delta\omega_x kT) - j\sin(\Delta\omega_x kT)) \end{aligned} \tag{2.158}$$

The real part and the imaginary part of the X-polarization output are, respectively,

$$\begin{aligned} V_{fout\_k\_x\_r} &= X'_{I,k} \cdot \cos(\Delta\omega_x kT) + X'_{Q,k} \cdot \sin(\Delta\omega_x kT) \\ V_{fout\_k\_x\_i} &= X'_{Q,k} \cdot \cos(\Delta\omega_x kT) - X'_{I,k} \cdot \sin(\Delta\omega_x kT) \end{aligned} \tag{2.159}$$

Similarly, the real part and the imaginary part of the Y-polarized state output is, respectively,

$$\begin{aligned} V_{fout\_k\_y\_r} &= Y'_{I,k} \cdot \cos(\Delta\omega_y kT) + Y'_{Q,k} \cdot \sin(\Delta\omega_y kT) \\ V_{fout\_k\_y\_i} &= Y'_{Q,k} \cdot \cos(\Delta\omega_y kT) - Y'_{I,k} \cdot \sin(\Delta\omega_y kT) \end{aligned} \tag{2.160}$$

2.7 Phase Compensation Algorithm

2.7.1 V-V Based Phase Compensation Algorithm

Viterbi–Viterbi algorithm is a classical compensation algorithm. The principle block diagram is as follows (Fig. 2.40).

Set the phase of the received signal as $\theta_k = \theta_S(k) + \Delta\omega kT_i + \theta_n + \theta_{ASE}$, in which $\Delta\omega kT_i$ has been removed in the previous frequency offset estimation. Then, do fourth power operation for the rest signal, $V^4(k) = \exp\{j4\theta_s(k)\} \cdot \exp\{j4\theta_n\} \cdot \exp\{j4\theta_{ASE}\}$. Assuming $\theta_S = \{0, \frac{\pi}{2}, \pi, -\frac{\pi}{2}\}$, $V^4(k)$ can remove the signal phase. The phase offset estimation phase can be obtained by adding and averaging N_P $V^4(k)$s, removing the phase noise, and extracting argument. This can be expressed as follows:

$$\theta_e = \frac{1}{4} \cdot \arg\left[\sum_{i=1}^{N} V^4(i)\right] \tag{2.161}$$

Algorithm decomposition,

1. **The fourth power operation (It can be divided into two square operations**.):
   X. pol:

$$V_{p\_2\_x} = V_{p\_2\_x\_r} + jV_{p\_2\_x\_i}$$

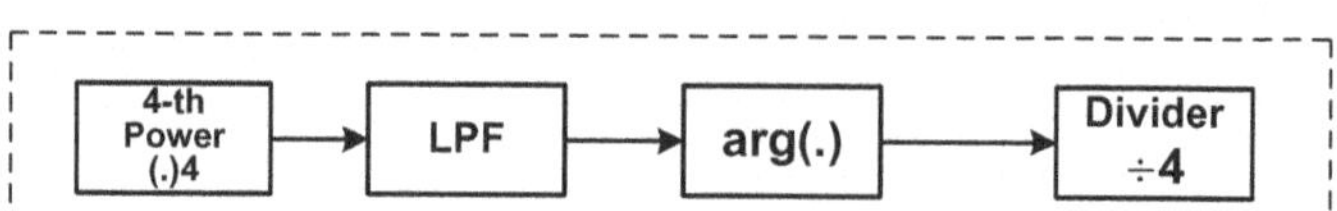

Fig. 2.40 Schematic diagram of Viterbi–Viterbi phase offset estimation algorithm

$$
\begin{aligned}
&= (V_{\mathrm{fout\_k\_}x\_r} + jV_{\mathrm{fout\_k\_}x\_\mathrm{i}})^2 \\
&= V^2_{\mathrm{fout\_k\_}x\_r} - V^2_{\mathrm{fout\_k\_}x\_\mathrm{i}} + 2jV_{\mathrm{fout\_k\_}x\_r} \cdot V_{\mathrm{fout\_k\_}x\_\mathrm{i}}
\end{aligned} \tag{2.162}
$$

$$
\begin{aligned}
V_{p\_4\_x} &= V_{p\_4\_x\_r} + jV_{p\_4\_x\_i} \\
&= (V_{\mathrm{p}\_2\_x\_r} + jV_{\mathrm{p}\_2\_x\_i})^2 \\
&= V^2_{\mathrm{p}\_2\_x\_r} - V^2_{\mathrm{p}\_2\_x\_i} + 2jV_{\mathrm{p}\_2\_x\_r} \cdot V_{\mathrm{p}\_2\_x\_i}
\end{aligned} \tag{2.163}
$$

Y. pol:

$$
\begin{aligned}
V_{p\_2\_y} &= V_{p\_2\_y\_r} + jV_{p\_2\_y\_i} \\
&= (V_{\mathrm{fout\_k\_}y\_r} + jV_{\mathrm{fout\_k\_}y\_\mathrm{i}})^2 \\
&= V^2_{\mathrm{fout\_k\_}y\_r} - V^2_{\mathrm{fout\_k\_}y\_\mathrm{i}} + 2jV_{\mathrm{fout\_k\_}y\_r} \cdot V_{\mathrm{fout\_k\_}y\_\mathrm{i}}
\end{aligned} \tag{2.164}
$$

$$
\begin{aligned}
V_{p\_4\_y} &= V_{p\_4\_y\_r} + jV_{p\_4\_y\_i} \\
&= (V_{\mathrm{p}\_2\_y\_r} + jV_{\mathrm{p}\_2\_y\_i})^2 \\
&= V^2_{\mathrm{p}\_2\_y\_r} - V^2_{\mathrm{p}\_2\_y\_i} + 2jV_{\mathrm{p}\_2\_y\_r} \cdot V_{\mathrm{p}\_2\_y\_i}
\end{aligned} \tag{2.165}
$$

2. **Averaging**: After summation, four values can be obtained:

$$
\sum_{Np} V_{p\_4\_x\_r}, \sum_{Np} V_{p\_4\_x\_i}, \sum_{Np} V_{p\_4\_y\_r}, \sum_{Np} V_{p\_4\_y\_i} \tag{2.166}
$$

3. **Taking argument**: Get the phase estimation. First, calculate the angle according to the following formula, and judge the quadrant based on the sign of $\sum_{Nf} V_{p\_4\_x\_r}$, $\sum_{Nf} V_{p\_4\_x\_i}$, $\sum_{Nf} V_{p\_4\_y\_r}$, $\sum_{Nf} V_{p\_4\_y\_i}\frac{dy}{dx}$. Taking X. pol as an example:
   a. if the real part $\sum_{Np} V_{p\_4\_x\_r} < 0$, and the imaginary part $\sum_{Nf} V_{p\_4\_x\_i} < 0$, the argument is $PE_{x,4} = PE_{x,1,4} - \pi$;
   b. if the real part $\sum_{Np} V_{p\_4\_x\_r} < 0$, and the imaginary part $\sum_{Nf} V_{p\_4\_x\_i} > 0$, the argument is $PE_{x,4} = PE_{x,1,4} + \pi$;
   c. Otherwise, if the real part $\sum_{Np} V_{f\_4\_x\_r} > 0$, the argument is $PE_x = PE_{x,1}$.
   d. Otherwise, if the real part is zero, when the imaginary part is larger than zero, $PE_{x,4} = \frac{\pi}{2}$; when the imaginary part is smaller than zero, $PE_{x,4} = -\frac{\pi}{2}$.

$$
PE_{x,1,4} = arctg\left(\frac{\sum_{Np} V_{p\_4\_x\_i}}{\sum_{Np} V_{p\_4\_x\_r}}\right), PE_{y,1,4} = arctg\left(\frac{\sum_{Np} V_{p\_4\_y\_i}}{\sum_{Np} V_{p\_4\_y\_r}}\right) \tag{2.167}
$$

Then divide the corrected phase by 4 to get the phase deviation value:

$$PE_x = \frac{1}{4} \cdot PE_{x,4}, PE_y = \frac{1}{4} \cdot PE_{y,4} \tag{2.168}$$

4. **Phase offset correction**: Multiply the values of the two polarization states after frequency offset correction by e^{-jPE}.

$$\begin{aligned} VP_{k\_x} &= VP_{pout\_k\_x\_r} + jVP_{pout\_k\_x\_i} \\ &= (V_{fout\_k\_x\_r} + jV_{fout\_k\_x\_i}) \cdot (\cos(PE_{x\_k}) - j\sin(PE_{x\_k})) \end{aligned} \tag{2.169}$$

The real part and the imaginary part of the X-polarization output are, respectively,

$$VP_{pout\_k\_x\_r} = V_{fout\_k\_x\_r} \cdot \cos(PE_{x\_k}) + V_{fout\_k\_x\_i} \cdot \sin(PE_{x\_k})$$

$$VP_{pout\_k\_x\_i} = V_{fout\_k\_x\_i} \cdot \cos(PE_{x\_k}) - V_{fout\_k\_x\_r} \cdot \sin(PE_{x\_k}) \tag{2.170}$$

Similarly, the real part and the imaginary part of the Y-polarization output are, respectively,

$$VP_{pout\_k\_y\_r} = V_{fout\_k\_y\_r} \cdot \cos(PE_{y\_k}) + V_{fout\_k\_y\_i} \cdot \sin(PE_{y\_k})$$

$$VP_{pout\_k\_y\_i} = V_{fout\_k\_y\_i} \cdot \cos(PE_{y\_k}) - V_{fout\_k\_y\_r} \cdot \sin(PE_{y\_k}) \tag{2.171}$$

2.7.2 Feedforward Based Phase Rotation Algorithm

For high-order modulation, the phase compensation generally adopts the phase rotation algorithm based on feedforward, and its principle is shown in Fig. 2.41.

For 16QAM and 64QAM modulation formats, the maximum likelihood-based phase rotation estimation algorithm can be used to compensate for the phase deviation.

For each sample value in the π/2 phase range, rotate them according to the unit of ϕ = n/B. *π/2, n−{0, 1, 2…B − 1}, B represents the rotation sample value used for the test (B is a constant of 32.64). For all the rotated sample values, do maximum likelihood decision, E(k) = min(abs(Xin.*ϕb-ideal_signal)), with the ideal high-order modulation signals (16QAM, 64QAM, if considering the SDO compatibility, QPSK can also use this BPS algorithm). Select the smallest e (k) and decide which ϕ should be selected. Then phase shift the signal by ϕ to realize phase compensation (Fig. 2.42).

When the signal is 16QAM modulation format, the ideal signal for maximum likelihood decision is {−3 −1 1 3} + 1J * {−3 −1 1 3}.

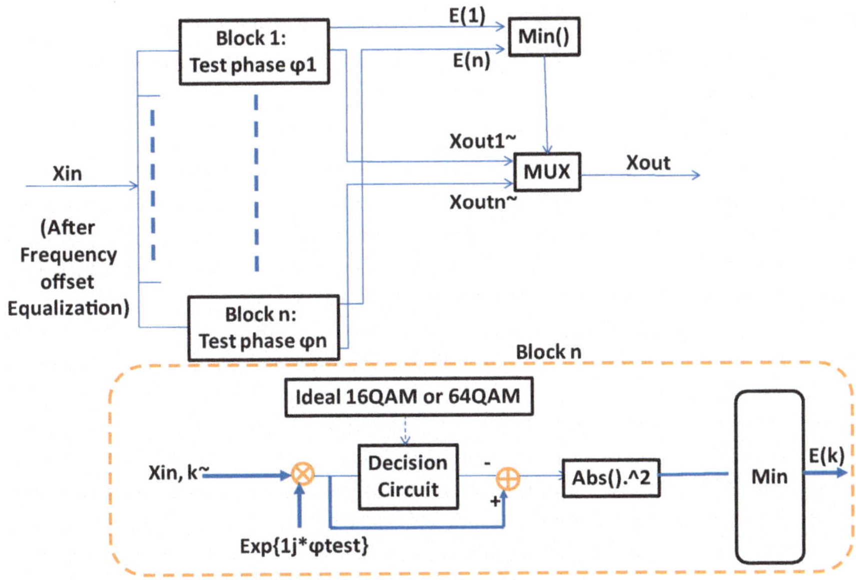

Fig. 2.41 Phase rotation algorithm based on feedforward

Fig. 2.42 16QAM decision rule

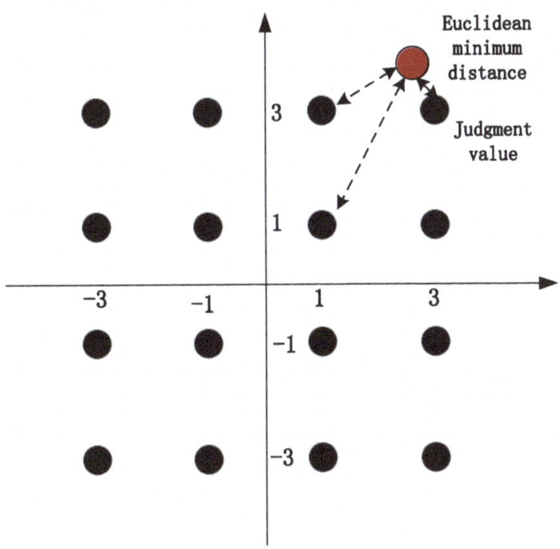

2.7.3 A Recovery Algorithm Based on the Combination of Modified V-V Phase Bias Algorithm and Maximum Likelihood

From the above introduction, we can find that both frequency offset and phase offset recovery algorithms based on classical V-V algorithm are only for constant modulus signals. However, for high-order QAM signals, there are amplitude changes besides phase changes. If we directly use quartic operation, it will be difficult to completely remove the signal phase and realize carrier phase estimation.

Here, taking 16QAM and 9QAM as examples, a recovery algorithm based on the combination of modified V-V phase offset algorithm and maximum likelihood is introduced [37].

Figure 2.43 shows the constellations of 16QAM and 9QAM. In the algorithm, we divide 16QAM constellation into three rings C1, C2, and C3. We can see that the two rings C1 and C3 are constellations of standard QPSK signals, but they have different OSNR. This means we can use the v-v-based phase offset recovery algorithm for the constellation of these two QPSK signals. Similarly, when we observe the constellation diagram of 9QAM signal, the two rings C2 and C3 are four-point signals. The only difference is that the four-point signals on C2 ring are not QPSK signals. After phase rotation processing, they can be processed the same as 16QAM signals.

Figure 2.44 shows the system diagram of the whole algorithm. Taking 16QAM signal as an example, we first divide the signal into three rings according to the amplitude of the signal data. Take the data on C1 and C3 for the modified V-V phase offset recovery algorithm to get the first phase estimation value θ_n^{est1}, which can be obtained from the following formula:

$$\theta_n^{est1} = \frac{1}{4} \cdot \left\{ \sum_{i:s_i=C_1} \frac{s_i^4}{\left|s_i^4\right|} + W_1 \cdot \sum_{i:s_i=C_3} \frac{s_i^4}{\left|s_i^4\right|} \right\} \tag{2.172}$$

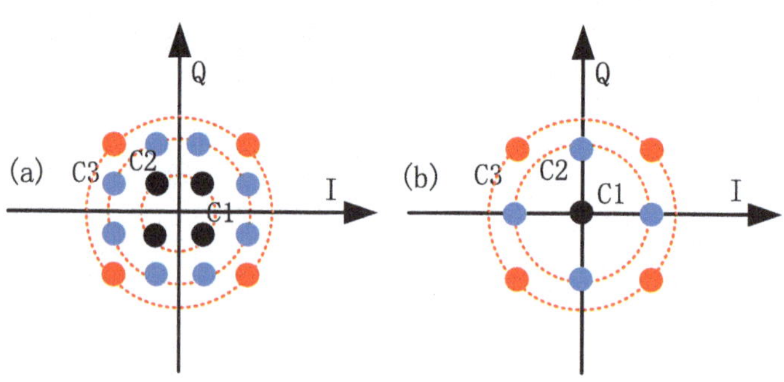

Fig. 2.43 Constellations of 16QAM and 9QAM

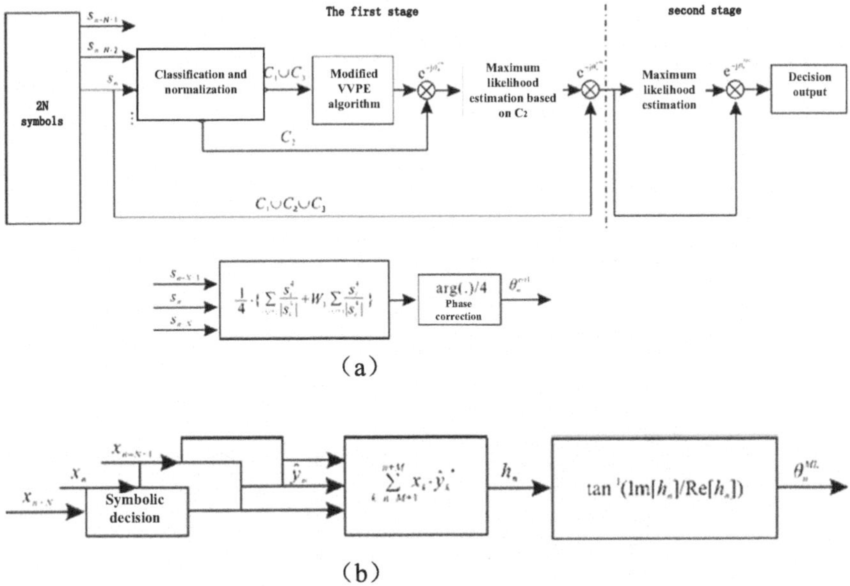

Fig. 2.44 **a** System diagram of algorithm **b** modified V-V algorithm **c** maximum likelihood estimation

Considering OSNR of the constellation points in the outer circle C3 are higher than those in the inner circle C1, the weighting coefficient W1 is introduced to make the phase noise estimated by the outer circle constellation points better than that of the inner circle constellation points.

On the basis of the existing rough estimation of phase noise θ_n^{est1}, the symbols in the middle circle C2 are compensated (multiplied $\mathrm{e}^{-\mathrm{j}\theta_n^{est1}}$) according to θ_n^{est1}. The compensated constellation points and the corresponding symbol decision results are put into the first maximum likelihood phase estimator to obtain the residual phase noise θ_n^{est2}. More accurate phase noise can be obtained.

$$\theta_n^{est} = \theta_n^{est1} + W_2\theta_n^{est2} \tag{2.173}$$

The result of Eq. (2.173) is taken as the output value of the first phase noise estimation. W2 is a weight coefficient related to the proportion of constellation points in C2 and the corresponding OSNR of three circles of constellation points. At the end of the first stage, all constellation points distributed on C1, C2, and C3 are multiplied by $e^{-j\theta_n^{est}}$ for phase correction to obtain x_n.

In order to obtain a more accurate and reliable phase noise estimation, then put the first stage modified result x_n into the second-stage maximum likelihood phase estimator, whose structure is shown in Fig. 2.43b, and the obtained second-stage phase noise estimate is

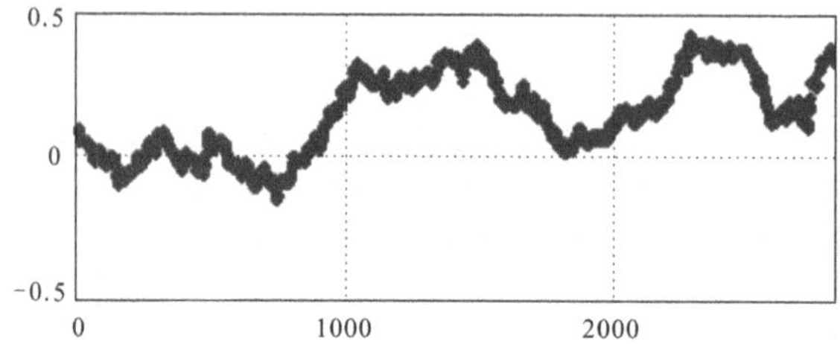

Fig. 2.45 The phase noise tracking process

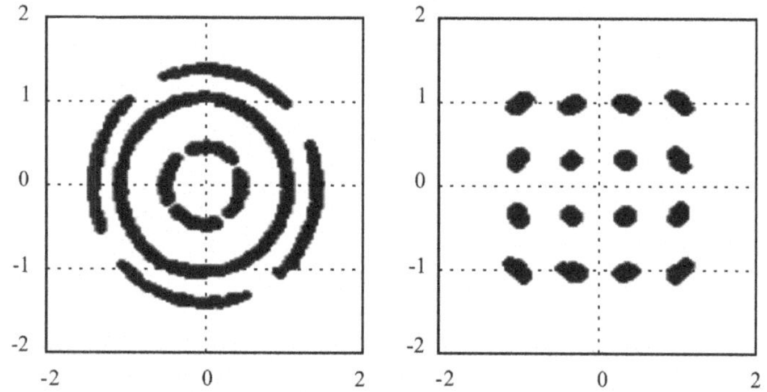

Fig. 2.46 16QAM constellations before and after modification by the new segmentation algorithm

$$\theta_n^{ML} = \tan^{-1}(\mathrm{Im}[h_n]/\mathrm{Re}[h_n]), \quad h_n = \sum_{k=n-M+1}^{n+M} x_k \cdot \hat{y}_k^* \tag{2.174}$$

Here $\hat{y}_k^*$ is the decision of three input symbol x_n, and the symbol length involved in the second stage of MLE is 2\M.

Therefore, the final output of the carrier recovery module after phase noise compensation is $x_n' = x_n e^{-j\theta_n^{ML}}$. As shown in Figs. 2.45 and 2.46, the phase noise tracking process and the constellations of 16QAM signal recovered with the above algorithm are presented, respectively. By mapping the x_n' to the 01 bit sequence by decision circuit, the received signal can be completely recovered.

2.7.3.1 Blind Phase Search Algorithm and Its Improvement

The classical V-V phase compensation algorithm and angle based phase estimation algorithm require the signal as constant mode signal. In order to improve the linewidth tolerance of high-order QAM, the blind phase search (BPS) algorithm is proposed for carrier frequency and phase estimation. In Sect. 2.7.3, the modified Viterbi–Viterbi phase estimation algorithm is used to estimate and compensate the phase noise of 16QAM signal.

At present, for the arbitrary M-order QAM modulation format, we mainly introduce the blind phase search algorithm (BPS) [38] and the improved blind phase search algorithm (BPS/ML) [39].

The basic block diagram of BPS algorithm is presented in Fig. 2.47, which adopts a feedforward structure.

Firstly, rotate the sampling symbol r_k with B test phases ϕ_b. The test phase φ_b B is

$$\varphi_b = \frac{b}{B} \cdot \frac{\pi}{2} \tag{2.175}$$

where b is taken from $-B/2$ to $B/2$-1 to comprehensively correct the influence of the phase noise. Parameter B directly affects the accuracy of the algorithm. A large parameter B will make the phase noise estimation more accurate, but at the same time, it will lead to more computation and algorithm complexity. Generally, 32 is a suitable B value for 16QAM signal.

Then the rotated symbol is subsequently input into the decision circuit, which outputs the ideal constellation point $\hat{y}_{k,b}$ closest to the input symbol. The square distance $\left|d_{k,b}\right|^2$ between the rotated constellation with phase noise and the ideal constellation can be calculated. In order to further filter out other additional noises in the receiver, we calculate the sum of the square distance of 2N consecutive constellation points (the best value of N depends on the quotient of laser linewidth and symbol rate, generally n is 6, 7… 10):

$$s_{k,b} = \sum_{n=-N}^{N} \left|d_{k-n,b}\right|^2 \tag{2.176}$$

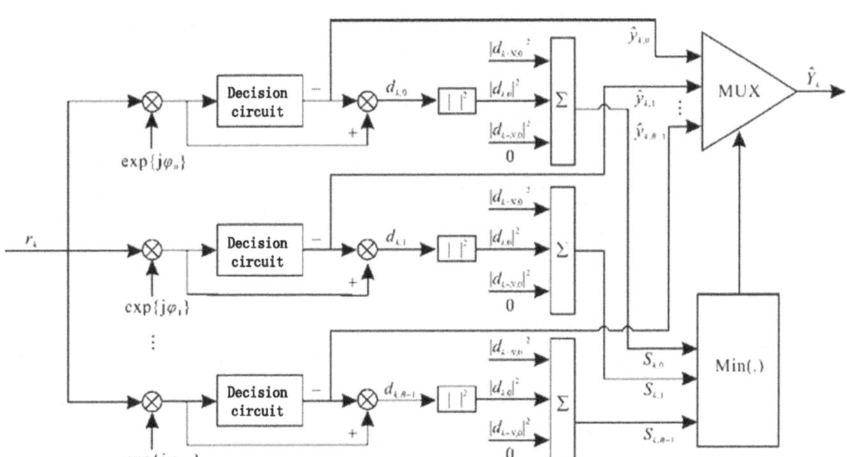

Fig. 2.47 Block diagram of 16QAM blind phase search algorithm (BPS)

Finally, the minimum value is selected from the sum of the square distance of B constellation points, and the corresponding value $\hat{y}_{k,b}$ is selected by switch control as the decision of the transmitted signal $\hat{Y}_k$, so as to eliminate the influence of phase noise.

Although BPS algorithm can effectively compensate the phase noise of 16QAM signal and can be flexibly applied to higher order QAM modulation, it has an obvious disadvantage of high computational complexity. Therefore, a BPS/ML algorithm with two-level cascaded structure is proposed in Ref. [39], and its algorithm structure is shown in Fig. 2.48.

In the first stage, BPS algorithm is still used to roughly estimate the best phase angle of constellation. Because of the low accuracy requirement of the first level rough estimation, the number of test phases ϕ_b can be reduced for lower computational complexity. In the second stage, a maximum likelihood phase estimator is introduced to improve the accuracy of the first stage. The originally received symbol r_k and the decision result output by the first stage processing $\hat{Y}_k^{(1)}$ are input into the maximum likelihood phase estimator to obtain the second-stage phase noise estimation value.

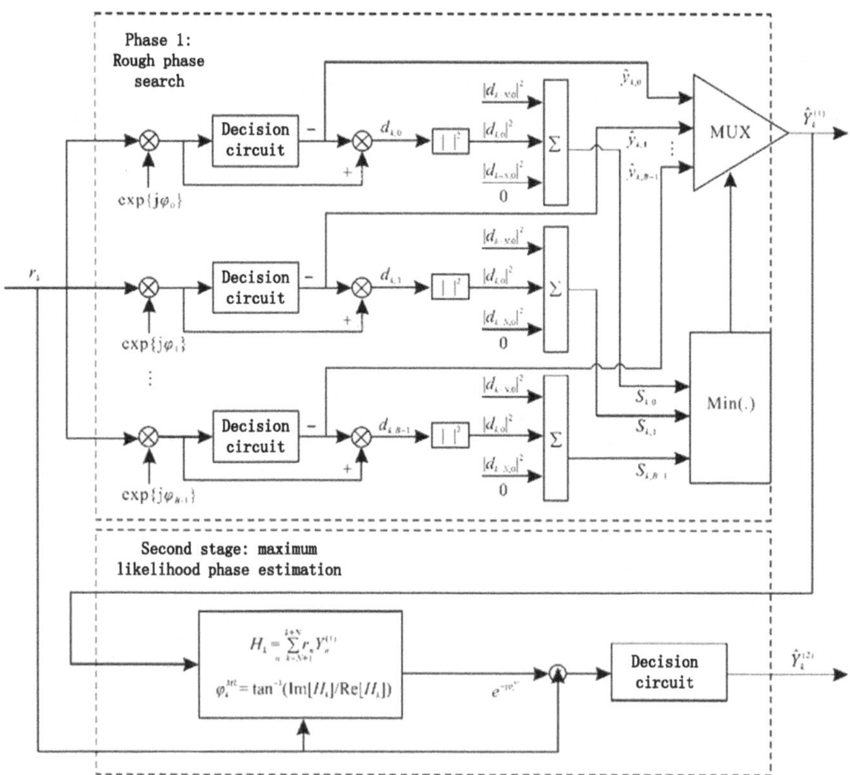

Fig. 2.48 Improved BPS/ML block diagram

$$\varphi_k^{ML} = \tan^{-1}(\mathrm{Im}[H_k]/\mathrm{Re}[H_k]), \quad H_k = \sum_{n=k-N+1}^{k+N} r_n \hat{Y}_n^{(1)} \tag{2.177}$$

The phase-corrected signal $r_k e^{-j\varphi_k^{ML}}$ is then input into the decision circuit to obtain a second-level decision output $\hat{Y}_k^{(2)}$.

Of course, in order to obtain higher estimation accuracy, multiple cascaded MLE phase estimation units can be added following the second level of MLE to approach the best estimation value infinitely.

In terms of algorithm complexity, it is proved in Ref. [39] that the computation cost of every additional cascaded maximum likelihood phase estimator is equivalent to the computation cost of adding two new test phases in pure BPS algorithm. However, in the first stage of rough estimation, the computational complexity of test phases can be significantly reduced by nearly 50%. Therefore, BPS/ML algorithm continues the performance of BPS algorithm, moreover, the computation cost can be effectively reduced (Figs. 2.49 and 2.50).

In reference [37], the performance of the traditional segmentation algorithm [15], BPS, BPS/ML, and the above new segmentation algorithm are compared under different laser linewidth conditions. The simulation results show that the new segmentation algorithm combined with maximum likelihood estimation can improve the tolerance

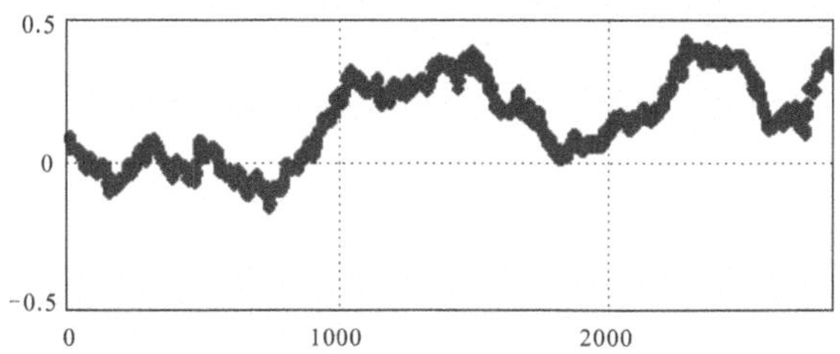

Fig. 2.49 The phase noise tracking process

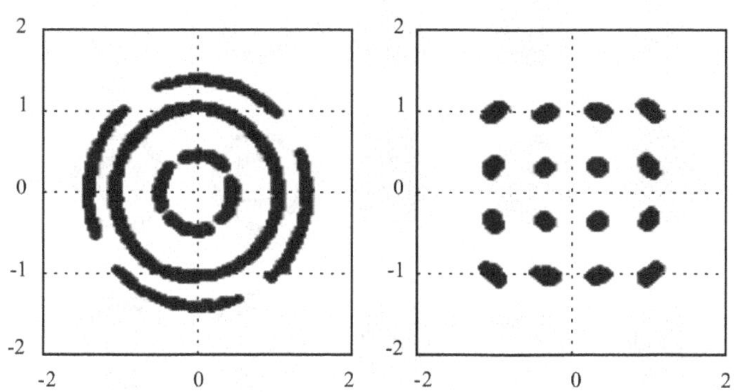

Fig. 2.50 16QAM constellations before and after modification by the new segmentation algorithm

of phase noise $\Delta v \cdot T_s$ by 1 dB at the $BER = 10^{-3}$. When the phase noise $\Delta v \cdot T_s$ is larger than 2×10^{-3}, the performance of the new segmentation algorithm is better than that of the other three algorithms. The performance of traditional segmentation algorithm is the worst of the four algorithms. In addition, the computational complexity of the new segmentation algorithm isanalyzed in detail in Ref. [37]. Assuming that the number of rotation test phases used in BPS and BPS/ML algorithm is 14 and 32, respectively, the computational complexity of the new segmentation algorithm compared with BPS/ML is only about half of the latter, and compared with BPS algorithm, the complexity is only one quarter of the latter. Therefore, the 16QAM carrier recovery algorithm with low computational complexity and feedforward structure is more suitable for high-speed coherent optical system than other algorithms.

2.7.4 Algorithm Demonstration and Summary

The above describes a series of equalization algorithms for the damage in optical communication system. According to the nature of the damage, these algorithms operate in a certain order, which is shown in Fig. 2.1. In this section, the digital signal processing flow chart of 32 Gbaud polarization multiplexing QPSK, 8QAM, and 16QAM signals with coherent reception is given. The result constellations after the recovery algorithms are shown in Fig. 2.51.

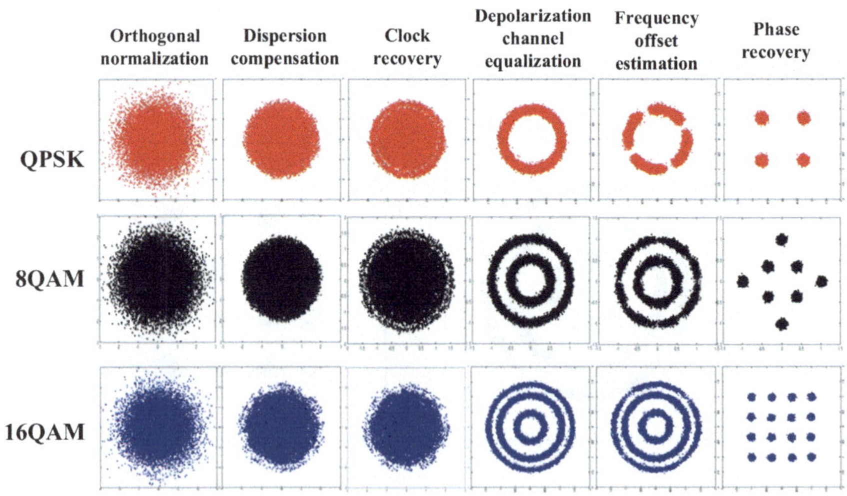

Fig. 2.51 The result constellations after DSPs

After the above-mentioned basic DSP algorithms in this section, the signal is orthogonalized and normalized; the dispersion of optical fiber link is compensated; the clock component of the signal is extracted; the sampling error is eliminated; the channel estimation and equalization are carried out simultaneously to get the independent signals in each polarization; then the frequency offset of signal light and local oscillator light is estimated, and the phase noise is estimated after the frequency offset estimation. Finally, the constellation of the originally transmitted signal can be recovered. These basic digital signal processing algorithms interact with each other and are indispensable. They also form the basis of the follow-up work in this chapter. The following work of this chapter aims at some damage of the system, or do some innovation and supplement of these algorithms, or propose new treatment methods.

References

1. J.F. Cardoso, maximum likelihood source separation: equivariance and adaptivity, in *Proceedings of Sysid'97 Ifac Symposium on System Identification*, pp. 1063–1068 (1997)
2. A. HyväRinen, Fast and robust fixed-point algorithms for independent component analysis. IEEE Trans. Neural Netw. **10**(3), 626–634 (1999)
3. X. Zhou, J. Yu, Multi-level, multi-dimensional coding for high-speed and high-spectral-efficiency optical transmission. J. Lightwave Technol. **27**(16), 3641–3653 (2009)
4. J. Yang, J. Werner, A.D. Guy, The multimodulus blind equalization and its generalized algorithms. IEEE J. Sel. Areas Commun. **20**(5), 997–1015 (2002)
5. E. Lach, W. Idler, Modulation formats for 100G and beyond. Optical Fiber Technol. **17**(5), 377–386 (2011)
6. L. Tao, Y. Ji, J. Liu, A.P.T. Lau, N. Chi, C. Lu, Advanced modulation formats for short reach optical communication systems. IEEE Netw. **27**(6), 6–13 (2013)
7. A. Nafta, P. Johannisson, M. Shtaif, Blind equalization in optical communications using independent component analysis. J. Lightwave Technol. **31**(12), 2043–2049 (2013)
8. J. Yu et al., Coherent optical communication technology based on digital signal processing. People's Posts and Telecommunications Press, Beijing 70 (2013)
9. W. Shieh, I. Djordjevic, *OFDM for Optical Communications*. Academic Press, pp. 156–159 (2010)
10. C.D. Poole, R.W. Tkach, A.R. Chraplyvy et al., Fading in lightwave systems due to polarization-mode dispersion. IEEE Photon. Technol. Lett. **3**(1), 68–70 (1991)
11. D.N. Godard, Self-recovering equalization and carrier tracking in two-dimensional data communication systems. Commun. IEEE Trans. **28**(11), 1867–1875 (1980)
12. M. Nakazawa, K. Kikuchi, T. Miyazaki et al., high spectral density optical communication technologies. Springer, Berlin Heidelberg **36–39**, 42 (2010)
13. J. Yu, X. Zhou, Ultra-high-capacity DWDM transmission system for 100G and beyond. IEEE Commun. Mag. **48**(3), S56–S64 (2010)
14. S.J. Savory, Digital filters for coherent optical receivers. Opt. Express **16**(2), 804–817 (2008)
15. X. Zhou, J. Yu, P. Magill, Cascaded two-modulus algorithm for blind polarization de-multiplexing of 114-Gb/s PDM-8-QAM optical signals. Optical Fiber Communication—incudes post deadline papers, 2009. OFC 2009. Conference on. IEEE, 2009:1–3
16. H. Louchet, K. Kuzmin, A. Richter. Improved DSP algorithms for coherent 16-QAM transmission[C]// Optical Communication, 2008. ECOC 2008. 34th European Conference on. IEEE, 2008:1–2

17. M. Tseytlin, Ritterbush O, Salamon A. Digital, endless polarization control for polarization multiplexed fiber-optic communications[C]// Optical Fiber Communications Conference, 2003. OFC 2003. IEEE, 2003
18. R. Noe, PLL-free synchronous QPSK polarization multiplex/diversity receiver concept with digital I&Q baseband processing. IEEE Photonics Technol. Lett. **17**(4), 887–889 (2005)
19. H. Yan, L. Guifang, Coherent optical communication using polarization multiple-input-multiple-output. Optics Express, 2005, **13**(19):7527
20. K. Kikuchi, Polarization-demultiplexing algorithm in the digital coherent receiver. LEOS Summer Topical Meeting. 2008:101–102
21. P. Johannisson, H. Wymeersch, M. Sjodin et al. Convergence comparison of CMA and ICA for blind polarization demultiplexing of QPSK and 16-QAM signals, in *Optical Communication (ECOC), 2010 36th European Conference and Exhibition on*. IEEE, 2010:1–3
22. J. Tang, J. He, J. Xiao, et al. Low-Complexity and Fast-Converging Polarization Demultiplexing by ICA for Coherent Optical Receivers. Signal Processing in Photonic Communications. Optical Society of America, 2013: JT3A. 4
23. A.J. Viterbi, A.M. Viterbi, Nonlinear Estimation of PSK-modulated carrier phase with application to burst digital transmission. IEEE Trans. Inf. Theory **29**(4), 543–551 (1982)
24. Y. Cao, S. Yu, J. Shen et al., Frequency estimation for optical coherent MPSK system without removing modulated data phase. Photonics Technology Letters, IEEE **22**(10), 691–693 (2010)
25. J.F. Cardoso, B.H. Laheld, Equivariant adaptive source separation. IEEE Trans. Signal Process. **44**(12), 3017–3030 (1997)
26. J.F. Cardoso, Blind signal separation: statistical principles. Proc. IEEE **86**(10), 2009–2025 (1998)
27. X. Shi, Blind Signal Processing: Theory and Practice. Springer Berlin Heidelberg, 2011:13,48–51,60,71-74
28. A.J. Bell, T.J. Sejnowski, An Information-Maximization Approach to Blind Separation and Blind Deconvolution. Neural Comput. **7**(6), 1129–1159 (1995)
29. A. Cichocki, R. Unbehauen, E. Rummert, Robust learning algorithm for blind separation of signals. Electron. Lett. **30**(17), 1386–1387 (1994)
30. D.T. Pham, P. Garrat, C. Jutten, Separation of a mixture of independent sources through a maximum likelihood approach, in *Proc. EU-SIPCO*. 1992:771–774
31. H. Zhang, Z. Tao, L. Liu, et al. Polarization demultiplexing based on independent component analysis in optical coherent receivers, Optical Communication, 2008. ECOC 2008. 34th European Conference on. IEEE, 2008:1–2
32. T. Oktem, A.T. Erdogan, A. Demir, Adaptive Receiver Structures for Fiber Communication Systems Employing Polarization-Division Multiplexing. J. Lightwave Technol. **27**(23), 5394–5404 (2010)
33. X. Xie, F. Yaman, X. Zhou et al., Polarization Demultiplexing by Independent Component Analysis[J]. IEEE Photonics Technol. Lett. **22**(11), 805–807 (2010)
34. P. Johannisson, H. Wymeersch, M. SjöDin et al., Convergence Comparison of the CMA and ICA for Blind Polarization Demultiplexing. J. Optical Commun. Netw. **3**(6), 493–501 (2011)
35. M. Novey, T. Adali, Complex Fixed-Point ICA Algorithm for Separation of QAM Sources using Gaussian Mixture Model, in *IEEE International Conference on Acoustics*, pp. 445–448 (2009)
36. Y. Cao, S. Yu, Y. Chen, et al. Modified frequency and phase estimation for M-QAM optical coherent detection[C]//Proc. Eur. Conf. Optical Commun. (ECOC) (2010)
37. R. Peveling et al., Multiplier-Free Real-Time Phase Tracking for Coherent QPSK Receivers. IEEE Photon. Technol. Lett. **21**(3), 137–139 (2009)
38. Y. Gao, A.P.T. Lau, C. Lu, et al., Low-complexity two-stage carrier phase estimation for 16-QAM systems using QPSK partitioning and maximum likelihood detection. Optical Fiber Communication Conference. Optical Society of America, OMJ6 (2002)

39. P. TimoPfau. Hard-efficient coherent digital receiver concept with feedforward carrier phase recovery M-QAM constellations. J. Lightwave Technol. 27(8): 989–998 (2009)
40. L. Liu, Z. Tao, W. Yan, et al. Initial Tap Setup of Constant Modulus Algorithm for Polarization De-Multiplexing in Optical Coherent Receivers[C]// Optical Fiber Communication—incudes post deadline papers, 2009. OFC 2009. Conference on. IEEE, 2009:1–3

Chapter 3
Quasi-linear Coherent Optical Transmission System and Digital Signal Processing

3.1 Introduction

In this chapter, we introduce and analyze the coherent optical transmission system. We focus on the quasi-linear coherent optical detection system based on advanced digital signal processing technology, including front-end linear pre-equalization algorithm and back-end nonlinear processing algorithm. The bandwidth limitations of the device and the nonlinear damage of the fiber link have always been two important factors limiting the transmission of high-speed optical signals. The former limits the bandwidth and baud rate of the signal generation, while the latter limits the transmission distance of high-speed signals. The coherent optical transmission system can effectively improve the receiving sensitivity of the system, and at the same time can utilize the digital signal processing algorithm to equalize and compensate for various impairments of the system link and the device. This chapter first introduces the system model and damage mechanism of digital coherent optical transmission, then discusses a new front-end digital time-domain pre-equalization scheme, investigates the implementation technology of the scheme, and evaluates the performance of the scheme through theory, simulation, and experimental research. The key factors, including system bandwidth, adaptive equalizer tap length, and signal-to-noise ratio, have successfully achieved 40 GBaud polarization-multiplexed QPSK/8QAM/16QAM signal generation and long-distance transmission.

Early in the early twentieth century, theoretical models and low-speed experimental devices for coherent optical communication were proposed and confirmed [1, 2]. Although high sensitivity was recognized at the time, high-speed digital signal processing was difficult to achieve due to the need for complex devices such as phase-locked loops. And it did not win in the competition with the wavelength division multiplexing system based on direct detection. Now, with the continuous improvement of the signal rate requirements of optical transmission systems, the continuous adoption of high-order optical modulation technology, and the maturity of high-speed digital signal processing technology and chips, coherent optical

J. Yu and N. Chi, *Digital Signal Processing In High-Speed Optical Fiber Communication Principle and Application*,
https://doi.org/10.1007/978-981-15-3098-2_3

communication has once again become the development trend of ultra-high-speed large-capacity long-distance optical transmission.

3.2 Theoretical Model and Damage Mechanism of Coherent Optical Transmission System

Figure 3.1 shows the basic structure of a basic polarization multiplexing coherent optical transmission system and the damage mechanism diagram of each part. The transmitter is a standard polarization-multiplexed QPSK or 16QAM transmitter received by phase diversity and polarization diversity coherent detection after optical transmission. A standard phase diversity and polarization diversity coherent detection receiver includes a local oscillator laser, a polarizing beam splitter, a 90° optical mixer, four balanced photodetectors, and a four-channel sampling and analog-to-digital converter. In a homodyne detection system based on phase diversity and polarization diversity, a polarization beam splitter splits the received signal into two optical signals of orthogonal polarization states. The local oscillator laser LO is also divided into two channels of completely orthogonal polarization states by a polarization beam splitter, and each polarized signal light and local oscillator light source realize phase diversity reception of a single polarization state through a 90° optical mixer. The coherently detected signal is converted into an electrical signal by a balanced photodetector, and then subjected to digital signal processing by sampling and analog-to-digital conversion. High-speed optical signals are subject to a series of linear or nonlinear impairments of the device and link during transmission, which can cause signal distortion and degradation, thereby degrading system performance.

On the device side, since the output of the amplifier is limited, the amplification gain is usually nonlinear, resulting in that the signals have different gains at different amplitudes. At the same time, the amplifier and the modulator have a certain operating bandwidth, corresponding to high-order multilevel signals. The high-frequency

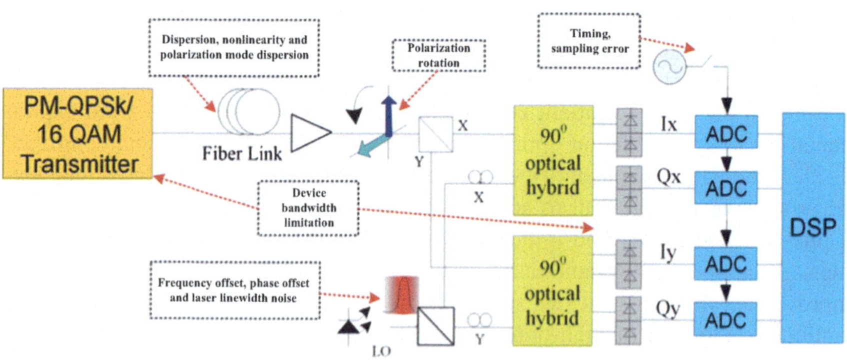

Fig. 3.1 Basic structure and damage mechanism of coherent optical transmission system

part of the signal is suppressed, and the reaction introduces distortion into the waveform in the time domain. On the other hand, modulator bias offset and manufacturing process variations, path imbalance introduced by the receiver at the receiving end, etc., can cause the received signal to be unbalanced, introducing phase and amplitude errors. In addition, narrow-band filtering effects are also introduced during transmission, resulting in inter-channel interference and inter-symbol interference, and the problems caused by narrow-band filtering effects that need to be overcome. In practical systems, there is a need to overcome device impairments in coherent optical transmission systems, including linear and nonlinear impairments induced by modulators, amplifiers, filters, receiver mixers, and other system devices in time and frequency domains. It is also necessary to explore the equilibrium and compensation mechanisms.

Linear damage of fiber links includes chromatic dispersion, polarization mode dispersion, linear phase noise, and frequency difference due to laser or local oscillator laser linewidth. High-speed high-order optical signals are very sensitive to dispersion, polarization mode dispersion, and phase noise in the transmission chain. Dispersion introduces severe inter-symbol interference, and polarization mode dispersion has a coupling effect in two polarization directions, further aggravating the ISI caused by symbol broadening. Linear phase noise causes phase aliasing on the constellation diagram, and the signal cannot be recovered correctly without processing. Therefore, it is necessary to conduct in-depth research on the dispersion, polarization mode dispersion, and phase noise of the system. On the one hand, it is necessary to model these linear damages with different mechanisms and different sources, and to explore their theoretical models. On the other hand, based on the mechanism of linear damage and phase noise formation, the mechanism and method of specific dispersion compensation, polarization mode dispersion compensation and phase noise suppression and recovery are proposed. Therefore, in practical systems, it is necessary to jointly consider the effects of various linear impairments including chromatic dispersion, polarization mode dispersion (first-order and high-order) effects, and laser linewidth and frequency difference, and summarize the action law. It is necessary to explore the corresponding theoretical model and build a theoretical basis for digital signal processing algorithms. The mechanism of linear damage and phase noise formation is analyzed. Based on this analysis, algorithms involving a comprehensive dispersion compensation, polarization mode dispersion compensation, and phase noise suppression and recovery are proposed.

On the other hand, the nonlinear damage of the fiber channel is the main constraint limiting the transmission capacity and transmission distance of the fiber. In order to achieve long-distance transmission, especially for high-order signals, advanced nonlinear compensation is needed to improve the transmission performance of the system. Nonlinear impairments include self-phase modulation, cross-phase modulation, and four-wave mixing. For a single channel, the main nonlinear impairment is self-phase modulation; for high-speed WDM signals, the cross-phase modulation between the channel and the channel and the four-wave mixing nonlinear effect will affect each other, introducing a nonlinear phase. Shift degraded signal quality. Therefore, in the actual system, in view of the nonlinear effects of high-speed high-order optical signals in the optical transmission process, it is necessary to study the mechanism of various nonlinear damage formations, explore the ways to suppress nonlinearity, and provide a theoretical basis for nonlinear compensation algorithms.We make an effort to establish a theoretical model of nonlinearity between channel and channel, analyze the influence of different channel spacing, channel modulation pattern, and channel bandwidth on nonlinear effects, study the nonlinear compensation mechanism of single channel and multi-channel, and improve the transmission capacity and distance of the system.

3.3 Research on Time-Domain Digital Pre-equalization Technology with Band Limit Signal

Bandwidth limitation is a common challenge in high-speed optical communication and a key cause in signal quality degradation [3–5]. Due to the limited bandwidth of the device, high-speed signals are compressed by the bandwidth-limited spectrum during the generation process, causing time-domain broadening and inter-symbol interference. Therefore, device bandwidth is an important reason for limiting the generation of high-speed optical communication signals. On the one hand, with the maturity of high-speed digital-to-analog converter DAC technology, any signal-generating technology based on DAC has attracted wide research interest due to its flexibility and software controllability [3–7]. The DAC-based signal generation also makes the pre-equalization technique at the transmitter end possible. The experimental verification of the dispersion of the emitter end region is reported in the literature [8].

On the other hand, the 3 dB bandwidth of the currently reported DAC is generally less than half of the sampling rate, which means that the high-speed signal generated by the DAC is limited by the DAC bandwidth and the signal is degraded; at the same time, other devices, such as the driver amplifier, are considered. With the bandwidth limitations of devices such as I/Q modulators, the spectrum of high-speed signals is further compressed, and system performance is degraded by inter-symbol interference. In earlier reports, frequency-domain pre-equalization techniques were commonly adopted [3, 5]. This pre-equalization technique first needs to convert the transmitted signal and the received signal into the frequency domain by FFT variation, and then obtains the inverse transfer function of the DAC and other devices, and then uses the transfer function of the frequency domain to perform signal pre-equalization when the signal is generated [3–5]. However, this method requires a large number of signals to be calculated. In order to minimize the influence of noise, it is usually necessary to calculate hundreds of signals [5]; at the same time, in order to obtain an accurate transfer function, strict clock synchronization and symbol synchronization are required to achieve [3]. In order to improve the accuracy, the training signal needs to be specially processed [5].

In order to overcome the shortcomings of frequency-domain pre-equalization, in fact, the time-domain pre-equalization technology is also a good method to mitigate bandwidth limitation impairments [4, 9, 10]. In theory, in a system without inter-symbol interference, the linear adaptive equalizer at the receiver is itself a good channel estimation tool. In fact, the adaptive filter will respond to the response of the band-limited channel. Convergence is the inverse of the channel, thereby compensating for channel bandwidth limitations [4, 9, 10]. This technology is also widely used in cable digital television and wireless communication networks [9, 10].

The advantages of time-domain pre-equalization are illustrated as follows: First, the channel estimation tool can be directly implemented by the adaptive filter module ubiquitous at the receiver end without additional module overhead, thus eliminates the extra FFT calculation module for frequency-domain equalization; secondly, the adaptive filter itself is blind equalization, does not require additional symbol synchronization, but only requires to achieve clock synchronization, which is simpler than the frequency-domain pre-equalization scheme. This section will study and introduce the theoretical analysis, numerical simulation, and experimental verification of digital time-domain pre-equalization based on receiver-side adaptive equalizer.

3.3.1 Digital Time-Domain Pre-equilibrium Principle Based on Receiver-Side Adaptive Equalizer

The digital time-domain pre-equalization principle based on the receiver-side adaptive equalizer is shown in Fig. 3.2. It mainly consists of two steps: the first step is channel estimation, that is, transmitting the unbalanced signal $X(t)$ as the training signal. After receiving the signal of $H(t)$, the receiver receives the signal $Y(t)$ degraded by the band-limited channel $H(t)$. $H(t)$ can be regarded as the end-to-end transfer function of the actual system, including the DAC, driver, and modulator at the transmitter end, as well as the receiver-side balance detector and ADC. It is

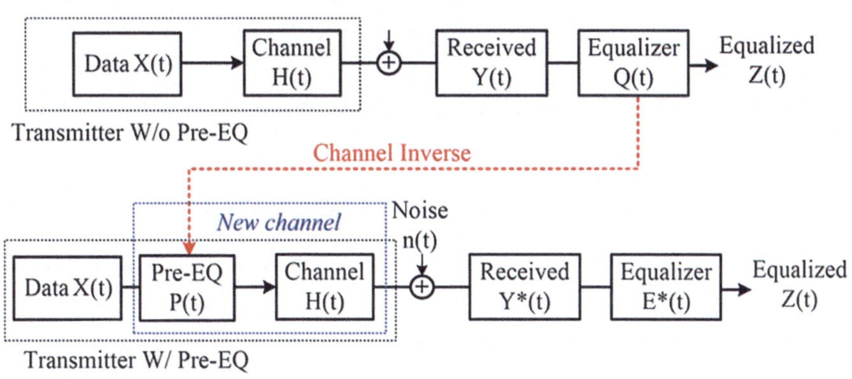

Fig. 3.2 Digital time-domain pre-equalization based on receiver-side adaptive equalizer

worth noting that $H(t)$ or channel is used here to describe all devices including the transmitter hardware, not just devices and links other than the transmitter. But here, it is assumed that the receiver is ideal and the bandwidth is large enough.

It is assumed here that high-speed white noise is added after the transmitter, then the received signal can be expressed by

$$Y(t) = X(t) * H(t) + n(t) \tag{3.1}$$

In order to eliminate the inter-symbol interference of the equalized signal, a linear adaptive filter is used at the receiving end to equalize the received signal.

$$Z(t) = Y(t) * Q(t) = X(t) * H(t) * Q(t) + n(t) * Q(t) \tag{3.2}$$

where $Z(t)$ is the equalized signal and $Q(t)$ is the adaptive linear equalizer for the signal and channel. Usually $Q(t)$ can be obtained by a zero-forcing algorithm or a minimum mean square error algorithm (MMSE). In practical systems, the MMSE algorithm is more common because the zero-forcing algorithm will over-amplify the noise. In coherent optical communication, the basic algorithms of polarization demultiplexing and channel equalization such as CMA, CMMA, and DD-LMS are all MMSE algorithms [3, 8]. The MSE is used to characterize the error level of the signal. If the system is free of noise, the recovered signal satisfies the formula.

$$Z(t) \cong X(t) * H(t) * Q(t) = X(t) \tag{3.3}$$

Therefore, it is obvious that the equalizer filter transfer function at this time satisfies $Q(t) = H(t)^{-1}$, which means that the transfer function of the equalizer is the inverse function of the channel response. Considering the CMA and DD-LMS algorithms used in actual systems, such as the finite impulse response (FIR) filter, which is T/2 interval in time, the actual update and calculation ISI are done at the T symbol interval, which makes the actual recovered signal as the T symbol sampling interval, that is,

$$Z(t) = X(kT) * X_N(t) \tag{3.4}$$

Here $X_N(t)$ is a Nyquist pulse shaping function, then according to Eqs. (3.3) and (3.4),

$$Q(t) \cong H(t)^{-1} * X_N(t) \tag{3.5}$$

Then the frequency-domain transfer function of the equalization filter can be expressed as follows:

$$Q(f) \cong 1/H(f) \quad |f| < 1/2T \tag{3.6}$$

That is, when the noise is negligible, the transfer function and frequency-domain response of the equalizer are in the Nyquist frequency range, which is the reciprocal function of the channel $H(f)$. In this way, under the condition of high SNR, the CMA algorithm or DD-LMS algorithm with T/2 symbol interval can accurately estimate the channel response, so that it can be used as a time-domain pre-equalization function to achieve time-domain pre-equalization of signals.

It is worth noting that the above analysis assumes that the system is high signal-to-noise ratio and the noise is negligible. In practical systems, noise always exists; on the other hand, the adaptive equalizer on the receiver side is itself affected by the tap length of the filter, so if these factors are taken into account, the receiver-side linear adaptive equalizer actually passes. The resulting pre-equalization function should be expressed as a function of channel response, noise level, and tap length.

$$Q(f) = F[H(f), N_0, L] \tag{3.7}$$

Here, N0 is the noise power spectral density and L is the adaptive linear equalizer length. According to the analysis of the literature, after channel pre-equalization, there can be a smaller MSE than the post-equalization, that is, the signal recovery quality will be higher. Assuming that the length of the adaptive filter is constant, the equalization algorithm of the MMSE can be used to equalize the response of the equalizer itself.

$$Q(f)_{\mathrm{MMSE}} = 1/(N_0 + H(f)) \ |f| < 1/2T \tag{3.8}$$

Then, after the post-equalization adaptive filter, the signal has the smallest MSE.

$$MSE_{min\_post-eq} = T \int_{-f/2}^{f/2} N_0/[N_0 + H(f)]df \tag{3.9}$$

Therefore, after the signal is equalized, the minimum MSE is determined by the noise power and the response of the channel. It can be seen from the above equation that when H(f) is small, that is, when the bandwidth of the signal is greatly limited, the MSE of the equalized signal will still be large due to the balance of noise. If the pre-stage pre-equalization is introduced, the above problem can be solved. Assuming that the response of the channel can be estimated by (3.6), the new channel phase will satisfy $H_{\mathrm{pre}}(f) = Q(f)H(f) = 1$, then the linearly adaptive signal is pre-equalized. The MSE after the filter satisfies the formula

$$MSE_{min\_Pre-eq} = T \int_{-f/2}^{f/2} N_0/[N_0 + 1]df \tag{3.10}$$

Comparing the above two Eqs. (3.9) and (3.10), it can be clearly seen that at the same noise level, for a narrow-band-limited signal, the pre-equalized signal is the same as the unpre-equalized signal. The back-end linear adaptive equalization has a smaller MSE, and the signal quality after equalization is improved. For systems with more severe bandwidth limitations, the system performance improvement gain brought by pre-equalization is more significant. This is the significance of studying the pre-equalization technique in this section of the paper.

3.3.2 Linear Digital Pre-equalization Implementation Method for Coherent Optical Communication System

For a coherent optical communication system, channel estimation can be implemented by a method as shown in Fig. 3.3, based on a receiver-side adaptive equalizer such as DD-LMS. First, the DAC is utilized to generate the mQAM signal without any processing as a pre-equalized training signal. Since the bandwidth limitation of the system is mainly determined by, for example, DAC, electric drive, and I/Q modulator, these factors are independent of polarization, so when doing channel estimation, only one polarized signal is needed. At the receiver end, in order to avoid the effects such as frequency difference, and phase difference, signal light can also be received with the local oscillator self-timer frequency. In this way, commonly used receiver-end post-equalization algorithms, such as CMA and DD-LMS, can be used for channel estimation. Thus, the amplitude–frequency response of the resulting CMA or DD-LMS is the reciprocal function of the system transfer function. Taking DD-LMS as an example, after the system is stable, the taps of the filter will converge to a steady value. This value can be used as a function of front-end pre-equalization by normalizing and frequency-symmetric processing to generate a time-domain FIR filter. On the other hand, as shown in the analysis we performed in the previous section, in order to obtain an accurate channel response, the noise should be minimized, and the higher the signal-to-noise ratio at the time of measurement, the better the performance. Since the channel estimation is a conventional adaptive blind equalization

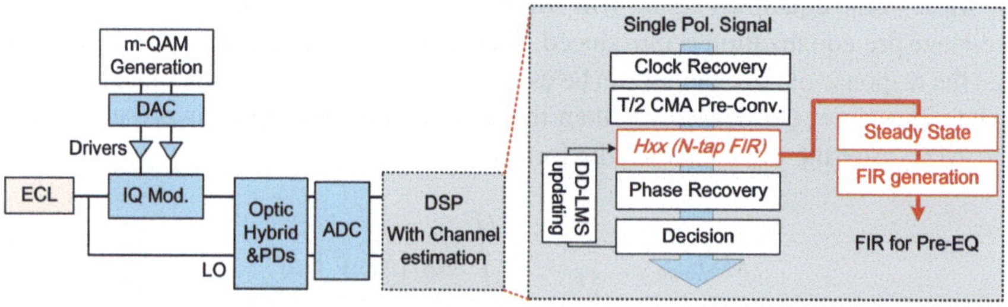

Fig. 3.3 Schematic diagram of DD-LMS-based adaptive filter channel estimation implementation method for coherent optical communication system

algorithm, the receiver only needs to use the existing blind equalization module, and does not need additional modules for channel estimation, which is simpler than the frequency-domain pre-equalization method and easy to implement.

3.3.3 *Analysis of Time-Domain Digital Pre-equalization Simulation Results for Coherent Optical Communication*

In order to study the pre-equalization performance of time-domain digital pre-equalization technology for band-limited optical communication systems, the following statements will be verified through a series of numerical simulations, and will be studied under different system conditions, such as different channel filtering bandwidths, different lights, different optical signal–to–noise–ratio (OSNR) and different adaptive equalizer tap lengths.

1. Simulation system model

Figure 3.4 shows the model of the digital pre-equalization numerical simulation system. The simulation system is based on commercial optical communication simulation software. The transmitter end contains four sets of data, which are X and Y polarized I and Q signals. The simulation system consists of two parts, one is channel estimation simulation and the other is pre-equalization simulation. When doing channel estimation, the generated mQAM signal is directly obtained by QAM mapping without any processing, thereby performing conventional back-end digital signal processing through the self-coherent beat frequency of the receiver end, and the adaptive equalization module through the back-end digital signal processing. The DD-LMS gets a pre-equalized function. The I/Q modulator is an ideal modulator with two Mach Zed modulators, both offset by 0 and operating in the linear region. After modulation, the signals in the two polarization states are combined by polarization combining to obtain a polarization-multiplexed signal. In order to justify the time-domain digital pre-equalization scheme proposed in this paper, two pairs of low-pass filters (4) are used in the front end of the modulator drive to simulate the bandwidth limiting effect of the device. These four low-pass filters are ideal

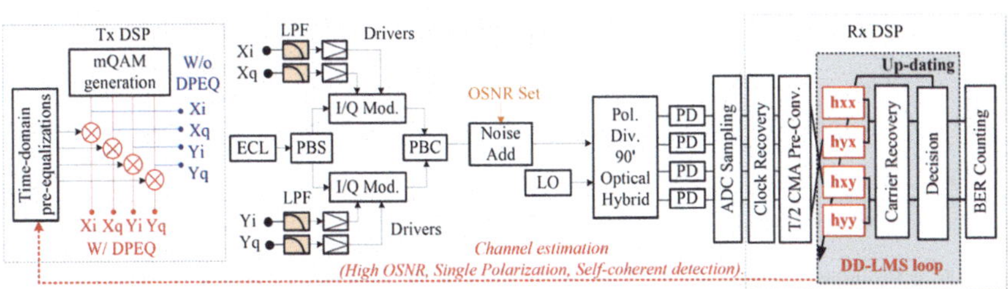

Fig. 3.4 Digital pre-equalization simulation model of coherent optical communication system

bandwidth-adjustable 3rd-order Bessel filters. The laser linewidth is 100 kHz, and other devices are ideally wide enough for the device bandwidth. The noise is added after polarization multiplexing and the system OSNR value is adjustable.

At the receiver end, the polarization and phase diversity coherent detection signals are used. A laser with the same frequency as the signal source is used as the local oscillator to coherent beat frequency. The line width of the local oscillator is also 100 kHz, and the response coefficient of the balanced photodetector is 1 A/W, and the thermal noise figure is $10*10^{-12}$ A/(Hz)$^{1/2}$. The ideal ADC operates on analog-to-digital conversion at a sampling rate of twice the baud rate. The digital signal processing module at the receiver side includes basic algorithm modules as described in Sect. 3.2, namely clock recovery, channel equalization based on T/2 symbol interval CMA and DD-LMS cascade, carrier recovery, and the like. The baud rate of the signal is 32 GBaud, while the measured bandwidth of OSNR is 0.1 nm. The channel estimation scheme shown in Fig. 3.5 is used for pre-equalization, and then the pre-equalization is accomplished by using the obtained pre-equalized FIR filter.

2. Digital pre-equalization performance under different filter bandwidths

Figure 3.5a is the frequency response diagram of the low-pass filter with a 3 dB bandwidth of 7 GHz in the simulation system, and Fig. 3.5b is the channel estimation algorithm developed by the paper in this section, i.e., based on the receiver. The channel estimation frequency response obtained by the adaptive equalization DD-LMS algorithm includes the frequency response of the DD-LMS and the spectral response of the generated FIR pre-equalization function, and the ideal channel inverse function response is compared with them. It is evident that the channel estimation response obtained by the DD-LMS algorithm is in good agreement with the ideal channel response. Figure 3.5c, d is the amplitude and phase frequency response of the DD-LMS. It can be seen that the phase of the DD-LMS is linear over the entire spectrum. It is obvious that the FIR filter generated by the DD-LMS has an "M" type frequency response, as analyzed in Sect. 3.3.1. Although for the T/2 DD-LMS, the spectrum range covers the entire 2 × Nyquis band, since the actual tap coefficient update and the MSE update are at the T symbol interval, the actual tap coefficients

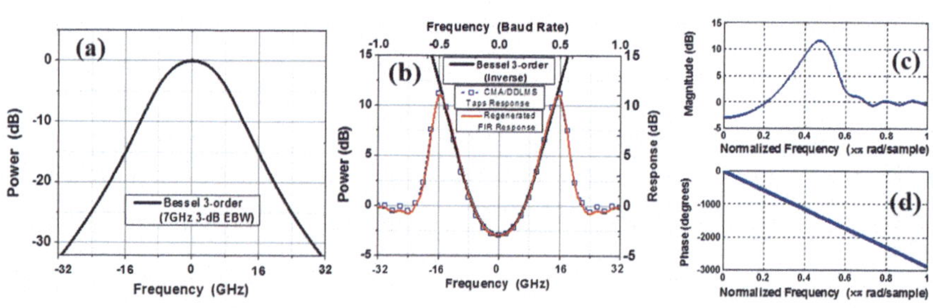

Fig. 3.5 **a** Frequency response of the low-pass filter; **b** Frequency response of the DD-LMS and the spectral response of the generated FIR pre-equalization function and the ideal channel inverse function response; **c** and **d** are DD- Frequency response of amplitude and phase of LMS

are only in Nyquist. Within the bandwidth, the equalizer response is guaranteed to compensate for channel bandwidth limitations.

Figure 3.6a shows the 32 GBaud polarization-multiplexed QPSK signal with pre-equalization and no pre-equalization BER and OSNR curves in different filtering bandwidths (3 dB bandwidth at 7 and 9 GHz). It can be seen that under the unpre-equalized signal at 7 GHz and 9 GHz, the OSNR performance of the pre-equalized signal can be increased by 4.5 and 2.5 dB, respectively. The illustrations (i) and (ii) are constellations with equalized and unbalanced signal recovery at a 7 GHz filter bandwidth with an OSNR of 16 dB, respectively. Figure 3.6b shows the OSNR cost of a 32 GBaud polarization-multiplexed QPSK signal without equalization and equalization in the case of a bit error rate of $1*10^{-2}$. At this time, the bandwidth of the filter is changed from 4 GHz to 16 GHz. The inset (iii) is the frequency response of the front-end pre-equalized FIR filter generated by the DD-LMS at different filter bandwidths. It is evident that by adopting the proposed time-domain digital pre-equalization scheme, the cost of OSNR is always less than 0.5 dB when the bandwidth of the low-pass filter is greater than 6 GHz, and the OSNR cost is even higher when the unbalanced signal has a filtering bandwidth of 6 GHz. The simulation results demonstrate that the pre-equalization scheme proposed in this section can greatly improve the performance of the band-limited system.

3. Adaptive equalizer length

Figure 3.7a–d illustrates the frequency response of the DD-LMS at different tap lengths at 7 GHz with the ideal channel inverse function. It can be seen that with adaptive equalization DD-LMS as the length of the tap increases, the frequency response of the tap is also getting closer to the ideal channel inverse function, especially at the high-frequency portion near the Nyquist frequency. At 14 dB OSNR, the bit error rate performance of the pre-equalized signal at different tap lengths and different filter bandwidths is shown in Fig. 3.7e. The tap frequency response of the DD-LMS here is calculated under the OSNR condition of 45 dB, and the update step size of the DD-LMS is set to $5*10^{-4}$. It can be seen that for a 7 GHz filter bandwidth,

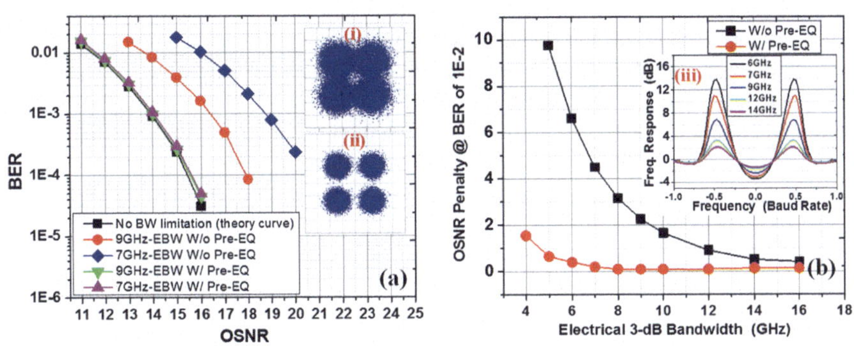

Fig. 3.6 **a** Bit error rate and OSNR curve of 32 GBaud PDM-QPSK; **b** OSNR cost curve at different bandwidths with bit error rate of $1*10^{-2}$

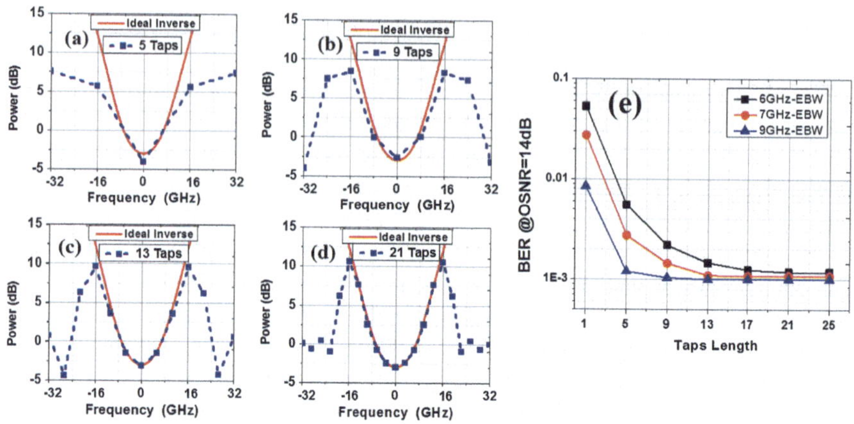

Fig. 3.7 Comparison of the frequency response of the DD-LMS filter with the ideal channel inverse function for different tap lengths: **a** 5 taps; **b** 9 taps; **c** 13 taps; **d** 21 taps **e** Error rate of pre-equalized signal at different tap lengths and different filter bandwidths at 14 dB OSNR

the required DD-LMS length needs to be greater than 13, and for a 9 GHz filter bandwidth, the required DD-LMS length needs to be greater than 9, and for 6 GHz, it needs to be greater than 17 taps. It can be seen that the stronger the bandwidth limit of the system, the larger the required tap length.

4. ONSR affects channel estimation

As analyzed by Eq. (3.7), in order to obtain an accurate channel response, it is necessary to perform channel estimation using an adaptive equalization algorithm module under low noise conditions. Unfortunately, in practical systems, noise is always present. According to the Eq. (3.8), this means that the channel calculated by the adaptive equalization MMSE algorithm module is always different from the ideal channel response inverse function. In this section, the impact of the noise level OSNR on the channel estimate will be investigated.

As shown in Fig. 3.8a, the result is that the frequency response of the FIR filter generated based on DD-LMS under different OSNR conditions is compared with the ideal channel inverse function, when the DD-LMS has a tap length of 33, and when the update step size is $5*10^{-4}$, if the OSNR of the channel estimation is 13 dB, the estimated channel at $\pm$ 16 GHz will differ from the ideal channel by more than 6 dB. It can be seen that the larger the noise, the farther the estimated channel is from the ideal channel inverse function, and the greater the difference, so the estimated channel is less accurate. This result is in good agreement with the conclusions analyzed by Eq. (3.8). Figure 3.8b studies the front-end digital pre-equalization of the band-limited signal at different OSNR for different channel bandwidths. The OSNR of the bit error rate measurement is 14 dB. The results in Fig. 3.8 reveal the following points. First, the higher the OSNR at channel estimation, the more accurate the estimated channel. Second, the stronger the bandwidth limit for the signal (the narrower the filter bandwidth), the higher the required channel estimation OSNR.

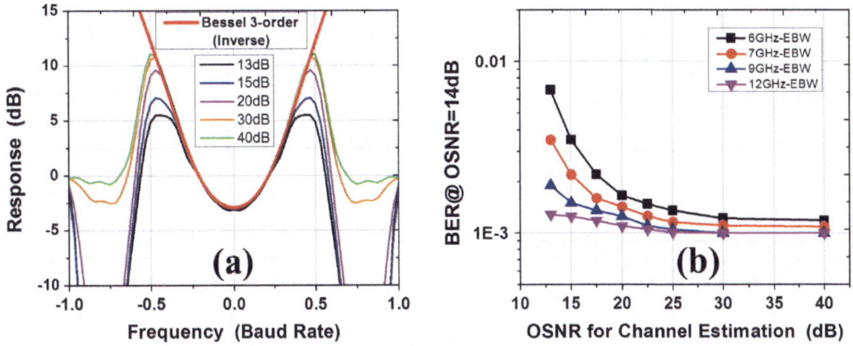

Fig. 3.8 **a** Comparison of frequency response and ideal channel inverse function of FIR filter based on DD-LMS under different OSNR conditions; **b** Pre-equalized signal with the influence of OSNR of channel estimation under different filtering bandwidths

4. Performance of digital time-domain pre-equalization in Nyquist signals and super Nyquist wavelength division multiplexing systems

In the above section, we focus on the performance analysis of signal pre-equalization of a single channel, but not evaluate the performance of pre-equalization in a wavelength division multiplexing system. Particularly in practical systems, more and more Nyquist wavelength division multiplexing (N-WDM) systems with approximate rectangular spectrum compression are used. On the other hand, the channel spacing is even compressed to less than the single-channel baud rate to achieve super Nyquist wavelength division (SN-WDM). **Reuse system**. In these two systems, time-domain pre-equalization will improve the resistance of the single channel itself to the band-limited system and reduce the ISI. On the other hand, crosstalk between channels may be caused by the enhancement of the high-frequency part (ICI). Therefore, it is necessary to study the performance of time-domain pre-equalization in N-WDM and SN-WDM systems.

Figure 3.9 shows the spectrum of the simulated signal. Figure 3.9a shows the ideal 32 GBaud Nyquist QPSK signal, and Fig. 3.9b shows the 32 GBaud Nyquist QPSK signal with a low-pass filtering bandwidth of 7 GHz. Using the channel estimation method proposed in this paper, after time-domain digital pre-equalization, Fig. 3.9c

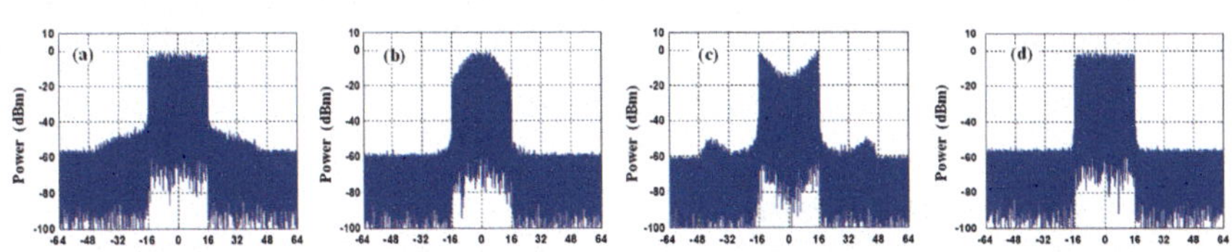

Fig. 3.9 Spectrum of the simulated signal: **a** ideal 32 GBaud Nyquist QPSK signal; **b** low-pass filtering bandwidth of 32 GBaud Nyquist QPSK signal at 7 GHz; **c** pre-equalized 32GBaud Nyquist QPSK signal filtered at 7 GHz **d** pre-equalized 32 GBaud Nyquist QPSK signal after 7 GHz filtering

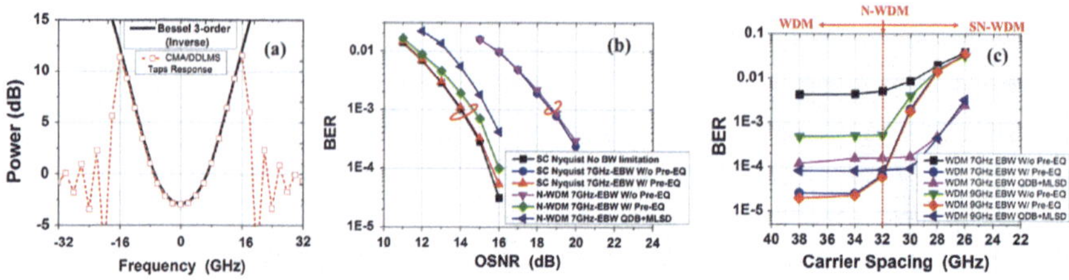

Fig. 3.10 **a** The tap frequency response obtained by using DD-LMS for the Nyuqit signal is compared with the ideal channel response inverse function; **b** the bit error rate OSNR curve of the back-to-back of different signals in the single channel and N-WDM systems; **c** bit error rate of WDM signals with different channel spacings under 7 and 9 GHz narrow-band filtering

is the pre-equalized 32 GBaud Nyquist QPSK signal in the 7 GHz pre-filter spectrum, and Fig. 3.9d is the pre-equalized 32 GBaud. The Nyquist QPSK signal is filtered at 7 GHz.

Figure 3.10a shows the frequency response of the DD-LMS tap at 45 dB OSNR with a tap length of 33. We can see that the tap response matches well with the ideal channel response inverse function. Therefore, channel estimation can be conducted in the same way as the method used in this section of the paper. Figure 3.10b shows the OSNR curve of the 32 GBaud Nyquist QPSK signal at different filter bandwidths (7 and 9 GHz), single channel or N-WDM with equalization and unbalanced bit error rate. Here, the channel spacing of the N-WDM system is maintained at 32 GHz with a roll-off factor of zero. The results show that the performance of the N-WDM system can be improved by pre-equalization.

Figure 3.10c shows the bit error rate performance of WDM, N-WDM, and SN-WDM systems with different channel spacings. It is clear that the errored performance deteriorates rapidly when the channel is separated, and the pre-equalization can be performed well for common WDM and N-WDM systems. When the channel spacing is less than the baud rate, ICI has become the dominant factor, which has deteriorated system performance. At this time, although pre-equalization reduces ISI, it is difficult to improve performance due to the increase in high-frequency components and crosstalk between channels. Therefore, new algorithms, such as orthogonal bipolar signal processing and maximum likelihood sequence estimation algorithms, are required to improve system performance. The discussion on this aspect will be specifically developed in the next chapter of Super Nyquist digital signal processing.

3.3.4 Experimental Results

To verify the time-domain pre-equaliztion scheme of this paper, an eight-channel 40 GBaud QPSK/8QAM/16QAM time pre-equalization transmission experiment is constructed as shown in Fig. 3.11. The experiment uses a high-speed DAC as a

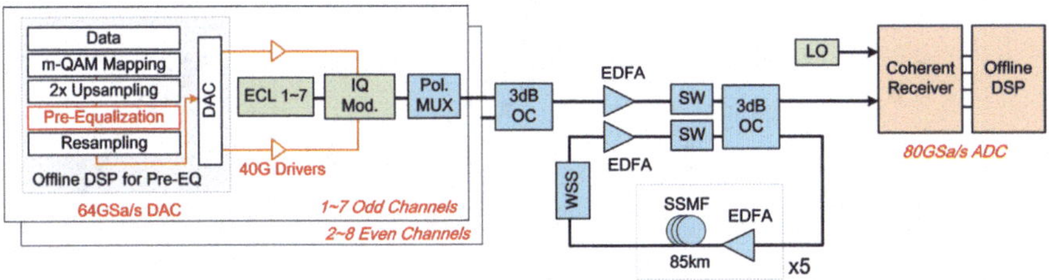

Fig. 3.11 Transmission experiment of eight-channel 400GBaud QPSK/8QAM/16QAM time-domain pre-equalization

signal generator to generate 40 GBaud QPSK/8QAM/16QAM signals, respectively. The 3.dB bandwidth of the DAC is only 11.3 GHz and operates at a sampling rate of 64GSa/s. First, when doing channel estimation, the DAC produces a signal of m-QAM (m is 4/8/16) without equalization and then drives the I/Q modulator. At the receiving end, the same laser is used for self-beat frequency, and the obtained signal is subjected to digital signal processing of the back-end offline after passing through the sampling oscilloscope. The sampling oscilloscope has a bandwidth of 30 GHz and a sampling rate of 80 GSa/s.

The results of the channel estimation are shown in Fig. 3.12. The tap response of the DD-LMS of QPSK without equalization is calculated at the receiver, and the result is shown in Fig. 3.12a. Since the bandwidth limitation is mainly determined by the DAC, the driver amplifier, and the I/Q modulator, a single polarization, coherent laser self-coherent mixing is used to avoid the effects of frequency offset, to reduce polarization crosstalk. Figure 3.12b shows the response of the front-end pre-equalized FIR filter obtained by normalizing and symmetrizing the 33-tap DD-LMS frequency response.

Figure 3.13a–c is the curve of BTB BER versus OSNR of 40 GBaud PDM-QPSK signals, PDM-8QAM signals, and PDM-16QAM signals with or without equalization. Here, the channel estimation uses a 33-tap DD-LMS. It can be seen

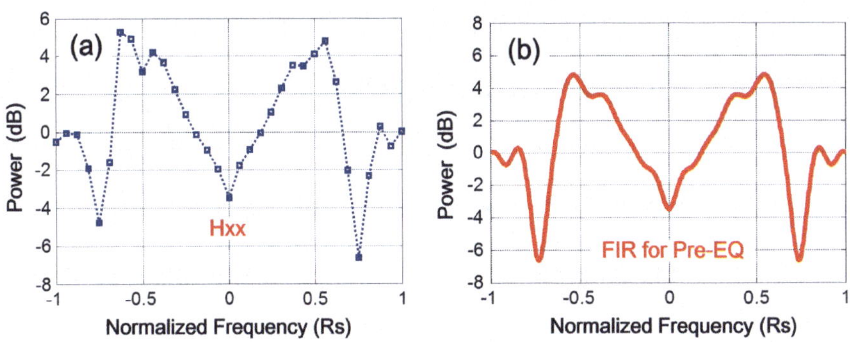

Fig. 3.12 **a** DD-LMS tap frequency response of channel estimation in experiments; **b** Generated front-end pre-equalized FIR filter tap response

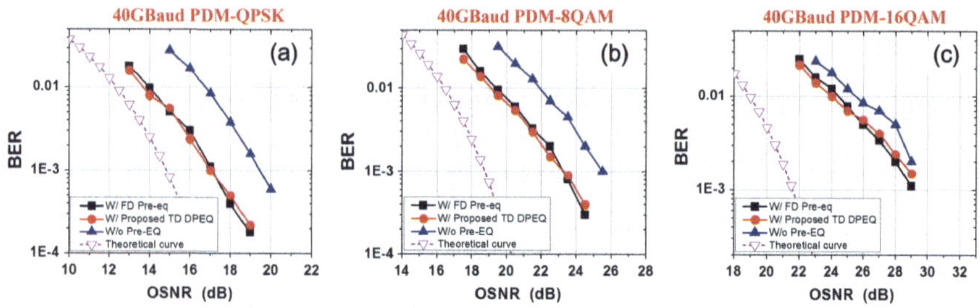

Fig. 3.13 BTB BER versus OSNR with or without pre-equalization **a** 40 GBaud PDM-QPSK signal; **b** 40 GBaud PDM-8QAM signal; **c** 40 GBaud PDM-16QAM signal

that compared with QPSK, 8QAM, and 16QAM signals, the OSNR performance can be improved by 3.5 dB, 2.5 dB, and 1.5 dB, respectively, by the digital time-domain pre-equalization scheme proposed in this section when the bit error rate is 1×10^{-3}. Since the advanced modulation format is more sensitive to ISI and channel bandwidth limitations, the improvement of pre-equalization in high-order modulation format is less significant, and the performance improvement of pre-equalization in QPSK signal is maximized.

Figure 3.14a shows the BTB BER curve versus the tap length of PDM-QPSK/8QAM/16QAM signals at a certain OSNR. Here, the training signal modulation format of channel estimation is maintained at QPSK. The OSNR of the PDM-QPSK/8QAM/16QAM signals is 16, 21.5, and 25 dB, respectively. The results suggest that QPSK is the least sensitive to the length of the tap, and only 9 taps are needed to achieve the best pre-equalization effect. While 8QAM and 16QAM require more tap lengths for channel estimation, 17 and 21 taps are required, respectively.

Figure 3.14b shows the transmission performance comparison of equalized and unequalized signals. The experimental setup of eight channels of WDM signals with a wavelength interval of 50 GHz is shown in Fig. 3.11. The effects of long-distance

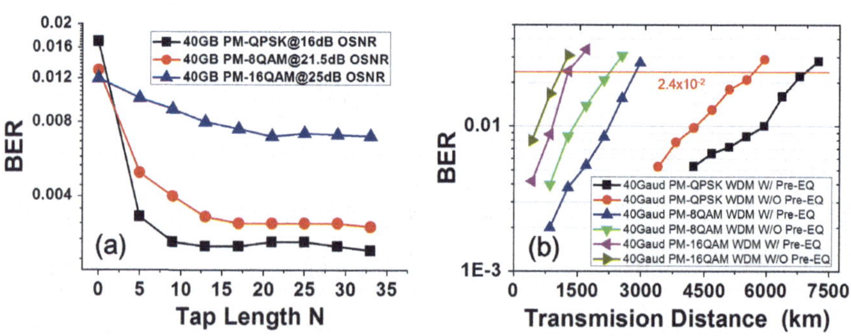

Fig. 3.14 Experimental results: **a** At a certain OSNR, the BTB BER versus the tap length of different signals; **b** in the case of WDM, the BER versus transmission distance of PDM-QPSK/8QAM/16QAM signals with or without equalization

transmission are simulated by circulating fiber loops, each of which contains five 85-km long standard single-mode fibers with a fiber loss of 18.5 dB and a dispersion coefficient of 17 ps/km/nm. In the link, the optical amplifier is an erbium-doped fiber amplifier. To suppress the accumulation and expansion of ASE noise, a programmable optical bandpass filter is used to suppress out-of-band noise. Finally, the transmission performance of 40 GBaud PDM-QPSK/8QAM/16QAM is shown in Fig. 3.15b. The results show that through the time-domain digital pre-equalization proposed in this section, the signal can transmit longer distances and displays better transmission performance.

Finally, the experiment also verified the pre-equalization effect of the Nyquist signal. The result is shown in Fig. 3.15. Figure 3.15a, b illustrates the FFT spectrum of the 40GBaud Nyquist QPSK signal after sampling the oscilloscope with or without equalization, where the roll-off factor is set to zero. It can be seen that in the absence of equalization, the high-frequency portion of the signal is compressed due to the bandwidth limitations of the DAC; by equalization, the high-frequency portion is boosted to form a complete Nyquist signal. Figure 3.15c shows the BTB BER versus OSRN for an equalized and unequalized 40 GBaud polarization-multiplexed Nyquist QPSK signal over single channel and N-WDM multi-channel. It has been observed that the best agreement between the results of simulation and experimental data would achieve. Through the time-domain digital pre-equalization proposed in this section, the BER and OSNR performance of Nyquist signal can be improved in both single-channel and N-WDM systems.

3.4 Conclusion

We present the research of quasi-linear coherent optical detection systems based on advanced digital signal processing techniques, including front-end linear pre-equalization algorithms and back-end nonlinear processing algorithms in this chapter. Bandwidth limitations of devices and nonlinear damage to fiber links are the two most important factors affecting high-speed optical signal transmission. The former limits the bandwidth and baud rate generated by the signal, while the latter limits the transmission distance of high-speed signals. The coherent optical transmission system can effectively improve the receiving sensitivity of the system, and at the same time can utilize the digital signal processing algorithm to equalize and compensate for various damages in the system link. Through the electrical domain pre-compensation technology at the transmitter end, the problems faced by advanced modulation signals can be well overcome. This chapter first introduces the basic coherent optical communication digital signal processing, and then based on this, introduces a new front-end digital time-domain pre-equalization scheme, and studies the implementation technology of the scheme. The key factors affecting the performance of the scheme, including system bandwidth, adaptive equalizer tap length, and signal-to-noise ratio, are studied by theory, simulation, and experiment. Through implementing this scheme, the 40GBaud polarization multiplexing QPSK/8QAM/16QAM signal generation and

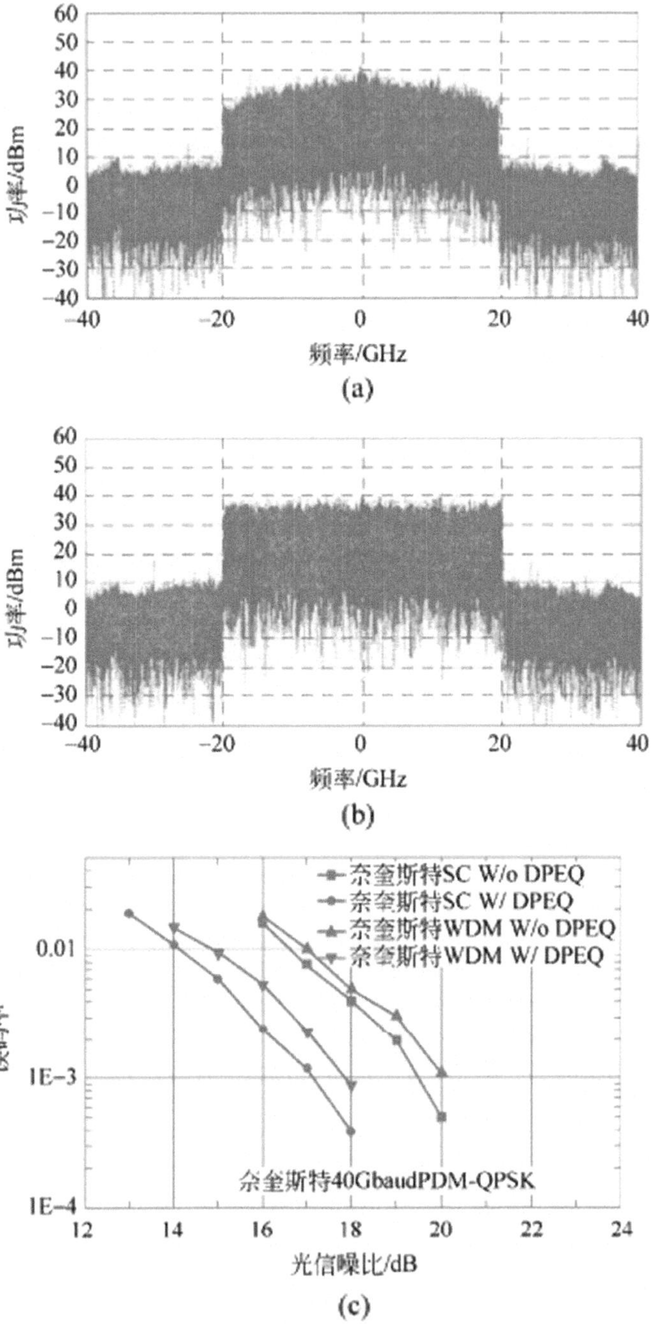

Fig. 3.15 The FFT spectrum of the 40 GBaud Nyquist QPSK signal after oscilloscope sampling **a** without equalization **b** with equalization **c** BER versus OSNR curve of 40 GBaud Nyquist QPSK signal in single channel and N-WDM

long-distance transmission are successfully realized. System performance has been significantly improved through front-end pre-equalization.

References

1. J.J. Yu, Z. Dong, H.C. Chien, et al., *30-Tb/s (3x12.84-Tb/s) signal transmission over 320 km using PDM 64-QAM modulation*. Los Angeles, CA, pp. 1–3 (2012)
2. J.J. Yu, Z. Dong, H.C. Chien, et al., 7Tb/s (7 × 1.284 Tb/s/ch) Signal transmission over 320 km using PDM-64QAM modulation. Photon. Technol. Lett. IEEE **24**(4), 264–266 (2012)
3. G.F. Li, Coherent optical communication, in *Optics InfoBase Conference Papers*. Shanghai, China: Optical Society of America (2008)
4. K. Kazuro, Digital coherent optical communication systems: fundamentals and future prospects. IEICE Electron. Expr. **8**(20), 1642–1662 (2011)
5. X. Zhou, J.J. Yu, Digital signal processing for coherent optical communication, in *WOCC 2009—18th Annual Wireless and Optical Communications Conference*. New York, NJ, United states: IEEE Computer Society (2009)
6. J.S. Seb, Digital signal processing for coherent optical communication systems, in *The 18th Opto Electronics and Communications Conference Held Jointly with 2013 International Conference on Photonics in Switching* (2013)
7. J.W. Zhang, J.J. Yu, Z.S. Jia et al., 400 G transmission of super-Nyquist-filtered signal based on single-carrier 110-GBaud PDM QPSK With 100-GHz grid. J. Lightwave Technol. **32**(19), 3239–3246 (2014)
8. J.W. Zhang, J.J. Yu, Z. Dong, et al. *Transmission of 20x440-Gb/s super-Nyquist-filtered signals over 3600 km based on single-carrier 110-GBaud PDM QPSK with 100-GHz grid*. San Francisco, CA (2014), pp. 1–3
9. J.J. Yu, J.W. Zhang, Z. Dong et al., Transmission of 8 480-Gb/s super-Nyquist-filtering 9-QAM-like signal at 100 GHz-grid over 5000-km SMF-28 and twenty-five 100 GHz-grid ROADMs. Opt. Express **21**(13), 15686–15691 (2013)
10. C.J. Xie, R. Gregory, Digital PLL based frequency offset compensation and carrier phase estimation for 16-QAM coherent optical communication systems, in *European Conference and Exhibition on Optical Communication* (2012)

Chapter 4
Super-Nyquist Wavelength Division Multiplexing System

4.1 Introduction

With the wide application of wavelength division multiplexing (WDM) systems, high spectral efficiency optical transmission is the key point for high-speed and large-capacity optical communication systems, whose available bandwidth is limited [1, 2]. On the one hand, high spectral efficiency optical transmission can be achieved by advanced modulation formats. However, due to the high SNR requirement, the transmission distance for advanced modulation formats is very limited. On the other hand, the spectrum compression technology proposed in recent years has become a hot topic, since it will not increase the number of system equipment or increase the SNR requirement [1, 2]. Compared to quadrature phase shift keying (QPSK) signals in 100 G and 200 G coherent optical communications, quadrature duobinary (QDB) spectral compression techniques have attracted a lot of attention due to their nearly doubled spectral efficiency and greater tolerance to channel crosstalk and dispersion (CD). The transmission of 100 G signals (including the rate of 107–112 Gb/s for forward error correction (FEC) overhead) in the optical links designed for 50 GHz or even 25 GHz channel spacing is a large challenge. The reason is that the available optical bandwidth per channel is limited and the signal baud rate is greater than the designed channel spacing. QDB has been proven a spectral compression technique with strong tolerance of narrowband optical filter, and its spectral efficiency is greater than 4 bit/s/Hz. Therefore, it is an effective method which can achieve the Nyquist limit at a certain baud rate or even can realize super-Nyquist transmissions.

Figure 4.1 shows the development trend of the WDM system and the comparison for different channel multiplexing modes, including conventional WDM (symbol bandwidth is smaller than channel spacing), Nyquist WDM (symbol bandwidth equals channel spacing) and super-Nyquist WDM system (symbol bandwidth is smaller than channel spacing) [1, 2]. In the conventional WDM scheme, there is a guard band between channels, which can completely avoid channel crosstalk and inter-symbol interference. However, due to the existence of guard bands, this scheme

J. Yu and N. Chi, *Digital Signal Processing In High-Speed Optical Fiber Communication Principle and Application*,
https://doi.org/10.1007/978-981-15-3098-2_4

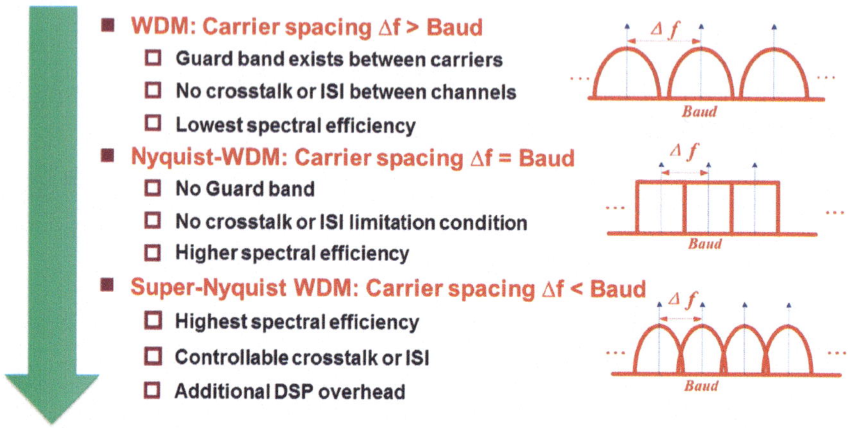

Fig. 4.1 Comparison of development trends and different multiplexing methods of WDM systems

has the lowest spectral efficiency. Nyquist WDM uses time-domain orthogonal pulses whose signal spectrum is shaped in the frequency domain such that the channel baud rate is equal to the channel spacing, theoretically the limit of zero-channel crosstalk and zero-symbol interference. However, if the overhead of the FEC is considered, it will be difficult to transmit 100 G or 400 G signals in the current optical transmission system based on the ITU frequency interval (25/50 GHz). Since the signal baud rate after considering FEC will be larger than the available bandwidth of the channel, the additional bandwidth overlap will cause severe channel crosstalk. On the other hand, super-Nyquist WDM spectrum compression can be achieved by narrowband optical filtering or electrical pre-filtering. Such algorithms can compress the signal spectrum into a certain bandwidth, which is greater than the channel spacing, and introduces controllable inter-channel crosstalk (ICI) and inter-symbol interference (ISI). It has the highest spectral efficiency in theory. However, in order to achieve good system performance, additional algorithms are needed to equalize ISI. At the same time, due to the filtering effect, after using the traditional recovery algorithm, the system performance is seriously degraded by the influence of channel crosstalk. In order to solve the problem of severe channel crosstalk and ISI, and achieve higher spectral efficiency, the new digital signal processing and recovery algorithms are urgently needed [1, 2].

In this chapter, we investigate the super-Nyquist wavelength division multiplexing system. Firstly, a new multi-modulus blind equalization algorithm for super-Nyquist signals is introduced. It can effectively suppress inter-channel crosstalk, and can effectively equalize inter-symbol interference by multi-symbol equalization decision, so that good system performance can be achieved for the signal transmission in a super-Nyquist WDM system. Moreover, the performance of the new multi-modulus blind equalization algorithm and conventional algorithms and other equalization algorithms are compared by experimental verification. Subsequently, the generation of the digital super-Nyquist signal realized by the front-end filtering is introduced, and the 400 G optical transmission with four carriers are realized. Finally, the experiment of super-high-speed PDM-QPSK coherent transmission with a speed of 110 GBaud is introduced. By using this system, we successfully create the world's first single-carrier super-Nyquist-based 400 G optical transmission system, and create the record of the world's highest baud rate of 110 GBaud signal transmission. The carrier spacing is compressed to 100 GHz, the transmission capacity is up to 8.8 Tb/s, and the spectral efficiency is greater than 4 bits/s/Hz (after removing 7% of the FEC overhead). In addition, we also confirmed the high anti-filtering performance of the 9QAM super-Nyquist signal. Based on this scheme, we successfully transmitted ten channels of 440 Gb/s/channel QPSK signals over 3000 km ULAF with 10 cascaded 100 GHz reconfiguration optical add and drop switch (ROADMS).

4.2 Multi-modulus Blind Equalization Algorithms for Super-Nyquist Signals

For signals with strong spectrum compression, the conventional channel equalization algorithm, such as CMA or DD-LMS, could cause the enhancement of the high-frequency part due to the filtering effect, which can refer to the channel estimation effect of the adaptive filter in the previous chapter. On the one hand, the enhancement of the high-frequency part will cause the noise in this band to be strengthened as well. On the other hand, considering the super-Nyquist WDM system, ICI is near the high-frequency band. Therefore, ICI will also be strengthened. In order to achieve high-quality signal recovery, a post-filter is cascaded after adaptive equalization for noise and crosstalk compression. Finally, multi-symbol equalization detection algorithms, such as maximum likelihood sequence estimation (MLSE), are used to equalize ISI so that the super-Nyquist signal can be recovered with high performance.

However, the traditional constant modulus blind equalization (CMBE) algorithm for polarization-multiplexed QPSK coherent detection is not compatible due to the zero-point effect caused by the filtering effect. The cascaded multi-modulus algorithm (CMMA) proposed and adopted in the PM 8QAM system shows a good modulo decision performance [3], which offers a possibility to utilize multi-modulus blind equalization (MMBE) in the QDB spectrum compression super-Nyquist PM-QPSK system.

4.2.1 Principle of Multi-modulus Blind Equalization Algorithm

Figure 4.2a shows the principle of QDB super-Nyquist spectrum compression for the PM-QPSK signal. Generally, the spectral shaping can be performed by either two

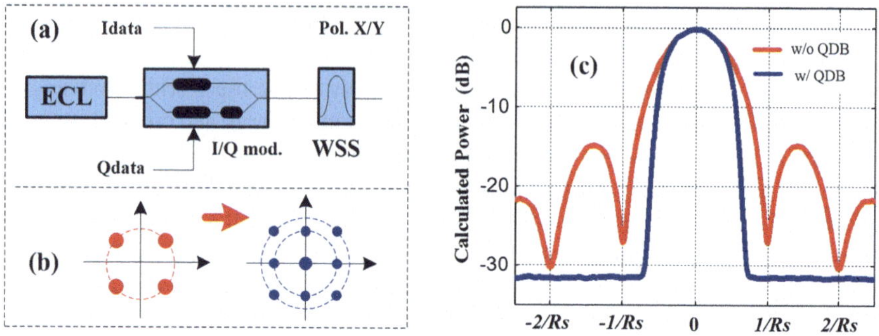

Fig. 4.2 **a** The principle of QDB spectrum shaped PM-QPSK by WSS or wave-shaper for one polarization; **b** Constellation of PM-QPSK signal before and after QDB spectrum shaping; **c** Calculated spectrum comparing PM-QPSK signal with/without QDB spectrum shaping

electrical low-pass filters on the two orthogonal electrical signals or an optical band-pass filter after the optical QPSK modulation. As shown in Fig. 4.2a, we can use a waveshaper or wavelength selective switch (WSS) with 3 dB pass bandwidth less than Rs for spectral shaping. Figure 4.2b shows the PM-QPSK signal constellations before and after QDB spectrum shaping. After QDB spectrum shaping, the 4-point QPSK signal becomes a 9-point duobinary QPSK signal with zero point in constellation due to the filtering effect. The 9 points of QDB QPSK signals are located on three circles with different radii. Compared with QPSK, the spectrum of the QDB shaped signal is significantly narrower, and the spectral side lobes are also greatly suppressed, which are presented in Fig. 4.2c. These spectral features of QDB spectrum shaped signals lead to the tolerance on WDM crosstalk, narrow optical filtering, and chromatic dispersion. It provides a practical means for achieving or even exceeding the Nyquist limit at a certain baud rate.

4.2.1.1 Cascaded Multi-modulus Algorithm for Polarization De-multiplexing and Channel Equalization

For QDB spectrum shaped PM-QPSK, the 9-point signal does not present constant symbol amplitude so that the classic CMA is not well compatible. It not only leads to extra noise after equalization, but also causes a problem with the frequency response of filter taps. Thus, we use the CMMA, which is proposed in PM 8QAM systems with good modulus decision performances, for blind polarization demultiplexing. Figure 4.3 shows the principle of CMMA for QDB spectrum shaped PDM-QPSK signals. It also contains four butterfly-configured adaptive digital equalizers. Here, $\varepsilon_{x,y}$ is the feedback error function for filter tap updating. The corresponding updating of the filter tap weight is given as follows:

$$h_{xx}(k) \rightarrow h_{xx}(k) + \mu\varepsilon_x(i)e_x(i)\hat{x}(i-k)$$

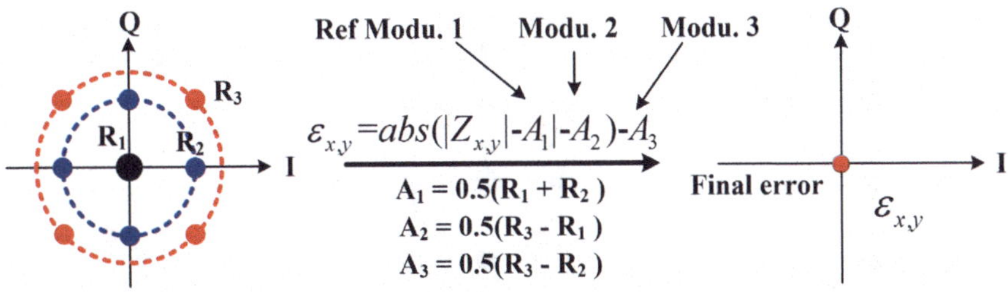

Fig. 4.3 The principle of CMMA for QDB spectrum shaped PDM-QPSK

$$\begin{aligned} h_{xy}(k) &\rightarrow h_{xy}(k) + \mu\varepsilon_x(i)e_x(i)\hat{y}(i-k) \\ h_{yx}(k) &\rightarrow h_{yx}(k) + \mu\varepsilon_y(i)e_y(i)\hat{x}(i-k) \\ h_{yy}(k) &\rightarrow h_{yy}(k) + \mu\varepsilon_y(i)e_y(i)\hat{y}(i-k) \end{aligned} \tag{4.1}$$

$$e_{x,y}(i) = \text{sign}(||Z_{x,y}(i)| - A_1| - A_2) \cdot \text{sign}(|Z_{x,y}(i)| - A_1) \cdot sign(Z_{x,y}(i)) \tag{4.2}$$

Here, $\hat{x}$ and $\hat{y}$ denote the complex conjugates of the received signals x and y, respectively. *Sign (x)* is the sign function and μ is the convergence parameter. By introducing three reference circles A_1–A_3, the final error can approach zero for ideal QDB signal. R_1–R_3 are the radii of the three modulus in QDB PDM-QPSK signal and $Z_{x,y}$ is the output of the equalizer. As a result, it is clear that the regular CMA error signal will not approach zero even for an ideal 9-point signal.

4.2.1.2 Joint Dual-Polarization Frequency Offset Estimation Algorithm with QPSK Partitioning

The partition scheme has been presented in some references for frequency offset estimation (FOE) in a 16QAM coherent system. In this way, the regular m-power algorithm can be also used for FOE in the 9-point QDB spectrum shaped signal with partitioning. On the other hand, for polarization-multiplexed coherent system, the same transmitter and LO are used for the signals in two polarizations. So both polarizations are affected by the same frequency offset. To address this issue, we propose a joint dual-polarization QPSK partitioning algorithm for FOE.

Figure 4.4 shows the principle of QPSK partition and rotation for the 9-point QDB spectrum shaped signal. The processing consists of 3 steps. Since the amplitude of symbols in R_1 is zero, the frequency offset and carrier phase noise have no impact on the R_1 symbols. Thus, we only need to use the symbols in R_2 and R_3 for FOE and phase recovery. The input symbols are first partitioned into three rings based on their amplitudes. Then, only the symbols in R_2 are rotated with $-\pi/4$. Finally, the symbols in R_2 and R_3 are normalized according to their amplitudes. In this way, these two rings are combined to one ring with QPSK 4-point constellation.

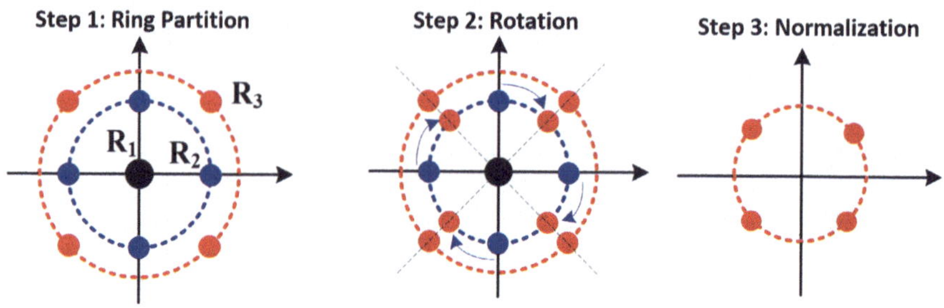

Fig. 4.4 The principle of QPSK partition and rotation for the 9-point QDB spectrum shaped signal

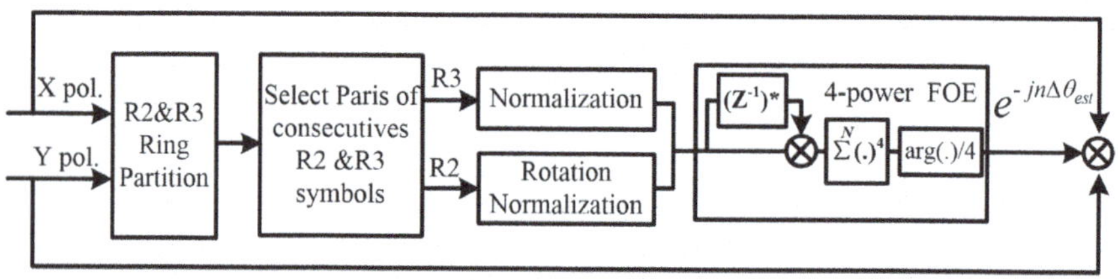

Fig. 4.5 The block diagram for joint-polarization QPSK partitioning FOE

Figure 4.5 shows the block diagram for the proposed joint dual-polarization QPSK partitioning algorithm for FOE. After polarization demultiplexing by CMMA, the incoming X and Y polarized symbols are first portioned into three groups with different circle radius. Here, only the symbols in R_2 and R_3 are selected for estimation to reduce the complexity of the algorithm. Then, symbols in R_2 are firstly rotated with $-\pi/4$ angle and then normalized. R_3 symbols are only needed to be normalized. After that, the two symbol groups can combine with a "QPSK"-like constellation. In this way, the 4-power frequency estimation for QPSK can operate now. For N pairs of symbols in R_2 and R_3, the estimation of the phase offset caused by frequency offset is

$$\Delta\theta_{\text{est}} = 2\pi \Delta f_{est} T_s = \frac{1}{4} \arg \sum_{1}^{N} (S_{k+1} \cdot S_k^*)^4 \tag{4.3}$$

where S_k is the combination of the normalized symbols in R_2 and R_3 groups, T_s is the symbol duration and Δf_{est} is the estimated frequency offset. Then the frequency offset can be compensated by $e^{-jn\Delta\theta_{est}}$ for the nth received symbol in both X and Y polarizations. The frequency offset Δf_{est} can be estimated in the range of $[-1/(8\ T_s), +1/(8\ T_s)]$ by the proposed fourth power operation.

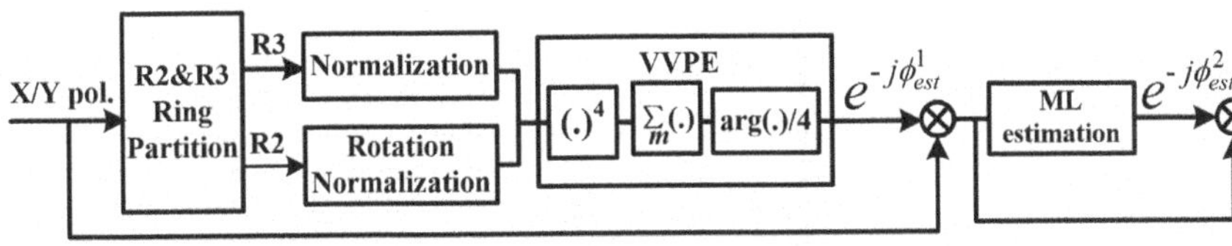

Fig. 4.6 The block diagram for two stages phase estimation based on QPSK partitioning/ML

4.2.1.3 Carrier Phase Recovery with QPSK Partitioning Cascaded Maximum Likelihood Estimation Algorithm

As analyzed above, the QPSK partitioning scheme can be also used in the carrier phase recovery (CPR) for QDB spectrum shaped signals. On the other hand, the maximum likelihood (ML) algorithm shows good improvement and low complexity for the 16QAM phase estimation [4]. In this way, we propose a two-stage phase recovery for the 9-point QDB spectrum shaped signal based on QPSK Partition/ML as shown in Fig. 4.6.

The principle of R_2 and R_3 ring partition, rotation, and normalization are the same as shown in Fig. 4.4. Here we also only need R_2 and R_3 symbols for phase recovery. In practice, the partition steps for FOE and phase recovery can be merged. After that, the symbols in R_2 are first rotated with $-\pi/4$ angle and then normalized, while R_3 symbols are only normalized and then combined with R_2 symbols. Here, all the combined symbols can be used in groups and m is the number of symbols in each group. In this way, the first stage phase ϕ_{est}^1 can be eliminated by Viterbi and Viterbi phase estimation (VVPE) as

$$\phi_{est}^1=\left(\sum_m S_k^4\right)/4 \tag{4.4}$$

The second stage of phase estimation ϕ_{est}^2 based on ML can be expressed as

$$h=\sum_m S_k\cdot D_k^*,\phi_{est}^2=\tan^{-1}(\mathrm{Im}[h]/\mathrm{Re}[h]) \tag{4.5}$$

Here, D_k is the decision of S_k after the first stage phase recovery. The second stage phase recovery is implemented before the final output.

4.2.2 Analysis of Simulation Results of Multi-modulus Blind Equalization Algorithm

The simulation is carried out based on the proposed DSP scheme with 112 Gb/s PM-QPSK signals. The QDB spectrum shaping is operated by a fourth-order Gaussian

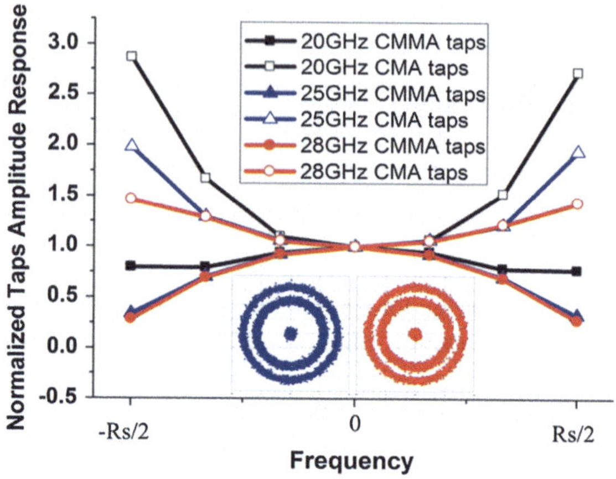

Fig. 4.7 Normalized taps amplitude–frequency response for CMMA and CMA under different spectrum shaping bandwidth

optical band-pass filter with different 3 dB bandwidth. The final output is detected by MLSD for BER measurement. In our simulation, the performances of CMMA compared with CMA for different spectrum shaping bandwidth are studied. The performances of FOE and CPR are also investigated with different spectrum shaping bandwidth, group size, and laser linewidth. Here, in the simulation we use PRBS with a length of 2^{10} and the BER is calculated over 32×1024 symbols.

Figure 4.7 shows the normalized taps amplitude frequency for CMMA and CMA taps in response to the change of Nyquist spectrum shaping bandwidth from 20to 28 GHz. Here Rs is the symbol rate. It shows that CMMA has better frequency response performance in noise compression compared with CMA. For spectrum shaping, the noise around $\pm Rs/2$ can be significantly enhanced for Nyquist WDM channels. However, CMMA taps are compressed at $\pm Rs/2$. The constellations after CMMA are shown as insets in Fig. 4.7.

Figure 4.8a shows the effect of the number of blocks on BER performance in the FOE algorithm in different QDB spectrum shaping bandwidth and OSNR. Here, we

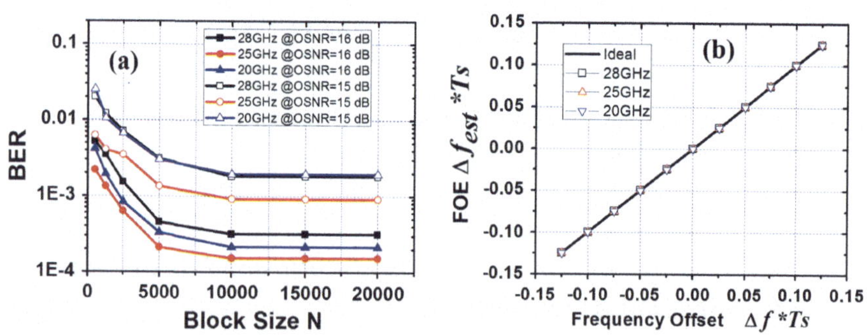

Fig. 4.8 **a** Impact of block size *N* on the performance of FOE; **b** performance of proposed FOE algorithm for different frequency offsets

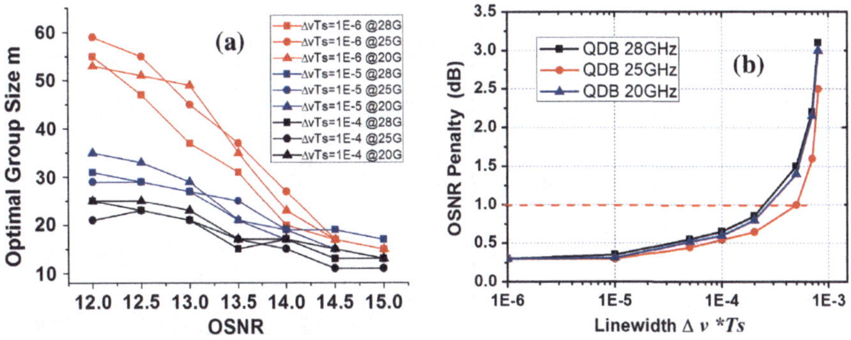

Fig. 4.9 **a** Optimum group size m varying with OSNR for different linewidth and QDB bandwidth; **b** OSNR penalty at BER of 1×10^{-3} varying with linewidth for different QDB bandwidth

keep the offset frequency $\Delta f \cdot T_s = 0.1$ and linewidth of the signal source and LO with $\Delta v = 100$ kHz. We can see that $N = 10{,}000$ is the optimal block size for FOE under different QDB spectrum shaping bandwidth and OSNR. The frequency offset between the transmitter laser and the LO was assumed to be constant in the simulation. Figure 4.8b shows the performance of FOE results for different frequency offset in the whole FOE range under different QDB spectrum shaping bandwidth with OSNR at 16 dB. The proposed FOE algorithm shows good estimation accuracy within the whole estimation range for different QDB spectrum shaping bandwidth.

Figure 4.9a shows the optimum group size *m* varying with OSNR for different linewidth and QDB spectrum shaping bandwidth in the proposed two-stage CPR algorithm. It shows that the optimum *m* decreases with the increase of OSNR and linewidth. The optimal *m* becomes larger and more sensitive to ASE noise when ASE noise is dominant and the linewidth is small. On the other hand, when ASE noise gets larger with smaller OSNR, larger group size *m* is needed to reduce the effect of the noise for phase estimation and recovery. Figure 4.9b shows simulation results of the OSNR penalty varying with linewidth $\Delta v{*}T_s$. (at BER of 1×10^{-3}) for different QDB spectrum shaping bandwidth. The QDB spectrum shaping bandwidth of 25 GHz shows the best performance. At the same time, it shows that our proposed algorithm for QDB spectrum shaping bandwidth of 25 GHz can tolerate linewidth $\Delta v T_s$ of 5×10^{-4} with an OSNR penalty of 1 dB.

Figure 4.10 shows the simulation results of back-to-back BER performance varying with QDB spectrum shaping bandwidth for the conventional constant modulus algorithms (CMBE), and our proposed multi-modulus algorithms (MMBE) for both single carrier (SC) and NWDM condition. Here we keep the OSNR at 16 dB. For NWDM channels, the channel spacing is set as 25 GHz. It can be shown that the proposed MMBE scheme has better tolerance for strong QDB spectrum shaping and crosstalk from other channels. For SC case, the MMBE scheme shows better BER performance when the QDB spectrum shaping bandwidth is less than 27 GHz. On the other hand, the MMBE scheme shows significant BER improvement over conventional CMBE for the NWDM case. The conventional CMBE for NWDM has

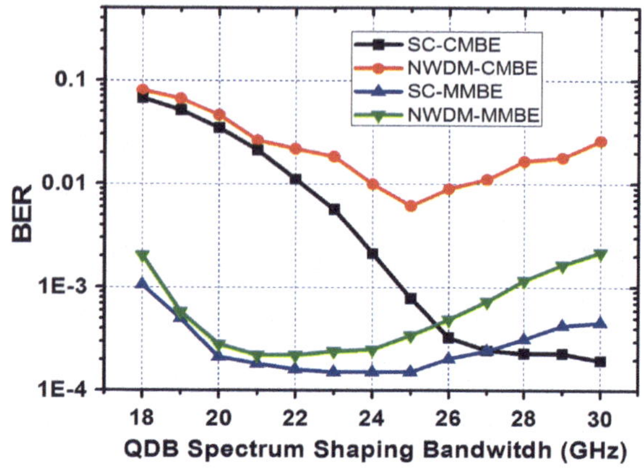

Fig. 4.10 The simulation results of BER performance varying with QDB spectrum shaping bandwidth for different DSP schemes

the worst performance due to crosstalk and strong shaping. It also shows that the optimal QDB spectrum shaping bandwidth for our proposed MMBE scheme is from 23 to 25 GHz for SC case and 21 to 23 for the NWDM case. It is worth noting that, our scheme has two additional steps compared with regular QPSK signals for the proposed 9-point QDB, including the QPSK partitioning and rotation. These two additional steps require additional computation. However, the QDB spectrum shaped signals have a higher spectrum efficiency (nearly doubled) and tolerance to channel crosstalk with respect to QPSK signals. Thus, we believe there is a trade-off when considering the spectrum efficiency and real-time implementation complexity.

4.2.3 Experimental Comparative Study of Multi-modulus Blind Equalization Algorithm

To verify the performance of the multimode blind equalization algorithm proposed in this section in the super-Nyquist WDM system, an eight-channel PM-QPSK super-Nyquist WDM signal transmission experiment with 28 Gbaud/s QDB filtering is demonstrated, as shown in Fig. 4.11. The tolerance of the proposed MMEQ compared with the CMEQ on noise and crosstalk suppression is measured in the experiment. Eight tunable external cavity lasers (ECLs) are used as eight sub-channels in our system with the linewidth less than 100 kHz and the output power of 14.5 dBm from 1555.61 to 1557.02 nm. The odd and even channels are implemented with two sets of polarization-maintaining optical couplers (PM-OCs) before the independent in-phase and quadrature (I/Q) modulation. For QPSK modulation, the 28-Gbaud binary electrical signals are generated from an electrical two-channel pulse pattern generator (PPG) with a word length of 2^{15}–1. The I/Q modulator is biased at the null point. The bandwidth of I/Q modulators is 32 GHz and the bandwidth of the modulator driver amplifier is 45 GHz (at 20 dB gain). The polarization multiplexing of the signal is realized by the polarization multiplexer, which comprises a PM-OC

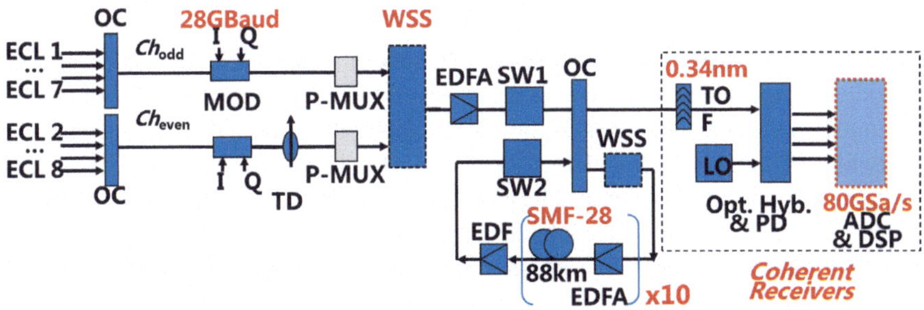

Fig. 4.11 Experimental setup for 8 × 112-Gb/s QDB filtering signals generation, transmission and receiving (TOF: tunable optical filter; TD: time delay; P-MUX: polarization multiplexing; WSS: wavelength selective switch)

to halve the signal, an optical delay line to provide a delay of 150 symbols, and a polarization beam combiner to recombine the signal. The even and odd channels are modulated and polarization-multiplexed individually. After that, they are combined and QDB filtered by a commercial WSS with 25-GHz grid.

The generated 8 × 112-Gb/s, 25 GHz-grid channel signals are then launched into a re-circulating transmission loop, which consists of 10 spans of 88-km conventional SMF-28 with average loss of 18.5 dB and chromatic dispersion (CD) of 17 ps/km/nm, loop switches (SWs), optical coupler (OC), and erbium-doped fiber amplifier (EDFA)-only amplification without optical dispersion compensation. The noise figure of EDFA is about 5.5 dB. Another WSS is placed in the loop, which is programmed to work as an optical band-pass filter to suppress the ASE noise. This WSS has the fourth-order Gaussian spectral profile with a 3 dB bandwidth of 2.2 nm. For the loop transmission, there is no obvious difference using the WSS before or after fiber in the loop. Figure 4.12 shows the optical spectrum of the eight-channel WDM signals before and after 2640-km SMF transmission. We can see the clear ASE noise in the two sides after transmission.

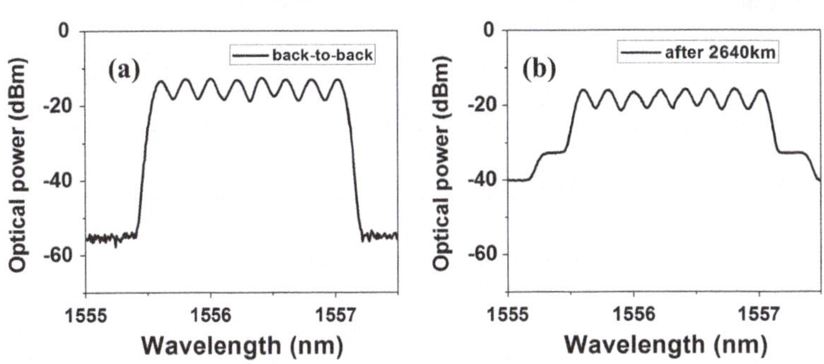

Fig. 4.12 Optical spectrum of eight channels in 25 GHz grid: **a** back to back; **b** after 2640 km fiber (0.1 nm resolution)

At the receiver, one tunable band-pass filter with a 3 dB bandwidth of 0.34 nm is employed to choose the measured channel. Polarization and phase diversity coherent homodyne detection is employed at the receiver. Here, the linewidth of ECL at the transmitter and local oscillator at the receiver is both around 100 kHz. The 3 dB bandwidth of balanced photodetectors is 42 GHz. The powers of received signal and LO with pre-amplification before the optical hybrid are 3 dBm and 20 dBm, respectively. The analog-to-digital conversion is realized in the digital oscilloscope with the sample rate of 80 GSa/s and 30-GHz bandwidth. The data is first resampled to 56 GSa/s with CD compensation, and then processed by the regular CMEQ with post-filtering and the proposed MMEQ algorithm for polarization demultiplexing, carrier frequency offset estimation, and phase recovery before BER measurement.

Figure 4.13 shows the experiment error-free results of signal constellation plots of X (upper) and Y (bottom) polarizations for QDB spectrum shaped PM-QPSK data in MMBE DSP. Figure 4.13a is the received signals before polarization demultiplexing and Fig. 4.13b is the result after CMMA. Figure 4.13c, d is the results after the proposed multi-modulus QPSK partitioning FOE and two stages of CPR with ML phase estimation, respectively.

Figure 4.14a shows the measured back-to-back BER performance for NWDM versus OSNR based on regular CMBE and proposed MMBE schemes as a comparison. For the regular CMBE scheme, the polarization is demultiplexed by CMA and carrier frequency offset estimation and phase recovery are done by fourth power Viterbi-Viterbi algorithm. For MMBE scheme, algorithms proposed in this chapter are performed. It shows that our scheme based on the multi-modulus scheme MMBE shows better BER performance compared with the conventional CMBE scheme. The

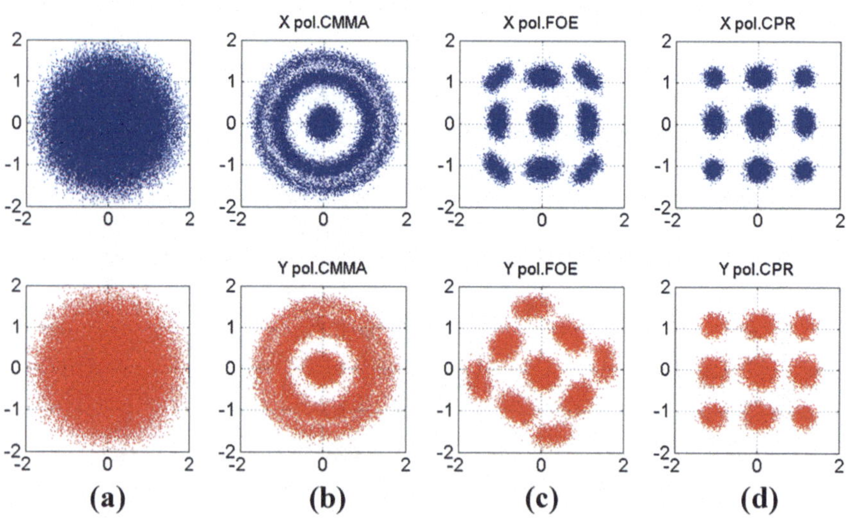

Fig. 4.13 Signal constellation plots of X (upper) and Y (bottom) polarizations for QDB spectrum shaped PM-QPSK data in MMBE DSP **a** before polarization demultiplexing, **b** after polarization demultiplexing, **c** after FOE, and **d** after carrier phase recovery

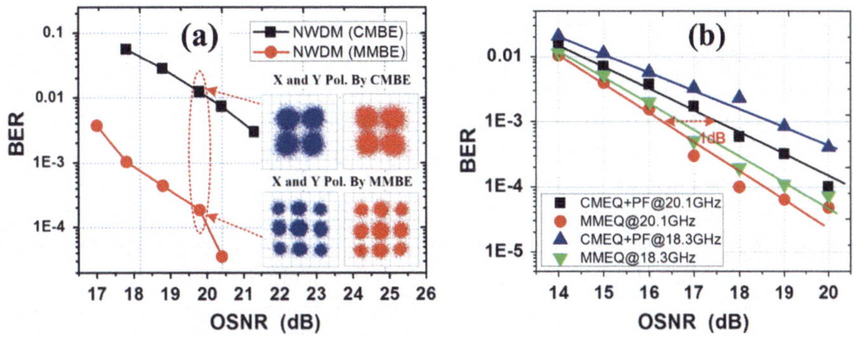

Fig. 4.14 **a** The back-to-back (BTB) BER performance versus OSNR for CMBE and MMBE; **b** The BTB BER performance of the two processing schemes versus the OSNR with filter bandwidth of 20.1 and 18.3 GHz at 25 GHz-grid

required OSNR for BER at 1×10^{-3} for NWDM channels based on our MMBE scheme is about 17.8 dB. The constellations of X and Y polarization obtained by the two schemes at OSNR of 19.8 dB is also inserted in Fig. 4.14a. We can see an enhanced suppression of noise for the 9-point constellation obtained by our scheme.

The measured BERs of channel 4 with a filter bandwidth of 20.1 and 18.3 GHz based on different processing schemes as a function of OSNR are shown in Fig. 4.14b. The OSNR is measured in a 0.1-nm noise bandwidth. It is shown that the MMEQ scheme has better BER performance compared with CMEQ with the post-filtering scheme. We believe that it is because the MMEQ scheme has an improved filter tolerance with noise and crosstalk suppression. For 20.1-GHz filter bandwidth, the required OSNR for BER at 1×10^{-3} based on the MMEQ scheme is around 16.5 dB, which has 1 dB improvement compared with the regular CMEQ with the post-filtering scheme.

We also measure the BER performance of QDB channels with different carrier spacing under different processing schemes as shown in Fig. 4.15a. Here we keep

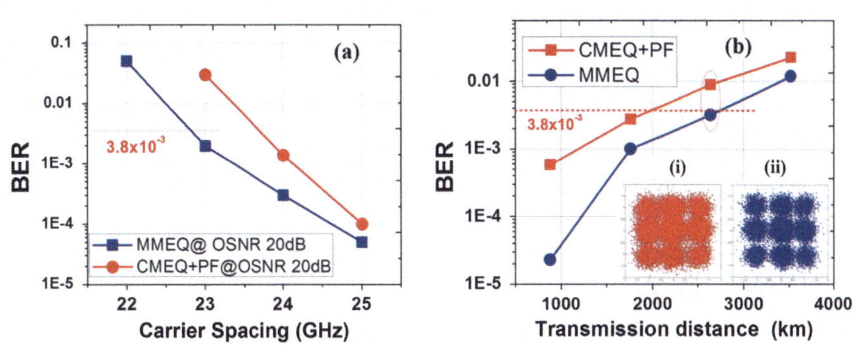

Fig. 4.15 **a** The BTB BER performance versus the carrier spacing; **b** The BER performance of QDB channels at 25 GHz grid versus transmission distance under different processing schemes

the filter bandwidth at 20.1 GHz and OSNR at 20 dB, and reduce the carrier spacing from 25 to 22 GHz to increase inter-channel crosstalk. The carrier spacing is changed by changing both the ECLs wavelength and the WSS grid. It is shown that the MMEQ scheme has better BER performance compared with the CMEQ with the post-filtering scheme. It is because the MMEQ has improved tolerance to the crosstalk. Finally, the transmission BER performance of the QDB channels with the 25-GHz grid as a function of the transmission distance under different processing schemes is shown in Fig. 4.15b. Here, we measure the BER of channel 4 under optimal input power at 1 dBm per channel. The total input power of eight channels is 8 dBm. By using the MMEQ scheme with improved filtering tolerance to noise and crosstalk, the maximum transmission distance of the 25-GHz QDB channel is 2640 km. The measured BER is 3.5×10^{-3}, which is below the soft-defined FEC limit of 3.8×10^{-3}. However, for the regular CMEQ with the post-filtering scheme, only about 2000-km maximum transmission distance is achieved with a BER below the FEC limit. The other seven channels have identical BER performances and no obvious BER variations among these channels are measured. The constellations of the processed 9QAM-like signals after the CMEQ with post-filtering and the MMEQ are shown in the insets (i) and (ii) of Fig. 4.8, respectively. We can see a clearer constellation in the inset (ii). Therefore, a better transmission performance for the MMEQ can be obtained compared with regular CMEQ with the post-filtering scheme with increased transmission distance by about 32% at the BER level of 3.8×10^{-3}.

4.3 Four-Carrier Digital Super-Nyquist Signal Generation and 400 G Transmission

The above super-Nyquist spectrum compression uses a pre-optical filter scheme. In practical systems, with the gradual maturity of devices such as high-speed DACs, it is possible to consider to achieve super-Nyquist signals by performing electrical domain filtering of the digital method. This section introduces a new type of digital super-Nyquist signal generation method, which can effectively simplify the system structure and avoid the use of optical filters. The spectrum of the digitally generated super-Nyquist 9QAM signal is more compact than the spectrum of conventional Nyquist QPSK signals. Therefore, in the process of super-Nyquist WDM channel multiplexing, only the optical coupler is needed.

4.3.1 The Principle of Digital Super-Nyquist Signal Generation

Figure 4.16 shows the principle of the proposed novel DAC-based super-Nyquist

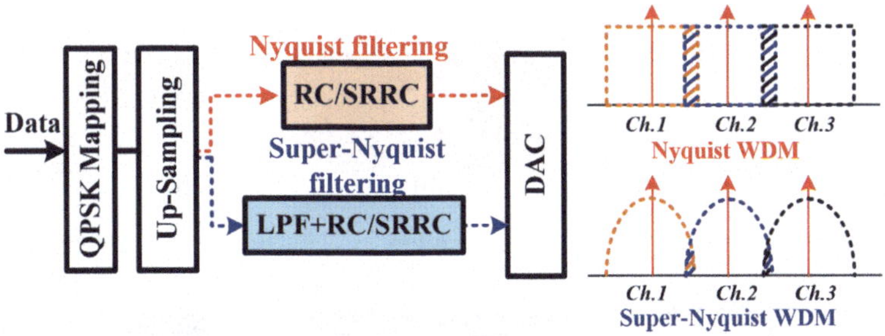

Fig. 4.16 The principle of DAC-based Nyquist and super-Nyquist 9QAM signal generation and the crosstalk impairments in SN WDM

9QAM signal generation compared with the regular Nyquist QPSK signal. For regular Nyquist filtering, a raised cosine (RC) or square root raised cosine (SRRC) filter with a roll-off factor of 0 is used for Nyquist pulse generation. However, as shown in Fig. 4.16, when the channel spacing is less than the baud rate, the excess bandwidth causes severe crosstalk. To realize the super-Nyquist transmission, an additional low-pass filter (LPF) is added for super-Nyquist pulse generation. In this way, the signal spectrum is further suppressed to reduce the channel crosstalk.

In our scheme, the low pass filter can be simply realized by the quadrature duobinary (QDB) delay—insertion filter, of which the transfer function in z-transform is given by

$$H_{QDB}(z) = 1 + z^{-1} \tag{4.6}$$

It can be simply implemented by a two-tap finite impulse response (FIR) digital filter with good performance and turn the QPSK to the 9QAM signal. The super-Nyquist digital filter in the time domain by cascading the QDB and SRRC filters is

$$h_{SN}(t) = h_{QDB}(t) \otimes h_{srrc}(t) \tag{4.7}$$

where $h_{ssrc}(t)$ is the typical time-domain impulse response of SRRC filter. The $h_{QDB}(t)$ is the impulse response of QDB filter HQDB described in Eq. (4.6). When the roll-off factor is 0, the SRRC filter has the same pulse and frequency response with RC. Therefore, the Nyquist filtering can also be implemented with the RC filter here.

Figure 4.17a, d shows the time-domain impulse response of the regular Nyquist filter based on SRRC and the super-Nyquist filter based on cascading QDB and SRRC filters, respectively. Here, the roll-off factor of SRRC is set as 0. Fewer oscillations and faster convergence can be observed for the super-Nyquist digital filter compared with Nyquist filter. Figure 4.17b, e shows eye diagrams of the generated Nyquist QPSK 2-level baseband signal and the super-Nyquist 9QAM 3-level baseband signal, respectively. The electrical power spectra of Nyquist QPSK and super-Nyquist 9QAM signals are shown in Fig. 4.17c, f, respectively. We can see that the power

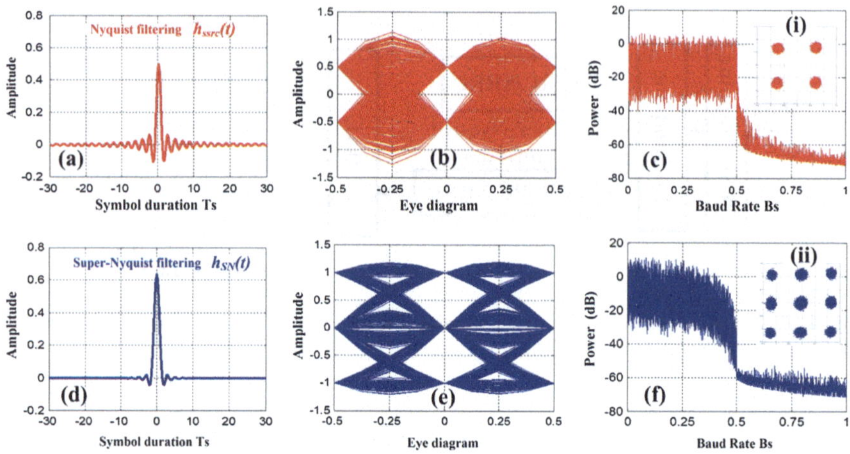

Fig. 4.17 The impulse response of **a** the Nyquist filter, **d** the super-Nyquist filter; The eye diagrams of the **b** Nyquist QPSK signal and **e** the super-Nyquist 9QAM signal; The FFT spectrum of the generated **c** Nyquist QPSK signal and **f** super-Nyquist 9QAM signal

spectrum of the super-Nyquist signal is significantly suppressed compared with the Nyquist signal, and the spectral side lobes are also greatly suppressed. The 3 dB bandwidth is less than 0.5 baud rate. Insets (i) and (ii) show the constellations of Nyquist QPSK and super-Nyquist 9QAM signal, respectively. In this way, a super-Nyquist signal is generated by the Nyquist filtering of the QDB 9QAM signal.

4.3.2 Signal Generation and Transmission Experiment for 4 × 512 Gb/s 4-Carrier Super-Nyquist Channels

Figure 4.18 shows the experimental setup for 4 × 512 Gb/s 4-carrier super-Nyquist signal generation, transmission, and coherent detection. In each 512-Gb/s, 100-GHz-wide super-Nyquist channel, four sub-channels with carrier spacing of 25 GHz are used, each carrying 128-Gb/s data. In this way, 16 tunable external cavity lasers (ECLs) with a linewidth less than 100 kHz are used as sixteen sub-channels in our system, the output power is 14.5 dBm, and carrier spacing is 25-GHz from 1548.20 to 1551.20 nm. The odd and even channels are implemented with two sets of polarization-maintaining optical couplers (PM-OC) before the independent in-phase and quadrature (I/Q) modulations. The super-Nyquist signals are generated by a 64 GSa/s DAC, in which the I and Q data is generated by procedures described in Fig. 4.16. The data after QPSK mapping is upsampled by 2 times, then passes through the super-Nyquist filter as described in Eq. (4.7). The roll-off factor of the SRRC filter is 0. The 3-level baseband signal is of 32 GBaud, with a word length of 2^{15}. The odd and even channels are independently modulated by the 4 port outputs

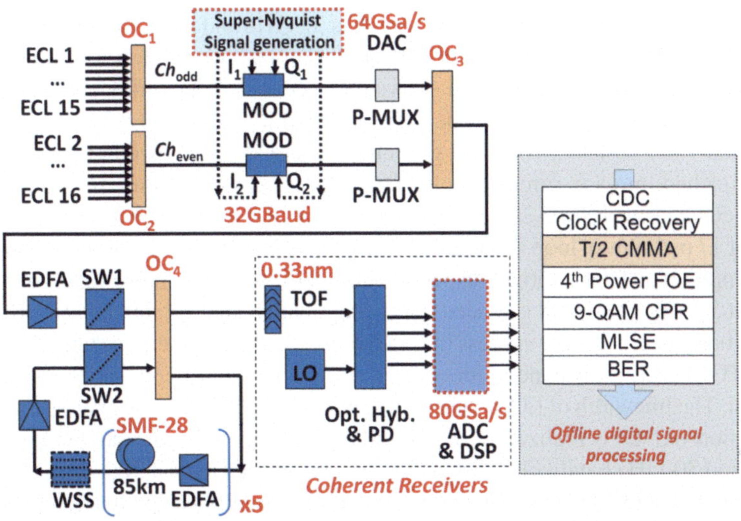

Fig. 4.18 Experimental setup of 4 × 512 Gb/s quad-carrier super-Nyquist channels signal generation, transmission and receiving. (MOD: modulator; OC: optical coupler; P-MUX: polarization multiplexing; WSS: wavelength selective switch; TOF: tunable optical filter)

of the DAC. Different from the regular optical spectrum shaping schemes, in our experiment, the even and odd channels are combined by a 2 × 1 optical coupler without using a WSS or array waveguide grating (AWG).

Figure 4.19a shows the optical spectrum of a generated single-channel (SC) 32-GBaud PDM-QPSK, Nyquist PDM-QPSK, PDM-9QAM (with QDB filter only), and proposed super-Nyquist PDM-9QAM signal. We can see that the PDM-QPSK signal without any operation occupies the largest bandwidth. The bandwidth of Nyquist

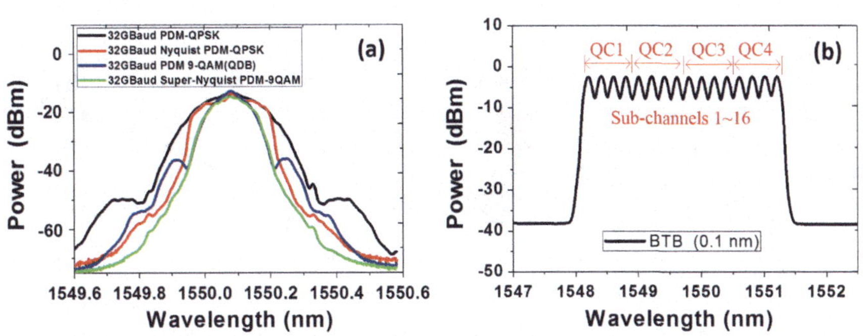

Fig. 4.19 **a** The optical spectrum of single-channel 32-Gbaud PDM-QPSK, Nyquist PDM-QPSK, PDM-9QAM and super-Nyquist PDM-9QAM signal; **b** The back-to-back optical spectrum of four-channel 4-carrier quad-carrier super-Nyquist PDM-9QAM with sixteen sub-channels

PDM-QPSK is equal to the baud rate, which is 32 GHz and exceeds the 25-GHz carrier spacing. The 9QAM signal with only QDB filer has two sidelobes. Our proposed super-Nyquist PDM-9QAM signal shows the narrowest bandwidth, compared with the other three types of signals. The 3 dB bandwidth is less than 0.5 baud rate. The generated 4×512-Gb/s, 100-GHz grid channel signals, as shown in Fig. 4.19b, are then launched into a re-circulating transmission loop, which consists of 5 spans of 85-km conventional SMF-28 with average loss of 18.5 dB and chromatic dispersion (CD) of 17 ps/km/nm, loop switches (SWs), optical coupler (OC), and erbium-doped fiber amplifier (EDFA) optical dispersion compensation. The noise figure of EDFA is about 5.5 dB. One WSS is placed in the loop, which is programmed to work as an optical band-pass filter to suppress the ASE noise. At the receiver, one tunable optical filter (TOF) with 3 dB bandwidth of 0.33 nm is employed to choose the measured channel. The linewidth of LO at the receiver is around 100 kHz. The analog-to-digital conversion (ADC) is realized in the digital oscilloscope with the sample rate of 80 GSa/s and 30-GHz bandwidth. The data is first resampled to 64-GSa/s with CD compensation and then processed by the proposed multi-modulus equalization (MMEQ) algorithms and MLSE. The QDB 9QAM signal is directly recovery by using the MMEQ scheme. The frequency offset estimation and carrier phase recovery are also based on the 9QAM constellation. After phase recovery, the 9QAM signal is converted to the QPSK signal by the multi-symbol equalization and detection algorithm MLSE. The FEC encoding is applied to the original QPSK signal.

Figure 4.20a shows the back-to-back (BTB) bit error ratio (BER) performance of 32-GBaud Nyquist PDM-QPSK and super-Nyquist PDM-9 QAM signals versus the optical signal-to-noise ratio (OSNR) in single-channel case and 25-GHz grid WDM case. In single-channel case, the Nyquist signal shows the best BER performance. Due to the narrow digital QDB filtering, the super-Nyquist 9QAM signal shows about 1.5 dB OSNR penalty at BER of 1×10^{-3} compared with SC Nyquist QPSK. However, for the 25-GHz grid WDM case, the 32-GBaud Nyquist signal cannot be recovered due to large crosstalk from adjacent channels. There is an error floor at 4

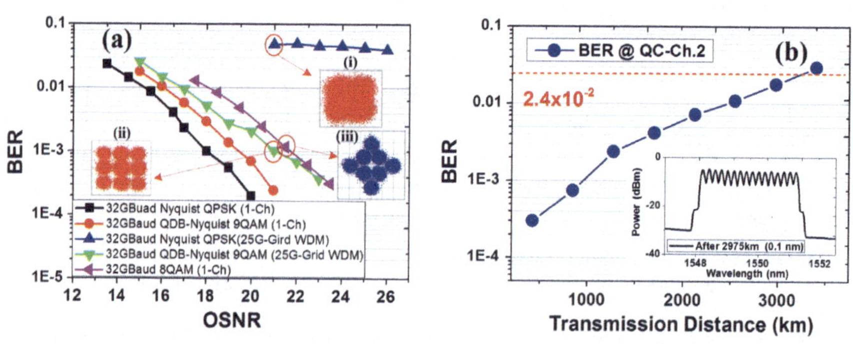

Fig. 4.20 **a** The BTB BER performance of 32-GBaud Nyquist-QPSK, super-Nyquist 9QAM, and 32 GBaud 8QAM signals versus OSNR in single-channel case and WDM case; **b** The BER of QC channel 2 versus the transmission distance

$\times$ 10^{-2} for Nyquist QPSK signal even under large OSNR condition. Only about 1.5 dB OSNR penalty is observed for the 32-GBaud super-Nyquist signal in the 25-GHz grid WDM case compared with SC case. Insets (i) and (ii) show the constellations of Nyquist QPSK and super-Nyquist 9QAM signal in the WDM case with OSNR at 21 dB, respectively. We also measure the BER of SC 32-GBaud PDM-8QAM using the same experimental setup. It shows about 2.5 dB OSNR penalty compared with the QDB 9QAM signal. Thus, 9QAM signal has better BER performance compared with 8-QAM, we believe there are two reasons. First, the minimum Euclidean distance of the QDB 9QAM signal is larger than that of the 8QAM signal; second, a multi-symbol equalization and decision based on MLSE is used for 9QAM signal. Figure 4.20b shows the measured BER of channel 2 of the 512-Gb/s four-channel versus transmission distance ranging from 425 to 3400 km. The measured BER of QC channel 2 after 2975-km transmission is 1.8 $\times$ 10^{-2}. The optical spectrum of the WDM signal after 2975-km transmission is inset in Fig. 4.20b. It is worth noting that the 3 dB analog bandwidth of the DAC used in our experiment is only 11.3 GHz. The performance of generated Nyquist and super-Nyquist signals is affected by the bandwidth limitation effect.

The BER performance of sub-channel 7 after 2975-km transmission versus the fiber input power per channel is shown in Fig. 4.21a. It shows that optimal input power per sub-channel is −1 dBm. The measured BER of all 16 WDM sub-channels of 32-GBaud 128-Gb/s and the averaged BER of all 4-carrier 512-Gb/s channels after 2975-km transmission are presented in Fig. 4.21b. After 2975-km transmission, the BER for all super-Nyquist-WDM channels is below the 2.4 $\times$ 10^{-2} BER threshold for 20% soft decision FEC. The constellations of the received signal in X and Y polarization of channel 9 after transmission processed by the proposed MMEQ are shown in Fig. 4.21b as insets. It shows that our proposed super-Nyquist signal has good performances for long-haul 400 G transmission.

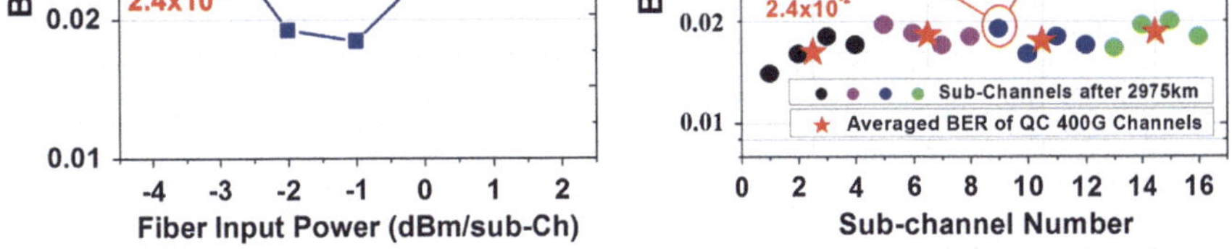

Fig. 4.21 **a** The BER of sub-channel 7 versus the input power per sub-channel; **b** The BER of all sub-channels and averaged BER of all QC channels

4.4 400 G Transmission Experiment of Single-Carrier 110-GBaud Super-Nyquist Filtered Signal

Here we introduce the research on a 400 G transmission system based on ultra-high-speed super-Nyquist signal. The high baud rate electronic time-division multiplexing (ETDM)-based super-Nyquist filtered signal has become the world's first single-carrier super-Nyquist-based optical transmission system and created the world's highest record of 110 GBaud signal transmission. The carrier spacing is compressed to 100 GHz, the transmission capacity is 8.8 Tb/s, and the signal spectrum efficiency is greater than 4 bits/s/Hz [4]. At the same time, the performance of the super-Nyquist filtered signal in multiple resettable optical add/drop multiplexers (ROADMs) is also studied experimentally. By using the system, we successfully transmitted 100-GHz-grid, twenty channels single-carrier 440-Gb/s super-Nyquist 9QAM-like signal over 3600-km ultra-large effective-area fiber (ULAF) at a net SE of 4b/s/Hz (after excluding the 7% hard decision FEC overhead). In addition, we also confirmed the high anti-filtering performance of the 9QAM super-Nyquist signal. Based on this scheme, we successfully transmitted ten channels 440 Gb/s QPSK signals over 3000-km ULAF in 10 cascaded 100 GHz ROADMs.

4.4.1 20 × 440 Gb/s Super-Nyquist WDM Long Distance Transmission Experiment

Figure 4.22 shows the experimental setup of the twenty channel, 100 GHz-grid, 440-Gb/s single-carrier SN filtering 9QAM-like signal generation and transmission. At the transmitter, 20 external cavity lasers (ECLs) with less than 100-kHz linewidth and a frequency space of 100 GHz are divided into two groups as the odd and even channels. The odd or even channels are multiplexed by one 200-GHz arrayed waveguide grating (AWG). The two pairs of 110-GBaud in-phase (I) and quadrature (Q) signal is

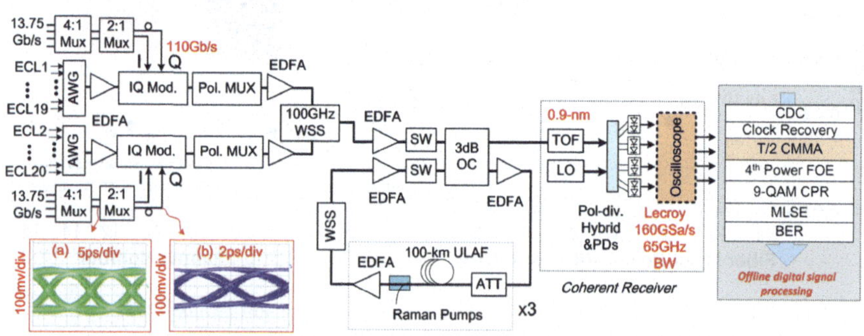

Fig. 4.22 Experimental setup of 20 × 440 Gb/s super-Nyquist WDM long-distance transmission

generated by a two-stage ETDM with 4:1 and 2:1 electrical multiplexer (MUX) and four-channel 13.75-GBaud binary pseudo-random binary sequence (PRBS) signals. The 4:1 MUX is a 56-Gb/s 4:1 broadband multiplexer module and 2:1 MUX is a 100-Gb/s 2:1 broadband multiplexer module. In our case, the 4:1 MUX works well with 55 Gb/s output and the 2:1 MUX also works well with 110 Gb/s output. The output of the 4:1 MUX has a 500 mV Vpp and the output of 2:1 MUX has a 300 mV Vpp. For data decorrelation, the delay between the two 55 Gb/s data sequences is about 55 bits for multiplexing and the delay between the two 110 Gb/s data sequences is about 35 bits for modulation. The odd and even channels are modulated independently by using two I/Q modulators driven by the generated 110-GBaud PRBS signals. The I/Q modulator is an integrated $LiNbO_3$ dual-drive modulator and the 3 dB BW is 33 GHz. Driven by the output 110-Gb/s data of the 2:1 MUX, the total insertion loss is about 27 dB. Insets (a) and (b) in Fig. 4.3 show the electrical eye diagrams of 55 and 110-GBaud binary signals, respectively. The vertical scale of these two eye diagrams is 100 mV/div and the time scales of (a) and (b) are 5 and 2 ps/div, respectively. For QPSK modulation, the I/Q modulator is biased at the null point. After modulation, the polarization multiplexing of each path is realized via the polarization-multiplexer. Then the odd and even channels are spectrally filtered to achieve the 9QAM-like constellation signals and combined by using a programmable wavelength selective switch (WSS) with 100-GHz fixed grid and 94.8-GHz 3 dB BW. The optical spectral filtering by WSS only includes the band-pass filtering with a flat-top shape. Since the channel BW is less than the signal symbol rate, SN channels can be obtained after the WDM multiplexing.

The twenty 100 GHz-grid, SN-WDM channels are generated as shown in Fig. 4.23, where the back-to-back (BTB) spectrum of twenty channels from 1544.32 to 1559.64 nm is measured. In this way, the 8.8 Tb/s (20 × 440 Gb/s) with a net SE of 4bit/s/Hz after removing the 7% HD-FEC overhead is obtained based on the single-carrier 400 G PDM-QPSK signals within a 100 GHz grid. The generated 20 × 440 Gb/s, 100 GHz-grid SN-WDM channel signals are then launched into a re-circulating transmission loop, which includes three 100-km spans of ULAF (21.1 ps/nm/km, 132 mm^2) with an average loss of 21 dB. Each span of the 100-km

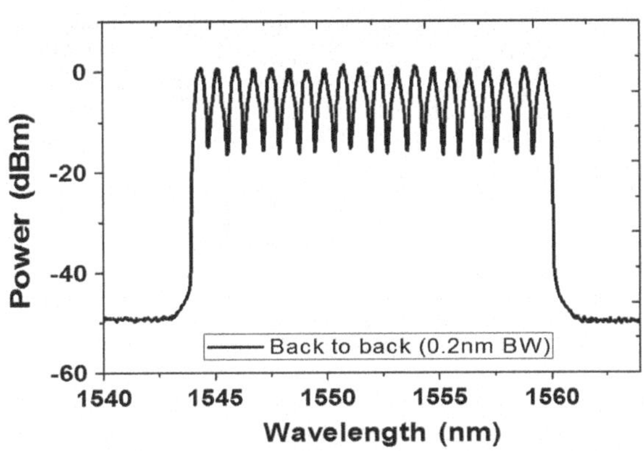

Fig. 4.23 The optical spectrum of 20 × 440 Gb/s, 100 GHz-grid SN-WDM channel signals

ULAF is followed by a hybrid post-Raman amplifier and an EDFA to compensate for the fiber loss. The ON–OFF Raman gain is 9 dB per span from 1450 nm pumps. One attenuator is used before each span of fiber to control the launch power per channel. An equalization filter with adjustable amplifier gain is used in the loop to flatten the gain slope and one WSS is employed as a band-pass filter to suppress the ASE noise. At the coherent receiver, one tunable optical filter with 3 dB BW of 0.9 nm is employed to select the desired sub-channel. An ECL with a linewidth less than 100 kHz is utilized as a local oscillator (LO). In addition, a polarization-diverse 90-degree hybrid is used for polarization and phase diversity coherent detection. The BW of the balanced detector is 50 GHz. The sampling and digitization (A/D) are realized by the real-time digital oscilloscopes with 160-GSa/s sample rate and 65-GHz electrical BW. After the ADC with an oversampling rate of about 1.45 for the 110 GBaud SN signal, the offline DSP is then applied to the 160 GSa/s sampled data sequence. CD compensation is conducted, the data is resampled to 220 GSa/s, and then processed by the MMEQ algorithms with MLSE as described in Fig. 4.22.

Figure 4.24a shows the BTB BER results of 440-Gb/s channel as a function of OSNR (0.1 nm resolution) under different spectrum shaping filter BW. The theory BER/OSNR curve of a single-channel 110 GBaud PDM-QPSK signal based on additive white Gaussian noise is also plotted in Fig. 4.24a. The required OSNR for that odd and even 440-Gb/s sub-channel in SN WDM at the BER of 3.8×10^{-3} (7% HD FEC limit) is about 22.5 dB/0.1 nm, which has 0.5 dB OSNR penalty compared to the single-channel case after 100 GHz-grid WSS. Compared with the theory curve, the single-channel 110-GBaud PDM-QPSK has about 4 dB penalty for experimental implementation due to the low-pass filtering effects caused by the electrical and optical components used. In addition, we also verified that all other channels exhibit similar performance except that the side channel has 0.5 dB better OSNR tolerance. Inset (i) and (ii) in Fig. 4.24a show the error-free constellation diagrams of X and Y polarization signals recovered under high OSNR. The BER performance of 100-GHz-grid channel 4 after 3600-km ULAF transmission versus the launch power per channel is shown in Fig. 4.24b. The input power is controlled by the attenuator at the input of each span of fiber. The optical input power per channel is about 4 dBm.

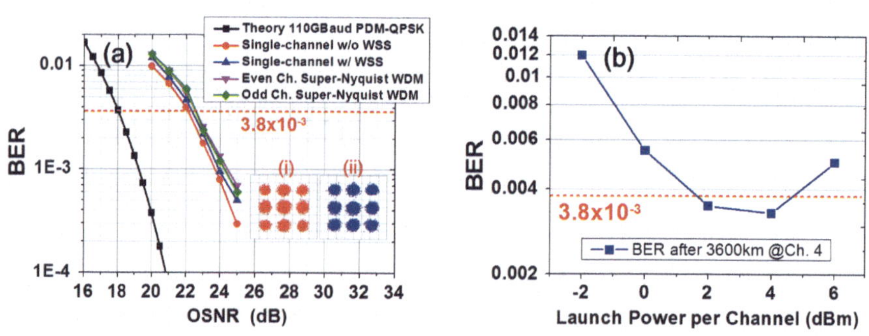

Fig. 4.24 **a** The BTB BER results versus the OSNR for the 110 GBaud PDM-QPSK signals. **b** The BER performance after 3600-km transmission of channel 4 versus the launch power per channel

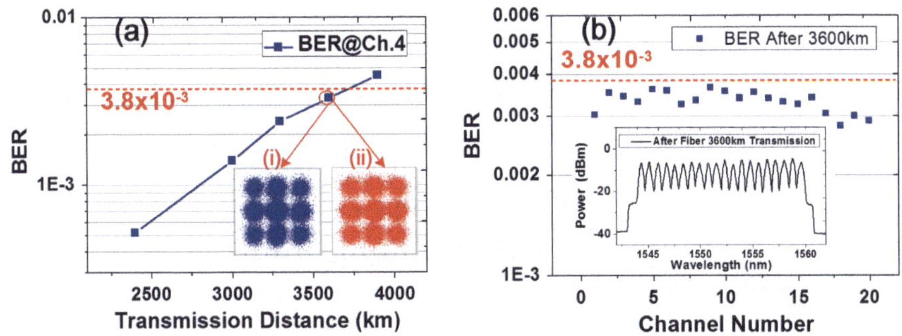

Fig. 4.25 **a** The BER performance of channel 4 versus the transmission distance **b** The BER of all tewnty channels after 3600-km transmission

The BER degrades at power larger than 4 dBm per channel due to the fiber nonlinear impairments. However, the BER also degrades fast at power less than 3 dBm per channel due to OSNR decrease.

Figure 4.25a shows the measured BER of channel 4 versus the transmission distance ranging from 2400 to 4000 km under the optimal launch power. The measured BER of channel 4 after the 3600-km transmission is 3.3×10^{-3}, which is below the 7% HD-FEC limit.

For all the 400 G SN-WDM channels, the measured BER after the 3600-km transmission is shown in Fig. 4.25b. The OSNR after transmission over 3600-km ULAF is about 25 dB/0.1 nm. The optical spectrum is inserted in Fig. 4.25b. After 3600-km ULAF transmission, the BER of all the SN-WDM channels is below the 3.8×10^{-3} BER threshold of the 7% HD-FEC.

4.4.2 Experiment of Ten Channels Super-Nyquist WDM Signal Transmission over ROADMs Link

The performance of the ETDM high-speed signal under super-Nyquist filtering after filtering by multiple ROADM nodes in the long-distance transmission is studied. The following experiments demonstrate the high anti-filtering performance of the 9QAM Super-Nyquist signal.

The experimental setup of the ten-channel, 100 GHz-grid, 440-Gb/s single-carrier SN signal generation and transmission over 3000-km ULAF and 10 ROADMs is shown in Fig. 4.26. The setup of the generation of single-carrier 110-Gbaud ETDM PDM-QPSK signal is the same as that in Fig. 4.22. At the transmitter, 10 ECLs with a frequency space of 100 GHz are divided into two groups as the odd and even channels. The odd and even channels are modulated independently. The odd and even channels are spectrally filtered to achieve the 9QAM-like constellation signals and combined by a programmable WSS1 with 100-GHz fixed grid and tunable 3 dB BW. Since

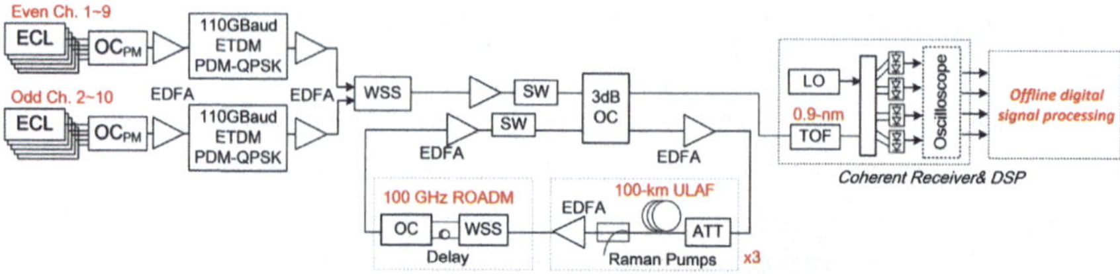

Fig. 4.26 Experimental setup of ten channels, 100-GHz-grid, 440 G-b/s single-carrier super-Nyquist signal transmission over multiple ROADMs link

the channel BW is less than the signal symbol rate, SN channels can be obtained after the WDM multiplexing. The generated ten-channel 440 Gb/s, 100 GHz-grid SN-WDM signals are then launched into a re-circulating transmission loop, which contains three spans of 100-km ULAF and a ROADM based on 100 GHz-grid WSS. One EDFA is used before the first fiber link. As shown in Fig. 4.22, each span of the 100-km ULAF is followed by a hybrid post-Raman amplifier and an EDFA to compensate for the fiber loss. The ULAF, Raman amplifier, and EDFA are the same as those used in Fig. 4.22. After fiber loop re-circulation, the signal is coherently detected and recovered by the offline DSP. The coherent receiver and offline DSP are also the same as that in Fig. 4.22.

In order to study the tolerance of the 110 GBaud SN signal against the cascade narrowing filter effect and inner-band crosstalk caused by ROADMs along the transmission path, one WSS (WSS2) is placed in the loop. The even and odd channels are filtered and routed to port 1 and port 2 after the WSS2, respectively. Then, both signals were recombined again by using a 50:50 OC with one arm delayed for decorrelation. The delay is about 5 ns using a fiber delay line. The transmittances of the two ports are shown in Fig. 4.27a. The 3 dB BW of WSS is 94.8 GHz. Here, we only consider the simple ROADM structure with only one filter element inside. Thus, the ten channels pass one ROADM per round trip with a 300-km distance.

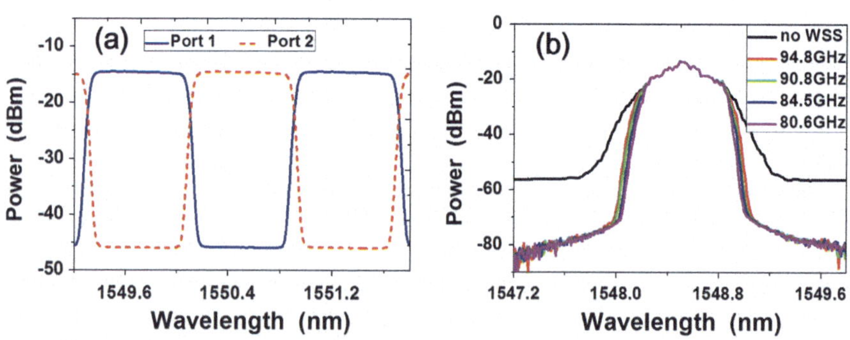

Fig. 4.27 **a** The transmittances of the two ports of the 100-GHz grid WSS in the ROADM. **b** The spectrum under different bandwidths after WSS filtered

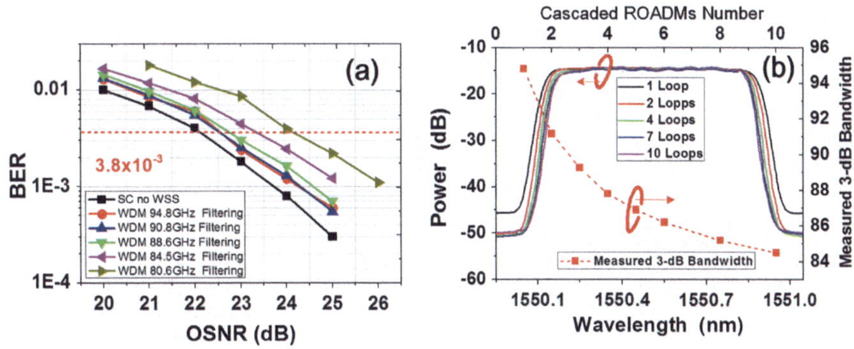

Fig. 4.28 **a** The BTB BER results versus the OSNR for the 110 GBaud PDM-QPSK signals under different filtering BW. **b** The spectrum of the cascaded ROADMs after re-circulation loops and the 3 dB bandwidth of cascaded ROADMs after transmission versus the loop re-circulating times

We first measure the BTB performance of the 440-Gb/s channel as a function of OSNR (0.1 nm resolution) under different filtering BW. The 3 dB BW of WSS1 for channel multiplexing is changed from 94.8 to 80.6 GHz. The optical spectrum of a single-channel 110-Gbaud QPSK signal under different filtering BW after the WSS is shown in Fig. 4.27b.

The measured BTB BER results of 440-Gb/s channel as a function of OSNR (0.1 nm resolution) under different filtering BW is shown in Fig. 4.28a. The required OSNR for the SN-WDM 440 Gb/s channel under 94.8-GHz filtering (100 GHz grid WSS) at the BER of 3.8×10^{-3} (7% HD FEC limit) is about 22.5 dB/0.1 nm, which has 0.5 dB OSNR penalty compared to the single-channel case after 100 GHz-grid WSS. Negligible OSNR penalty for BW larger than 88.6 GHz and less than 0.8 dB penalty are observed for the SN 110 GBaud signal for even 84.6-GHz filtering at the BER of 3.8×10^{-3}. Therefore, our proposed SN signal scheme is highly tolerant of the filtering effect. We measure the transfer function of the cascaded ROADMs after different re-circulation loops transmission and the 3 dB BW of cascaded ROADMs versus the loop re-circulating times as shown in Fig. 4.28b. Each ROADM has the same 3 dB BW of about 94.8 GHz. We can see that the BW is narrowed down with the increasing of the loop re-circulating times. The 3 dB BW of the 10 cascaded ROADMs after 10 loops has been reduced to 84.5 GHz.

The BER of channel 4 versus the transmission distance under the optimal launch power without and with 1 ROADM per 300 km is shown in Fig. 4.29a. Without ROADM, the maximum transmission distance can be 3600 km. With 1 ROADM per 300 km, the BER of channel 4 after 3000-km and 10 ROADMs is 3.4×10^{-3}, which is below the 7% HD-FEC limit. For all the 400 G SN-WDM channels, the measured BER after 3000-km transmission and 10 ROADMs with 100 GHz-grid is shown in Fig. 4.29b. The optical spectrum after 3000-km ULAF transmission and 10 ROADMs is inserted in Fig. 4.29b. After transmission, the BERs of all the SN-WDM channels are below 3.8×10^{-3}. The above results show that the 9QAM super-Nyquist filtered signal generation and reception scheme proposed in this chapter have good

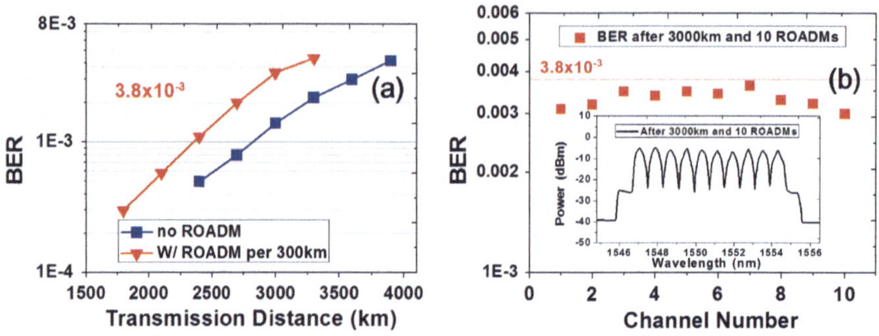

Fig. 4.29 **a**The BER performance of channel 4 as a function of the transmission distance without and with ROADMs. **b** The BER of all ten channels after 3000 km transmission with 10 cascaded ROADMs

anti-filtering and anti-channel interference performance. It has important potential in the future ultra-high-speed, large-capacity, high spectral efficiency and ultra-dense super-Nyquist wavelength division multiplexing system.

4.4.3 Transmission Experiment of Single-Carrier 400 G Signal Based on 128.8 GBaud PDM-QPSK in the Terrestrial Fiber Link

So far, a single-carrier 400 G submarine system transmission over a 7200-km fiber link is reported as the longest distance [5]. The transmission was done by using 64-GBaud 16QAM format with a net SE of 5.33b/s/Hz over cascaded 50-km fiber spans. As for terrestrial systems with 80 to 100-km fiber spans, a single-carrier 400 G transmission over 4800 km is reported [6]. It was realized with a net SE of 3.64 bit/s/Hz based on 107 GBaud QPSK signal. Also, 3600-km transmission with a net SE of 4 bit/s/Hz based on 110-GBaud QPSK, which has been the highest ETDM baud rate so far, is reported in [6].

We successfully increased the limit of ETDM-based optoelectronic schemes by employing the symbol rate of 128.8-GBaud to over record-breaking distances. Namely, we successfully transmitted single-carrier 515.2-Gb/s PDM-QPSK signals over 10130-km link consisting of 100 km-spans of terahertz wave SLA + fiber. To the best of our knowledge, it is the highest symbol rate and the longest WDM transmission distance with single-carrier 400 G signals reported so far. We have also demonstrated the transmission of single-carrier 128.8-GBaud filtered QPSK signals in 100 GHz-grid over 6078 km, which has the SE distance product of 31,314 (b/s/Hz) km.

Figure 4.30 shows the experimental setup of the system based on 515.2-Gb/s single-carrier generation on 128.8-GBaud PDM-QPSK format. The same setup was

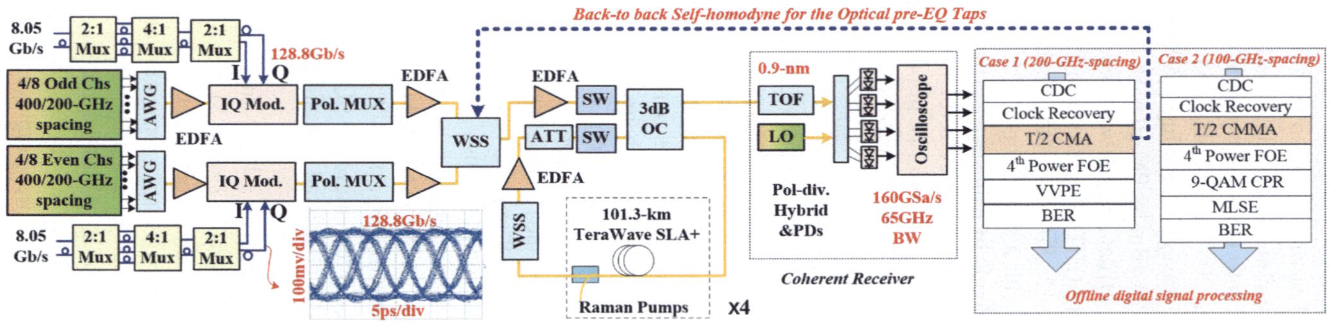

Fig. 4.30 Experimental setup. (ECL: external cavity lasers; MUX: time-division multiplexer; IQ Mod: IQ modulator; Pol. MUX: polarization multiplexer; BW: bandwidth; WSS: wavelength selective switch; OC: optical coupler; SW: switch; ATT: attenuator; TOF: tunable optical filter; LO: local oscillator; ADC: analog-to-digital converter)

used to generate 9QAM-like filtered QPSK signal that effectively doubles the spectral efficiency. At the transmitter side, 8 (for PDM-QPSK) or 16 (for 9QAM-like filtered QPSK) external cavity lasers (ECLs) with less than 100-kHz linewidth have been used. The carrier frequency space is 200 or 100 GHz while the output power is 14.5 dBm. As such, they were divided into two groups as the odd and even channels to form the WDM channel setup in 200- or 100 GHz-grid. The odd/even channels were multiplexed by two arrayed waveguide gratings (AWGs). After AWGs, the power of signals was amplified to 23 dBm by PM-EDFAs. The two-pairs of 128.8-GBaud in-phase (I) and quadrature (Q) data signals were generated by three-stage all-electronic time-division multiplexing (ETDM) blocks with 2:1, 4:1, and 2:1 electrical multiplexing ratios, thus producing 128.8-GBaud rate from the 8.05-GBaud pseudo-random binary sequence (PRBS) with a word length of 2^{15}–1. The odd and even channels were modulated independently by using two I/Q modulators directly driven by the generated 128.8-GBaud PRBS signals. The electrical eye diagram of the 128.8-GBaud binary signal is shown in Fig. 4.30 (inset i). In our case the employed 4:1 MUX was a 56-Gb/s 4:1 broadband multiplexer module, while the 2:1 MUX was a 120 Gb/s 2:1 broadband multiplexer module. As we confirmed, the 4:1 MUX performs quite well with 64.4-Gb/s output and the 2:1 MUX also works well with the 128.8 Gb/s output. The obtained output of the 4:1 MUX has a peak-to-peak value Vpp of 500 mV, while the output of 2:1 MUX has a Vpp of 400 mV. For the QPSK modulation format, the I/Q modulator was biased at the null point.

We first measured the long-distance transmission performances for the single-carrier 400 G transmission with transmitter-side optical pre-equalization using 200 GHz-grid channels. For eight 200-GHz grid channels, the transmitter-side pre-equalization by was conducted by enhancing the power of high-frequency components while attenuating the low-frequency components [5, 6]. In this case, the transfer function of WSS is designed via the obtained channel response. The selection of taps number and the frequency value of WSS impulse response is based on the similar scheme we proposed for digital pre-equalization using receiver-side channel estimation under back-to-back self-homodyne coherent detection in Ref. [5]. After that, sixteen 100-GHz-grid super-Nyquist WDM (SN-WDM) channels without any pre-equalization were tested. For the 100-GHz-grid SN-WDM channels, the odd and even channels were spectrally filtered to achieve the 9QAM-like constellation signals and combined by the WSS at 100-GHz fixed grid and 94.8-GHz 3 dB bandwidth (BW).

The fiber transmission was conducted over a 405.2-km re-circulating loop, which consisted of four 101.3-km spans of terahertz wave SLA + fiber with an average effective area A_{eff} of 122 μm^2, an attenuation coefficient of 0.185 dB/km (20 dB span loss including connectors) and a chromatic dispersion coefficient of 20.0 ps/(nm km) at 1550 nm. One backward-pumped Raman amplifier with about 20 dB ON–OFF gain was used for each span to compensate for the signal loss. The average power of the Raman pumps was about 950 mW. One attenuator was used to control the launch power per channel. Also, one WSS was employed to flatten the gain slope band-pass filter. At the coherent receiver side, a tunable optical filter (TOF) with 3 dB bandwidth of 0.9 nm was employed to select the desired sub-channel. An ECL with a linewidth less than 100 kHz was utilized as a local oscillator (LO) with the input to a polarization-diverse 90° hybrid. The bandwidth of the balanced detector that we used was 50 GHz. Next, the sampling and digitization (A/D) was realized by the real-time digital oscilloscopes with 160-GSa/s sampling rate and 65-GHz electrical bandwidth, which was followed by offline DSP applied to four-channel sampled data sequence. The data was first resampled to 257.6-GSa/s with CD compensation applied, and then processed by either regular constant modulus equalization (CMEQ) algorithm for QPSK or the multi-modulus equalization (MMEQ) algorithm with maximum likelihood sequence estimation (MLSE) for a 9QAM-like signal. The total errors were counted over 12 million bits. The line bit rate of 515.2 Gb/s also implies that some overhead was added to 400 G data. It would be ≥20% SD-FEC enabling error-free transmission at BER $\leq 2.4 \times 10^{-2}$.

We have initially measured the performance of the WSS-based pre-equalization scheme. Figure 4.31a shows the frequency response of the WSS with different filter alpha factors (we use the alpha factor to adjust the pre-emphases strength; no pre-equalization when alpha is 0, and full pre-equalization when alpha is 1), while Fig. 4.31b shows the QPSK signal spectrum after passing WSS. The back-to-back (BTB) BER of signals with different filtering alpha factors is shown in Fig. 4.31c. We can see the optimal alpha factor for QPSK recovery is 0.75. Finally, Fig. 4.31d shows the BTB BER results of 128.8-GBaud signals as a function of OSNR (0.1 nm resolution) under different transmission cases for a single channel. The required OSNR at the BER of 2×10^{-2} for the 200 GHz-grid channels with pre-equalization is about 19 dB, and it increases to 20.5 dB for the 100 GHz-grid SN-WDM channels. In addition, we also verified that all other channels exhibit similar performance (please see Fig. 4.33b). The corresponding error-free constellation diagrams of 9QAM and QPSK signals are shown by insets (i) and (ii) in Fig. 4.31d.

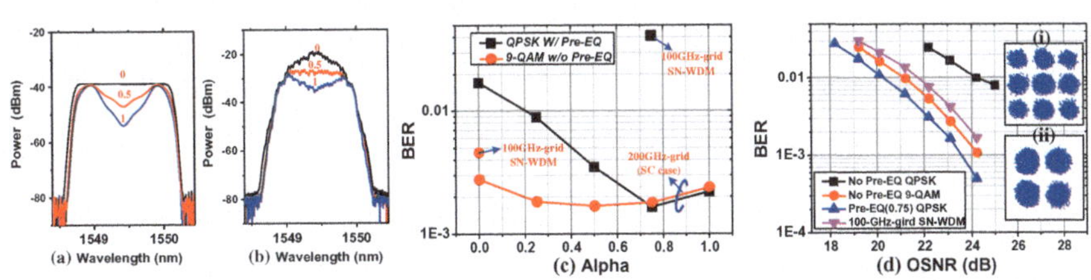

Fig. 4.31 **a** The optical spectrum of WSS, **b** the generated QPSK signals after WSS with pre-equalization. The back-to-back BER results versus the OSNR for the 128.8-GBaud PDM-QPSK signals

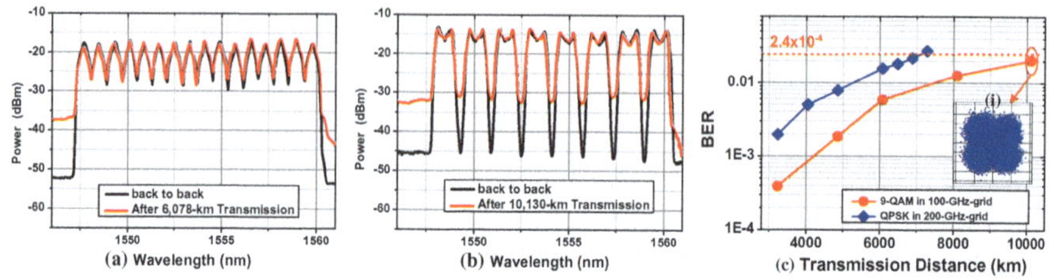

Fig. 4.32 **a** The spectrums of the 16 SN-WDM channels in 100-GHz-grid before and after 6,078-km transmission, **b** the eight channels in 200-GHz-grid before and after the 10,130-km transmissions, **c** The BER performance versus the transmission distance

The spectrum of the 16 SN-WDM 100 GHz-grid channels before and after the 6078-km transmission is shown in Fig. 4.32a, while Fig. 4.32b shows a spectrum of the eight channels in 200-GHz-grid before and after the 10130-km transmission. As an illustration, Fig. 4.32c shows the measured BER of channel 7 inside the SN-WDM channels as well as the measured BER of channel 3 inside 200 GHz-grid channels as a function of transmission distance under the optimal launch power. The measured BERs of 100 GHz-grid channel 7 after 6078-km transmission and 200 GHz-grid channel 3 after 10130 km are 1.6×10^{-2} and 2.15×10^{-2}, respectively, which are all below the SD-FEC limit. The constellation diagram of channel 3 after 10,130-km transmission is also shown in Fig. 4.32c (inset i).

We also measured the BER performance of 100-GHz-grid channel 7 after 6,078-km terahertz wave SLA + fiber transmission as a function of the launch power per channel, which is shown in Fig. 4.33a. The optimal input power per channel is ~0.75 dBm. The BER degrades at the launched powers higher than 0.75 dBm due to nonlinear impairments. We have also confirmed that the BER of all WDM channels in 100 GHz or 200 GHz grid after 6078-km or 10130-km transmission are below the BER threshold, respectively (Fig. 4.33d).

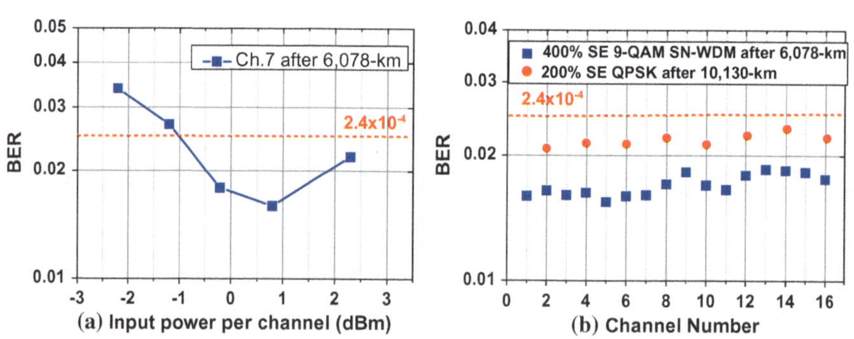

Fig. 4.33 **a** The BER performance after 6078-km transmission of channel 7 versus the launch power per channel; **b** The BER of all sixteen channels after 6078-km transmission and eight channels after 10130-km transmission

4.5 Conclusions

This chapter first introduces a new multi-modulus blind equalization algorithm for super-Nyquist spectrum compression signals, which can effectively suppress noise and inter-channel crosstalk, thus achieving super-Nyquist WDM signal transmission. At the same time, the experimental results show that the proposed algorithm has better anti-noise, anti-filtering, and anti-channel interference performance than the common processing equalization algorithm. Through this algorithm, we realize the digital generation of super-Nyquist signals based on front-end filtering, and the 400 G optical transmission of four carriers. This chapter also introduces an experiment of a 100-GBaud ultra-high-speed PDM-QPSK coherent transmission system. By using this system, we successfully created the world's first single-carrier super-Nyquist-based 400 G optical transmission system, and created the world's record of the highest baud rate of 110 GBaud signal transmission. The carrier spacing is compressed to 100 GHz, the transmission capacity is up to 8.8 Tb/s, and the spectral efficiency is greater than 4 bits/s/Hz. At the same time, the performance of the super-Nyquist filtered signal in multiple resettable optical add/drop multiplexers (ROADMs) is also studied experimentally. We also confirmed the high filter resistance of the 9QAM super-Nyquist signal in experiments. Based on this scheme, we successfully transmitted ten-channel 440 Gb/s QPSK signals in more than 3000-km ULAF and 10 cascaded 100-GHz ROADMs. We successfully transmitted single-carrier 515.2-Gb/s PDM-QPSK/9QAM signals over 10130 km/6078-km, respectively, over 100-km spans of terahertz wave SLA + fiber. As far as we know, this is the first time to realize single-carrier 400 G signal transmission over distances above 10,000 km in a terrestrial transmission environment. We have also demonstrated the coherent transmission of single-carrier 128.8-GBaud 9QAM-like filtered QPSK signals in 100-GHz-grid over 6078 km, with line spectral efficiency-distance product that exceeds 30000 (b/s/Hz) km.

References

1. J.W. Zhang, B. Huang, X.Y. Li, Improved quadrature duobinary system performance using multi-modulus equalization. IEEE Photon. Technol. Lett. **25**(16), 1630–1633 (2013)
2. J.W. Zhang, J.J. Yu, N. Chi, Generation and transmission of 512-Gb/s quad-carrier digital super-Nyquist spectral shaped signal. Opt. Expr. **21**(25), 31212–31217 (2013)
3. Z.S. Jia, H.C. Chien, J.W. Zhang et al., Super-Nyquist shaping and processing technologies for high-spectral-efficiency optical systems, in *Proceedings of SPIE—The International Society for Optical Engineering*. San Francisco, CA, United states: SPIE, 2014: The Society of Photo-Optical Instrumentation Engineers (SPIE)
4. J.W. Zhang, J.J. Yu, N. Chi et al., Multi-modulus blind equalizations for coherent quadrature duobinary spectrum shaped PM-QPSK digital signal processing. J. Lightwave Technol. **31**(7), 1073–1078 (2013)
5. M.R. Rios, J. Renaudier, P. Brindel et al., Spectrally-efficient 400-Gb/s single carrier transport over 7 200 km. J. Lightwave Technol. **33**(7), 1402–1407 (2015)

6. H.C. Chien, J. Zhang, Y. Xia et al., *Transmission of 20 × 440-Gb/s Super-Nyquist-Filtered Signals over 3600 km based on Single-Carrier 110-GBaud PDM QPSK with 100-GHz Grid[M]* (CA, San Francisco, 2014), pp. 1–3

Chapter 5
All-Optical Nyquist Signal

5.1 Introduction

The bandwidth requirements of network data services are driven by video streaming, cloud computing, social media, and mobile internet transmission, growing at a rate of 30–60% per year [1–4]. Considering the cost, high-speed optical transmission interfaces and high-spectrum efficiency technologies are becoming more and more important. High-order modulation formats and orthogonal multiplexing techniques can effectively reduce system bandwidth and improve system spectral efficiency [1]. Orthogonal multiplexing technologies can be realized by the generation of optical or electrical waveforms, including orthogonal OFDM technology [2–10] in the frequency-domain and orthogonal Nyquist pulse technology [3–6] in the time domain. So far, OFDM with high-order modulation formats has been widely studied, and OFDM signal generation based on electrical and optical has been reported [2, 3], especially optical OFDM. Hillerkuss et al. have implemented 26 Tb/s superchannel all-optical OFDM transmission based on 16QAM and coherent detection [1].

The time–frequency relationship between OFDM and Nyquist is shown in Fig. 5.1, and the performance of both in the time domain and frequency domain is reversed. The OFDM signal is orthogonal in frequency domain and rectangular in time domain; the Nyquist signal is a rectangular spectrum in frequency domain and an orthogonal pulse sequence in time domain. Compared with OFDM, Nyquist signal transmission has several unique advantages, such as low receiver complexity, low receiver bandwidths, and low peak-to-average ratios for improving fiber nonlinear damage [4–6]. Hillerkuss et al. have demonstrated the single-laser 32.5 Tbit/s Nyquist WDM transmission system based on 12.5 GBaud PDM-16QAM and 325 optical carriers [1]. However, most of the reported Nyquist pulse-shaping solutions are realized by the electronics devices with limited sampling rate and limited processor capacities.

The Nyquist signal has the theoretical minimum signal bandwidth and can achieve high-spectral efficiency optical transmission by using high-order signal modulation.

J. Yu and N. Chi, *Digital Signal Processing In High-Speed Optical Fiber Communication Principle and Application*,
https://doi.org/10.1007/978-981-15-3098-2_5

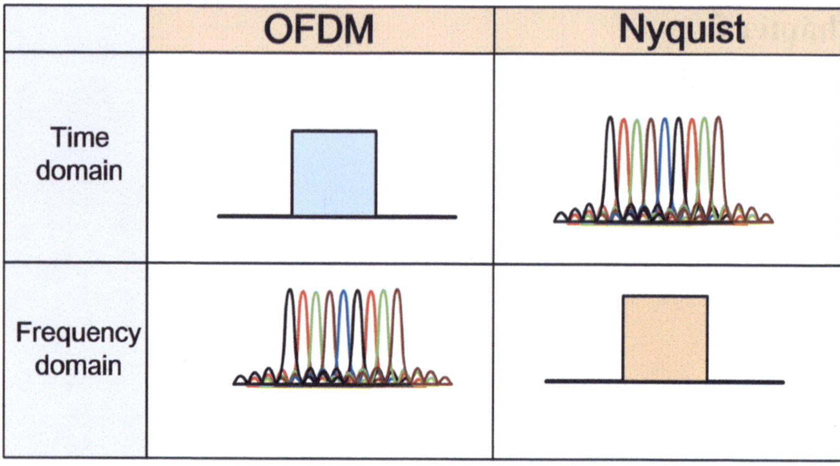

Fig. 5.1 Time–frequency relationship diagram between OFDM signal and Nyquist signal

In general, the Nyquist signal needs to be implemented using high-speed analog-to-digital conversion electronic devices. This electrical Nyquist signal is limited by the bandwidth of the electronic devices, making it difficult to achieve high-speed signal generation. All-optical signal processing can break the bandwidth limitation of electronic devices and obviously improve the signal generation and processing rate. Therefore, the key theories and technologies for the generation, modulation and coding, transmission, and signal processing of all-optical Nyquist signals have attracted extensive attention and it has become a hot topic in the optical transmission research community [4–6].

This chapter introduces the generation and processing of all-optical Nyquist signal, such as the all-optical Nyquist pulse generation, signal modulation and multiplexing, and detection and signal processing. Firstly, the principle of all-optical Nyquist signal pulse generation, signal modulation, and multiplexing are introduced. This chapter will also introduce the theoretical basis of optical frequency comb based on single MZM to achieve frequency lock, linear phase, and equal amplitude. Then, the all-optical Nyquist signal generation and coherent detection system will be described. It utilizes a full-light frequency comb to generate and coherently detect all-optical Nyquist QPSK signals up to 62.5, 75, and 125 GBaud. Finally, polarization-multiplexed all-optical Nyquist signal long-distance transmission is demonstrated, including the generation, transmission, and full-band coherent detection of 37.5 and 62.5-GBaud all-optical Nyquist PDM-QPSK and 16QAM signals. The all-optical Nyquist 37.5 and 62.5-GBaud 16QAM signals have successfully demonstrated transmission over 1200 km and 850 km SSMF.

5.2 The Generation and Signal Multiplexing Principle of All-Optical Nyquist Pulse

5.2.1 The Basic Theory of High-Quality Sinc-type Nyquist Pulse Generation

The all-optical scheme of high-quality Sinc-type Nyquist pulses can be generated by a linear phase and frequency-locked optical comb. This scheme was first proposed by Soto et al. in Nature Communication in 2013 [4]. The time–frequency transforms for Sinc-shaped Nyquist pulses and the modulated Nyquist signals are shown in Fig. 5.2. On one hand, the ideal Sinc pulse of the rectangular spectrum is difficult to generate by electronic devices because its time-domain impulse response is infinite, and the digitally generated Sinc pulse filter taps are always limited; on the other hand, the high-speed Sinc-type Nyquist Pulses require a large sampling rate to ensure their pulse shape, which places further demands on the processing speed of the device.

It can be known from the basic concept of Fourier transforms that the spectrum of a single Sinc-type time-domain pulse is rectangular in frequency domain. Furthermore, if it is a Sinc-type pulse sequence, its spectrum is frequency-locked, phase-continuous, and linear flat light comb. Then, compared with generating a high-quality Sinc-type pulse from the time domain, a simple and effective method is to generate a frequency-locked and phase-continuous optical comb from the frequency domain, and the optical comb in the time domain is naturally the Sin-type Nyquist optical pulses.

Assuming that the frequency of the *N* carriers is locked to the optical comb, the center frequency is f_0, the carrier spacing is Δf, and the equivalent amplitude is E_0/N, then the time-domain expression of the optical field of the comb can be expressed as:

$$E_N(t) = \frac{E_0}{N}\sum_{-(N-1)/2}^{(N-1)/2} \exp[j2\pi(f_0 + n\Delta f)t + j\phi + jn\varphi_0]$$

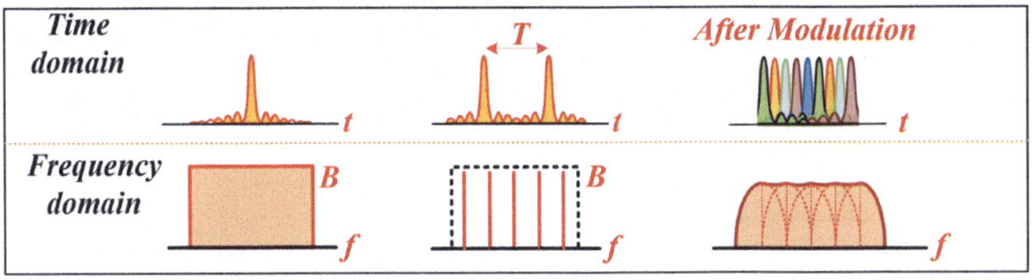

Fig. 5.2 The basic principle of time–frequency transforms for Sinc-shaped Nyquist pulses and the modulated Nyquist signals

$$=E_0 \exp(j2\pi f_0 + j\phi)\frac{\sin[\pi N \Delta f(t+\tau_0)]}{N\sin[\pi \Delta f(t+\tau_0)]}$$
$$=E_0 \exp(j2\pi f_0 + j\phi)\sum_{-\infty}^{+\infty}(-1)^{(N-1)n}\sin c[N\Delta f(t-nT+\tau_0)] \quad (5.1)$$

Here T $= 1/\Delta f$ is the pulse repetition period. We define the time delay τ_0, and $\varphi_0 = 2\pi\,\Delta f\tau_0$, which is the phase condition that the pulse sequence needs to satisfy, that is, the linear continuous phase. Thus for a linear phase of the N carrier and a frequency-locked comb of equal amplitude, the time-domain optical signal is a periodic Sinc-shaped Nyquist pulse with a period of T and a zero-crossing pulse width of 2T/N. Thus, the problem of pulse generation for all-optical Nyquist is achieving a frequency-locked, linear phase, and flat amplitude optical comb.

5.2.2 *Principle of All-Optical Nyquist Signal Multiplexing and Coherent Detection*

Figure 5.3 shows the principle of all-optical Nyquist pulse generation, high-order modulation based on I/Q modulators, multiplexing, and digital signal processing. Flexible Nyquist pulses can be generated by a frequency-locked linear phase comb cascading with an adjustable bandpass filter. The frequency-locked optical comb can be generated by a radio frequency driven cascaded MZM modulator, and each subcarrier can be guaranteed to have a linear phase relation. The subcarrier frequency spacing and the total bandwidth of the optical comb are tunable by adjusting the frequency of the RF signal and the passband of the bandpass filter. This proposed scheme is developed from then scheme of [4], mainly by adding a tunable bandpass filter. This change has two main advantages. First, the undesired sidebands of the bandpass filter can be greatly suppressed to reduce crosstalk. The results in [4] show

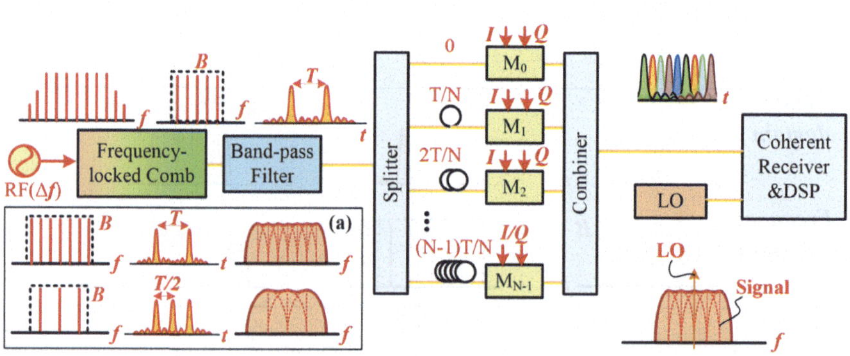

Fig. 5.3 The principle of all-optical Nyquist signal multiplexing and coherent detection system

that the power suppression of the frequency components of the sidebands is only 21–27 dB. However, by using a bandpass filter, the suppression of unwanted high-order sidebands can be as high as 40 dB. Second, the number of time division channels can be easily changed without changing the bias and drive voltage by changing the passband bandwidth of the bandpass filter. As analyzed in [4], bias and driven voltage are very sensitive and difficult to match each other.

Assuming B is the bandwidth of BPF, the number of optical comb carriers is $N = B/\Delta f$. Changing the BPF bandwidth or the carrier spacing of the optical comb can change the pulse width and repetition period, as shown in Fig. 5.3a. The periodic Sinc-shaped Nyquist pulse sequence is divided into N branches for individual modulation, and each branch performs an exact time delay $t_n = nT/N + KT$, where $n = 0–N - 1$, and K is an arbitrary nonnegative integer. The symbol period of the multiplexed Nyquist signal is T/N. After combining, a modulated Nyquist signal with a baud rate of B or $N\Delta f$ can be obtained. For the Nyquist pulse before modulation, the rectangular comb spectrum is with a roll-off factor of zero. However, as analyzed in [74], since the modulation of the time domain corresponds to the convolution of the frequency domain, the spectrum of the modulated Sinc-shaped pulses is given by the convolution of the frequency comb and the frequency representation of the modulating signal. Therefore, the modulated Nyquist signal spectrum is actually spread after modulation, and the spectral roll-off and overflow bandwidth are no longer zero.

It is worth noting that the actual overflow bandwidth is the ratio of the actual bandwidth occupied by the signal to the Nyquist frequency bandwidth, which is not determined by the roll-off factor of the unmodulated pulse. Because the pre-modulation coefficient is zero, the Nyquist bandwidth of the signal is

$$B_{\text{Nyquist}} = B = N\Delta f \tag{5.2}$$

The modulation baud rate for each pulse is Δf, as shown in Fig. 5.3, and the effective bandwidth after modulation is $B_{effective} = (N + 1)\,\Delta f$. The effective overflow bandwidth and effective frequency roll-off factor after modulation can be obtained:

$$\beta_{\text{effective}} = \frac{B_{\text{effective}}}{B_{\text{Nyquist}}} - 1 = \frac{(N+1)\Delta f}{N\Delta f} - 1 = \frac{1}{N} \tag{5.3}$$

Therefore the effective roll-off factor (Note: the effective roll-off factor here needs to be distinguished from the roll-off factor of the commonly used raised cosine signal, we use the effective roll-off factor to describe the overflow bandwidth generated by the Nyquist signal modulation) or the overflow bandwidth of modulated Nyquist signal is $\beta_{\text{effective}} = 1/N$. The relationship between the effective roll-off factor and the number of carriers after modulation is shown in Fig. 5.4. We can find that when the optical comb is 5 or 10 carriers, the effective roll-off coefficients after modulation are 0.2 and 0.1, respectively.

A coherent receiver can be used at the receiver side, and a free-running continuous-wave local oscillator laser operating at the center frequency of the optical comb

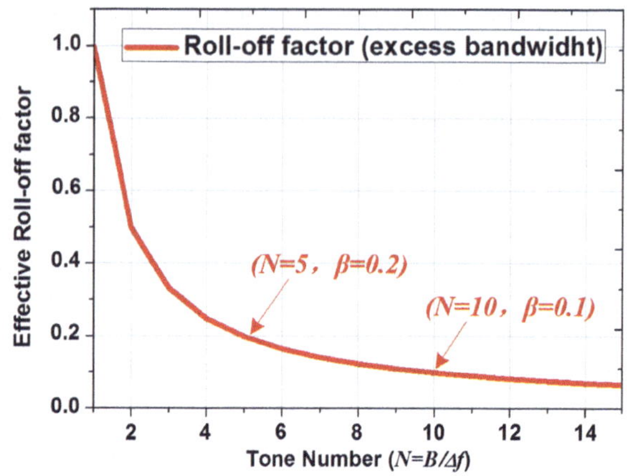

Fig. 5.4 The relationship between the effective roll-off coefficient of all-optical Nyquist signal and the number of optical comb carriers

is used for coherent homodyne beating. As mentioned earlier, full-band coherent receivers are simpler than traditional OTDM receivers [9, 10]. Compared to OTDM optical receivers, the simpler coherent receiver mainly shows in two aspects: the reduction in the number of receivers and the free-running LO. On the one hand, we propose to use a wide-bandwidth full-band receiver compared to a large number of narrowband receivers for OTDM optical receivers. The receiver size can be reduced, the structure can be simplified, and the integration is easier. On the other hand, as described in [9, 10], in order to obtain the correct time slot, the receiver terminal LO should be a strictly synchronized pulse source. In an actual system, it is difficult to achieve such synchronization. However, in our proposed system, a commercial free-running continuous-wave ECL laser is used, which is very easy to implement. Therefore, for coherent receiver described in this Chapter, there is no need for strict synchronization pulse demultiplexing or optical comb local oscillator, while the coherent detection and digital signal processing are used. Digital signal processing is based on conventional algorithms and subsystems, as described in Sects. 5.3 and 5.4 of this chapter.

5.2.3 Principle of All-Optical Nyquist Pulse Generation Based on MZM and Bandpass Filter

From the above, the key to generating an all-optical Nyquist pulse is to produce a linear comb, a frequency lock, and an equal amplitude optical comb spectrum. In the all-optical Nyquist signal generation scheme proposed in this Chapter, we use one single MZM and a PM-TOF to generate optical combs of equal amplitude and linear phase. The following conditions are used to achieve equal amplitude and linear phase when generating five optical carriers. In order to achieve the frequency lock, phase linearity, and equal amplitude conditions given in Sect. 5.2.1, we must adjust the RF drive signal and the DC bias voltage to get the best output. For a zero-chirped MZ intensity modulator, assuming that the voltage of the RF driven signal is $aV_{pi}cos(2\pi \Delta ft)$ and the voltage of the DC bias is bV_{pi}, then the output of the MZM can be expressed as

$$E(t) = \cos\left\{\frac{\pi}{2}[a\cos(2\pi\Delta ft) + b]\right\} = \cos\left\{\frac{\pi}{2}\left[a\cos(2\pi\Delta ft) + \frac{\pi}{2}b\right]\right\}$$
$$= \cos\left[\frac{\pi}{2}a\cos(2\pi\Delta ft)\right]\cos\left(\frac{\pi}{2}b\right) - \sin\left[\frac{\pi}{2}a\cos(2\pi\Delta ft)\right]\sin\left(\frac{\pi}{2}b\right) \quad (5.4)$$

It can be expanded by the Jacobi–Anger extension:

$$E(t) = \sum_{n=-\infty}^{+\infty}(-1)^n\left\{\cos\left(\frac{\pi}{2}b\right)J_{2n}\left(\frac{\pi}{2}a\right)\cos[2\pi(f_0 + 2n\Delta f)t]\right.$$
$$\left. + \sin\left(\frac{\pi}{2}b\right)J_{2n-1}\left(\frac{\pi}{2}a\right)\cos[2\pi(f_0 + (2n-1)\Delta f)t]\right\} \quad (5.5)$$

Here J_n is the n-th order first-class Bessel function, V_{pi} is the half-wave voltage of the modulator, and a and b are the normalized driven voltage and DC bias. Therefore, there are two degrees in the MZM-based all-optical Nyquist pulse generation, which is the modulation index and the bias index, and they can be used to equalize the amplitude and phase of each carrier frequency of the generated optical comb. In order to obtain a pulse of Sinc-shaped, the optical comb should satisfy the conditions of equal amplitude and linear phase given by Eq. 5.1. From Eq. 5.4, we can see that the phase of each carrier is related to the sign of the amplitude, which is 0 or π. Therefore, assuming that the phase of the center carrier is 0, the phase of the five carriers can be {0, 0, 0, 0, 0} or {0, π, 0, π, 0}. Even carriers should have the same phase. Based on Eq. 5.4, by changing the normalized amplitude and DC bias, we can plot the power difference and phase relationship of the five carriers of the optical comb, as shown in Fig. 5.5a, b.

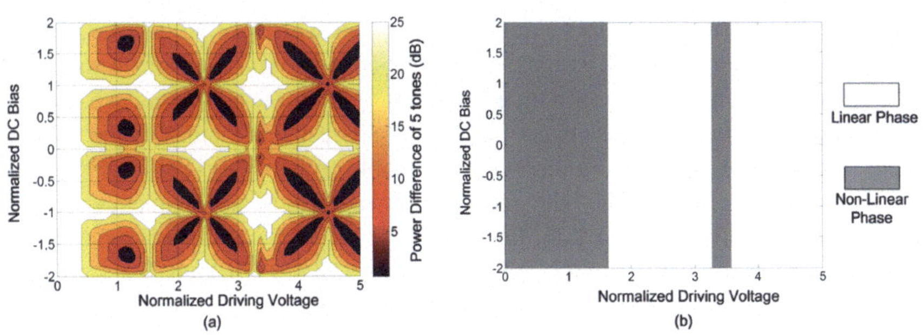

Fig. 5.5 **a** The relationship of power difference of 5 tones, normalized DC bias and normalized driving voltage; **b** The phase relationship of the driving voltage and the DC bias

From Fig. 5.5a, b, several conclusions can be drawn. First, there are several optimal regions for the RF drive voltage and DC bias (the black region in Fig. 5.5a). A flat 5 carrier comb with the smallest power difference can be obtained, and the power difference is determined by the drive voltage and DC bias. Second, we can find that the linear phase relationship of the generated carrier is determined only by the driving voltage, and the linear phase relationship of the five carriers can be easily obtained by a suitable driving voltage within a large adjustment range. In addition, the optimized region of Fig. 5.5a and the linear phase region of Fig. 5.5b can be satisfied at the same time because there are several overlapping portions. However, we have found that the optimal area of a flat comb with the smallest power difference is much smaller than the linear phase region. Compared to the phase relationship, the power difference of the optical comb is more sensitive to the driving voltage and the DC bias. In an actual system, the power balance condition can also be alleviated by a power equalization device, such as a filter or a wavelength selective switch.

5.3 Generation and Coherent Detection Experiment of Single-Polarized 125 GBaud All-Optical Nyquist QPSK Signal

Based on the above basic principles of all-optical Nyquist pulse generation, modulation, multiplexing, and coherent detection, the following is an experimental study and verification of the ultra-high-speed all-optical Nyquist signal generation and coherent detection system [11, 12]. Here we present a conceptual verification experimental setup for all-optical Nyquist pulse generation, high-order formats signal modulation, and full-band coherent detection, as shown in Fig. 5.6. An ECL having a wavelength of 1549.50 nm, a line width of less than 100 kHz, and output power of 14.5 dBm was used as the laser source. An RF source driven MZM is used to generate a frequency-locked optical comb. Here we choose 6.25 and 12.5 GHz RF and a 1:2 frequency multiplier to produce optical combs with 12.5 and 25 GHz spacing. A power amplifier is used to amplify the RF signal. It is worth noting that the voltage and DC bias of the driving RF signal should be adjusted simultaneously to match each other to obtain a flat carrier with a linear phase. In our experiments, we chose a drive voltage of $2V_{pi}$ and a DC bias of 0.52 V_{pi}, where V_{pi} is the half-wave voltage of the modulator. A polarization-maintaining tunable optical filter (PM-TOF) is used to select the carrier. PM-TOF is the commercial wavelength and frequency tunable optical bandpass filter (Alnair Lab, BVF-200). The filter bandwidth is from 0.1 to 15 nm and the center wavelength is adjustable from 1525 to 1610 nm. It has an ideal flat-top response and a steep roll-off of >150 dB/nm with an out-of-band rejection of approximately 50 dB. Its dispersion is less than 0.1 ps/nm and is almost negligible. The TOF does not compensate the phase relation because the linear phase can be very easily obtained. However, we can use this filter to suppress the second-order

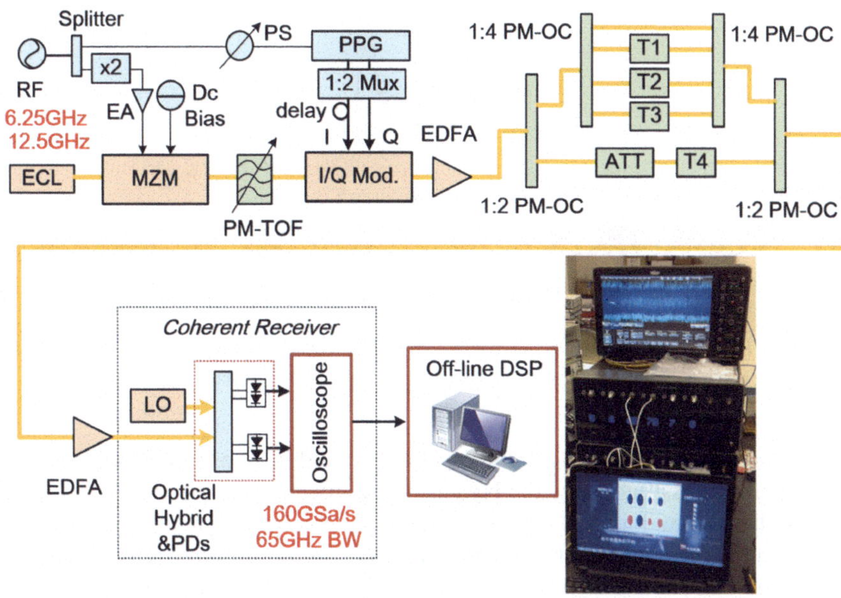

Fig. 5.6 Conceptual verification experimental device for all-optical Nyquist pulse generation, high-order modulation format signal modulation, and full-band coherent detection of a receiver

sidebands to equalize the power and achieve a flatter comb. The power difference between the five carriers generated after the filter is less than 0.3 dB.

Figure 5.7a, b shows the spectra of the resulting optical combs with a spacing of 12.5 GHz before and after the PM-TOF. The bandwidth of the PM-TOF here is 62.5 GHz (5 × 12.5 GHz). Figure 5.7c shows the spectrum of the Nyquist signal after modulation. We can observe that the spectrum of the full Nyquist signal is close to a rectangular shape. Since there are 5 peaks in the 5 time-domain subchannels in this optical comb, the overflow bandwidth or roll-off factor is 0.2.

As previously analyzed, a Sinc-shaped Nyquist pulse can be generated after the PM-TOF. Figure 5.8a shows the time-domain waveform of a 5 × 12.5 GHz optical comb. We can see the Nyquist pulse of the Sinc-shape and a total of 4 zero crossings

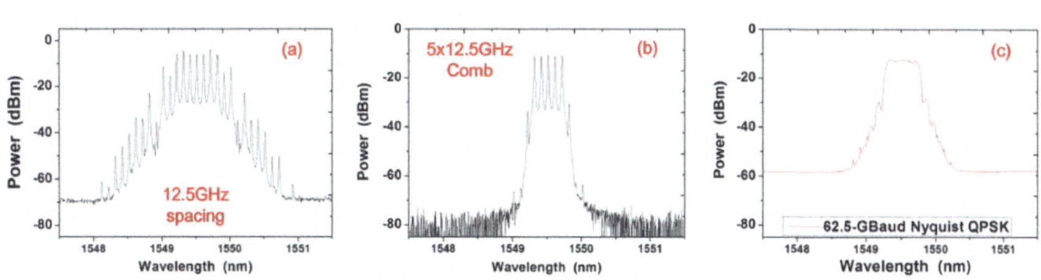

Fig. 5.7 Optical Spectra (**a**) and (**b**) the optical comb with 12.5-GHz carrier spacing before and after PM-TOF; **c** the optical comb after modulation

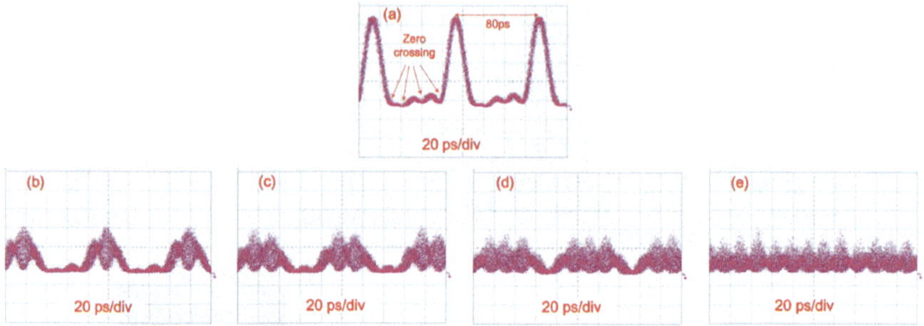

Fig. 5.8 **a** Time-domain waveform of a 5 × 12.5 GHz optical comb; **b–e** is a time-domain multiplexing process diagram of Nyquist pulses, from 2 pulses to 5 pulses

between every two pulses. The pulse repetition period is 80 ps. For Nyquist signal modulation, we first modulate the Nyquist pulse directly, and then multiplex it on the time division multiplexed channel through the polarization-maintaining optical coupler and the polarization-maintaining tunable optical delay line (PM-TDL, T1–T4). Noting that in our experiments, each branch carries the same data, so to ensure orthogonality and decorrelate the slots, the delay line must be long enough. The PM-TDL in our experiment consists of a fixed delay line (polarization-maintaining fiber jumper, about 5 ns m^2 delay) and a manually adjusted delay module (0–300 ps adjustable time delay). For QPSK modulation, a 12.5 GBaud binary I/Q signal is generated by a programmable pulse generation (PPG) followed by a cascade of 2:1 multiplexers. The I/Q modulator is biased at zero and driven across the full range to achieve zero chirp and phase modulation from 0 to π. An electric phase shifter is used for the clock signal of the PPG to separate the RF source for the comb production. It synchronizes the I/Q modulated signal with the Nyquist pulse to ensure that the peak of the Nyquist pulse is within the duration of the QPSK symbol.

To generate a five-channel time division multiplexed Nyquist pulse, two pairs of 1:2 PM-OC and 1:4 PM-OC are used as four polarization-maintaining delay lines. Each Nyquist pulse branch delay is long enough to decorrelate, while the delay length is an integer multiple of the symbol period (T/N, 16 ps for a 62.5 GBaud Nyquist signal). For a 5 × 12.5 GHz optical comb, T1 is 10 ns + 16 ps, T2 is 20 ns + 32 ps, T3 is 30 ns + 48 ps, and T4 is 40 ns + 64 ps. Each branch has a 125-symbol delay for decorrelation. Using this transmitter device, we generate a 62.5-GBaud Nyquist QPSK signal with a time-domain waveform as shown in Fig. 5.8a. It can be seen that the symbol period of the 62.5-GBaud Nyquist QPSK signal is 16 ps. Figure 5.8b–e clearly shows the process of obtaining a 62.5 GBaud Nyquist signal by time-domain multiplexing of a five-channel Nyquist pulse. There are 2, 3, 4, and 5 branches in Fig. 5.8b–e, respectively. Since all Nyquist pulses are multiplexed at zero crossings, orthogonal overlap in the time domain without ISI can be achieved. Therefore, the modulated time-multiplexed modulated Nyquist pulse shows minimal ISI and maximum spectral efficiency. In the waveform measured in Fig. 5.8, the pulses are not strictly symmetrical. We believe that the asymmetry of

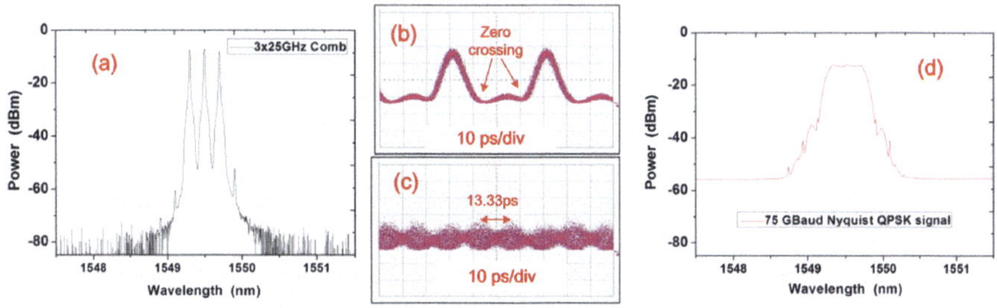

Fig. 5.9 **a** 3 × 25 GHz optical comb spectrum after the PM-TOF; **b** the time-domain Nyquist pulse waveform of the 3 × 25-GHz comb; **c** the waveform of the multiplexed 75-GBaud Nyquist QPSK signal; **d** the modulated and multiplexed 75-GBaud Nyquist QPSK signal

the shape of the waveform in this measured pulse is due to the overbalance problem of the sampling oscilloscope during pulse down and rise. If you have a sampling oscilloscope with a high-performance, large sampling bandwidth for fast pulse-to-noise, the measurement results will be better.

Higher baud rate Nyquist signals can be generated by optical combing with more carriers or larger carrier spacing. Here we successfully generated 75 and 125-GBaud Nyquist QPSK signals with a 25-GHz spaced optical comb. Using a 12.5 GHz RF source, we set the BPF bandwidth to 75 GHz to produce a 3 × 25-GHz comb, as shown in Fig. 5.9a. At this time, the waveform of the Sinc-shaped Nyquist pulse is also changed. Figure 5.9b shows the time-domain Nyquist pulse waveform of a 3 × 25 GHz comb. We can see that there are 2 zero crossings between each Nyquist pulse. The pulse repetition period is 40-ps, which is only half of the 5 × 12.5-GHz comb. Therefore, only the delay branch of the integer multiple of the three symbol periods is used here. For a 3 × 25 GHz optical comb, T1 and T2 are approximately 5 ns + 13.33 ps and 10 ns + 26.66 ps, respectively. Each branch also has a 125-symbol delay to decorrelate. The waveform of the multiplexed 75-GBaud Nyquist QPSK signal is shown in Fig. 5.9c, and its symbol period is 13.33-ps. The spectrum of the Nyquist pulse after signal modulation and multiplexing is shown in Fig. 5.9d. The effective roll-off factor after modulation is 0.33, because the optical comb has only 3 carriers.

In addition to the tunability of the Nyquist pulse generated by different RFs, we also verified that the same RF is maintained but the BPF bandwidth is changed to flexibly generate pulses. Using the same 12.5-GHz RF source, we increased the bandwidth of the PM-TOF to 125 GHz to produce a 5 × 25-GHz optical comb, as shown in Fig. 5.10a. Figure 5.10b shows the time-domain waveform of a 5 × 25-GHz optical comb. Compared to the 5 × 12.5-GHz optical comb, we found that the pulse repetition period dropped to 40 ps. There are 4 zero crossings in each cycle. The waveform of the modulated and multiplexed 125-GBaud Nyquist signal is shown in Fig. 5.10c. For the 125-GBaud Nyquist signal, the symbol period is 8 ps. For a 5 × 25 GHz optical comb, T1–T4 are 5 ns + 8 ps, 10 ns + 16 ps, 15 ns + 24 ps, and 20 ns + 32 ps, respectively. Approximately 125 symbol delays are used

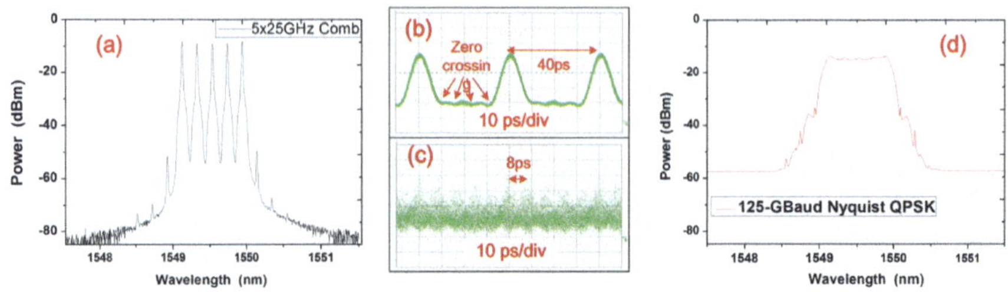

Fig. 5.10 **a** 5 × 25 GHz optical comb spectrum after the PM-TOF; **b** the time-domain Nyquist pulse waveform of the 5 × 25-GHz comb; **c** the waveform of the multiplexed 125-GBaud Nyquist QPSK signal; **d** the modulated and multiplexed 125-GBaud Nyquist QPSK signal

for decorrelation between branches. The spectrum of the modulated and multiplexed 125-GBaud Nyquist signal is shown in Fig. 5.10d. The effective overflow bandwidth or roll-off factor after modulation is also 0.2.

For the receiver side of the full-band signal coherent detection, a free-running ECL with a linewidth of less than 100 kHz is used as the LO, as shown in Fig. 5.6. A light 90° mixer is used for phase diversity coherent detection. The 3 dB bandwidth of the balanced detector is 50 GHz. Sampling and digitization are achieved with a high-speed real-time digital oscilloscope with a sampling rate of 160-GSa/s with an electrical bandwidth of 65-GHz, which is sufficient for a 125-GBaud Nyquist signal. After the ADC, the offline DSP is used to process the 160-GSa/s sampled data sequence for the four channels. The offline DSP is similar to the flow of a conventional coherent system, as shown in Fig. 5.11a, including I/Q imbalance compensation, clock recovery, linear equalization, frequency offset estimation, and carrier phase recovery. The data is first resampled to two samples per symbol and then processed

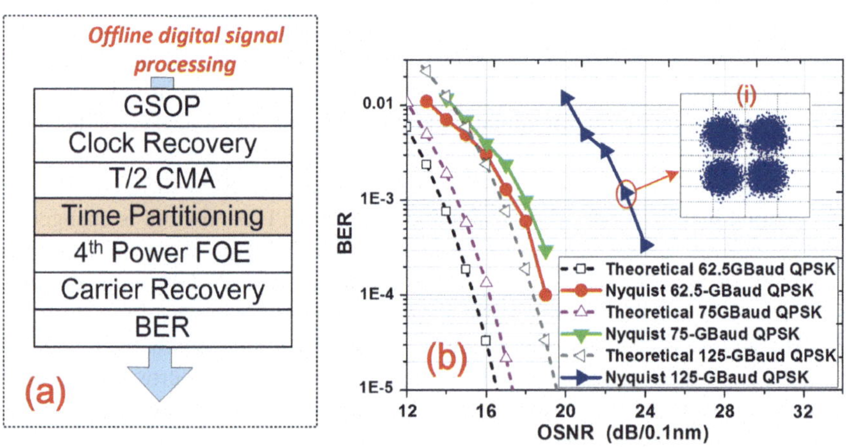

Fig. 5.11 **a** The offline DSP blocks for the Nyquist QPSK signal recovery; **b** BER versus OSNR for back-to-back coherent detection

by the improved QPSK digital signal processing module. Since the phase between each time slot is unknown, we need to perform time grouping before the classical linear CMA equalization, bias estimation, and phase recovery.

In order to measure the performance of coherent detection of full-band signals, we tested the back-to-back BER performance of the resulting 62.5, 75, and 125-GBaud all-optical Nyquist QPSK signals (OSNR takes a 0.1 nm reference bandwidth). As shown in Fig. 5.11b, there is a 4 dB OSNR penalty for 62.5 and 75-GBaud Nyquist QPSK signals compared to the theoretical BER curve, and a 6 dB penalty for 125-GBaud Nyquist QPSK signals, mainly due to inaccurate delays. These theoretical curves are the BER performance limits (best performance) for different baud rate signals. We believe that the inconsistency between theoretical and experimental results comes mainly from two reasons: The first is the inaccuracy of delay in signal multiplexing. Because the fixed fiber patch cord is commercial grade and non-adjustable, the manually adjusted delay line is used for precise pulse position adjustment. The resolution of this module during the move is 0.05 mm, which corresponds to a 0.33-ps delay variation. However, because the delay line is manually adjustable, we must use a light sampling oscilloscope for pulse orthogonal time division multiplexing in the experiment. There is a large delay error in manually adjusting the delay line compared to the programmable delay line or integrated optical delay. This delay error reduces the orthogonality of the time-domain pulses and causes ISI. Therefore, the generated Nyquist signal is deteriorated by the ISI introduced by the clock error in multiplexing. The second reason we believe is the bandwidth limitation of optoelectronic devices in the system. The 3 dB bandwidth of the balanced PD for coherent detection in our experiments is 50 GHz, which imposes an additional penalty. We believe that better performance can be achieved if integrated broadband optoelectronic devices are used for signal generation and modulation. In summary, the above experiments successfully achieved high-speed all-optical Nyquist signal generation, modulation, multiplexing, and full-band coherent detection based on Sinc-shaped pulse [12].

5.4 Polarization-Multiplexed All-Optical Nyquist Signal Long-Distance Transmission Experiment

The third section introduces the results of single-polarized all-optical Nyquist signal generation and back-to-back coherent detection experiments, and successfully achieves 125 GBaud high-speed all-optical Nyquist signal generation, modulation, multiplexing, and full-band coherent detection based on Sinc-shaped pulse [12]. This section will further introduce the experimental results of coherent detection of all-optical Nyquist signal generation, long-distance transmission, and polarization diversity using polarization multiplexing. Figure 5.12 shows the schematic diagram of a polarization-multiplexed all-optical Nyquist signal generation, fiber transmission, and full-band coherent detection based on a Sinc-type Nyquist pulse. An ECL

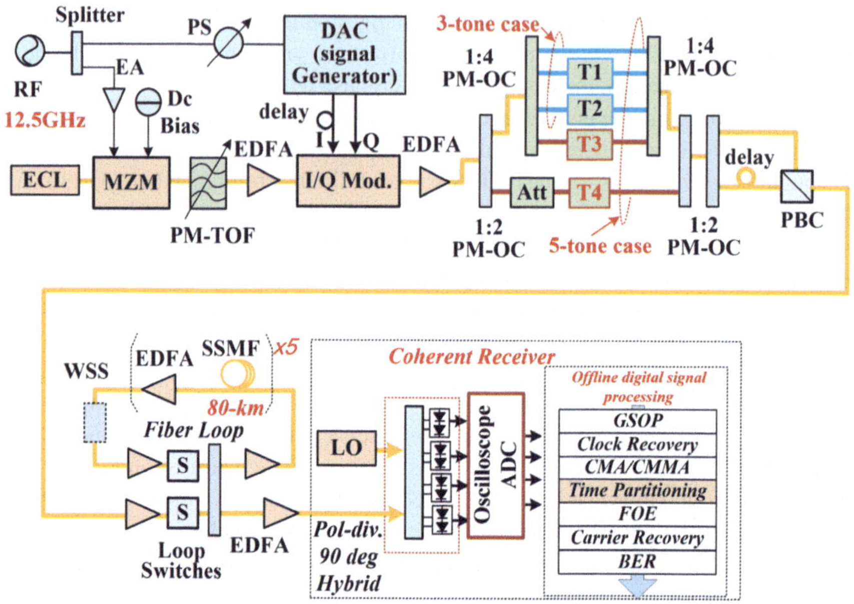

Fig. 5.12 Experimental setup of polarization-multiplexed all-optical Nyquist signal generation and long-distance transmission

is used as a light source with a wavelength of 1549.50 nm, a linewidth of less than 100 kHz, and output power of 14.5 dBm. A frequency-locked optical comb of 12.5-GHz carrier spacing is generated by MZM driven by a 12.5 GHz RF source. A power amplifier is used to amplify the amplitude of the RF signal. It is worth noting that the RF driving signal and the DC bias voltage need to be adjusted simultaneously to match each other to obtain a linear phase and flat amplitude optical comb. In our experiments, we chose a drive voltage of $2V_{pi}$ and a DC bias of $0.52V_{pi}$, where V_{pi} is the half-wave voltage of the modulator. A PM-TOF is used to select the optical comb carrier. PM-TOF is a commercially available wavelength and bandwidth tunable optical bandpass filter with an ideal flat-top response and steep roll-off > 150 dB/nm with an out-of-band rejection of approximately 50 dB. So we can use this filter to suppress high-order sidebands and get a flat comb.

Figure 5.13a is the spectrum of the optical comb that produces the 12.5-GHz spacing before PM-TOF, while Fig. 5.13b, c shows the flat light of 3 × 12.5-GHz and 5 × 12.5-GHz combs. The power difference between the three carriers generated by the filter and the five carrier optical combs is less than 0.3 dB. The phase of these optical comb carriers is linearly locked by the RF drive voltage. In the time domain, the Sinc shaping pulses of the 3 carrier and 5 carrier combs are shown in Fig. 5.13d, e). We can find that for the 3 × 12.5-GHz optical comb, the resulting Sinc-shaped Nyquist pulse has two zero crossings between each Nyquist pulse, and the pulse repetition period is 80 ps. The repetition period of the 5 × 12.5 GHz optical comb

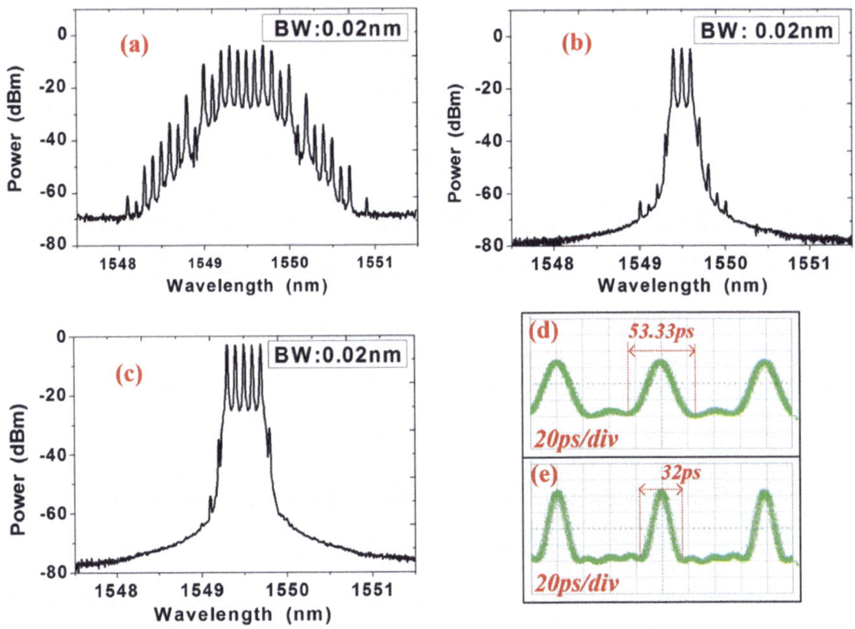

Fig. 5.13 **a** The spectrum of the optical comb that produces the 12.5-GHz spacing before PM-TOF; **b** 3 × 12.5 GHz flat comb; **c** 5 × 12.5 GHz flat comb; **d** and **e** 3 × 12.5 GHz and 5 × 12.5 GHz flat time-domain waveform of Sinc-shaped Nyquist pulse of optical comb

is also 80 ps, but there are 4 zero crossings between each pulse. The duration of the zero crossing pulse in Fig. 5.13d, e is 53.33 and 32 ps.

For Nyquist signal generation, we first modulate the Sinc-shaped Nyquist pulse directly and then multiplex these time division channels with PM-OC and PM-TDL in the experiment. For QPSK/16QAM modulation, 12.5-GBaud in-phase and quadrature data signals are generated by a programmable DAC. The I/Q modulator biases the signal at zero. An electrical phase shifter is used between the DAC clock signal and the RF source to generate an optical comb that synchronizes the I/Q signal modulation with the Nyquist pulse to ensure that the peak of the Nyquist pulse is within the symbol duration. In order to ensure orthogonality between slots and decorrelation, the delay line for each slot must be adjustable and long enough. In our experiments, the adjustable delay line consisted of a fixed length of fiber (polarization-maintaining fiber jumper, 5 ns per meter delay) and a manually adjustable delay module (0 to 300 ps adjustable time delay). For three-channel Nyquist pulse multiplexing, we use a pair of 1:4 PM-OC and two PM-TDL. For five-channel Nyquist pulse time division multiplexing, two 1:2 PM-OC, two 1:4 PM-OC, and four PM-TDL are used. The delay time is an integer multiple of the half-zero duration (37.5 GBaud Nyquist signal is 26.67 ps, 62.5 GBaud Nyquist signal is 16 ps) to ensure that the pulses are multiplexed in the zero-forcing slots of adjacent pulses. For decorrelation, each

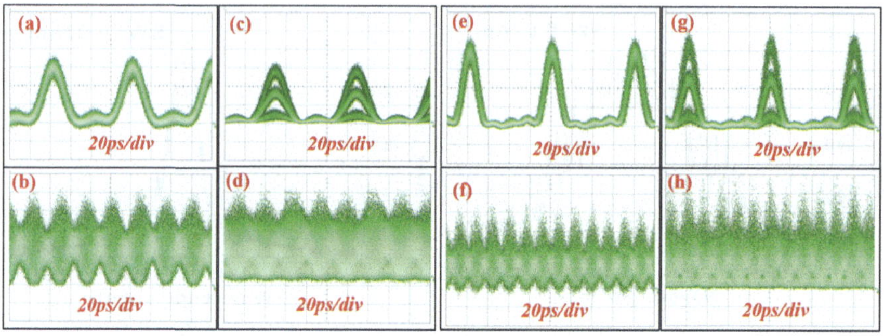

Fig. 5.14 The time-domain Nyquist pulse waveforms of the modulated Sinc-type Nyquist pulse and time division multiplexed all-optical Nyquist signals

branch has a 125-symbol delay. The PDM is then implemented by PM-OC split signals, one PM-TDL providing precise pulse alignment over 100 symbol durations and recombining with another PBS. Thus 37.5 and 62.5-GBaud PDM all-optical Nyquist QPSK/16QAM signals are generated.

Figure 5.14 shows the time-domain Nyquist pulse waveforms of the modulated Sinc-type Nyquist pulse and time division multiplexed all-optical Nyquist signal. The time scale of all these figures is 20 ps/div. Figure 5.14a, c is Sinc-type Nyquist pulses of 3 × 12.5-GHz optical combs modulated by 12.5-GBaud QPSK and 16QAM, respectively. Figure 5.14e, g is modulated 5 × 12.5-GHz Nyquist pulse, the eye pattern of the modulated pulse is very clear. Figure 5.14b, d, f, and g is waveforms of time division multiplexed 37.5-GBaud QPSK, 37.5-GBaud 16QAM, 62.5-GBaud QPSK, and 62.5-GBaud 16QAM. Since all Nyquist pulses are multiplexed at zero crossings, the time domain can be orthogonally superimposed without ISI. The multiplexed Nyquist signal shows an overbalance of amplitude between the symbol centers (that is, the zero crossings of adjacent pulses). The generated PDM all-optical signals are injected into a recirculating fiber loop, which consists of 5 spans of 80-km SSMF with chromatic dispersion (CD) of 17 ps/km/nm and average loss of 18.5 dB, loop switches (SWs), optical coupler (OC), Erbium-doped fiber amplifier (EDFA), and no optical dispersion compensation A wavelength selective switch (WSS) is also placed in the loop, and WSS is programmed to operate as an optical bandpass filter that rejects ASE noise.

At the receiver end, a free-running ECL with a line width of less than 100 kHz is used as the LO. A polarization-division 90° optical mixer and 4 balanced PDs are used for coherent detection. The bandwidth of the balanced PD is 50 GHz. Sampling and digitization are achieved by a high-speed real-time digital sampling oscilloscope with a sampling rate of 160-GSa/s and an electrical bandwidth of 65 GHz. After the ADC, the sampled data sequence of the four channels is subjected to offline digital signal processing. The data is first resampled to 2 samples per symbol and clocked and then processed using the improved QPSK and 16QAM digital signal processing algorithms. The classical equalization algorithms of CMA cascaded CMMA are used for QPSK and 16QAM, respectively. Since the phase between each time slot

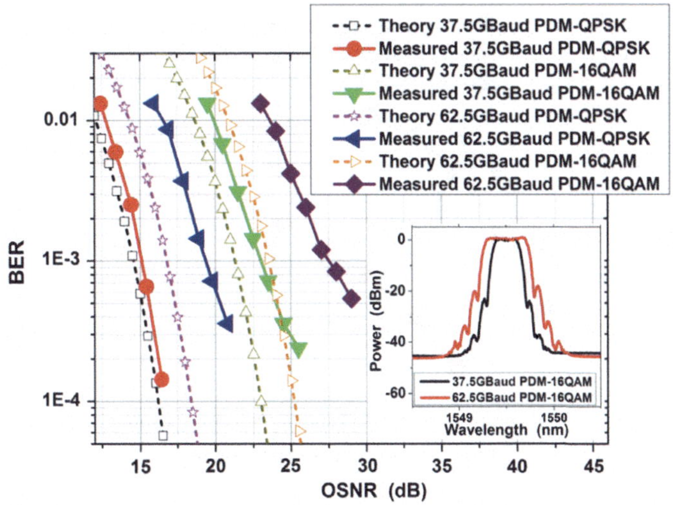

Fig. 5.15 The relationship of Back-to-back bit error rate of PDM all-optical Nyquist 37.5 GBaud QPSK/16QAM and 62.5 GBaud QPSK/16QAM signals with OSNR

is unknown, we must divide the time after polarization demultiplexing. Frequency offset estimation and phase recovery are performed separately for each TDM time slot. After equalization, each slot bit is demodulated and all bits together calculate the bit error rate.

Figure 5.15 shows the relationship of the back-to-back bit error rate and OSNR of PDM all-optical Nyquist 37.5 GBaud QPSK/16QAM and 62.5 GBaud QPSK and 16QAM signals. The theoretical curves for the BER and OSNR relationships of these signals are also drawn. For the 37.5-GBaud PDM-QPSK and 16QAM signals, the OSNR penalty at a 1×10^{-3} bit error rate is 0.8 dB and 1.8 dB, respectively. However, 62.5-GBaud PDM-QPSK and 16QAM cost 2.5 dB and 4 dB at the same BER level. We believe that the inconsistency between experiment and theory is mainly due to the inaccurate delay between TDM caused by the manually adjustable delay line. These time errors reduce the time-domain orthogonality of the pulse and introduce ISI. Obviously, high baud rate and high-order modulation signals are more sensitive to this. It is believed that the integrated signal generation and modulation device will have better performance. The inset (i) in Fig. 5.15 shows the spectrum at 37.5 and 62.5 GBaud PDM-16QAM (0.02-nm resolution). A better way to generate signals is to use high-speed DACs, but the bandwidth of commercial DACs is still limited.

Finally, we also measured the relationship between BER performance and transmission distance of PDM all-optical Nyquist 37.5-GBaud 16QAM, 62.5-GBaud QPSK, and 16QAM signals, as shown in Fig. 5.16. The 62.5-GBaud PDM all-optical Nyquist QPSK signal can achieve a 1×10^{-2} bit error rate over a 4000 km transmission distance. At the same BER level, the 37.5 and 62.5-GBaud 16QAM signals can transmit 1200 km and 850 km. The above experimental results have successfully verified the generation of PDM all-optical Nyquist signals, high-order modulation, fiber

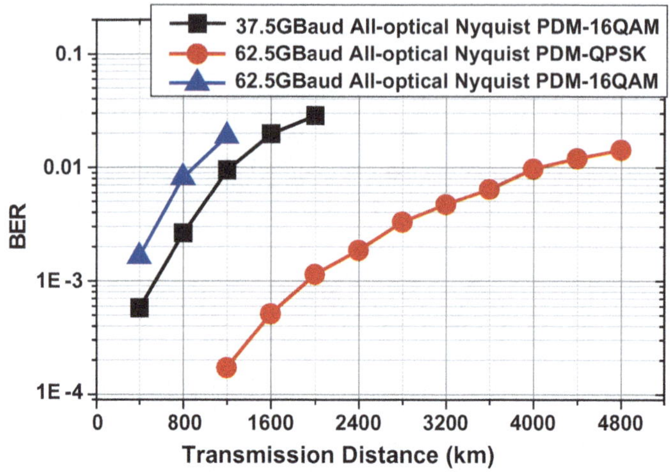

Fig. 5.16 The relationship of BER performance and transmission distance of PDM all-optical Nyquist signals

transmission, and full-band coherent detection. This solution provides high-quality ways of long-distance transmission of all-optical Nyquist.

5.5 Summary

In this chapter, the research work on the generation and processing of all-optical Nyquist signals is introduced from the aspects of all-optical Nyquist pulse generation, signal modulation and multiplexing, and detection and signal processing. First, the principle of all-optical Nyquist signal pulse generation, signal modulation, and multiplexing are introduced. The theoretical basis of frequency-locking, linear phase, and equal amplitude optical comb based on a single MZM are also given. Then, the first complete all-optical Nyquist signal generation and coherent detection system that utilizes a full-light comb spectrum to generate and coherently detect all-light Nyquist QPSK signals up to 62.5, 75, and 125 GBaud have been successfully demonstrated. Finally, the research work on polarization-multiplexed all-optical Nyquist signal long-distance and high-order modulation formats are introduced, including the results of 37.5 and 62.5-GBaud all-optical Nyquist PDM-QPSK and 16QAM signal generation, transmission, and full-band coherent detection. By using coherent detection, all-optical Nyquist's 37.5 and 62.5-GBaud 16QAM signals were successfully transmitted over 1200 km and 850 km SSMF.

References

1. D. Hillerkuss, R. Schmogrow, T. Schellinger, et al., 26 Tbit s-1 line-rate super-channel transmission utilizing all-optical fast Fourier transform processing. Nat. Photon. **5**(6), 364–371 (2011)
2. G. Bosco, A. Carena, V. Curri et al., Performance limits of Nyquist-WDM and CO-OFDM in high-speed PM-QPSK systems. IEEE Photon. Technol. Lett. **20**(15), 1129–1131 (2010)
3. D. Hillerkuss, R. Schmogrow, M. Meyer, et al. Single-laser 32.5 Tbit/s Nyquist WDM transmission. Opt. Commun. Netw. IEEE/OSA J **4**(10), 715–723 (2012)
4. Marcelo A. Soto, Mehdi Alem, Mohammad Amin Shoaie, et al. Optical sinc-shaped Nyquist pulses of exceptional quality[J]. Nature Communications, 2013, 4
5. M Orteza Ziyadi, M.R. Chitgarha, A. Mohajerin Ariaei, et al., Experimental demonstration of optical Nyquist generation of 32-Gbaud QPSK using a comb-based tunable optical tapped-delay-line FIR filter: OSA, 1–3 (2014)
6. J. Wang, C. Xie, Z. Pan, Generation of spectrally efficient Nyquist-WDM QPSK signals using DSP techniques at transmitter. IEEE 1–3 (2012)
7. A. Sano, E. Yamada, H. Masuda et al., No-guard-interval coherent optical OFDM for 100-Gb/s Long-Haul WDM transmission. J. Lightwave Technol. **25**(16), 3705–3713 (2009)
8. B. Zhu, X. Liu, S. Chandrasekhar, et al. Ultra-long-haul transmission of 1.2-Tb/s multicarrier no-guard-interval CO-OFDM superchannel using ultra-large-area fiber. Photon. Technol. Lett. IEEE **22**(11), 826–828
9. M. Nakazawa, T. Hirooka, P. Ruan et al., Ultrahigh-speed "orthogonal" TDM transmission with an optical Nyquist pulse train. Opt. Express **20**(2), 1129–1140 (2012)
10. H. Hao, D. Kong, E. Palushani et al., 320 Gb/s Nyquist OTDM received by polarization-insensitive time-domain OFT. Opt. Express **22**(1), 110 (2014)
11. J. Zhang, J. Yu, Y. Fang, Nan Chi, High speed all optical Nyquist signal generation and full-band coherent detection. Scientif. Reports **4**, 6156. http://www.nature.com/srep/2014/140821/srep06156
12. J. Zhang, J. Yu, N. Chi, Transmission and full-band coherent detection of polarization-multiplexed all-optical Nyquist signals generated by Sinc-shaped Nyquist pulses. Scientif. Reports **5**, 13649 (2015)

Chapter 6
Nonlinear Compensation in Optical Fiber

6.1 Introduction

Compared with the wireless channel, the biggest difference of fiber channel is the nonlinear effect. In the current high-speed optical communication system, the nonlinear damage of optical fiber is the most important factor which limits the long-distance transmission of high-speed optical signals [1–11]. In order to realize high-speed signal transmission with large capacity and long distance, fiber nonlinear compensation must be considered. With the maturity of a coherent optical communication system and the development of related devices such as ADC sampling analog-to-digital conversion and digital signal processing chip technology, the digital fiber nonlinear compensation algorithm has attracted extensive attention; there are a lot of researches at home and abroad [1–11]. Around the same time, the team represented by Li Guifang and Ezra Ip proposed digital backward propagation (DBP) algorithm [2, 4]. It uses the nonlinear Schrödinger equation of the optical fiber channel to inversely deduce the propagation process of an optical signal in optical fiber from the received optical signal, thus the nonlinear dispersion and equilibrium are compensated. By using the fractional Fourier transform, the DBP algorithm continuously differentiates all the linear and nonlinear processes that the signal goes through in the time and frequency domain in the optical fiber, and then inverts until it comes back to the transmitter to obtain the original signal.

The DBP algorithm has attracted extensive attention since it was proposed. A large number of experiments, including single-polarization and polarization multiplexing systems, have verified its good nonlinear compensation performance [7–9, 11]. In addition, similarly, another time-domain nonlinear algorithm based on the Volterra series has also been reported by solving the frequency-domain nonlinear response of fiber channel [10]. This section will introduce the research on the nonlinear compensation algorithm of fiber channel, from the aspects of algorithm principle, high-speed long-distance experiment, and algorithm improvement.

J. Yu and N. Chi, *Digital Signal Processing In High-Speed Optical Fiber Communication Principle and Application*,
https://doi.org/10.1007/978-981-15-3098-2_6

6.2 Principle of DBP Fiber Nonlinear Compensation Algorithm

For ordinary single-mode fiber, the transfer function under single polarization can be expressed by the nonlinear Schrödinger equation as follows [1, 2, 4]:

$$\frac{\partial E(z,t)}{\partial z} = -\frac{\alpha}{2}E(z,t) + j\frac{\beta_2}{2}\frac{\partial^2 E(z,t)}{\partial t^2} - \frac{\beta_3}{6}\frac{\partial^3 E(z,t)}{\partial t^3} - j\gamma|E(z,t)|^2 E(z,t) \tag{6.1}$$

where $E(z,t)$ is a pulsed field intensity on the time and distance, α is the loss coefficient, β_2 and β_3 are the second-order and third-order dispersion, and γ is the optical fiber nonlinear coefficient. The nonlinear Schrödinger equation describes the various effect of propagation of the single-polarization light pulse in optical fiber, mainly including three parts: the first on the right formula expresses the loss of the optical pulse, the second and the third describe the optical fiber dispersion, and the fourth describes the phase change caused by the Kerr nonlinear effect of the fiber (including the self-phase modulation, cross-phase modulation, and four-wave mixing effect). The above equation can be simply written as

$$\frac{\partial E}{\partial z} = -\frac{\alpha}{2}E + j\frac{\beta_2}{2}\frac{\partial^2 E}{\partial t^2} - \frac{\beta_3}{6}\frac{\partial^3 E}{\partial t^3} - j\gamma|E|^2 E = \left(\widehat{N} + \widehat{D}\right)E \tag{6.2}$$

where $\widehat{D} = -\frac{\alpha}{2} + j\frac{\beta_2}{2}\frac{\partial^2}{\partial t^2} - \frac{\beta_3}{6}\frac{\partial^3}{\partial t^3}$ and $\widehat{N} = -j\gamma|E|^2$, here $\widehat{D}$ is called linear operator, $\widehat{N}$ is called the nonlinear operator. The whole nonlinear Schrödinger equation is dominated by these two operators.

If formula (6.1) only considers the linear part, the analytic solution can be obtained by the Fourier transform to the frequency domain, and the result is as follows:

$$\widetilde{E}(z+h,\omega) = \widetilde{E}(z,\omega)\exp[-j\left(\frac{\beta_2}{2}\omega^2 + \frac{\beta_3}{6}\omega^3\right)h]\exp(-\alpha h) \tag{6.3}$$

where $\widehat{E}$ is for the Fourier transform of $E(z,t)$. It can be seen from the above equation that the linear part has a very simple transfer function in the frequency domain, through which the influence of the dispersion and loss of the signal on the signal can be obtained. On the other hand, if formula (6.1) only considers the nonlinear part, then the analytical solution of the equation can be obtained in the time domain. The nonlinear solution is as follows:

$$E(z+h,t) = E(z,t)\exp(j\gamma h|E(z,t)|^2) \tag{6.4}$$

Therefore, the nonlinear damage is related to the power of the signal, and a phase torsion related to the power will be introduced in the time domain. Because the intensity of the signal itself is constantly changing in the fiber channel due to dispersion

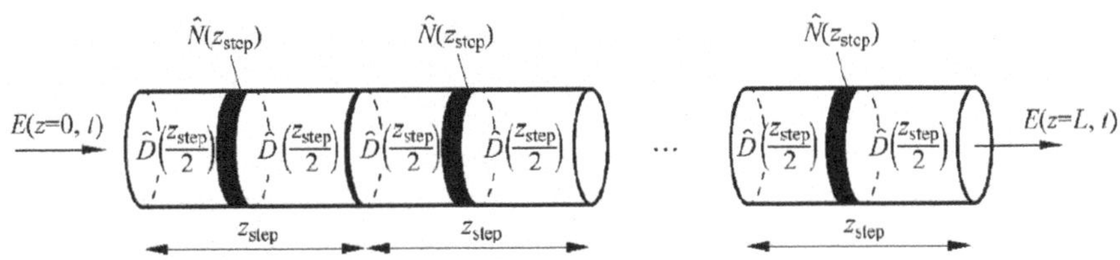

Fig. 6.1 Signal propagation process of the fractional Fourier algorithm for solving the nonlinear Schrödinger equation of optical fiber

and loss, constant power or intensity cannot be used to solve the phase. At the same time, in the WDM system, the signal impairments also include the influence of adjacent channels, such as cross-phase modulation and four-wave mixing. Therefore, the combined effects of linear and nonlinear operators need to be comprehensively considered in order to accurately describe the influence of fiber channel.

According to the classical nonlinear optical fiber literature [12], a suitable description method is to adopt the fractional Fourier algorithm, as shown in Fig. 6.1. Fiber channel can be divided into several small pieces, and the signal power of each piece can be considered constantly, so each piece of the optical fiber can be regarded as three processes: transmitted to frequency domain by Fourier transform and performing linear operator processing, then adopting Fourier inverse transform to the time domain and conducting nonlinear operator processing. Next, continue to conduct Fourier transform to transmit the signals to the frequency domain and perform linear operator processing, and finally conduct the inverse transform back to the time-domain signal. In this way, the nonlinear transmission of the signal in the fiber channel can be simulated by processing each segment.

Based on the forward propagation in Fig. 6.1, it is logically easy to deduce the optical channel model of reverse propagation, as shown in Fig. 6.2, which is the basic principle of the DBP algorithm. The DBP algorithm or the improved DBP algorithm reported in various pieces of the literature [1–5, 8–11] are all based on this schematic diagram. Similarly, the optical fiber is divided into many small segments. In each segment the signal power can be considered as constant, and each segment can be regarded as going through two inverse processes, including negative linear operator and negative nonlinear operator. At the same time, the propagation direction of the

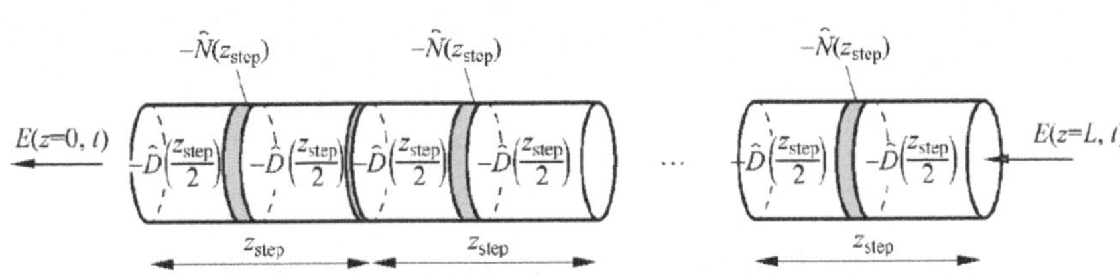

Fig. 6.2 Schematic diagram of digital backward propagation algorithm (DBP) analog optical signal backward propagation in optical fiber

optical fiber is changed from the receiving end to the transmitting end. Therefore, by calculating from the received optical signal, the original signal of the transmitting end can be obtained through continuous promotion.

Finally, for polarization multiplexing systems, small changes need to be made to Eqs. 6.1 and 6.2 [1, 4]. Due to the birefringence of optical fiber, the polarization state of optical signal changes constantly during the transmission of optical fiber, and its correlation length is even less than 10 m. On the other hand, because the polarization correlation length is far less than the effective nonlinear calculation length, the nonlinear calculation process can be averaged on the two polarization states in practical systems. Based on the Manakov equation, the improved DBP method can obtain the following relation after the averaged power and phase of the two polarization states:

$$\frac{\partial E_{x,y}}{\partial z} = -\frac{\alpha}{2}E_{x,y} - j\frac{\beta_2}{2}\frac{\partial^2 E_{x,y}}{\partial t^2} + j\frac{8}{9}\gamma(|E_{x,y}|^2 + |E_{y,x}|^2)E_{x,y} = (\hat{D} + \hat{N})E_{x,y} \quad (6.5)$$

$$\hat{D} = -\frac{\alpha}{2} - j\frac{\beta_2}{2}\frac{\partial^2}{\partial t^2}, \quad \hat{N} = j\tfrac{8}{9}\gamma(|E_{x,y}|^2 + |E_{y,x}|^2) \quad (6.6)$$

where $E_{x,y}$ represents the light field of the signal in X or Y polarization.

6.2.1 Nonlinear Compensation Experiment Based on 4 × 160.8 Gb/s Wavelength Division Multiplexing PDM-QPSK Signal

This section introduces the experimental verification of nonlinear compensation of high-speed optical transmission system. In the experiment, WDM 4 × 160.8 Gb/s PDM-QPSK signal with channel spacing of 50 GHz is transmitted over 1300-km standard single-mode fiber SMF-28. In the experiment, the effects of digital nonlinear compensation and channel crosstalk suppression are studied. Using DBP-based nonlinear compensation and maximum likelihood sequence estimation crosstalk suppression, the bit error rate of 4 × 160.8 Gb/s WDM PDM-QPSK signals after 1300-km SMF-28 fiber transmission is reduced from 1.0×10^{-3} to 3.5×10^{-4}.

Figure 6.3 shows the experimental devices of heterodyne coherent detection and nonlinear compensation for 4 × 160.8-Gbit/s WDM PDM-QPSK signal with 50-GHz channel spacing transmitted over 1300-km SMF-28. At the transmitter, four external cavity lasers (ECL) with an output power of 14.5 dBm and a linewidth of less than 100 kHz are used for the four WDM channels.

The operating wavelength of ECL1 to ECL4 ranges from 1549.0 nm to 1550.2 nm with an interval of 50 GHz. The odd channel (ECL1 and ECL3) and even channel (ECL2 and ECL4) are implemented by two polarization-maintaining optical couplers (PM-OC) before independent I/Q modulation. Each I/Q modulator is driven by a 40.2-GBaud electrical binary signal with a peak-to-peak value of 0.5 V, the signal

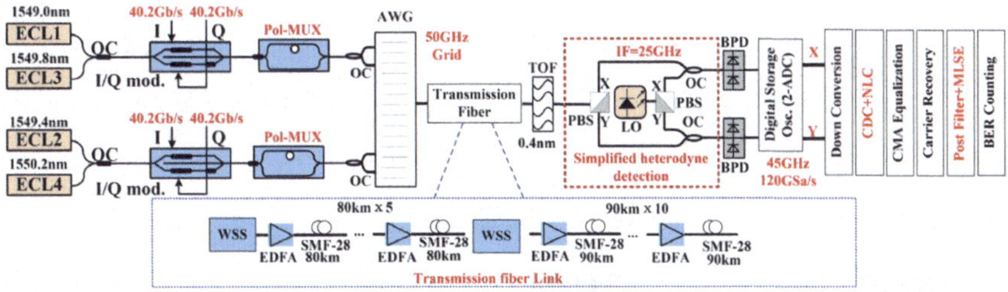

Fig. 6.3 Experimental device diagram of nonlinear compensation for 4 × 160.8-Gbit/s WDM PDM-QPSK signal with 50 GHz channel spacing transmitted over 1300-km SMF-28 based on heterodyne detection

is a pseudo-random binary sequence (PRBS) with a length of $(2^{13} - 1) \times 4$. 40.2 GBaud is generated by a 10.05-GBaud binary signal through a 4 × 1 electrical multiplexer. After polarization multiplexing, odd and even channels are merged and amplified before optical fiber transmission. 40.2-GBaud WDM PDM-QPSK signal passes through the array waveguide grating (AWG) with a frequency interval of 50 GHz for spectrum for filtering and merging, while the 3, 10, and 20 dB bandwidths of AWG are 26.7, 44.9, and 59.6 GHz, respectively.

The linear transmission link consists of 5 segments of 80-km single-mode fiber and 10 segments of 90-km single-mode fiber. In the link, there is only an erbium-doped fiber amplifier (EDFA) amplification and no optical dispersion compensation module. Two bandpass filters with a bandwidth of about 2 nm (implemented with a wavelength selector switch) are inserted into the link to suppress the accumulation of erbium-doped fiber amplified spontaneous emission (ASE) noise. In order to study the nonlinear damage, we adjusted the output power of each EDFA. Figure 6.4a, b, respectively, shows the spectra of the four channels of 160.8-Gbit/s PDM-QPSK signal generated by 50-GHz AWG and transmitted over 1300 km. The resolution of Fig. 6.4a, b is 0.1 nm. At the receiving end of heterodyne coherent detection, a tunable optical filter (TOF) with a 3 dB bandwidth of 0.4 nm was used to select the tested channel in the experiment. The linewidth of the local oscillator (LO) is less than 100 kHz, and there is a frequency difference of 25 GHz between the received optical signal and LO. In the experiment, only two LOs were used. One was coherent with channel 1 (1549.0 nm) and channel 2 (1549.4 nm) at 1549.24 nm, and the other was coherent with channel 3 (1549.8 nm) and channel 4 (1550.2 nm) at 1550.04 nm. Before being detected by a 50-GHz balanced photodetector (PD), two polarization beam splitters and OC and LO realized polarization diversity of the received optical signals in the optical domain. The analog/digital conversion is realized by a real-time digital storage oscilloscope with a 120-GSa/s sampling rate and 45-GHz electrical bandwidth. After that, the signal recovery is realized by the offline digital signal processing module, including fiber nonlinear compensation.

The offline digital signal processing module after the mode/number transformation is shown in Fig. 6.5. First, the received signal is converted down to the baseband in 25 GHz and resampled to 2 sampling/symbol. Then, nonlinear compensation (NLC)

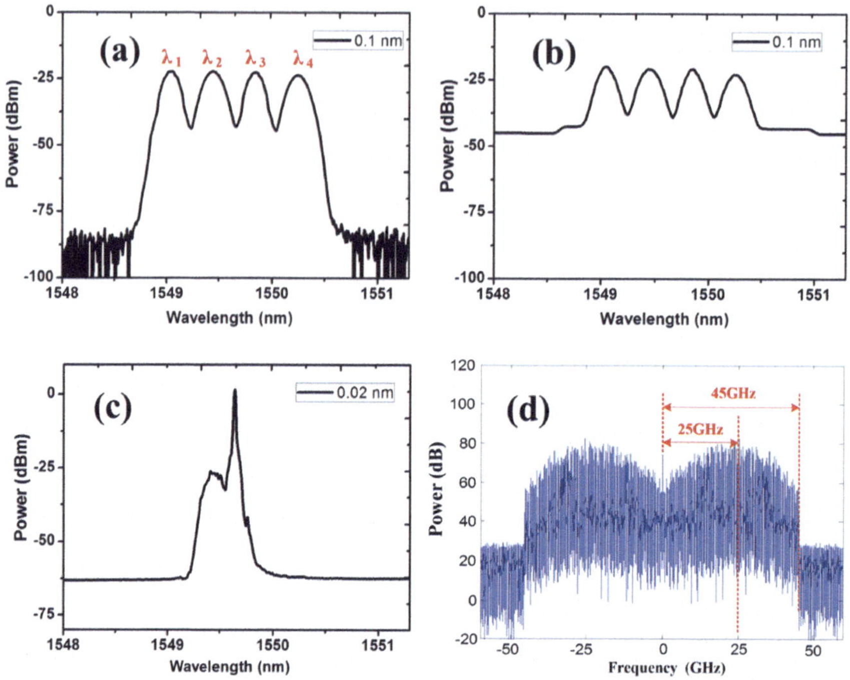

Fig. 6.4 Spectra of 4 × 160.8-Gbit/s PDM-QPSK signal in the experiment: **a** spectrum after AWG filtering; **b** spectrum after 1300-km transmission; **c** spectrum of heterodyne beats of 160.8-Gbit/s PDM-QPSK signal and LO before PD; **d** electrical spectrum of IF signal received after PD and 45-GHz ADC

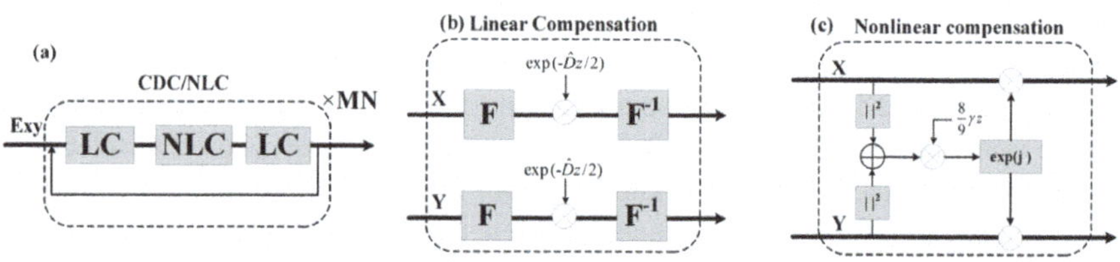

Fig. 6.5 Offline digital signal processing module after analog/digital transformation: **a** main module diagram of nonlinear compensation; **b** dispersion linear compensation processing module; **c** processing module of nonlinear compensation

and dispersion compensation (CDC) are implemented by the DBP method based on solving the Manakov equation through standard single-mode fiber (SSMF). In this paper, the average loss of each fiber segment is about 0.26 dB/km, the dispersion is 17 ps/km/nm, and the nonlinear parameter γ is 1.5 w^{-1} km^{-1}. After dispersion compensation and nonlinear compensation, channel equalization and polarization demultiplexing of PDM-QPSK signals are realized by adaptive FIR filters based on two complex fields of classical CMA with 13 taps and T/2 interval. Next is carrier recovery, and the Viterbi-Viterbi algorithm described above is used for residual frequency offset recovery and phase recovery. After that, the dual binary signal is converted by delay addition filter, and the maximum likelihood sequence estimation (MLSE) is used to realize noise and ISI equalization. Finally, the bit error rate is calculated, and the length of the data used to calculate the bit error rate in the experiment is greater than 10×10^6 bit.

Among them, the DBP algorithm described above is adopted to compensate for the nonlinear damage of optical fiber, and its step-by-step processing module is shown in Fig. 6.5a. We compensate for the linear fiber dispersion and nonlinear loss by solving the nonlinear Schrödinger equation in reverse. In each step size z, we first compensate for the linear dispersion and fiber loss of the fiber with a length of z/2 in the frequency domain through fast Fourier transform (FFT), as shown in Fig. 6.5b. Then we calculate the total power of the signal and compensate for the nonlinear phase shift in the time domain, as shown in Fig. 6.5c. Finally, the dispersion and loss of the remaining half-step z/2 are compensated in the frequency domain. For the fiber with N segments, each segment needs M steps of compensation, and the total dispersion compensation and nonlinear compensation are calculated as MN steps.

In the experiment, we first tested the bit error rate performance versus OSNR of back-to-back single-channel and four-channel WDM signals under coherent heterodyne detection, as shown in Fig. 6.6a. In the case of 23.7 dB OSNR, the constellation diagram of channel 2 in the WDM system before and after filtering is respectively shown in the inset (i) and (ii) of Fig. 6.6a, and the QPSK constellation diagram of 4 points is converted into the 9QAM constellation diagram of binary signals through the post-filter. At the bit error rate of 1×10^{-3}, more than 2 dB sensitivity gain is obtained by using post-filtering and MLSE, indicating that inter-channel crosstalk and ISI are suppressed. Due to the influence of other channels, the performance of the WDM signal here is 0.5 dB worse than that of single channel. Figure 6.6b, c shows the results of nonlinear compensation, where Fig. (b) shows the single-channel bit error rate performance at different input powers after transmission over 850-km single-mode fiber. The results show that the optimal fiber input power can be improved

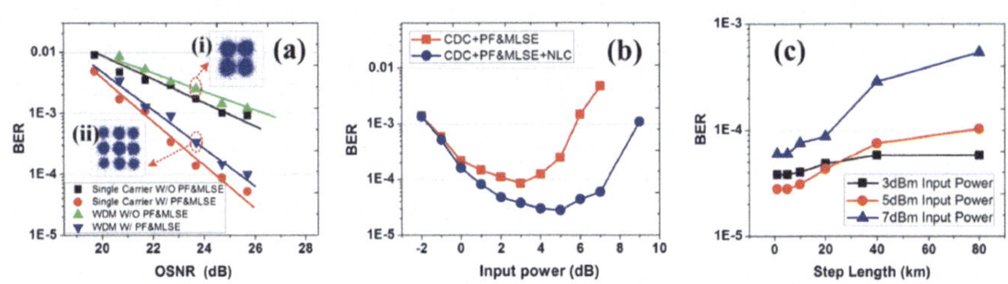

Fig. 6.6 Experimental results: **a** back-to-back bit error rate performance versus OSNR in the case of single-channel and WDM channels; insets (i) and (ii) show the constellation of X polarization under different conditions; **b** single-channel bit error rate performance of different processing schemes at different optical powers into fiber after transmission of 850-km single-mode fiber; **c** bit error rate performance with nonlinear compensation calculation step size under different fiber input powers

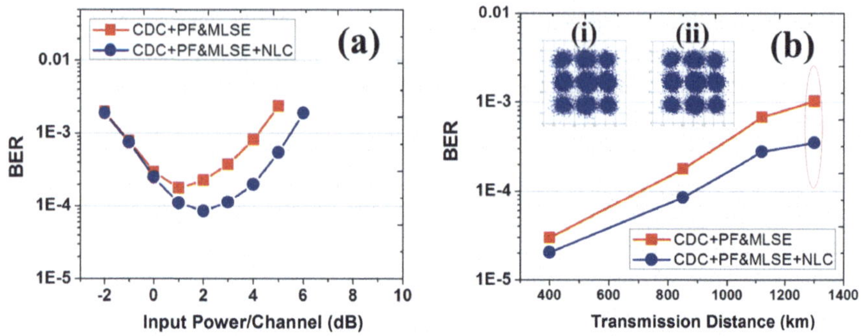

Fig. 6.7 Experimental results: **a** the change of bit error rate of 4 × 160.8-Gbit/s WDM signal with input power after 850-km transmission; **b** the bit error rate of 4 × 160.8-Gbit/s WDM signal with the transmission distance from 400 to 1300 km

from 3 dBm to 5 dBm, and the bit error rate performance can also be improved after using the DBP nonlinear compensation algorithm. Therefore, dispersion compensation and nonlinear compensation, as well as post-filtering and MLSE can achieve the best performance. Figure 6.6c shows the relation between the bit error rate performance of the DBP nonlinear compensation algorithm and the calculated step size from 1 to 80 km under different fiber input powers. The smaller the step size is, the better the DBP nonlinear compensation performance. At the same time, the higher the input power is, the more sensitive the nonlinear compensation is to the step size.

Figure 6.7a shows the variation of bit error rate with fiber input power after 4 × 160.8-Gbit/s WDM signal transmission of 850 km in the WDM system. Here, channel 2 is selected as the test channel. The input power of each fiber changes at the same time, and the DBP nonlinear compensation step length remains at 5 km. The results also show that the nonlinear compensation can improve the system performance and improve the optimal fiber power. On the other hand, the optimal fiber input power of 4 × 160.8-Gbit/s WDM is 2 dBm/channel. Compared with the single-channel results in Fig. 6.7b, the optimal fiber input power is reduced, because, in the case of WDM, nonlinear damage between channels such as cross-phase modulation and four-wave mixing will also occur.

Finally, the bit error rate of 4 × 160.8-Gbit/s WDM system is measured experimentally with and without nonlinear compensation, and the BER performance versus transmission distance is shown in Fig. 6.7b. For the test, the fiber input power is optimized input power and the DBP nonlinear compensation step length is 5 km. Insets (i) and (ii) respectively show the X polarization constellation results without and with the nonlinear compensation algorithm after 1300-km transmission. The results show that by using DBP-based nonlinear compensation and MLSE-based crosstalk suppression, the bit error rate of 4 × 160.8-Gbit/s WDM PDM-QPSK signal is reduced from 1.0×10^{-3} to 3.5×10^{-4} in heterodyne coherent detection after the transmission of 1300-km single-mode fiber SMF-28. The above results show that digital nonlinear loss compensation and channel crosstalk suppression can significantly improve the error performance of the heterodyne detection WDM system.

6.2.2 Improved Digital Nonlinear Compensation Algorithm Based on the Logarithmic Step Size

The above two sections introduce the DBP nonlinear compensation algorithm based on the fractional Fourier method to solve the nonlinear Schrödinger equation, including its principle and experimental verification of high-speed signal transmission. Many nonlinear compensation researches based on the DBP algorithm have also been reported in the relevant literature, including single-channel signals, polarization-multiplexed single-carrier signals, multi-carrier signals such as PDM-OFDM signals [1–11]. However, the above DBP methods are all based on the fixed step size fractional Fourier method, that is, the step size is fixed in every step calculation of SSFM. Thus, the effect of nonlinear compensation is determined by the calculated step size or step number of each fiber segment. Therefore, reducing the number of DBP steps per fiber segment can effectively reduce the complexity of the algorithm. On the other hand, some researchers have carried out simulation analysis on the non-fixed step size method, and compared with the fixed step size, the non-fixed step size is more accurate in the nonlinear damage estimation of the signal [12]. In addition, studies have confirmed that the logarithmic step method can effectively reduce the number of steps under the premise of guaranteeing the nonlinear compensation performance in the same channel [12]. However, the optimal logarithmic step distribution has not been analyzed in the above study, and the effect of asynchronous step distribution on nonlinear compensation between channels in multi-channel systems has not been studied.

This section will introduce a new DBP nonlinear compensation algorithm based on improved logarithmic step distribution, and analyze its compensation effect in the Nyquist wavelength division multiplexing (N-WDM) system. Different from the traditional method of fixed step size, the scheme uses logarithmic step size distribution in DBP and introduces adjustment parameter k to improve the traditional logarithmic step size distribution. Simulation results show that the improved logarithmic step distribution can reduce the complexity of the algorithm and improve the compensation effect. In addition, we also carried out experimental verification, using EDFA and SMF-28, 3×50-Gbit/s NWDM PDM-QPSK signal was transmitted over 1120 km. Experimental results show that compared with single-channel fixed step nonlinear compensation, the improved algorithm can obviously improve the bit error rate performance and Q value of the system, and greatly reduce the complexity of the nonlinear compensation algorithm.

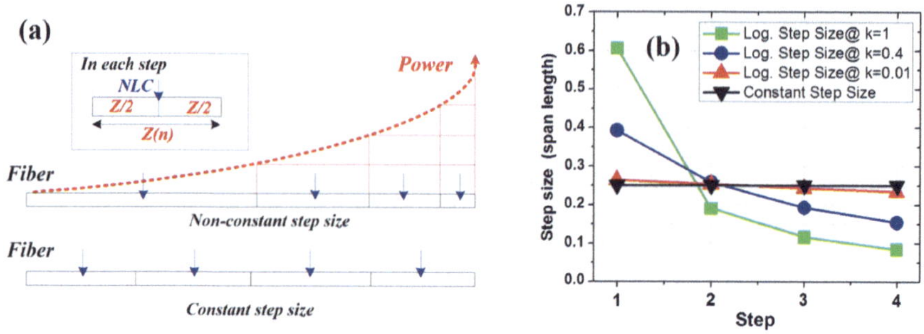

Fig. 6.8 DBP method based on SSMF: **a** principle of improved logarithmic step DBP algorithm; **b** comparison of logarithmic step size distribution with different k values and constant step size distribution

1. Principle of improved logarithmic step DBP algorithm

Similarly, the DBP method based on nonlinear compensation assumes that the length of each fiber segment is L, and the DBP calculation of each fiber segment is divided into N steps. As shown in Fig. 6.8a, for the traditional SSMF-based DBP method, the step size calculated by SSMF is fixed as $z(n) = L/n$ for each step. Thus, the performance of nonlinear compensation depends heavily on the step size and the number of steps per fiber segment. Therefore, reducing the DBP step number of each fiber segment can effectively reduce the algorithm complexity. However, considering optical fiber attenuation, the power distribution is not linear. At the beginning of transmission, the light energy is very strong and the nonlinear effect is obvious. At the end of transmission, the nonlinearity weakens, as shown in Fig. 6.8a. Therefore, a reasonable method is to reduce the step size to improve the accuracy of signal damage estimation when the light power is strong. Here, we introduce the adjustment factor k to improve the logarithmic step formula given in the literature:

$$z(n) = -\frac{1}{k\alpha}\ln\left[\frac{1-(N-n+1)\delta}{1-(N-n)\delta}\right], n = 1, N, \delta(1-e-k\alpha L)/N \tag{6.7}$$

where $z(n)$ is the step length of DBP step n and α is fiber loss. By adjusting factor k, the step size distribution calculated by DBP may be changed. However, it is worth noting that fiber loss does not change with k value in each DBP calculation step. The reasons for introducing the adjustment factor k are explained in detail below.

When $k = 1$, $z(n)$ is the inverse logarithmic distribution; when k approaches 0, the logarithmic step size becomes a fixed step size:

$$\begin{aligned}\lim_{k\to 0} z(n) &= \lim_{k\to 0}\{-\frac{1}{k\alpha}\ln[\frac{1-(N-n+1)\delta}{1-(N-n)\delta}]\}\\ &= L/N,\ n = 1,\ldots,N\end{aligned} \tag{6.8}$$

As the value of k changes from 0 to 1, the step size distribution changes between the fixed step size and the logarithmic step size. Figure 6.8b shows the change of length step by step under the condition of asynchronous length distribution, and each fiber segment is compensated in 4 steps. Under the condition of fixed step size, the step size of each step is 0.25 L. As the value of k increases, the difference between adjacent steps becomes bigger and bigger. When k approaches 0 ($k = 0.01$), logarithmic step size approaches fixed step size. When $k = 1$, the first step grows larger than 0.6 L while the length of the last step becomes smaller than 0.1 L. As the value of k increases, the step size gap becomes larger and larger. Since the number of steps is limited, such a step size gap is too large for DBP nonlinear compensation. Moreover, the ASE noise of DBP also reduces the accuracy of signal recovery. In conclusion, the optimal value can be found between fixed step size and logarithmic step size by adjusting k value.

2. Multi-channel nonlinear joint compensation principle of N-WDM system

Figure 6.9a shows the principle of Nyquist multi-channel joint detection, where broadband multi-channel detection is adopted to achieve nonlinear compensation.

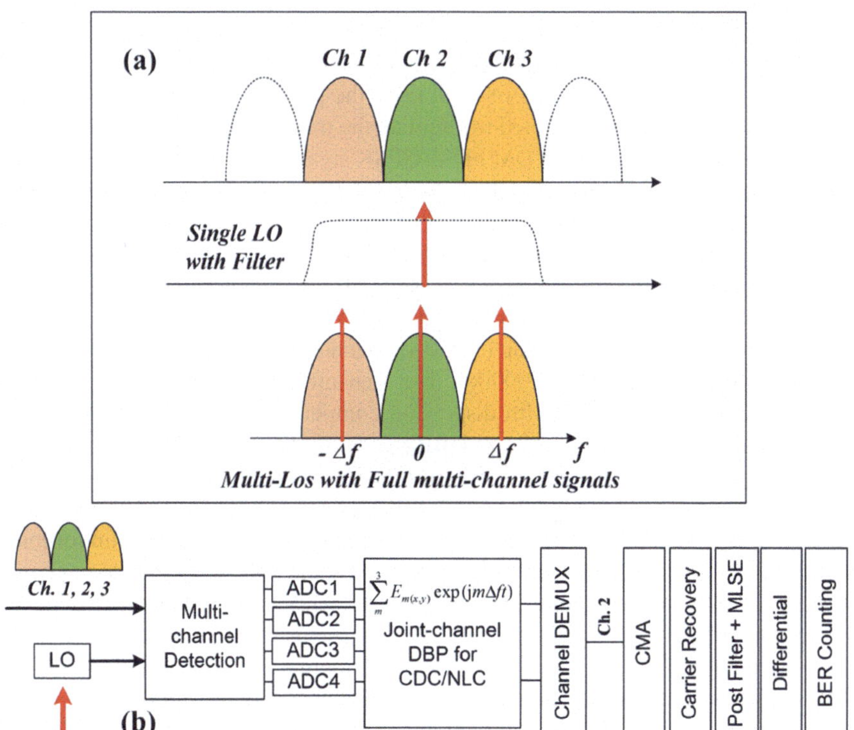

Fig. 6.9 Multi-channel nonlinear joint compensation principle of N-WDM system: **a** multi-channel signal joint detection of N-WDM system; **b** coherent detection and digital signal processing module of multi-channel nonlinear joint compensation

In N-WDM systems, channel spacing is equal to baud rate, and nonlinear damage between channels, such as cross-phase modulation (XPM) and four-wave mixing (FWM), is more serious in N-WDM systems. Wideband multi-channel detection can compensate for the nonlinear effect between channels, which requires the bandwidth of ADC to be at least greater than the bandwidth of three sub-channels, so as to realize simultaneous detection and processing. In particular, because the two sideband information of channel 2 can be retained, the multi-channel DBP method can compensate for the nonlinear damage between channels. Figure 6.9b shows the flow chart of the multi-channel detection and digital signal processing module, which includes joint channel-nonlinear compensation (JC-NLC), dispersion compensation (CDC), channel demultiplexing, and the subsequent equalization algorithm. The sum of the three channel signals can be expressed as

$$E_{x,y} = \sum_{m=1}^{3} E_{m(x,y)} \exp(jm\Delta f t) \tag{6.9}$$

where $E_{m(x,y)}$ denotes the optical field of the mth channel, Δf denotes channel spacing.

3. Simulation results

In order to verify the nonlinear compensation based on the improved log-step DBP (LS-DBP) method and the effect of k value on the performance, different processing algorithm schemes are adopted to simulate the transmission performance of the three-channel 12.5-Gbaud N-WDM PDM-QPSK signal in the 10-disk 100-km SMF fiber. In this simulation, 3 × 50-Gbit/s N-WDM PDM-QPSK signals were modulated on three subcarriers with an interval of 12.5 GHz. Each channel carries 1024 QPSK symbols per polarization. Each fiber span is connected to a 20 dB EDFA, the noise parameter is 4 dB, the fiber loss is 0.2 dB/km, the dispersion coefficient is 16 ps/km/nm, and the nonlinear coefficient is 1.5 W^{-1} km^{-1}.

Figure 6.10a shows the Q value of WDM channel 2 varying with the input optical power into each span after 1000-km fiber transmission using different schemes, including the single channel with dispersion compensation only (SC-CDC only), single-channel with constant step size DBP algorithm (SC-CS-DBP), single channel with logarithmic step size DBP algorithm (SC-LS-DBP), joint multi-channel with dispersion compensation only (JC-CDC only), constant step size DBP algorithm for joint multi-channel (JC-CS-DBP), logarithmic step size DBP algorithm for joint multi-channel (JC-LS-DBP) and so on. For the scheme only using dispersion compensation, the dispersion is compensated in the frequency domain. For single-channel dispersion compensation or NLC/CDC, channel 2 is first demultiplexed from three channels and then processed by a single-channel nonlinear compensation algorithm. For joint channel nonlinear compensation and dispersion compensation, three channels are compensated together and then demultiplexed, as shown in Fig. 6.9b. The step number for DBP of SC-NLC and JC-NLC is 4, and the adjustment factor k of logarithmic DBP is 0.4.

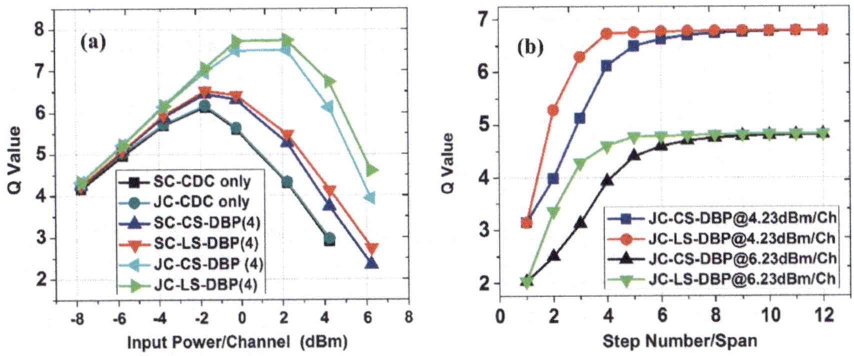

Fig. 6.10 Simulation results: **a** the change of signal Q value with the optical power into the fiber under different processing algorithms after 1000-km transmission; **b** the change of signal Q value with the number of steps calculated for each span of fiber under different processing algorithms after 1000-km transmission

According to the results shown in Fig. 6.10a, the following conclusions can be drawn. First, compared with SC-CDC-only and JC-CDC-only schemes, the single-channel NLC/CDC scheme based on SC-DBP or JC-DBP can improve bit error rate performance to some extent. However, because single-channel processing can only compensate for nonlinear damage in the channel (such as self-phase modulation (SPM)), such improvement is not obvious. Joint channel NLC/CDC improved BER performance even more significantly. Secondly, both JC-CS-DBP and JC-LS-DBP have better performance than the single-channel scheme, because the joint channel nonlinear compensation can compensate for the nonlinear damage in and between channels at the same time. Finally, compared with the fixed step size, the logarithmic step size scheme has better performance in single-channel and multi-channel processing.

Figure 6.10b shows the relationship between Q value and the number of steps per span when using the DBP method with fixed and logarithmic step size, respectively. The adjustment factor k was 0.4. Both fixed step size and logarithmic step size DBP can improve the performance of Q value by increasing the number of steps. As the number of steps increases and the step size decreases, the estimation of the phase shift caused by nonlinear damage becomes more accurate. But at the same time, the computational complexity increases. When the number of DBP steps increases to a certain number, the Q value stops increasing and gradually approaches the maximum value. When the number of DBP steps is very large, the logarithmic step size does not outperform the fixed step size. However, compared with CS-DBP, the optimal number of steps in LS-DBP is reduced to half. When the input power of each channel is 4.23 dBm, the optimal step number of CS-DBP is 4, while the optimal step number of CS-DBP is 8. Therefore, compared with CS-DBP, the computational complexity of CS-DBP is significantly reduced to 50%. LS-DBP performs better than CS-DBP when the number of steps is small and the input power is the same. When the input power is different, the optimal step number is also different. At the same time, no

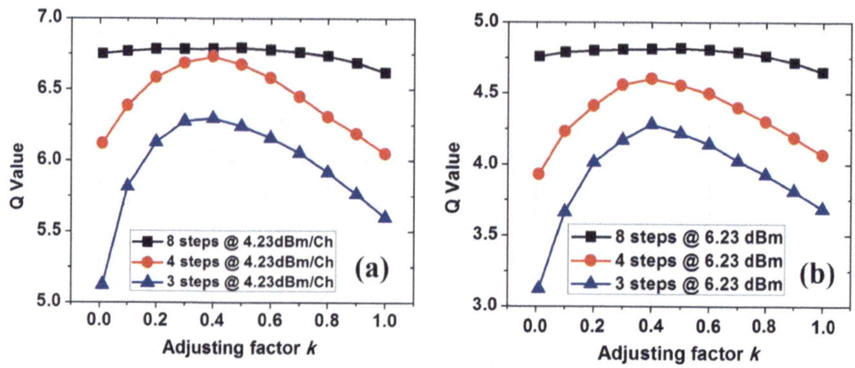

Fig. 6.11 Simulation results: The Q value performance of JC-LS-DBP scheme versus the adjusting factor k with different numbers of steps at **a** 4.23 and **b** 6.23 dBm per channel input power after 1000-km transmission

matter the CS or LS DBP method is used, the optimal step number at 4.23 dBm input power is less than 6.23 dBm input power. Because more input power will lead to more serious nonlinear damage, more steps are needed.

Figure 6.11a, b shows the effect of adjustment factor k on the nonlinear compensation performance of the JC-LS-DBP method for different step numbers at 4.23 and 6.23 dBm per channel input power, respectively. During the simulation, k value changes from 0.01 to 1. When k = 0.01, the step size distribution was close to the fixed step size. When k = 1, the step size distribution is the total logarithm distribution. It can be seen that when the number of steps is small, the adjustment factor k has an obvious influence on the Q value (when the number of steps is 3, k has a great influence on the Q value). On the contrary, when the number of steps is large, k has little influence on Q value. On the other hand, the optimal adjusting factor k is around 0.4, where Q value reaches the maximum. Therefore, the optimal step size distribution should be between the fixed distribution and the logarithmic distribution, which is called the improved logarithmic step size distribution. From Figs. 6.11 and 6.12, it can be seen that the performance can be improved by adjusting factor k for the following two reasons: first, compared with the fixed step size, the improved logarithmic step size distribution matches better with the optical power distribution; secondly, since the number of steps we set is limited (less than 4), the step size distribution can be changed by adjusting k, so as to find the optimal value of k.

Figure 6.12 shows the change curve of adjustment factor k with different transmission parameters: (a) EDFA noise, (b) input power, (c) transmission distance, (d) optical fiber attenuation, and (e) length of each fiber span. Here, the input power of each channel is 2.23 dBm. In Fig. 6.12a, b and d, the transmission link is composed of 10 spans of 100-km optical fiber. In Fig. 6.12c, the length of each fiber span is 100 km, and an improved logarithmic step JC-NLC method with step number of 4 is used. Noise, input power, and transmission distance have no obvious influence on the optimal k value. As the optical fiber attenuation increases, the optimal k value

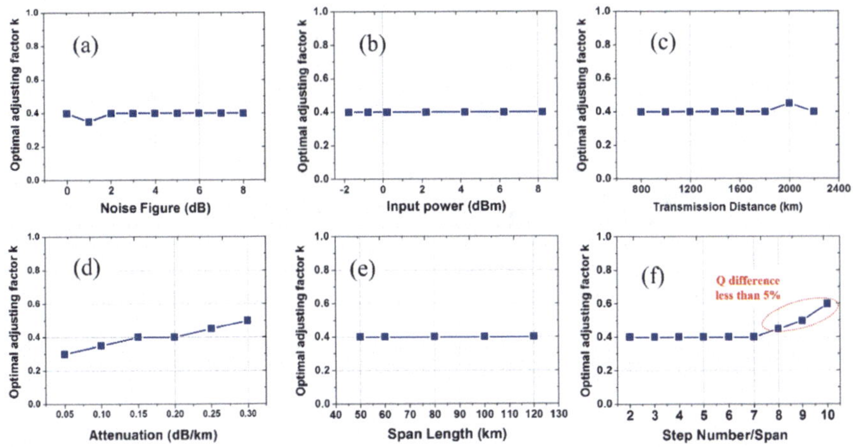

Fig. 6.12 Simulation results: the optimal adjusting factor k versus **a** noise figure; **b** input optical power; **c** optical fiber transmission distance; **d** optical fiber attenuation; **e** span length of the 1200-km optical fiber; **f** the calculation step number per span

will increase slightly, because the optical fiber attenuation changes the power distribution in the optical fiber. When the optical fiber attenuation is less than 0.15 dB/km, the optimal k value is less than 0.4. However, the influence of k value on the performance of the system is not significant, because under the specific optical fiber attenuation, k value can only change the step size distribution without affecting the optical fiber loss. In Fig. 6.12e, we divide the total transmission distance of 1200 km into segments with different span lengths. The results show that the length of each fiber span has no significant influence on the optimal k value, because the length of each fiber span does not affect its power distribution. Figure 6.12f shows the relationship between the optimal k value and the number of steps per span. The input power of each channel is 2.23 dBm, and JC-NLC uses an improved logarithmic step size. The results show that when the step number of each span is larger than 7, the optimal k value increases with the increase of the step number. The reason is that when the step size is smaller and smaller (the number of steps is more and more), the backward propagation of the nonlinear compensation is close to the full logarithmic distribution, and the optimal k value increases accordingly. Since only three WDM channels are considered in the simulation, four-steps per span is enough to achieve the optimal performance. When the distribution number is larger than 7, the increase of Q value is less than 5%. However, when the number of channels is larger, more computational steps are needed, and this effect is obvious.

In addition, we also analyzed the relationship between the optimal calculation step number (the minimum step number reaching the maximum Q value) and the optimal k value, dispersion value and baud rate, as shown in Fig. 6.13a, b. Similarly, in the N-WDM system, the input power of each channel is 2.23 dBm, and the transmission link consists of 10 spans of 100-km optical fiber. For Fig. 6.13a, the baud rate is 12.5 GBaud and channel spacing is 12.5 GHz. As for Fig. 6.13b, with

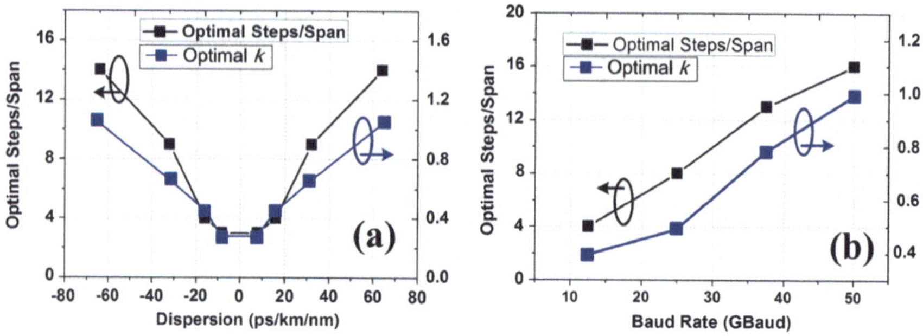

Fig. 6.13 The number of optimization steps per span and the optimal k value versus **a** fiber dispersion coefficient and **b** signal baud rate

the increase of baud rate, the carrier spacing remains the same as the baud rate, and the dispersion parameter is 16 ps/nm/km. The simulation uses JC-NLC, CMA, and carrier recovery algorithm based on improved logarithmic step. The optimal step number increases with the increase of dispersion due to the pulse broadening caused by dispersion. When the dispersion is large, more steps are needed to ensure the accuracy of compensation. Similarly, when the baud rate is large, larger dispersion influence is suffered and a larger step number is required. Therefore, as a larger step number is required, the optimal k value increases with the increase of dispersion and baud rate. When the step size becomes smaller, the optimal k value becomes larger, and the backward propagation in nonlinear compensation in optical fiber is more close to the logarithmic distribution.

The simulation further increases the number of channels to 9 for nonlinear compensation. At this time, 9 × 50-Gbit/s NWMD PDM-QPSK signal is transmitted with 12.5-GHz channel spacing and 12.5-Gbaud baud rate. After 10 spans of 100-km SMF transmission, the performance impact of the JC-NLC method based on improved logarithmic step size on the 5th channel is simulated, and the input power is 1.23 dBm per channel. Figure 6.14a shows the relationship between the Q value performance of the 5th channel using the JC-LS method and the step number per span when a different number of sub-channels are used for calculation. The results show that, for the NWDM systems with many channels, the single-channel nonlinear compensation has very limited performance. Compared with the scheme only using dispersion compensation, the SC-LC nonlinear compensation only improves the Q value by less than 0.5. When JC-NLC is used, the performance can be significantly improved by adding more adjacent channels. At the same time, the computational complexity increases. Figure 6.14b shows the relationship between the optimal step number of each optical fiber and the number of sub-carriers used for calculation. It is possible to see that for both JC-LS and JC-CS schemes, if two more sub-carriers are used in the calculation of JC-NLC, the optimal step number will be doubled. In order to calculate more subcarriers, it is necessary to increase the step number of each optical fiber, so that the optimal k value will increase with the increase of the number of sub-carriers. Similarly, when the number of steps becomes larger and the

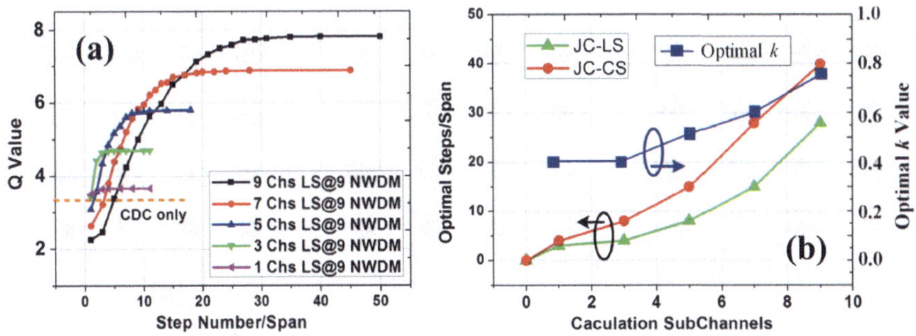

Fig. 6.14 Simulation results: **a** the relationship between the signal Q value and the number of steps of each fiber segment after the nonlinear compensation of joint multi-channel logarithmic distribution under a different number of computational channels; **b** the relationship between the optimal number of steps and the optimal k factor under a different number of computational channels

step length becomes smaller, the k value becomes larger, and the transmission of nonlinear compensation gradually tends to the full logarithmic distribution.

Finally, we study the optimal k values under different pulse shapes and modulation formats. Figure 6.15a shows the relationship between the Q value of PDM-QPSK signal and the adjustment factor k when using different pulse shapes (NRZ and RZ with a duty cycle of 67%) based on the improved logarithmic step size nonlinear compensation. The optimal k value exists for both the signals, which are 0.3 and 0.4, respectively. When k is optimized, the improved logarithmic step size performs better than both constant step size (k = 0.01) and full logarithmic step size (k = 1). Here, a single-channel signal with 12.5-Gbaud baud rate is used and the transmission link consists of 10 spans of 100-km SMF, and the number of computational steps of each optical fiber span is 4.

Figure 6.15b shows the relationship between the Q value of the 12.5-GBuad PDM-16QAM signal and the adjustment factor k when the step number per span is 4 and 2, respectively. The results show that the optimal k value is about 0.4 in both cases.

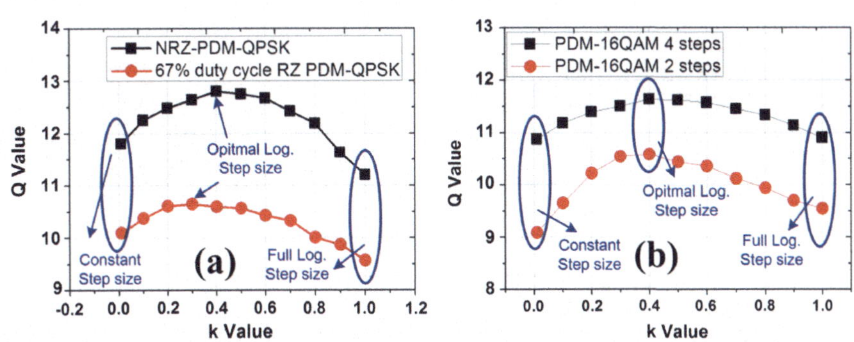

Fig. 6.15 Relation between Q value of nonlinear compensation and adjusting factor k value for **a** different pulse shapes; **b** polarization multiplexing 16QAM

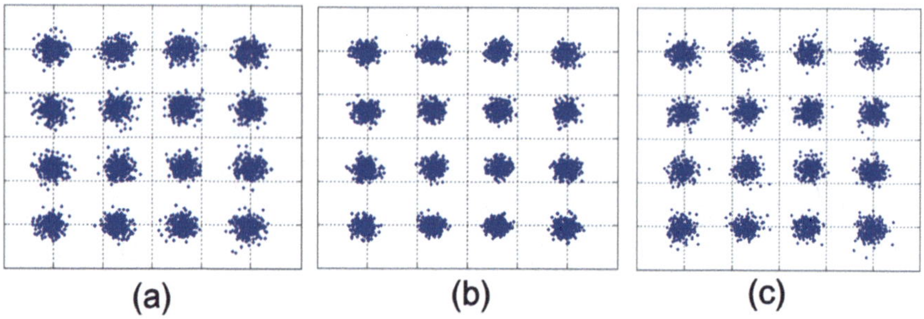

Fig. 6.16 Constellation diagram of 16QAM signal obtained after transmission of 1000 km by using the improved logarithmic step nonlinear compensation and carrier recovery: **a** when k value is 0.01, **b** when k value is 0.4, and **c** when k value is 1

When k is optimized, the improved logarithmic step size performs better than both constant step size (k = 0.01) and full logarithmic step size (k = 1). The constellation after DSP and NLC based on the improved logarithmic step nonlinear compensation is shown in Fig. 6.16a–c, when k = 0.01, k = 0.4, and k = 1, respectively. Here, the number of steps per span is 2. As can be seen from Fig. 6.16, the constellation diagram is the clearest when k takes the optimal value of 0.4.

6.3 Experimental Setup and Results

In order to verify the proposed JC-NLC scheme, the experimental setup is shown in Fig. 6.17. The transmitted signal is the multi-channel 3 × 50-Gbit/s NWDM PDM-QPSK signal, the SMF-28 fiber transmission distance is 1120 km, the link only uses EDFA as the amplifier, and the multi-channel joint coherent detection is adopted at the receiving end. At the transmitter end, ECL with a central wavelength of 1549.42 nm and linewidth less than 100 kHz is adopted, and its output optical

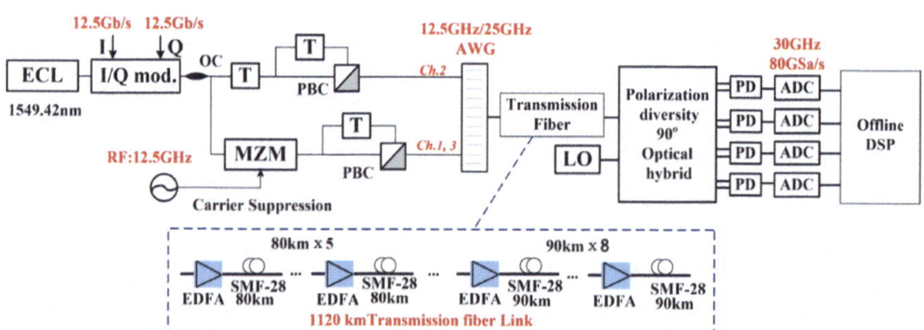

Fig. 6.17 Multi-channel experimental setup based on improved logarithmic step nonlinear compensation algorithm

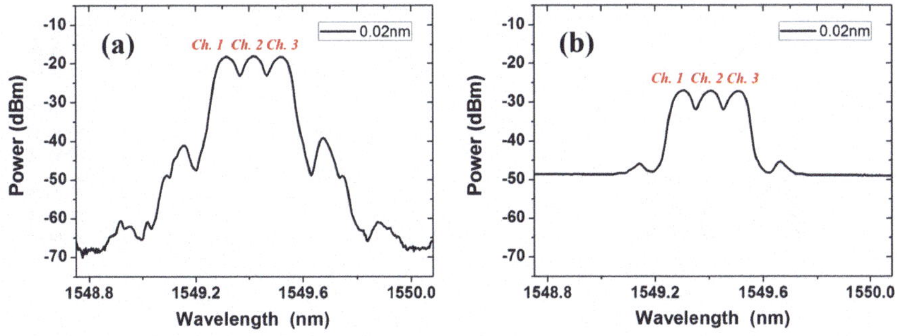

Fig. 6.18 **a** Spectrum of channel 3 N-WDM signal before transmission, and **b** spectrum after 1120-km transmission

power is 14.5 dBm. After I/Q modulation, 25-Gbit/s optical QPSK signal is divided into two beams: one beam is used for polarization multiplexing to generate channel-2 PDM-QPSK signal; another beam is injected into MZM to generate another two odd channels by carrier suppression. MZM is driven by a 12.5-GHz RF signal with a bias point of 0. The polarization multiplexing of the QPSK signal is combined into three channels of the N-WDM PDM-QPSK signal. The polarization multiplexer consists of a polarization-preserving optical coupler (PM-OC), an optical delay line that provides a delay of 150 symbols, and a polarization beam combiner to recombine signals. In the N-WDM channel combination, a 12.5/25-GHz array waveguide grating is used to combine odd channels and even channels. The transmission link consists of 5 spans of 80-km SMF-28 and 8 spans of 90-km SMF-28, and there is no optical dispersion compensation module in the link. In order to study the nonlinear damage, the output power of each EDFA is controllable. Figure 6.18a, b is the spectra of 3 × 50-Gbit/s N-WDM signal before and after 1120-km fiber transmission (0.02-nm resolution).

At the coherent receiver, an ECL with a linewidth less than 100 kHz is used as the local oscillator, and the center frequency matches channel 2. Polarization diversity and phase diversity are realized by using a 90-degree hybrid with polarization diversity. Analog-to-digital conversion is achieved by a digital oscilloscope with a sampling rate of 80 GSa/s and a bandwidth of 30 GHz. Detailed offline DSP is shown in Fig. 6.9b, including joint multi-channel dispersion compensation and nonlinear compensation. The average attenuation of the fiber link is 0.26 dB/km, the dispersion is 17 ps/km/nm, and the nonlinear coefficient is 1.5 $W^{-1}\cdot km^{-1}$. After the joint channel dispersion compensation and nonlinear compensation, the signals of the three channels are demultiplexed by using the fifth-order Bessel filter. After channel demultiplexing, two 13-order, T/2-spaced, complex adaptive FIR filters based on the CMA algorithm are used to recover the modulus of the PDM-QPSK signal and realize polarization demultiplexing. The same method was used for each carrier recovery, that is, the fourth-order Viterbi-Viterbi algorithm was used for frequency offset estimation and phase recovery. The binary signal is then converted into a binary signal using a delay-add post-filter, followed by a 1-bit MLSE symbol decision. Finally, the bit error rate of channel 2 is calculated by using a differential decoder to remove the

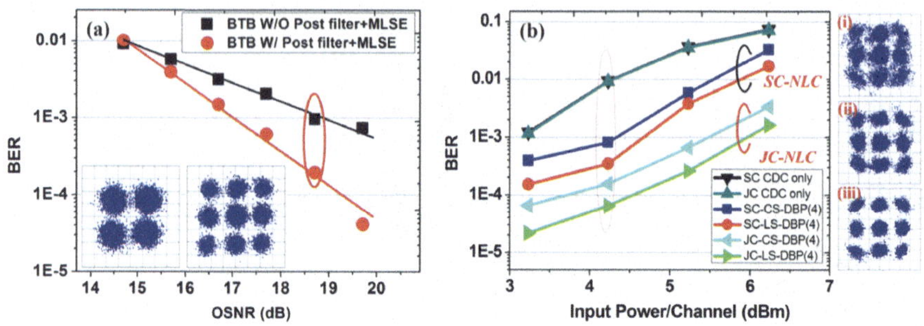

Fig. 6.19 **a** The relationship between the back-to-back bit error rate and OSNR of multi-channel N-WDM systems. **b** The BER performance versus fiber input power after 1120-km transmission for different processing algorithms

uncertainty of the $\pi/2$ phase. It is worth mentioning that the central frequency of LO used by the receiver is the same as the central frequency of channel 2. In other words, the center frequency of distance channel 1 is -12.5 GHz, and the center frequency of distance channel 3 is $+12.5$ GHz. Therefore, the bandwidth of the photodetector and ADC is large enough to realize the joint detection of the three channels.

In the multi-channel N-WDM system, the relationship between the back-to-back BER and OSNR of channel 2 is shown in Fig. 6.19a under the condition of using or not using post-filtering and MLSE. Figure 6.19a also shows the constellation of QPSK signals when OSNR is 18.7 dB with or without post-filtering. As can be seen, after the post-filtering, the four-point QPSK signal is shaped into a 9QAM-like signal. The inter-channel crosstalk and ISI can be suppressed by the post-filter and MLSE. When the bit error rate is 1×10^{-3}, 2 dB OSNR gain is obtained. Thus, the channel 2 signal can be recovered better by using post-filtering and MLSE.

Figure 6.19b shows the relationship between the bit error rate of channel 2 and the input power per span under different nonlinear compensation algorithms after the transmission of 1120 km SMF-28. The nonlinear compensation schemes include single-channel dispersion compensation, joint channel dispersion compensation, single-channel nonlinear compensation (SC-CS-DBP, SC-LS-DBP) based on fixed and logarithmic step DBP, and joint channel nonlinear compensation (JC-CS-DBP, JC-LS-DBP) based on fixed and logarithmic step DBP. To keep the transmission in the nonlinear region, the input power is increased from 3.23 dBm per channel to 6.23 dBm. For single-channel dispersion compensation and nonlinear compensation, channel 2 signal is demultiplexed and then processed by single-channel nonlinear compensation. In contrast, for joint channel nonlinear compensation, three channels are simultaneously compensated and then demultiplexed, as shown in Fig. 6.9b. Single-channel nonlinear compensation based on CS-DBP and LS-DBP has limited improvement in performance. Compared with the single-carrier scheme, JC-CS-DBP and JC-LS-DBP have better performance, because the nonlinear impairments between channels and within channels can be compensated effectively. Similarly, logarithmic step size is better than the fixed step size in both single and joint channel methods. The insets (i), (ii), and (iii) in Fig. 6.19b show the X-polarized signal constellation using JC-CDC, SC-CS-DBP (step is 4), and JC-LS-DBP (step is 4) methods when the channel input power is 4.23 dBm, respectively. JC-LS-DBP can significantly improve constellation performance, and JC-LS-DBP performs best in all NLC schemes. Compared with the CDC-only scheme and the traditional SC-CS-DBP scheme, the JC-LS-DBP scheme can improve the input power by 3 dB and 2 dB, respectively, when the bit error rate is 1×10^{-3}. The above results are very consistent with the simulation results.

Considering the computational complexity of the algorithm, Fig. 6.20a shows the relationship between the bit error rate of the system after 1120-km optical fiber transmission and the calculation step number under different input power and different DBP schemes conditions, where the adjustment factor remains at 0.4. The results in Fig. 6.20a are also in good agreement with the simulation analysis. Compared with CS-DBP, the maximum step number of LS-DBP is reduced to nearly half, which means the computational complexity is reduced by about 50%. When the input power per channel is 5.23 dBm, the optimal step number for LS-DBP is 4, while the optimal step number for CS-DBP is 8. Therefore, compared with CS-DBP, the computational complexity of LS-DBP is obviously reduced. In addition, when the number of steps is small, the performance of LS-DBP is better than that of CS-DBP at both 4.23 and 5.23 input powers.

Finally, we study the relationship between the bit error rate performance of the system and the adjustment factor k (0.01–1) of JC-LS-DBP when the number of steps is different. The results are shown in Fig. 6.20b, where the input power of each channel is 5.23 dBm. In three different cases, k value has an impact on bit error rate, especially when the number of steps is small. When the number of steps is large, the k value has little influence on the bit error rate. The results in Fig. 6.20b show that the improved logarithmic step nonlinear compensation has the best performance when the adjusted factor k is about 0.4, which is consistent with the simulation results.

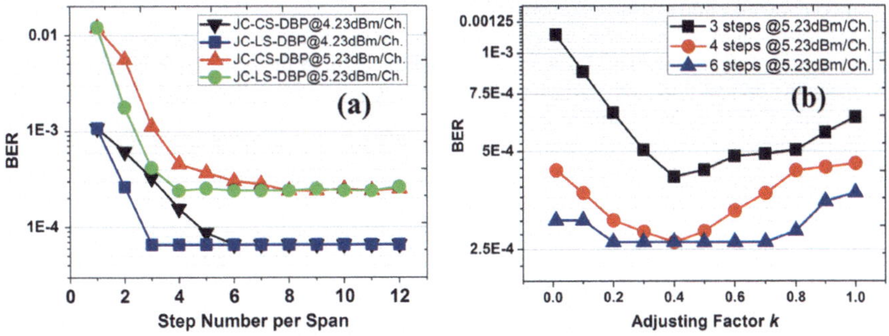

Fig. 6.20 **a** The relation between the bit error rate and the number of steps calculated after different nonlinear compensation algorithms are adopted for signals with different input optical powers. **b** The impact of the regulator k on signal bit error rate under different calculation steps

6.4 Summary

In this chapter, a new DBP nonlinear compensation algorithm based on improved logarithmic step distribution is introduced, and its compensation effect in the Nyquist wavelength division multiplexing (N-WDM) system is analyzed. Different from the traditional fixed step size method, the scheme uses logarithmic step size distribution in DBP, and improves the traditional logarithmic step size distribution by introducing adjustment parameter k. Simulation results show that the improved logarithmic step distribution can reduce the complexity of the algorithm and improve the compensation effect. In addition, we also carried out experimental verification. Using EDFA and SMF-28, 3×50-Gb/s NWDM PDM-QPSK signal was transmitted over 1120 km. Experimental results show that compared with single-channel fixed step nonlinear compensation, the improved algorithm can significantly improve the bit error rate performance and Q value of the system, greatly simplify the nonlinear compensation algorithm, and reduce the computational complexity by about 50%.

References

1. F. Yaman, G. Li, Nonlinear impairment compensation for polarization-division multiplexed WDM transmission using digital backward propagation. IEEE Photon. J. **1**(2), 144–152 (2009)
2. G. Li, E. Mateo, L. Zhu, Compensation of nonlinear effects using digital coherent receivers, in *Optics InfoBase Conference Papers*. Los Angeles, CA, United states: Optical Society of America (2011)
3. I.P. Ezra, Complexity reduction algorithms for nonlinear compensation using digital backpropagation, in *2012 IEEE Photonics Conference, IPC 2012*. Burlingame, CA, United states: IEEE Computer Society, pp. 388–389
4. Ezra Ip, Nonlinear compensation using backpropagation for polarization-multiplexed transmission. J. Lightwave Technol. **28**(6), 939–951 (2010)
5. Kazuro Kikuchi, Electronic post-compensation for nonlinear phase fluctuations in a 1000-km 20-Gbit/s optical quadrature phase-shift keying transmission system using the digital coherent receiver. Opt. Express **16**(2), 889–896 (2008)
6. G.P. Agrawal, *Nonlinear fiber optics*. Elsevier/Academic Press, p. 629 (2013)
7. O.V. Sinkin, R. Holzlöhner, J. Zweck et al., Optimization of the split-step Fourier method in modeling optical-fiber communications systems. J. Lightwave Technol. **21**(1), 61–68 (2003)
8. R. Asif, C.-Y. Lin, M. Holtmannspoetter, et al. Logarithmic step-size based digital backward propagation in N-channel 112 Gbit/s/ch DP-QPSK transmission. IEEE, pp. 1–4 (2011)
9. Irshaad Fatadin, S. J. Savory. Compensation of Frequency Offset for 16-QAM Optical Coherent Systems Using QPSK Partitioning[J]. Photonics Technology Letters, IEEE, 2011, 23(17): 1246–1248
10. Y. Gao, A.P.T. Lau, L. Chao et al., *Low-complexity two-stage carrier phase estimation for 16-QAM systems using QPSK partitioning and maximum likelihood detection* (CA, Los Angeles, 2011), pp. 1–3
11. M Orteza Ziyadi, M.R. Chitgarha, A.M. Ariaei, et al. Experimental demonstration of optical Nyquist generation of 32-Gbaud QPSK using a comb-based tunable optical tapped-delay-line FIR filter. OSA, pp. 1–3 (2014)
12. Junyi Wang, Chongjin Xie, Z. Pan. Generation of spectrally efficient Nyquist-WDM QPSK signals using DSP techniques at transmitter: IEEE,2012:1–3

13. Akihide Sano, Eiichi Yamada, Hiroji Masuda et al., No-guard-interval coherent optical OFDM for 100-Gb/s Long-Haul WDM transmission. J. Lightwave Technol. **25**(16), 3705–3713 (2009)
14. Benyuan Zhu, Xiang Liu, Sethumadhavan Chandrasekhar, et al. Ultra-Long-Haul Transmission of 1.2-Tb/s Multicarrier No-Guard-Interval CO-OFDM Superchannel Using Ultra-Large-Area Fiber[J]. Photonics Technology Letters, IEEE, 2010, 22(11): 826–828
15. Junwen Zhang, Xinying Li, Ze Dong, Digital nonlinear compensation based on the modified logarithmic step size. J. Lightwave Technol. **31**(22), 3546–3555 (2013)
16. J. Zhang, J. Yu, N., Ze Dong, X. Li, Nonlinear compensation and crosstalk suppression for 4 $\times$ 160.8 Gb/s DWDM PDM-QPSK signal with heterodyne detection. Optics Expr. **21**, 9230–9237 (2013)

Chapter 7
Probabilistic Shaping

7.1 Introduction

Faced with the increasing demand for transmission capacity and flexibility of communication systems, the probabilistic shaping (PS) technique, as a typical modulation format optimization technology, has attracted more and more attention and become a promising technology due to its high transmission capacity and low complexity.

PS can provide unparalleled flexibility for optical communication systems without increasing the complexity of the system. In particular, the optical communication system channel is subject to nonlinear power limitations, making the PS more tailor-made for optical communication. Under the premise of not increasing the transmission power, obtaining high spectral efficiency and high transmission capacity is the advantage of PS. PS is really a good choice for optical communication systems to solve power limitation problems and increase transmission capacity.

PS has started to attract researchers' attention since September 2016 when Nokia Bell Labs et al. achieved 1-Tbit/s transmission using the 4-carrier super-channel in the German backbone network under the help of probabilistically shaped constellations, which has achieved unprecedented transmission capacity and can be seen as a breakthrough in optical communications [1]. Besides, in October 2016, Alcatel-Lucent et al. announced that they have achieved an amazing breakthrough in implementing 65 Tbit/s data transmission over 6600 km single-mode fiber in laboratory tests using PS technology [2]. So far, the research work of PS has focused on single-carrier fiber transmission with coherent detection. In a comparison between a uniformly distributed QAM format and a PS-QAM format, it has been found that in a nonlinear channel, the PS can provide a sensitivity gain greater than 1.53 dB [3]. The improvement of the transmission distance of PS for optical communication has also been confirmed in the relevant researches [4].

J. Yu and N. Chi, *Digital Signal Processing In High-Speed Optical Fiber Communication Principle and Application*,
https://doi.org/10.1007/978-981-15-3098-2_7

As a breakthrough technology, PS will become an epoch-making technique that will allow faster transmission speed and greater flexibility in the future of optical communication. The application of PS technique in optical communication systems is undoubtedly a significant advancement for optical wireless convergence systems that focus on power limitation, spectral efficiency, and transmission capacity. PS technology will be a promising candidate for improving optical communication performance in the future. However, PS has not been further developed, so we have started to explore the new modulation format optimization technology of probabilistic shaping in optical communication.

This section introduces the research of probabilistic shaping technology from several aspects, including the basic principle of probabilistic shaping technology, simulation, and experimental exploration of performance improvement on the communication system by PS technology.

7.2 Principle of Probabilistic Shaping

Just when people think that optical communication has reached a bottleneck, probabilistic shaping technology has begun to enter people's realization. As a typical example of modulation format optimization technology, probabilistic shaping technology has attracted more and more attention and become a promising technology due to its advantages of higher transmission capacity and lower system complexity according to many researches and verification [5–12]. Here, we introduce the principle of PS by taking 16QAM modulation format as an example.

In the general optical fiber communication system, which is shown in Fig. 7.1a, the probability of 16 constellation points is evenly distributed for 16QAM. While in Fig. 7.1b, a distribution matcher (DM) is applied before the FEC encoder, and a distribution de-matcher is added after the FEC decoder. Probabilistic shaping is just added to the communication system through the above change.

By now, there have been many implementation schemes of PS. While one method that is widely regarded is the probability amplitude shaping (PAS) scheme using low-density parity check (LDPC) codes. The role of the distribution matcher is to

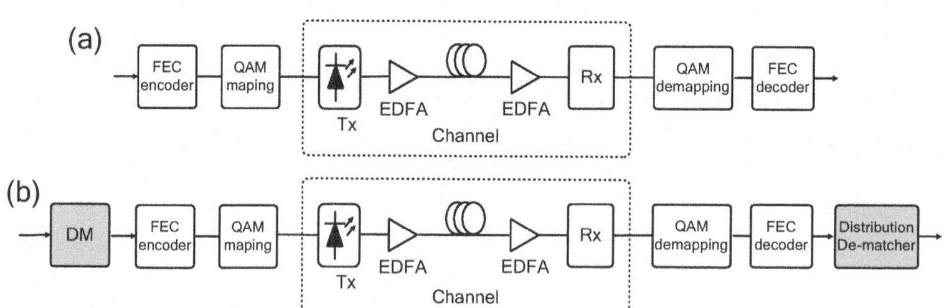

Fig. 7.1 Principle of optical fiber communication systems **a** without and **b** with PS technique

perform an external encoding before the FEC encoding, in order to subject the probability of occurrence of the 16QAM symbols after encoding and mapping to our preset probability distribution. And this probability distribution is determined by the signal-to-noise ratio (SNR). Compared with the uniformly distributed 16QAM, the probability of occurrence of four constellation points in the inner circle is increased, and the probability of occurrence of four constellation points in the outer circle is reduced, thereby improving bit error rate (BER) performance and saving transmission power.

Figure 7.2a–c shows the probability distribution of uniformly distributed 16QAM constellation points, and the probability distribution of the 16QAM constellation points probabilistically shaped when the signal-to-noise ratio is 15 dB and 10 dB, respectively. In theory, each SNR has an optimal corresponding probability distribution that allows the system performance to be optimally improved. In general, the worse the SNR, the greater the degree of probabilistic shaping, that is, the difference between the probability of a signal with a large probability and the probability of a signal with a small probability is more disparity.

There have been studies that found the robustness of probabilistic shaping [10]. The SNR of the shaping optimization cannot always match the SNR of the channel. However, the penalty due to this mismatch is small. That is to say, a certain SNR can share a probability distribution with the SNR that is closer to it. Reference [10] found that 4 probability distributions can just cover SNR in the range of 5–25 dB, as shown in Fig. 7.3 (citing Fig. 4 of Ref. [10]). The dashed line in the figure represents perfect shaping, while different colors indicate the penalty in comparison to perfectly matched shaping, and green corresponds to an acceptable penalty caused by the mismatch.

As we know, the 16 constellation points of 16QAM contain three amplitudes, which are considered three circles here. As shown in Fig. 7.4a, the distance between the inner ring and the second ring is 1.75, while the distance between the second ring and the outer ring is only 0.80, which results in the outermost ring becoming the error-prone area. The second ring and the third ring after CMMA in Fig. 7.4b are indistinguishable, which can also illustrate the above viewpoint. In fact, when we use 16QAM as the modulation format, a large part of errors come from the four

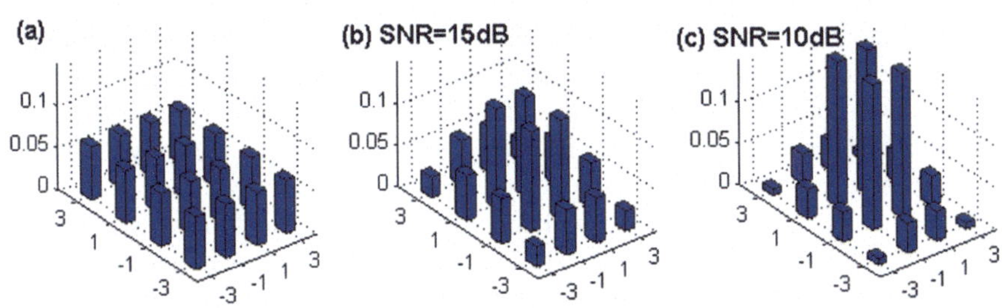

Fig. 7.2 Probability distribution of **a** regular-16QAM and PS-16QAM when **b** SNR = 15 dB and **c** SNR = 10 dB

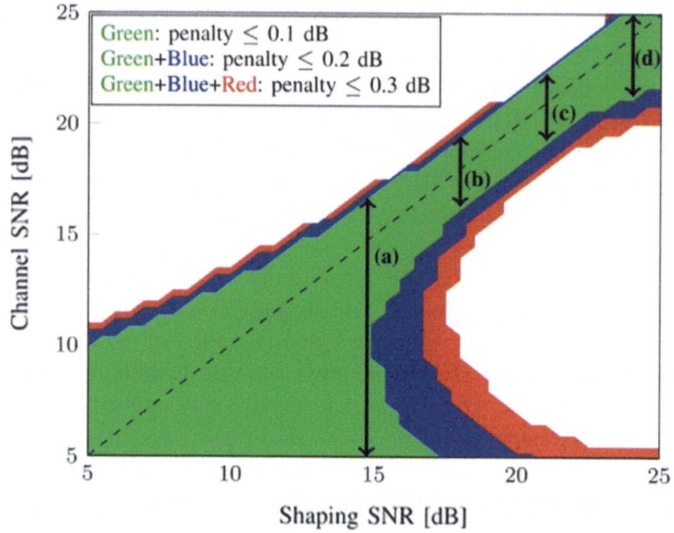

Fig. 7.3 Shaping robustness for 64QAM over the AWGN channel [11]

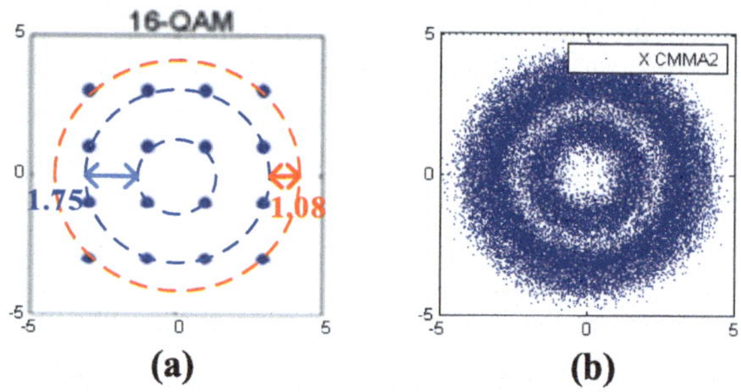

Fig. 7.4 **a** 16QAM constellation; **b** 16QAM constellation after CMMA equalizer during DSP

constellation points of the outer ring. On the other hand, the constellation points of the outermost circle are also most affected by nonlinearities.

Probabilistic shaping increases the probability of occurrence of four constellation points in the inner circle of 16QAM, and reduces the probability of occurrence of four constellation points in the outer circle. It just increases the probability of constellation points with good error performance, and reduces the probability of occurrence of constellation points with poor error performance. The above is one of the reasons why probabilistic shaping can reduce the bit error rate.

Probabilistic shaping, on the other hand, increases the probability of occurrence of four constellation points in the inner circle, which also means increasing the probability of occurrence of constellation points with smaller amplitude. Reducing the probability of occurrence of 4 constellation points in the outer ring also means reducing the probability of occurrence of larger amplitude points. Therefore, the average power of the 16QAM signal after probabilistic shaping is lower than that of the regular 16QAM signal, thereby saving the transmission power and solving the power limitation problem caused by the nonlinearity of the optical communication. Under the same signal-to-noise ratio, the noise power of PS signals will be lower, therefore the bit error rate will be reduced. The above is the second reason why probabilistic shaping can improve BER performance.

As we know, probabilistic shaping is to increase the probability of occurrence of four constellation points in the inner circle and reduce the probability of four points in the outer circle. Specifically, the probabilistic shaping signal is subject to a power-efficient Maxwell Boltzmann distribution, as shown in Eq. (7.1).

$$P_X(x_i) = \frac{\exp(-v(\mathrm{Re}(x_i)^2 + \mathrm{Im}(x_i)^2))}{\sum_{k=1}^{M} \exp(-v(\mathrm{Re}(x_k)^2 + \mathrm{Im}(x_k)^2))} \tag{7.1}$$

where x_i denotes the ith complex QAM constellation point, and v is a crucial parameter, which is determined by SNR and can be used to control the degree of shaping. In theory, each SNR can be matched to an optimum v to maximize mutual information (MI) gain. The different probability distributions shown in Fig. 7.2 are derived from different parameters v. In addition, Eq. (7.2) is a calculation formula of MI.

$$I(X:Y) = H(X) - H(X|Y) \tag{7.2}$$

where X is the transmitted symbol and Y is the received symbol. $H(X)$ represents the source entropy, and $H(X|Y)$ represents the uncertainty that X still exists after receiving Y, which is the reduction of the amount of information obtained by the receiver at the receiving end due to various interferences. Therefore, mutual information can represent the information capacity of the communication system. So we can use mutual information to measure the optimization that probabilistic shaping brings to a communication system.

Probabilistic shaping increases the probability of occurrence of a signal with a small amplitude, and reduces the probability of occurrence of a signal with a large amplitude, which results in a decrease in the average power of the signal. However, this will make the PS and regular-QAM signal not comparable, so in order to make the average power of the probability shaping signal the same as that of the original signal, it is necessary to introduce a scaling factor Δ to expand the probabilistic shaping constellation diagram so that the average power of the PS signal is equal to the average power of the uniformly distributed signal.

$$E\left[|\Delta X|^2\right] = E\left[|X_0|^2\right] \tag{7.3}$$

where X represents the probabilistic shaped signal and X_0 represents the uniformly distributed signal. In this way, Eq. (7.1) can also be changed to

$$P_{\Delta X}(x_i) = \frac{\exp(-v(\mathrm{Re}(x_i)^2 + \mathrm{Im}(x_i)^2))}{\sum_{k=1}^{M} \exp(-v(\mathrm{Re}(x_k)^2 + \mathrm{Im}(x_k)^2))} \tag{7.4}$$

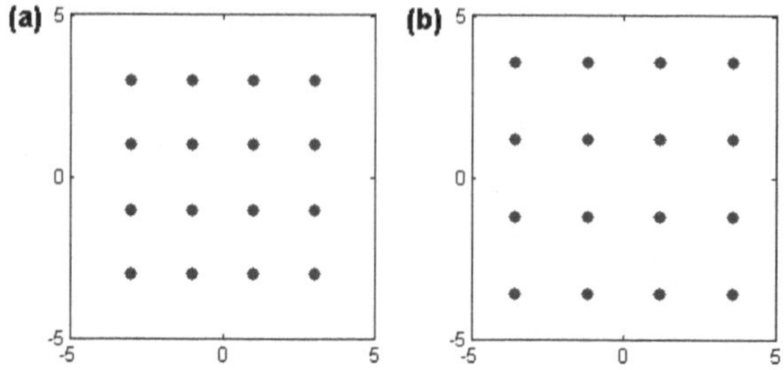

Fig. 7.5 Constellation of **a** regular-16QAM; **b** PS-16QAM

We take PS-16QAM as an example, here we take $v = 0.1$. According to Eq. (7.4), the probability distribution of each constellation point can be obtained as

$$P(\Delta X) = \begin{vmatrix} 0.0240 & 0.0535 & 0.0535 & 0.0240 \\ 0.0535 & 0.1190 & 0.1190 & 0.0535 \\ 0.0535 & 0.1190 & 0.1190 & 0.0535 \\ 0.0240 & 0.0535 & 0.0535 & 0.0240 \end{vmatrix} \quad (7.5)$$

$$\overline{P} = \sum_i |X_i|^2 \cdot P(X_i) \quad (7.6)$$

The average power of the 16QAM signal after probabilistic shaping can be calculated by Eq. (7.6) to be 6.96. The average power of the original 16QAM signal is 10, which is 1.4367 times that of PS-16QAM. The amplitude of each symbol of PS-16QAM should be increased by 1.1986 times, that is, $\Delta = 1.1986$. The original 16QAM constellation diagram is changed from Fig. 7.5a, b. In this way, under equal average power condition, the minimum distance between the constellation points is improved and the fault tolerance is enhanced.

7.3 Simulation of Probabilistic Shaping

7.3.1 Simulation Setup

The structure of the simulation setup is shown in Fig. 7.6. The external cavity laser (ECL) 1 that produces the continuous-wave (CW) lightwave provides the optical input of the I/Q modulator, and the regular and PS-16QAM signal generated by the Tx DSP are used to drive the I/Q modulator. After the Gaussian white noise source is added, the resulting regular and PS-16QAM optical signal is sent to the 90° optical hybrid. The other input to the 90° optical hybrid is the CW lightwave generated from ECL 2, as a local oscillator with a 15 GHz frequency difference from the transmitted optical signal. In addition, the linewidth of the two ECLs is ideally set to be zero. Then, the balanced photodetector (BPD) is used for up-conversion at the I output of

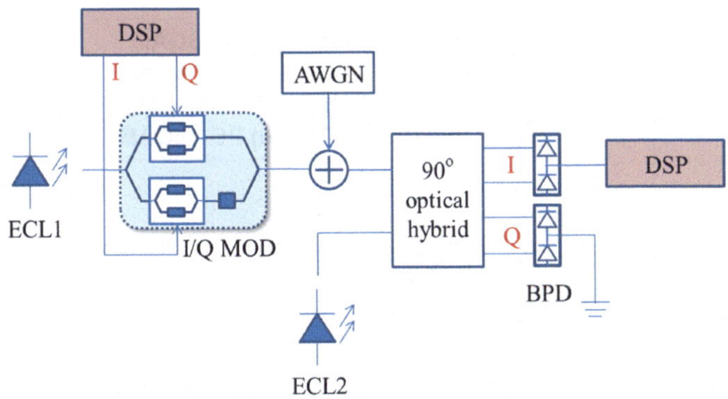

Fig. 7.6 Simulation setup

the 90° optical hybrid to obtain a 15 GHz regular (or PS) 16QAM millimeter-wave signal. The channel noise in the channel comes from the thermal and shot noise of the BPD. The Q output of the 90° optical hybrid is grounded because we do not need it. Afterwards, the 15 GHz 16QAM millimeter-wave signal is processed by the DSP on the receiver side. The operation in the optical domain is implemented by the VPI platform. The operation of the electrical domain is implemented by MATLAB. The data can be recovered by MATLAB programming, including IF down-conversion, symbol decision, 16QAM de-mapping, and other processes. The bit error rate (BER) calculation is performed based on the comparison of the recovered data with the original transmitted data.

7.3.2 Simulation Result and Discussion

The calculated BER performance is shown in Fig. 7.7. The curves of BER versus the optical signal-to-noise ratio (OSNR) input to optical hybrid are given in four different situations: (1) 20 Gbit/s (5 Gbaud) regular-16QAM, (3) 20 Gbit/s (20/3.4 Gbaud) PS-16QAM, (4) 20 Gbit/s (20/3 Gbaud) PS-16QAM, and (5) 20 Gbit/s (20/2.6 Gbaud) PS-16QAM. Here, the denominator of each baud rate in the brackets represents the number of bits carried by one symbol. In order to obtain the same bit rate, the baud rate should be different for the PS scheme with different transmission rates. In the second to fourth situations, we set parameter v of PS as 0.0792, 0.1373, and 0.2100, respectively. From the simulation results shown in Fig. 7.7, we can conclude that, in the case of the same bit rate, PS-16QAM has a better BER performance than regular-16QAM. In addition, the fewer bits one symbol carries, the better BER performance will be achieved for PS-16QAM. What is more, Fig. 7.8 shows the recovered constellations of the 20 Gb/s regular-16QAM and PS-16QAM signals at 15 GHz corresponding to a certain point of different cases in Fig. 7.7, respectively. Figure 7.8a is measured at 17.26 dB OSNR input to an optical hybrid in the case of

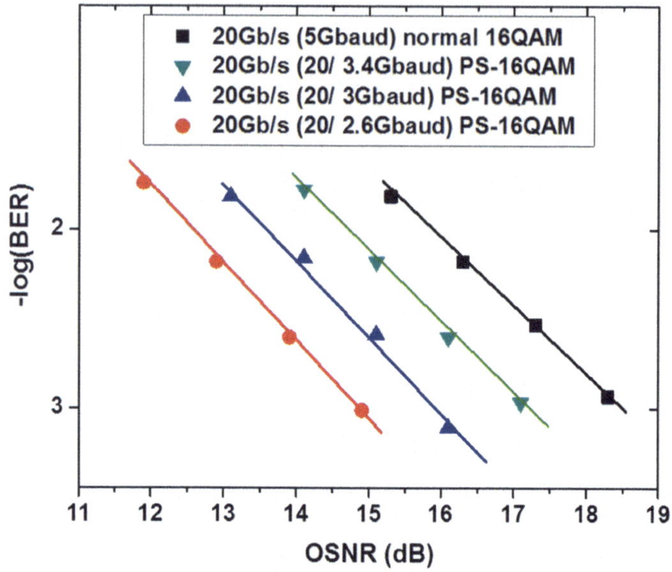

Fig. 7.7 BER versus OSNR input to optical hybrid

regular-16QAM and gets a BER of 3.3×10^{-3}. Figure 7.8b is measured at 16.07 dB OSNR in the case of PS-16QAM with an entropy of 3.4 bit/symbol (called PS-16QAM-3.4 here) and gets a BER of 2.8×10^{-3}. Figure 7.8c is calculated at 15.1 dB OSNR input to an optical hybrid in the case of PS-16QAM-3 and gets a BER of 2.9 $\times 10^{-3}$. Figure 7.8d is measured at 13.9 dB OSNR in the case of PS-16QAM-2.6 and gets a BER of 2.8×10^{-3}. In addition, Fig. 7.9 illustrates the captured 15 GHz spectra at different locations for the situation in Fig. 7.8c. The optical spectrum in Fig. 7.9a is calculated before BPD and the electrical spectrum in Fig. 7.9b is calculated after BPD when the signal OSNR is 15.1 dB.

In summary, the PS-16QAM modulation format is superior to the regular-16QAM modulation format in terms of BER performance. Ideally, different SNRs require different PMFs, i.e., different parameter v values. In a realistic system, we can use the same v over an SNR range because of the tolerance for the mismatch between channel SNR and the SNR corresponding to our PMF, which makes the implementation of PS easier. Through this simulation, we can demonstrate the feasibility of probabilistic shaping in coherent optical communication systems. Next, we will continue to explore through specific experiments.

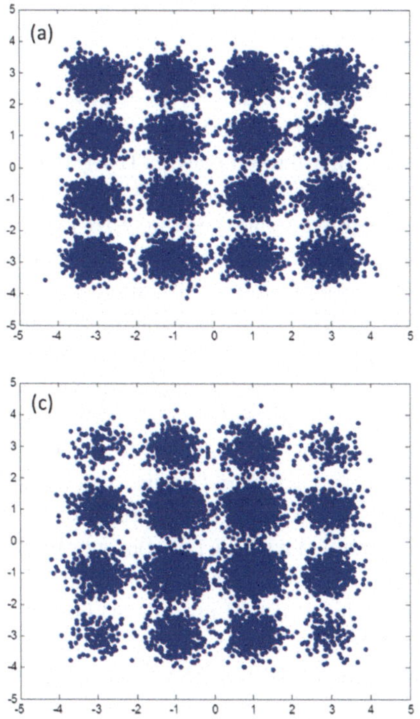

Fig. 7.8 The recovered constellations of the 20 Gb/s regular-16QAM and PS-16QAM signal at 15 GHz. **a** 17.26 dB OSNR, 4 bit/symbol, BER = 3.3×10^{-3}, **b** 16.07 dB OSNR, 3.4 bit/symbol, BER = 2.8×10^{-3}, **c** 15.1 dB OSNR, 3 bit/symbol, BER = 2.9×10^{-3}, and **d** 13.9 dB OSNR, 2.6 bit/symbol, BER = 2.8×10^{-3}

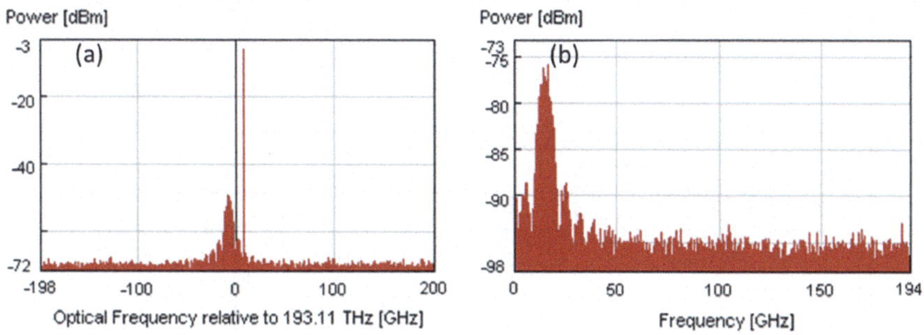

Fig. 7.9 Captured 15GHz spectra at different locations for the situation in Fig. 7.8c. **a** Before BPD, and **b** after BPD

7.4 Principle of PAS Scheme

It is universally acknowledged that PS is an excellent technology to achieve rates close to Shannon's limit and probabilistic amplitude shaping (PAS) [6, 9] has become a standard scheme, which carries out shaping by a distribution matcher (DM) before FEC encoding. The principle of combining DM with FEC encoder based on PS-64QAM with an entropy of 5.5-bit/symbol/polarization is shown in Fig. 7.10. Firstly, the binary data source is split into two parts consisting of $U_1 = 56678$ and $U_2 = 11435$ bits, respectively. U_1 bits are fed into a constant composition distribution matcher (CCDM) and $V_1 = 32400$ amplitude symbols $\{1, 3, 5, 7\}$ are generated according to the designated probability distribution. Inset (a) of Fig. 7.10 offers the probability distribution of amplitude symbols generated by CCDM. Then the amplitude symbols are mapped into $2V_1$ binary bits, and these mapped bits are fed together with the U_2 bits into a low-density parity-check coded (LDPC) encoder with 27.5% overhead, which generates parity bits. These parity bits together with the U_2 bits, which both follow a uniform distribution, are used as sign bits according to the mapping ($0 \rightarrow +1$ and $1 \rightarrow -1$). These mapped signs are then combined with the previous amplitude symbols to form $\{+1, +3, +5, +7\}$ or $\{-1, -3, -5, -7\}$ symbols, which are just the generated PS-PAM8 symbols as shown in the inset (b) of Fig. 7.10. At last, two paths of PS-PAM8 symbols with an entropy of 2.75 bit/symbol can be combined into PS-64QAM with an entropy of 5.5 bit/symbol (called PS-64QAM-5.5 here).

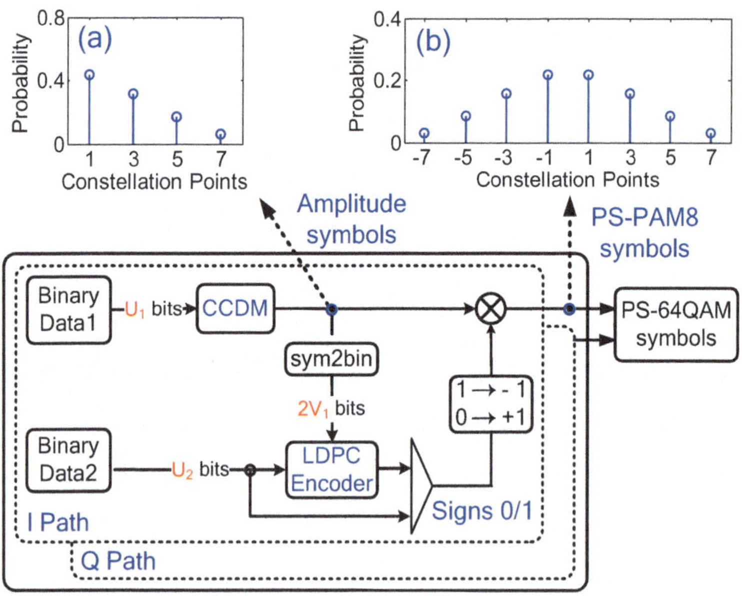

Fig. 7.10 Principle of PAS transmitter system

When considering overheads from CCDM and FEC, we transmit a total of V_1 64QAM symbols, which can only carry $2(U_1 + U_2)$ desired bits. Therefore, the actual transmission rate can be expressed as

$$R_{\text{actual}} = \frac{2(U_1 + U_2)}{V_1} \tag{7.7}$$

In this way, the actual transmission rate of PS-64QAM-5.5 is only 4.2 bit/symbol. Our actual transmission rate is exactly consistent with the theoretical rate R for probabilistic shaping based on 64QAM constellations, which is obtained by Eq. 7.3 of Ref. [12]:

$$R = H - (1 - r) \cdot m \tag{7.8}$$

where H is entropy (5.5 bit/symbol), r is FEC code rate (40/51), and m is 64QAM constellation bits per symbol (6 bit/symbol). In this way, the calculated theoretical transmission rate R is also 4.2 bit/symbol. Therefore, the net bit rate per channel can be calculated as 48 Gsymbol/s × 2 polarization × 4.2 bit/symbol/polarization = 403 Gbit/s.

7.5 Comparison of PS and Hybrid-QAM

Similar to PS, hybrid-QAM [13, 14] is another modulation format optimization technique, which can also yield OSNR sensitivity gain and transmission distance improvement. Different from regular-QAM, hybrid-QAM, which consists of two or more regular-QAMs with different source entropies that are time-interleaved at various ratios within a frame, can realize any entropy that falls between the entropies of the two regular-QAMs. In this way, the gap between regular-32QAM and regular-64QAM can be filled smoothly by hybrid-32/64QAM, which can yield a better compromise between spectral efficiency and OSNR sensitivity compared with regular-32QAM and regular-64QAM. As for the PS technique, a power-efficient Maxwell–Boltzmann distribution is used in order to decrease the gap between the capacity of communication systems and the Shannon limit. It is therefore interesting to compare hybrid-QAM with PS technology to explore a better modulation format for 400G WDM transmission based on the 50-GHz grid.

As we know, regular 2^n-ary QAM can only realize integer entropy, while a recently proposed hybrid-QAM can achieve arbitrary entropy by assigning different time slots to two or more regular 2^n-ary QAMs with different entropies and fill the gap between regular-QAMs smoothly. To take the hybrid-32/64QAM with a hybrid ratio of 1:1 (with an entropy of 5.5-bit/symbol/polarization) as an example, regular-32QAM symbols (mapped to 5 bits) and regular-64QAM symbols (mapped to 6 bits) appear in turns following the ratio of 1:1, as shown in Fig. 7.11a. The corresponding constellation is illustrated in Fig. 7.11b.

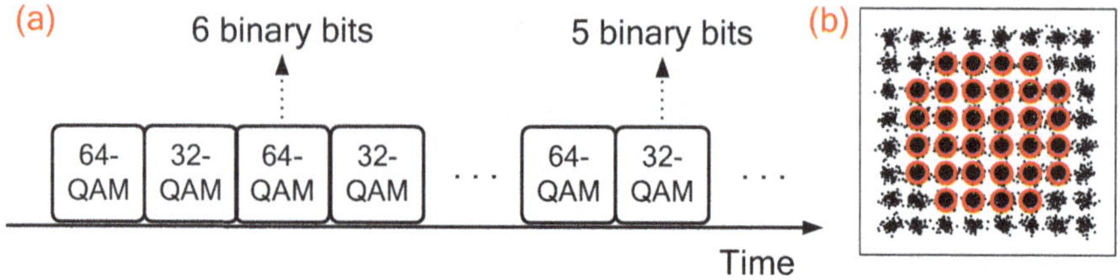

Fig. 7.11 **a** Symbol mapping and **b** constellation of a time-domain hybrid-32/64QAM with an entropy of 5.5-bit/symbol/polarization

Different from hybrid-QAM, PS uses a power-efficient Maxwell–Boltzmann distribution shown in Eq. (7.1) to increase the probability of constellation points with smaller amplitude but decrease the probability of constellation points with larger amplitude and therefore it can realize a lower average power at the same entropy and obtain a shaping gain.

$$P_X(x_i) = \frac{\exp(-v(\mathrm{Re}(x_i)^2 + \mathrm{Im}(x_i)^2))}{\sum_{k=1}^{M} \exp(-v(\mathrm{Re}(x_k)^2 + \mathrm{Im}(x_k)^2))} \tag{7.9}$$

where x_i denotes the ith complex M-QAM constellation point and v denotes the shaping parameter. Moreover, hybrid-QAM is made up of different regular-QAMs and its components maintain their individual numbers of mapped bits per symbol, while PS technique realizes the designated probability distribution of symbols by introducing redundancy and therefore the original number of mapped bits per symbol is maintained. We compared the performance of PS-64QAM with hybrid-32/64QAM with the same entropy (5.5 bit/symbol) in the transmission experiment shown in Sect. 7.6.2.

Figure 7.12 shows the NGMI versus post-FEC BER after LDPC decoding when using LDPC with 40/51 and 55/72 code rates, corresponding to 27.5 and 30.9% overheads. As shown in Fig. 7.12, post-FEC error-free performance can be achieved when the NGMI reaches up to 0.8 and 0.784 for the case with 27.5% and 30.9% overheads, respectively. In other words, 0.8 and 0.784-NGMI can be proven to be the SD-FEC threshold with 27.5% and 30.9% overheads, respectively. Here, the LDPC with 27.5% overhead is used on PS-64QAM-5.5 signal. While in order to get the same net bit rate, the LDPC with 30.9% overhead is used on regular-64QAM and hybrid-32/64QAM-5.5 signals.

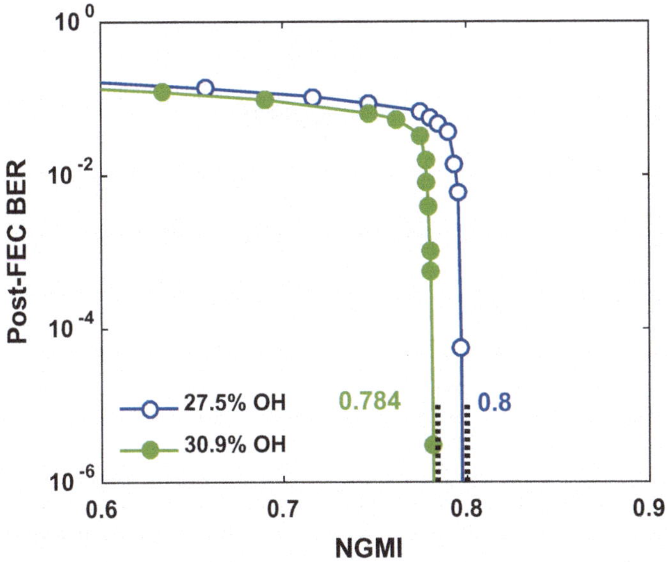

Fig. 7.12 Post-FEC BER as a function of NGMI with 27.5% and 30.9% SD-FEC overheads for PS-64QAM-5.5 and hybrid-32/64QAM-5.5

7.6 Experiment of PS Transmission

7.6.1 Experimental Setup

The experimental setup and DSP principle of the generation and transmission of eight-channel 48-Gbaud WDM PS-64QAM (or hybrid-32/64QAM) signals are given by Figs. 7.13 and 7.14. In the transmitter DSP, hybrid-32/64QAM and PS-64QAM symbols, both with an entropy of 5.5 bit/symbol/polarization, were generated, respectively. For hybrid-32/64QAM, as shown in inset (a) of Fig. 7.14, 32QAM and 64QAM

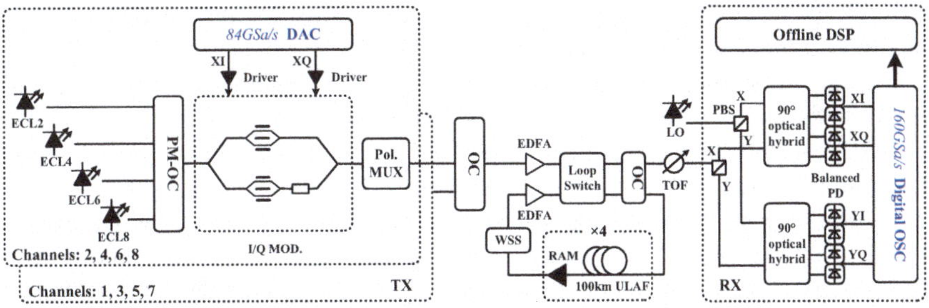

Fig. 7.13 Experimental setup of eight-channel WDM transmission over ULAF with Raman amplifier based on 48-Gbaud PM PS-64QAM or hybrid-32/64QAM

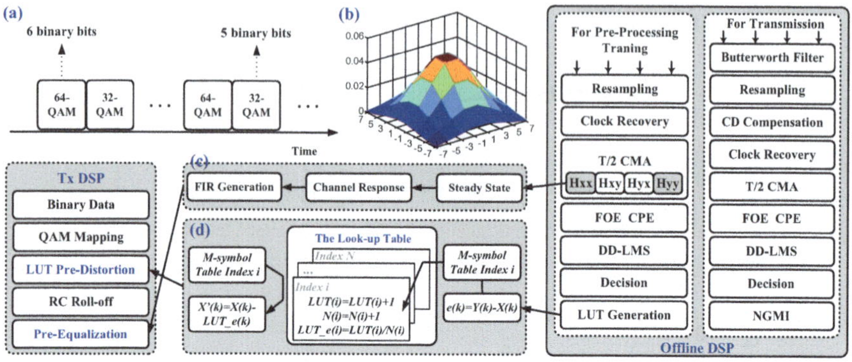

Fig. 7.14 Principle of Tx DSP, preprocessing, and offline DSP after transmission. **a** Symbol mapping of hybrid-32/64QAM; **b** Probability distribution of PS-64QAM signals

signals both account for half of the time slots and appear alternately. As for PS-64QAM, we generated symbols with 5.5-bit/symbol/polarization entropy with designated Maxwell–Boltzmann distribution, and the probability distribution of generated PS-64QAM signals are illustrated in inset (b) of Fig. 7.14. After QAM-mapping, the generated PS-64QAM (or hybrid-32/64QAM) symbols were fed into an LUT pre-distortion and pre-equalization section based on a training sequence used for preprocessing training. Meanwhile, the raised cosine (RC) roll-off with a roll-off factor of 0.04 was applied, as shown in the transmitter DSP part of Fig. 7.14. The pre-equalization finite impulse response (FIR) with 1024 filter length was calculated according to the 21-tap coefficients of constant modulus algorithm (CMA) filter after convergence under the back-to-back (BTB) case, which is shown in Fig. 7.14c. The LUT-based pre-distortion shown in Fig. 7.14d is carried out based on the training sequence after pre-equalization. The pattern dependent lookup table with 9-symbol memory is firstly generated by comparing the transmitted signals with the corresponding recovered ones. The average difference between the transmitted signals and the corresponding recovered ones for each pattern in the table index was calculated. Here around 35000 patterns were covered through several sets of 216 length training sequences. The corresponding constellation of pre-distortion was generated lastly by the calculated average differences. Afterwards, the preprocessed signals were loaded into a digital-to-analog converter (DAC).

The transmitter block consisted of a commercial DAC with 20-GHz analog bandwidth and 84-GSa/s sampling rate, a 100-kHz-linewidth external cavity laser (ECL), a 32-GHz-bandwidth I/Q modulator, two 65-GHz modulator drivers, and a polarization multiplexer (Pol. MUX). For optical signal modulation, the light source at 1553.126-nm from the ECL was fed into the I/Q modulator driven by the preprocessed signals. After optical modulation, a Polarization MUX was utilized to realize polarization multiplexing. At the transmitter side, eight-channel 50 GHz-WDM PM signals, which were divided into odd-channel group (ch. 1, 3, 5, and 7) and even-channel group (ch. 2, 4, 6, and 8), were combined by a polarization-maintaining optical coupler (PM-OC) and launched into the re-circulating ULAF loop with backward-pumped Raman amplifier, which consisted of several spans of 400-km TeraWaveTM + fiber with an average effective area of 125-μm^2, an attenuation coefficient of 0.182-dB/km and a chromatic dispersion coefficient of 20-ps/(nm·km) at 1550 nm. In the re-circulating loop, a backward-pumped Raman amplifier with 20-dB ON-OFF gain was utilized to compensate for signal loss for each span. The average power of the Raman pumps is around 950-mW. What is more, we used an attenuator (ATT) to control the optical power and a WSS was used to flatten the gain slope of the band-pass filter.

At the receiver end, a tunable optical filter (TOF) with 3-dB bandwidth of 0.9 nm was used to select the desired sub-channel from the eight-channel WDM signals. In our experiment, the 4th sub-channel is selected by the TOF. Then we implemented the polarization- and phase-diversity operation for the selected WDM signal and the 100-kHz-linewidth optical local oscillator (LO) using two polarization beam splitters (PBSs) and two 90° optical hybrids. Afterwards, the outputs of hybrids were fed into four balanced photodetectors (PDs), each with a bandwidth of 70-GHz. Finally, the digitization and sampling were realized by the digital oscilloscope (OSC) with 65-GHz bandwidth and 160-GSa/s sampling rate, which was followed by the offline DSP shown in Fig. 7.14. The offline DSP consisted of Butterworth digital low-pass filter, resampling, chromatic dispersion (CD) compensation, clock recovery, T/2 CMA, frequency offset estimation (FOE), carrier phase estimation (CPE) based on blind phase search (BPS), decision-directed least mean square (DD-LMS) equalization, decision, and NGMI calculation.

7.6.2 Comparison of PS-64QAM and Hybrid-32/64QAM

Firstly, the theoretical and experimental NGMI results versus BTB OSNR for single-channel hybrid-32/64QAM and PS-64QAM signals with an entropy of 5.5-bit/symbol/polarization are illustrated in Fig. 7.15. PS-64QAM and hybrid-32/64QAM signals incur around 2.8-dB and 3.4-dB OSNR penalty compared with theory results, respectively. PS-64QAM signals incur lower penalty because the probability distribution of outermost constellation points is lower and the performance in overcoming nonlinearity impairment is better for PS-64QAM compared with hybrid-32/64QAM.

The WDM experiment results of measured NGMI versus BTB OSNR for regular-64QAM, hybrid-32/64QAM, and PS-64QAM are given in Fig. 7.16. There is around 0.9-dB OSNR penalty for both 48-Gbaud WDM PM hybrid-32/64QAM-5.5 and PS-64QAM-5.5 signals compared with the single-channel case. The required OSNR for WDM PS-64QAM signals (528-Gbit/s line rate and 403-Gbit/s net bit rate) at the SD-FEC threshold with 27.5% overhead of 0.8 NGMI is around 22.7-dB, which obtains around 2.6-dB OSNR gain compared with 44-Gbaud regular-64QAM (528-Gbit/s line rate and 403-Gbit/s net bit rate) at 30.9% SD-FEC threshold of 0.784

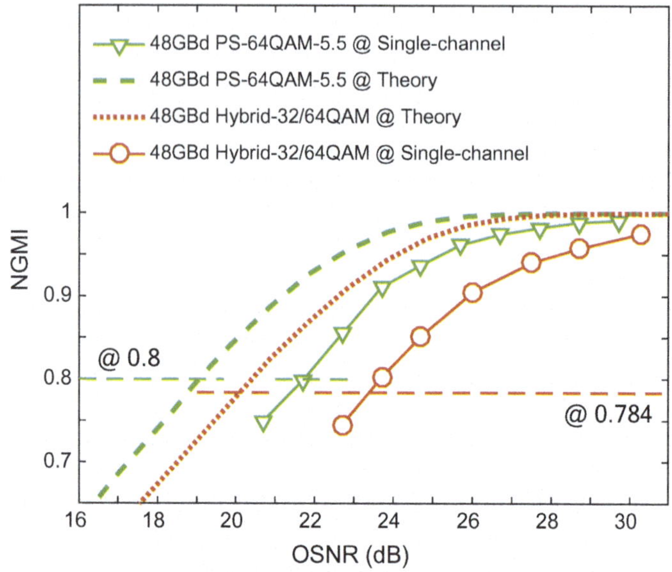

Fig. 7.15 Theoretical and experimental NGMI results versus BTB OSNR

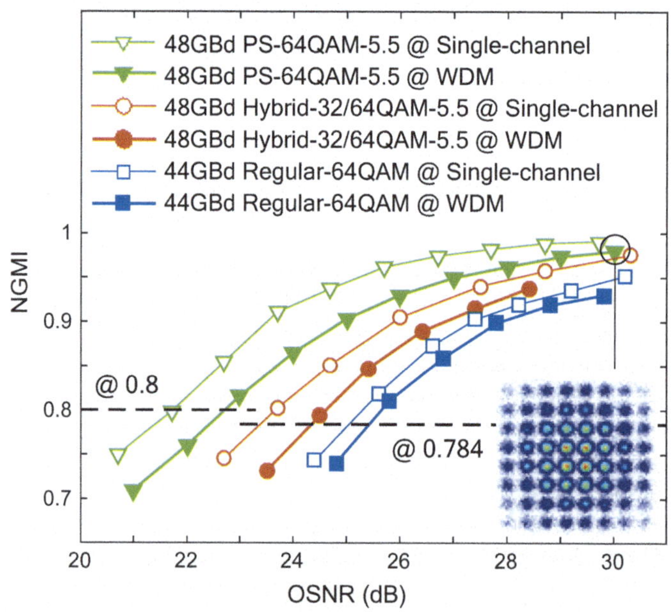

Fig. 7.16 NGMI results versus BTB OSNR for regular-64QAM, hybrid-32/64QAM, and PS-64QAM signals

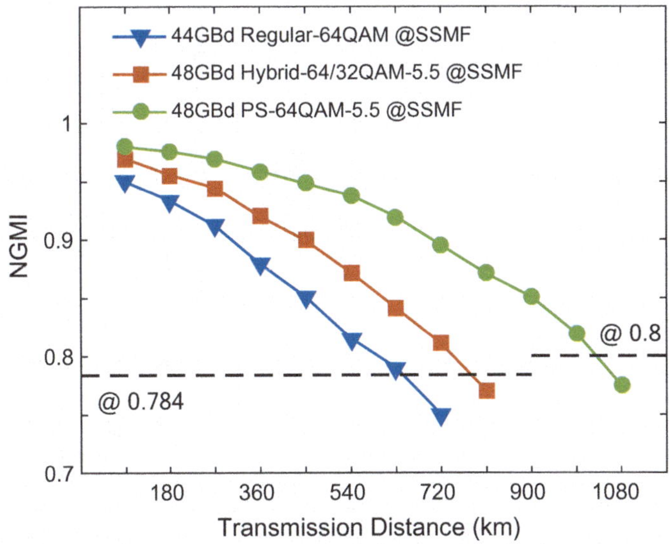

Fig. 7.17 NGMI versus SSMF transmission distance for channel-4 WDM regular-64QAM, hybrid-32/64QAM, and PS-64QAM signals

NGMI. For hybrid-32/64QAM with the same baud rate and bit rate as PS-64QAM, however, only 1.1-dB OSNR gain can be obtained compared with regular-64QAM.

Afterwards, we carry out an eight-channel 50-GHz WDM transmission experiment over several spans of 90-km EDFA-amplified standard single-mode fiber (SSMF) and several spans of 400-km ULAF (TeraWave-SLA + fiber, 122-μm2 average effective area) with backward-pumped Raman amplifier, respectively. The transmission results over SSMF and ULAF of the 4th WDM sub-channel is illustrated in Figs. 7.17 and 7.18a, respectively. It can be demonstrated that the SSMF transmission of WDM PM PS-64QAM signals with a net bit rate of 403-Gbit/s over up to 990-km can be fulfilled with measured NGMI above the 27.5% SD-FEC threshold of 0.8 NGMI, leading to 37.5% SSFM transmission distance improvement compared with hybrid-32/64QAM (720-km). Simultaneously, 403-Gbit/s WDM PS-64QAM signals can be transmitted over 3,600-km Raman-amplified ULAF, obtaining apparent transmission distance improvement (more than 250%) compared with EDFA-amplified SSMF. Moreover, WDM PS-64QAM signals can obtain 80% ULAF transmission distance improvement compared with regular-64QAM (2,000-km), while transmission over 2,400-km ULAF can be realized based on WDM hybrid-32/64QAM, leading to only 20% reach improvement compared with regular-64QAM. For the 50-GHz WDM system with 400G per channel, PS-64QAM has 50% transmission distance improvement compared with hybrid-32/64QAM. In addition, we have summarized the above experimental results in Table 7.1. Besides, Figs. 7.18b, c show the optical spectra of eight-channel WDM PS-64QAM signals before and after 3,600-km ULAF transmission, respectively.

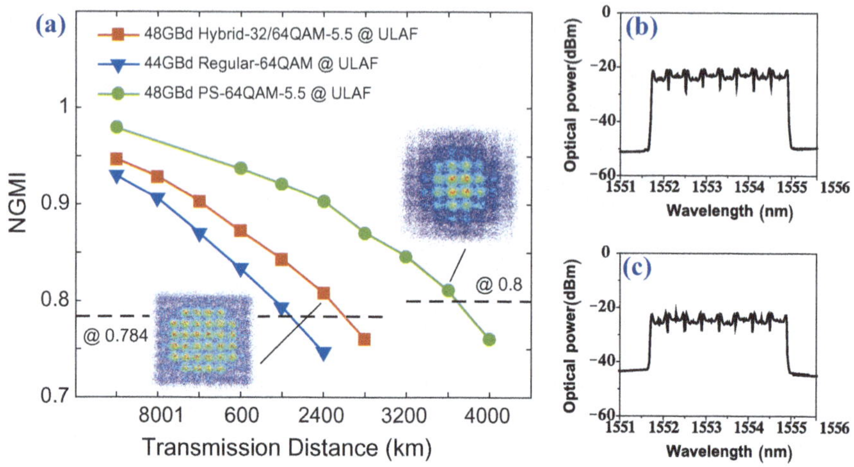

Fig. 7.18 **a** NGMI versus ULAF transmission distance for channel-4 WDM regular-64QAM, hybrid-32/64QAM, and PS-64QAM signals; optical spectra of eight-channel WDM PS-64QAM signals **b** before, and **c** after 3600-km ULAF transmission

Table 7.1 Experimental results of WDM regular-64QAM, hybrid-32/64QAM, and PS-64QAM

Modulation formats	Baud rate per channel (Gbaud)	SD-FEC overhead (%)	Net bit rate per channel (Gbit/s)	SE (b/s/Hz)	OSNR sensitivity (dB)	SSMF reach (km)	ULAF reach (km)
Regular-64QAM	44	30.9	403.3	8.06	25.3	630	2,000
Hybrid-32/64QAM-5.5	48	30.9	403.3	8.06	24.2	720	2,400
PS-64QAM-5.5	48	27.5	403.7	8.07	22.7	990	3,600

What is more, the NGMI results versus optical input power for the 4th sub-channel of WDM PS-64QAM signals after 3,600-km ULAF transmission is illustrated in Fig. 7.19 and the optimal input optical power per channel is about 0.5-dBm. When optical power is too low, the OSNR is not enough for reliable transmission, while if the power is too high, the photodiode will be saturated. The reliable input optical power range is around 0–1.5-dB for 3,600-km WDM PS-64QAM transmission. In addition, Fig. 7.20 shows that the measured NGMI values of all eight WDM channels after 3,600-km ULAF transmission are all above 0.8, which is 27.5% SD-FEC threshold. It can be concluded that the eight-channel WDM transmission with a net bit rate of 403-Gbit/s over 3,600-km Raman-amplified ULAF based on PM PS-64QAM can be fulfilled.

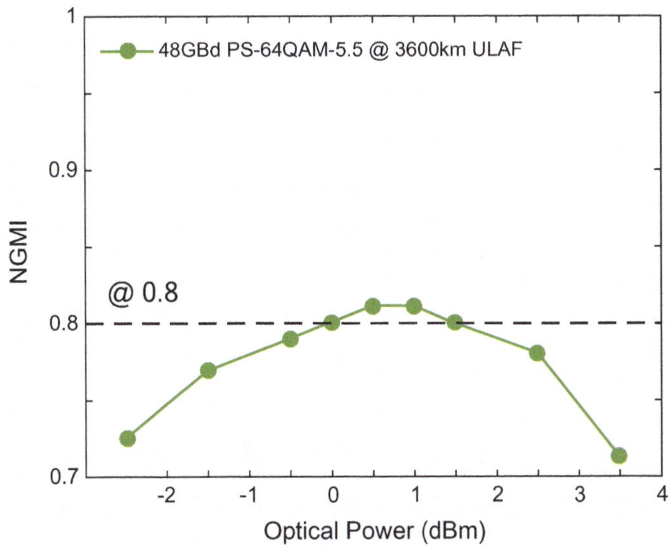

Fig. 7.19 NGMI versus optical power for channel-4 WDM PS-64QAM signals after 3,600-km ULAF transmission

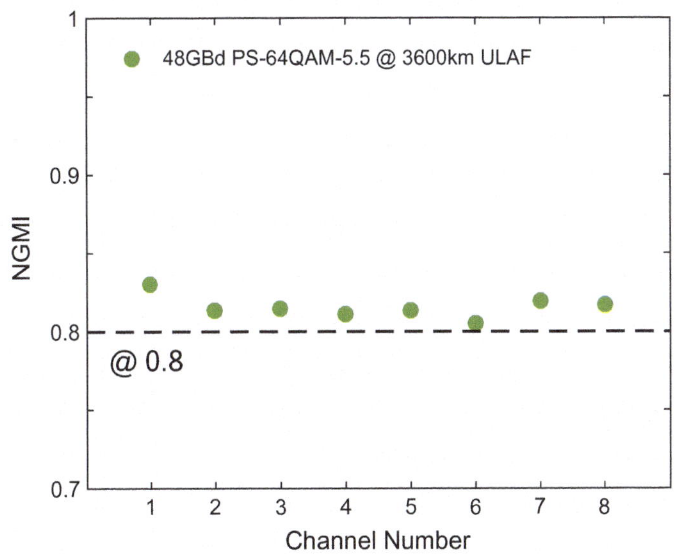

Fig. 7.20 NGMI results of all eight channels for PS-64QAM signals after 3,600-km ULAF transmission

7.7 Conclusion

The PS technology, pre-equalization, LUT-based pre-distortion, and ULAF with Raman amplification have allowed us to experimentally realize the eight-channel 50-GHz WDM coherent transmission over 3,600-km ULAF based on 403-Gbit/s net-bit-rate PM PS-64QAM at 48-Gbaud when considering the 27.5% SD-FEC threshold of 0.8 NGMI, which obtains 80% transmission distance improvement compared with the regular-64QAM and outperforms hybrid-32/64QAM by 50% reach improvement in the 400G per channel 50-GHz WDM system. The application of DM and Maxwell–Boltzmann distribution makes PS-QAM more power-efficient than hybrid-QAM and other modulation formats for high-speed and long-reach coherent optical transmission.

We have compared PS-64QAM with hybrid-32/64QAM signals with an entropy of 5.5-bit/symbol/polarization in 48-Gbaud 50-GHz WDM transmission. It can be concluded that PS-64QAM signals can obtain around 1.5-dB WDM sensitivity gain and around 50% ULAF transmission distance improvement compared with the hybrid-32/64QAM. In addition, the application of Raman-amplified ULAF offers apparent transmission distance improvement (more than 250%) compared with EDFA-amplified SSMF.

References

1. W. Idler, F. Buchali, L. Schmalen, et al., Field demonstration of 1 Tbit/s super-channel network using probabilistically shaped constellations, in *European Conference on Optical Communication, 2016*
2. J. Colin. Nokia's super-fast subsea data cable torpedos the competition. http://newatlas.com/record-fiber-optic-transmission-alcatel-nokia/45889, Accessed 13 Oct 2016
3. U. Wachsmann, R.F.H. Fischer, J.B. Huber, Multilevel codes: theoretical concepts and practical design rules. IEEE Trans. Inf. Theor. **45**(5), 1361–1391 (1999)
4. T. Fehenberger, LDPC coded modulation with probabilistic shaping for optical fiber systems, in *Optical Fiber Communications Conference and Exhibition*. IEEE, pp. 1–3 (2015)
5. P. Schulte et al., Constant composition distribution matching [J]. IEEE Trans. Inf. Theory **62**, 430 (2015)
6. L. Schmalen et al. *Probabilistic constellation shaping: challenges and opportunities for forward error correction*. OFC 2018, Paper M3C.1
7. T. Fehenberger et al., Sensitivity gains by mismatched probabilistic shaping for optical communication systems. IEEE Photon. Technol. Lett. **28**(7), 786–789 (2016)
8. F.R. Kschischang et al. Optimal nonuniform signaling for Gaussian channels. IEEE Trans. Inform. Theory **39**(3) (1993)
9. G. Böcherer, F. Steiner, P. Schulte, Bandwidth efficient and rate-matched low-density parity-check coded modulation. IEEE Trans. Commun. **63**(12), 4651–4665 (2015)
10. T. Fehenberger et al., On probabilistic shaping of quadrature amplitude modulation for the nonlinear fiber channel. J. Lightwave Technol. **34**(21), 5063–5073 (2016)
11. J. Cho et al. Trans-atlantic field trial using probabilistically shaped 64-QAM at high spectral efficiencies and single-carrier real-time 250- Gb/s 16-QAM. OFC 2017, Paper Th5B.3
12. M. Kong et al., High spectral efficiency 400 Gb/s transmission by different modulation formats and advanced DSP. J. Lightwave Technol. **37**(20), 5317–5325 (2019)

13. X. Zhou et al., High spectral efficiency 400 Gb/s transmission using PDM time-domain hybrid 32-64 QAM and training-assisted carrier recovery. J. Lightwave Technol. **31**(7), 999–1005 (2013)
14. X. Zhou et al., *4000 km transmission of 50 GHz spaced 10 × 494.85 Gb/s hybrid- 32–64 QAM using cascaded equalization and training-assisted phase recovery*. OFC-NFOEC 2012, paper PDP5C. 6

Chapter 8
High Baud Signal Transmission

8.1 Introduction

To increase bit rate per channel is the trend of optical fiber communication [1–10]. As shown in Fig. 8.1, the bit rate of an optical fiber communication system can be scaled from two aspects [11–21]. The first one is to increase the number of OFDM or Nyquist WDM carriers [11–16]. The second one is to increase the bit rate per channel, which can be achieved by increasing the symbol rate or spectral efficiency. The spectral efficiency here refers to the number of bits carried by each symbol. In order to increase the number of bits per symbol, 64QAM, and even modulation formats with higher order have been applied [5–10]. However, the required OSNR increases almost exponentially with the increase in the number of constellation points, as shown in Fig. 8.2. Compared with dual-polarization QPSK with 4 bits per symbol, the dual-polarization 16QAM with 8 bits per symbol incurs more than 6-dB OSNR sensitivity penalty [3, 4, 9, 13, 19–21]. To make matters worse, as the modulation order continues to increase, the sensitivity penalty will increase faster [1, 5]. Therefore, it is desirable to seek a better compromise between spectral efficiency and OSNR tolerance.

In order to cope with the exponential growth of Internet protocol (IP) traffic, 400 Gbit Ethernet (400 GbE) is expected to become the basis of the next-generation transmission standard, so the 400 Gbps wavelength division multiplexing (WDM) transmission per channel in the telecommunication standardization sector of the international telecommunications union (ITU-T) network has attracted many researchers [1–10]. On the network transmission line, one promising solution is 400 Gb/s per channel. Although the limits of optoelectronic devices and modules will be reached, the complexity of the system can be reduced [5–10]. The use of optical devices and the cost of transponders can be reduced by realizing the highest multiplexing rate and the minimum number of subcarriers in the electrical domain. There are two main ways for optical modulation. On the one hand, high-order modulation formats can

J. Yu and N. Chi, *Digital Signal Processing In High-Speed Optical Fiber Communication Principle and Application*,
https://doi.org/10.1007/978-981-15-3098-2_8

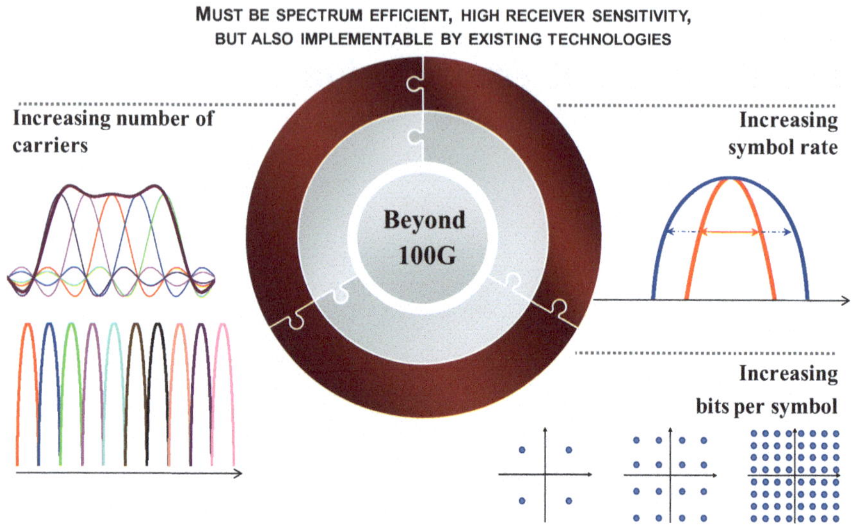

Fig. 8.1 Methods of scaling the bit rate of the optical signals

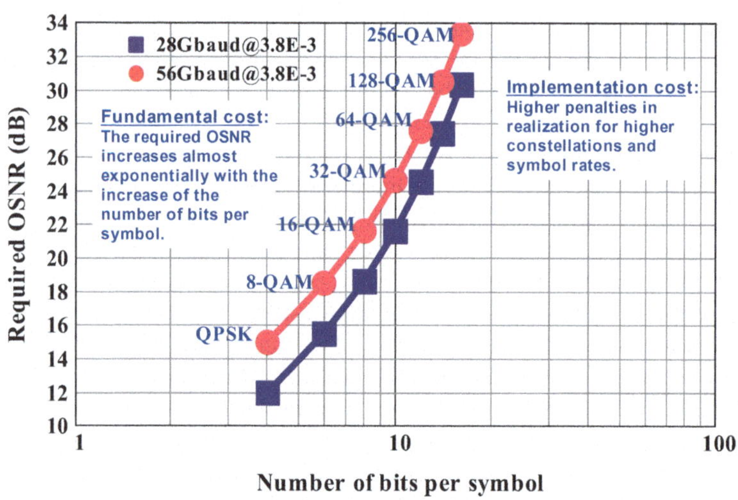

Fig. 8.2 Challenge in increasing constellation points and baud rate

be used, such as 16QAM and higher order modulation, which will require a higher optical signal-to-noise ratio (OSNR) as shown in Fig. 8.2. Also, the transmission distance is limited, meanwhile increasing the complexity of the system and the power consumption of the coherent receiver DSP module [5, 10]. On the other hand, a modulation scheme with a lower order but better noise resistance can be used, and the baud rate is improved, such as PMD-QPSK, which can realize long-distance transmission, and DSP processing is relatively simple [7–9]. We believe that in the future large-capacity optical transmission network, the transmission of single-carrier 400 Gbps per WDM channel by increasing the baud rate is a more promising and cost-effective way, and can provide a basis for a multi-carrier channel with terabit capacity. At present, the single-carrier 400 G submarine transmission system has realized 7200 km optical fiber link transmission, which creates the longest distance. In the cascaded 50 km fiber transmission experiment, the 16GAM modulation format is used for 64 Gbaud rate signal transmission, and the spectral efficiency can reach 5.33 b/s/Hz. In terrestrial transmission systems with 80–100 km cascading fiber, single-carrier 400 Gbaud transmission has created a distance of 4800 km [8]. The transmission experiment is based on 107 Gbaud QPSK with a spectral efficiency of 3.6 4bit/s/Hz. At the same time, the transmission experiment based on 110 Gbaud QPSK achieves a transmission distance of 3600 km, and the spectral efficiency is 4 bit/s/Hz [9]. In our research, we break the transmission distance record with the 128 Gbaud code rate and improved the capacity limit of the ETDM-based optical-electric transmission [22]. We successfully increased the rate of the single-carrier PDM-QPSK signal to 515.2 Gb/s, and transmitted 10,130 km in the link composed of 100 km cascading fiber [23]. We also used a filtered QPSK signal for single-carrier 128.8 Gbaud transmission experiments over 6,078 km in a 100 GHz network with a spectral efficiency of 5.152 b/s/Hz [23].

In order to achieve an increase for a single-channel data rate, the conventional method is to increase the symbol rate on a single carrier [1–8]. Electrical domain time-division multiplexing (ETDM) is an important scheme to increase the telecommunication number rate [2–8]. In the past few years, the optical communication industry has also made this technology a priority [3]. At present, advanced ETDM technology has been able to achieve transmission rates of more than 100 Gbaud. The literature in [2] describes the generation of the 107 Gbaud QPSK signal and carries out the transmission experiment of 4800 km distance. Afterwards, 107 Gbaud PMD-QPSK signal back-to-back transmission experiment proved the feasibility of this scheme [3]. In the previous research work, we sent a 128.8 Gb baud rate QPSK signal and transmitted it over long distances in 100 and 200 GHz terrestrial network cables [4]. Recent studies have shown that WDM transmission of 120 Gbaud PMD-16QAM optical signals using ETDM in a 150 GHz network can achieve a spectral efficiency of 5.33 b/s/Hz. At present, the data rate of single-carrier PDM-64QAM signals with ETDM reaches 1.08 Tbit/s [6]. High-rate signals are generated by combined multiplexing of multi-band signals [7], and digital band interleaving is also a technique that can be used to further increase the rate. Due to the high-rate signal generation technology, the Ethernet transmission rate standard is expected to exceed 1 Tb/s, and the number of channel subcarriers can be minimized, thereby reducing the use of optical devices and reducing the cost of transponders [2–8]. In Ref. [28], we generated and transmitted the 128 Gbaud PMD-QPSK signal based on the photoelectric method, which further improved the rate limit of ETDM. Thanks to the optical pre-emphasis at the transmitting end, the partial response (PR) at the receiving end, and maximum likelihood sequence estimation (MLSE) detection techniques, the narrowband filtering effect caused by the limited bandwidth of the optoelectronic device is reduced, thereby greatly improving the system performance. We successfully performed a sixteen-channel WDM transmission experiment on a 32 Gz optical network with a signal rate of 1,024 Gbps per channel (data rate 800 Gb/s + 20% FEC extra bit overhead), and only used EDFA for relay amplification over 320 km standard single-mode fiber transmission. The high-speed data interface achieves a spectral efficiency of 6.06 bit/s/Hz, which is the largest known ETDM 16QAM signal rate and has a relatively high spectral efficiency [22].

8.2 110-GBaud Polarization-Multiplexed QPSK Signal for 3000 Km Transmission

8.2.1 Experimental Setup

Figure 8.3 shows a schematic diagram of a twenty-channel, 100 GHz channel spacing, single-carrier 440 Gb/s transmission experiment. The transmitted signal is a 9QAM-like modulated signal with super-Nyquist filtering, based on a 110 Gbaud PMD-QPSK signal. The transmitter uses 20 external cavity lasers (ECLs) with a linewidth less than 100 kHz to generate signals with a spectral separation of 100 GHz. The output power of 14.5 dBm is divided into two groups, odd channels and even channels. The transmission of odd and even channels is multiplexed through a 200 GHz arrayed waveguide grating (AWG). After passing through the AWG, the signal is amplified by PM-EDFA to 23 dBm. The 110 Gbaud signals of I and Q are generated by the second-order ETDM. The signal source is a binary pseudo-random sequence (PRBS) with a word length of 215 − 1. The 4-way 13.75 Gbaud's signal is combined by the 4:1 and 2:1 electrical multiplexers. The 4:1 MUX is a 56 Gb/s 4:1 multiplexer and the 2:1 MUX is a 100 Gb/s 2:1 multiplexer. In the experiment, the 4:1 MUX achieved an output rate of 55 Gb/s, and the 2:1 MUX remained stable at an output rate of 110 Gb/s. Vpp of 4:1 MUX output is 500 mV, and Vpp of 2:1 MUX output is 300 mV. In order to recover the data, there is a 55-bit delay between the two 55 Gb/s signals and a

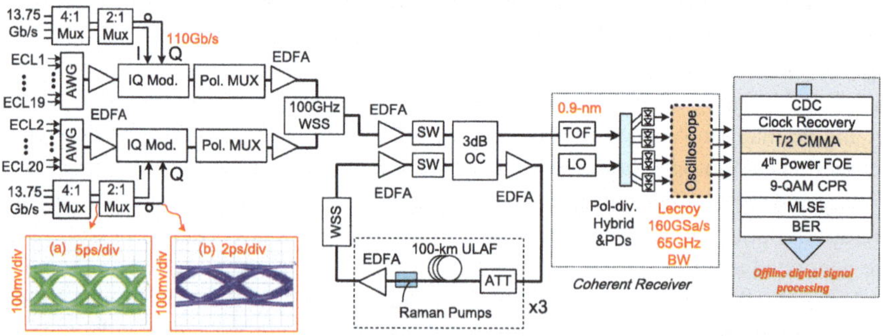

Fig. 8.3 Experimental setup

35-bit delay between the two 110 Gb/s signals. The odd channel and the even channel are modulated independently and are generated by a 110 Gbaud PRBS signal driven I/Q modulator. The I/Q modulator consists of a two-arm $LiNbO_3$ modulator with a 3-dB bandwidth of 33 GHz and is driven by a 110 GHz signal generated by a 2:1 MUX with an overall insertion loss of 27 dB. The subgraphs (a) and (b) of Fig. 8.3 are the OOK signal eye diagrams of 55 Gbaud and 110 Gbaud, respectively, with a vertical axis scale of 100 mV/div and horizontal axis scales of 5 ps/div and 2 ps/div, respectively. The I/Q modulator is biased to zero to modulate the QPSK signal. Each modulated signal is polarization-multiplexed by a polarization multiplexer consisting of a polarization-maintaining coupler (PM-OC), optical delay lines (D1 and D2), and a polarization beam combiner (PBC). The PM-OC is used to separate signals, optical delay lines are used to provide delays in excess of 100 symbol periods, and PBC is used to recombine signals. The signals of the odd and even channels are filtered in frequency domain to produce a signal similar to the 9QAM constellation modulation, and then combined via a programmable wavelength selective switch (WSS) with a fixed frequency spacing of 100 GHz and a 3 dB bandwidth of 94.8 GHz. After optical filtering by WSS, the signal spectrum becomes bandpass and the top becomes flat. WDM multiplexing enables super-Nyquist transmission because the channel bandwidth is less than the signal symbol rate.

In the experiment, the twenty-channel 100 GHz ultra-Nyquist WDM channel is shown in Fig. 8.4. The spectrum in the range of 1555.32–1559.64 nm is measured by back-to-back (BTB) experiments with a transmission signal rate of 8.8 Tb/s (20 × 440 Gb/s). The spectral efficiency of the single carrier 400 Gbaud PDM-QPSK signal is 4 bits/s/Hz after removing 7% of the hard decision (HD) FEC bit overhead.

The generated 20 × 440 Gb/s super-Nyquist signal has a spectral separation of 100 GHz and is transmitted in a fiber loop consisting of three 100-km ULAF fibers (21.1 ps/nm/km, 132 mm^2) with an average loss of 21 dB. As shown in Fig. 8.3, each segment of the ULAF is connected to a Raman amplifier and an EDFA to compensate for fiber loss, with one EDFA being accessed before the first segment of the fiber. When the pump source is 1450 nm, the switching Raman amplification has a gain of 9 dB on each segment of the fiber. The attenuator is used to adjust the signal transmission power of each sub-channel. The adjustable amplifier gain equalization

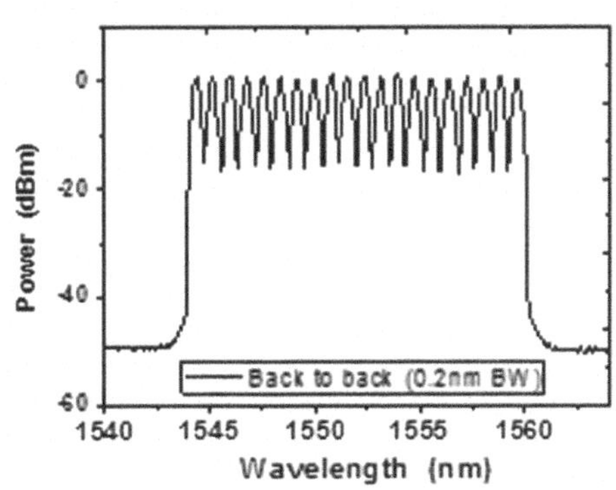

Fig. 8.4 20 × 440 Gb/s, 100 GHz interval super-Nyquist WDM channel back-to-back transmission

filter is used to eliminate the linear change of the gain and flatten its slope. The WSS acts as a band-pass filter to reduce the amplified spontaneous emission (ASE) noise. At the coherent receiver, the sub-channel is selected by a tunable optical filter with a 3 dB bandwidth of 0.9 nm. The local oscillator (LO) is an ECL laser with a linewidth of less than 100 kHz. A 90° polarization multiplexer is used for polarization and phase diversity coherent detection. The balanced detector at the receiver end has a bandwidth of 50 GHz. The sampling and digitization of the received signal are performed by a digital oscilloscope (LeCroy, LabMaster 10 Zi) with a sampling rate of 160 GSa/s and a bandwidth of 65 GHz.

The 110 Gbaud super-Nyquist signal is first oversampled 1.45 times by the ADC at a rate of 160 GHz, followed by offline digital signal processing. The sampled data is resampled at a sampling rate of 220 GSa/s and subjected to dispersion compensation, and then recovered by a multimode equalization (MMEQ) algorithm and a maximum likelihood sequence estimation (MLSE) algorithm.

8.2.2 *Experiment Results and Analysis*

Figure 8.5 shows the BER optical signal-to-noise ratio curve after the filter shaping of different bandwidths for 440 Gb/s channel back-to-back transmission. The figure also shows the theoretical BER/OSNR (0.1 nm resolution) curve of the 110 Gbaud PMD-QPSK signal under additive white Gaussian noise (AWGN). In the super-Nyquist WDM channel, the odd and even sub-channels for 440 Gb/s require an OSNR of 22.5 dB/0.1 nm to achieve a 3.8×10^{-3} error rate threshold (7% HD FEC), which costs more 0.5 dB OSNR penalty than single-channel transmission under the same conditions. Compared with the theoretical curve, the 110 Gbaud PMD-QPSK signal transmitted by a single channel in the experiment has a power penalty of 4 dB, which is caused by the low-pass filtering effect of the optoelectronic component. At the same time, we also confirmed that all sub-channels have similar error performance

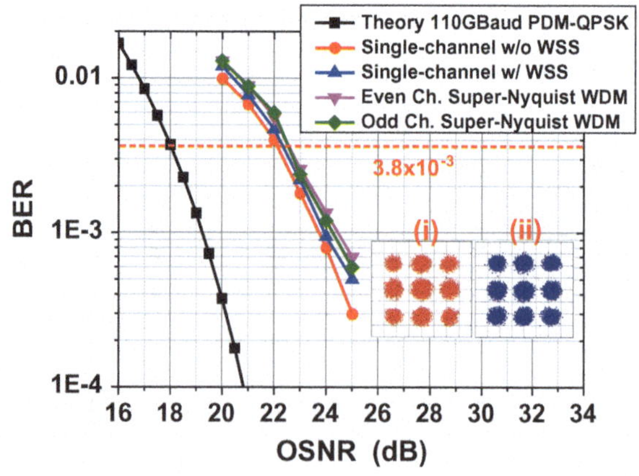

Fig. 8.5 BER-OSNR of back-to-back transmission of 110GBaud PDM-QPSK

except for the edge channel with 0.5 dB OSNR tolerance. Subgraphs (i) and (ii) show constellations of X and Y polarization directions when the recovered signal reaches zero error at high OSNR.

Figure 8.6 shows the error rate curve of channel 4 transmission over 3600 km ULAF fiber, and the sub-channel spectrum interval is 100 GHz. The input power is controlled by the attenuator at the input end of the fiber. The input optical power per channel is about 4 dBm. When there is a nonlinear effect, the increase of input power will lead to a decrease of BER. When the power per channel is less than 3 dB, the BER rapidly declines with the increase of OSNR.

Figure 8.7 shows the error rate versus transmission distance. The transmission distance ranges from 2400 to 4000 km, and the optical power is at the most advantageous. As can be seen from the figure, when the transmission distance is 3600 km, the BER of channel 4 is 3.3×10^{-3}, which is still lower than the 7% HD-FEC threshold [12]. It can be seen from sub-pictures (i) and (ii) of Fig. 8.7 that the received signal is a 9QAM constellation point in the X and Y polarization directions, and further third-order post-equalization can recover the signal therefrom.

Figure 8.8 shows the BER of all 400 Gb/s ultra-Nyquist WDM signals after 3600 km transmission, where the OSNR at the receiver is approximately 25 dB/0.1 nm. The results show that the BER of all channels after the 3600 km transmission fiber is lower than the 7% HD-FEC threshold of 3.8×10^{-3}.

8.2.3 Performance of the Multi-link System

1. Experimental schematic

We further study the high-speed super-Nyquist-filtered signal based on ETDM, and generate a 9QAM signal through super-Nyquist filter. This signal can be used for

Fig. 8.6 BER of fourth channel versus transmitted power of every channel over 3600 km transmission power

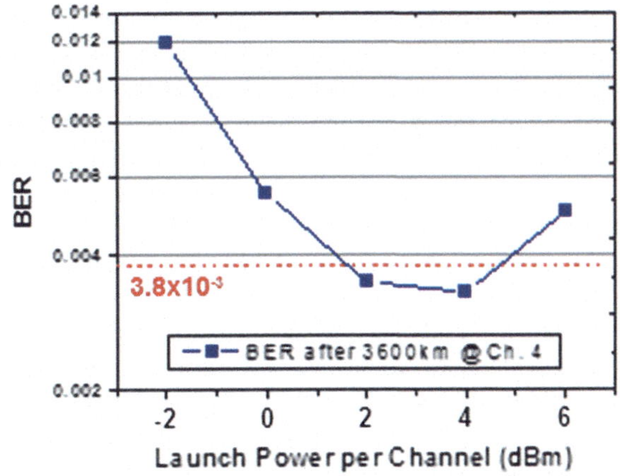

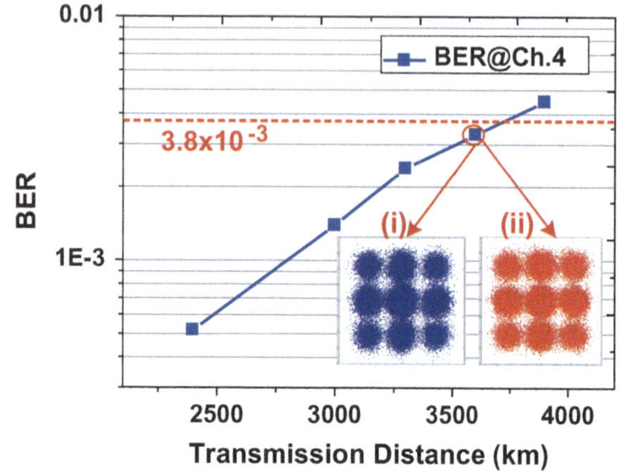

Fig. 8.7 BER of fourth channel versus transmission distance

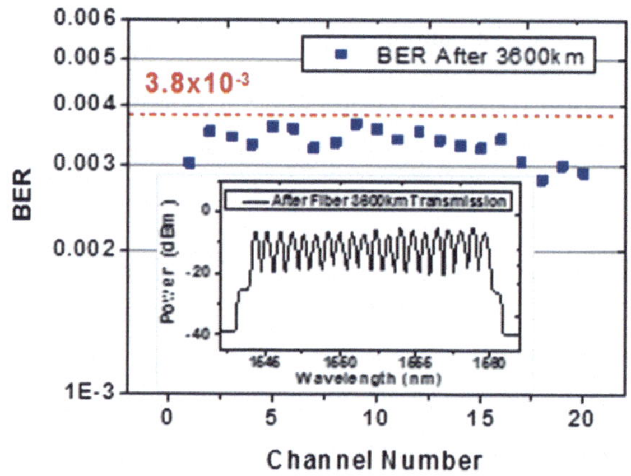

Fig. 8.8 BER of 3600 km transmission on twenty channels

multiple reusable optical add-drop multiplexers in long-distance optical fiber link transmission. The strong filtering effect caused by the (ROADM) node has better robustness. Later experiments will verify this feature.

As shown in Fig. 8.9, ten channels are set up with a channel spacing of 100 GHz. The single-carrier 440 Gb/s super-Nyquist signal is generated based on the 110 Gbaud PDM-QPSK signal. The transmission distance is 3000 km, and 10 ROADMs are used. The PDM-QPSK signal of the single-carrier 110Gbaud is generated in the same principle as shown in Fig. 8.3. At the transmitting end, 10 ECLs with an optical frequency interval of 100 GHz are divided into two groups and are separately modulated as odd channels and even channels. The QPSK signals of each channel are then filtered to produce 9QAM constellation points, and the signals are combined by a 100-GHz spaced programmable wavelength selective switch (WSS). Since the bandwidth of each channel is smaller than the signal rate, the signal after WDM multiplexing is super-Nyquist.

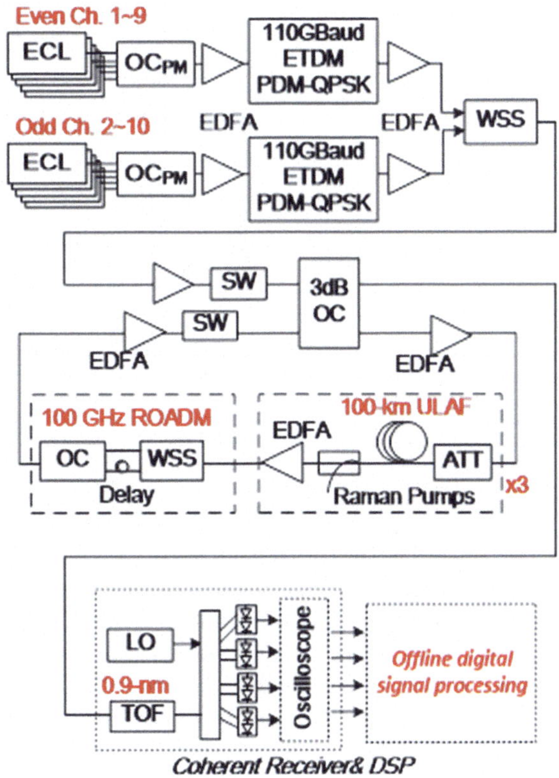

Fig. 8.9 Ten channel, 100 GHz interval 440 Gb/s single-carrier super-Nyquist signal transmitted through ROADM

The signal rate after multiplexing and combining is 440 Gb/s, which is transmitted in the fiber link ring. The transmission ring is composed of three segments of 100-km ULAF and 100 GHz interval ROADM. The signal is amplified by the EDFA before entering the first segment of the fiber. The power loss is compensated by the back-end Raman amplifier and EDFA after each segment of the fiber. The ULAF, Raman amplifier, and EDFA are the same as in Fig. 8.3. After transmission in the fiber loop, the receiver performs coherent detection and offline DSP recovery on the signal, and the process is the same as that shown in Fig. 8.3.

In order to analyze tolerance of the 110 Gbaud super-Nyquist signal to the cascading narrowband filtering effect and in-band crosstalk caused by the ROADM device, a WSS is used in the link loop to simulate a single ROADM node and study its filtering effect. The odd channel and the even channel are filtered by WSS and then input to port 1 and port 2, respectively, and then combined into one channel using 50:50 OC. In order to meet the needs of the receiver, there is a 5 ns delay between the two arms of the OC. The transmission characteristics of the two input ports are shown in Fig. 8.10, where the 3-dB bandwidth of the WSS is 94.8 GHz. If the ROADM model is simplified with only one internal filter, it can be considered that each 300 km loop transmission passes through a ROADM.

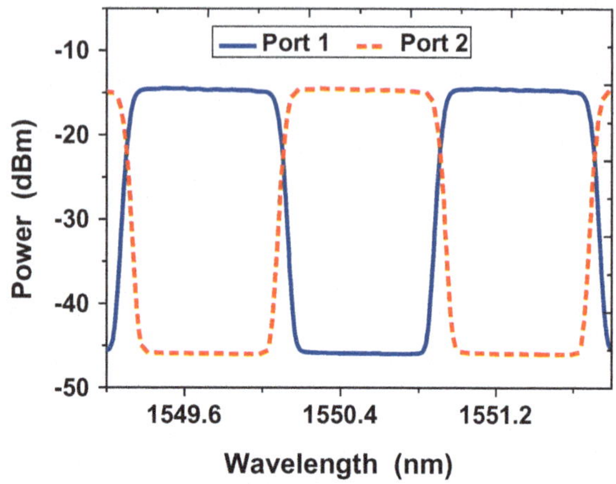

Fig. 8.10 Two-port 100 GHz frequency-domain WSS transmission curve in ROADM

2. Experiment results and analysis

We first measured the BER-OSNR curve (0.1 nm resolution) for back-to-back transmission and compared the error performance after filtering with different bandwidth filters. In the experiment, the 3-dB bandwidth of WSS is 94.8–80.6 GHz, and the spectrum of the 110-Gbaud QPSK signal of a single channel filtered by different bandwidth WSS is shown in Fig. 8.11.

Figure 8.12 shows the BTB BER-OSNR curve (0.1 nm resolution) of a 440 Gb/s WDM signal after different bandwidth filters. When the bandwidth of the WSS filter is 94.8 GHz, the BER of 3.8×10^{-3} (7% HD-FEC threshold) requires an OSNR of approximately 22.5 dB/0.1 nm, which costs 0.5-dB OSNR penalty higher than that for single-channel transmission. When the channel bandwidth is wider than 88.6 GHz, the OSNR penalty is negligible. For the 110 Gbaud super-Nyquist signal, even if the channel bandwidth is 84.6 GHz, the OSNR penalty is less than 0.8 dB to reach

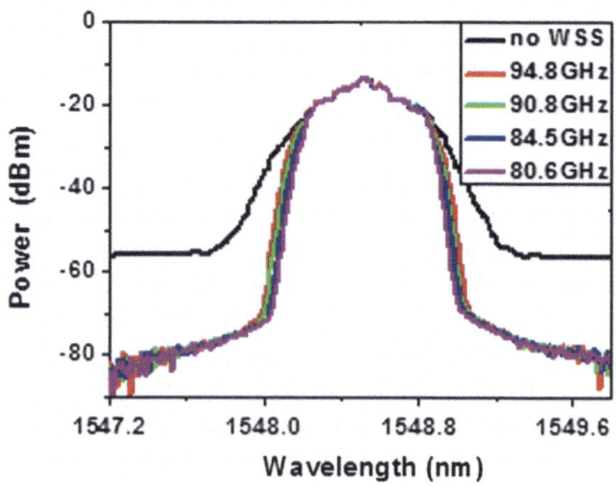

Fig. 8.11 Spectrum of 110-Gbaud QPSK signal after WSS filtering with different bandwidth

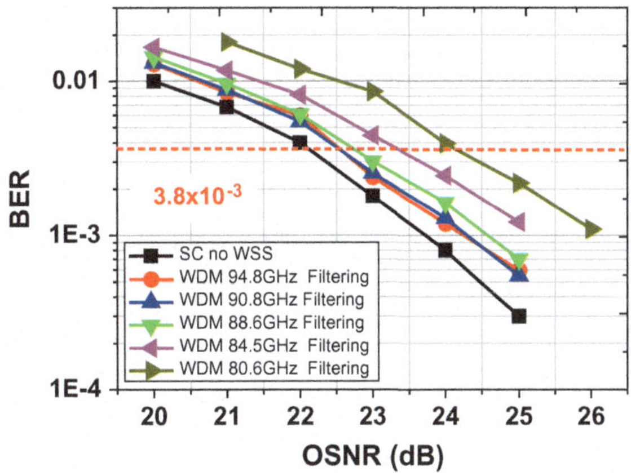

Fig. 8.12 BER-OSNR curve of 110-Gbaud QPSK signal after WSS filtering with different bandwidths

the BER threshold of 3.8×10^{-3}. Therefore, the super-Nyquist signal transmission scheme has a good tolerance for filtering effects.

The transfer function of the cascaded ROADM under different times of fiber loop and the corresponding 3-dB bandwidth are measured in the experiments, as shown in Fig. 8.13. Each ROADM has a bandwidth of 93.8 GHz, and the bandwidth of the cascaded ROADM system decreases as the loop time increases. After 10 loops transmission, it is equivalent to 10 ROADM cascades, and the system bandwidth is 84.5 GHz, which is reduced by about 10 GHz. The BER of channel 4 that varies with the transmission distance is shown in Fig. 8.14. The transmitted optical power is at the best advantage. The BER curves for ROADM-free and one ROADM per 300 km are measured, respectively. It can be seen from the figure that, in order to achieve the error threshold, the transmission distance limit without ROADM is 3600 km, and in the case of one ROADM per 300 km, the transmission of 3000 km can still reach the BER of 3.8×10^{-3}, which is lower than the threshold.

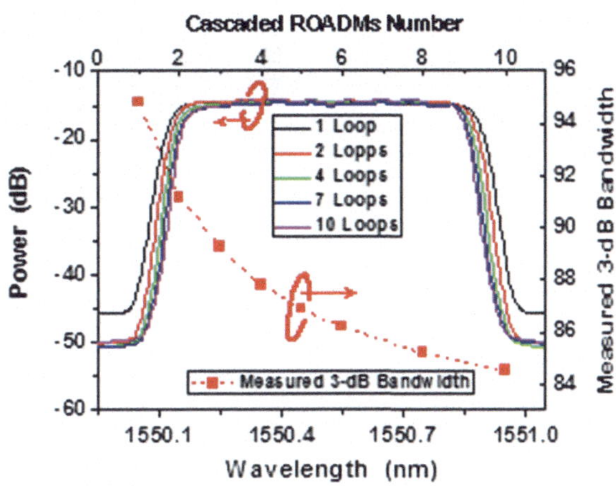

Fig. 8.13 Cascaded ROADM spectrum characteristics in the fiber loop and 3-dB bandwidth of different loop transmission times

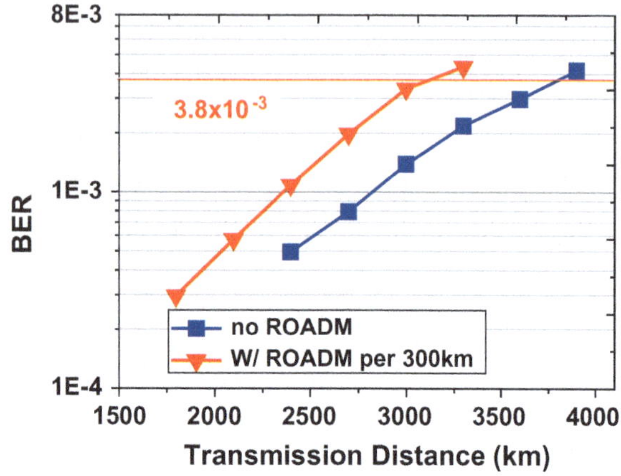

Fig. 8.14 BER of the fourth channel versus transmission distance with and without ROADM

We measured the error characteristics of a wavelength division multiplexed 400G super-Nyquist signal over 3,000 km optic fiber transmission and 10 ROADMs with 100 GHz channel spacing. The spectrum after transmission through the fiber and ROADM is shown in Fig. 8.15. The BER of the signal after the transmission is less than 3.8×10^{-3}. Note that each of our ROADM passes only one filter, which means that only the direct communication channel is considered, and the upper and lower voice channels are not considered. If it is considered, this signal needs to be filtered twice in each ROADM, and the number of ROADMs passed in this case will be greatly reduced.

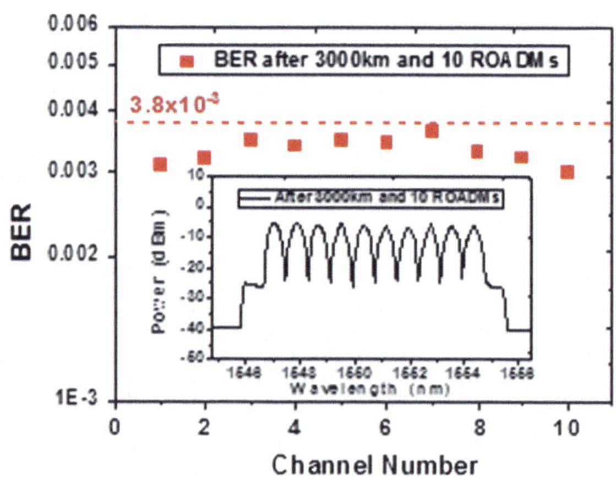

Fig. 8.15 BER of 10 wavelength signals over 3,000 km with 10 cascaded ROADMs

8.3 128 GBaud Polarization Multiplexing QPSK Signal Transmission Over 10,000 Km

8.3.1 *Experimental Setup*

Figure 8.16 shows the experimental schematic of the 515.2 Gb/s single-carrier transmission system, in which the transmission signal is generated based on the 128.8 Gbaud PDM-QPSK. At the transmitting end, 8 or 16 ECLs are combined to generate a WDM channel, wherein the PDM-QPSK signal requires 8 ECLs, and the 9QAM filtered signal is 16. The subcarrier frequency interval is 200 or 100 GHz, and the output power is 14.5 dBm. It is divided into two groups of the odd channel and even channels, which are respectively multiplexed and combined by two arrayed waveguide gratings (AWGs). The combined signals are amplified to 23 dBm by PM-EDFA. The in-phase and quadrature component of 128.8 Gbaud are generated in pairs by electrical domain time-division multiplexing (ETDM). After third-order multiplexing (2:1, 4:1, 2:1), 8.5Gbaud pseudo-random binary sequence (PRBS) signals are combined into a 128.8 Gbaud signal as a drive signal for the I/Q modulators of the odd channel and even channel. The subgraph (a) of Fig. 8.16 shows the eye diagram of the 128.8 Gbaud drive signal. The signals of different polarization modes after modulation are polarization-multiplexed using a polarization multiplexer, and finally the signals of the odd channel and the even channel are combined into a WDM signal by WSS.

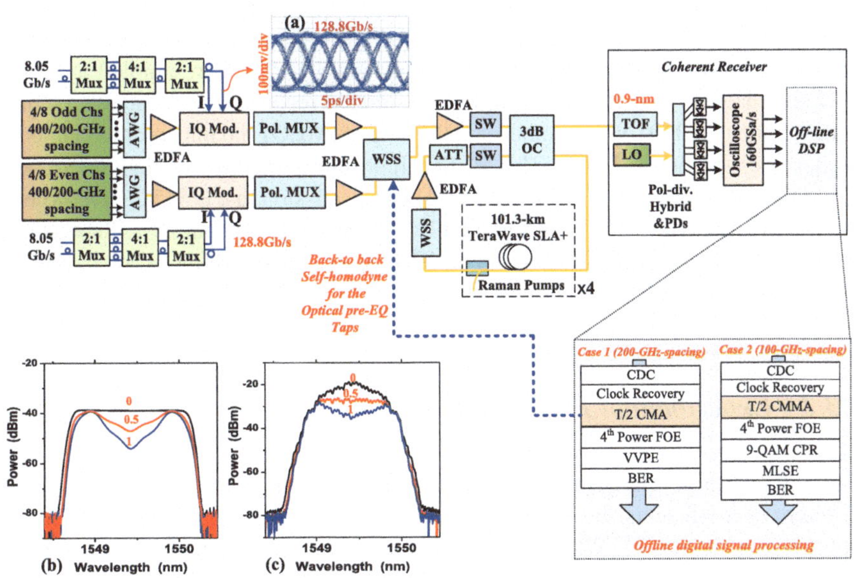

Fig. 8.16 Experimental setup

Two WDM channel arrangements were used in the experiment: (i) The space among eight channels is 200 GHz. Spectral efficiency is 2.64 b/s/Hz. And pre-equalization is used at the transmitter; (ii) 16 super-Nyquist channels are arranged at 100 GHz space. The transmitting end is not pre-equalized, and the spectral efficiency is 5.15 b/s/Hz. In the first mode, the pre-equalized WSS filter enhances the high-frequency components in the optical signal spectrum while attenuating the low-frequency components; the transmission function of the WSS is designed according to the channel response. By changing α,pre-equalization can be designed for different intensities. Figure 8.16b shows the filter frequency response for different values of α. When α is 0, there is no pre-equalization, and when α is 1, it is fully pre-equalized. Figure 8.16c shows the spectrum of the QPSK signal after WSS. In the second mode, the signals are not pre-equalized, and the transmission signals of the odd channel and the even channel are respectively subjected to frequency-domain filtering to form a 9QAM modulated signal, and then combined by WSS.

The total length of the fiber loop is 405.2 km, which is connected by 4 segments of 101.3 km TeraWave-SLA + fiber. The effective area is 122 μm^2, the attenuation coefficient is 0.185 dB/km, and the dispersion coefficient at 1550 nm wavelength is 20.0 ps/(nm-Km). Each segment of the fiber uses a 20-dB switching gain back-pumped Raman amplifier to compensate for power loss. The average power of the Raman pump is about 950 mW. An attenuator is used for each channel for power control, and WSS is used to compensate for the in-band gain unevenness of the filter. At the coherent receiver, a tunable optical filter (TOF) is used to select the channel. An ECL is used as the local oscillator (LO) with a linewidth of less than 100 kHz. It is an input of the polarization diversity 90° optical mixer and beats the received signal to produce the baseband signal. Then, the baseband signals are captured using a real-time oscilloscope at the sampling rate of 160 GSa/s. The sampled signal is processed offline by the DSP algorithm. The sampled data is resampled at 257.6 GSa/s, then dispersion compensated, and the signal is recovered by constant modulus equalization (CMEQ) or multimode equalization (MMEQ) and maximum likelihood sequence estimation (MLSE). The MLSE algorithm is used to equalize inter-code crosstalk (ISI) caused by filtering effects. The transmission rate is 515.2 Gb/s, where the information rate is 400 Gb/s and the FEC bit overhead is about 20%. In practice, in order to achieve the BER threshold of SD-FEC 2.4×10^{-2}, the FEC bit overhead is not less than 20%.

8.3.2 Experimental Result and Analysis

Figure 8.17a shows the results of pre-equalized BTB transmission using WSS with different α coefficients. It can be seen from the figure that for QPSK signals, the optimal α is about 0.75. When α is less than 0.75, ISI effect is large, and when the signal is too large, the signal is distorted, which will reduce the BER performance. For the 9QAM signal, the effect of α is not obvious, because MLSE is implemented to equalize the ISI. On the other hand, under SN-WDM conditions, 9QAM signals

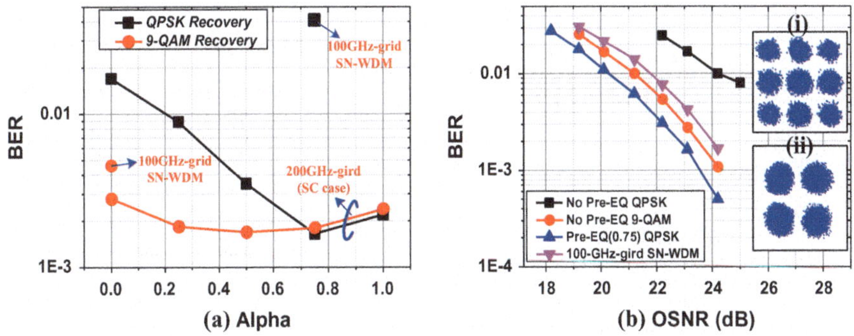

Fig. 8.17 **a** BER of QPSK and 9QAM signals for back-to-back transmission with different alpha values, **b** BER-OSNR of 128.8 GBaud PDM-QPSK signals for back-to-back transmission

can achieve good BER performance by minimizing inter-channel crosstalk (ICI) and receiver-side equalization even without pre-equalization. Figure 8.17b shows the BER-OSNR curve (0.1 nm resolution) of a single-carrier 128.8 Gbaud signal during BTB transmission, comparing different modulation formats and pre-equalization conditions. For a 200 GHz spaced channel, the OSNR is 19 dB to achieve a BER threshold of 2.4×10^{-2} after pre-equalization, and when the SN-WDM channel spacing is 100 GHz, the required OSNR increase by 20.5 dB. Experiments have shown that all other channels exhibit similar characteristics (see Fig. 8.19b), and the corresponding QPSK and 9QAM constellations are shown in subgraphs (i) and (ii) of Fig. 8.17.

Figure 8.18a shows the spectrum of the sixteen-channel SN-WDM signal before and after the 6,078 km fiber transmission, with a channel spacing of 100 GHz. Figure 8.18b shows the eight-channel WDM signal before and after 10,130 km fiber transmission. The spectrum of the channel is 200 GHz. Figure 8.16c shows the error rate-transmission distance curve of the seventh channel and the third channel transmitted by the experiment. The channel spacing is 200 and 100 GHz, and the transmission distance is 6,078 km and 10,130 km, respectively. The optical power is at the most advantage. After transmission, the BERs measured on channels 7 and 3 are 1.6×10^{-2} and 2.15×10^{-2}, respectively, both below the SD-FEC threshold. The sub-picture (i) of Fig. 8.18c is a constellation diagram of the third channel received signal.

The transmission error rate of the seventh channel varies with the optical power in Fig. 8.19a. It is found that the optimal transmitted optical power is about 0.75dBm. When the power is greater than 0.75 dBm, the enhancement of nonlinear effect makes BER performance decrease. Figure 8.19b shows the transmission error rate of each channel. It can be seen that the transmission BER of all channels is lower than the SD-FEC threshold of 20%.

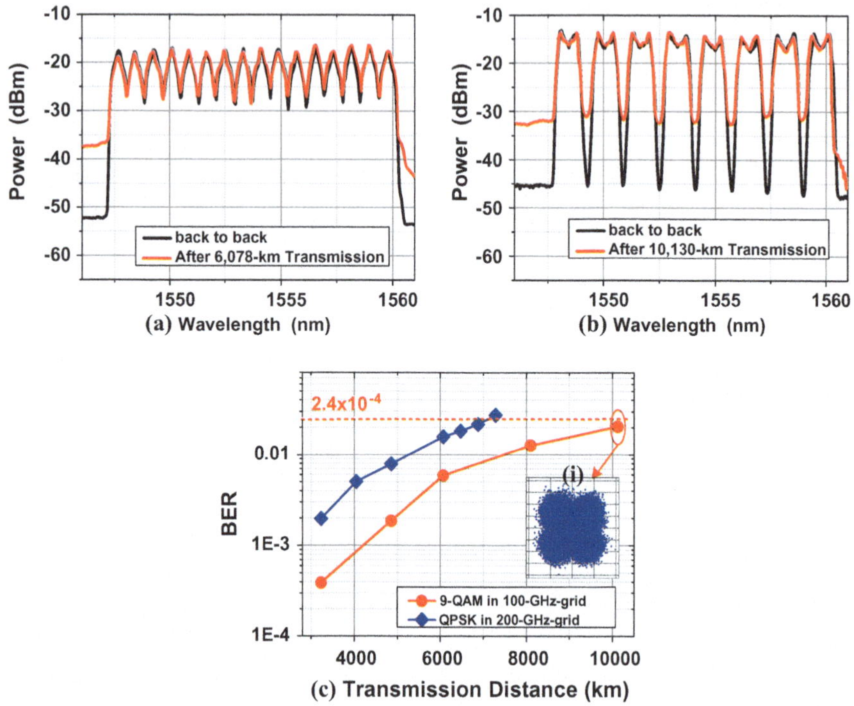

Fig. 8.18 **a** Spectrum before and after 6,078 km fiber over 16 SN-WDM channels at 100 GHz; **b** Spectrum before and after 10,130 km fiber at eight channels spaced 200 GHz; **c** BER versus transmission distance

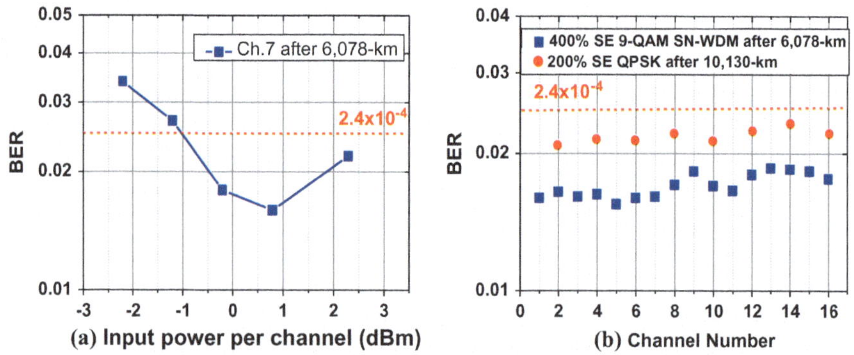

Fig. 8.19 **a** Transmission power-bit error rate curve after channel 7 over 6,078 km fiber; **b** BER of sixteen channels over 10,130 km fiber and BER of eight channels over 6,078 km fiber

8.4 Long-Distance Transmission of 128-GBaud Polarization-Multiplexed 16QAM Signal

8.4.1 Experimental Setup

The structure of the WDM transmission experiment is shown in Fig. 8.20. The transmitting side (Tx) uses 16 ECLs to generate sixteen sub-channels with an interval of 132 GHz, each channel carrying a 1,024-Gb/s (128-Gbaud) PDM-16QAM signal. The channel is divided into two groups of odd channels and even channels. Each group is polarization-multiplexed by a 1 × 8 polarization-preserving optical coupler (PM-OC). The signal after multiplexing and combining is amplified by the PM-EDFA to 23 dBm. After 2:1, 4:1 and 2:1 ETDM multiplexing, 16-Gbaud, 64-Gbaud, 128-Gbaud binary electrical signals are generated in turn, and two 128-Gbaud NRZ signals are combined to generate a 4-level signal. As the driving signal of the I/Q modulator, the signal power of one 128-Gbaud is attenuated by 6 dB. The ETDM signals are demodulated by setting the delay, and there is also a 43-bit delay between the I and Q paths to create independence. The odd channel and the even channel are independently modulated by an I/Q modulator, and the I/Q modulator is at a zero bias point when performing 16QAM modulation. The bandwidth limitation of the optoelectronic device will have a filtering effect, and so WSS is used for pre-equalization.

In order to reduce the influence of filtering effect and improve performance, the system adopts the optical frequency pre-equalization (OPEQ) at the transmitting side and the PR-MLSE algorithm at the receiving side. The signal processing at the receiving end is based on the adaptive equalization algorithm. The processing of the coherent receiver is the same as the QPSK signal of the previous 128-GBaud polarization multiplexing. The signal is recovered using the cascaded multimode algorithm (CMMA) and the DD-LMS algorithm for equalization and polarization demultiplexing. After carrier recovery, the PR filter and the 4-state MLSE are used to further compensate for residual filtering effects and improve signal quality. Both the WSS transfer function and the optimal OPEQ parameters are determined based on the receiver's channel estimate. Figure 8.20a is the frequency-domain response of the

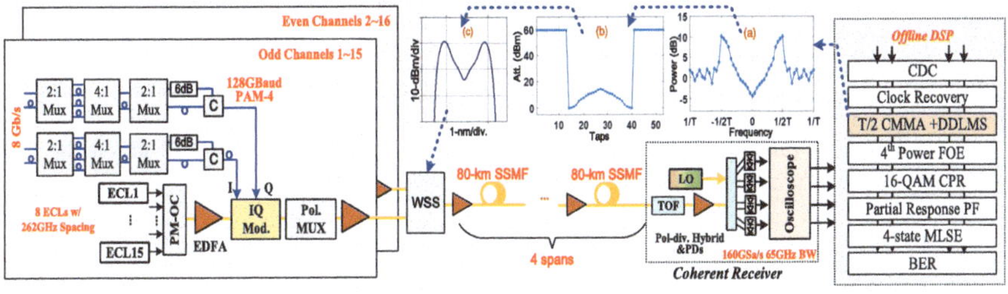

Fig. 8.20 Experimental setup

adaptive equalizer, which is the inverse function of the overall transmission channel response. According to the obtained channel response, the WSS can be programmed to attenuate differently with different frequency components, as shown in Fig. 8.20b. Figure 8.20c shows the output of WSS when matching the channel function. The signals of all channels pass OPEQ before transmission, and the transmission link consists of four 80-km SSMFs with a loss factor of 0.2 dB/km and a dispersion coefficient of 17.0 ps/(nm-km) at 1550 nm.

In the experiment, the transmitting side OPEQ and the receiving side PR-MLSE are used for signal processing, and the transmitting side adjusts the pre-equalization intensity by the α coefficient; the WSS transmission function becomes

$$H(f_{\text{WSS}}) = \alpha H(f_{\text{FIR}}) \text{ (in dB)}$$

where $H(f_{\text{WSS}})$ and $H(f_{\text{FIR}})$ represent the frequency response of the WSS and the adaptive FIR filter, respectively. At the receiving side, the response of the PR equalizer is controlled by the tap coefficient β, and its transfer function can be expressed as $H(z) = 1 + \beta z - 1$. By adjusting the tap coefficient β ($0 \leq \beta \leq 1$), the PR equalizer and the next 4-level MLSE module with a stored depth are optimally achieved. In the literature [5], when β is 0, it corresponds to the case of hard decision decoding, and MLSE has no memory depth. When β is not 0, the PR equalizer has a memory depth, and MLSE performs a soft decision. when β is 1, the PR is a double binary partial response. Figure 8.19a shows the BER performance of the system with different combinations of alpha and beta parameters. As can be seen from the figure, the optimal (α, β) region is around (0.8, 0.6).

8.4.2 Experimental Results and Analysis

Figure 8.21b shows the BER-OSNR curve (0.1 nm resolution) for back-to-back transmission of 128 Gbaud PDM-16QAM signals. As can be seen from the figure, the error performance has been greatly improved with OPEQ. With the PR-MLSE,

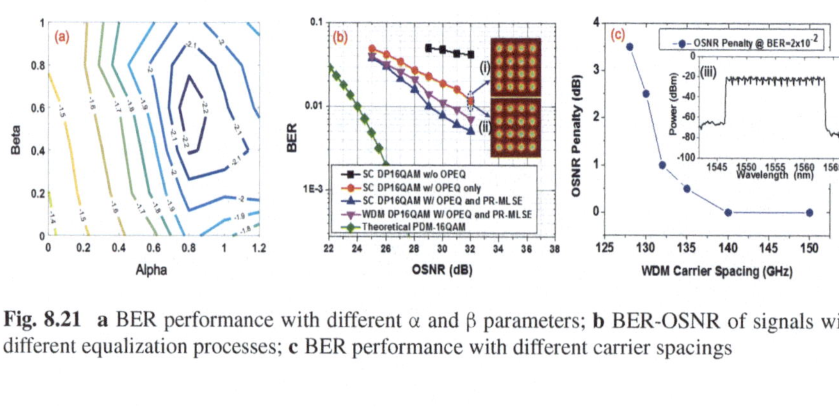

Fig. 8.21 **a** BER performance with different α and β parameters; **b** BER-OSNR of signals with different equalization processes; **c** BER performance with different carrier spacings

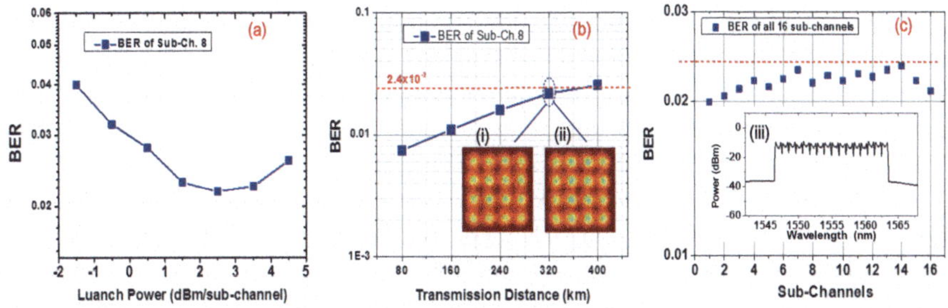

Fig. 8.22 **a** BER of channel 8 versus transmit power over 320 km; **b** BER of channel 8 versus transmission distance; **c** BER performance of 16 WDM channels over 320 km

the system BER performance can be improved by 2 dB to 2×10^{-2}. For WDM channels, the OSNR is 28 dB/0.1 nm to achieve BER of 2.4×10^{-2}. Subgraphs (i) and (ii) of Figs. 8.19 are received signal constellations in the X and Y polarization directions with an OSNR of 32 dB/0.1 nm. The experiment also compares the OSNR cost under different WDM channel spacings when the BER is 2×10^{-2}. When the subcarrier spacing is 132 GHz, the OSNR cost does not exceed 1 dB. But when the subcarrier spacing is less than 130 GHz, the OSNR cost is large to achieve the required BER, which is about 2.5 dB. Other channels exhibit similar characteristics, as shown in Figs. 8.21b, c.

Figure 8.22a–c shows the experimental results of sixteen-channel 1,024 Gb/s WDM signal transmission with a channel spacing of 132 GHz and a transmission fiber length of 320 km. Figure 8.22a shows the transmit power-bit error rate curve of the intermediate channel (channel 8). It can be found that the optimal transmit power is 2.5 dBm/channel. Figure 8.22b shows the BER of different distances is transmitted on the optimal power, and a constellation after 320 km is also displayed. Figure 8.22c shows the BER of a 320 km fiber transmitted on each channel. It can be seen that the BER of all channels is below the threshold of 2.4×10^{-2}.

8.5 Conclusion

We experimentally study 110 Gbaud's ultra-Nyquist high-rate coherent system, and successfully transmit twenty-channel super-Nyquist filtered 9QAM signals on a single carrier. The data rate is 440 Gb/s, channel spacing is 100 GHz, and transmission fiber length is 3600 km. The spectral efficiency has reached 4b/s/Hz (with 7% HD-FEC bit overhead removed). At the same time, the 9QAM transmission scheme with good tolerance to filtering effect is also studied. Using this scheme, a single-carrier ten-channel ETDM multiplexed 110 Gbaud QPSK signal is successfully transmitted on a 3000 km fiber link with a channel spacing of 100 GHz with 10 cascaded ROADMs. The receiving end uses coherent detection for demodulation. For a 400 G system, the data rate of 110 Gbaud allows for a 10% error correction bit overhead.

Since the 10% FEC bit overhead is not the actual standard used, we used 7% FEC bit overhead and 5% protocol bit overhead standard in the result analysis. In the higher rate transmission of 112 Gbaud, the requirements for optoelectronic devices will also be higher. The experimental results show that the single-carrier 400 G transmission technology based on PDM-QPSK is a promising solution.

In 512.5 Gb/s PDM-QPSK/9QAM single-carrier transmission, PDM-QPSK and 9QAM transmit 10,130 km and 6,078 km TeraWave SLA + fiber, respectively, which is known as the first single-carrier 400 G signal experiments with exceeding 10,000 km on land. In another single-carrier 128.8 Gbaud 9QAM transmission scheme, the signal is transmitted over 6,078 km in a 100 GHz spaced channel, and the spectral efficiency-distance product exceeds 30,000 (b/s/Hz) km.

We studied the generation of 128-GBaud ETDM PDM-16QAM signals and WDM transmission. Due to the use of OPEQ at the transmitting end and the use of the PR-MLSE processing algorithm at the receiving end, sixteen channels successfully transmitted 320 km SSMF in the channels of 132 GHz interval. The data rate reaches 1,024 Gb/s (800 Gb/s data rate + 20% FEC bit overhead) per channel, and the spectral efficiency reaches 6.06 b/s/Hz. This is the highest known ETDM 16QAM signal rate with high spectral efficiency.

References

1. X. Zhou, L.E. Nelson, P. Magill, B. Zhu, and D. W. Peckham, "8x450-Gb/s,50-GHz-spaced,PDM-32QAM transmission over 400 km and one 50 GHz-grid ROADM," in Proc. OFC 2011, paper PDPB3
2. H. Zhang, J.-X. Cai, H. G. Batshon, M. V. Mazurczyk, O. Sinkin, D. G. Foursa, A. Pilipetskii, G. Mohs, and N. S. Bergano, "200 Gb/s and dual wavelength 400 Gb/s transmission over transpacific distance at 6.0 b/s/Hz spectral efficiency," in Proc. OFC 2013, Paper PDP5A.6
3. J. Yu, J. Zhang, Z. Dong, Z. Jia, H.-C. Chien, Y. Cai, X. Xiao, X. Li, Transmission of 8 × 480-Gb/s super-Nyquist-filtering 9-QAM-like signal at 100 GHz-grid over 5000-km SMF-28 and twenty-five 100 GHz-grid ROADMs. Opt. Express **21**, 15686–15691 (2013)
4. J. Zhang, J. Yu and N. Chi., "Generation and transmission of 512-Gb/s quad-carrier digital super-Nyquist spectral shaped signal," Opt. Express 21, 31212–31217 (2013)
5. P. J. Winzer, A. H. Gnauck, S. Chandrasekhar, S. Draving, J. Evangelista, and B. Zhu, "Generation and 1, 200-km transmission of 448-Gb/s ETDM 56-Gbaud PDM 16-QAM using a single I/Q modulator," in Proc. ECOC 2010, paper PDP 2.2
6. J. Renaudier, W. Idler, O. Bertran-Pardo, F. Buchali, K. Schuh, H. Mardoyan, P. Tran, G. Charlet, S. Bigo, Long-Haul WDM Transmission of 448-Gb/s Polarization-Division-Multiplexed 16-ary Quadrature-Amplitude Modulation using Coherent Detection. IET Elec. Letters **47**(17), 973–975 (2011)
7. G. Raybon, A. Adamiecki, P. J. Winzer, C. Xie, A. Konczykowska, F. Jorge, J.-Y.Dupuy, L. L. Buhl, S. Chandrasekhar, S. Draving, M. Grove, K. Rush, B. Zhu, and D. W. Peckham, "All-ETDM 107-Gbaud (214-Gb/s) single-polarization QPSK transmitter and coherent receiver," in Proc. ECOC 2012, Paper We3 A.2
8. G. Raybon, A. Adamiecki, P. Winzer, C. Xie, A. Konczykowska, F. Jorge, J. Dupuy, L. L. Buhl, C. Sethumadhavan, S. Draving, M. Grove, and K. Rush, "Single-carrier 400G interface and 10-channel WDM transmission over 4800 km using all-ETDM 107-Gbaud PDM-QPSK," in Proc. OFC 2013, PDP5A.5

9. J. Zhang, Z. Dong, H. Chien, Z. Jia, Y. Xia, and Y. Chen, "Transmission of 20 × 440-Gb/s Super-Nyquist-Filtered Signals over 3600 km based on Single-Carrier 110-GBaud PDM QPSK with 100-GHz Grid," in Proc. OFC 2014, paper Th5B.3
10. R. Rios-Müller, J. Renaudier, P. Brindel, A. Ghazisaeidi, I. Fernandez, P. Tran, C. Simonneau, L. Schmalen, G. Charlet, Spectrally-Efficient 400-Gb/s Single Carrier Transport Over 7 200 km. J. Lightwave Technol. **33**, 1402–1407 (2015)
11. X. Liu, S. Chandrasekhar, B. Zhu, P. J. Winzer, A. H. Gnauck, and D. W. Peckham, "Transmission of a 448-Gb/s reduced-guard-interval CO-OFDM signal with a 60-GHz optical bandwidth over 2000 km of ULAF and five 80-GHz-grid ROADMs," Proc. OFC/NFOEC 2010, San Diego, California, paper PDPC2
12. J. Zhang, Z. Dong, H. Chien, Z. Jia, Y. Xia, and Y. Chen, "Transmission of 20 × 440-Gb/s Super-Nyquist-Filtered Signals over 3600 km based on Single-Carrier 110-GBaud PDM QPSK with 100-GHz Grid," in OFC 2014, paper Th5B.3
13. J. Zhang, J. Yu, N. Chi, Z. Dong, J. Yu, X. Li, L. Tao, Y. Shao, Multi-modulus blind equalizations for coherent quadrature duobinary spectrum shaped PM-QPSK digital signal processing. J. Lightwave Technol. **31**, 1073–1078 (2013)
14. J. Zhang, B. Huang, X. Li, Improved quadrature duobinary system performance using multi-modulus equalization. Photonic Technology Letters **25**(16), 1630–1633 (2013)
15. R.-J. Essiambre, G. Kramer, P.J. Winzer, G.J. Foschini, B. Goebel, Capacity Limits of Optical Fiber Networks. J. Lightwave Technol. **28**, 662–701 (2010)
16. A. Sano, H. Masuda, E. Yoshida, T. Kobayashi, E. Yamada, Y. Miyamoto, F. Inuzuka, Y. Hibino, Y. Takatori, K. Hagimoto, T. Yamada, and Y. Sakamaki, "30x100-Gb/s all-optical OFDM transmission over 1300 km SMF with 10 ROADM nodes," ECOC 2007, PDS 1.7 (2007)
17. V. Sleiffer, D. van den Borne, V. Veljanovski, M. Kuschnerov, M. Hirano, Y. Yamamoto, T. Sasaki, S. L. Jansen, and H. Waardt, de, "Transmission of 448-Gb/s dual-carrier POLMUX-16QAM over 1230 km with 5 flexi-grid ROADM passes," inOptical Fiber Communication Conference, OSA Technical Digest (Optical Society of America, 2012), paper OW4C.3
18. G. Raybon, A. Adamiecki, P. J. Winzer, M. Montoliu, S. Randel, A. Umbach, M. Margraf, J. Stephan, S. Draving, M. Grove, and K. Rush, "All ETDM 107-Gbaud PDM-16QAM (856-Gb/s) transmitter and coherent receiver," in Proc. ECOC 2013, Paper PD 2.D.3
19. J. Zhang, J. Yu, N. Chi, H.-C. Chien, Time-domain digital pre-equalization for band-limited signals based on receiver-side adaptive equalizers. Opt. Express **22**, 20515–20529 (2014)
20. J. Li, E. Tipsuwannakul, T. Eriksson, M. Karlsson, P.A. Andrekson, Approaching Nyquist Limit in WDM Systems by Low-Complexity Receiver-Side Duobinary Shaping. J. Lightwave Technol. **30**, 1664–1676 (2012)
21. H. Chien, J. Yu, Z. Jia, Z. Dong, X. Xiao, Performance Assessment of Noise-Suppressed Nyquist-WDM for Terabit Superchannel Transmission. J. Lightwave Technol. **30**, 3965–3971 (2012)
22. J. Zhang, J. Yu, and H. Chien, "WDM Transmission of 16-Channel Single-Carrier 128-GBaud PDM-16QAM signals with 6.06 b/s/Hz SE," in *Optical Fiber Communication Conference*, OSA Technical Digest (online) (Optical Society of America, 2017), paper Tu2E.5
23. J. Zhang, J. Yu, B. Zhu, F. Li, H.-C. Chien, Z. Jia, Y. Cai, X. Li, X. Xiao, Y. Fang, Y. Wang, Transmission of single-carrier 400G signals (515.2-Gb/s) based on 128.8-GBaud PDM QPSK over 10,130- and 6,078 km terrestrial fiber links. Opt. Express **23**, 16540–16545 (2015)

Chapter 9
Advanced Modulation Code Optical Signal Transmission Technology

9.1 Introduction

The single-carrier 400 G transmission network (SC-400 G), as a solution for future metropolitan area networks (metro) and data center networks (DCI), is expected to quadruple the network capacity while reducing the cost per bit [1, 2]. Because of the many options available for the 400 G network, the optical internetworking forum (OIF) members are considering developing detailed SC-400 G network standards to meet the availability and manageability requirements of existing optical transceiver equipment. The current alternative modulation formats are PM-16QAM, PM-32QAM, and PM-64QAM, with frequency intervals of 50 GHz, 62.5 GHz, and 75 GHz, respectively [3]. Among them, PM-256QAM/37.5 GHz is the most promising, because it can make use of the existing components of the 100 G PM-QPSK network, but some engineering problems need to be solved [3]. Table 9.1 summarizes the current SC-400 G network potential solutions. In this chapter, we propose and experimentally verify the generation of 34 Gbaud PM-256QAM, and consider the encoding overhead and bandwidth limitation. In this scheme, the coding overhead of soft decision forward error correction (SD-FEC) code is up to 30%, can obtain more coding gains and coding strategies [4], and can still effectively utilize the 37.5 GHz channel interval.

Table 9.1 Alternative solutions for a single-carrier 400 G network

Channel spacing (GHz)		75	62.5	50	37.5
Net bit rate	PM-16QAM	400			
	PM-32QAM		400		
	PM-64QAM			400	
	PM-256QAM				400

J. Yu and N. Chi, *Digital Signal Processing In High-Speed Optical Fiber Communication Principle and Application*,
https://doi.org/10.1007/978-981-15-3098-2_9

9.2 Trade-Offs Between Coding Overhead and Bandwidth Limitations

Figure 9.1 shows theoretical curves of required optical signal-to-noise ratio (ROSNR) and modulation efficiency for a single-carrier 400 G signal with 15 and 25% SD-FEC coding overhead. For example, the modulation efficiency of PM-256QAM signal is 16 bits/symbol. Considering 15 and 25% of SD-FEC coding cost, the symbol rate of 400 G PM-256QAM signal is about 30 and 32.75 Gbaud. The added coding cost can obtain 1.6 dB ROSNR gain at a 400 G net bit rate. Here, under the 15% and 25% SD-FEC coding overhead, the Q^2 factors are 6.34 and 4.95 dBQ, respectively [5]. It can be seen from Fig. 9.1 that when the modulation efficiency increases, the ROSNR gain will also increase, but the coding overhead will affect the signal bandwidth, and the ROSNR gain brought by the coding will be offset by the cost caused by the bandwidth limitation effect. Therefore, for a single-carrier 400G signal, the trade-off between bandwidth-limited effect and coding gain needs to be considered. For example, 25% SD-FEC encoding overhead is not suitable for SC-400 G PM-128QAM signal (37.4 Gbaud) with a channel interval of 37.5 GHz, and other modulation formats can meet the requirements (as shown in Table 9.1).

Figure 9.2 illustrates the 34-Gbaud PM-256QAM signal generation scheme in a back-to-back situation, with a total transmission rate of 544 Gbps. This scheme can be applied in four user channels, each with a channel rate of 104.79 Gbps and a coding overhead of 30%, of which 25% is for SD-FEC coding. The transmitter (Tx) part includes an external cavity laser (ECL1) with a working wavelength of 1550.3 nm and a linewidth of 400 Hz, a 26 GHz I/Q modulator, two 6 dB attenuators, a sampling rate of 87.04 GSa/s, a digital-to-analog converter (DAC) with an analog bandwidth of 20 GHz, and a polarization multiplexer (pol-mux). The receiver (Rx) section includes 2 EDFA amplifiers, 1 integrated coherent receiver (ICR), 1 40-GHz balanced photodetector, 1 analog-to-digital converter (ADC) with a sampling rate of

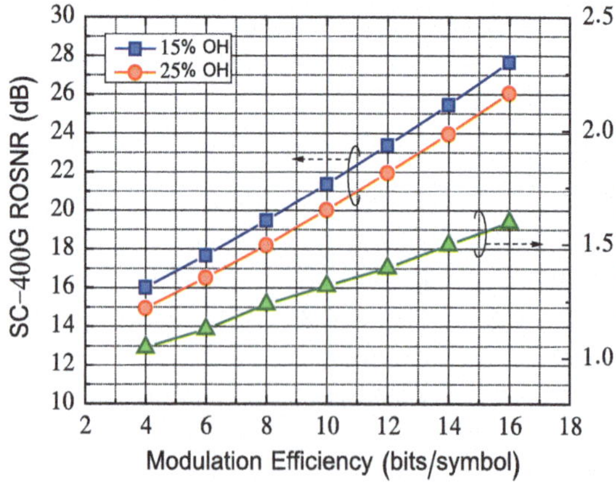

Fig. 9.1 Theoretical curves of OSNR and modulation efficiency under different coding overhead 9.3. 34 GBaud PM-256QAM signal generation

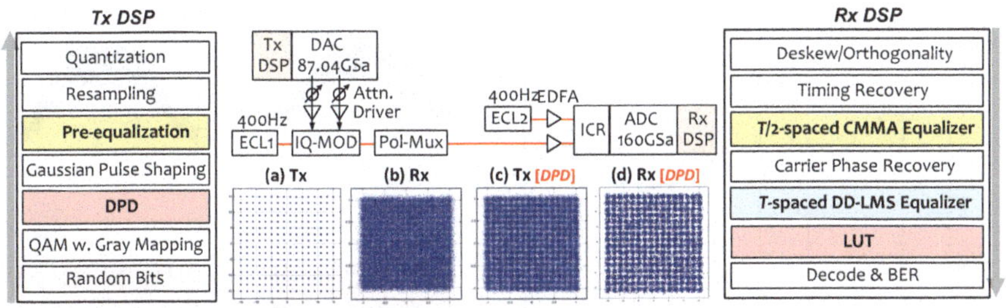

Fig. 9.2 Experimental setting of 34 GBaud single-carrier 400 G pm-256QAM signal generation

160 GSa/s and a digital bandwidth of 19.28 GHz. An external cavity laser (ECL2) with a working wavelength of 1550 nm acts as a local oscillator (LO).

In the transmitter signal processing (DSP) part, the gray mapping 256QAM symbol with a length of 25600 is generated first, then the digital pre-distortion module (DPD) is entered to pre-compensate the nonlinear damage, and the digital pre-distortion module (DPD) is the 9-symbol lookup table (LUT) generated based on the receiver [6]. The index address of the lookup table (LUT) is determined by the sending symbol sequence of different modes, and its content stores the distortion of different symbol modes. The data of the lookup table under different addresses are averaged at the receiving end, and the final lookup table (LUT) is set at the sending end. Subsequently, Gaussian pulse shaping, double over-sampling, pre-equalization, and symbol sequence resampling were conducted to match the sampling frequency of DAC, in which the real part (I channel) and the imaginary part (Q channel) were quantized by 8 bits, respectively. The digital pre-equalization filter is obtained by inverting the channel impulse response obtained by the receiver blind equalizer. In the receiving end signal processing (DSP) section, IQ orthogonalization, resampling, square timing clock recovery, T/2 interval multimode cascade algorithm (CMMA), carrier phase recovery based on blind phase search (BPS), t-interval decision feedback minimum mean square error algorithm (DD-LMS), and lookup table generator (LUT) are included. Among them, the tap number of CMMA is 33 and the step length is 2^{-18}. In the polarization multiplexing high-order QAM system, the CMMA algorithm has slower convergence speed and larger mean square error than DD-CMA and DD-LMS algorithms [7]. However, CMMA can correct the frequency offset so that no additional frequency offset compensation algorithm is required. In addition, the DD-LMS algorithm of 193-tap T interval is used to match the channel impulse response, and considerable processing gain can be obtained [3, 7]. In Fig. 9.2, insets (i) and (iii) are respectively the constellation diagram of 256QAM on the sending end when DPD is adopted and when DPD is not adopted, and insets (ii) and (iv) are the constellation diagram of 256QAM on the receiving end when DPD is adopted and when DPD is not adopted.

Figure 9.3 illustrates the influence of pre-distortion module (DPD) on the system performance of 34 Gbaud PM-256QAM signal in the back-to-back case, where the crude bit rate is 544 Gb/s, and the overhead bit rate of SD-FEC coding with the

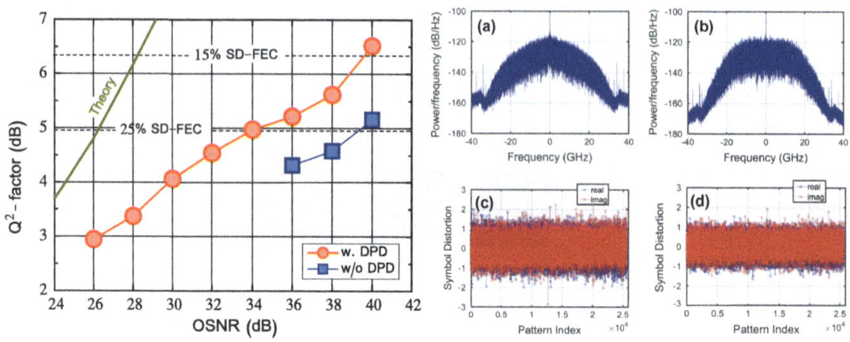

Fig. 9.3 BER performance of 34 Gbaud PM-256QAM signals in a back-to-back situation. Figures (a) and (b) are the spectra of the electrical signal with and without pre-equalization, and Figures (c) and (d) are the lookup table when DPD is used and when DPD is not used

removal of 15% and 25% is 473 Gb/s and 435.2 Gb/s, respectively. With and without DPD, the required ROSNR of 25% SD-FEC encoding Q^2-factor 4.95 dBQ is 34 dB and 39.2 dB, respectively, and the cost is 7.8 dB and 13 dB, respectively. In Ref. [8], 15% SD-FEC coding overhead was considered, and 30-GBd PM-256QAM signal back-to-back system surplus was 1.04 dBQ. Compared with that, the system surplus was increased to 1.6 dBQ under 25% SD-FEC coding overhead. Figure (a) and (b) represents the spectrum of 34 Gbaud PM-256QAM electrical signals when DPD is adopted and when DPD is not adopted. Figure (c) and (d) represents the receiver lookup table (LUT) data when DPD is adopted and when DPD is not adopted. Symbol distortion is suppressed after DPD of 9 symbol length.

9.3 Single-Carrier 400 G PM-256QAM Signal Transmission Experiment

Figure 9.4 illustrates the 34 Gbaud PM-256QAM signal wavelength division multiplexing (WDM) experiment. Each wavelength can transmit 4 user channels, each channel rate 104.79 Gbps, with 30% coding overhead, 25% for SD-FEC coding, 5% for probability coding, and information entropy of 7.64 bits/symbol. In Fig. 9.4, Tx includes an external cavity laser (ECL1) with a working wavelength of 1550.3 nm and a linewidth of 400 Hz, an orthogonal polarization multiplexing optical modulator (PMQ) with a bandwidth of 23 GHz, an integrated four-channel 32-GBd linear drive, a digital-to-analog converter (DAC) with a sampling rate of 87.04 GSa/s and 20 GHz analog bandwidth, and four 30-GHz balance transformer (Balun) for differential single-ended output and to suppress the clock leakage. The same transmitter assembly is used to simultaneously modulate the seven-channel 100-KHz external cavity laser (ECL) to form a single-carrier 400 G channel with an eight-channel 100 GHz interval, coupled to a single signal by an optical coupler (OC). The transmission link

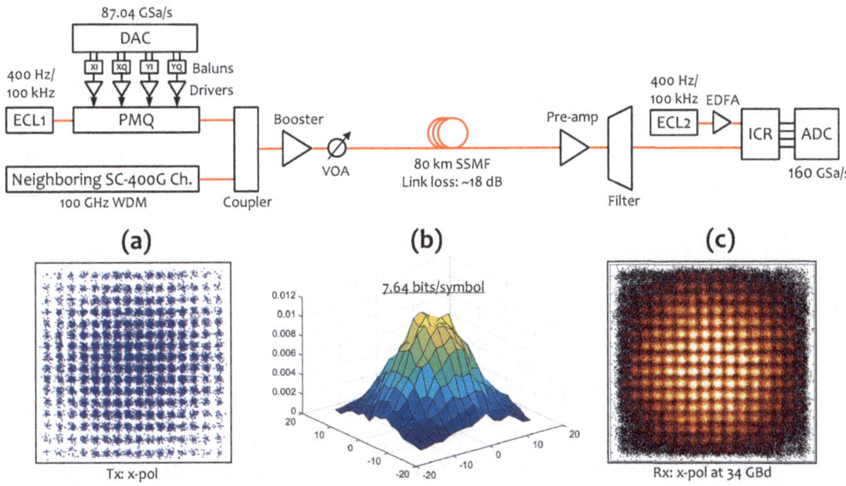

Fig. 9.4 34 Gbaud SC-400G PM-256QAM signal wavelength division multiplexing (WDM) experiment, **a** constellation point of 256QAM with PS and DPD polarization direction, **b** two-dimensional probability mass function (PMF) with PS-256 QAM source entropy of 7.64 bits/symbol, **c** constellation point of 256QAM with X polarization direction

includes an 18 dB loss 80-km standard single-mode fiber (SSMF), a 23 dBm booster, a tunable optical attenuator (VOA), a 23 dBm amplifier, and a tunable optical filter for channel selection. The receiver (Rx) section includes an EDFA amplifier, an integrated coherent receiver (ICR), a 40-GHz balanced photodetector, and an ADC with a sampling rate of 160 GSa/s and a digital bandwidth of 19. 28 GHz. An external cavity laser (ECL2) with a working wavelength of 1550 nm acts as a local oscillator (LO).

In the transmitting end signal processing (DSP) section, the gray mapping 256QAM symbol with a length of 25,600 is first generated, in which the I and Q channels are subject to the defined probability mass function (PMF) [0.0185, 0.0284, 0.0410, 0.0557, 0.0712, 0.0856, 0.0967, 0.1028, 0.1028, 0.0967, 0.0856, 0.0712, 0.0284, 0.0185]. The inset (ii) in Fig. 9.4 illustrates the probability distribution function of the designed 256QAM signal. Given a code table, it is easy to carry out probability shaping (PS) simulation by MATLAB without knowing the mapping relationship between bits and symbols. Specifically, although PS-QAM signal information entropy is 7.64 bits/symbol, the receiver can still take PS-QAM signal as a general QAM signal and decode it according to 8 bits/symbol [9]. How to decode a PS-QAM signal is not the focus of this chapter. The PS-QAM signal enters the digital pre-distortion module (DPD) to pre-compensate the nonlinear damage. The digital pre-distortion module (DPD) is based on the 9-symbol lookup table (LUT) generated by the receiver. After 2 times of oversampling, 33-tapped FIR filter is used for post-equalization, symbol sequence resampling is conducted to match the DAC sampling frequency, then output 34 GBd 256QAM signal. Figure 9.4 (i) illustrates the constellation diagram of the 256QAM signal after PS and DPD. The DSP at the receiving end is like Ref. [8] on a single-carrier 400 G PM-256QAM, except that the DD-LMS algorithm with the enhanced 193-t time interval is used to match the channel impulse response and reduce the signal crosstalk caused by the imbalance between I channel and Q channel.

Figure 9.5 illustrates the BTB results of the traditional PM-256QAM signal and PS7.64 shaping, the symbol rate of the two cases are 34 GBd, and the total bit rate is 544 Gbit/s. However, the PM-256QAM—PS7.64 signal has a bit rate of 420 Gb/s before 25% SD-FEC and 5% PS, while the traditional PM-256QAM signal does not require PS, and the SD-FEC encoding rate is 440 Gb/s. In the 25% SD-FEC coding Q^2-factor threshold of 4.95 dBQ, with and without PS, the required optical signal-to-noise (ROSNR) is 28 dB and 31.7 dB, respectively [9]. Note that the actual probability shaping gain is less than 3.5 dB when the same bit rate is considered. The actual execution cost of conventional PM-256QAM signal is about 5.5 dB, which is 2.3 dB lower than the previous Ref. [8], due to the use of more advanced photoelectric equipment and signal processing technology (DSP). Theoretically, the execution cost of PM-256QAM—PS7.64 signal is about 2 dB. Since the distribution of the theoretical curve is assumed uniform, the actual execution cost will be greater than 2 dB.

Figure 9.4 (iii) illustrates the constellation point of 256QAM in the direction of X polarization at the receiver when OSNR is 40 dB. In addition, figures (i) and (ii) in Fig. 9.5 respectively illustrate the data of PS7.64 and LUT of conventional 256QAM signals in the direction of X polarization. LUT mode distortion of the PS symbol is lower than that of the conventional QAM signal. Figure 9.6a illustrates the spectrum before and after 80 km fiber transmission. The transmitted power of channel 5 is 1dBm, and the received OSNR is 29.3 dB. Figure 9.6b illustrates the relationship between channel 5 Q^2 factor and the transmitted power of a single channel (single channel and WDM channel). It can be seen that the optimal transmission power of single channel is 3 dBm, while for WDM channel, the optimal transmission power

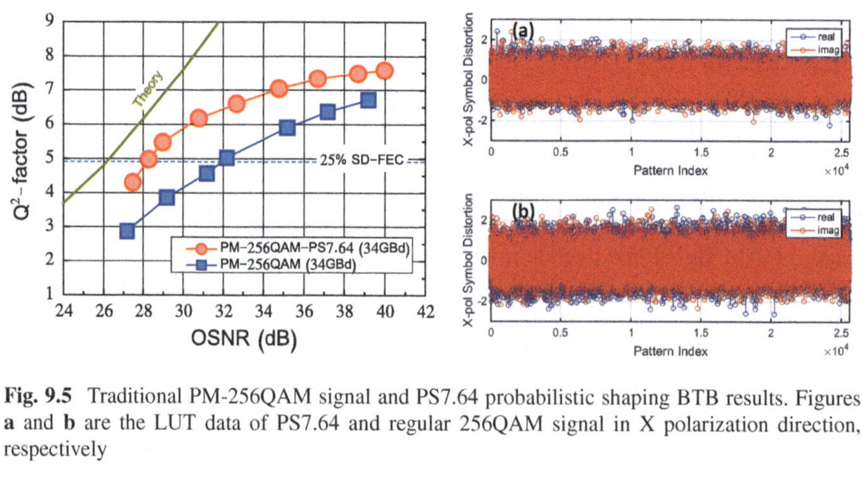

Fig. 9.5 Traditional PM-256QAM signal and PS7.64 probabilistic shaping BTB results. Figures **a** and **b** are the LUT data of PS7.64 and regular 256QAM signal in X polarization direction, respectively

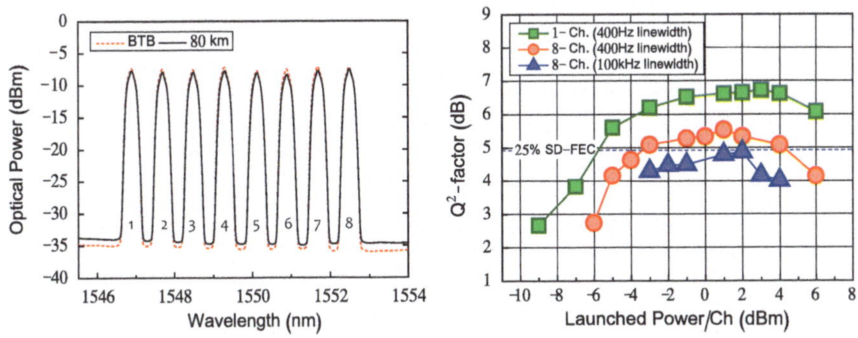

Fig. 9.6 **a** Spectrum before and after 80 km fiber transmission, **b** relation between channel 5 Q^2 factor and transmitted power of a single channel (single channel and WDM channel)

(400 Hz linewidth) is 1 dBm/Ch, the minimum transmission power is about −3.4 dBm, the dynamic range is about 7 dB, and the optimal transmission power (100 Hz linewidth) is 2 dBm/Ch.

9.4 Conclusion

In the first part, we experimentally verified the single-carrier 400 G 34 Gbaud PM-256QAM signal generation scheme, and the selected baud rate was used to obtain higher FEC coding gain. The encoding cost increases from 15 to 25%, and combined with LUT pre-distortion the BER threshold of precoding will be reduced. When ROSNR is 34 dB, we achieve 544 Gb/s ultra-high-rate transmission, currently, the execution cost is 7.8 dB, and the system surplus is 1.6 dBQ. With the continuous improvement of the sampling rate of the ADC/DAC interface, we believe that the single-carrier 400 G scheme with a modulation efficiency of 16 bits/symbol will be gradually commercialized in the point-to-point large-capacity transmission system in the future. In the second part, we verify the WDM experiment of 34 Gbaud single-carrier 400 G PM-256QAM signal with 30% coding overhead, of which 25% is used for SD-FEC coding and 5% is used for probability coding. The experiment shows that the minimum transmitting power of a single channel is about −3.4 dBm, and the dynamic range is about 7 dB, which makes it possible to gain coding gain by introducing probability coding in the transmission experiment.

References

1. F. Buchali et al., Implementation of 64QAM at 42.66 GBaud using 1.5 samples per symbol DAC and demonstration of up to 300 km fiber transmission. OFC, M2A.1 (2014)
2. J. Zhang et al., Transmission of single-carrier 400G signals (515.2-Gb/s) based on 128.8-GBaud PDM QPSK over 10,130- and 6,078 km terrestrial fiber links. Opt. Expr. 23(13), 16540–16545 (2015)
3. Z. Jia et al., Flex coherent DWDM transmission framework. Opt. Internetw. Forum OIF2015.339.06 (2015)
4. J. Geyer et al., Practical implementation of higher order modulation beyond 16-QAM, OFC, Th1B.1 (2015)
5. N. Swenson et al., Flexible coding and modulation techniques for next generation DSP-based coherent systems, OFC W4A.1 (2016)
6. K. Sugihara et al., A Spatially-coupled type LDPC code with an NCG of 12 dB for optical transmission beyond 100 Gb/s. OFC, OM2B 4 (2013)
7. J. Ke et al., 400 Gbit/s single-carrier and 1 Tbit/s three-carrier superchannel signals using dual polarization 16-QAM with look-up table correction and optical pulse shaping. Opt. Expr. **22**(1), 71–83 (2013)
8. H. Chien et al., On single-carrier 400G line side optics using PM-256QAM. ECOC 28–30 (2016)
9. H. Chien et al., Single-carrier 400G PM-256QAM generation at 34 GBaud trading off bandwidth constraints and coding overheads. OFC, W1J.3 (2017)

Chapter 10
Carrierless Amplitude and Phase Modulation

10.1 Introduction

With the rapid development of new services and new technologies represented by high-definition video, mobile Internet, cloud computing, and the Internet of Things, the bandwidth requirements of access networks and short-distance optical transmissions are increasing rapidly year by year [1–8]. On the one hand, with the advent of the cloud computing era and the popularity of big data centers, large-capacity and high-speed data centers are gradually becoming the basis of next-generation Internet services, and there is an urgent need for short-distance high-speed optical interconnections for data centers [6, 7]. On the other hand, as the bandwidth requirements of the client data service and the mobile Internet continue to increase, the transmission rate of the optical access network as the "last mile" is also increasing [4, 5]. How to realize short-distance high-speed optical transmission through technological innovation to adapt to the bandwidth requirements of data center optical interconnection and access networks such as passive optical networks (PON) has become a popular research field [1–8].

The short-distance high-speed optical transmission is oriented to the access network and the data center optical interconnection [9–19]. The application scenario is shown in Fig. 10.1. The transmission distance of the data center is about 0.5–10 km, and the optical access network transmission distance is about 20–50 km. Different from the bearer network for long-distance transmission, its technical development also needs to consider the cost, scheme complexity, and system power consumption. Considering cost, power consumption, and complexity, combining intensity modulation and direct detection (IM/DD) with high-order modulation formats is a more practical approach [1–8]. Overall, IM/DD-based carrierless amplitude and phase modulation (CAP) has proven to be an uncomplicated and well-performing solution that enables the use of low-cost optical components to achieve relatively high data rates, such as direct modulation lasers. Low-cost optoelectronic devices such as

J. Yu and N. Chi, *Digital Signal Processing In High-Speed Optical Fiber Communication Principle and Application*,
https://doi.org/10.1007/978-981-15-3098-2_10

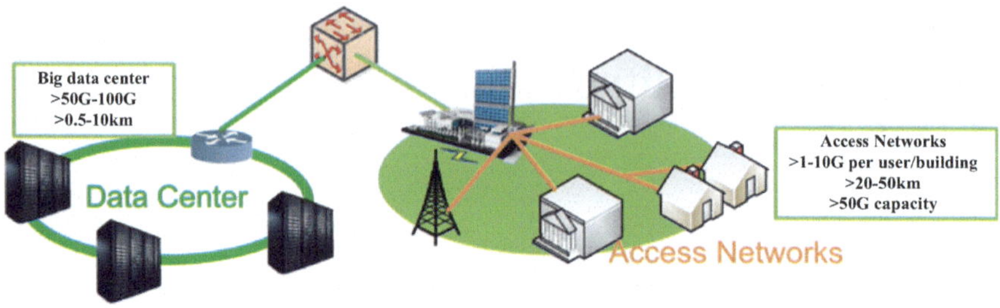

Fig. 10.1 High-speed optical transmission application scenarios for data centers and access networks

(DML) or vertical cavity surface emitting lasers (VCSELs) [4–8]. Compared to subcarrier modulation (SCM) and orthogonal frequency division multiplexing (OFDM), CAP requires neither an electrical complex conversion device, a complex mixer, nor a radio frequency (RF) source or optical in-phase/orthogonal (I/Q) modulator [1, 3, 7]. It has been pointed out that CAP has the advantages of low power consumption and low cost [4]. Compared to other modulation formats such as non-return-to-zero code reception (NRZ), pulse amplitude adjustment (PAM), or OFDM, CAP has great potential in short-range optical transmission.

This chapter will introduce the new modulation format–carrierless amplitude and phase modulation (CAP). First, the modulation and demodulation principles and implementation methods of CAP will be introduced, including the generation and reception of single-band CAP-mQAM, and the generation and reception of multi-band multi-level CAP signals. Then, the WDM-CAP-PON multi-user access network based on multi-level multi-band CAP modulation will be introduced through experimental research, and its high-speed access performance will be verified. Next, this chapter introduces the wireless access network transmission system based on high-order modulation CAP-64QAM, and experimentally verifies the performance of 24 Gb/s CAP-64QAM fiber transmission 40 km and wireless transmission access system. Finally, the direct modulation and direct detection high-speed CAP-64QAM system based on DML and digital signal processing will be introduced. With DD-LMS equalized CAP-64QAM, 60-Gb/s CAP-64QAM based on 10 G DML direct modulation is successfully transmitted for 20 km under direct detection conditions.

10.2 Modulation and Demodulation Principles of CAP

10.2.1 Modulation and Demodulation Principles of Single-Band CAP

As shown in Fig. 10.2, it is a schematic diagram of the modulation and demodulation principle of a single-band CAP-mQAM. At the transmitting end, the original data bit sequence is first mapped to the complex symbol of m-QAM (*m* is the order of

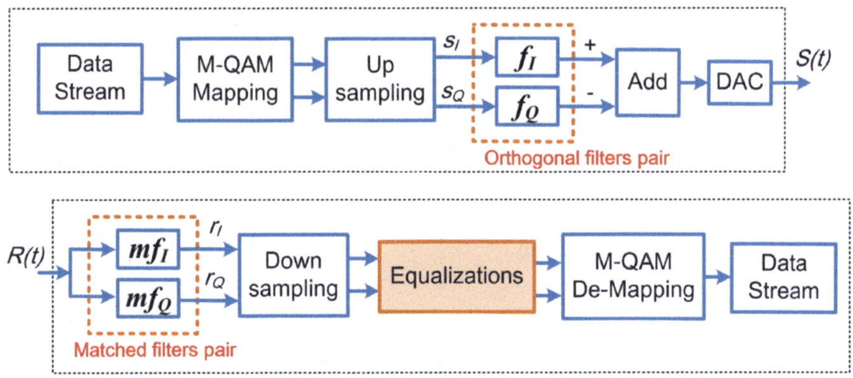

Fig. 10.2 Modulation and demodulation principles of single-band CAP

QAM), and then the mapped symbol is upsampled to match the sampling rate of the subsequent shaping filter. After the data is upsampled, the orthogonal signal is obtained by a pair of orthogonal shaping filters, and then the orthogonal filter outputs are added to obtain a modulated single-band CAP-mQAM signal. Usually, a digital-to-analog converter (DAC) is used to generate an output waveform, and a subsequent signal can drive a modulator, a DML, or a VCSEL. At the receiving end, the signal obtained after direct detection can be recovered by digital signal processing after passing through an analog-to-digital converter (ADC). The received signal is first fed to a pair of matched filters to separate the in-phase and quadrature components, and the resulting signal is then downsampled to employ a linear equalizer for channel equalization. Finally, the equalized signal is decoded to obtain the original bit sequence.

In the transceiver module for CAP generation and demodulation, the filters and matched filters used for orthogonal shaping and separation are called Hilbert filter pairs, including $f_I(t)$, $f_Q(t)$, $mf_I(t)$, and $mf_Q(t)$. In general, the two sets of Hilbert orthogonal filter pairs can be constructed by multiplying the square root raised cosine pulse by the sine and cosine functions. Assuming that $s_I(t)$ and $s_Q(t)$ are orthogonal signals obtained after QAM mapping, the CAP signal $S(t)$ superimposed and output by the orthogonal filter can be expressed as

$$S(t) = [s_I(t) \otimes f_I(t) - s_Q(t) \otimes f_Q(t)] \tag{10.1}$$

At the receiving end, the matched filter usually satisfies $mf_I(t) = f_I(-t)$ and $mf_Q(t) = f_Q(-t)$. For the demodulation of CAP, the I and Q signals passing through the matched filter can be expressed as

$$r_I(t) = R(t) \otimes mf_I(t), \quad r_Q(t) = R(t) \otimes mf_Q(t) \tag{10.2}$$

where R(t) is the received signal, and $r_I(t)$ and $r_Q(t)$ are the output signals of the matched filter.

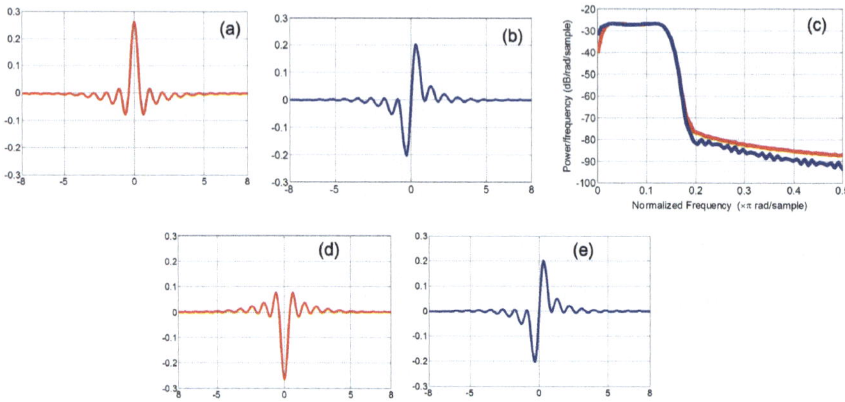

Fig. 10.3 The single-band CAP signal Hilbert filter's time-domain impulse response: **a** is $f_I(t)$, **b** is $f_Q(t)$, and **c** is the frequency response of the two filters, **d** and **e** is $mf_I(t)$ and $mf_Q(t)$

As shown in Fig. 10.3, it is the Hilbert filter and the time-domain impulse response of the matched filter and its frequency response. It can be seen that due to the orthogonality of the filters, in-phase and quadrature data can be obtained by orthogonal matched filters. At the same time, since the filter has a square root raised cosine pulse with the Nyquist filtering effect, the signal bandwidth is compressed, and high spectral efficiency modulation is achieved. On the other hand, as can be seen in Fig. 10.3, since the peaks between matched filters are required at the center of the pulse, clock synchronization is very important in demodulation in order to properly recover in-phase and quadrature data. However, in practical systems, the sampling time is not fixed, so the sampling time offset will result in severe intersymbol interference, and proper linear channel equalization is required for the in-phase and quadrature components that are subsequently demodulated.

10.2.2 Modulation and Demodulation Principles of Multi-band CAP

Unlike single-band CAP signals, multi-band CAP can more effectively utilize the bandwidth of the DAC, while reducing the bandwidth rate and bandwidth of each sub-band, making signal equalization and processing easier. The principle of generating multi-band CAP signals and the demodulation principle is shown in Fig. 10.4. At the transmitting end, different bands carry different data, and each sub-band has the same data processing manner. First, the bit sequence is mapped into a complex symbol of m-QAM and after the data is upsampled, the filtered orthogonal signals are then obtained by a pair of orthogonal shaping filters, the signals of each sub-band pass through the shaping filters of the corresponding sub-bands, and finally

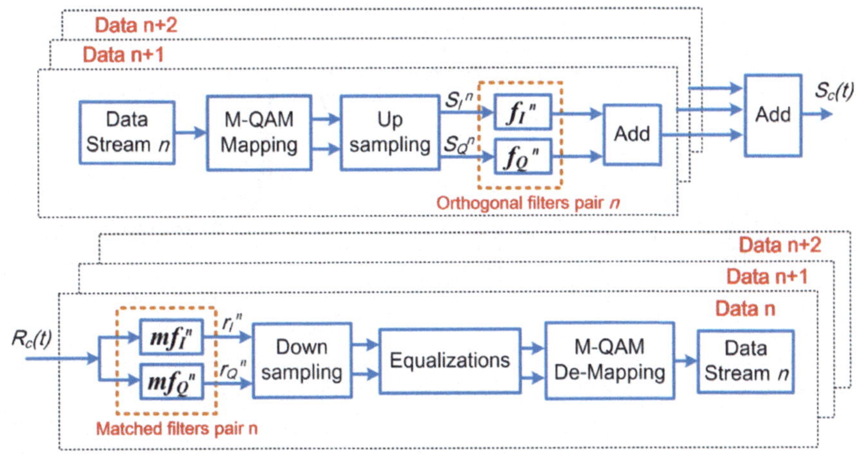

Fig. 10.4 Modulation and demodulation principles of multi-band CAP

the output results of all the sub-band filters are superimposed to obtain a multi-band CAP signal.

Similarly, at the receiving end, the signal obtained after direct detection can be recovered by digital signal processing after passing through an analog-to-digital converter (ADC). Different matched filters can be used to obtain the filtered signals of different sub-bands. After the in-phase and quadrature components are separated, the obtained orthogonal signals are downsampled, and then the linear equalizer is used for channel equalization. Finally, the equalized signal is decoded to obtain the original bit sequence. It is worth noting that different sub-bands correspond to different matched filters. In this way, a specific matched filter can only obtain a specific subcarrier signal, which ensures the confidentiality and integrity of the signal to a certain extent.

Assuming that the multi-band CAP signal has N sub-bands, the nth (n takes $1-N$) sub-band orthogonal filter pairs can be expressed as

$$f_I^n(t) = \frac{\sin\left[\pi(1-\beta)\frac{t}{T_s}\right] + 4\beta\frac{t}{T_s}\cos\left[\pi\frac{t}{T_s}(1+\beta)\right]}{\pi\frac{t}{T_s}\left[1-\left(4\beta\frac{t}{T_s}\right)^2\right]} \sin\left[\pi(2n-1)(1+\beta)\frac{t}{T_s}\right] \tag{10.3}$$

$$f_Q^n(t) = \frac{\sin\left[\pi(1-\beta)\frac{t}{T_s}\right] + 4\beta\frac{t}{T_s}\cos\left[\pi\frac{t}{T_s}(1+\beta)\right]}{\pi\frac{t}{T_s}\left[1-\left(4\beta\frac{t}{T_s}\right)^2\right]} \cos\left[\pi(2n-1)(1+\beta)\frac{t}{T_s}\right]. \tag{10.4}$$

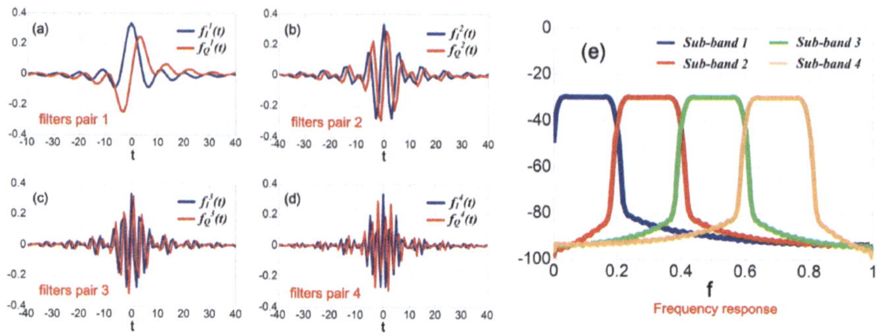

Fig. 10.5 Hilbert filter combination of four sub-bands: **a** first sub-band filter; **b** second sub-band filter; **c** third sub-band filter; **d** fourth sub-band filter; **e** Spectrogram of four sub-band filters

Here T_s is the symbol interval, and β is the roll-off factor, which is between 0 and 1. Assuming that there are 4 subcarrier CAP signals generated, the time-domain response and spectrum of each sub-band filter are shown in Fig. 10.5. Here the upsampling rate is 10, and the roll-off factor is taken as 0.2. As can be seen from Fig. 10.5, different sub-band filters have different time-domain responses, and these filters are located over different frequencies.

Similarly, assuming that the I and Q signals of the nth sub-band can be represented as $s_I^n(t)$ and $s_Q^n(t)$ after QAM mapping, the total N bands of CAP signals can be expressed as

$$S_c(t) = \sum_{n=1}^{N} [s_I^n(t) \otimes f_I^n(t) - s_Q^n(t) \otimes f_Q^n(t)] \tag{10.5}$$

At the receiver, generally, we have the matched filters $mf_I^n(t)$ and $mf_Q^n(t)$ of the nth sub-band. The demodulated I and Q signals can be expressed as

$$r_I^n(t) = R_c(t) \otimes mf_I^n(t), \quad r_Q^n(t) = R_c(t) \otimes mf_Q^n(t) \tag{10.6}$$

Here $R_c(t)$ is the received multi-band CAP signal, and $r_I^n(t)$ and $r_Q^n(t)$ are the result of passing the matched filter of the nth sub-band.

10.3 Multi-band Multi-level CAP for WDM-PON Access Network

Using the sub-band independence of the multi-band CAP signal proposed in the previous section, we propose and experimentally demonstrate a new WDM-CAP-PON multi-user access network based on multi-band multi-stage CAP modulation. This section introduces the multi-band multi-level CAP for WDM-PON access networks.

Figure 10.6 shows the downlink structure of WDM-CAP-PON based on multi-band CAP modulation. At the central office, each optical line termination (OLT) transmitter modulated at the *i*th wavelength carries the multi-band multi-level CAP (MM-CAP) signals for multiple users. The MM-CAP signal generation principle is shown in Fig. 10.4. Different sub-bands carry different information, and different orthogonal filters are used to generate multi-band CAP signals. At the optical network unit (ONU) side, each user can recover the data in each sub-band using a pair of matched filters without any interference. In this way, N CAP sub-bands can be assigned to N users. For WDM-CAP-PON, K wavelengths with N sub-bands can totally be assigned to K $\times$ N users. For upstream signals, the system can use OOK or other low-level modulation formats due to the low signal rate.

As a proof of concept, the experimental setup of 11 $\times$ 5 $\times$ 9.3-Gb/s WDM-CAP-PON downstream with 55 users over 40-km SMF is shown in Fig. 10.7. In the experiment, 11 multi-carriers generated by two intensity-modulated MZMs driven by radiofrequency are used as the light source of eleven channels, and the odd and even channels are, respectively, modulated and merged by WSS of two channels of 28 GHz frequency spacing. The 5 bands MM-CAP-16 signals are generated by the high-speed DAC with its sample rate of 30 GSa/s. Five different sub-band sequences are firstly mapped to the 16QAM I and Q signals and 12 $\times$ upsampled to perform shape filtering.

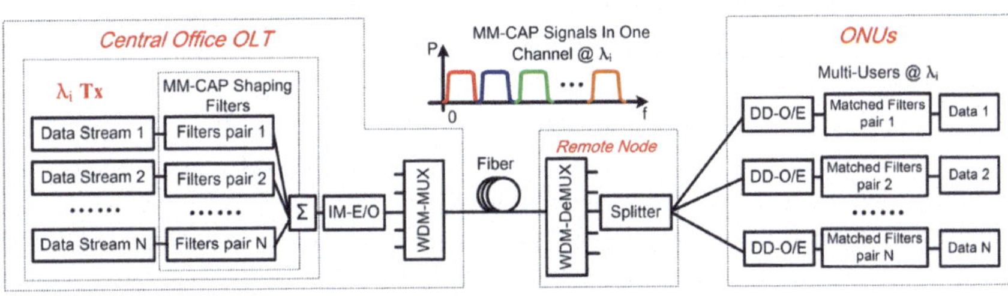

Fig. 10.6 Schematic diagram of multi-band multi-stage CAP-WDM-PON

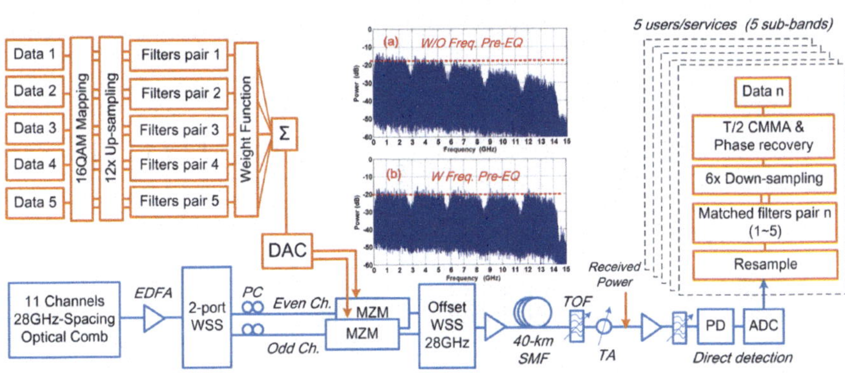

Fig. 10.7 $\times$ 5 $\times$ 9.3-Gb/s WDM-CAP-PON experimental setup

The 5 sub-bands are achieved with 5 pairs of different filters with a roll-off factor of 0.2 and an excess bandwidth of 15%. The 50-Gb/s MM-CAP-16 signal is generated by adding the outputs of the 5 filter pairs. It is worth noting that, the frequency response of the output of the DAC is non-flat, with a 3 dB bandwidth less than 13-GHz. Therefore, it is necessary to improve the power of the high-frequency band by frequency-domain equalization. Inset (a) and (b) in Fig. 10.7 depicts the spectrum of output signals of the DAC with and without frequency-domain equalization. In order to achieve longer distance optical transmission, we shift the WSS center carrier with 14 GHz frequency offset to achieve optical single-sideband filtering. The single-wavelength channel and the 11-wavelength channel WDM optical single-sideband multi-band CAP signal are shown in Fig. 10.8.

At the receiver, the channel is first filtered out by a tunable filter (TOF) with 0.3-nm bandwidth. After direct detection, the signals are sampled by a digital oscilloscope at 50-GSa/s for offline signal processing. A 0.9 nm TOF is used before PD front-end to remove excess ASE noise. The digital signal is first upsampled to 30GSa/s, and then the signals of I and Q are recovered by the matched filter. After that, the signal is downsampled to 2 Sa/Symbol for linear equalization recovery. Each sub-band is with 2.5 GHz bandwidth and 10-Gb/s data rate.

Figure 10.9 shows the bit error rate (BER) of each sub-band versus the received optical power for back-to-back (BTB) transmission. We can see that, with frequency-domain pre-equalization by adding weights on each sub-band, identical BER performance for sub-band 1–3 and negligible power penalty for sub-band 4 and 5 at BER below the FEC limit of 3.8×10^{-3} for 7% hard decision FEC are obtained. We also measure the BER of sub-band 5 without pre-equalization, which shows very poor performance. Therefore, it is necessary to perform pre-equalization processing.

Figure 10.10a, b shows the received MM-CAP signal spectrum over 40 km of single-mode fiber for double sideband (DSB) and optical single sideband (OSSB) signals. We can see that, without OSSB, sub-band 3 and 4 are degraded due to the power fading effect caused by CD and direct detection. Almost 40% of data transmission is cut-off and the power fading can be avoided by OSSB as shown in Figure 10.10b.

Figure 10.11a shows the measured BER of each sub-band in channel 4 versus

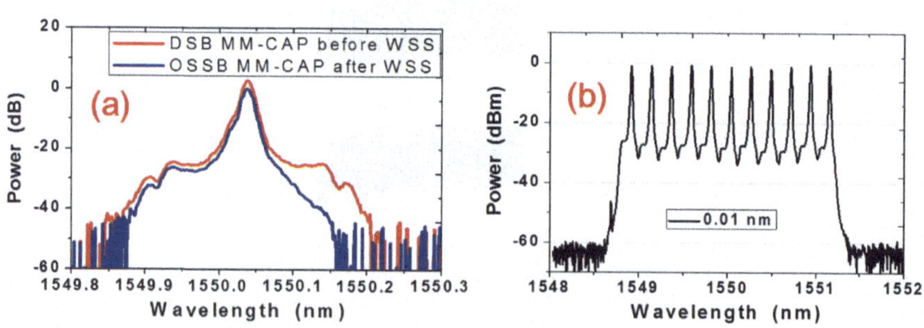

Fig. 10.8 a Optical spectrum of a single-wavelength channel; **b** optical spectrum of 11 wavelength channels

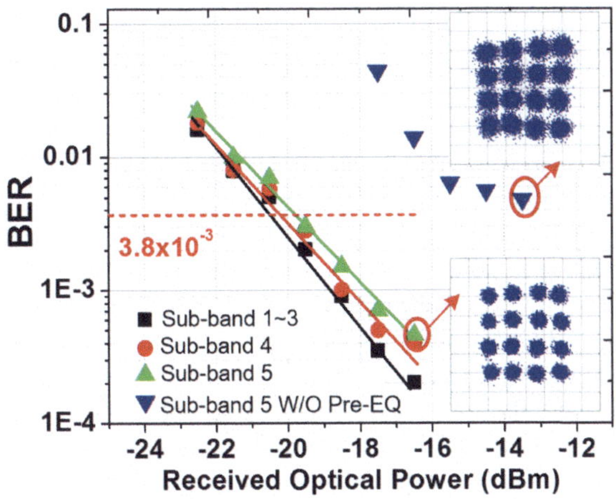

Fig. 10.9 The back-to-back BER results versus receiver optical power for each sub-band

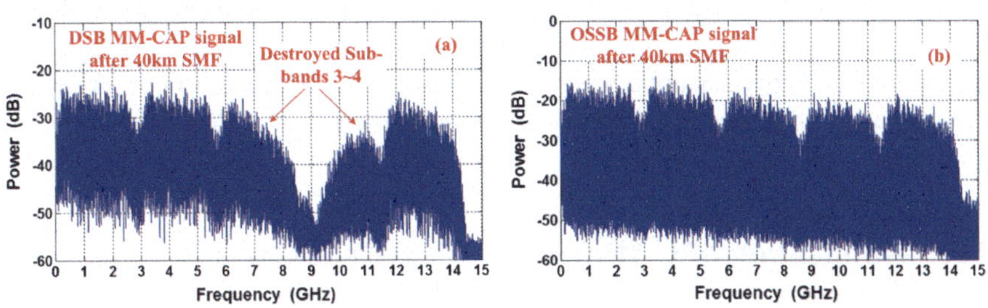

Fig. 10.10 Spectrum of MM-CAP after 40-km SMF for DSB and OSSB signals

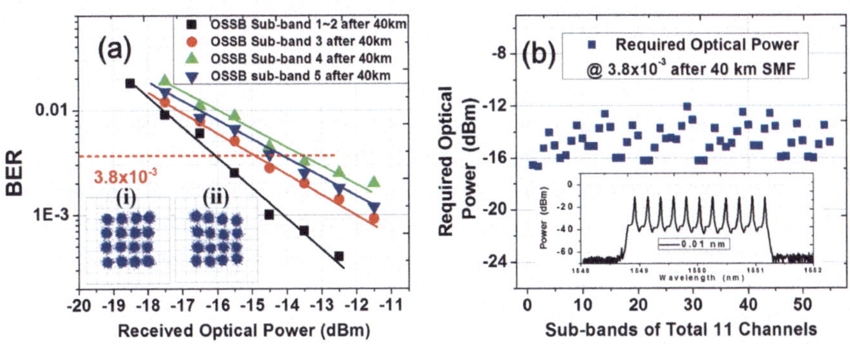

Fig. 10.11 **a** The BER of each sub-band in channel 4 versus received optical power after 40-km transmission; **b** The required optical power for total 55 sub-bands in eleven channels at the BER of 3.8×10^{-3} after 40 km SMF transmission

received optical power after 40-km SMF transmission. Inset (i) and (ii) show the constellation of sub-band 1 and 5 at the received power of −15 dBm. Sub-bands 1 and 2 have the same performance with about 4.5 dB power penalty compared to the BTB case. However, sub-bands 3–5 have a larger power penalty of about 5.5 dB, 6.5 dB, and 6 dB, respectively. It is due to the residual power fading effect under the imperfect OSSB filtering. Figure 10.11b shows the required optical power for the total 55 sub-bands in eleven channels at the BER of 3.8×10^{-3} for the 7% hard decision FEC limit after 40-km SMF transmission. The downstream data rate for each wavelength is 50 Gb/s. After removing the FEC overhead, the total rate of 55 users is $11 \times 5 \times 9.3$ Gb/s. Inset shows the optical spectrum of the received eleven channels. Considering that in a typical WDM-PON, the input power of each channel is about 10 dBm, and the loss of the fiber can be optimized by about 8 dB. The insertion loss of the commercial wavelength demultiplexer (such as AWG) is less than 4 dB, and in our case the required received power of the worst WDM channel for MM-CAP is about −12 dBm. Therefore, there is a power margin of 10 dB per wavelength. Since only 5 users are considered here, this power margin is sufficient for each user. On the other hand, the wavelength interval and number of users should be optimized in practical applications. However, if more users or longer distances are required, a highly sensitive receiver (such as a coherent receiver) or an optical amplifier needs to be used in the local or remote switch.

10.4 Application of CAP-64QAM in Wireless Access Network

With the advent of various multimedia data services for end users, the demand for wireless access rates is also increasing. To meet this capacity and spectral efficiency requirements, fiber-optic wireless systems have evolved from on-off keying (OOK) to more advanced modulation formats such as subcarrier modulation (SCM)-based quadrature amplitude modulation (QAM), orthogonal frequency division multiplexing (OFDM), etc. In this section, we propose the carrierless amplitude and phase (CAP) modulation based on intensity modulation and direct detection. As a candidate technology it can provide good system performance and low complexity. Recently, it has been widely studied and used in short-distance optical access networks. Compared with OFDM, CAP has great potential in high power efficiency and low cost in short-distance optical transmission applications. Therefore, considering the technical advantages of large capacity, high spectral efficiency, etc. CAP is worthy of study in fiber-optic wireless access systems.

As a proof of concept, the experimental setup of the 24-Gb/s CAP-64QAM ROF system over 40-km SMMF and 1.5-m 38-GHz wireless transmission is shown in Fig. 10.12. An external cavity laser (ECL) with a wavelength of 1550.10 nm is used as a CW source for carrier modulation and signal modulation via an I/Q modulator. The I/Q modulator consists of two parallel MZMs, one on each of the two arms. The

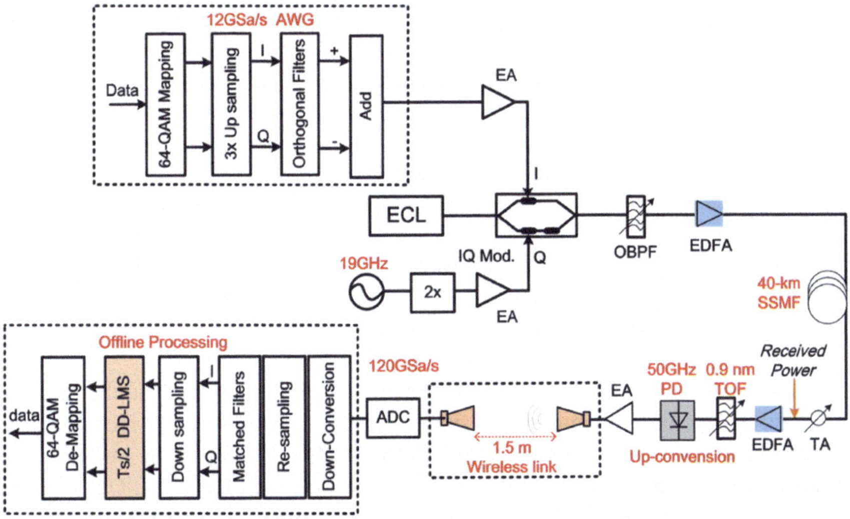

Fig. 10.12 The experimental setup of the 24-Gb/s CAP-64QAM ROF system

upper arm MZM is driven by the 4-Gbaud CAP-64QAM signal, which is generated by the 12 GSa/s commercial arbitrary waveform generator (AWG). The data sequence is first mapped to the 8-level 64QAM I and Q signals with 16×2^{11} symbols. Then, the 16 sets of 8-level I and Q sequences are upsampled to 3Sa/symbol and filtered by the orthogonal Hilbert filters pair. The filters are finite impulse response (FIR) filters with a length of 10 symbols each. The roll-off coefficient is 0.2 and the excess bandwidth is set to 15%. In this way, the 4-Gbaud baseband CAP-64QAM signal is generated by the AWG. The lower arm which is used for mm-wave tones generation is operated at OCS condition. The lower arm MZM2 is biased at null-point, and driven by the 38-GHz RF signal. One OBPF with 0.5-nm 3 dB bandwidth is used for the baseband signal and first-order tone filtering.

The spectrum of the lower arm MZM_2 millimeter-wave carrier output and the output after I/Q modulator with both baseband signal and mm-wave tones are shown in Fig. 10.13a, b. From Fig. 10.13a, we can see an obvious carrier suppression result with more than 50 dB first-tone to carrier power ratio. Figure 10.13c shows the result after OBPF, where only baseband signal and the left sideband first-order mm-tone are kept with carrier spacing of 38-GHz. The ratio of carrier to baseband power (first-order millimeter-wave in the left band) is approximately 18 dB. After that, the filtered optical baseband signal and the first-order sideband signal are input to the 40 km SSMF, which has about 10 dB loss and 17-ps/km/nm chromatic dispersion (CD) at 1550 nm without optical dispersion compensation. One erbium-doped fiber amplifier (EDFA) is used before the fiber to compensate for the fiber loss.

For wireless transmission, the optical baseband signal is up-converted by the optical beating at the 50-GHz PD. The generated 38-GHz mm-wave is firstly amplified

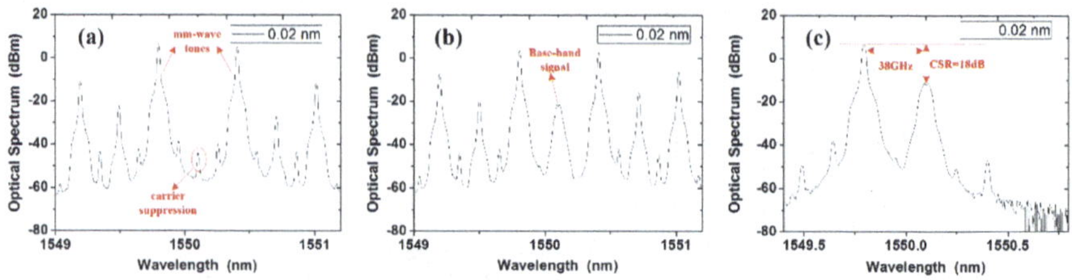

Fig. 10.13 The optical spectrum of **a** the output of lower arm MZM_2 with mm-wave tones, **b** the output after I/Q modulator with both baseband signal and mm-wave tones, and **c** the output after OBPF with only baseband signal and the left sideband first-order mm-tone

by an electrical amplifier (EA) with a bandwidth of about 10-GHz and then broadcasted through a horn antenna with a gain of 25 dBi. At the wireless receiver side, the broadcasted wireless signal is received by another 40 GHz horn antenna with 25 dBi gain. The received signal is then directly sampled by a 120-GSa/s high-speed oscilloscope with 45-GHz bandwidth for analog-to-digital conversion (ADC). Figure 10.14a shows the FFT spectrum of a 38-GHz millimeter-wave signal passing through the ADC. It can be seen that the millimeter-wave signal carrying 24 Gb/s CAP-64QAM signal with a center frequency of 38-GHz occupies less than 9-GHz bandwidth.

After the ADC, the signal is recovered using offline digital signal processing. First, the down-conversion of the signal is performed by multiplying by the cosine or sine function of the synchronization. Then, the down-converted signal is resampled to 16 GSa/s. Figure 10.14b shows the FFT spectrum of the down-converted baseband CAP 64QAM signal with 16-GSa/s. It should be pointed out that we use a high-speed ADC for digital down-conversion to demonstrate the ROF system. For practical use, an analog down-conversion can be employed using RF local oscillator (LO) to reduce the requirements for high-speed ADCs. The resampled signal is then generated by orthogonal matched filters to generate the I and Q path signals. The matched filter operates at 4 Sa/symbols and has a length of 10 symbols. The I and Q signals are then downsampled to 2 Sa/symbol before linear equalization. Here, four real-valued,

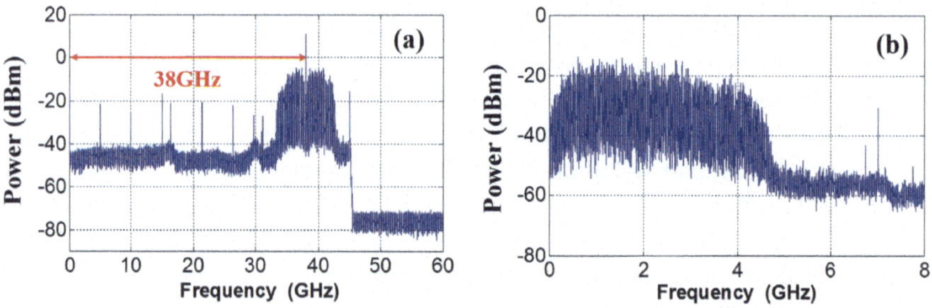

Fig. 10.14 The FFT spectrum of **a** the received 24-Gb/s CAP-64QAM millimeter-wave signal at 38-GHz carrier after ADC and **b** the down-converted baseband 24-Gb/s CAP-64QAM signal

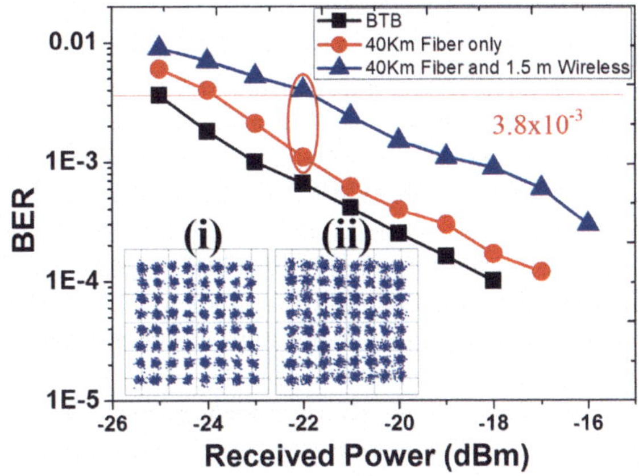

Fig. 10.15 The BER results versus receiver optical power for BTB case, after 40-km fiber transmission only and after 40 km fiber and 1.5 m wireless transmission

31-order, T/2-spaced butterfly adaptive FIR filters based on decision-directed least mean squares (DD-LMS) algorithm are used for signal equalization and recovery.

Figure 10.15 shows the bit error ratio (BER) results of the 24-Gb/s CAP 64QAM signal versus the received optical power in back-to-back (BTB) case without fiber or wireless transmission, after 40-km SMMF transmission and after 40-km SMMF and 1.5-m wireless transmission. The optical power is measured after the adjustable attenuator (TA). The 0.9 nm TOF is used to remove additional ASE noise after the EDFA. It can be concluded that the required received optical power at BER of 3.8×10^{-3} for the 7% hard decision FEC limit in BTB case is about −25 dBm. Less than 1 dB power penalty after 40-km fiber transmission is observed at the BER of 3.8×10^{-3} compared to the BTB case. However, a power loss of approximately 2 dB is introduced by 1.5 m wireless transmission due to power loss and impairments of wireless transmission. The constellations of the processed CAP 64QAM signal at −22 dBm received power after 40-km fiber transmission without and with 1.5-m wireless transmission are shown in the insets (i) and (ii) of Fig. 10.15, respectively. As a proof of concept, we only measured the results of a 1.5-m wireless link. We believe that longer distance wireless transmission can be achieved by obtaining larger received optical power or by using an electrical amplifier on the wireless transmitter and receiver side. In summary, these results clearly demonstrate the feasibility of the proposed high spectral efficiency CAP-64QAM in fiber-optic wireless access systems.

10.5 DML-Based 60 Gb/s CAP-64QAM Transmission Experiment

The IM/DD-based CAP transmission system has many advantages for achieving relatively high data transmission rates using low-cost optical components such as low-cost optoelectronic devices, directly modulated lasers (DML) or vertical cavity surface emitting lasers (VCSELs). This section will introduce the research work on short-distance optical transmission in high-speed CAP, and demonstrate the high-speed CAP-64QAM transmission based on DML-based direct detection and digital equalization technology. Modified decision-directed least mean squares (DD-LMS) are used to equalize the CAP-64QAM. Compared with CMMA, reduced complexity and improved performance are obtained. Using this scheme, we successfully generate and transmit up to 60-Gb/s CAP-64QAM over 20-km standard single-mode fiber (SSMF) based on the 10G-class DML and direct detection. To the best of our knowledge, it is the first demonstration and investigation for CAP-64QAM based on DML with such a high speed up to 60 Gb/s.

Figure 10.16 shows the experimental setup of 60-Gb/s CAP-64QAM generation, transmission, and detection based on 10G-class DML with direct detection and digital signal processing. The DML has a working wavelength of 1295.14 nm and is driven by a CAP-64QAM signal. In the experiment, a DAC of programmable 30-GSA/s is used to generate a signal of 10-Gbaud CAP-64QAM. The data sequence is first mapped to the 8-level 64QAM with a length of 16×2^{11} symbols. Then, the 8-level in-phase and quadrature data are upsampled to 3 Sa/symbol and filtered by the orthogonal Hilbert filters pair. The filters are FIR filters with a length of 10 symbols each. The roll-off coefficient is 0.2 and the excess bandwidth is set to 15%. In this way, the 10-Gbaud baseband CAP-64QAM signal is generated by the DAC. The output power of the DML is 8 dBm. After modulating by the DML, the optical signal is transmitted over 20-km SMF for direct detection.

At the receiving end, the signal is directly detected by the PD and then sampled by the digital oscilloscope at 40-GSa/s and then digitally processed offline. The sampled signal is first demodulated by the two orthogonal matched filters. After the matched filters, the I and Q signals are then downsampled to 2-Sa/symbol before the linear equalization Here, four real-valued, 31-order, T/2-spaced butterfly adaptive FIR filters based on decision-directed least mean squares (DD-LMS) algorithm are used for signal equalization and recovery. The BER is measured after signal de-mapping. Figure 10.17a shows the FFT spectrum of the received 10-Gbaud CAP-64QAM signal at a 40-GSa/s sampling rate.

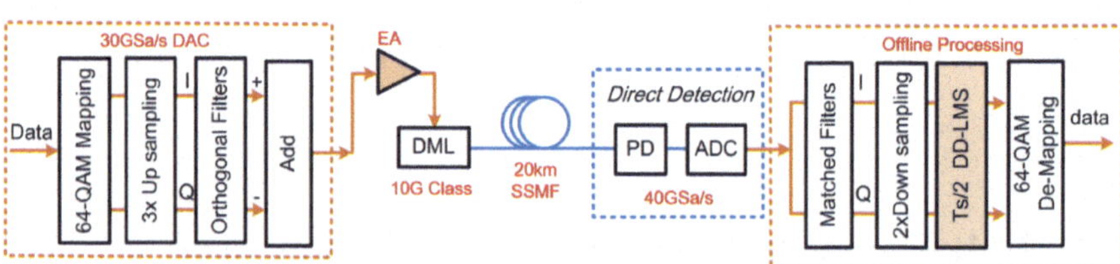

Fig. 10.16 Experimental setup of 60-Gb/s CAP-64QAM transmission based on 10G class DML with direct detection

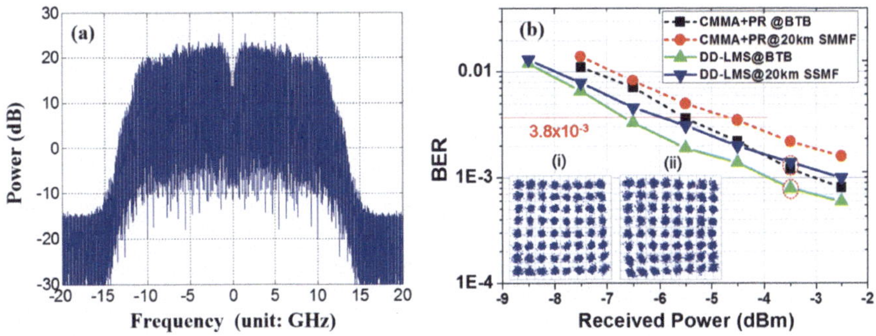

Fig. 10.17 a The FFT spectrum of received 10 Gbaud CAP-64QAM signal; **b** The BER curves of 60-Gb/s CAM-64QAM signal versus the received optical power before and after 20-km SMF

Figure 10.17b shows the BER performance of 60-Gb/s CAM-64QAM signal versus the received optical power before and after 20-km SMF using different equalization methods. We can see that for both back-to-back (BTB) and 20-km SMF transmission cases, the DD-LMS equalization shows better performances compared with the CMMA plus phase recovery method. About 1.5 dB and 2.5 dB received power sensitivity improvement can be obtained for BTB and 20-km SMF transmission case at the BER of 3.8×10^{-3}, respectively. The constellations of the processed CAP-64QAM signal at -3.5 dBm received power before and after 20-km SMF transmission using DDLMS are shown in the insets (a) and (b) of Fig. 10.17b, respectively. These results clearly demonstrate the feasibility of the proposed CAP-64QAM system based on DML and direct detection with modified DD-LMS for equalization.

In previous works [7], two-stage equalizations, ISI equalization, and phase recovery (PR) are used to equalize the CAP signal based on CMMA. However, it is difficult for high-order CAP-QAM signal equalization using CMMA, since the ring spacing in QAM is generally smaller than the minimum symbol spacing. On the other hand, since CMMA is based on the radius of the circles of the symbol, it is a phase-independent algorithm. This method is not sensitive to phase, so additional phase recovery after CMMA is required. In this work, we propose a one stage equalizer based on modified DD-LMS for CAP-QAM signal ISI and crosstalk equalization. Figure 10.18a shows the structure of the DD-LMS algorithm. Different from the regular DD-LMS used in a coherent optical system based on four complex-valued FIR filters, a four real-valued, T/2-spaced butterfly configured adaptive digital FIR filters structure is used for CAP signal equalization as shown in Fig. 10.18a. In this way, the in-phase and quadrature signals are independently equalized by the filters of h_{ii} and h_{qq} and the crosstalk is equalized by the coupling filters h_{iq} and h_{qi}.

The two benefits of using modified DD-LMS for CAP-mQAM signal equalization are depicted as shown in Fig. 10.18a, c with simulation results of the CAP-64QAM signal. Figure 10.18b shows the simulation results of the phase rotation versus sampling offset after using CMMA and DD-LMS. The upsampling ratio here is 8. We can see that the phase rotation caused by the sampling offset cannot be removed

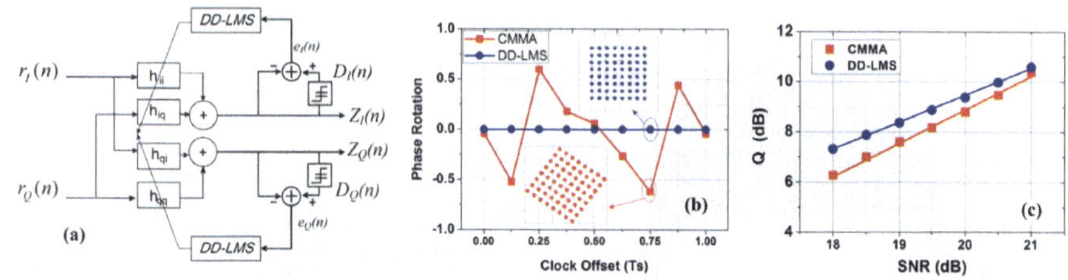

Fig. 10.18 a Structure of DD-LMS algorithm; **b** The phase rotation versus the sampling offset; **c** the Q performance versus different SNR for CAP-64QAM

by CMMA, which requires the additional phase recovery after CMMA. However, the phase is simultaneously recovered by the modified DD-LMS after equalization since the DD-LMS is phase-sensitive. Figure 10.18c shows the relationship between Q performance and different SNRs for CAP-64QAM under different equalization schemes. We can see that DD-LMS shows better Q performance for CAP-64QAM compared with CMMA, especially under low SNR conditions. More than 1 dB improvement is obtained when SNR is less than 19.5 dB.

Finally, we measure the BER performance with different FIR filter lengths. Figure 10.19a shows the BTB BER performance versus the DD-LMS filter tap length under different transmitter orthogonal filter length (T-OFL) for CAP-64QAM signal generation. We can see that larger T-OFL has better BER performance. The optimal filter length for CAP-64QAM generation is about 10Ts. Figure 10.19b shows the BER curve with the DD-LMS filter tap length under different matched filter lengths at the receiver (R-MFL). We can see that the R-MFL has less impact on the received BER performance. Figure 10.19a, b show that BER performance is improved when the DD-LMS tap length is ranging from 1 to 41. The optimum tap length for the improved DD-LMS is approximately 31. Here, the received optical power is set at −6.5 dBm. These results clearly demonstrate the feasibility of the proposed CAP-64QAM system based on DML and direct detection with modified DD-LMS for

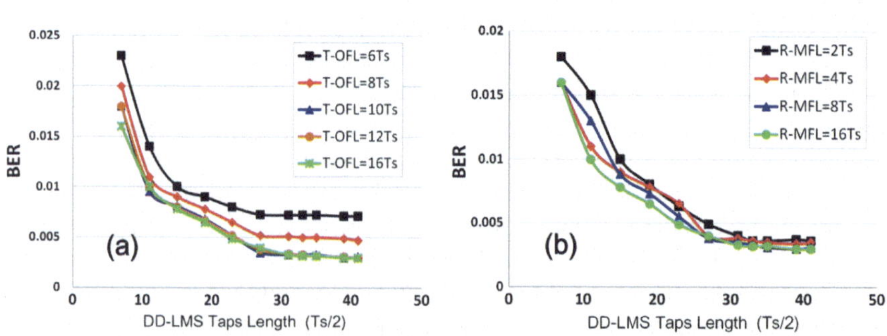

Fig. 10.19 a and **b** The BTB BER performances versus the DD-LMS filter tap length under different T-OFL and R-MFL

equalization. Using this scheme, a record 60-Gb/s CAP-64QAM over 20-km SSMF based on the DML and direct detection is demonstrated.

10.6 100 G CAP Long-Distance Transmission with Dispersion Compensation

Recently, the demand for ultra-high data rate optical transmission has been continuously growing in optical transport networks, metro networks and access networks [20]. Wavelength division multiplexing (WDM) and even ultra-dense WDM (UDWDM) with advanced modulation formats are widely used in coherent systems to realize the most promising solutions for 400 Gb/s and 1 Tb/s transmission [21]. Especially for metro networks, both transmission distance and cost should be considered in the architecture to achieve 100 Gb/s per lane. Compared with coherent receivers, direct detection (DD) optical transmission is considered as a more attractive and feasible solution in terms of system construction cost, computation complexity, and power consumption [22].

Carrierless amplitude and phase modulation (CAP) is an advanced single-carrier modulation format with low cost and bandwidth-limited optical components. Bitrates of 221 and 336 Gb/s are successfully transmitted over 225 and 451 km SSMF employing PDM multi-band CAP and a coherent transceiver [23]. And 100 Gb/s CAP over 15 km fiber with a direct detection quadrature amplitude modulation (QAM) receiver has been reported in [24].

Although many researchers have investigated PAM, CAP, and OFDM modulation formats for metro networks. There is not any 100 G transmission reported over 400 km SSMF using low-cost direct detection. The major reason is that long-haul transmission suffers from chromatic dispersion (CD) penalties. Applying SSB or vestigial sideband (VSB) is one way to overcome the CD limitation in systems with direct detection. 100 Gb/s SSB DMT over 80 km fiber [25] and 110.3 Gb/s VSB DMT over 100 km fiber [26] are achieved. However, employing SSB will introduce 3 dB signal-to-noise (SNR) loss compared with double sideband (DSB) signals [27]. Pre-CD compensation is another way to suppress CD distortion. 336 Gb/s PDM-64 QAM has been experimentally demonstrated with IQ modulator over 40 km SSMF [32]. Recently, 56 Gb/s DMT over 320 km SSMF [28] and 100 Gb/s DMT over 80 km SSMF [27] with dual-drive Mach–Zehnder modulator (DDMZM) have been realized.

In this section, we experimentally demonstrate a low-cost CAP-16 transmission with direct detection. It is realized at the bit rate of 112 Gb/s/λ with commercial optical components. We also compare the system performance between dispersion compensating fiber (DCF) and pre-CD compensation over 80 km SSMF. And the transmission performance of SSB and pre-CD signal with DDMZM or IQ modulator

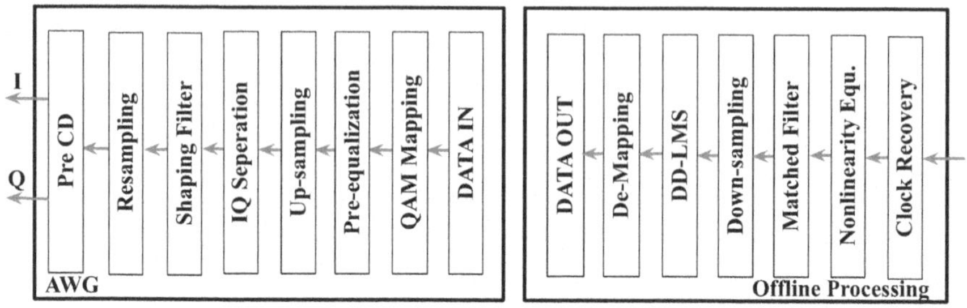

Fig. 10.20 Block diagram of CAP-16

is also evaluated. To the best of our knowledge, a single-wavelength 100 G transmission over 480 km SSMF is experimentally studied for the first time with CAP-16 based on direct detection in metro networks.

10.6.1 Digital Signal Processing of CAP

10.6.1.1 CAP-16

Figure 10.20 shows the block diagrams of a CAP-16 signal. At the transmitter, the data is first mapped into complex symbols of the 16QAM signal. After pre-equalization, the data is upsampled by a factor of 4. IQ separation is used to form a Hilbert pair and a square root raised cosine shaping filter with a roll-off factor of 0.1 is used as a shaping filter. The center frequency is set as 15.6 GHz, while the baud rate of CAP-16 is 28 GBaud. The bit rate is still 112 Gb/s and the same pre-CD process is carried out. After resampling, the signal is pre-distorted with the inverse of the phase delay caused by CD [27]. Because of the process of pre-CD, the signal becomes a complex signal. Real and imaginary signals are both fed into the upper and lower arms of DDMZM or IQ modulator.

During the offline process, the signal is sent into the matched filter to separate the in-phase and quadrature components after timing recovery and the nonlinearity equalizer. The BER performance of the final data is measured after the DD-LMS and the de-mapping process.

10.6.1.2 Pre-CD Method

The main factor which limits the transmission distance for the DSB signal is the power fading issue caused by chromatic dispersion [30]. The general frequency-domain channel response of CD is

$$H(w) = \exp\left(-j\frac{DL\lambda^2}{4\pi c}w^2\right) \tag{10.7}$$

where D is the dispersion parameter, L is the fiber length, λ is the carrier wavelength, and c is the speed of light. Its corresponding time-domain expression is

$$h(t) = \sqrt{\frac{c}{jDL\lambda^2}}\exp\left(j\frac{\pi c}{DL\lambda^2}t^2\right) \tag{10.8}$$

According to Eq. (10.8) and the square-law detection, we only present the final formula as in [31]

$$I_{PD}^2(t) \propto cos^2\left[\frac{\pi DL\lambda^2}{c}f^2\right] \tag{10.8}$$

When the phase sum of the signal is $\frac{\pi}{2} + N * \pi$ (N is integer), the signal will suffer the destructive power fading. So the bandwidth of the first lobe is expressed as

$$f_{\text{bandwidth}} = \frac{c}{2DL\lambda^2} \tag{10.9}$$

In order to solve the serious power fading, the modulated signal should be pre-distorted by the inverse of the CD channel response. However, due to the pre-CD method, the signals will carry the phase information in the meantime. That's why we used DDMZM and IQ modulator in this experiment.

10.6.2 Generation of SSB Signal

SSB signal is another way to avoid power fading caused by CD. Recently, much work about SSB signals generated by a low-cost DDMZM is published [25, 26]. A DDMZM consists of two parallel PMs and they are driven with a bias difference of $V_\pi/2$. The output of the DDMZM can be expressed as [29]:

$$E_{out} = \frac{\sqrt{2}}{2}E_{in} * \left\{e^{j*\left[\frac{\pi}{V_\pi}I(t)-\frac{\pi}{2}\right]} + e^{j*\left[\frac{\pi}{V_\pi}Q(t)\right]}\right\}$$

$$= \frac{\sqrt{2}}{2}E_{in} * \left\{-j * e^{j*\left[\frac{\pi}{V_\pi}I(t)\right]} + e^{j*\left[\frac{\pi}{V_\pi}Q(t)\right]}\right\}$$

$$\approx \frac{\sqrt{2}}{2}E_{in} * \left\{-j * \left[1 + j * \frac{\pi}{V_\pi}I(t)\right] + \left[1 + j * \frac{\pi}{V_\pi}Q(t)\right]\right\}$$

$$= \frac{\sqrt{2}}{2} E_{in} * \left\{ \frac{\pi}{V_\pi} * [I(t) + j * Q(t)] + 1 - j \right\} \tag{10.10}$$

From Eq. (10.10), it is observed that the electrical signal $I(t) + j * Q(t)$ is linearly converted to the optical domain. For the generation of SSB signal, we set the electrical signal $I(t)$ as a real signal x and the signal $Q(t)$ as its Hilbert pair x. The output of $x + \hat{x} * j$ is the analytic signal of x and is a single-band signal. Then, the optical domain expression could be

$$E_{out} = E_{in} * \left(x + j * \hat{x}\right) \tag{10.11}$$

The output of Eq. (10.11) will be an optical single-band signal.

10.6.3 Experimental Setup and Results

The setup is shown in Fig. 10.21. We generate the drive signals using an 81.92 GSa/s DAC with 20 GHz bandwidth and an offline MATLAB® program. Both DDMZM

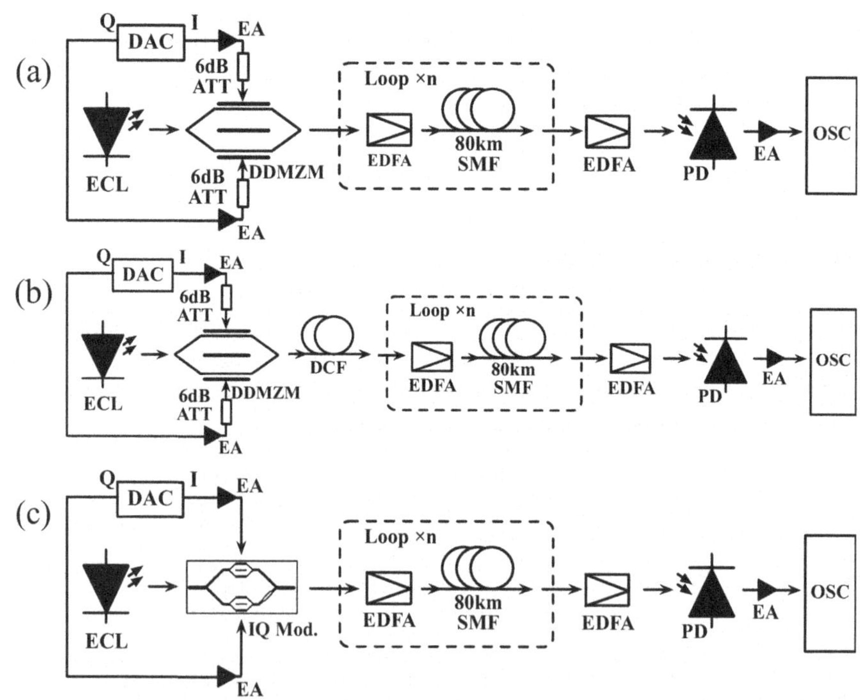

Fig. 10.21 Experimental setup of optical transmission with direct detection, **a** DDMZM with Pre-CD, **b** DDMZM with DCF and **c** IQ modulator with Pre-CD

(35 GHz bandwidth) and IQ modulator (30 GHz bandwidth) are used in this experiment. The bias of the two parallel PMs in the DDMZM is driven with a bias difference of $V_{\pi}/2$ to achieve the function of IQ modulation [29], and IQ modulator is driven at the quarter point. Before driving the upper and lower arms of the modulator, the signals are amplified by electrical amplifiers (EA, 32 GHz bandwidth and 20 dB gain). 6 dB and 0 dB electrical attenuators for DDMZM and IQ modulator, respectively, are used to fit the linear region of the modulators. Continuous-wave light of 1542.9 nm is input to the modulator. The fiber transmission loop includes an erbium-doped fiber amplifier (EDFA) and an 80 km SSMF fiber. After the EDFA amplifies the signal, it detects the signal with a 50 GHz photodetector. Finally, the signal is sampled with a real-time digital oscilloscope with a sampling rate of 80 GSa/s and an electrical bandwidth of 33 GHz.

10.6.4 Comparison Between Pre-CD and DCF with DDMZM

Firstly, we investigate the BER performance of CAP-16 at BTB and 80 km SSMF cases. We use the dispersion compensating fiber (DCF) to compensate for the CD caused by 80 km fiber in Fig. 10.22.

To investigate the effect of the peak-to-average power ratio (PAPR), we used the different DSP procedures shown in Fig. 10.23 to evaluate the PAPR of the CAP signal. The figure shows the relationship between complementary cumulative distribution function (CCDF) and PAPR. We can see that after pre-equalization, PAPR will increase. If pre-dispersion is used, PAPR will be much higher than other methods.

Then, we compared the BER performance with pre-CD and dispersion-compensating fiber. According to Fig. 10.24, using the pre-CD method, the receiver sensitivity of the CAP-16 can achieve a gain of approximately 2 dB under the HD-FEC threshold. Finally, we also investigate the BER performance utilizing pre-CD and SSB with DDMZM at 240 km case. Their performance is similar (Fig. 10.25).

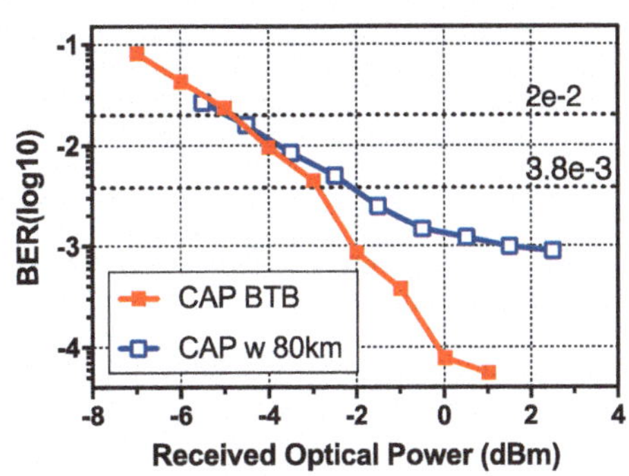

Fig. 10.22 BER versus received optical power of CAP-16 at BTB and 80 km cases with DDMZM

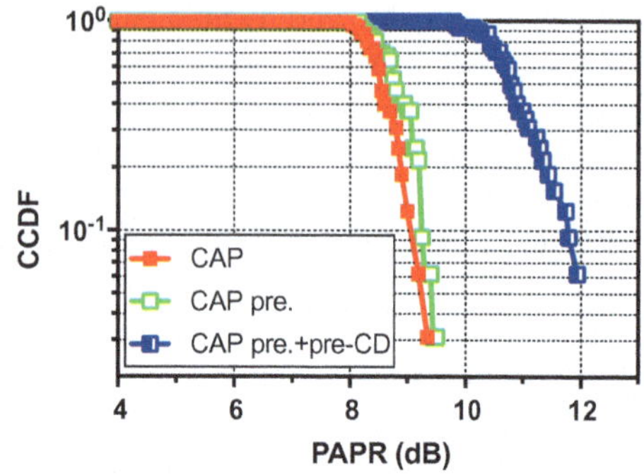

Fig. 10.23 CCDF versus PAPR of CAP-16 without pre-equalization, with pre-equalization and with both pre-equalization and pre-CD

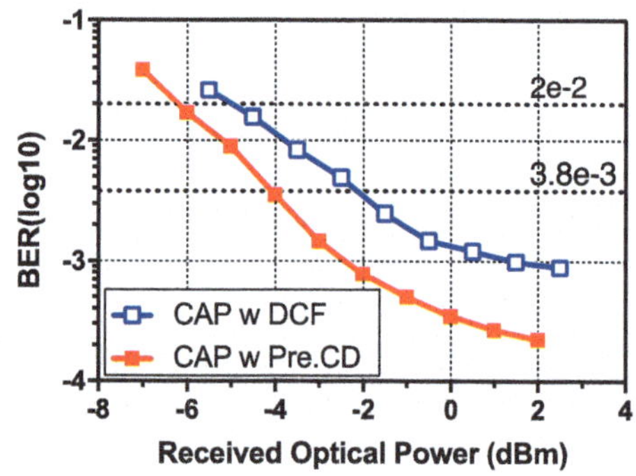

Fig. 10.24 BER versus received optical power of CAP-16 utilizing pre-CD and DCF with DDMZM at 80 km case

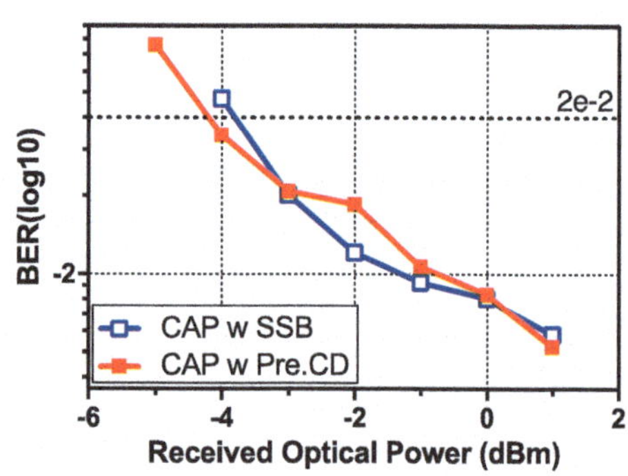

Fig. 10.25 BER versus received optical power of CAP-16 utilizing pre-CD and SSB with DDMZM at 240 km case

10.6.5 Comparison Between Pre-CD and SSB with DDMZM

SSB is another way to overcome the CD limitation in the systems with direct detection. We also want to compare the performance between applying pre-CD and SSB for three modulation formats. The results are firstly obtained over 240 km SSMF transmission as shown in Fig. 10.8. SSB signals slightly outperform pre-CD signals. However, employing SSB will introduce 3 dB SNR loss compared with DSB signal

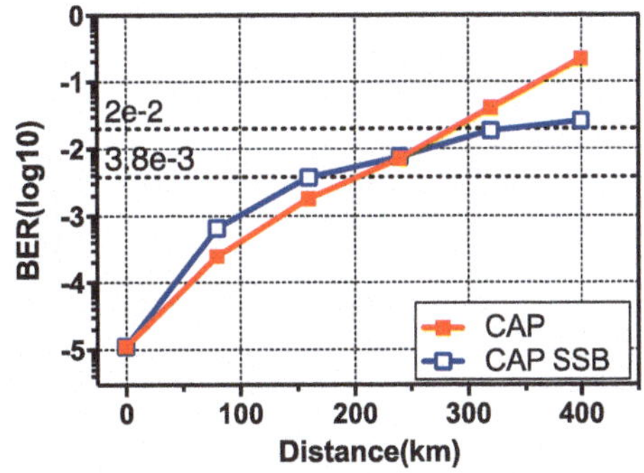

Fig. 10.26 BER versus transmission distance of CAP-16 utilizing pre CD and SSB with DDMZM

in [27], where DDMZM over 80 km SSMF is utilized. Our results do not seem to match with their conclusion.

Therefore, we continue to study the performance comparison between pre-CD and SSB at different fiber lengths, as shown in Fig. 10.26. When the transmission length is short (about less than 240 km), the pre-CD signal is better than the SSB signal, which matches with the conclusion of the reference. When the transmission distance increases, SSB will have a better performance. The reason is the nonlinearity of the DDMZM [29]. The effect of the nonlinearity will sharply deteriorate as the transmission distance increases.

10.6.6 Comparison Between DDMZM and IQ Modulator

As DDMZM is roughly similar to the IQ function [29], we also investigate the performance of the IQ modulator in this system. Figure 10.27 shows the BER performance versus received optical power of CAP-16 utilizing DDMZM and IQ modulators in the BTB case. Employing IQ modulator will get about 2 dB receiver sensitivity gain compared with DDMZM at the HD-FEC threshold.

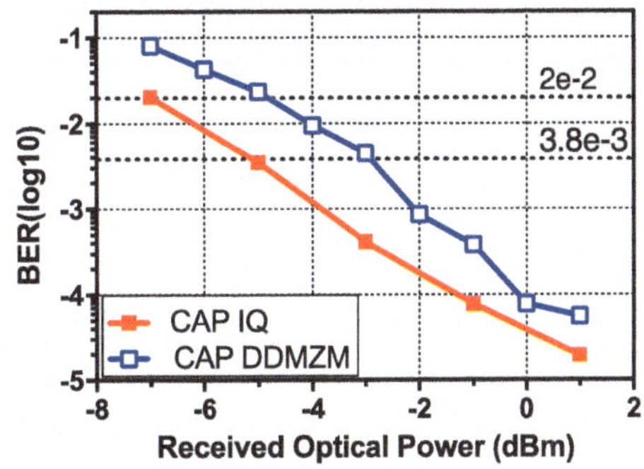

Fig. 10.27 BER versus received optical power of CAP-16 utilizing DDMZM and IQ modulator at BTB case

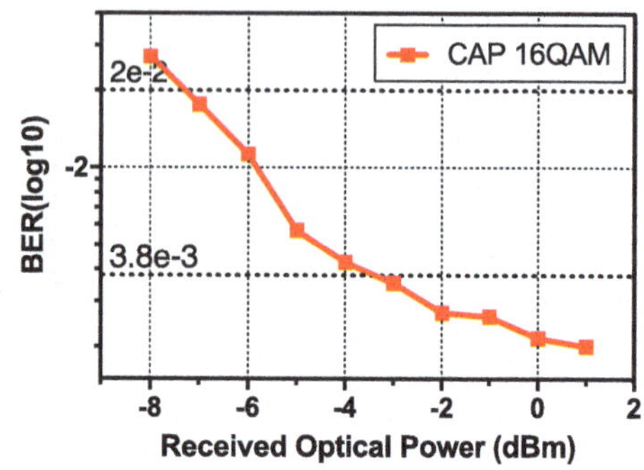

Fig. 10.28 BER versus received optical power of CAP-16 utilizing IQ modulator at 400 km case

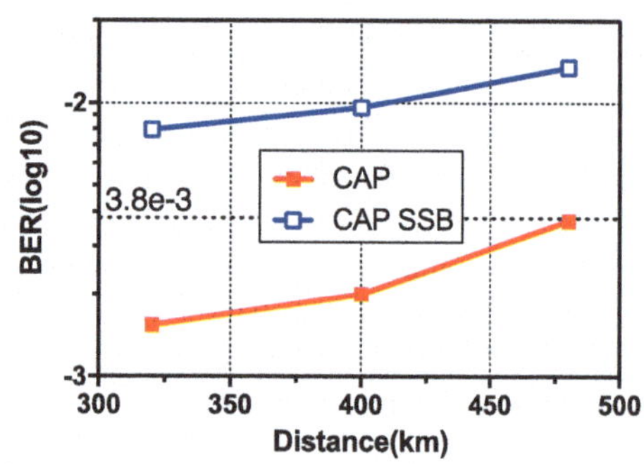

Fig. 10.29 BER versus transmission distance of CAP-16 utilizing pre-CD and SSB with IQ modulator

First, we measure BER performance versus the received optical power using the pre-CD method and the IQ modulator over 400 km SSMF transmission, as shown in Figs. 10.28.

Then, we investigate the BER performance versus the transmission distance utilizing pre-CD method and SSB with the IQ modulator as shown in Fig. 10.29. Unlike the phenomenon of DDMZM, DSB signals are always better than SSB signals with the IQ modulator. Finally, a 480 km transmission is experimentally demonstrated utilizing CAP-16 under the HD-FEC threshold of 3.8×10^{-3}. To our knowledge, this is the longest transmission distance for 100 G signal per lane with direct detection.

The optical spectra of DSB and SSB signals for three modulation formats over 400 km and 480 km SSMF are shown in Fig. 10.30. According to the above results, we believe that applying an IQ modulator could achieve better performance and acceptable cost in the metro networks.

10.6.7 Conclusion

We experimentally demonstrate a 100G transmission experiment of the metro optical transmission system based on CAP-16 low-cost direct detection. Among three types of CD compensation methods including pre-CD, SSB, and DCF, DCF has the worst

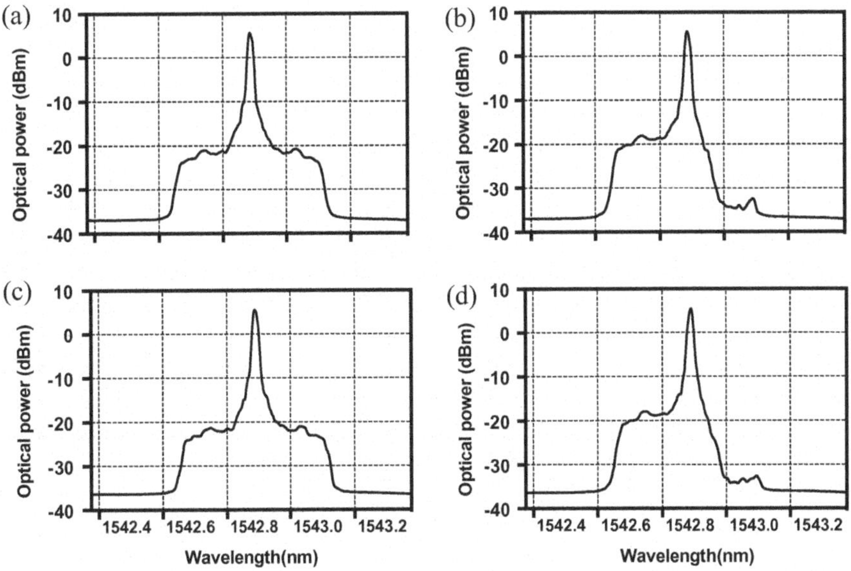

Fig. 10.30 Optical spectra (0.1 nm resolution) **a** CAP-16 DSB,400 km SSMF,**b** CAP-16 SSB,400 km,**c** CAP-16 DSB, 480 km, **d** CAP-16 SSB, 480 km

performance but no extra DSP and it is suitable for low-cost single-drive MZM. Pre-CD and SSB are both generated by DDMZM or IQ modulator. Utilizing pre-CD method, CAP-16 can get about 2 dB receiver sensitivity gain at the HD-FEC threshold compared with DCF. And pre-CD signal is always better than SSB signal for IQ modulator, while it is different for DDMZM due to the nonlinearity of the modulators. For DDMZM, when the transmission length is short (about less than 240 km), pre-CD signal is better than SSB signal. When the transmission distance increases, SSB will have a better performance.

In the whole experiment, the pre-CD method with an IQ modulator shows the best performance in the medium distance transmission system with direct detection. To our knowledge, it is the first time to achieve a 100G single-wavelength over 480 km SSMF transmission with the modulation format of CAP-16 based on direct detection in metro networks.

10.7 Summary

This chapter studies the carrierless amplitude and phase modulation (CAP). Firstly, the principle and implementation method of CAP modulation and demodulation are introduced, including the generation and reception of single-band CAP-mQAM, and the generation and reception of multi-band multi-level CAP signals. Then, a

WDM-CAP-PON multi-user access network based on multi-band multi-level CAP modulation is introduced, and its high-speed access performance is verified. The experiment successfully demonstrates 11 WDM channels, 55 sub-bands, for 55 users with 9.3-Gb/s per user (after removing 7% overhead for forward error correction (FEC)) in the downstream over 40-km SMF. To the best of our knowledge, this is the first demonstration for MM-CAP used for WDM-CAP-PON with such a high speed. Then, the transmission experiment of wireless access networks based on high-order modulation CAP-64QAM is introduced, which realizes a 24-Gb/s CAP-64QAM radio-over-fiber (ROF) system over 40-km stand single-mode fiber (SMMF) and 1.5-m 38-GHz wireless transmission. Finally, we propose and experimentally demonstrate a high-speed CAP-64QAM system using DML based on direct detection and digital equalizations. Modified decision-directed least mean squares (DD-LMS) are used to equalize the CAP-64QAM. Using this scheme, we successfully generate and transmit up to 60-Gb/s CAP-64QAM over 20-km standard single-mode fiber (SSMF) based on the 10G-class DML and direct detection.

References

1. D. Breuer, F. Geilhardt, R. Hülsermann, M. Kind, C. Lange, T. Monath, E. Weis, Opportunities for next-generation optical access. IEEE Commun. Mag. **49**(2), s16–s24 (2011)
2. G.K. Chang, A. Chowdhury, Z. Jia, H.C. Chien, M.F. Huang, J. Yu, G. Ellinas, Key technologies of WDM-PON for future converged optical broadband access networks. J. Opt. Commun. Netw. **1**(4), C35–C50 (2009)
3. J.M. Kang, S.K. Han, A novel hybrid WDM/SCM-PON sharing wavelength for up- and down-link using reflective semiconductor optical amplifier. IEEE Photon. Technol. Lett. **18**(3), 502–504 (2006)
4. A. Wiberg, B. Olsson, P. Andrekson, Single cycle subcarrier modulation, in *Proceedings of OFC 2009*, Paper OTuE1
5. J. Karout, M. Karlsson, E. Agrell, Power efficient subcarrier modulation for intensity modulated channels. Opt. Express **18**(17), 17913–17921 (2010)
6. J. Karout, E. Agrell, K. Szczerba, M. Karlsson, Optimizing constellations for single-subcarrier intensity-modulated optical systems. IEEE Trans. Inf. Theory **58**(7), 4645–4659 (2012)
7. B. Liu, X. Xin, L. Zhang, J. Yu, Q. Zhang, C. Yu, A WDM-OFDM-PON architecture with centralized lightwave and PolSK-modulated multicast overlay. Opt. Express **18**(3), 2137–2143 (2010)
8. N. Cvijetic, M. Cvijetic, M.F. Huang, E. Ip, Y.K. Huang, T. Wang, Terabit optical access networks based on WDM-OFDMA-PON. J. Lightwave Technol. **30**(4), 493–503 (2012)
9. J.L. Wei, D.G. Cunningham, R.V. Penty, I.H. White, Study of 100 Gigabit ethernet using carrierless amplitude/phase modulation and optical OFDM. J. Lightwave Technol. **31**(9), 1367–1373 (2013)
10. J. D. Ingham, R. Penty, I. White, and D. Cunningham, "40 Gb/s carrierless amplitude and phase modulation for low-cost optical data communication links," in Proc. OFC2011, paper OThZ3
11. R. Rodes, M. Wieckowski, T.T. Pham, J.B. Jensen, J. Turkiewicz, J. Siuzdak, I.T. Monroy, Carrierless amplitude phase modulation of VCSEL with 4 bit/s/Hz spectral efficiency for use in WDM-PON. Opt. Express **19**(27), 26551–26556 (2011)
12. M.B. Othman, X. Zhang, L. Deng, M. Wieckowski, J. Jensen, I.T. Monroy, Experimental investigations of 3D/4D-CAP modulation with DM-VCSELs. IEEE Photon. Technol. Lett. **24**(22), 2009–2012 (2012)

13. M. I. Olmedo, T. Zuo, J. B. Jensen, Q. Zhong, X. Xu, and I. T. Monroy, "Towards 400GBASE 4-lane solution using direct detection of multiCAP signal in 14 GHz bandwidth per lane," in Proc. of OFC2013, paper PDP5C.10
14. J. Wei, L. Geng, D. G. Cunningham, R. V. Penty, and I. White, "100 Gigabit Ethernet transmission enabled by carrierless amplitude and phase modulation using QAM receivers," in Proc. of OFC2013, paper OW4A.5
15. L. Tao, Y. Wang, Y. Gao, A.P. Lau, N. Chi, C. Lu, Experimental demonstration of 10 Gb/s multi-level carrier-less amplitude and phase modulation for short range optical communication systems. Opt. Express **21**(5), 6459–6465 (2013)
16. G.H. Im, D.D. Harman, G. Huang, A.V. Mandzik, M.H. Nguyen, J.J. Werner, 51.84 Mb/s 16-CAP ATM LAN standard. IEEE J. Sel. Areas Comm. **13**(4), 620–632 (1995)
17. J. Yu, M.F. Huang, D. Qian, L. Chen, G.K. Chang, Centralized lightwave WDM-PON employing 16-QAM intensity modulated OFDM downstream and OOK modulated upstream signals. IEEE Photon. Technol. Lett. **20**(18), 1545–1547 (2008)
18. J. Zhang, J. Yu, L. Tao, Y. Fang, Y. Wang, Y. Shao, N. Chi, Generation of coherent and frequency-lock optical subcarriers by cascading phase modulators driven by sinusoidal sources. J. Lightwave Technol. **30**(24), 3911–3917 (2012)
19. Junwen Zhang, Yu. Jianjun, Fan Li, Nan Chi, Ze Dong, Xinying Li, 11 × 5 × 9.3 Gb/s WDM-CAP-PON based on optical single-side band multi-level multi-band carrier-less amplitude and phase modulation with direct detection. Opt. Express **21**, 18842–18848 (2013)
20. R.P. Davey and D.B. Payne, "The future of optical transmission in access and metro networks – an operator's view," in ECOC 2015 Proceeding, vol. 5, Symposium We 2.1.3
21. H. Rohde, E. Gottwald, A. Teixeira et al., Coherent ultra dense WDM technology for next generation optical metro and access networks. J. Lightwave Technol. **32**(10), 2041–2052 (2014)
22. J. C. Rasmussen, T. Takahara, T. Tanaka, et al., "Digital signal processing for short reach optical links," in Proc. IEEE Eur. Conf. Opt. Commun. (ECOC), Sep. 2014, pp. 1–3
23. J. Estaran, M. Iglesias, D. Zibar, et al., "First experimental demonstration of coherent CAP for 300-Gb/s metropolitan optical networks," OFC, 2014: Th3K. 3
24. J. Wei, Q. Cheng, D.G. Cunningham et al., 100-Gb/s hybrid multiband CAP/QAM signal transmission over a single wavelength. J. Lightwave Technol. **33**(2), 415–423 (2015)
25. S. Randel, D. Pilori, S. Chandrasekhar, et al., "100-Gb/s discrete-multitone transmission over 80-km SSMF using single-sideband modulation with novel interference-cancellation scheme," ECOC 2015. European Conference on. IEEE, 2015: 1–3
26. R. Okabe, B. Liu, M. Nishihara, et al., "Unrepeated 100 km SMF Transmission of 110.3 Gbps/lambda DMT signal," ECOC, 2015. European Conference on. IEEE, 2015: 1–3
27. J. Zhou, L. Zhang, T. Zuo, et al., Transmission of 100-Gb/s DSB-DMT over 80-km SMF Using 10-G class TTA and Direct-Detection, ECOC 2016; 42nd European Conference on Optical Communication; Proceedings of. VDE, 2016: 1–3
28. C.Y. Wong, S. Zhang, L. Liu, et al., "56 Gb/s direct detected single-sideband DMT transmission over 320-km SMF using silicon IQ modulator," Optical Fiber Communication Conference. Optical Society of America, 2015: Th4A. 3
29. J. Shi, Y. Zhou, Y. Xu et al., 200-Gbps DFT-S OFDM Using DD-MZM-Based Twin-SSB with a MIMO-Volterra Equalizer. IEEE Photonics Technol. Lett. **29**(14), 1183–1186 (2017)
30. A. Lebedev, J.J.V. Olmos, M. Iglesias et al., A novel method for combating dispersion induced power fading in dispersion compensating fiber. Opt. Express **21**(11), 13617–13625 (2013)
31. H. Schmuck, Comparison of optical millimetre-wave system concepts with regard to chromatic dispersion. Electron. Lett. **31**(21), 1848–1848 (1995)
32. N. Kikuchi, R. Hirai, Y. Wakayama. High-speed optical 64QAM signal generation using InP-based semiconductor IQ modulator, Optical Fiber Communications Conference and Exhibition (OFC), 2014. IEEE (2014)

Chapter 11
PAM4 Signal Modulation and Digital Signal Processing-Based Detection Technology

11.1 Introduction

To meet the rapidly growing demand for data center traffic, flexible and low-cost 400 Gbit/s transmission schemes have been proposed as alternatives for next-generation data center interconnect applications [1]. Recently, intensity modulation/direct detection systems (IM/DD) based on advanced modulation formats, such as pulse amplitude modulation (PAM), carrierless phase modulation (CAP), discrete multi-tone (DMT), as a low-cost data center interconnection scheme have attracted the interest of many researchers [1]. In order to support 400 Gbit/s transmission, one of the promising solutions is to use PAM modulated four-channel $\times$ 100 Gbit/s/λ transmission, which can reduce the transceiver design complexity and energy consumption [1]. In addition, some single-channel 100 Gbit/s short-distance access experiments have been published. These experiments use PAM4, PAM8 internal modulation or external modulation scheme [2–5]. It is worth noting that the internal modulation scheme using an electro-absorption modulated laser (EML) [3, 4] or a direct modulated laser (DML) [5] is less expensive and simpler to design than a Mach–Zehnder modulator (MZM) [1, 2] based on external modulation. However, there are two bottlenecks for these schemes that limit the performance of the system: modulation bandwidth limitations of optoelectronic devices and nonlinear damage during modulation and demodulation. Many digital signal processing (DSP) methods have been proposed to address these two limitations, such as decision feedback equalization [1], nonlinear Volterra equalization [5], which require high computational complexity at the receiving end. As an alternative, this chapter proposes a simple design of a lookup table (LUT) pre-distortion nonlinear compensation scheme [6], and experimental verification in the PAM4 system.

J. Yu and N. Chi, *Digital Signal Processing In High-Speed Optical Fiber Communication Principle and Application*,
https://doi.org/10.1007/978-981-15-3098-2_11

11.2 Principle and Related Algorithms of PAM4 Modulation

11.2.1 Principle of PAM4 Modulation

The amount of data in today's era is showing a spurt of growth, which not only means more data, but also means faster data transmission rates. Traditional non-return to zero (NRZ) modulation schemes are no longer suitable for development needs. We need to explore more efficient point-to-point transmission schemes, which can be used in fiber transmission and chip-level transmission. One of the most popular modulation methods is pulse amplitude modulation (PAM). We first discuss the basic concepts of PAM4 modulation and then analyze the experimental results and the challenges faced.

For a long time, the non-return to zero coding scheme has been the mainstream data transmission modulation method. In NRZ, we encode the bitstream information into a series of level values, such as 001100, low level represents 0, high level represents 1 (as shown in Fig. 11.1). It is assumed that the bit rate is 28 Gbit/s.

In Fig. 11.1, PAM4 is modulated by 4-level amplitude, and each level value can carry 2 bits of information at the expense of being more sensitive to noise. If we observe the eye diagram of the NRZ signal, if the bit period is T and the amplitude is A, then the channel bandwidth is the reciprocal of the bit period ($1/T$). The higher the bit rate, the smaller the bit period and the larger the signal bandwidth. There is usually a requirement for signal-to-noise ratio (SNR), which is related to the signal amplitude. From a vertical perspective, the smaller the amplitude of the eye diagram opening, the more difficult it is to distinguish the original signal from the receiving end with a fixed (SNR).

Fig. 11.1 PAM4 signal modulation principle

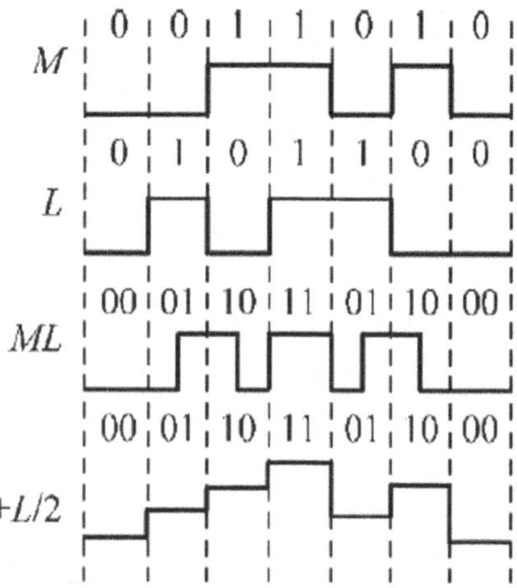

In general, one of the ways to double the point-to-point transmission rate is to use two channels. On another channel, different bit streams are transmitted, such as 0101100. However, this method also has obvious disadvantages, generally requiring two transmitters, two receivers, and two channels. We may not want to pay extra space or energy consumption, so we need to find other solutions.

Is there any way to double the bit rate? One method is to serialize two bit streams and replace two 28 Gbit/s channels with one 56 Gbit/s channel. Thus, at the original 28 Gbit/s rate, the current rate is now 56 Gbit/s (Fig. 11.1). As can be seen from the eye diagram of the ML signal, its amplitude is still A, but the period becomes $T/2$. If the reciprocal of the bit period is calculated, the signal bandwidth $2/T$ will be obtained. The SNR remains unchanged with A, but the signal bandwidth is doubled. Therefore, considering the SNR and bandwidth, this scheme has its own advantages and disadvantages.

We need a solution that doubles the bit rate without increasing bandwidth, which is the advantage of PAM4. The PAM4 signal maps the L (least significant bit) and the M (most significant bit) to a 4-level amplitude, each level representing 2 bits of information. The PAM4 signal is equivalent to M + L/2 in Fig. 11.1. The level from low to high represents 00, 01, 10, and 11, and PAM4 represents 4-level pulse amplitude modulation.

The eye diagram of the PAM4 is unusual. There are 3 open eyes and 4 amplitudes in the longitudinal direction, and the symbol period is T. However, each eye has an opening amplitude of A/3 and a corresponding bandwidth requirement of 1/T. In this way, we get a 56 Gbit/s signal, which is the same as the bandwidth of the 28 Gbit/s single signal M or L, but the SNR is related to A/3, so PAM4 has a trade-off between SNR and signal bandwidth. Many serial links are bandwidth limited, so it is difficult to increase 28 Gbit/s by shortening the bit period. However, when there is an SNR tolerance space, a PAM4 scheme that sacrifices a portion of the SNR exchange rate multiple times would be a good choice.

11.2.2 DD-LMS Algorithm

The decision-direct minimum means square error (DD-LMS) algorithm is a widely used blind equalization algorithm. The steady-state error of the algorithm is small, but only when the bit error rate drops to the order of 10^{-1}, 10^{-2}, and the eye diagram is basically open, the algorithm can show the effect [7].

Figure 11.2 shows the algorithmic schematic of the DD-LMS equalizer. Since the DD-LMS algorithm is performed after the inverse discrete Fourier transform (IDFT), the DD-LMS equalizer can be regarded as a time-domain equalization.

The relationship between an L-tap DD-LMS equalizer input x(k) and an output $y(k)$ is

$$y(k) = w(k)^H x(k) \tag{11.1}$$

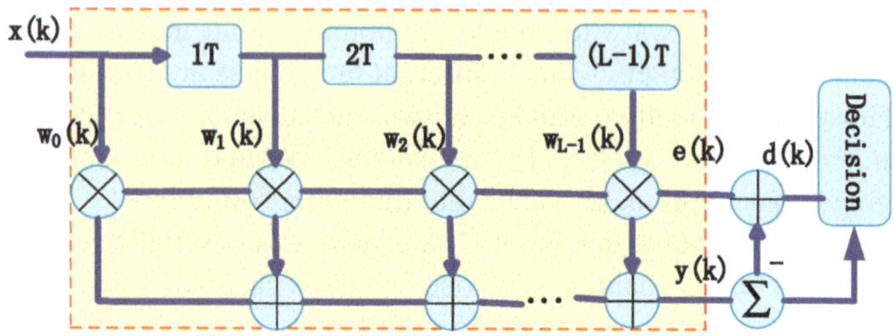

Fig. 11.2 Schematic diagram of the DD-LMS algorithm

where $w(k)$ represents the tap coefficients at time k, $x(k)$ and $y(k)$ represent the input and output vectors at time k, and $()^H$ represents the conjugate transpose of the matrix. The expressions for $x(k)$ and $w(k)$ are

$$x(k) = [x(k), x(k-1), x(k-2), \ldots x(k-L+1)]\,\overline{T|} \tag{11.2}$$

$$w(k) = [w_0(k), w_1(k), w_2(k), \ldots, w_{L-1}(k)]^T \tag{11.3}$$

The output value $y(k)$ at time k is sent to the decider for symbol decision, and $d(k)$ is the standard constellation point output by the decider. The error between the two is

$$e(k) = d(k) - y(k) \tag{11.4}$$

The tap coefficient is updated at the next moment according to the error value $e(k)$. The iterative process expression is

$$w(k+1) = w(k) + \mu e(k)^* x(k) \tag{11.5}$$

In the formula (11.5), μ is called the step size, and determines the speed at which the error converges. When the tap coefficient is updated at $k + 1$, the new sequence $x(k)$ is sent, and iteratively continues to complete the update of the output sequence $y(k)$. As a stochastic gradient descent algorithm, the DD-LMS algorithm does not depend on the statistical characteristics of the output symbols, but on the symbol decision process [8].

11.2.3 Principle of Dispersion Pre-compensation

Electrical dispersion compensation (EDC) is considered a flexible and efficient way to improve the performance of optical transmission systems. In the direct detection system, the phase information of the signal is lost due to the square-law detection of the photodetector (PD) at the receiver end, so that the use of the EDC at the receiving end is limited. An alternative is to use EDC at the transmitter [9] to pre-distort the amplitude and phase of the transmitted signal.

The transfer function in the fiber can be considered linear without considering the nonlinearity of the single-mode fiber [10]. In this case, the frequency-domain response of the fiber dispersion can be modeled as

$$H(\omega) = \exp(j\frac{\lambda^2}{4\pi c}DL\omega^2) \tag{11.6}$$

where λ is the wavelength of light, D is the dispersion coefficient, L is the transmission distance, c is the speed of light, and the dispersion effect can be compensated by the reciprocal of the fiber's frequency-domain response

$$H^{-1}(\omega) = \exp(-j\frac{\lambda^2}{4\pi c}DL\omega^2) \tag{11.7}$$

It can be known from Eq. (11.7) that the electrical dispersion pre-compensation can effectively attenuate the dispersion effect at the transmitting end.

11.2.4 Lookup Table Algorithm

Lookup table (LUT) pre-distortion is primarily used to reduce mode impairments associated with transmitted sequences in high-speed systems [11]. It has the advantages of low computational complexity and flexible configuration, and can be applied to correct nonlinear damage problems in intensity modulation/direct detection (IM/DD) systems.

Figure 11.3 illustrates the schematic diagram of the LUT algorithm in the PAM4 transmission system. The transmission symbol sequence $X(k - M: k: k + M)$ is 2M+1 in length, referring to a pattern in the PAM4 signal, where $X(k) \in \{\pm 1, \pm 3\}$. In the initial state, the data in the LUT is all set to zero. The sliding window selects 2M+1 symbols in the transmission sequence each time, and calculates the address of this mode, that is, the LUT index i. $Y(k - M: k: k + M)$ represents the recovery

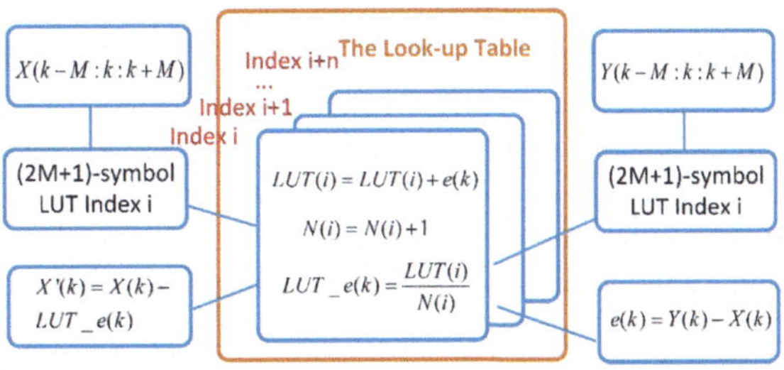

Fig. 11.3 Schematic diagram of the LUT algorithm

sequence obtained at the receiving end, and the center symbol of the transmission sequence and the reception sequence is subtracted to obtain $e(k)$, that is

$$e(k) = Y(k) - X(k) \tag{11.8}$$

As the sliding window moves forward, more and more data is stored in the LUT, and the parameter $N(i)$ is used to record the number of data stored in the LUT index i. The final data in each LUT index i is the average of all the differences $e(k)$. The specific calculation process is

$$LUT(i) = LUT(i) + e(k) \tag{11.9}$$

$$N(i) = N(i) + 1 \tag{11.10}$$

$$LUT\_e(k) = \frac{LUT(i)}{N(i)} \tag{11.11}$$

LUT_e(k) is the final value in each LUT index i. Once the LUT based on various possible modes is established, it can be arranged to the transmitting end for transmitting pre-distortion of data. The specific pre-distortion process is as follows

$$X'(k) = X(k) - LUT\_e(k) \tag{11.12}$$

11.3 PAM4 High-Speed Transmission System

11.3.1 4-Lane IM/DD 112.5 Gbit/s PAM4 Transmission System

Figure 11.4 illustrates the schematic and signal processing flow for a 4-lane IM/DD 112.5 Gbit/s PAM4 transmission system [12]. As can be seen from the figure, 4 independent electro-absorption modulated lasers (EMLs) are used as laser sources with a channel spacing of 0.8 nm, a 3 dB bandwidth of 40 GHz, an optimum bias voltage of −1.4 V, and a drive current of 100 mA. At the transmitter end, a high-speed digital-to-analog conversion (DAC) module with a sampling rate of 80 GSa/s and a 3 dB bandwidth of 20 GHz produces four channels of 56.25 Gbaud PAM4 signals. The four channels data is first mapped to the PAM4 symbol, followed by a lookup table (LUT) pre-distortion and pre-equalization process to reduce the nonlinear damage of the system. At the receiver end, it is converted by a 3 GHz photodetector (PD) with a bandwidth of 40 GHz. The signal is sampled and processed offline by a digital storage oscilloscope (DSO) with a sampling rate of 80 GSa/s and 3 GHz bandwidth of 33 GHz. The received signal is first resampled to 112.5 GSa/s, then recovered by

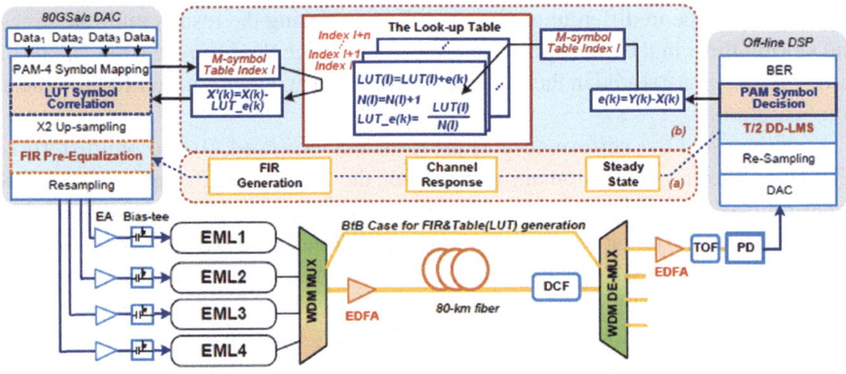

Fig. 11.4 Schematic diagram of the 4-lane IM/DD 112.5 Gbit/s PAM4 transmission system and signal processing flow. **a** Pre-equalization (Pre-EQ) based on the receiver equalizer; **b** Pre-distortion (Pre-DT) based on LUT

a DD-LMS equalizer with 21 tap coefficients and $T/2$ interval. And the bit error rate (BER) is calculated after PAM4 symbol decision.

For an 80-km single-mode fiber link, 4 PAM4 signals are injected into the fiber channel after passing through a WDM multiplexer and an erbium-doped fiber amplifier (EDFA). The optimum light emission power was determined experimentally. After passing through the tunable optical filter (TOF), the receiving end is connected to another EDFA to suppress the amplified spontaneous emission (ASE) noise. To compensate for the dispersion, a matched dispersion compensating fiber (DCF) is cascaded behind the single-mode fiber.

Before the fiber is transmitted, the LUT pre-distortion and pre-equalization process is first performed. In order to obtain good channel quality for the LUT correction process, we use the training sequence to estimate the channel condition in an end-to-end case. The parameters of the pre-equalization filter at the transmitting end are determined by the tap coefficients of the DD-LMS adaptive filter at the receiving end (as shown in Fig. 11.5a). Figure 11.5b illustrates the generation process of the LUT, mainly by comparing the difference between the transmitted sequence and the

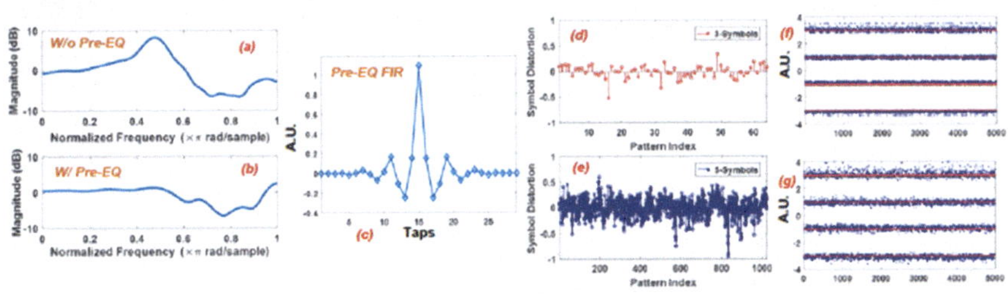

Fig. 11.5 DD-LMS equalizer frequency-domain response **a** before and **b** after Pre-EQ; **c** FIR filter tap coefficients; **d** 3 symbols and **e** 5 symbols LUT data; PAM4 signal constellation after **f** 3 symbols and **g** 5 symbols LUT

received sequence in different modes, and then averaging the results multiple times and storing them in the corresponding LUT index. Once the compensation values in different modes are stored in the LUT, it can be placed at the transmitter for nonlinear compensation.

We first tested the performance of pre-equalization (Pre-EQ) and pre-distortion (Pre-DT). The frequency-domain response of the receiver end adaptive DD-LMS equalizer is shown in Fig. 11.5a, b. A flat frequency-domain response is obtained after pre-equalization. The tap coefficients of the pre-equalized FIR filter are generated by an adaptive DD-LMS equalizer (Fig. 11.5c). Figure 11.5d, e shows the 3 symbols (64 modes) and 5 symbols (1024 modes) LUT processes. Larger distortion values can be corrected using longer symbols. Figure 11.5f, g shows the constellation points of the PAM4 signal after pre-distortion through the 3 symbols and 5 symbols LUTs.

Back-to-back BER performance is shown in Fig. 11.6. We tested the effect of the number of FIR filter taps and LUT symbols on the system BER (Fig. 11.6a, b). It can be obtained that the 21-tap FIR filter and the 7 symbols LUT can well satisfy the Pre-EQ and Pre-DT processes. Figure 11.6c shows the performance curves of received power and BER. There is a bit error bottleneck when Pre-EQ and Pre-DT are not used, and a receiver sensitivity gain of 10 dB is obtained at a hard decision forward error correction (HD-FEC) threshold of 3.8×10^{-3} after using Pre-EQ. An additional 2 dB sensitivity gain can be obtained from the LUT Pre-DT. When the two are applied at the same time, there is a big performance improvement. The illustrations (i), (ii), and (iii) respectively show the PAM4 constellation diagram when the received power is −5 dBm with a BER of 3.9×10^{-3}, 4.0×10^{-4}, and 0.

The 4-lane 400 Gbit/s WDM PAM4 signal spectrum is shown in Fig. 11.7a. Figure 11.7b shows the relationship curve between BER and input optical power of channel 2 after transmitting 80-km fiber in a 4-lane 400 Gbit/s system. It can be seen from the different transmission schemes in the figure that the best system performance can be obtained by using Pre-EQ and Pre-DT together, and the input power range exceeding 8 dB can meet the system error threshold requirement. The illustrations (i) and (ii) show the PAM4 constellation diagram when only Pre-EQ is used, Pre-EQ and Pre-DT are used in combination with over 80 km fiber. The BER

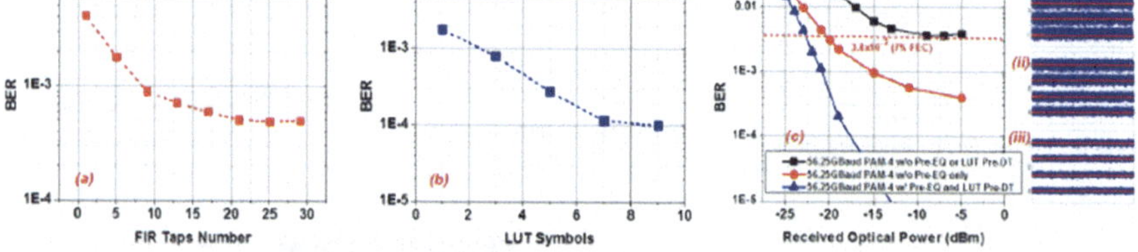

Fig. 11.6 Back-to-back error rate performance. **a** Relationship between BER and FIR taps; **b** Relationship between BER and LUT symbols; **c** Relationship between BER and received signal power under different algorithms. Illustration (i)–(iii) is the PAM4 signal recovery constellation point at −5 dBm received power

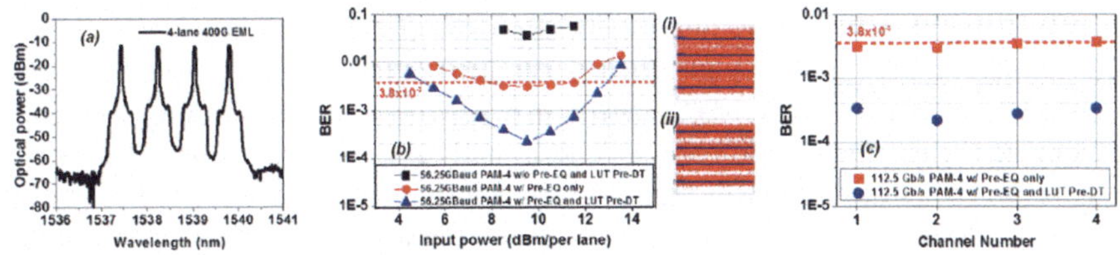

Fig. 11.7 Experimental results. **a** four channels WDM power spectrum; **b** BER and input power relationship over 80 km SMF + DCF of channel 2; **c** BER for each channel at 9.5 dBm input power

of the 112.5 Gbit/s PAM4 signal four channels at an input power of 9.5 dB is shown in Fig. 11.7c. Similarly, better system performance is verified.

11.3.2 50 Gbit/s and 64 Gbit/s PAM4 PON Downlink Transmission System

Figure 11.8 illustrates the 50 Gbit/s and 64 Gbit/s/λ PAM4 PON downlink transmission systems. At the optical line termination (OLT), a 1550 nm laser diode is used as the light source of the system. A dual-drive Mach–Zehnder modulator (MZM) with a 3 dB bandwidth of 23 GHz is biased at the quadrature position for complex modulation of the signal. Digital-to-analog conversion (DAC) modules with sampling rates of 80 GSa/s and 81.92 GSa/s are used to generate 25 Gbaud and 32 Gbaud PAM4 signals, respectively. The 3 dB bandwidth of the DAC modules is 16 GHz. The flow of sending data for I and Q channels is shown in Fig. 11.8. The bit information is mapped to the PAM4 symbol, followed by nonlinear Pre-DT and linear Pre-EQ to reduce system nonlinear damage. For the 20 km fiber transmission, the transmitter adds the dispersion pre-compensation module, which is based on the dispersion compensation in the frequency domain. The specific process is shown in the illustration (iii) in Fig. 11.8.

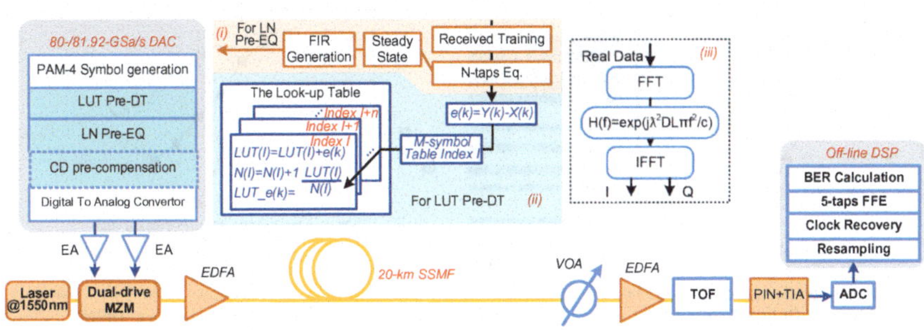

Fig. 11.8 Experimental setup of 50 Gbit/s and 64Gbit/s/λ PAM4 PON downlink transmission system

Before the fiber is transmitted, the lookup table (LUT) pre-distortion and pre-equalization process are first performed ((i) and (ii) in Fig. 11.8). In order to obtain good channel quality for the LUT correction process, we use the training sequence to estimate the channel condition in an end-to-end case. The parameters of the pre-equalization filter at the transmitting end are determined by the tap coefficients of the DD-LMS adaptive filter at the receiving end. Figure 11.8(ii) illustrates the generation process of the LUT, mainly by comparing the difference between the transmitted sequence and the received sequence in different modes, and then averaging the results multiple times and storing them in the corresponding LUT index. Once the compensation values in different modes are stored in the LUT, it can be placed at the transmitter for nonlinear compensation.

At the optical network unit (ONU) end, a variable optical attenuator (VOA) is used to measure the sensitivity of the receiver. The receiver includes an erbium-doped fiber amplifier (EDFA) that guarantees an output optical power fixed at 0dBm and a tunable optical filter (TOF) to eliminate out-of-band amplified spontaneous emission (ASE) noise. In addition, the photodiode (PIN + TIA) is used for signal detection. The PAM4 signal is sampled and processed offline by a digital storage oscilloscope (DSO) with a sampling rate of 80GSa/s and 3 GHz bandwidth of 33 GHz. The sampled signal is first resampled to 2 samples/symbol, then recovered to 1 sample/symbol by clock, and finally subjected to a 5-tap feedforward equalizer (FFE) for bit error rate (BER) calculation.

Figure 11.9a shows the relationship between the BER and the received optical power in the case of back-to-back transmission. As can be seen from the figure, the best receiver sensitivity can be obtained by using pre-equalization (Pre-EQ) and pre-distortion (Pre-DT) only in the case of a 5-tap FFE. The illustrations (i), (ii), and (iii) indicate the 64 Gbit/s PAM4 constellation diagrams without pre-digital signal processing (Pre-DSP), only Pre-EQ, and both Pre-EQ and Pre-DT. Obvious linear and nonlinear damage suppression effects can be seen from the inset. When only Pre-EQ is used, the system has approximately 1.5 dB of power loss due to nonlinear damage. Figure 11.9b shows the receiver sensitivity after 20-km fiber transmission.

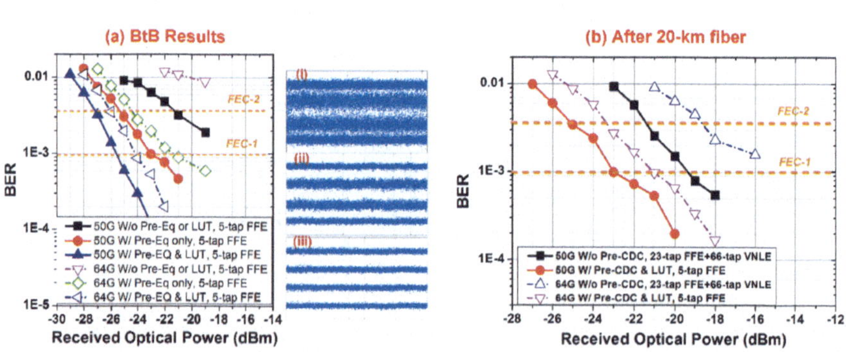

Fig. 11.9 BER curve of the experiment. **a** Relationship between BER and received power of PAM4 signal; **b** Relationship between BER and received power of PAM4 signal after 20 km fiber

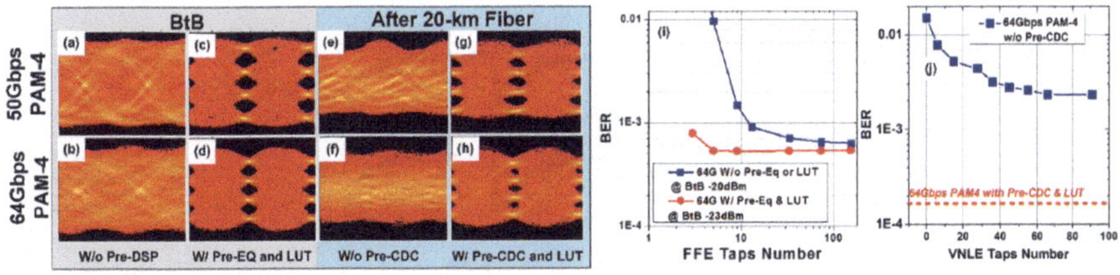

Fig. 11.10 Eye diagram of 50 Gbit/s and 64 Gbit/s PAM4 signals in different situations. (i) Relationship between BER and FFE tap number of 64 Gbit/s PAM4 signal under back-to-back case; (ii) Relationship between BER and VNLE tap number of 64 Gbit/s PAM4 signal after 20 km fiber

We compare the effects of dispersion pre-compensation and Pre-DT on system BER. When dispersion pre-compensation is not used, nonlinear compensation is critical for signal recovery. A 66-tap Volterra nonlinear equalizer (VNLE) and a 23-tap FFE are used in the experiment. However, this joint nonlinear compensation scheme is still not superior to the Pre-DSP scheme. By using a combination of dispersion pre-compensation, LUT Pre-DT, and 5-tap post-equalizer, 2 dB and 4 dB sensitivity gains are obtained for 50 Gbit/s and 64 Gbit/s PAM4 signals, respectively.

To better illustrate the superior performance of OLT digital signal processing, Fig. 11.10a–h shows the 50 and 64 Gbit/s PAM4 signal eye diagrams at the receiver end under different conditions. Even before the FFE, you can still see a clear eye (Fig. 11.10c, d, g, h). We also compared the performance of VNLE and FFE with different tap numbers. Figure 11.10(i) illustrates the effect of different tap numbers on the BER for 64 Gbit/s PAM4 signals. The joint sender signal processing (Pre-EQ and LUT) can only meet the BER requirement by using 5 taps FFE, and if there is no signal processing at the transmitting end, at least 73 taps FFE are needed. Figure 11.9(ii) illustrates the effect of different tap numbers of VNLE on the BER of a 64 Gbit/s PAM4 signal at −18 dBm received power. VNLE is not required in the case of joint transmitter signal processing (Pre-CDC and LUT), and conversely requires at least 66 taps VNLE.

11.3.3 4-Lane IM/DD 112 Gbit/s PAM4 Dispersion Pre-compensation Transmission System

Figure 11.11 shows a 4-lane IM/DD 112 Gbit/s PAM4 dispersion pre-compensation transmission system with a transmission distance of 400 km. At the transmitting end, four tunable laser diodes (LD) have a wavelength range of 1554.92–1557.35 nm, an output power of 14.5 dBm, and a wavelength interval of 100 GHz. The four LDs are divided into an odd channel group and an even channel group. The same group of LDs is multiplexed and I/Q modulated by a polarization-maintaining optical coupler (PM-OC), and the I/Q modulator has a 3 dB bandwidth of 37 GHz. The high-speed digital-to-analog conversion (DAC) module with a sampling rate of 81.92 GSa/s and

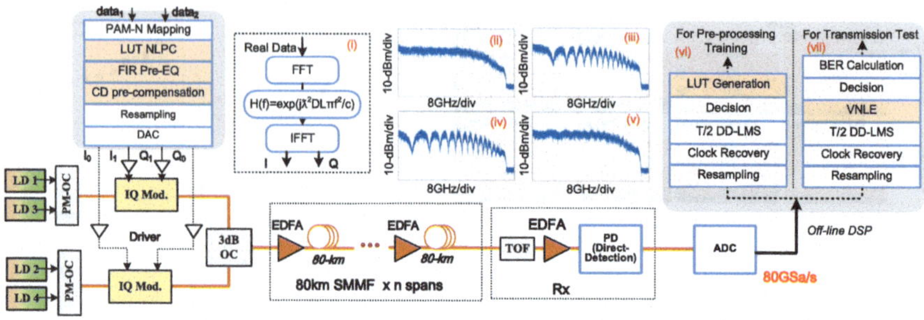

Fig. 11.11 4-lane IM/DD 112 Gbit/s PAM4 dispersion pre-compensation transmission system. (i) Frequency-domain dispersion pre-compensation; (ii)–(v) Spectra of the BTB, 80 km fiber without dispersion compensation, before and after 80 km fiber with dispersion compensation

a 3 dB bandwidth of 20 GHz produces four channels of 56.25 Gbaud PAM4 signals. The flow of the *I* and *Q* data is shown in Fig. 11.11. The data of four channels is first mapped to the PAM-4 symbol, followed by the lookup table (LUT) pre-distortion (Pre-DT) and pre-equalization (Pre-EQ) process to reduce the nonlinear damage of the system. For back-to-back (BTB) systems, dispersion compensation is not required, and the flow generation process of I and Q data is similar. For the 400 km fiber transmission, the transmitter adds the dispersion pre-compensation module, which is based on the dispersion compensation in the frequency domain. The specific process is shown in the illustration (i) in Fig. 11.11.

At the receiver end (RX), a tunable optical filter (TOF) is used to filter out the desired channel, directly detected by a photodetector (PD) with a 3 dB bandwidth of 40 GHz. The signal is sampled and processed offline by a digital storage oscilloscope (DSO) with a sampling rate of 80 GSa/s and 3 dB bandwidth of 33 GHz. The signal spectrum after direct detection is shown in Fig. 11.11(ii)–(v), which is the spectrum of the BTB, 80-km fiber without dispersion compensation, before and after 80-km fiber with dispersion compensation. By using dispersion pre-compensation, the power fading after direct detection is overcome. For offline processing, the received signal is first resampled to 56 GSa/s, followed by a 33-tap coefficient, a T/2-interval DD-LMS equalizer signal recovery. Finally, the PAM4 symbol decision is made and the BER is calculated.

Before the BTB and wavelength division multiplexing (WDM) experiments, the LUT-based Pre-DT [12] and Pre-EQ processes were first used to overcome the nonlinear damage of the channel. In order to obtain good channel quality for the LUT correction process, we use the training sequence to estimate the channel condition in an end-to-end case. The coefficient of the FIR filter at the transmitting end is determined by the tap coefficient of the DD-LMS adaptive filter at the receiving end. The generation process of the LUT is mainly by comparing the difference between the transmission sequence and the reception sequence in different modes, and then averaging the results multiple times and storing them in the corresponding LUT index. Once the compensation values in different modes are stored in the LUT, they can be placed at the transmitting end for nonlinear compensation. A 7-symbol LUT was used in the test, and the final DAC output data was pre-distorted and dispersion pre-compensated. For BER testing, after the DD-LMS algorithm, a Volterra nonlinear equalizer (VNLE) was used to further improve system performance (Fig. 11.11).

For a 400-km single-mode fiber link, four PAM4 signals are injected into the fiber channel through a 3 dB coupler and an erbium-doped fiber amplifier (EDFA). The optimum light emission power was determined experimentally. The EDFA is used to compensate for the power loss in front of each segment of 80 km fiber. The other EDFA after TOF is used to suppress the amplified spontaneous emission (ASE) noise and ensure that the optical power before inputting the PD is maintained at 0 dBm.

Figure 11.12a illustrates the relationship between the BER and the optical signal-to-noise ratio (OSNR) (dB/0.1 nm) of the 56Gbaud PAM4 signal BTB transmission. The system has a large BER bottleneck when no pre-digital signal processing (Pre-DSP) is used, and a significant OSNR gain is obtained when Pre-EQ is used. An additional 3 dB gain at the BER threshold of 3.8×10^{-3} (7% HD-FEC) is available through the Pre-DT process. When using LUT-based Pre-DT, the OSNR required for a single-channel PAM4 signal at the BER threshold of 3.8×10^{-3} is approximately 32.5 dB/0.1 nm. We also observed a slight improvement in system performance when using the receiver VNLE. When combined with Pre-EQ and LUT-based Pre-DT, significant BER performance improvement can be observed when the received power is large enough. The illustrations (i), (ii), and (iii) respectively show the PAM4 constellation diagram when the BER is 2.1×10^{-3}, 5.0×10^{-4}, and 0 under −48 dB OSNR.

We also tested the performance of the system after 80-km fiber (Fig. 11.12b). In the test, the OSNR is measured by changing the input power of the transmitter end, so when the OSNR becomes large, the BER increases due to the nonlinear effect of the optical fiber. Through experiments, we compared the effects of chromatic dispersion pre-compensation (Pre-CDC), Pre-DT, and Pre-EQ on the BER. The results again show that the system nonlinear damage can be significantly reduced by using a series of pre-equalization and post-equalization techniques.

The signal transmission result of the 4-lane 400 Gbit/s WDM PAM4 is shown in Fig. 11.13. Figure 11.13a shows the relationship between the BER and the input

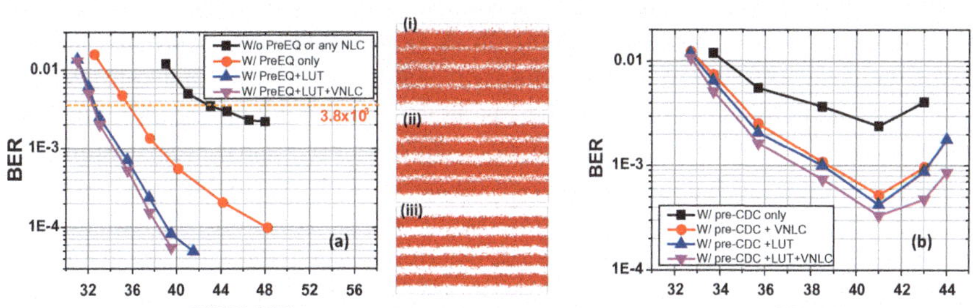

Fig. 11.12 Experimental BER curve **a** Relationship between BER and OSNR of 56 Gbaud PAM4 signal BTB transmission under different schemes; **b** Relationship between BER and OSNR of a single-channel 56 Gbaud PAM4 signal after 80 km fiber under different schemes

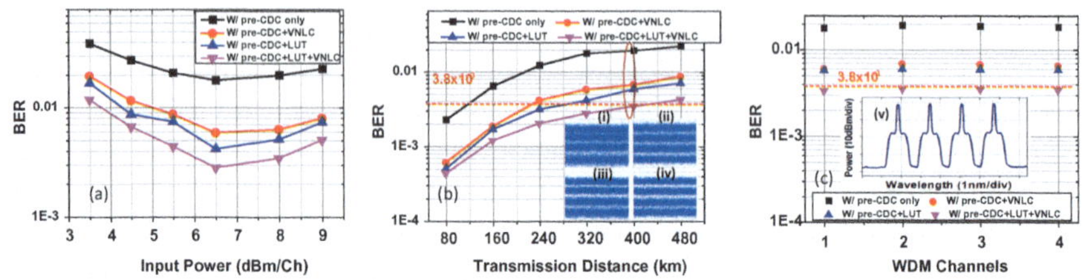

Fig. 11.13 Experimental BER graph. **a** BER versus input power for channel 2 after 320-km fiber; **b** BER versus transmission distance; **c** BER of four channels; (i)–(iv) PAM4 constellation after 400-km fiber; (v) four-channel WDM spectrum

optical power of channel 2 after transmitting 320 km fiber in a 4-lane 400 Gbit/s system. It can be seen from the different transmission schemes in the figure that the LUT-based Pre-DT and VNLE-based post-equalization used in combination give the best system performance. The optimal transmit power of each 100 Gbit/s channel is 6.5 dBm, and the BER of the channel 2 at the optimal transmit power with the transmission distance is shown in Fig. 11.13b. The illustrations (i)–(iv) respectively represent the PAM4 recovery constellation over 400-km fiber when only the Pre-CDC is used, the Pre-CDC and VNLE are used in combination, the Pre-CDC and the LUT are used in combination, and the three are used simultaneously. Channel 2 has a BER of 3.6×10^{-3} after transmission of 400-km fiber below the BER threshold of 3.8×10^{-3} (7% HD-FEC). The BER of the 4-lane 112 Gbit/s PAM4 signal after 400-km fiber is shown in Fig. 11.13c. The BER is lower than the threshold of 3.8×10^{-3} (7% HD-FEC). The 4-lane 400 Gbit/s WDM PAM4 signal spectrum is shown in illustration (v).

11.3.4 400 Gbit/s PDM-PAM4 Signal Generation and Coherent Detection

In order to cope with the increasing bandwidth demand, the data rate of the network side and the client side has reached 400 Gbit/s [13–20]. There are currently several ways to achieve 400 Gbit/s rates, including wavelength division multiplexing (WDM) to increase the number of channels, increasing the signal baud rate or increasing the modulation order [15–20]. Considering the complexity, cost, and power consumption issues, the number of channels should be reduced as much as possible to achieve efficient device integration of the optical interface [21–23]. Therefore, the data transfer rate can only be increased from the perspective of increasing the signal baud rate and the modulation order. In addition, pulse amplitude modulation (PAM) has attracted the interest of many researchers in the field of short-range optical communication and metropolitan area networks [15] due to its simple setup, low cost, and high spectrum efficiency. In order to further increase the single-channel data rate, the PAM4 signal baud rate is increasing year by year. The 56 Gbaud PAM4 intensity modulation/direct detection (IM/DD) system has recently been proposed, which is the highest baud rate currently with a data rate of 112 Gbit/s [15]. One way to achieve 400 Gbit/s is to increase the number of channels, such as a four-channel 112 Gbit/s PAM4 signal [15] or a two-channel 224 Gbit/s polarization division multiplexing (PDM) PAM4 signal joint MIMO transceiver [18]. Furthermore, for the 56 Gbaud PAM4 signal, at least two channels are required to reach 400 Gbit/s, and even polarization diversity is considered. Therefore, increasing the single-channel baud rate to achieve 400 Gbit/s is an efficient and promising next-generation network solution. In addition, the adoption of PDM will make the direct detection system receiver more complicated [18], and a better solution is to use coherent detection to bring better performance [19, 20].

The 120-Gbaud PDM-PAM4 signal coherent detection system is shown in Fig. 11.14. The PAM4 signal is generated by three-stage electrical time-division multiplexing (ETDM). First, send two 7.5 Gbit/s uncorrelated pseudo-random binary sequences (PRBS, 215-1) to form a 15 Gbit/s binary sequence through a 2:1 multiplexer (MUX), and then form a 60 Gbit/s binary sequence through a 4:1 MUX, as shown in Fig. 11.14(i); and then form a 120Gbaud NRZ signal through another 2:1 MUX, as shown in Fig. 11.14(ii). In this experiment, the 4:1 MUX working voltage Vpp is 500 mv, which can get 60 Gbit/s binary signal output, and the 2:1 MUX working voltage Vpp is 400 mv, which can get 120 Gbit/s binary signal output. Finally, the PAM4 signal is composed of two uncorrelated NRZ signals, one of which is superimposed by a 6 dB attenuator, as shown in Fig. 11.14(iii).

For PAM4 signal modulation, an external cavity laser with a wavelength of 1549.44 nm is used as a light source with a linewidth of less than 100 kHz and output power of 13.5 dBm. The electrical domain PAM4 signal is optically modulated by a Mach–Zehnder modulator (MZM). Due to the use of coherent detection, the MZM is biased at zero. After modulation, the signal is connected to an erbium-doped fiber amplifier (EDFA) for power amplification. The polarization multiplexer is composed of a polarization-maintaining optical coupler (PM-OC), an optical delay line, and a polarization beam combiner (PBC). In the experiment, the MZM 3 dB bandwidth is 40 GHz, so the PAM4 signal will be subject to low-pass filtering damage

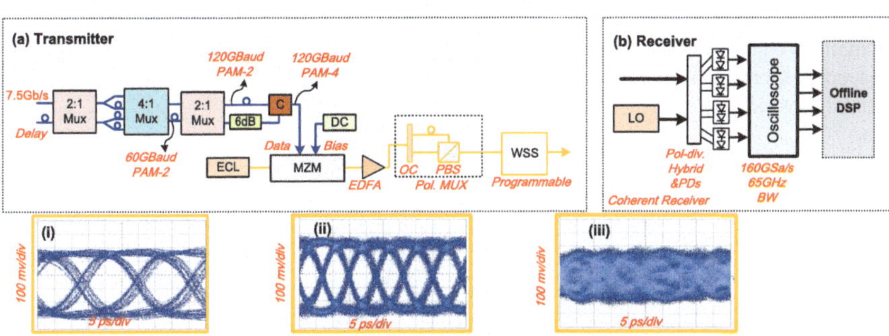

Fig. 11.14 GBaud PDM-PAM4 signal coherent detection system. Illustrations (i), (ii), (iii) are eye diagrams of 60 Gbaud NRZ signal, 120 Gbaud NRZ signal, and 120 Gbaud PAM4 signal, respectively

due to bandwidth limitation. At the same time, other optoelectronic devices, such as photodetectors (PD) and analog-to-digital converters (ADC), also have bandwidth limitations, further compressing the spectrum, resulting in impairments such as inter-symbol interference (ISI) and signal crosstalk.

In order to overcome the filtering effect caused by the bandwidth limitation, we use the wavelength selective switch (WSS) to perform the pre-equalization of the transmitter end, which is achieved by reducing the energy of the low-frequency component and increasing the energy of the high-frequency component. This pre-equalization scheme is widely used in high-speed systems with obvious effects [21, 23, 24]. Here we adopt a similar pre-equalization scheme [24], that is, the channel spectrum response is obtained by channel estimation. Figure 11.15a illustrates the principle and experimental method of channel estimation, in which the 120 Gbaud PAM2 signal is used as the training sequence. At the receiving end, the DD-LMS algorithm converges to obtain 49 adaptive tap coefficients. Figure 11.15b illustrates the adaptive filter spectral response, which is also the reciprocal of the channel response. By adaptive tap coefficients, we can design WSS to get a matching transfer function, as shown in Fig. 11.15c.

At the receiver end, another ECL is used as the local oscillator source, and the other 90° optical mixer is mainly for phase diversity reception of a single-polarization state, and the bandwidth of the balanced detector is 50 GHz. The real-time digital oscilloscope has a sampling rate of 160 GSa/s and a 3 dB bandwidth of 65 GHz. After the ADC has an oversampling rate of 1.33, the 120-Gbaud PDM-PAM4 signal is obtained. Then four channels of 160 GSa/s sampled signals are processed offline, resampled to 240 GSa/s, and passed through the adaptive filter as channel estimation, as shown in Fig. 11.15a. Since the PAM4 signal constellation points have two module values, the *T*/2 interval cascaded multi-modulus algorithm (CMMA) is introduced to pre-converge, and the 21-tap DD-LMS algorithm is further used to improve the decision effect. It is worth noting that the phase rotation due to the linewidth of the

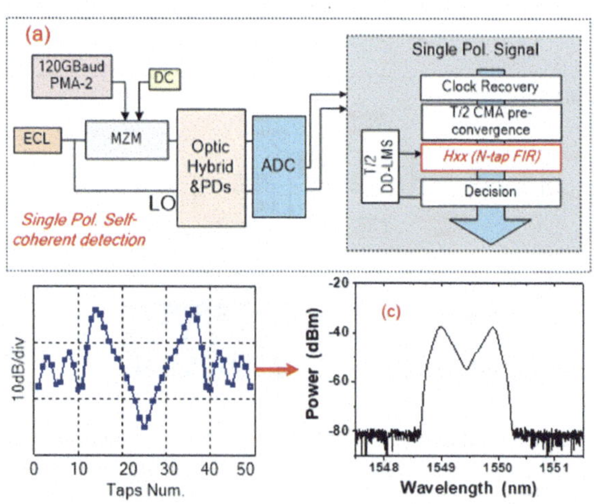

Fig. 11.15 Experimental results. **a** Channel estimation experimental setup and schematic; **b** adaptive filter frequency response; **c** WSS optical filter spectrum

laser and the frequency offset of the coherent detection still exists. The phase rotation and phase noise caused by the frequency offset can be corrected by the clock recovery algorithm, and the DD-LMS algorithm is performed after the clock recovery. Since the baud rate exceeds 100Gbaud, it is extremely sensitive to dispersion caused by optical fibers.

The signal processing results at the receiver end are shown in Fig. 11.16. The figure shows the 120 Gbaud PDM-PAM4 signal with an OSNR greater than 30 dB. Figure 11.16a, b illustrates the constellation of X and Y polarization directions after the first-stage CMMA and the amplitude information of the signal is separated by the CMMA algorithm. The second-stage DD-LMS algorithm is used to further improve performance. Figure 11.16c, d illustrates the PAM4 signal constellation after the DD-LMS algorithm. The amplitude information is clearly separated. We also counted

Fig. 11.16 Receiver signal processing results

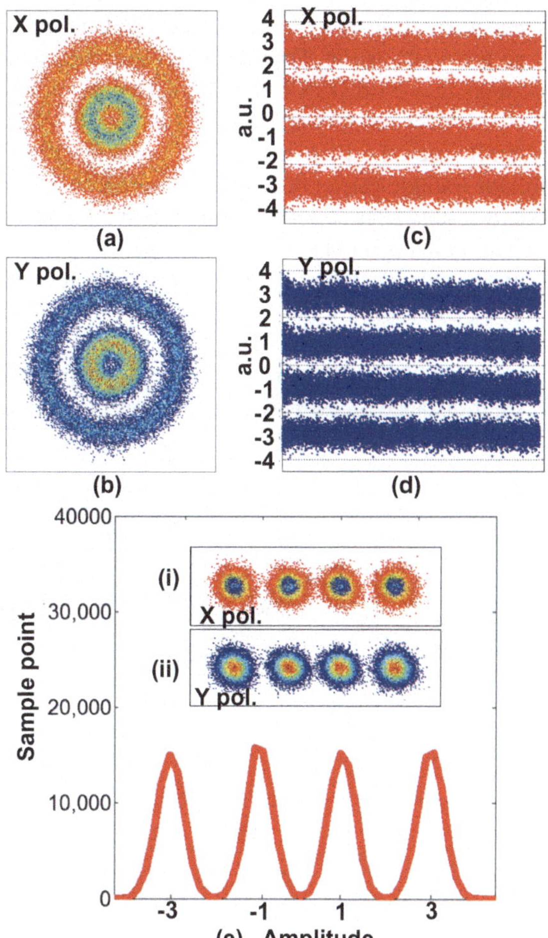

the histogram of the recovered signal amplitude information, which can be derived from the Gaussian distribution from Fig. 11.16e. The illustrations (i) and (ii) are constellations of the X and Y polarization directions of the 120 Gbaud PAM4 signal after signal processing.

To overcome the effects of filtering and receiver noise, we use the transmitter pre-equalization technique. First, the relationship between BER and pre-emphasis intensity in back-to-back (BTB) conditions is measured, which is shown in Fig. 11.17a. Here we use the α factor to adjust the pre-emphasis intensity, and when it is 0 there is no pre-equalization. It can be seen from Fig. 11.17a that for the PAM4 signal, the optimal α factor is 0.8, and the performance is slightly decreased when α is above 0.8. Figures 11.17b, c shows the spectra of the 120 Gbaud PAM4 signal with and without pre-equalization. The high-frequency component of the signal is significantly enhanced after pre-equalization.

Finally, Fig. 11.18 measures the relationship between BER and OSNR for different schemes. It can be seen from the figure that there is a large BER bottleneck when there is no pre-equalization. When using pre-equalization, the 120 Gbaud PAM4 signal has more than 5 dB OSNR gain. Here the pre-emphasis intensity is 0.8. Considering 20% soft decision forward error correction (SD-FEC) coding, the OSNR is about 24.8 dB when the BER is 2×10^{-2}, and the 120 Gbaud PDM-PAM4 signal can provide a single-carrier 400 Gbit/s information rate. We also measured the BER and OSNR results after 20 km of single-mode fiber, and there was no significant OSNR damage compared to the BTB condition.

We have studied the single-carrier 400 G Ethernet solution, and realized the 120 Gbaud PDM-PAM4 signal through external intensity modulator and coherent detection. The data rate reaches 480 Gbit/s, and the net bit rate can still reach 400 Gbit/s considering 20% FEC overhead. These show that it has broad application prospects in short-distance access networks and metropolitan area networks.

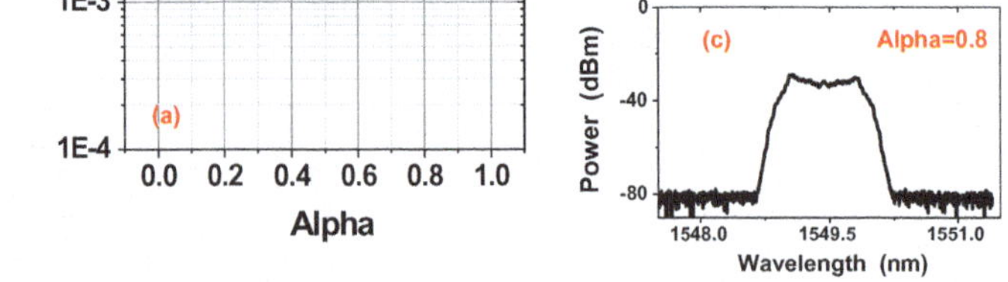

Fig. 11.17 Experimental results

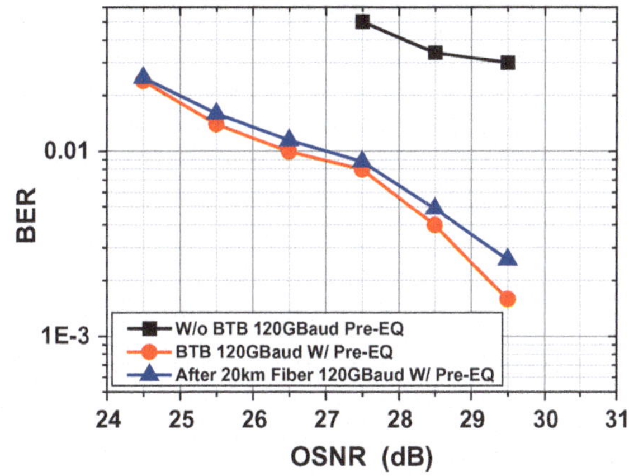

Fig. 11.18 Relationship between BER and OSNR in different schemes

11.4 Summary

This chapter mainly introduces the PAM4 signal modulation and the corresponding digital signal processing algorithms. The introduction in Sect. 11.1 illustrates PAM4 as a simple, low-cost modulation method that has attracted extensive research interest in short-range communications. Section 11.2.1 analyzes its advantages in signal-to-noise ratio (SNR) requirements and signal bandwidth from the basic principles of PAM4, and can cope well with bandwidth-limited channels. However, when there is room for SNR, a PAM4 scheme that sacrifices a fraction of the SNR for multiple improvements of rate is a good choice. Section 11.2.2–11.2.4 introduces some PAM4 signal processing algorithms, firstly the DD-LMS algorithm, which is a widely used blind equalization algorithm that can converge constellation points and further improve decision performance; then dispersion pre-compensation algorithm that can pre-compensate at the transmitter to cancel the dispersion effect in the fiber transmission. Finally, the lookup table algorithm is a simple operation algorithm. By pre-distortion at the transmitter end, it can resist various nonlinear damage in the direct detection system and reduce system BER. Section 11.3 analyzes various PAM4 high-speed transmission systems from an experimental perspective. Section 11.3.1 analyzes a 4-lane IM/DD 112.5 Gbit/s PAM4 transmission system. The system uses pre-equalization and lookup table algorithms to reduce system linear and nonlinear damage. Section 11.3.2 introduces 50 Gbit/s and 64 Gbit/s/λ PAM4 PON downlink transmission systems, adding dispersion pre-compensation based on Sect. 11.3.1, and compares and analyzes the difference in computational complexity between the Volterra nonlinear equalizer and lookup table algorithm. Section 11.3.1 introduces a 4-lane IM/DD 112 Gbit/s PAM4 system, which combines a series of algorithms such as dispersion pre-compensation, pre-equalization, lookup table, and DD-LMS to achieve a 400 Gbit/s high-speed system. Section 11.3.4 is a polarization division multiplexing 400 G PAM4 signal coherent system. The system overcomes the bandwidth limitation of the optoelectronic device by time-division multiplexing and polarization division multiplexing, and realizes the 120 Gbaud PDM-PAM4 signal. The data transmission rate reaches 480 Gbit/s, and the net bit rate of 400 Gbit/s can still be achieved considering 20% FEC overhead. These show that it has broad application prospects in short-distance access networks and metropolitan area networks.

References

1. A. Dochhan, et al., "Solutions for 400 Gbit/s Inter Data Center WDM Transmission", ECOC 2016, pp. 680–682
2. N. Eiselt, et al., "First Real-Time 400G PAM-4 Demonstration for Inter-Data Center Transmission over 100 km of SSMF at 1550 nm," OFC, 2016, paper W1K.5
3. D. Sadot, Single channel 112Gbit/sec PAM4 at 56Gbaud with digital signal processing for data centers applications. Opt. Exp ress **23**, 991–997 (2015)
4. X. Pang, "Evaluation of High-Speed EML-based IM/DD links with PAM Modulations and Low-Complexity Equalization", ECOC 2016, pp. 872–874
5. Y. Gao, "112 Gb/s PAM-4 Using a Directly Modulated Laser with Linear Pre-Compensation and Nonlinear Post-Compensation", ECOC2016, pp. 121–123
6. Z. Jia, et al., "Experimental Demonstration of PDM-32QAM Single-Carrier 400G over 1200-km Transmission Enabled by Training-assisted Pre-equalization and Look-up Table," OFC 2016, paper Tu3A.4
7. J. Zhang et al., "Joint CMA + DDLMS blind equalization algorithm [J]. Journal of Tsinghua university: nature science **10**, 1681–1683 (2009)
8. Wang Y, Yu J, Chi N. Demonstration of 4 × 128-Gb/s DFT-S OFDM Signal transmission over 320-km SMF with IM/DD[J]. Photonics Journal, IEEE, 2016, 8(2)
9. J. Zhou, L. Zhang, T. Zuo, et al., "Transmission of 100-Gb/s DSB-DMT over 80-km SMF Using 10-G class TTA and Direct-Detection," presented at the Eur. Conf. Exhib. Opt. Commun., Dusseldorf, Germany, 2016, Paper Tu3F.1
10. S.J. Savory, Digital filters for coherent optical receivers. Opt. Express **16**(2), 804–817 (2008)
11. J.H. Ke, Y. Gao, J.C. Cartledge, 400 Gbit/s single-carrier and 1 Tbit/s three-carrier superchannel signals using dual polarization 16-QAM with look-up table correction and optical pulse shaping. Opt. Express **22**(1), 71–84 (2014)
12. J, Zhang, J. Yu, et al.," EML-based IM/DD 400G (4x112. 5-Gbit/s) PAM-4 over 80 km SSMF Based on Linear Pre-Equalization and Nonlinear LUT Pre-Distortion for Inter-DCI Applications," presented at the Opt. Fiber Commun. Conf. Exhib., Los Angeles, California United States, USA, 2017, Paper W4I. 4
13. M. Iglesias Olmedo, Z. Tianjian, J. Bevensee Jensen, Z. Qiwen, X. Xu, and I. T. Monroy, "Towards 400GBASE 4-lane solution using direct detection of MultiCAP signal in 14 GHz bandwidth per lane," presented at the Opt. Fiber Commun., Nat. Fiber Opt. Eng. Conf., Anaheim, CA, 2013, Paper PDP5C.10
14. J. Estaran, M. Iglesias, D. Zibar, X. Xu, and I. Tafur, "First Experimental Demonstration of Coherent CAP for 300-Gb/s Metropolitan Optical Networks," in Optical Fiber Communication Conference, OSA Technical Digest (online) (Optical Society of America, 2014), paper Th3K.3
15. L. Suhr, J. J. Vegas Olmos, B. Mao, x. xu, G. N. Liu, and I. Tafur Monroy, "Direct modulation of 56 Gbps duobinary-4-PAM," in Optical Fiber Communication Conference, OSA Technical Digest (online) (Optical Society of America, 2015), paper Th1E.7
16. K. Zhong, X. Zhou, T. Gui, L.Tao, Y. Gao, W. Chen, J. Man, L. Zeng, A. P. Tao Lau, and C. Lu, "Experimental study of PAM-4, CAP-16, and DMT for 100 Gb/s Short Reach Optical Transmission Systems," Opt. Express 23, 1176–1189 (2015)

17. X. Xu, E. Zhou, G.N. Liu, T. Zuo, Q. Zhong, L. Zhang, Y. Bao, X. Zhang, J. Li, Z. Li, Advanced modulation formats for 400-Gbps short-reach optical inter-connection. Opt. Express **23**, 492–500 (2015)
18. M. Morsy-Osman, M. Chagnon, M. Poulin, S. Lessard, D.V. Plant, "1λ × 224 Gb/s 10 km Transmission of Polarization Division Multiplexed PAM-4 Signals Using 1.3 ʒm SiP Intensity Modulator and a Direct-Detection MIMO-based Receiver", ECOC 2014, PD.4.4
19. C. Xie, S. Spiga, P. Dong, P. J. Winzer, A. Gnauck, C. Gréus, C. Neumeyr, M. Ortsiefer, M. Müller, and M. Amann, "Generation and Transmission of 100-Gb/s PDM 4-PAM Using Directly Modulated VCSELs and Coherent Detection," in Optical Fiber Communication Conference, OSA Technical Digest (online) (Optical Society of America, 2014), paper Th3K.2
20. C. Xie, S. Spiga, P. Dong, P. Winzer, M. Bergmann, B. Kogel, C. Neumeyr, and M.-C. Amann, "Generation and transmission of a 400-Gb/s PDM/WDM signal using a monolithic 2 × 4 VCSEL array and coherent detection," presented at the Opt. Fiber Commun. Conf., San Francisco, CA, USA, 2014, Paper PDP Th5C.9
21. G. Raybon, A. Adamiecki, P. Winzer, C. Xie, A. Konczykowska, F. Jorge, J. Dupuy, L. L. Buhl, C. Sethumadhavan, S. Draving, M. Grove, and K. Rush, "Single-carrier 400G interface and 10-channel WDM transmission over 4800 km using all-ETDM 107-Gbaud PDM-QPSK," in Proc. OFC 2013, PDP5A.5
22. J. Zhang, Z. Dong, H. Chien, Z. Jia, Y. Xia, and Y. Chen, "Transmission of 20 × 440-Gb/s Super-Nyquist-Filtered Signals over 3600 km based on Single-Carrier 110-GBaud PDM QPSK with 100-GHz Grid," in Proc. OFC 2014, paper Th5B.3
23. G. Raybon, A. Adamiecki, P. J. Winzer, M. Montoliu, S. Randel, A. Umbach, M. Margraf, J. Stephan, S. Draving, M. Grove, and K. Rush, "All ETDM 107-Gbaud PDM-16QAM (856-Gb/s) transmitter and coherent receiver," in Proc. ECOC 2013, Paper PD 2.D.3
24. J. Zhang, Y. Jianjun, N. Chi, H.-C. Chien, Time-domain digital pre-equalization for band-limited signals based on receiver-side adaptive equalizers. Opt. Express **22**, 20515–20529 (2014)

Chapter 12
Optical OFDM

12.1 Introduction

The modulation method of the optical communication system can be divided into direct modulation and external modulation, as shown in Fig. 12.1 [1, 2]. Direct modulation achieved optical intensity modulation by the injecting current of a semiconductor laser. This structure is simple, economical, and easily implemented. However, since the direct modulation semiconductor laser has limited bandwidth, low extinction ratio, and large linewidth, it is usually used in short-distance DDO-OFDM fiber communication systems. External modulation techniques are generally employed in DDO-OFDM with transmission distance over 100 km and large-capacity medium-long-distance CO-OFDM fiber communication systems. The external modulation method utilizes an external modulator which is independent of the laser source to achieve electro-optical conversion, and has the advantages of high bandwidth and high extinction ratio. The linewidth of the laser can be adjusted according to the application scene. At present, in optical fiber communication systems, the external modulators mainly include a Mach–Zehnder modulator (MZM) based on electro-optic effect and an Electro Absorption Modulator (EAM) based on electro-absorption effect. In this chapter, we mainly discuss the MZM modulation technique of DDO-OFDM system based on external modulation.

MZM uses the electro-optic effect of the material to realize the modulation of the optical signal by the electrical signal. It is divided into upper and lower electrodes. Its structure is shown in Fig. 12.2. Optical modulation changes the refractive index of the material by adjusting the applied voltage of the photovoltaic material loaded on the two electrodes, in order to control the light intensity of the output signal.

The relationship between the electric field of the input and output optical signals of MZM is

$$E_{\text{out}}(t) = \frac{1}{2}E_{\text{in}}(t)\left\{e^{j\pi\frac{V_1(t)}{V_\pi}} + \gamma e^{j\pi\frac{V_2(t)}{V_\pi}}\right\} \tag{12.1}$$

J. Yu and N. Chi, *Digital Signal Processing In High-Speed Optical Fiber Communication Principle and Application*,
https://doi.org/10.1007/978-981-15-3098-2_12

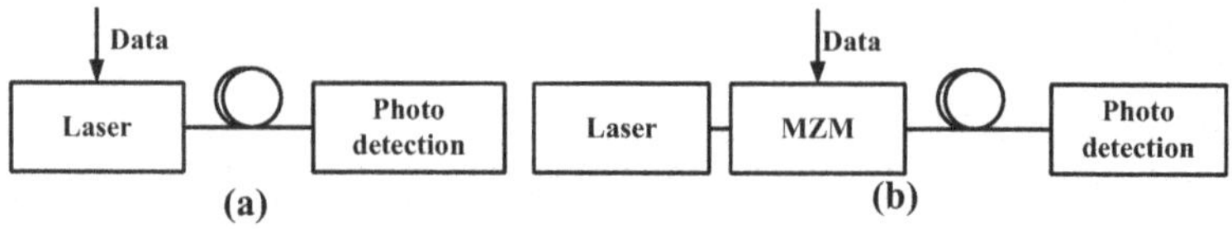

Fig. 12.1 Two different modulation methods for DDO-OFDM systems. **a** Direct modulation, **b** external modulation

Fig. 12.2 MZM structure diagram

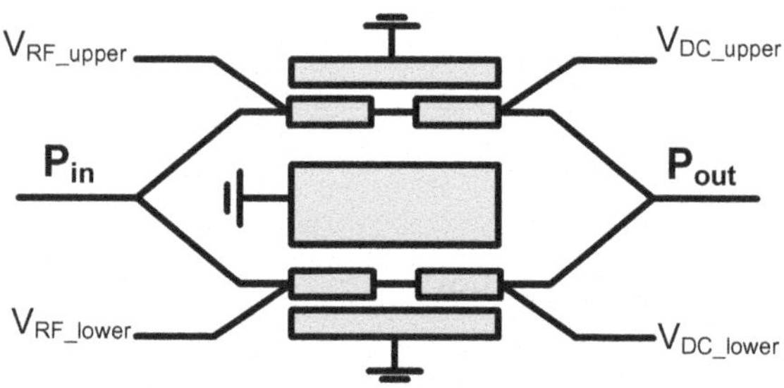

$E_{in}(t)$ is the input optical signal, $E_{out}(t)$ is the output optical signal, $V_1(t) = V_{RF\_upper}(t) + V_{DC\_upper}$ is the sum of the input modulation signal of the upper arm and the DC bias of the upper arm, $V_2(t) = V_{RF\_lower}(t) + V_{DC\_lower}$ is the sum of the input modulation signal of the lower arm and the DC bias of the lower arm, $\gamma = \frac{\sqrt{\delta}-1}{\sqrt{\delta}+1}$ is the symmetry factor of the upper and lower branches, δ is the extinction ratio of the modulator, and V_π is the half-wave voltage of the modulator. Ideal δ is more than 20 dB and γ is approximately equal to 1.

Bringing the input signal loaded on both arms into Eq. (12.1), we can get

$$\begin{aligned} E_{out}(t) &= \frac{1}{2}E_{in}(t)\left\{\cos\left(\pi\frac{V_1(t)}{V_\pi}\right) + j\sin\left(\pi\frac{V_1(t)}{V_\pi}\right) + \cos\left(\pi\frac{V_2(t)}{V_\pi}\right) + j\sin\left(\pi\frac{V_2(t)}{V_\pi}\right)\right\} \\ &= E_{in}(t)\cos\left\{\pi\frac{(V_1(t)-V_2(t)}{2V_\pi}\right\}\exp\left\{j\pi\frac{V_1(t)+V_2(t)}{2V_\pi}\right\} \end{aligned} \tag{12.2}$$

In Eq. (12.2), $\cos\left\{\pi\frac{(V_1(t)-V_2(t)}{2V_\pi}\right\}$ is an amplitude modulation component, and $\exp\left\{j\pi\frac{V_1(t)+V_2(t)}{2V_\pi}\right\}$ is a phase modulation component. When the MZM is operating in the complementary push-pull mode (when $V_1(t) + V_2(t) = 0$), the modulo of the modulator is zero. Let $V(t) = V_1(t) - V_2(t)$, when the direct detection method is adopted, the transfer function is the ratio of the output power to the input power, which can be expressed as

$$\frac{P_{out}}{P_{in}} = \frac{|E_{out}(t)|^2}{|E_{in}(t)|^2} = \cos^2\left\{\pi\frac{V_1(t)-V_2(t)}{2V_\pi}\right\} = \cos^2\left\{\pi\frac{V(t)}{2V_\pi}\right\} \tag{12.3}$$

When using the coherent detection method, the transfer function is the ratio of the output electric field to the input electric field, which can be expressed as

$$\frac{E_{out}(t)}{E_{in}(t)} = \cos\left\{\pi\frac{V_1(t) - V_2(t)}{2V_\pi}\right\} = \cos\left\{\pi\frac{V(t)}{2V_\pi}\right\} \tag{12.4}$$

The transfer function curves of direct detection and coherent detection represented by Eqs. (12.3) and (12.4) are shown in Fig. 12.3, and the MZM transmission curve is nonlinear. In order to make the MZM work efficiently with high performance, the modulation signal should be placed as far as possible in the region where the MZM linearity is high, which can be controlled by adjusting the peak-to-peak value of the modulated signal and the configuration voltage of the MZM. When the amplitude of the signal is in a region where the MZM transfer nonlinearly is high, the signal will be severely distorted, resulting in worse BER performance of the system. For DDO-OFDM systems, the best bias point should be chosen as the quadrature point. For CO-OFDM systems, the optimal bias point should be chosen to be a null point.

Depending on the detection mode of the receiver, the OFDM optical communication system can be divided into two categories: DDO-OFDM and CO-OFDM.

(1) Compared with CO-OFDM systems, DDO-OFDM systems do not need to provide high-cost devices such as hybrid and balance detector at the receiver, so that the cost will be controlled. However, the subcarrier-subcarrier mutual beat noise and the dispersion in the optical fiber transmission will cause the frequency selective fading, and the optical fiber transmission distance of the DDO-OFDM system is very limited. Therefore, it is mainly used for transmission between short-distance wired access networks and point-to-point data centers.
(2) The main advantage of the CO-OFDM system is that it can simultaneously achieve intensity and phase modulation, and then achieve ultra-high-rate signal

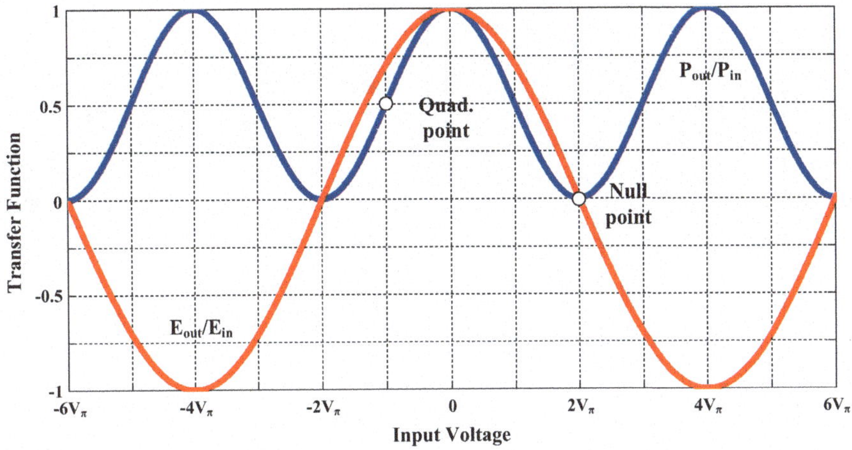

Fig. 12.3 Transmission curve when MZM adopts different detection methods

transmission. The SNR of the received signal can be changed by adjusting the power of the local oscillating signal, so the system can achieve ultra-long-distance transmission and the high receiving sensitivity of the receiver can be guaranteed. CO-OFDM is mainly used in ultra-high-rate and ultra-long-distance backbone transmission networks and in medium-range (<1000 km) metropolitan area transmission.

12.2 Basic Structure of Direct Detection Optical OFDM System

In the DDO-OFDM system, there are mainly five parts: OFDM transmitter, OFDM signal electro-optic modulation, transmission of optical OFDM signal, OFDM signal photoelectric conversion, and OFDM receiver. The schematic diagram of the DDO-OFDM system is shown in Fig. 12.4.

The OFDM transmitter mainly includes the following processes: (1) Pseudo Random Binary Sequence (PRBS) serial-to-parallel conversion; (2) inverse fast Fourier transform to realize signal conversion from frequency domain to time domain; (3) The training sequence is inserted at the beginning of signal, which is mainly used for symbol synchronization and channel estimation; (4) Parallel-to-serial conversion of the received signal; (5) The CP is inserted in the time domain of the OFDM signal to resist ISI and ICI.

The generated OFDM signal is converted into an analog signal by a DAC. After a low-pass filter (LPF), the signal is amplified by an amplifier and injected into an external modulator to realize electro-optical conversion. Another solution of photoelectric conversion can be achieved by direct modulation. After transmitted through

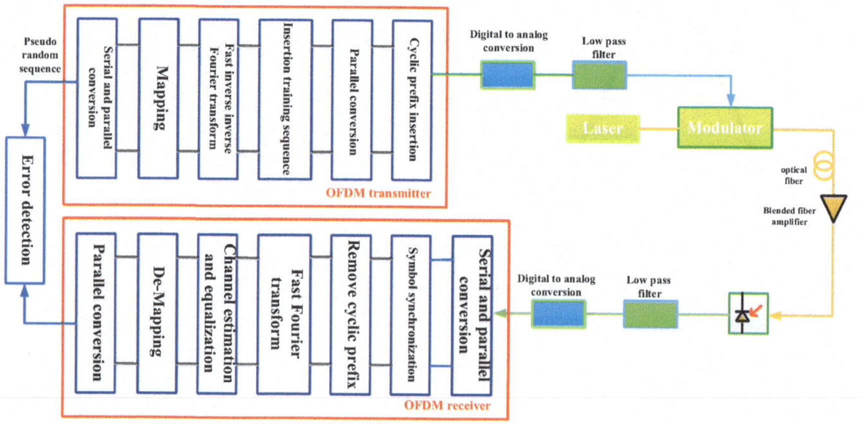

Fig. 12.4 System schematic of DDO-OFDM

the optical fiber, the optical OFDM signal is amplified by the EDFA before entering the photodiode (PD) to realize photoelectric conversion.

After photoelectric conversion, the electrical OFDM signal is firstly filtered by an LPF to remove the noise outside the signal band, and the ADC cascaded behind the LPF converts the OFDM signal into a digital OFDM signal. The digital signal at the OFDM receiver needs to perform the following digital signal processing to realize demodulation: (1) Serial-to-parallel conversion of the digital signal after the ADC; (2) Symbol synchronization is used to determine the beginning of the OFDM signal; (3) CP removal of OFDM signal; (4) Fast Fourier transform converts the signal from time domain to frequency domain; (5) Removal of the training sequence used for channel estimation and channel equalization; (6) Signal de-mapping according to the mapping law of the transmitter; (7) Converting the signal into parallel and calculating the bit error rate according to the original bits.

12.3 Structure and Basic Principles of Coherent Detection Optical OFDM Systems

In the early twentieth century, coherent optical communication was chased by many researchers because of its high sensitivity and large channel capacity [1–14]. However, with the widespread application of WDM systems and EDFA amplifiers in the 1980s, direct detection replaced coherent detection because of its simple structure and low implementation cost [15]. Therefore, coherent optical communication gradually faded out of the researcher's sight. However, with the requirement of higher transmission rates and the application of digital signal processing and high-order optical modulation technology, the coherent optical communication with high performance and high rate has become a research hotspot of modern optical communication [16–30]. Coherent optical communication not only greatly improves the spectral efficiency of the system, but also has better effects in the application of digital signal processing in terms of dispersion and nonlinear compensation. Introducing polarization multiplexing into the system of coherent detection can further improve the transmission capacity [16–25].

With the complexity of communication increasing, the required bandwidth of ethernet will reach 100 GHz or more. In the current simplified structure direct detection transmission system, since the maximum bandwidth of the existing photodetection device is about 100 GHz and only the intensity in the direct detection system can be used to modulate information, this will become a bottleneck for the improvement of the communication rate. In order to further increase the system capacity of optical fiber communication, researchers pay attention to coherent optical communication. In coherent optical communication, the amplitude and phase of the optical signal can be modulated. The use of a higher order amplitude/phase modulation format can greatly improve the spectral efficiency of the system and ensure the realization of a high-speed optical fiber communication system [21].

The fundamental reason coherent optical communication can ensure high signal rate is that coherent optical communication introduces vector modulation into the modulation format, which can greatly increase the spectral efficiency of the system [21]. Quadrature phase shift keying (QPSK) is the most common vector modulation format in coherent optical communication. In the QPSK modulation format, a symbol carries a signal of 2 bits. The information rate doubles over the same bandwidth. Similarly, when the M-QAM modulation format is applied to optical communication, the spectral efficiency is improved by M times relative to the OOK modulation format in the direct detection system. Similarly, in the CO-OFDM system, it can cause an improvement in spectral efficiency that the modulation format order of the signal carried on the subcarrier is increased.

In an optical fiber communication system for coherent detection, an optical signal modulated by a vector signal can be expressed as

$$E_s(t) = A_s(t)\exp(j\omega_s t) \tag{12.5}$$

In Eq. (12.1), $A_s(t)$ is the complex signal amplitude, and ω_s is the modulated optical signal angular frequency. Similarly, the LO of the receiver is defined as

$$E_{LO}(t) = A_{LO}(t)\exp(j\omega_{LO} t) \tag{12.6}$$

Analogous to Eq. (12.5), $A_{LO}(t)$ and ω_{LO} in Eq. (12.6) are the complex amplitude and angular frequency of the local oscillator signal, respectively. The modulated signal is correlated with the power and amplitude of the local LO, and is represented as $P_s = {|A_s|^2}/{2}$ and $P_{LO} = {|A_{LO}|^2}/{2}$, respectively.

Balance detection is commonly used in coherent optical communication systems, which suppresses the DC component while ensuring that the output photocurrent is the maximum. Figure 12.5 shows a block diagram of the balance detection method.

In this detection mode, the received signals E_S and E_{LO} local oscillation signals are guaranteed to have the same polarization state, so that the two optical signals entering the balanced receiver during detection can be expressed as

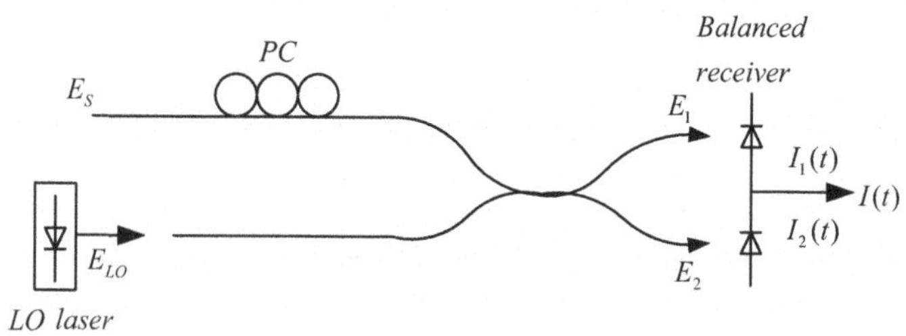

Fig. 12.5 Structure of the balance detection mode

$$E_1(t) = \frac{1}{\sqrt{2}}(E_S + E_{LO}) \tag{12.7}$$

$$E_2(t) = \frac{1}{\sqrt{2}}(E_S - E_{LO}) \tag{12.8}$$

After photodetection, the $I_1(t)$ and $I_2(t)$ output from the two PD are

$$\begin{aligned} I_1(t) &= R[\mathrm{Re}\{\frac{A_S(t)\exp(j\omega_S t) + A_{LO}(t)\exp(j\omega_{LO} t)}{\sqrt{2}}\}]^{ms} \\ &= \frac{R}{2}[P_S + P_{LO} + 2\sqrt{P_S P_{LO}}\cos\{\omega_{IF} t + \theta_{sig}(t) - \theta_{LO}(t)\}] \end{aligned} \tag{12.9}$$

$$\begin{aligned} I_2(t) &= R[\mathrm{Re}\{\frac{A_S(t)\exp(j\omega_S t) - A_{LO}(t)\exp(j\omega_{LO} t)}{\sqrt{2}}\}]^{ms} \\ &= \frac{R}{2}[P_S + P_{LO} - 2\sqrt{P_S P_{LO}}\cos\{\omega_{IF} t + \theta_{sig}(t) - \theta_{LO}(t)\}] \end{aligned} \tag{12.10}$$

In Eqs. (12.9) and (12.10), *ms* represents the square detection of the light intensity convert to current achieved by the photodiode in photodetection, where $\omega_{IF} = \omega_S - \omega_{LO}$ is the frequency difference between the received signal and the local LO. $\theta_{sig}(t)$ and $\theta_{LO}(t)$ are the phases of the received signal and the LO signal, respectively, and R is the responsiveness of the photodiode. The detailed expression is

$$R = \frac{e\eta}{\hbar\omega_S} \tag{12.11}$$

Responsivity indicates that in Eq. (12.11), e is the electron charge, η is the quantum efficiency of the photodiode, and $\hbar$ is the Planck's constant. According to the two currents, $I_1(t)$ and $I_2(t)$, in the balance detection diode from Eqs. (12.9) and (12.10), we can get the current output of the final balanced detection:

$$I(t) = I_1(t) - I_2(t) = 2R\sqrt{P_S(t)P_{LO}}\cos\{\omega_{IF} t + \theta_{sig}(t) - \theta_{LO}(t)\} \tag{12.12}$$

The power P_{LO} of the LO signal is a constant. According to the value of ω_{IF}, the coherent detection is divided into two types: homodyne coherent detection and heterodyne coherent detection. The homodyne coherent detection means that the frequency of the LO signal is exactly the same as the frequency of the received signal, that is $\omega_{IF} = 0$, the signal becomes the baseband signal after the balance detection. For the heterodyne coherent detection, the frequency of the LO signal and the frequency of the received signal are different with frequency difference ω_{IF}. The intermediate frequency is achieved after the balance detection, and the down-conversion of the electric domain is required to obtain the baseband signal.

Both homodyne coherent detection and heterodyne coherent detection have their own advantages. Compared with homodyne coherent detection systems, the

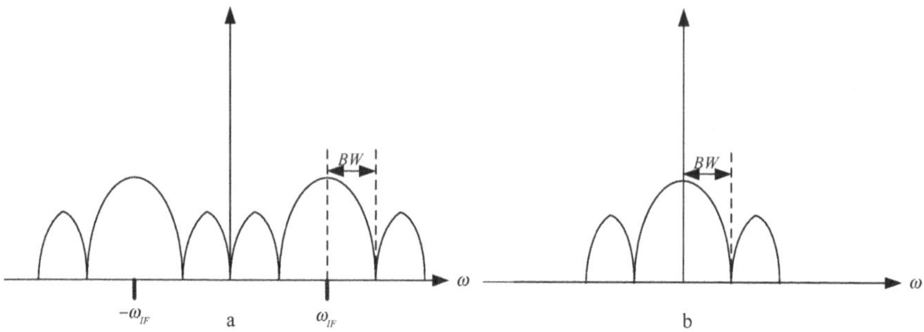

Fig. 12.6 Signal spectrum and bandwidth requirements for the two detection methods, **a** heterodyne detection, **b** homodyne detection

complexity of heterodyne coherent detection systems increases due to the down-conversion in the electric domain. In general, the homodyne detection is used in core transmission networks with its simplicity. In hybrid optical-wireless network, the heterodyne detection is widely utilized to generate and transmit the modulated microwave, millimeter wave, or even W-band signal. The specific bandwidth requirements for both are analyzed in Fig. 12.6. The bandwidth required by the homodyne coherent detection system is BW, which is the same as the bandwidth of the transmitted baseband signal. However, the bandwidth required of the heterodyne coherent detection system is $\omega_{IF} + BW$, which can be seen from Fig. 12.6a. In order to avoid aliasing and save as much bandwidth as possible, the frequency of the IF signal in the heterodyne coherent detection system must be set to $\omega_{IF} \geq BW$, so that the bandwidth of heterodyne coherent detection system is at least $2BW$.

In terms of the complexity of the system, since the modulation format selected in the coherent detection system is not limited to simple amplitude modulation, both amplitude modulation and phase modulation are usually adopted to improve spectral efficiency. Assuming the quadrature amplitude modulation (QAM) used in the system, the number of balanced detectors required to perform the homodyne and heterodyne coherent detection is different. One balanced detector is needed for heterodyne detection of the QAM format of a single-polarization state, but for homodyne detection we need two balanced detectors. In addition, in the heterodyne coherent detection system, a pre-filter is needed to filter out the virtual sideband of the signal, but this is not required for homodyne detection. A comparison of detailed requirements for both systems is given in Table 12.1.

This section focuses on how to extend the simple coherent detection system using a balanced receiver mentioned in the previous section to a complex coherent detection system using phase and polarization diversity reception. Firstly, the principle and implementation of phase diversity reception and polarization diversity reception are discussed separately. Finally, phase diversity reception and polarization diversity reception are integrated into the same coherent detection system to realize the coherent detection based on phase and polarization diversity reception.

Table 12.1 Comparison of requirements and settings for homodyne detection and heterodyne detection systems

Homodyne detection	• Two balanced detectors are required for quadrature amplitude modulation of one polarization state • System bandwidth is the baseband signal bandwidth BW • No preset amplifier is needed to remove the virtual sideband
Heterodyne detection	• A balanced detector is required for quadrature amplitude modulation of one polarization state • The bandwidth of the system is $\omega_{IF} + BW$, the minimum is $2BW$ • Pre-amplifier is required to remove the virtual sideband

For the sake of simplicity, only the homodyne detection system based on phase and polarization diversity reception is analyzed in this section. As for the heterodyne system, the analogy can be analyzed. According to the derivation of the previous section, $\omega_{IF} = \omega_S - \omega_{LO} = 0$ must be guaranteed to achieve homodyne coherent detection. In the homodyne coherent detection system, if both the LO and the component that produces 90° phase shift can be provided, the in-phase component and the quadrature component of the signal can be obtained through the two balanced detectors. The basic structure diagram is illustrated in Fig. 12.7.

The in-phase component $I_I(t)$ and the quadrature component $I_Q(t)$ obtained by photoelectric conversion from two pairs of balanced receivers can be expressed as

$$I_I(t) = I_{I1}(t) - I_{I2}(t) = R\sqrt{P_S P_{LO}} \cos\{\theta_{sig}(t) - \theta_{LO}(t)\} \tag{12.13}$$

$$I_Q(t) = I_{Q1}(t) - I_{Q2}(t) = R\sqrt{P_S P_{LO}} \sin\{\theta_{sig}(t) - \theta_{LO}(t)\} \tag{12.14}$$

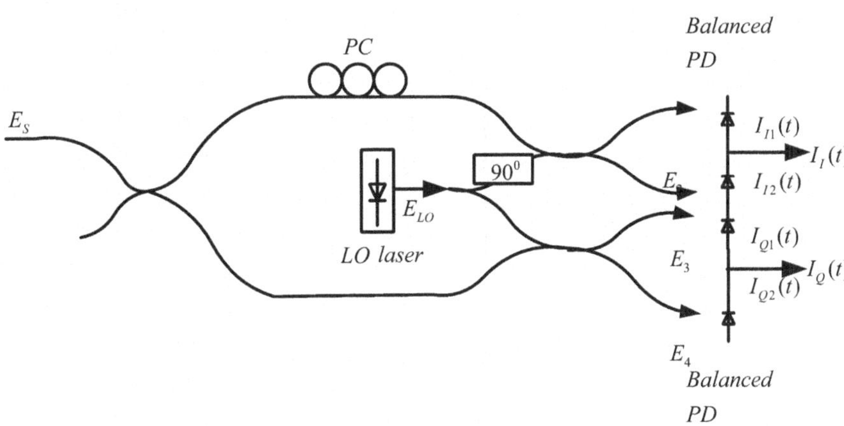

Fig. 12.7 Structure diagram of homodyne detection system for phase diversity reception

The in-phase component of Eq. (12.13) and the quadrature component of Eq. (12.14) together determine the final output signal:

$$I_C(t) = I_I(t) + jI_Q(t) = R\sqrt{P_S P_{LO}}\exp\{j[\theta_{sig}(t) - \theta_{LO}(t)]\} \quad (12.15)$$

The phase component of the output signal consists of its own phase and phase noise $\theta_{sig} = \theta_S + \theta_{sn}$. We define the demodulated phase noise θ_n as $\theta_n = \theta_{sn} - \theta_{LO}$. Put these two relations into Eq. (12.15), and the final output signal can be obtained:

$$I_C(t) = R\sqrt{P_S P_{LO}}\exp\{j[\theta_S(t) + \theta_n(t)]\} \quad (12.16)$$

Since the responsiveness of the photodiode R and the power P_{LO} of the LO are both constants, the only difference between the final received signal and the transmitted signal represented in Eq. (12.16) is an unknown phase $\theta_n(t)$. This unknown phase is the phase noise that is usually referred to. Many methods are proposed to estimate and eliminate phase noise. The most common method is digital signal processing based on the receiver.

To improve the spectral efficiency of the system, besides increasing the modulation order of the transmitted signals in the system, polarization multiplexing can also be used to achieve capacity expansion. The spectral efficiency of the system using polarization diversity reception is two times as much as the single-polarization direction. Next, how to implement polarization diversity reception in the homodyne coherent detection system will be discussed. With polarization diversity reception and phase diversity reception both taken into consideration, the schematic diagram of a general polarization diversity reception system is shown in Fig. 12.8.

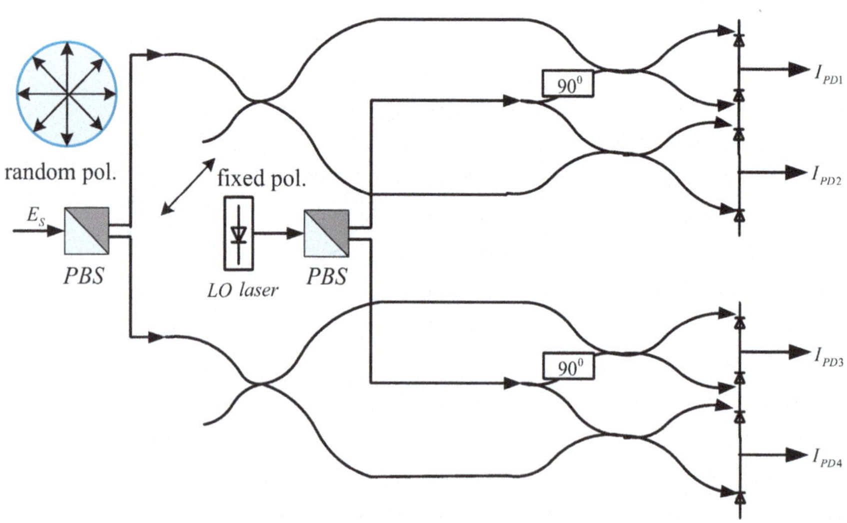

Fig. 12.8 Homodyne coherent detection system based on phase diversity and polarization diversity

In Fig. 12.8, the PBS is a polarization beam splitter that splits the signal into two orthogonal polarization states. The device in front of the photodiode is a 90° optical mixer, mainly for phase diversity reception in a single-polarization state. At the receiver side, the transmitted signal E_S is divided into two polarization states X and Y by the polarization beam splitter. The local oscillation signal LO is also divided into two completely orthogonal polarization signals by the polarization beam splitter. The signal obtained by the signal E_S after the polarization beam splitter is

$$\begin{bmatrix} E_{sx} \\ E_{sy} \end{bmatrix} = \begin{bmatrix} \sqrt{\alpha} A_s e^{j\delta} \\ \sqrt{1-\alpha} A_s \end{bmatrix} \exp(j\omega_s t) \tag{12.17}$$

The α in Eq. (12.17) is the ratio of the power assigned to the two polarization states, and δ is the phase difference between two polarization states. The two variables are related to the birefringence of the transmission fiber and change over time. The LO also obtains two polarization states after PBS.

$$\begin{bmatrix} E_{LOx} \\ E_{LOy} \end{bmatrix} = \frac{1}{\sqrt{2}} \begin{bmatrix} A_{LO} \\ A_{LO} \end{bmatrix} \exp(j\omega_{LO} t) \tag{12.18}$$

From Eq. (12.18), the signal is divided into two different polarization states by power, and the phase difference between the two polarization states is zero. Four pairs of balance detection photodiodes to be detected for electrical signals $E_{1,2\ldots8}$ are

$$E_{1,2} = \frac{1}{2}\left(E_{sx} \pm \frac{1}{\sqrt{2}} E_{LO}\right) \tag{12.19}$$

$$E_{3,4} = \frac{1}{2}\left(E_{sx} \pm \frac{j}{\sqrt{2}} E_{LO}\right) \tag{12.20}$$

$$E_{5,6} = \frac{1}{2}\left(E_{sy} \pm \frac{1}{\sqrt{2}} E_{LO}\right) \tag{12.21}$$

$$E_{7,8} = \frac{1}{2}\left(E_{sy} \pm \frac{j}{\sqrt{2}} E_{LO}\right) \tag{12.22}$$

The in-phase and quadrature components of different polarization states can be obtained after the balance detection of these signals. The currents after the photoelectric conversion of the four pairs of balanced receivers $PD_{1,2\ldots4}$ in Fig. 12.8 are

$$I_{PD1} = R\sqrt{\frac{\alpha P_S P_{LO}}{2}} \cos\{\theta_S(t) - \theta_{LO}(t) + \delta\} \tag{12.23}$$

$$I_{PD2} = R\sqrt{\frac{\alpha P_S P_{LO}}{2}} \sin\{\theta_S(t) - \theta_{LO}(t) + \delta\} \tag{12.24}$$

$$I_{PD3} = R\sqrt{\frac{(1-\alpha)P_S P_{LO}}{2}}\cos\{\theta_S(t) - \theta_{LO}(t)\} \tag{12.25}$$

$$I_{PD4} = R\sqrt{\frac{(1-\alpha)P_S P_{LO}}{2}}\sin\{\theta_S(t) - \theta_{LO}(t)\} \tag{12.26}$$

Signals of two polarization states can be expressed as

$$I_{xc}(t) = I_{PD1}(t) + jI_{PD2}(t) \tag{12.27}$$

$$I_{yc}(t) = I_{PD3}(t) + jI_{PD4}(t) \tag{12.28}$$

Equations (12.27) and (12.28) are signals of two polarization states of the signal. After subsequent processing, two signals with orthogonal polarization states can be separated.

To improve the performance of coherent optical communication system, complex digital signal processing (DSP) technology is added to signal detection and processing. Many key processes are implemented by DSP, which include

- Clock extraction and digital signals reshaping;
- Extraction and separation of signal polarization states;
- Channel equalization, compensation for linear effects of the channel (such as dispersion), nonlinear compensation, polarization mode dispersion compensation, etc.;
- Estimation of carrier phase;
- Identification of symbols when detection.

In the process of coherent detection, there are two alternative schemes depending on the modulation format of the transmission signal: a single-carrier coherent detection system and an OFDM coherent detection system. The biggest advantage of OFDM compared to single carrier is that the insertion of pilot tones and training sequences (TS) makes simpler DSP and lower computation complexity at the receiver. Moreover, its digital signal processing algorithms and the hardware do not have to be modified or adjusted, when the modulation format of the system changes. Moreover, the transparency of the system upgrade will greatly improve the system's flexibility.

In a single-carrier system with coherent detection, it is difficult to find the optimal sampling time with the increase of transmission rate in the system. Failure to find the best sampling time will greatly degenerate the single-carrier system performance. While for OFDM signals with coherent detection, there is no high requirement for the accuracy of the sampling time, as long as that the entire OFDM symbol is included in the time window. However, it is very sensitive to frequency offset and phase noise in the coherent OFDM systems. Therefore, frequency offset and phase estimation must be implemented accurately in CO-OFDM systems.

There are some key DSP technologies in CO-OFDM implementation, which include: (1) Symbol synchronization. Finding the exact position where the OFDM signal starts; (2) Frequency synchronization (frequency offset estimation, FOE). The OFDM signal is very sensitive to frequency offset. Since the linewidth of the laser is not zero and the center wavelength also fluctuates within a certain range, the frequency of the lasers at the transmitter and the receiver cannot be ensured completely consistent in the coherent detection system. After the photoelectric conversion, the frequency offset estimation and compensation of the received signal must be implemented by DSP; (3) Phase noise estimation. Phase noise will lead to the rotation and divergence of the OFDM signal constellation points, and a poor BER performance. The phase noise is mainly caused by the optical signal, which does not have a single frequency. So how to estimate phase noise accurately is also the key to realize the CO-OFDM system; (4) Channel estimation. Channel estimation in the CO-OFDM system requires insertion of a known training sequence at the transmitter. The insertion of the training sequence is to estimate the channel response after optical fiber transmission at the receiver and perform equalization according to the channel response; (5) The implementation of polarization multiplexing. Polarization multiplexing can be used to achieve capacity expansion of the transmission system. The difficulty in the implementation of a polarization multiplexing system is how to do polarization demultiplexing. In this section, the channel estimation and equalization algorithm based on the Jones matrix are used to implement polarization demultiplexing at the receiver in the CO-OFDM system.

In CO-OFDM systems, an OFDM signal is subjected to electro-optic modulation up-conversion, transmission in optical fiber, balanced detection, and down-conversion. The schematic diagram of the CO-OFDM system is shown in Fig. 12.9. It can be divided into 5 modules according to the function: OFDM transmitter, OFDM

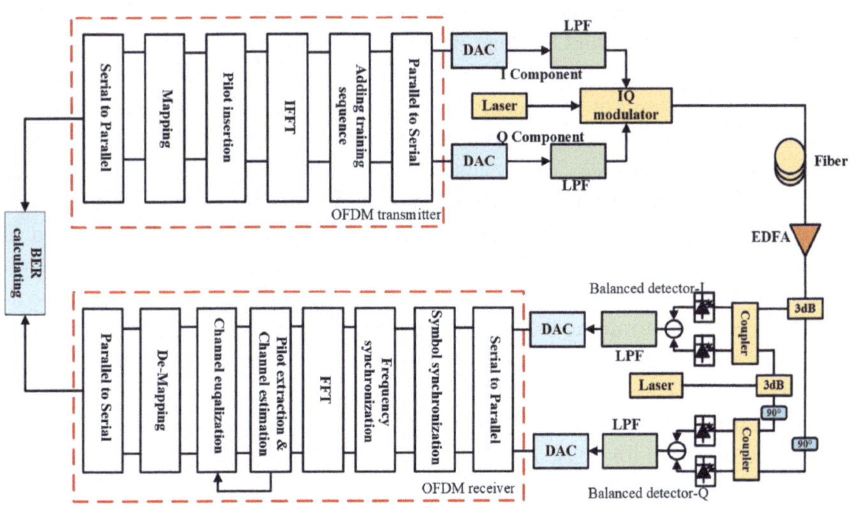

Fig. 12.9 System schematic diagram of CO-OFDM

electro-optic modulation module, optical fiber transmission, OFDM photoelectric detection module, and OFDM receiver. The process of generating OFDM at the transmitter mainly includes: (1) Firstly, the PRBS signal performs a serial-to-parallel conversion. (2) The serial-to-parallel converted signals are mapped according to different modulation formats. (3) Pilots are inserted in the frequency domain, mainly for phase noise estimation. (4) Inverse fast Fourier transform (IFFT) is performed to realize the conversion of the signal from the frequency domain to the time domain. (5) The training sequence is inserted at the beginning of the signal, which is mainly used for symbol synchronization, frequency synchronization, and channel estimation. (6) The obtained signal after the above steps is converted to parallel. The generated OFDM signal is converted to an analog signal by an ADC. After the low-pass filter and amplifier, the in-phase and quadrature components of the signal are amplified and injected into the I/Q modulator to realize the orthogonal modulation of the optical signal. The I/Q modulator consists of three two-arm MZM modulators, two of which modulate the signal, and the third modulator controls the phase difference between the in-phase and quadrature components of the optical signal to be $\pi/2$. The DC biases of the two modulators are adjusted, respectively, to ensure that the modulators operate at the minimum power point. The third modulator that controls the phase difference operates at the orthogonal point to ensure that there is a phase difference of $\pi/2$ between the two signals. Thus, the optical OFDM signal is obtained after passing through the I/Q modulator.

After the generated optical OFDM signal is transmitted through the optical fiber, the wavelength of the local laser source is determined according to the selected down-conversion method (homodyne reception and heterodyne reception). The down-conversion method of homodyne reception is generally selected in the backbone transmission network. The system schematic diagram based on homodyne reception is given in Fig. 12.13. The down-conversion method for heterodyne reception is very similar, except that two balanced receivers are not required at the receiver, but digital down-conversion should be done in the DSP. In the down-conversion mode of homodyne reception, the wavelength of the local laser is set to be the same as the wavelength of the laser at the transmitter. Two orthogonal in-phase and quadrature components of the OFDM signal are obtained after the photoelectric conversion of the balanced receiver.

The in-phase component and the quadrature component of the photoelectric converted OFDM signal become digital signals after A/D conversion. DSP is used to perform demodulation of the OFDM signal. The digital signal needs to be demodulated by the following DSPs: (1) Serial-to-parallel conversion. (2) Symbol synchronization for determining the start of the OFDM signal. (3) The frequency offset estimation and compensation for the received OFDM signal. (4) The fast Fourier transform converts the signal from the time domain to the frequency domain. (5) The training sequence is extracted for channel estimation and equalization. (6) The pilot is extracted and used for phase noise estimation and compensation. (7) The signal is converted to parallel and the BER is calculated according to the original signal.

In coherent optical OFDM reception, synchronization is a very important part of DSP. Synchronization can be mainly divided into three parts in the receiver of the coherent optical OFDM system. The first is time-domain synchronization, or symbol synchronization, which is used to find where the OFDM signal begins. The second is frequency-domain synchronization, that is, frequency offset estimation. The OFDM signal is very sensitive to the frequency offset. How to accurately estimate the frequency offset is also the key to correctly demodulate the OFDM signal. The third is the recovery of subcarriers, which mainly includes channel estimation, as well as the estimation and elimination of phase noise. The channel estimation is mainly implemented by training sequences, while the phase noise is mainly estimated by pilot extraction.

In the DSP of the CO-OFDM receiver, symbol synchronization is very important. If the signal start point cannot be found correctly, the DSP algorithm will not be completed. If the symbol synchronization is not accurate, it will cause inter-symbol interference and inter-carrier interference. The most famous algorithm in OFDM symbol synchronization was proposed by Schmidl and Cox in 1997. In this method, a training sequence containing two identical information streams in the time domain is inserted in front of the transmission information, which is mainly used for symbol synchronization. A schematic diagram of the signal format of a training sequence for symbol synchronization is shown in Fig. 12.10.

In the signal modulation of OFDM, the length of the FFT is defined as N. The training sequence satisfies the following equation in the time domain:

$$T_k = T_{k+N/2}, \quad k \in [1, N/2] \tag{12.29}$$

where k is the kth point in the specified FFT length N, which ranges from 1 to $N/2$, and T_k represents the time-domain value of the kth point. Assuming that this training sequence passes through a time-invariant channel (the fiber channel is very stable with a very slow-changed channel characteristic, and can be approximated as a time-invariant channel for a certain length of time), and its impulse response is $h(t)$, the received signal is

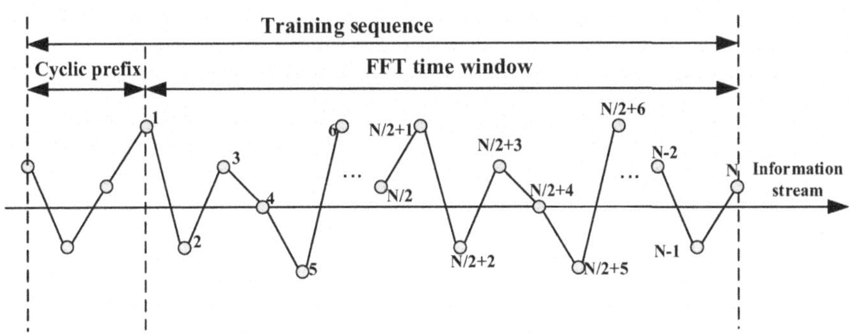

Fig. 12.10 Schematic diagram of the signal format of the training sequence used for symbol synchronization

$$r_m = r(mt_s/N) = T(mt_s/N) \otimes h(mt_s/N) \cdot e^{(j\omega_{off} mt_s/N + \Delta\varphi)} + n_m \tag{12.30}$$

where ω_{off} represents the center frequency difference between the laser at the transmitting end and the LO at the receiving end in a coherent optical OFDM system, and $\Delta\varphi$ represents the phase noise due to the linewidth of the laser (assuming that the phase noise is constant over the entire symbol range). The comparison shows that the two identical parts have only one phase shift after transmission and reception, meaning that there is only one phase shift between the two parts. Then the synchronization can adopt the method of correlating the two parts, with a sliding window whose length is *N/2*. Define the correlation calculation as

$$R_d = \sum_{m=1}^{N/2} r^*_{m+d} r_{m+d+N/2} \tag{12.31}$$

Define the power product of the two parts as

$$S_d = \sqrt{\left(\sum_{m=1}^{N/2} \left|r^2_{m+d}\right|\right)\left(\sum_{m=1}^{N/2} \left|r^2_{m+d+N/2}\right|\right)} \tag{12.32}$$

Normalize the calculated correlation function power:

$$M_d = \left|{R_d}/{S_d}\right|^2 \tag{12.33}$$

We obtain M_d by comparison and find the d corresponding to the largest M_d value, which is the optimal point of symbol synchronization, that is, the optimal time for the reception of the OFDM signal.

Frequency offset estimation is also a very important part of offline algorithms. The existence of a frequency offset will cause the orthogonality of OFDM subcarriers to be destroyed, which will lead to serious ICI. In this way, the frequency offset must be compensated during reception. There are also various estimation algorithms for the frequency offset, which are mainly divided into two categories: frequency offset estimation using training sequences and frequency offset estimation using pilots.

The frequency offset estimation using the training sequence is also the algorithm used by Schmidl and Cox. The optimal symbol synchronization point d_{opt} has been estimated above, and the autocorrelation function of the two parts before and after the point is

$$R_{d_{opt}} = \sum_{m=1}^{N/2} r^*_{m+d_{opt}} r_{m+d_{opt}+N/2}$$

$$= \sum_{m=1}^{N/2} \left(\begin{array}{r} (T((m + d_{opt})t_s/N) \otimes h((m + d_{opt})t_s/N) \\ \cdot e^{(j\omega_{off}(m+d_{opt})t_s/N+\Delta\varphi)} + n_{(m+d_{opt})})^* \\ \cdot(T((m + d_{opt} + N/2) \cdot t_s/N) \otimes h((m + d_{opt} + N/2) \cdot t_s/N) \\ \cdot e^{(j\omega_{off}(m+d_{opt}+N/2)\cdot t_s/N+\Delta\varphi)} + n_{(m+d_{opt}+N/2)}) \end{array} \right)$$

$$= \sum_{m=1}^{N/2} \left| \hat{r}_{m+d_{opt}} \right|^2 \cdot \exp(j \cdot \pi f_{off}/\Delta f) + \Theta(d_{opt}) \quad (12.34)$$

where $\Theta(d_{opt})$ represents the residual noise at that point and $\Delta f = {}^1/_{t_s}$ is the frequency spacing between each subcarrier. Take the phase of the autocorrelation function:

$$\text{phase} = \text{angle}(R_{d_{\text{opt}}}) = \pi f_{\text{off}}/\Delta f \quad (12.35)$$

then we can get the frequency offset:

$$f_{\text{off}} = \text{phase} \cdot \Delta f/\pi \quad (12.36)$$

Due to the presence of residual noise, when the SNR of the system is very low, the estimation may be inaccurate. Therefore, several training sequences can be inserted in front of the OFDM signal, and then estimated and averaged. When the frequency offset estimation is completed, the frequency offset compensation is needed. The compensation method is as follows:

$$r_c(t) = r(t) \cdot \exp(-j \cdot 2\pi f_{off} t) \quad (12.37)$$

Another feasible approach is to implement frequency offset estimation by inserting pilots. The principle of using pilot frequency offset estimation is as follows: inserting a pilot at a specific frequency point in the frequency domain, transforming the signal into the frequency domain, and finding the frequency point after transmission and demodulation. The frequency offset can be found by comparing it with the originally set frequency point. Since this pilot needs to be found obviously, its power must be larger than all subcarriers to make it easy to find this frequency point at the receiving end. This frequency point can be chosen at zero frequency, and zero frequency needs to be reserved for DC offset during OFDM modulation. When the DC offset is added, it is generally higher than the subcarriers for information transmission, so that it can be easily extracted at the receiving end (even if there is a frequency offset, the point with the highest original power is still the highest point). Figure 12.11 is a schematic diagram of pilot insertion for frequency offset estimation at zero frequency.

In Fig. 12.11, the solid line indicates the frequency-domain distribution of the OFDM signal subcarriers at the transmitting end, and the center frequency is reserved for DC offset. The DC offset is inserted during electro-optic modulation, the power of which is set higher than the other subcarriers to effectively estimate the frequency offset. The frequency offset is introduced after transmission and optical-electrical

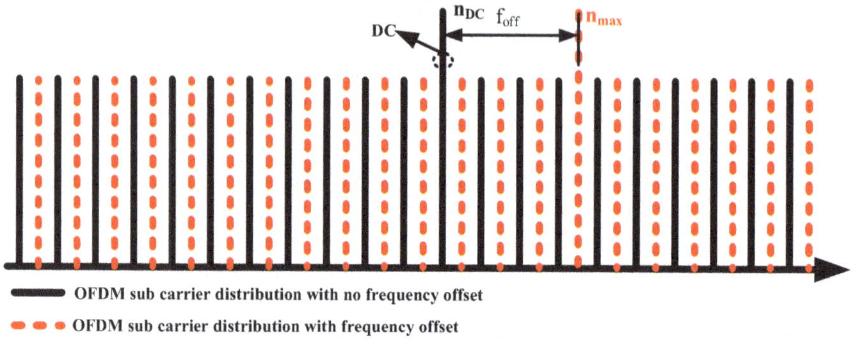

Fig. 12.11 Schematic diagram of frequency offset estimation and pilot insertion

demodulation. The frequency distribution of the frequency-shifted OFDM signal subcarriers is indicated by broken lines in Fig. 12.11. Similarly, the maximum value of the frequency point can be found in the spectrum of the signal with frequency offset. The difference between the position of the subcarrier where the maximum value is located and the position of the original DC offset subcarrier is

$$\Delta n = n_{\max} - n_{DC} \tag{12.38}$$

The corresponding frequency offset is

$$f_{off} = \begin{cases} \Delta n \Delta f & 0 \le \Delta n \le N/2 - 1 \\ (\Delta n - N)\Delta f & N/2 \le \Delta n \le N - 1 \end{cases} \tag{12.39}$$

Since all operations in the process are based on integer multiples of the subcarrier spacing, it can only accurately estimate the frequency offset of an integer multiple of the subcarrier spacing. The frequency offset of the fractional multiple of the subcarrier spacing can be obtained by a cyclic prefix. The specific process can refer to the algorithm for calculating the frequency offset using the training sequence. When both the integer frequency offset and the fractional frequency offset are obtained, the frequency offset can be compensated by the formula (12.37).

In fiber channel transmission of OFDM signals, the rotation and dispersion of constellation points can cause constellation points to be disorganized after optical-electrical conversion. There are three main factors that cause the chaos of the constellation: (1) The dispersion effect of the fiber channel in the entire spectrum. (2) The timing error of the FFT window. (3) Phase noise. The first two are slow-changing processes that can be estimated and compensated by inserting training sequences at equal intervals in the transmitted information. Phase noise is introduced due to the linewidth of the transmitting and receiving lasers. Since the operating state of the laser changes very quickly (up to the ns level), the phase estimation must be specific to each OFDM symbol, not the entire information frame. Therefore, the estimation

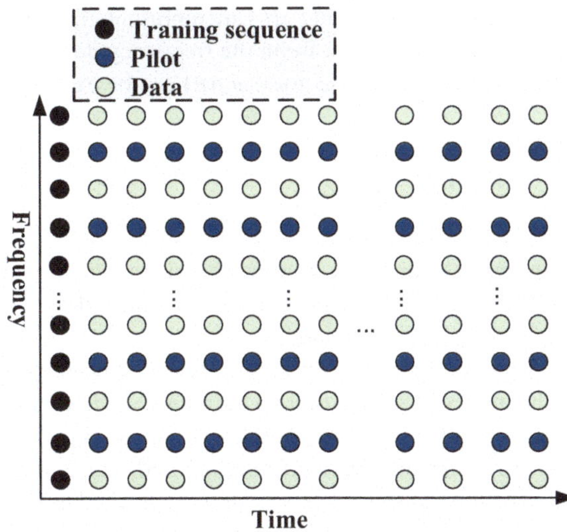

Fig. 12.12 Schematic diagram of training sequence and pilot insertion

of phase noise cannot be achieved by inserting training sequences at equal intervals in the transmitted information stream. In the channel estimation, it is necessary to discuss the estimation and compensation of the dispersion effect of the optical fiber and the timing error of the FFT window by using the inserted training sequence. Figure 12.12 is a schematic diagram of the training sequence and pilot insertion. The black color in the figure indicates the training sequence, which occupies the entire frequency domain at a certain time point. Its purpose is to estimate the channel response of all frequency points in the entire frequency domain. When the SNR is low or the channel estimation accuracy requirement of the system is very high, the influence of random noise on the training sequence-based channel estimation can be eliminated by inserting multiple training sequences in time and then averaging. For a transmission system of OFDM, the signal transmitted through the channel can be expressed as

$$r_m(n) = x_m(n) \otimes h_m(n) \cdot e^{j\varphi_m(n)} + \xi_m(n). \tag{12.40}$$

where $x_m(n)$, $h_m(n)$, and $\varphi_m(n)$ are expressed as the transmitted signal, channel response and phase noise, and $\xi_m(n)$ is the variance of the random noise in the channel. After removing the cyclic prefix and completing the FFT transform to the frequency domain, the channel transmission can be expressed as

$$\begin{aligned} R_m(k) &= X_m(k)H_m(k)I_m(0) \\ &\quad + \sum_{\substack{l=-N/2 \\ l \neq k}}^{N/2-1} X_m(l)H_m(l)I_m(l-k) + \zeta_m(k). \end{aligned} \tag{12.41}$$

where $X_m(k)$, $H_m(k)$, and $\zeta_m(k)$ are representations of the transmitted signal, channel response, and random noise in the frequency domain, respectively. The relationship between $I_m(i)$ and phase noise $\varphi_m(n)$ can be expressed as

$$I_m(i) = \frac{1}{N}\sum_{n=-N/2}^{N/2-1} e^{j2\pi ni/N} e^{j\varphi_m(n)}, \quad i = -\frac{N}{2}, \ldots, \frac{N}{2} - 1. \tag{12.42}$$

In order to properly recover the signal, $h_m(n)$ and $\varphi_m(n)$ must be accurately obtained, where the accurate $h_m(n)$ is achieved by channel estimation and $\varphi_m(n)$ is achieved by phase estimation.

Channel estimation is primarily based on training sequence. Among the methods of channel estimation based on training sequences, the maximum likelihood estimation is a method with low complexity and good performance. In channel estimation, if the influence of phase noise is ignored, the frequency-domain representation of the system's transfer function can be simplified as

$$R(k) = X(k)H(k) + \zeta_m(k). \tag{12.43}$$

where $\zeta_m(k)$ is the random noise in the fiber channel transmission. Since the fiber channel is assumed to be a time-invariant channel, the symbol index indicating the ith symbol of the OFDM signal in the time domain is omitted for simplicity. For a channel transmission function $H(k)$ of a certain channel that is unknown before the channel estimation, the received signal R and its joint probability density distribution are expressed as

$$\begin{aligned} &p(R(1), R(2), \ldots, R(N)|H(1), H(2), \ldots, H(N)) \\ &\propto \exp\pi\left(\sum_{k=1}^{N}\frac{(R(k) - H(k)X(k)) * (R(k) - H(k)X(k))}{2\delta^2}\right) \end{aligned} \tag{12.44}$$

where δ is the standard deviation of the random noise distributed on each subcarrier and assumes that the random noise distributed on each subcarrier is identical. The maximum likelihood criterion is to find an $H(k)$ that maximizes the joint probability density distribution function. The equivalent decision condition is to find an $H(k)$ to make the following function take the minimum value.

$$\Lambda(H(1), H(2), \ldots, H(N)) = \sum_{k=1}^{N}(R(k) - H(k)X(k))^*(R(k) - H(k)X(k)). \tag{12.45}$$

In the formula, $H(k)$ is a complex variable, and its conjugate $H^*(k)$ can be regarded as another single variable. In order to find a suitable $H(k)$ to make the equivalence likelihood function take the minimum value, the equivalent minimum likelihood function is derived for $H^*(k)$:

$$\frac{\partial \Lambda(H(1), H(2), \ldots, H(N))}{\partial H^*(k)} = X^*(k)(R(k) - H(k)X(k)). \quad (12.46)$$

In order to obtain the minimum value of the equivalent likelihood function, the value of the derivative must be satisfied to be zero. According to this criterion, the maximum likelihood estimation of the channel function is

$$\tilde{H}(k) = \frac{R(k)}{X(k)} = H(k) + \frac{n(k)}{X(k)} = H(k) + \tilde{n}(k). \quad (12.47)$$

In the OFDM channel estimation, the estimated channel characteristics include partial random noise and the actual channel characteristics. In addition to the maximum likelihood estimation criterion, there is a minimum mean square error criterion with higher computational complexity. This criterion can estimate the channel more accurately, but it is not discussed in this chapter. Next, we will discuss how to eliminate the random noise in channel estimation. Random noise leads to estimation errors in channel estimation. A common approach is to insert multiple training sequences in the time domain and then minimize the impact of this random noise on the channel estimation by time-domain averaging. However, this extra overhead will undoubtedly reduce the spectral efficiency of the system. In order to avoid reducing the spectral efficiency and eliminate the influence of random noise as much as possible, a method based on moving average in the frequency domain is proposed. In the method of moving average in the frequency domain, the final estimated value of a certain frequency point is jointly determined by the estimated values of its own and surrounding frequency points. The final channel estimation value can be expressed as

$$\hat{H}(k') = \frac{1}{\min(k_{\max}, k' + m) - \max(k_{\min}, k' - m)} \sum_{k=k'-m}^{k'+m} \tilde{H}(k). \quad (12.48)$$

where k and k' represent the frequency points before and after the moving average, respectively. Note that the number of samples at the beginning and end of the moving average will be reduced (the number of samples is m+1, while the number of samples at the intermediate frequency is 2m+1).

Since OFDM signals are very sensitive to phase noise, phase noise can corrupt the orthogonality between subcarriers of OFDM. The phase noise of a CO-OFDM system is classified into two categories according to its characteristics: common phase noise (CPN) and phase noise that causes ICI. The common phase noise is the same phase noise component present on each subcarrier in the entire frequency domain. This phase noise exhibits a characteristic that causes the rotation of the constellation of the signal. Another phase noise that causes ICI is irregular in the distribution of signal-modulated subcarriers, and this noise is smaller than the common phase noise, mainly leading to the dispersion of constellation points.

Phase noise cancelation can be done in both frequency and time-domain methods. The most common method of the frequency domain is based on the M-squares of the data points, which is commonly used in single-carrier systems for coherent detection. The advantage is that no additional bandwidth overhead is required. However, phase ambiguity is prone to occur, and phase ambiguity must be eliminated by differential coding, so that the implementation complexity is high. Another method is to reserve a certain number of subcarriers in each OFDM symbol for transmitting pilots. After the signal at the receiving end passes through the FFT, the signal on the corresponding subcarrier is compared with the original signal. Finding the phase change and averaging, we can get the phase noise in the frequency-domain estimation method. It can be expressed as

$$\phi_i = \frac{1}{N_p}\sum_{k=1}^{N_p}(\arg(Y_{ik}) - \arg(X_{ik})). \tag{12.49}$$

where X_{ik} represents the signal of the pilot inserted in the subcarrier reserved at the transmitting end, and Y_{ik} represents the received signal on the corresponding subcarrier at the receiving end. In the actual fiber transmission system, since the phase noise is introduced due to polarization mode dispersion and polarization loss and the distribution of the phase noise is random, the noise distribution using the previous average estimation method will lead to inaccuracy. In order to improve the accuracy of the phase noise estimation, a phase noise method based on the maximum likelihood estimation criterion is proposed. In this method, for a given ith OFDM symbol, assuming that the standard deviation of the noise distribution with Gaussian distribution on each subcarrier is δ_k, the likelihood function can be expressed as

$$\Lambda_i = \sum_{k=1}^{N_p}\frac{1}{\delta_k^2}\left|Y_{ki} - H_k X_{ki}\cdot e^{j\phi_i}\right|^2. \tag{12.50}$$

Let the likelihood function of Eq. (12.50) take the minimum value, then the phase noise estimate is

$$\phi_i = \arg\left(\sum_{k=1}^{N_p}\frac{1}{\delta_k^2}Y_{ki}H_k^* X_{ki}^*\right). \tag{12.51}$$

These two frequency-domain-based phase estimation schemes can only estimate common phase noise, but cannot estimate the phase noise that causes ICI. When the modulation order of the OFDM signal increases (16–2048QAM), it is very sensitive to phase noise, so the phase noise that causes ICI will seriously degrade the performance of the system. In order to overcome the influence of the phase noise that causes ICI on the decision of received OFDM signal, a time-domain-based phase noise estimation scheme is proposed. In the time-domain phase estimation scheme, the currently reported reception is based on heterodyne coherent detection. In the

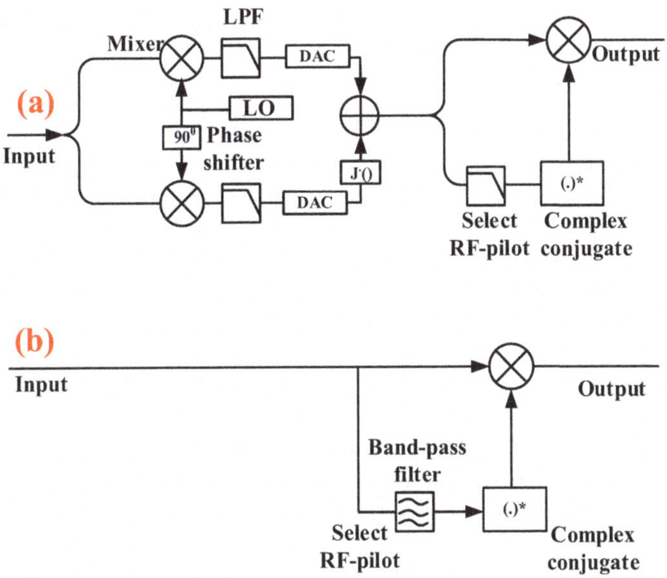

Fig. 12.13 Different RF-pilot extraction methods

modulation of OFDM, the first subcarrier is reserved for the insertion of the DC component in the time domain after the modulation of the signal, and the DC component is up-converted to the intermediate frequency by the electric IQ mixer. After electric up-conversion, the DC component inserted in the first subcarrier becomes a radio frequency pilot (RF-pilot). And after transmission over the fiber link, the RF-pilot is extracted by time-domain filtering and the phase noise is compensated. There are two ways to extract RF-pilot, as shown in Fig. 12.13.

Figure 12.13a is a way to extract RF-pilot and compensate phase noise by down-conversion and low-pass filtering. The biggest advantage of this method is that the bandwidth of the DAC is saved, which can greatly reduce the cost. Figure 12.13b is a direct band-pass filter to extract the intermediate frequency RF-pilot, which is used to achieve both down-conversion and phase noise compensation. Although this method can achieve down-conversion at the same time, it is related to the bandwidth of the DAC and the frequency of the RF-pilot. The bandwidth requirement of the DAC is much larger than the bandwidth requirement of the first scheme. In a coherent optical OFDM optical fiber transmission system based on RF-pilot compensation phase noise, an RF-pilot is extracted by filtering at the receiving end. Assuming that the impulse response of the M point of the filter bandwidth is h_m, the signal obtained after filtering is

$$\underset{k,i}{\tilde{x}} = \frac{1}{M}\sum_{k=1}^{M-1} x_{k+m,i} h_m. \tag{12.52}$$

where $x_{k,i}$ represents the OFDM signal in the time domain. The phase of the filtered signal is the phase noise:

$$\varphi_{k,i} = \arg\left(\widetilde{x_{k,i}} \Big/ \left|\widetilde{x_{k,i}}\right|\right). \tag{12.53}$$

The phase noise in Eq. (12.53) is the phase noise estimated in the time domain. In order to distinguish from the phase noise estimated in the frequency domain, it is represented by $\varphi_{k,i}$ here. Compared with the phase noise ϕ_i of the frequency-domain estimation, the phase noise of the time-domain estimation is continuously fluctuating at different time points. The estimated phase noise in the frequency domain is constant for a certain OFDM symbol period. This is also the reason why the time-domain estimation can compensate for phase noise that causes ICI but the frequency-domain estimation not.

In a single-mode fiber transmission system, both polarization states can be used to transmit signals. Polarization multiplexing means that the optical signals of the two polarization states are used to transmit information, respectively. The introduction of polarization multiplexing technology doubles the capacity of an optical fiber communication system. But when a polarization-multiplexed signal is transmitted in optical fiber, the two polarization states are randomly rotated in the fiber. This random polarization rotation is called polarization mode dispersion (PMD). In addition, polarization-dependent loss (PDL) can also cause signal distortion. Therefore, the effects of PDM and PDL on OFDM signals must be overcome during the implementation of polarization multiplexing.

In a polarization-multiplexed OFDM fiber transmission system, the polarization-multiplexed signal can be regarded as a 2 × 2 multiple-input multiple-output (MIMO) system. The OFDM signal at the transmitting end can be expressed as

$$\boldsymbol{t} = \begin{bmatrix} t_x \\ t_y \end{bmatrix}. \tag{12.54}$$

where t_x and t_y represent OFDM signals transmitted in the x, y polarization direction, respectively. After the signal is transmitted over the fiber link, it can be expressed as

$$\boldsymbol{s} = \boldsymbol{h} \otimes \boldsymbol{t} + \boldsymbol{n}. \tag{12.55}$$

where $\boldsymbol{n}$ is random noise, ⊗ represents channel convolution, and $\boldsymbol{s}$ and $\boldsymbol{h}$ represent the received signal and the impact response of the fiber channel, which are expressed in Eqs. (12.56) and (12.57), respectively.

$$\boldsymbol{s} = \begin{bmatrix} s_x \\ s_y \end{bmatrix} \tag{12.56}$$

$$\boldsymbol{h} = \begin{bmatrix} h_{xx} & h_{xy} \\ h_{yx} & h_{yy} \end{bmatrix} \tag{12.57}$$

Fourier transform is performed on Eq. (12.55), and the transmission expression in the time domain is converted to the frequency domain, which can be expressed as

$$\boldsymbol{S} = \boldsymbol{H} \cdot \boldsymbol{T} + \xi. \tag{12.58}$$

In (12.58), $\boldsymbol{S} = \begin{bmatrix} S_x \\ S_y \end{bmatrix}$, $\boldsymbol{H} = \begin{bmatrix} H_{xx} & H_{xy} \\ H_{yx} & H_{yy} \end{bmatrix}$, and $\boldsymbol{T} = \begin{bmatrix} T_x \\ T_y \end{bmatrix}$. $\boldsymbol{S}$, $\boldsymbol{H}$, $\boldsymbol{T}$, and ξ are the Fourier transform forms of the received signal, the fiber channel response, the transmitted signal, and the random noise, respectively. The time-domain convolution becomes the product of the frequency domain after the Fourier transform. Therefore, if the channel characteristic $\boldsymbol{H}$ of the polarization multiplexing system in the optical fiber can be correctly estimated, the demultiplexing and separation of the two polarization states can be realized. This matrix $\boldsymbol{H}$, which reflects the characteristics of the polarization channel, is called a Jones matrix.

After the Jones matrix $\boldsymbol{H}$ is calculated, it can be estimated that the transmitted information is

$$\tilde{\boldsymbol{T}} = \boldsymbol{H}^{-1}\boldsymbol{S} - \boldsymbol{H}^{-1}\xi. \tag{12.59}$$

Comparing $\tilde{\boldsymbol{T}}$ with the transmitted $\boldsymbol{T}$, even if the Jones matrix $\boldsymbol{H}$ can be estimated without any error, the random noise ξ will still affect the reception and recovery of the OFDM signal. To overcome the effects of random noise, maximum likelihood estimation can be used to eliminate the effects of random noise as much as possible. In addition, the effects of random noise can be reduced by using the channel estimation methods of time-domain average and frequency-domain moving average which are discussed in the previous part.

The key step of polarization demultiplexing in a polarization-multiplexed CO-OFDM system is to obtain a Jones matrix $\boldsymbol{H}$. For a conventional OFDM transmission system, a common method is to periodically insert a training sequence in the time domain, and then estimate the channel characteristics before equalization by comparing the received signal with the training sequence in the frequency domain. Since the polarization multiplexing system can be simply viewed as a 2×2 MIMO system, it is necessary to ensure that the training sequences in multiple input signals are orthogonal in the MIMO system. Then the receiving end can correctly find the training sequence. According to this criterion, a variety of orthogonal training sequence structures can be proposed for a polarization-multiplexed 2×2 MIMO system. The simplest way is to implement orthogonality of training sequences by means of time-division multiplexing. A training sequence based on time-division multiplexing is given in Fig. 12.14.

The two orthogonal sequences of time-division multiplexing can be expressed as

$$\boldsymbol{T}_1 = \begin{bmatrix} T_x \\ 0 \end{bmatrix}, \boldsymbol{T}_2 = \begin{bmatrix} 0 \\ T_y \end{bmatrix}. \tag{12.60}$$

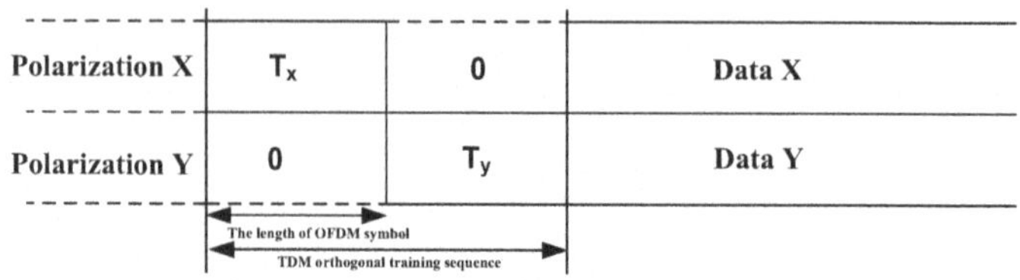

Fig. 12.14 Time-division multiplexed training sequence for MIMO channel estimation

where T_x and T_y are training sequences, respectively inserted in the x and y polarization states in the time domain. The training sequence needs to ensure good performance against system noise and consistency between the DSP of the training sequence and signal. Assuming that the effects of the fiber channel on the two consecutive pairs of training sequences in the time domain are the same and the effects of random noise are ignored, the received signals are

$$\boldsymbol{S}_1 = \begin{bmatrix} S_{1x} \\ S_{1y} \end{bmatrix} = \boldsymbol{H}\boldsymbol{T}_1 = \begin{bmatrix} H_{xx} & H_{xy} \\ H_{yx} & H_{yy} \end{bmatrix} \begin{bmatrix} T_x \\ 0 \end{bmatrix} = \begin{bmatrix} H_{xx}T_x \\ H_{yx}T_x \end{bmatrix} \quad \text{and} \quad \boldsymbol{S}_2 = \begin{bmatrix} S_{2x} \\ S_{2y} \end{bmatrix} = \boldsymbol{H}\boldsymbol{T}_2 = \begin{bmatrix} H_{xx} & H_{xy} \\ H_{yx} & H_{yy} \end{bmatrix} \begin{bmatrix} 0 \\ T_y \end{bmatrix} = \begin{bmatrix} H_{xy}T_y \\ H_{yy}T_y \end{bmatrix} \tag{12.61}$$

Solving the equation to get the Jones matrix $\boldsymbol{H}$:

$$\boldsymbol{H} = \begin{bmatrix} H_{xx} & H_{xy} \\ H_{yx} & H_{yy} \end{bmatrix} = \begin{bmatrix} S_{1x}/T_x & S_{2x}/T_y \\ S_{1y}/T_x & S_{2y}/T_y \end{bmatrix}. \tag{12.62}$$

According to the obtained Jones matrix, the signals of the two polarization states can be separated. Next, phase noise estimation and decision are made to obtain the received digital signal.

The schematic diagram of the polarization-multiplexed CO-OFDM system is shown in Fig. 12.15. The in-phase component and the quadrature component of the OFDM signal generated by the arbitrary waveform generator are jointly injected into the IQ modulator to implement electro-optical conversion of the OFDM signal. A quadrature modulated optical OFDM signal is produced by adjusting the DC bias of the IQ modulator. The optical OFDM signal is subjected to polarization multiplexing through an analog polarization multiplexing unit. The analog polarization multiplexing unit is implemented by a coupler, an optical delay line, an optical attenuator, and a polarization coupler. Two optical OFDM signals are separated by the coupler, and one signal is delayed by exactly one OFDM symbol length, while the power of the other signal is adjusted by an optical attenuator between the two channels. Then the two signals are coupled by a polarization coupler. The OFDM signals in the two polarization directions of x and y after the analog polarization multiplexing unit are shown in Fig. 12.15. This way of polarization multiplexing also ensures the

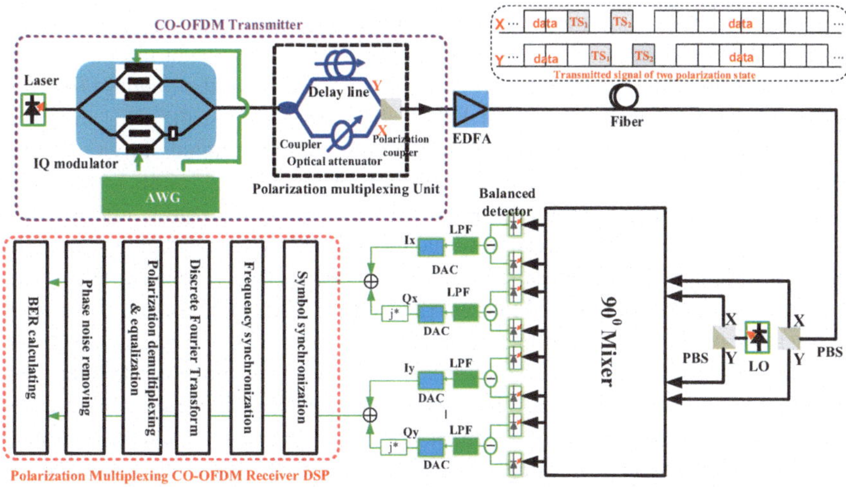

Fig. 12.15 Polarization-multiplexed coherent detection OFDM system diagram

orthogonality of the training sequences. After the polarization-multiplexed OFDM signal is amplified by EDFA and transmitted through a fiber of a certain length, the signal is photoelectrically converted by coherent detection. The four signals obtained by the balanced detector respectively pass through a low-pass filter and analog-to-digital conversion and then enter the DSP of the receiving end. The DSP at the receiving end mainly includes symbol synchronization, frequency synchronization, discrete Fourier transform, polarization demultiplexing and equalization, phase noise cancelation, and error calculation.

The main advantage of using a CO-OFDM system to achieve single-wavelength high-rate signal transmission compared with the single-carrier implementation is that inserting a CP of a certain length of time into each OFDM symbol can avoid using the DSP algorithm to compensate for dispersion at the receiving end. In addition, in the frequency-domain equalization of OFDM, the DSP algorithm is transparent to the modulation format of the signal. But for the single carrier, the methods of channel equalization changes with different signal modulation formats. When the system needs to increase the order of the modulation format to achieve system expansion, the advantage that OFDM is transparent to the signal modulation format during channel equalization is very obvious. In recent years, coherent OFDM long-distance transmission has achieved many record-making research results [26–42]. In 2007, William Shieh et al. from the University of Melbourne in Australia simulated the transmission of OFDM signals in a coherent detection system for the first time. In the absence of dispersion compensation, the 10 Git/s OFDM signal successfully transmitted over 3000 km through a single-mode fiber. This confirms that the OFDM signal can be transmitted over long distances without dispersion compensation. At the 2007 ECOC, William Shieh et al. reported that 8 Gbit/s OFDM was successfully transmitted over 1000 km through a single-mode fiber without dispersion compensation [33]. Since then, CO-OFDM has become a research hotspot in optical fiber communication. There are a lot of reports on CO-OFDM in the annual OFC and ECOC. In 2007, SL Jansen et al. of KDDI, Japan, reported at the OFC that they successfully realized the transmission of 20 Gbit/s OFDM signals in CO-OFDM systems and increased the distance to 4160 km over SSMF [27, 28]. In the same year of the ECOC, they reported the first successful use of polarization multiplexing technology to achieve the expansion of the CO-OFDM system, which achieved 16 optical wavelength WDM systems in the CO-OFDM system. And each light wavelength carries 52.5 G bit/s OFDM signal [28]. At the 2008 OFC, S. L. Jansen et al. successfully implemented a WDM-CO-OFDM system with a single-channel rate exceeding 100 Gbit/s and successfully transmitted 1000 km single-mode fiber [28]. Since then, more and more researchers have proposed CO-OFDM as an alternative to 100 G or even 400 G Ethernet optical interfaces. In order to achieve transmission of 400 Gbit/s signals in a single WDM channel, Bell Labs of the United States proposed a method based on generating multiple optical carriers in a single WDM channel and loading OFDM signals on different optical carriers, respectively [36]. Since the roll-off coefficient of the OFDM signal spectrum is particularly low, there is no need to maintain a guard interval between different optical OFDM sidebands, thus ensuring high spectral efficiency. By this technique, it is possible to transmit OFDM signals of 448 Gbit/s in the WDM channels with an interval of 80 GHz. In order to make the net bit rate over 400 Gbit/s, the ratio of overhead such as CP and pilot must be reduced. In this scheme, the electrical dispersion compensation is applied to eliminate the dispersion effect in fiber transmission.

In addition to ultra-high-rate signals transmission in the backbone transmission network, the CO-OFDM system is also applied to a large-capacity wireless access network realized by a hybrid network structure of optical fibers and wireless. The millimeter wave of the wireless link in the hybrid network is generated by optical beat frequency, and the millimeter wave in the OFDM-RoF system based on direct detection is generated by optical double frequency. Since there is no correlation between the two optical waves in the millimeter wave scheme using the optical beat frequency method, the DSP algorithm at the receiving end is more complicated (including additional frequency offset estimation and phase noise estimation). However, due to the use of coherent detection, the receiving sensitivity of the system is improved. The W-band (80–100 GHz) has been extensively studied in recent years because of its low transmission loss in air and the ability to provide a huge available bandwidth. In 2012, L. Deng et al. used a ROF network structure in which two uncorrelated optical waves generated a W-band (75–110 GHz) carrier signal by beat frequency, and each optical carrier carries three 8.3 Gbit/s OFDM signal sidebands [43]. In order to meet the consistency between the capacity of the wireless link transmission and the fiber link transmission in the next-generation ROF network, this ROF network using CO-OFDM will be a very feasible alternative. In 2013, we achieved a wireless distance of 5 m for W-band 40 Gbit/s signal transmission [44].

12.4 Summary

OFDM detection methods can be divided into direct detection and coherent detection. Considering the cost, direct detection is suitable for short-distance transmission systems, while coherent detection is suitable for long-distance transmission. This chapter briefly introduces their system architecture and basic algorithm principles and recent research progress.

References

1. T. Okoshi, K. Kikuchi, *Coherent Optical Fiber Communications* (Kluwer Academic Publishers, Boston, 1988)
2. Y. Yamamoto, T. Kimura, Coherent optical fiber transmission systems. IEEE J. Quantum Electron. **QE-17**(6), 919–935 (1981)
3. T. Okoshi, K. Kikuchi, Frequency stabilization of semiconductor lasers for heterodyne-type optical communication schemes. Electron Lett. **16**(4), 179–181 (1980)
4. F. Favre, D. LeGuen, High frequency stability of laser diode for heterodyne communication systems, Electron. Lett. **16**(18), 709–710
5. Y. Yamamoto, Receiver performance evaluation of various digital optical modulation-demodulation systems in the 0.5-IO-pm-wave- length region, IEEE J. Quantum Electron. QE-16(11)
6. T. Okoshi, K. Emura, K. Kikuchi et al., Computation of bit-error rate of various heterodyne and coherent-type optical communication schemes. J. Opt. Commun. **2**(3), 89–96 (1981)
7. F. Favre, L. Jeunhomme, I. Joindot et al., Progress towards heterodyne-type single-mode fiber communication systems, IEEE J. Quantum Electron., QE-17(6), pp. 897–906 (1981)
8. T. Okoshi, Heterodyne and coherent optical fiber communications: recent progress, IEEE Trans. Microwave Theory Tech., vol. MTT- 30, no. 8, pp. 1138–1149, Aug. 1982
9. D.W. Smith, I. W. Stanley, "The worldwide status of coherent optical fiber transmission systems, in Tech. Dig. ECOC'83, Ninth Europ. ConJ Opt. Commun. (Geneva, Switzerland), Oct. 23–26, 1983, pp. 263–266
10. R.C. Hooper, J.E. Midwinter, D. W. Smith et al., Progress in monomode transmission techniques in the United Kingdom, J. Lightwave Technol., vol. LT-1, no. 4, pp. 596–611, Dec. 1983
11. T. Okoshi, Recent progress in heterodynelcoherent optical fiber communications, in *Proc. OFC'84 Confi Opt. Fiber Commun.* (New Orleans, LA), Jan. 23–25, 1984, Pap. TUK1
12. T. Okoshi, Recent progress in heterodyneicoherent optical fiber communications. J. Lightwave Technol., vol. LT-2, no. 4, pp. 341–346, Aug. 1984
13. K. Kikuchi, T. Okoshi, M. Nagamatsu et al., "Degradation of bit-error rate in coherent optical communications due to spectral spread of the transmitter and local oscillator, J. Lightwave Technol., vol. LT-2, no. 6, pp. 1024–1032, Dec. 1984
14. K. Kikuchi, T. Okoshi, "Dependence of semiconductor-laser linewidth on measurement time: Evidence of predominance of l/f noise, Electron. Lett., vol. 21, no. 22, pp. 1011–1012, Oct. 24, 1985
15. I. Kaminow, T. Li (eds.), *Optical Fiber Telecommunications IV A +B* (Academic, New York, 2002)
16. J. Kahn, K.-P. Ho, Spectral efficiency limits and modulation/detection techniques for DWDM systems. IEEE J. Sel. Top. Quantum Electron. **10**(2), 259–272 (2004)
17. R. Griffin, A. Carter, "Optical Differential Quadrature Phase-Shift Key (oDQPSK) for High Capacity Optical Transmission, in Proc. OFC 2002, paper WX6, Anaheim, CA, USA, March 17–22, 2002

18. S. Shimotsu, S. Oikawa, T. Saitou et al., Single side-band modulation performance of a LiNbO3 integrated modulator consisting of four-phase modulator waveguides. IEEE Photonics Technol. Lett. **13**(4), 364–366 (2001)
19. S. Tsukamoto, D.-S. Ly-Gagnon, K. Katoh et al., Coherent Demodulation of 40-Gbit/s Polarization-Multiplexed QPSK signals with 16-GHz Spacing after 200-km Transmission, in Proc. OFC 2005, paper PDP29, Anaheim, CA, March 7–12, 2005
20. K. Kikuchi, Phase-diversity homodyne detection of multi-level optical modulation with digital carrier phase estimation. IEEE J. Sel. Top. Quantum Electron. **12**(4), 563–570 (2006)
21. S. Tsukamoto, K. Katoh, K. Kikuchi, Coherent demodulation of optical multilevel phase shift-keying signals using homodyne detection and digital signal processing. IEEE Photon. Technol. Lett. **18**(10), 1131–1133 (2006)
22. S. Tsukamoto, K. Katoh, K. Kikuchi, Unrepeated transmission of 20-Gbit/s optical quadrature phase-shift keying signal over 200-km standard single-mode fiber based on digital processing of homodyne-detected signal for group-velocity dispersion compensation. IEEE Photon. Technol. Lett. **18**(9), 1016–1018 (2006)
23. M. Taylor, Coherent detection method using DSP for demodulation of signal and subsequent equalization of propagation impairments. IEEE Photon. Technol. Lett. **16**(2), 674–676 (2004)
24. H. Bu¨low, Electronic Dispersion compensation, in Proc. OFC 2007, paper OMG5, Anaheim, CA, March 25–29, 2007
25. A. Leven, N. Kaneda, U. V. Koc et al., Coherent Receiver for Practical Optical Communication Systems, in Proc. OFC 2007, paper OThK4, Anaheim, CA, March 25–29, 2007
26. W. Shieh, C. Athaudage, Coherent optical orthogonal frequency division multiplexing. Electron. Lett. **42**, 587–589 (2006)
27. S.L. Jansen, I. Morita, T.C.W. Schenk et al., Coherent Optical 25.8-Gb/s OFDM Transmission over 4,160-km SSMF. J. Lightwave Technol. **26**(1), 6–15 (2008)
28. S.L. Jansen, I. Morita, H. Tanaka, "10x121.9-Gb/s PDM-OFDM Transmission with 2-b/s/Hz Spectral Efficiency over 1,000 km of SSMF," in Proc. Optical Fiber Communication (OFC) Conference 2008, PDP 2
29. A. Lobato, B. Inan, S. Adhikari et al., "On the efficiency of RF-Pilot-based nonlinearity compensation for CO-OFDM," in Proc. Optical Fiber Communication (OFC) Conference 2011, OThF2
30. D. van den Borne, V.A.J.M. Sleiffer, M.S. Alfiad et al., POLMUX-QPSK modulation and coherent detection: the challenge of long-haul 100G transmission. Proc. European Conference on Optical Communications (ECOC) **3**(4), 1 (2009)
31. S.L. Jansen, A. Al Amin, H. Takahashi et al., "132.2-Gb/s PDM-8QAM-OFDM Transmission at 4-b/s/Hz Spectral Efficiency," IEEE Photon. Tech. Lett., Vol. 21, No. 12, pp. 802–804, 2009
32. S.L. Jansen, I. Morita, T.C.W. Schenk et al., 121.9-Gb/s PDM-OFDM Transmission with 2 b/s/Hz Spectral Efficiency over 1,000 km of SSMF. J. Lightwave Technol. **27**(3), 177–188 (2009)
33. W. Shieh, X. Yi, Y. Tang, Transmission experiment of multi-gigabit coherent optical OFDM systems over 1000 km SSMF fibre. Electron. Lett. **43**, 134–136 (2007)
34. W.-R. Peng, K. Takeshima, I. Morita et al., Scattered pilot channel tracking method for PDM-CO-OFDM transmissions using polar-based intra-symbol frequency-domain average, in Proc. OFC 2011, pp. 1–3, paper OWE6
35. R. Dischler, F. Buchali, Transmission of 1.2 Tb/s Continuous Waveband PDM-OFDM-FDM Signal with Spectral Efficiency of 3.3 bit/s/Hz over 400 km of SSMF, OFC 2009, PDPC2
36. X. Liu, S. Chandrasekhar, P.J. Winzer et al., Single coherent detection of a 606-Gb/s CO-OFDM signal with 32-QAM subcarrier modulation using 4 × 80-Gsamples/s ADCs, in proc. ECOC 2010, PD2.6
37. D. Qian, M. Huang, E. Ip et al., 101.7-Tb/s (370 × 294-Gb/s) PDM-128QAM-OFDM Transmission over 3 × 55-km SSMF using Pilot-based Phase Noise Mitigation,OFC 2011, PDP B5
38. L. Tomba, On the Effect of Wiener Phase Noise in OFDM Systems. IEEE Trans. on Com **46**, 580–583 (1998)

39. S.L. Jansen, I. Morita, T.C.W. Schenk et al., Coherent Optical 25.8-Gb/s OFDM Transmission over 4,160-km SSMF. J. Lightwave Technol. **26**, 6–15 (2008)
40. X. Yi, W. Shieh, Y. Tang, Phase Estimation for Coherent Optical OFDM, IEEE Photon. Technol. Lett., vol. 19, 2007
41. X. Yi, W. Shieh, Y. Ma, Phase Noise Effects on High Spectral Efficiency Coherent Optical OFDM Transmission. J. Lightwave Technol. **26**, 1309–1316 (2008)
42. S. Randel, S. Adhikari, S.L. Jansen, Analysis of RF-Pilot-based Phase Noise Compensation for Coherent Optical OFDM Systems. IEEE Photon. Technol. Lett. **22**, 1288–1290 (2010)
43. L. Deng et al., Fiber wireless transmission of 8.3-Gb/s/ch QPSKOFDM signals in 75–110-GHz band. IEEE Photon. Technol. Lett. **24**(5), 383–385 (2011)
44. Fan Li et al., Fiber-wireless transmission system of PDM-MIMO-OFDM at 100 GHz frequency. J. Lightwave Technol. **31**(14), 2394–2399 (2013)

Chapter 13
Direct Detection OFDM

13.1 Introduction

According to our discussion in Sect. 13.2.1, due to its advantages of simple structure, low-cost, and low implementation complexity, DDO-OFDM is mainly applied in access network [1–12] and ultra-high-speed point-to-point optical fiber communication services [13–21]. DDO-OFDM system has some disadvantages, mainly including subcarrier–subcarrier mutual interference (SSMI) [6, 12], high frequency attenuation caused by fiber dispersion, and imperfect frequency response of devices [10]. In the short distance DDO-OFDM system, the optical fiber dispersion is not obvious, so the high frequency attenuation caused by dispersion can be ignored. The application scenarios of DDO-OFDM discussed and analyzed in this chapter are all short transmission distance, and imperfect frequency response will be the most important factor leading to high frequency attenuation. With the promotion and popularization of broadband communication service in daily life mentioned in Chap. 1, the transmission capacity of DDO-OFDM system that realizes access network and high-speed point-to-point optical fiber communication service will increase rapidly. It is a feasible scheme to improve the transmission performance of DDO-OFDM system with large capacity by using DSP to eliminate SSMI.

In addition, the high frequency attenuation caused by the imperfect frequency response of the device in the high-capacity DDO-OFDM system will be obvious. Generally, in order to improve the capacity of the system, the order of modulation format will be increased to improve the spectral efficiency, and the higher order modulation format will be more sensitive to ISI caused by such high frequency attenuation. In order to guarantee the transmission performance of the broadband DDO-OFDM system with high order modulation, it is necessary to study how to overcome the high frequency attenuation.

In the second section of this chapter, we propose a method to eliminate SSMI without extra overhead. In DDO-OFDM system, after the optical OFDM signal

J. Yu and N. Chi, *Digital Signal Processing In High-Speed Optical Fiber Communication Principle and Application*,
https://doi.org/10.1007/978-981-15-3098-2_13

passes the PD through square detection, the subcarrier–subcarrier mutual interference noise will also appear in the detected signal. In order to overcome the influence of SSMI on the OFDM signal, one scheme is to insert protection interval in the frequency domain, which is called OFDM of protection interval. Another scheme is called frequency-domain cross OFDM, in which the signal is modulated only on an even number of subcarriers. These two schemes have been proved to be effective in eliminating the influence of SSMI on the OFDM signal, but both schemes would reduce the overall spectral efficiency of the system by half. In a 64-ary quadrature-amplitude-modulation (64QAM) DDO-OFDM system, Cao et al. proposed interleave encoding and Turbo encoding technology to overcome SSMI and successfully transmit 64QAM-OFDM signal for 100 km [17]. This technique has been proved to be effective in overcoming SSMI, but the introduction of Turbo coding will still lead to a decrease in spectral efficiency, in addition, the introduction of forward error correction coding technology will lead to a great increase in the system implementation complexity. Another method proposed at present is to modulate signals of different modulation formats according to the signal-to-noise ratio of each subcarrier, so that the subcarrier at low frequency can carry a lower modulation format order due to the more serious SSMI. This scheme requires a complex calibration process at the early stage of system establishment in order to find out the distribution of SSMI, so it is not suitable for DDO-OFDM system with simple structure, low-cost, and low implementation complexity. In order to overcome the influence of SSMI without reducing the spectral efficiency and improving the system implementation complexity, the half-cycled DDO-OFDM scheme is proposed in Sect. 13.2 of this chapter. This scheme has been proved to be able to successfully resist SSMI without introducing additional overhead. The experimental results show that the receiver sensitivity of QPSK-OFDM and 16QAM-OFDM systems is improved by 2 dB and 1.5 dB, respectively, after 40-km transmission by standard single-mode fiber.

The third section of this chapter analyzes the difficulties in the implementation of the low-cost high order modulation format DDO-OFDM system and proposes the corresponding solutions. Simple structure and low-cost are the key advantages of the DDO-OFDM system for realizing short distance access network. Different from the long distance transmission backbone, the capital expenditures (CapEx), and the operational expenditures (OpEx) in the access network are required to be controlled in the range of minimum. Intensity modulation in DDO-OFDM system can be realized by directly modulating laser and external modulators. Compared with the external modulator, the cost of the directly modulated laser is its biggest advantage. Using direct modulation laser to realize electro-optic modulation in the access network will greatly reduce the input and expenditure of the access network system.

With the emergence of new high-bandwidth communications services (HDTV, cloud computing, interactive HD online games, etc.), the capacity of the next generation access network will exceed 10 Gb/s and even reach 100 Gb/s. In order to realize such a large capacity of signal transmission in DDO-OFDM system with limited device bandwidth, the most direct way is to increase the frequency spectrum efficiency of the system by increasing the order of OFDM signal modulation format.

In the DDO-OFDM system, 128QAM [4] is the most advanced modulation scheme reported. In the CO-OFDM system, 1024QAM-OFDM and 2048QAM-OFDM transmission can be realized by adopting large size FFT in OFDM signal modulation [3]. For a certain bandwidth OFDM signal, the increase of FFT size means the increase of resolution in the frequency domain, so the information that can be estimated by training sequence during channel estimation is more accurate. In addition, the OFDM signal of large-scale FFT can better resist ISI, but at the same time, the OFDM signal will be more sensitive to ICI. In order to eliminate the influence of ICI in the reported high order modulated CO-OFDM system with large-scale FFT, it is usually necessary to add complex frequency offset estimation and phase estimation algorithms. However, in the DDO-OFDM system with a short transmission distance of optical fiber, the ICI caused by phase noise and frequency deviation can be ignored. In a direct detection OFDM system, this increase in FFT size can effectively improve the overall error performance of the high order modulation OFDM signal. As the peak/average power ratio of the OFDM signal is proportional to the size of FFT, another problem is that the peak/average power ratio of the OFDM signal will increase when the size of FFT increases. The high order modulation OFDM signal is very sensitive to the nonlinear noise introduced by the high peak/average power ratio. In order to realize the high-performance transmission of the high order modulation OFDM signal, the peak/average power ratio of the OFDM signal must be controlled. In order to reduce the peak/average power ratio of OFDM signals, the discrete Fourier transform spread (DFT-spread) is added to the transmitter. This method can effectively reduce the peak/average power ratio of OFDM signals. In addition, in order to improve the accuracy of channel estimation, the intra-symbol frequency-domain averaging (ISFA) algorithm was used in channel estimation. The above DSP algorithm is adopted to realize the transmission of 2048 QAM-OFDM in DDO-OFDM adopting direct modulation laser to realize electro-optical modulation for the first time. The error performance of 31.7 Gb/s 2048QAM-OFDM signal after transmission through 20-km large effective area fiber (LEAF) is still lower than the error threshold (2.4×10^{-2}) of the soft judgment of forward error correction code at the cost of 20%.

The fourth section of this chapter is the research on the short distance transmission of DDO-OFDM signal with a large capacity based on DFT-spread. Direct modulation can greatly reduce the cost of the system in the DDO-OFDM system with short distance transmission. Therefore, direct modulation lasers are currently used in the short distance access network based on direct detection. Currently, advanced modulation technologies that can be applied to direct detection systems to improve the spectral efficiency of the system mainly include OFDM, carrierless amplitude phase modulation (CAP), pulse amplitude modulation (PAM), and 16QAM of half-cycle. The OFDM signal shows good resistance to dispersion and PMD, and the modulation format of the signal is transparent because of the frequency-domain equalization. Among these modulation techniques, the OFDM signal is the most competitive in realizing short distance optical fiber transmission. However, in OFDM, the problem that the peak/average power ratio of signals is too high must be solved, and the DFT-spread technology is proposed to reduce the peak/average power ratio of the

OFDM signals. This scheme does not cause distortion of the signal and the computational complexity is not high. This chapter mainly completes two aspects of work: DDO-OFDM optimization training sequence insertion scheme in DFT-spread; the performance comparison between pre-enhancement and DFT-spread in the DDO-OFDM system with large bandwidth DFT-spread in resistance to high frequency attenuation.

In the DDO-OFDM system based on DFT-spread, compared with the conventional DDO-OFDM system, a set of additional DFT/IDFT is added to the DSP of the transmitter end and the receiver end of OFDM. In the frequency-domain equalized OFDM system based on pilot frequency, it is usually necessary to ensure that the DSP algorithm of the OFDM symbol and training sequence is as consistent as possible in order to accurately estimate the channel characteristics. The extra DFT at the transmitting end will cause the signal going through OFDM modulation to change from the previous binary signal to the analog signal, which has no effect on the OFDM symbol except that the peak/average power ratio of the symbol after OFDM modulation will decrease, but this will make the training sequence change from the discrete digital signal to the continuous analog signal. This change will lead to a rapid deterioration of the performance of frequency-domain equalization based on training sequence. Therefore, for the training sequence, there is no need to add additional DFT/IDFT in the DSP of the transmitter and receiver. In Sect. 13.4 of this chapter, we experimentally investigate the optimal training sequence insertion scheme suitable for large capacity DDO-OFDM system. In the experiment, we found that the digital signal training sequence without adding additional DFT had better bit error performance than the analog signal training sequence without adding additional DFT. BPSK/QPSK shows the best error performance in the digital signal training sequence without additional DFT. When this optimal training sequence is adopted, 79.86-Gb/s DFT-spread 32QAM-OFDM signal has no power cost after being transmitted through the 20-km optical fiber and the signal bit error rate is far lower than the hard decision forward error correction (HD-FEC) Error threshold (3.8×10^{-3}).

In the DDO-OFDM system with large capacity and short distance transmission, high frequency attenuation must be considered and overcome. In order to realize the high-capacity DDO-OFDM system, the signal usually chooses the high order modulation format. When the OFDM signal of 100 Gb/s is transmitted by direct modulation, the insufficient bandwidth of the DML and DAC will lead to obvious HF attenuation. In addition, the amplifier, optical filter, and PD will also introduce HF signal attenuation. In order to eliminate the influence of high frequency attenuation on the OFDM signal, the traditional scheme is to adopt pre-emphasis. This method requires a complex-value channel estimation process using training sequence established at the transmitting side, to estimate the attenuation degree of each subcarrier channel and calculate the coefficient of pre-emphasis filter. This scheme is effective against high frequency attenuation but the whole process is relatively complex and less feasible. Another scheme we proposed is the application of DFT-spread in DDO-OFDM system. This scheme can also effectively resist high frequency attenuation and has much higher feasibility. In addition, this scheme also has a very big advantage that it can overcome high frequency attenuation and reduce the peak/average power ratio of the OFDM signals at the same time. In DFT-spread DDO-OFDM system all data in the subcarrier before going through the OFDM modulation will have an additional DFT, after the extra DFT the digital signal in the single frequency is split into all subcarrier in the OFDM signal frequency band, so after DFT-spread the OFDM signal will resist high frequency attenuation and the narrowband interference in subcarrier better. With this scheme, we successfully transmitted 100 Gb/s OFDM signal for 20 km LEAF. Compared with the pre-emphasis the OFDM signal, DFT-spread OFDM signal has a lower peak average power ratio and stronger resistance to high frequency attenuation and narrowband interference. The experiment result shows that the BER of 100 Gb/s DFT-spread 32QAM-OFDM signal with residual sideband transmitted by 20-km LEAF was far lower than the HD-FEC threshold (3.8 $\times$ 10^{-3}).

13.1.1 System Principle

Figure 13.1a shows three different OFDM signals in the DDO-OFDM system. The first is the traditional OFDM signal. After the square detection of PD, SSMI will exist in the frequency domain of the whole signal, so the bit error rate of the signal will be seriously deteriorated. The second is the OFDM signal with protection interval, which will reserve protection interval with the same size of signal bandwidth in frequency domain at low frequency. After the signal passes the PD through square detection, all the SSMI falls within the protection interval, so the signal error performance will be improved. However, the obvious drawback is that the overall spectral efficiency has fallen by half. The third is the OFDM with frequency-domain crossing. The data of such OFDM signals are modulated on even subcarriers, and the odd subcarriers are all set to zero. After the square detection of PD, the SSMI falls on the odd subcarrier, so the error performance of the signal will be improved. In this OFDM scheme with frequency domain crossover, when the data is only modulated on odd/even subcarriers, a single OFDM shows symmetry in the time domain. Based on this theory, a half-cycled DDO-OFDM scheme is proposed.

In OFDM modulation, the size of IFFT is defined as N, and the time length of a single OFDM symbol is T. After IFFT, OFDM signal can be expressed as

$$s(t) = \sum_{k=0}^{N-1} c_k \exp(j2\pi f_k t) \quad (1 \le t \le T) \tag{13.1}$$

In Eq. (13.1), k represents the exponent of the subcarrier, and f_k represents the carrier frequency corresponding to the kth subcarrier. In Eq. (13.2), it can be expressed as

$$f_k = k\Delta f = {}^{k}/_{T} \tag{13.2}$$

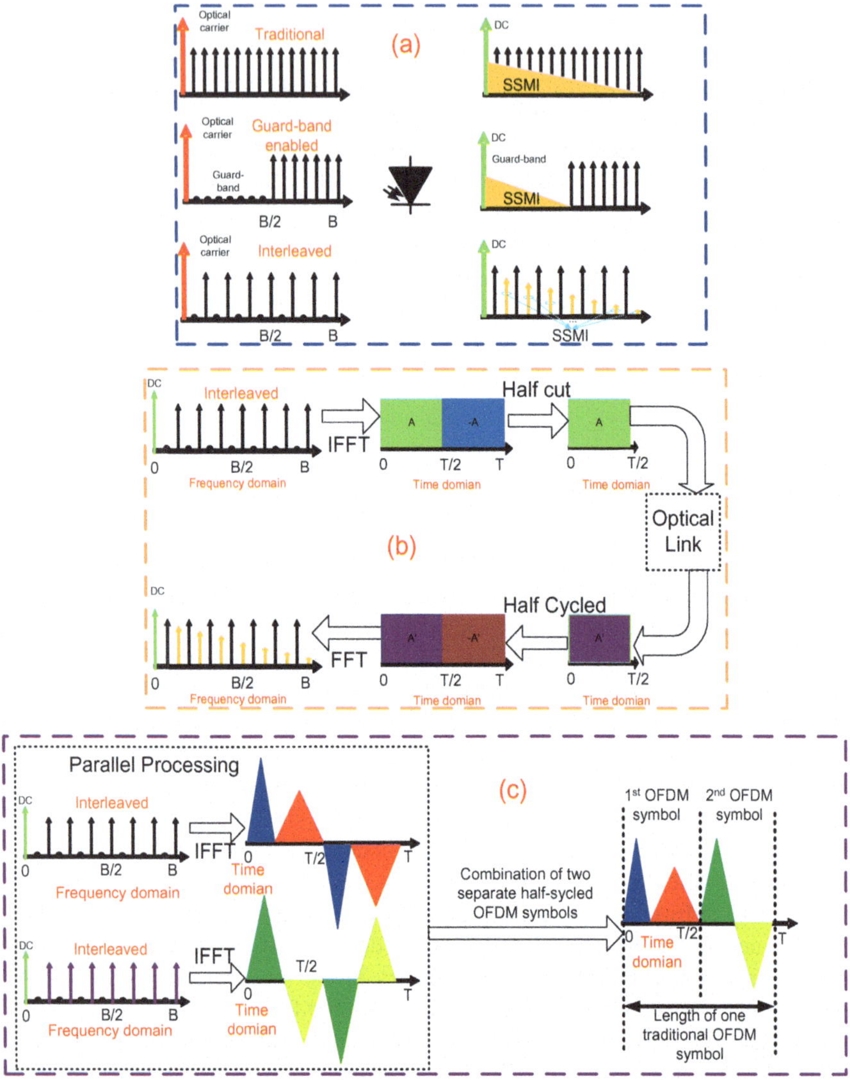

Fig. 13.1 **a** Three different types of direct detection of OFDM signals, **b** the proposed half-cycled DDO-OFDM structure, and **c** the symbol combination of two half-cycled OFDM symbols

In the time domain, the first half and the second half of a single OFDM symbol can be expressed as

$$s(t_1) = \sum_{k=0}^{N-1} c_k \exp(j2\pi f_k t_1) \quad (1 \le t_1 \le {}^{T}/_{2}) \tag{13.3}$$

$$s(t_2) = \sum_{k=0}^{N-1} c_k \exp(j2\pi f_k t_2) \quad \left({}^{T}\!/_{2} + 1 \le t_2 \le T\right) \tag{13.4}$$

By substituting $t_2 = t_1 + {}^{T}\!/_{2}$ into Eq. (13.4), the latter part of OFDM symbol evolves into

$$\begin{aligned} s(t_1 + {}^{T}\!/_{2}) &= \sum_{k=0}^{N-1} c_k \exp(j2\pi f_k (t_1 + {}^{T}\!/_{2})) \\ &= \sum_{k=0}^{N-1} c_k \exp(j2\pi f_k t_1 + jk\pi) \\ &= \sum_{k=0}^{N-1} c_k (\cos k\pi + j \sin k\pi) \exp(j2\pi f_k t_1) \\ &= \sum_{k=0}^{N-1} c_k \cos k\pi \exp(j2\pi f_k t_1) \qquad (1 \le t_1 \le {}^{T}\!/_{2}) \end{aligned} \tag{13.5}$$

According to the index of subcarriers, all subcarriers are divided into even subcarriers and odd subcarriers, in which the index of odd subcarriers and even subcarriers is expressed by m and n, respectively. The first half and the second half of OFDM symbols represented by Eqs. (13.3) and (13.5) can be expanded as follows:

$$s(t_1) = \sum_{n=1}^{N-1} c_n \exp(j2\pi f_n t_1) + \sum_{m=0}^{N-2} c_m \exp(j2\pi f_m t_1) \quad \left(1 \le t_1 \le {}^{T}\!/_{2}\right) \tag{13.6}$$

$$\begin{aligned} s(t_2) =& s\left(t_1 + {}^{T}\!/_{2}\right) \\ =& \sum_{k=0}^{N-1} c_k \cos k\pi \exp(j2\pi f_k t_1) \\ =& \sum_{n=1}^{N-1} c_n \cos n\pi \exp(j2\pi f_n t_1) + \sum_{m=0}^{N-2} c_m \cos m\pi \exp(j2\pi f_m t_1) \\ =& \sum_{n=1}^{N-1} -c_n \exp(j2\pi f_n t_1) + \sum_{m=0}^{N} c_m \exp(j2\pi f_m t_1) \quad \left(1 \le t_1 \le {}^{T}\!/_{2}\right) \end{aligned} \tag{13.7}$$

In the OFDM scheme with frequency domain crossover, the subcarrier with an odd exponent is set to zero to resist the SSMI. This means that in Eqs. (13.6) and (13.7) $c_m = 0$, the first half and the second half of the OFDM symbol can be simplified as follows:

$$s(t_1) = \sum_{n=1}^{N-1} c_n \exp(j2\pi f_n t_1) \quad \left(1 \le t_1 \le {}^{T}/_{2}\right) \tag{13.8}$$

$$s(t_2) = \sum_{n=1}^{N-1} -c_n \exp(j2\pi f_n t_1) = -s(t_1) \quad \left(t_2 = {}^{T}/_{2} + t_1\right) \tag{13.9}$$

By comparing the first half of the OFDM symbol in Eq. (13.8) with the second half of the OFDM symbol in Eq. (13.9), we find that the first half and the second half of the OFDM symbol in the time domain show odd symmetry about the middle point of the OFDM symbol. In order to improve the spectral efficiency, the latter part of the OFDM symbol is removed directly before transmission. After this operation, the time length of a single OFDM symbol is shortened from T to $T/2$, and the original time domain interval $[T/2\ T]$ can be placed with the first half of another OFDM symbol. Through this operation, the length of OFDM signals is doubled, so that the spectral efficiency of half-cycled OFDM signals is improved to be consistent with that of traditional OFDM signals.

Figure 13.1b gives the structure of the proposed half-cycled DDO-OFDM, which, like the OFDM with frequency domain crossing, only even number of subcarriers carry the signal. After the realization of OFDM modulation by IFFT, the symbol's period length is T. It has been discussed above that such OFDM signals show odd symmetry about the intermediate points of OFDM symbols. In the transmission process, the second half of the OFDM symbol is discarded directly, so that the symbol length of OFDM is reduced by half. After this treatment, the spectral efficiency was doubled, so that the spectral efficiency of half-cycled was the same as that of ordinary OFDM signals. Since the DDO-OFDM system can be approximately considered as a time-invariant communication system in a certain length of time, the latter half of the half-cycled OFDM signal can be copied through half-cycle replication after electro-optical conversion, optical fiber transmission, and photoelectric conversion. FFT at the receiving end is used to realize the demodulation of the OFDM signal. After this process, the SSMI falls on the odd subcarriers. The DDO-OFDM system with half-cycled odd number can eliminate the SSMI without reducing the spectral efficiency. Figure 13.1c shows the combination of two half-cycled OFDM symbols. After the two half-cycled OFDM symbols, respectively, discard the last half-cycle symbols in the time domain, the two remaining symbols with the period length of $T/2$ can be merged into a symbol with the period of T. After this processing, two half-cycled OFDM symbols can be transmitted in time length T. This will lead to the reduction of ISI capability caused by dispersion resistance of half-cycled OFDM signal. However, as the transmission distance of optical fiber in the access network based on DDO-OFDM is very short, this shortcoming is not something that must be overcome at present.

13.2 Study on Elimination of SSMI in DDO-OFDM Based on Half-Cycled Technique

13.2.1 Experimental Setup and Results

Figure 13.2 shows the experimental setup of the half-cycled DDO-OFDM system. In this experiment, a dual-band modulation DDO-OFDM system was used to verify the effectiveness of the half-cycled scheme. At the transmitter end, an external cavity laser with a wavelength of 1557.04 nm and a linewidth of less than 100-kHz is used as the laser source. The power of the optical signal emitted by the laser source is 14.5 dBm, which is input into an external MZM intensity modulator to realize electro-optical modulation. The modulation signal is an OFDM signal generated by an arbitrary waveform generator. When an OFDM signal is produced, the sampling rate of the arbitrary waveform generator is set as 12 GSa/s. In OFDM signal modulation, the size of FFT is 256, in which 200 subcarriers carry data but only 100 of them carry valid data and the other 100 are set to conjugate symmetry to produce real OFDM signals. The 55 subcarriers at high frequencies are set to zero to achieve oversampling, and the subcarriers at zero frequencies are also set to zero to facilitate subsequent DC offset adjustment. In this experiment, QPSK/16QAM signals are modulated to 100 effective subcarriers, respectively, for testing. After OFDM modulation is completed, a cyclic prefix with a length of 8 samples is inserted before each OFDM symbol, and the length of the OFDM symbol becomes 264 after the insertion of the cyclic prefix. In traditional OFDM, a single OFDM symbol transmits 264 samples, while in half-cycled, the number of alternative points of a single OFDM symbol

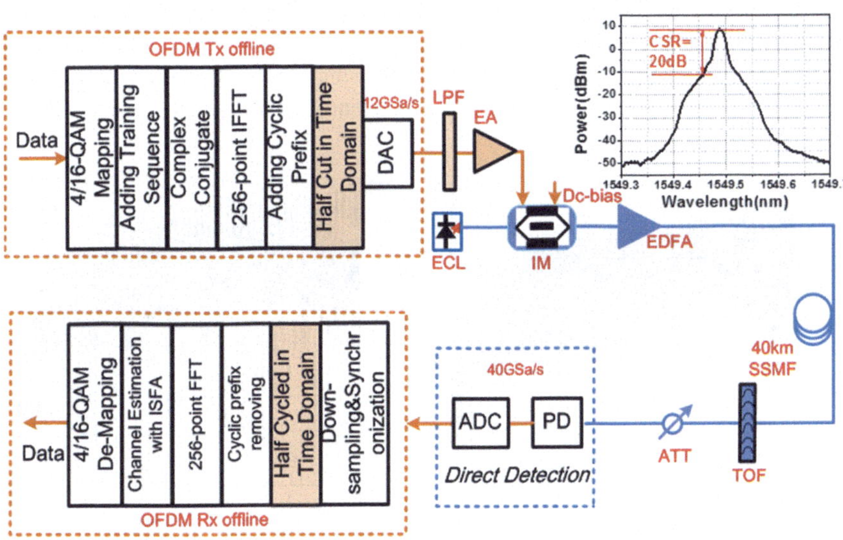

Fig. 13.2 Half-cycled DDO-OFDM system

becomes 132, including 128 data samples and 4 cyclic prefixes of samples. A training sequence is inserted into every 160 OFDM symbols for symbol synchronization and channel estimation. When QPSK and 16QAM modulation formats are adopted in OFDM modulation, the total transmission rate of the original signal is 9.1 Gb/s and 18.2 Gb/s, respectively. The OFDM signal generated by DAC in the arbitrary waveform generator is firstly filtered by a low-pass filter to eliminate the noise outside the sideband of the OFDM signal, and then the signal is injected into the MZM modulator to realize electro-optical modulation. The half-wave voltage of the MZM modulator was changed to 3.4 V. In the experiment, the DC bias was set to be near the normal intersection point of its power transmission curve (1.9 V). The resulting signal is fed into the EDFA, amplified to 8 dBm, and injected into the optical fiber. The optical signal modulated by OFDM is given in the illustration of Fig. 13.2, and the power of carrier and sideband signals is 20 dB at a resolution of 0.02 nm. After 40-km single-mode optical fiber transmission, the OFDM signal first goes through a 0.33 nm tunable optical filter to filter out the amplified spontaneous emission (ASE) noise. Before entering the receiver to realize photoelectric conversion, an optical attenuator is used to adjust the received power of the receiver, the 3-dB bandwidth of the receiver is 10 GHz. The converted signal is sampled and stored by a real-time oscilloscope with a sampling rate of 40 GSa/s. The collected signal is transmitted to the computer for off-line processing. Off-line DSP processing mainly includes symbol synchronization, half-cycled operation in time domain, removal of cyclic prefix, FFT, channel estimation using ISFA, channel equalization, QPSK/16QAM reflection, and bit error rate calculation. The bit error rate is calculated by counting the total error number of 320,000. In Fig. 13.3a, b, spectra of ordinary OFDM signals and half-cycled OFDM signals are given respectively.

In the experiment, we have tested the bit error rate and received power curves of three OFDM signals, which include traditional OFDM signals, frequency-domain cross OFDM signals, and half-cycled OFDM signals. The single symbol length of the first two OFDM signals is the same, while the single symbol length of half-cycled OFDM is reduced to half. To verify the effectiveness of the scheme, we tested the

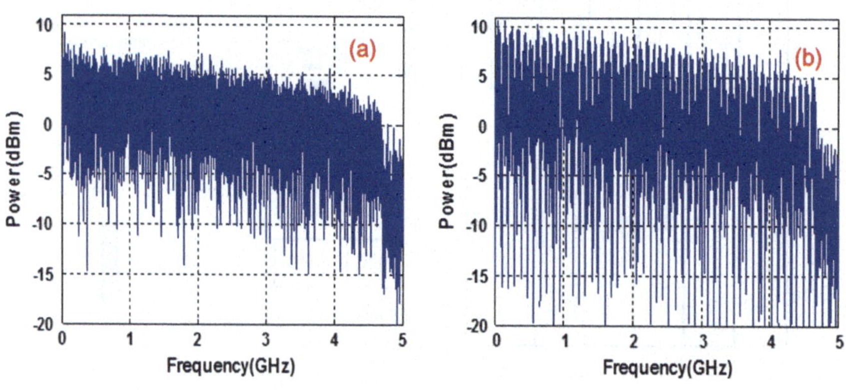

Fig. 13.3 **a** Spectrum of common OFDM signal and **b** half-cycled OFDM signal

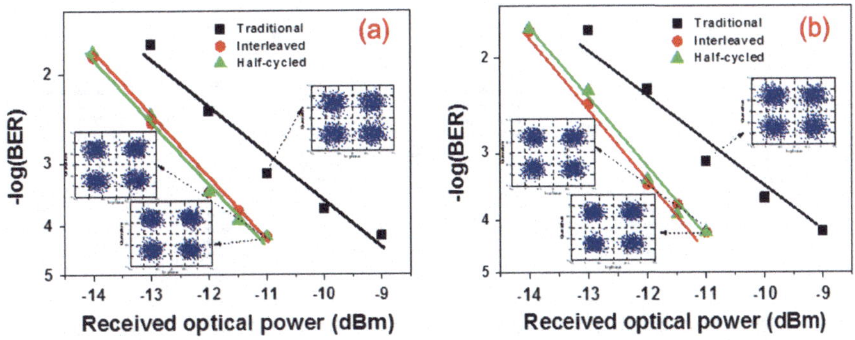

Fig. 13.4 Curve of bit error rate and received power of QPSK-OFDM signal: **a** Back-to-back, **b** transmission through 40 km of single-mode optical fiber

performance of the system under different modulation formats (QPSK and 16QAM). In the frequency-domain cross OFDM signal scheme, in order to compare with the half-cycled scheme, the second half of a single OFDM symbol in the time domain is also copied from the first part.

We have tested the bit error rate and received the power curve of QPSK-OFDM, in which the curve of light transmitted back-to-back and through 40 km single-mode fiber is given in Fig. 13.3a, b, respectively. The experimental results show that the sensitivity of the half-cycled OFDM signal is improved by 2 dB compared with the traditional OFDM signal after the transmission of the light back-to-back and the transmission of 40 km single-mode fiber, which is mainly because the half-cycled OFDM can effectively eliminate the SSMI. The sensitivity of half-cycled QPSK-OFDM signal and the frequency domain cross QPSK-OFDM signal is almost the same as that of the receiver, which also proves that it is feasible to use half-cycled OFDM to eliminate the SSMI. The constellation diagram with the received power of -11 dBm after the transmission of the three signals back-to-back and through 40 km single-mode optical fiber is given in the insets of Fig. 13.4a, b, respectively.

In 16QAM-OFDM transmission, we also measured the curve of bit error rate and received power. The curves of the back-to-back transmission and through 40 km single-mode fiber are given in Fig. 13.5a, b, respectively. Similarly, the sensitivity of the receiver is improved because half-cycled can overcome the SSMI. The experimental results show that the sensitivity of half-cycled 16QAM-OFDM signal is improved by 1.5 dB compared with the traditional 16QAM-OFDM signal after the transmission of back-to-back and 40 km single-mode fiber. Three types of 16QAM-OFDM signals receiving the power of -6 dBm after back-to-back and 40 km single-mode optical fiber transmission are given in the illustrations of Fig. 13.5a, b, respectively. It is found that the improvement of receiver sensitivity in QPSK-OFDM transmission using half-cycled technology is higher than that in 16QAM-OFDM signal transmission. Because the OFDM signal is sensitive to the error of symbol synchronization, the synchronization error will be introduced into ISI and ICI. The accuracy of traditional OFDM signals is higher than that of half-cycled OFDM signals, which is

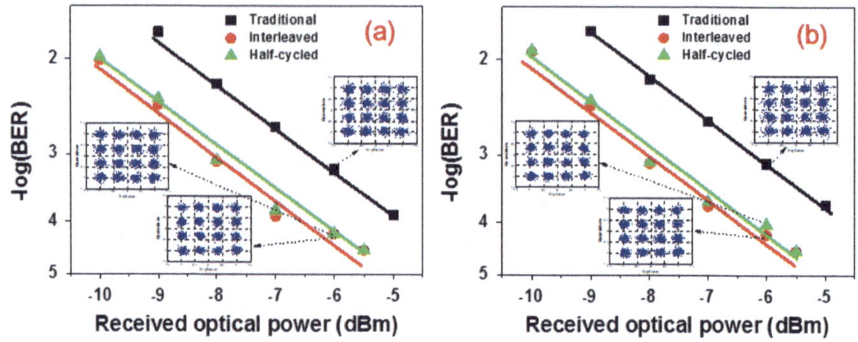

Fig. 13.5 Bit error rate and received power curve of 16QAM-OFDM signal. **a** Back-to-back, **b** transmission through 40 km of single-mode optical fiber

mainly because the number of training sequences used to calculate synchronization points in traditional OFDM signals is larger than that in half-cycled signals. However, the robustness of QPSK-OFDM for ISI and ICI caused by such synchronization error is higher than that of 16QAM-OFDM signal with the same bandwidth. Therefore, the adoption of half-cycled technology in QPSK-OFDM signal transmission will improve receiver sensitivity.

After 40 km optical fiber transmission, the distribution of bit error rate on the subcarriers of traditional 16QAM-OFDM signal and half-cycled 16QAM-OFDM signal when the received power is −7 dBm is shown in Fig. 13.6a, b, respectively. In the traditional OFDM symbol, the number of subcarriers carrying signal is 100, while in the half-cycled scheme, the number of subcarriers carrying signal is reduced to 50. By comparing Fig. 13.6a, b, it is found that after eliminating the SSMI through half-cycled technology, the BER on the subcarrier is significantly reduced, especially the BER of the low frequency subcarrier with high SSMI power is significantly improved.

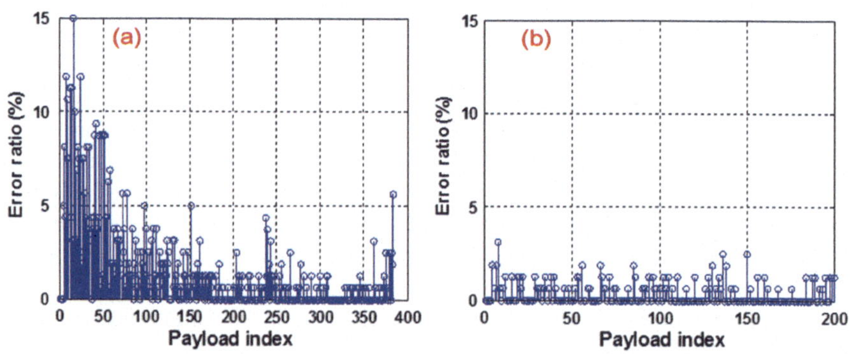

Fig. 13.6 Bit error rate distribution on subcarriers when the received power of 16QAM-OFDM signal is −7 dbm. **a** traditional 16QAM-OFDM, **b** half-cycled 16QAM-OFDM

13.2.2 *Summary*

In this chapter, experimental results show that half-cycled technology is effective to overcome the SSMI and this technology will not lead to the reduction of spectral efficiency. In this scheme, the latter half-cycle of OFDM symbol with frequency domain cross is discarded and a half-cycled OFDM symbol is formed. In this way, two half-cycled OFDM symbols can be transmitted in the traditional OFDM symbol cycle. In this way, the transmission rate and spectral efficiency of half-cycled are consistent with the traditional OFDM symbol, but the SSMI can be eliminated. The experimental results show that the receiver sensitivity of 9.1 Gb/s QPSK-OFDM signal and 18.2 Gb/s 16QAM-OFDM signal can be improved by using half-cycled technology after transmission of 40 km single-mode fiber.

13.3 Research on High Order QAM-OFDM Signal Modulation Transmission by Direct-Detect

13.3.1 *Experimental Setup*

A diagram of a high order QAM-OFDM signal experimental setup for direct detection is shown in Fig. 13.7. In OFDM systems with intensity modulation and direct detection, electro-optic modulation of OFDM signals is achieved by low-cost commercial direct modulation lasers. The center wavelength of the optical carrier generated by the direct modulation laser is 1537.92 nm. The modulation formats in the high order QAM-OFDM system implemented in this section include 256QAM, 521QAM, 1024QAM, and 2048QAM. The OFDM signal is first generated off-line by Matlab and then loaded into an arbitrary waveform generator (Tektronix 7122B) with a sampling rate set to 12 GSa/s. In the OFDM generation process, the size of the FFT is 8192. Among the 8192 subcarriers of OFDM, 2048 low frequency subcarriers at positive frequency are used to transmit signals, and 2048 subcarriers corresponding to negative frequencies are used to transmit signals with positive frequency Hermitian conjugate symmetry. In order to generate a real OFDM signal. The zero subcarrier of OFDM is reserved for DC bias and the other subcarriers are set to zero for oversampling. The cyclic prefix length of the OFDM signal is 14, and the complete OFDM symbol length is 8206 after adding 8192 samples of the OFDM symbol. In order to implement symbol synchronization and channel estimation, one training sequence symbol is inserted in front of every 25 OFDM symbols, and 25 OFDM symbols are included in the single frame OFDM signal. The bandwidth of the electrical OFDM signal generated by the arbitrary waveform generator in this experiment was 3 GHz ($2048/8192 \times 12$ GHz $= 3$ GHz). When the modulation format of the signal is 2048QAM, the original total information rate of the OFDM signal is

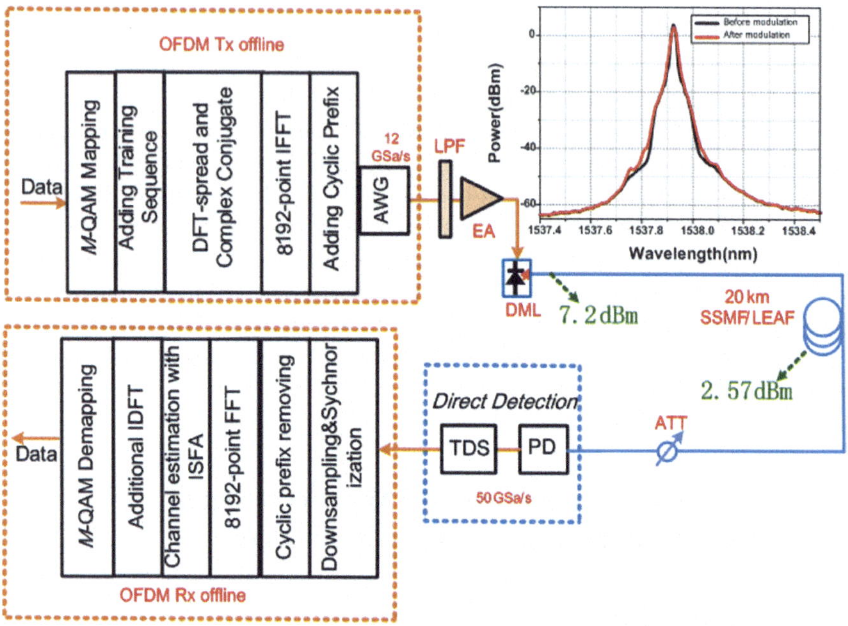

Fig. 13.7 Experimental setup of direct detection of high order QAM-OFDM signal

$$
\begin{aligned}
\text{Raw data rate} &= \frac{N_{\text{OFDM symbol}}}{N_{\text{Training}} + N_{\text{OFDM symbol}}} \times \frac{N_{\text{data\_ subcarrier}}}{N_{\text{total\_ subcarrier}} + N_{CP}} \\
&\quad \times \text{Sampling rate} \times {}^{\text{bits}}/_{\text{sample}} \\
&= 25/(25+1) \times 2048/(8192+14) \times 12 \times 11\,\text{Gbit/s} \\
&\approx 31.7\,\text{Gbit/s}
\end{aligned}
\tag{13.10}
$$

where N(.) represents the number of subcarrier types, symbols, and samples specified in parentheses. Since the modulation format of the signal is 2048QAM, the bit/symbol should be 11. A low-pass filter with a bandwidth of 3-GHz is connected in series behind the arbitrary waveform generator to filter out residual sideband signals. The OFDM signal filtering the vestigial sideband is amplified by an electric linear amplifier to 2.4 V, and then the amplified OFDM signal is injected into a distributed feedback direct modulation laser to achieve electro-optic modulation. In optical OFDM signal generation, a direct modulation laser with a 3-dB modulation bandwidth of 10 GHz is biased to 88 mA, and the power of the generated optical OFDM signal is 7.2 dBm. The spectral map of the optical carrier and the modulated optical carrier is indicated by the black and red lines in the insets in Figs. 13.7, respectively, with a resolution of 0.01 nm. The resulting optical OFDM signal is injected into a 20 km large effective area fiber/single-mode fiber. Because the distance traveled by the fiber is very short, there is no need to add an EDFA to the link. The main parameters of the two fibers at 1550 wavelength are given in Table 13.1. After transmission through the

Table 13.1 Main parameters of two types of fiber

Type of fiber	D (ps/km/nm)	S ($ps/nm^2/km$)	α (dB/km)
SSMF	17	0.06	0.2
LEAF	4	0.106	0.21

fiber, the power of the optical OFDM signal is attenuated to 2.57 dBm. An adjustable optical attenuator is used to adjust the power of the optical OFDM signal entering the receiver before the optical OFDM signal is input to the receiver for receiver sensitivity testing. The 3-dB bandwidth of the receiver used to achieve photoelectric conversion in the experiment was 10 GHz. The photoelectrically converted signal was acquired by a real-time oscilloscope (Tektronix DPO72004B) with a sampling rate of 50 GSa/s. The collected signal is sent to a computer to demodulate the OFDM signal using an off-line DSP. Off-line DSP mainly includes OFDM symbol synchronization, cyclic prefix removal, FFT, channel estimation, and channel equalization using ISFA, inverse mapping of high order QAM, and final bit error rate calculation. In this experimental test, a single 256QAM/512QAM/1024QAM/2048QAM-OFDM symbol contains 16384/18432/20480/22528 bits (2048 × 8/9/10/11 bits). When calculating the bit error rate, the sample of the signal is selected as a frame of the OFDM signal. The number of OFDM symbols in a frame OFDM signal is 25, so the number of bits used to calculate the bit error rate in 256QAM/512QAM/1024QAM/2048QAM-OFDM is 409600/460800/512000/563200 (25 × 6384/18432/20480/22528 bits = 409600/460800/512000/563200 bits).

13.3.2 Experimental Results and Analysis

Channel estimation is a very important procedure in OFDM communication. In DDO-OFDM systems, physical impairments such as dispersion and polarization mode dispersion in the fiber link can be eliminated by accurate channel estimation and equalization. In short-range DDO-OFDM systems, noise from the PD will be the dominant source of noise since EDFAs can be avoided. The ISFA technique is used in channel estimation of OFDM to improve the accuracy of channel estimation by eliminating received PD noise. The power of the noise decreases as the length of the ISFA window increases because the averaging method eliminates noise and as the length of the ISFA window increases, a sufficient number of samples can always be found to completely eliminate noise. Therefore, in order to minimize the power of the noise in the channel estimation, the window length of the ISFA needs to be selected as large as possible. However, there is a hypothesis in ISFA that the correlation of samples for the average frequency domain is very high, so the window length of this ISFA cannot be chosen too long because the sample in the average frequency domain is used as the window length increases. The relevance will be lower and lower. Therefore, when selecting the length of the ISFA window, it is necessary to consider these two factors comprehensively, so that an optimal value will appear

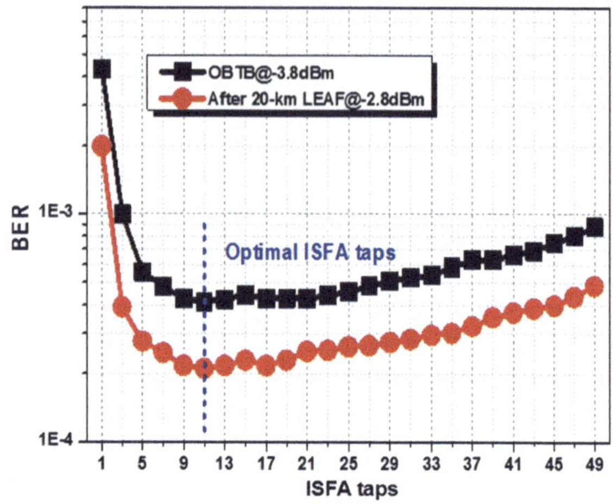

Fig. 13.8 256QAM-OFDM signal error rate curve with ISFA window length

when the length of the ISFA window is selected. In order to find this most significant ISFA window length value, the BER of the 256QAM-OFDM signal after optical back-to-back and 20 km of large effective area fiber transmission is tested with the ISFA window length curve and is shown in Fig. 13.8. It can be seen from Fig. 13.8 that the most ISFA window length is 11, and this optimal ISFA window length will be used in the channel estimation for the remainder of this section.

Another very important variable for implementing high order QAM-OFDM in a DDO-OFDM system is the number of subcarriers, i.e., the size of the FFT. In the modulation of OFDM, when the number of subcarriers is larger, the roll-off coefficient of the edge of the signal is smaller. ISI is caused when a signal is transmitted over a bandwidth-limited link. In the DDO-OFDM system, the electro-optical-electrical link can be viewed as a bandwidth-limited overall link because the bandwidth of the low-pass filter and DAC in the link is not sufficient for the OFDM signal. Insufficient bandwidth will result in ISI appearing in the time domain. When the system link is established, the overall system frequency response cannot be changed. At this time, the spectral shape of the signal will be a key factor to reduce ISI. In addition, the DDO-OFDM system can be considered as a time-invariant system for a certain length of time and the ICI due to dispersion can be ignored in ultra-short-range systems. In a DDO-OFDM system with high spectral efficiency and ultra-short transmission distance using a high order QAM modulation format, a large-sized FFT must be selected in order to achieve high-performance transmission. There are two main reasons for this: the OFDM signal generated by the large-scale FFT can better resist the ISI introduced due to insufficient link bandwidth; the large-scale FFT can improve the resolution of the frequency domain in the channel estimation based on the training sequence. Brings an increase in estimated accuracy. To verify this theoretical analysis, the bit error rate versus FFT size for the 512QAM-OFDM signal in back-to-back condition is given in Fig. 13.9. As can be seen from Fig. 13.9, the bit error rate performance of the OFDM signal is improved as the FFT size is increased. The

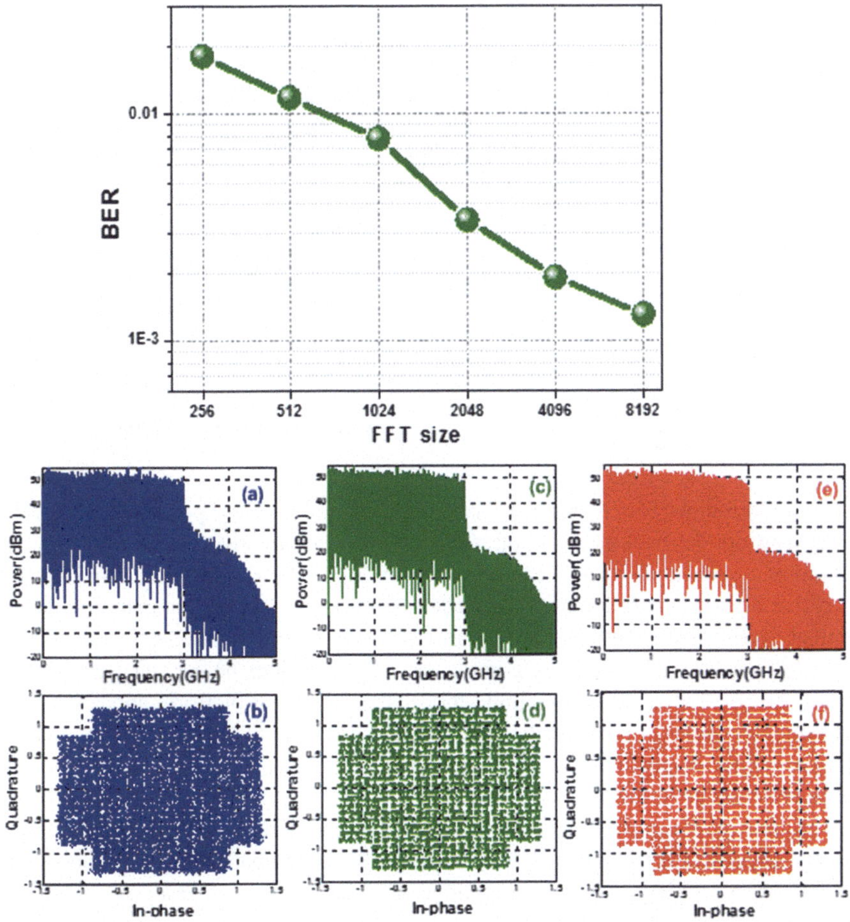

Fig. 13.9 Relationship between BER and FFT size of B2B 512QAM-OFDM **a** Spectrogram of an OFDM signal with an FFT size of 256, and **b** an OFDM signal constellation with an FFT size of 256 **c** Spectrogram of OFDM signal with FFT size 2048, **d** OFDM signal constellation with FFT size 2048 **e** OFDM signal spectrogram with FFT size 8192, **f** OFDM signal constellation with FFT size 8192

spectrograms of the received OFDM signals when the FFTs are 256, 2048, and 8192, respectively, are given in Fig. 13.9a, c, e, respectively. The corresponding constellation diagrams are given in the illustrations (b), (d), and (f) of Fig. 13.9. The OFDM signal with the FFT size of 8192 given in Fig. 13.9e is the fastest to roll-off at the edge of the spectrum compared to other signals, and its corresponding constellation in Fig. 13.9f is also performed. Out of the best performance.

The bit error rate and received power curves of the 256QAM-OFDM signal are shown in Fig. 13.10. In a DDO-OFDM system that uses a direct modulation laser

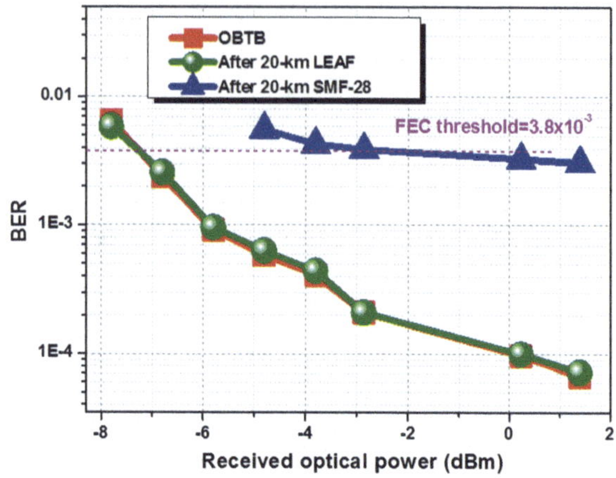

Fig. 13.10 Relationship of 256QAM-OFDM signal error rate and received power

for modulation, we tested the relationship of BER and received power of 256QAM-OFDM in optical back-to-back, after 20 km of large effective area fiber and after 20 km of single-mode fiber. It can be seen from Fig. 13.10 that after 20 km of large effective fiber transmission, there is basically no power penalty compared with the optical back-to-back, and the bit error rate is 3.8×10^{-3} compared with the optical back-to-back after 20 km of single-mode fiber. The 3 o'clock power penalty is higher than 4 dB. And after 20 km of single-mode fiber, the error platform appeared. The introduction of power cost and the emergence of error platforms are all due to severe ICI caused by dispersion. As shown in Table 13.1, the dispersion coefficient of a large effective area fiber is much smaller than that of a common single-mode fiber. Therefore, when transmitting the same 20 km length, the ICI caused by dispersion in a large effective fiber is much smaller. In a single-mode fiber, this is why the signal transmission of a large effective area fiber of 20 km does not find a power penalty. When the received power is 0.2 dBm, the constellation of the 256QAM-OFDM signal transmitted through the 20 km large effective area fiber and the 20 km single-mode fiber is, respectively, shown in Fig. 13.11a, b. Given. The apparent distortion of the constellation caused by ICI caused by dispersion can be seen from Fig. 13.11b.

The bit error rate and received power curves and constellation diagrams for the 512QAM-OFDM signal are shown in Fig. 13.12. DFT-spread can be used to reduce the peak-to-average power ratio of OFDM signals and is used in this section to reduce the peak-to-average power ratio. In order to verify the reduction of the peak-to-average power ratio of the OFDM signal, the BER performance of the OFDM signal can be improved. In the optical back-to-back transmission, the DFT-spread technique and the OFDM signal without the DFT-spread technique are used and the two OFDM signals are tested separately. Bit error rate curve. We also tested the BER curve of a 512QAM-OFDM signal using DFT-spread technology after 20 km of effective fiber transmission and is also shown in Fig. 13.12. The peak-to-average power ratio of the OFDM signal is evaluated by a Complementary Cumulative Distribution Function (CCDF), which represents a probability distribution in which the

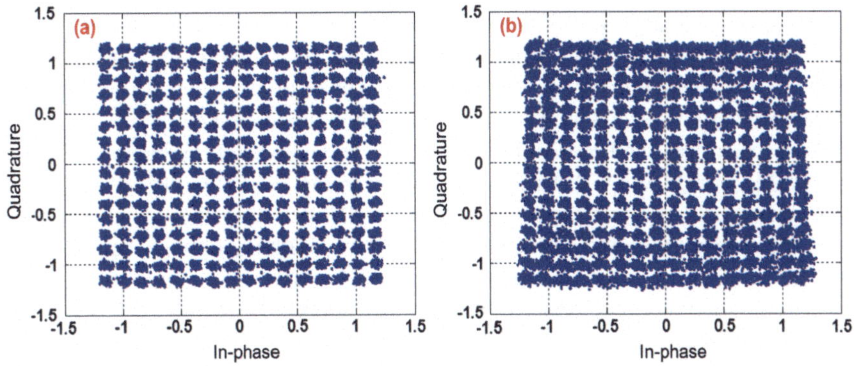

Fig. 13.11 256QAM-OFDM signal constellation. **a** After 20 km of effective area fiber transmission. **b** After 20 km of single-mode fiber transmission

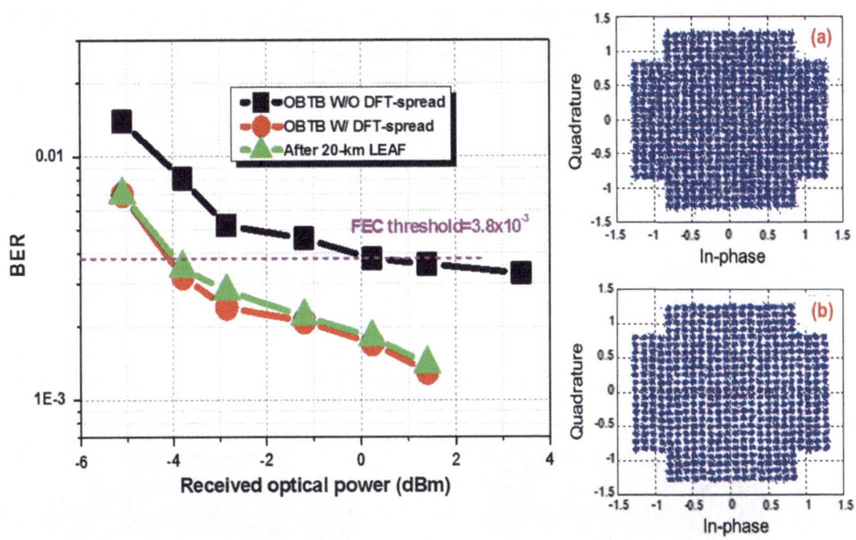

Fig. 13.12 Bit error rate and received power curve and constellation diagram of 512QAM-OFDM signal. **a** Traditional 512QAM-OFDM signal constellation diagram. **b** DFT-spread 512QAM-OFDM signal constellation diagram

peak-to-average power ratio of the OFDM signal exceeds a certain threshold. It is found by calculation that the peak-to-average power ratios of the DFT-spread OFDM signal and the conventional OFDM signal are 11.45 dB and 14.16 dB, respectively, when the distribution probability is 1×10^{-3}. The test results show that when the bit error rate is the error correction coding threshold (3.8×10^{-3}), the receiving sensitivity of the OFDM signal with DFT-spread reduced peak-to-average power ratio is improved by 4 dB, and is transmitted over 20 km. The power of the effective area does not introduce any power penalty. Figure 13.12a, b shows the constellation diagrams of the traditional 512QAM-OFDM signal and the DFT-spread 512QAM-OFDM signal at the receiving end at the same received power, respectively. In contrast, when the DFT-spread technique is used to eliminate the peak-to-average power ratio of the OFDM signal, the constellation of the signal is more concentrated.

The bit error rate and received power curves and constellation diagrams for the 1024QAM/2048QAM-OFDM signal are shown in Fig. 13.13. In the test results of these two signals, the transmission of the large effective area fiber for 20 km does not introduce an additional power penalty compared with the optical back-to-back. In order to achieve a bit error rate of less than 2.4×10^{-2}, the minimum received power required for 1024QAM-OFDM and 2048QAM-OFDM is −5.5 dBm and 0.23 dBm, respectively. After the 2048QAM-OFDM signal removes the overhead

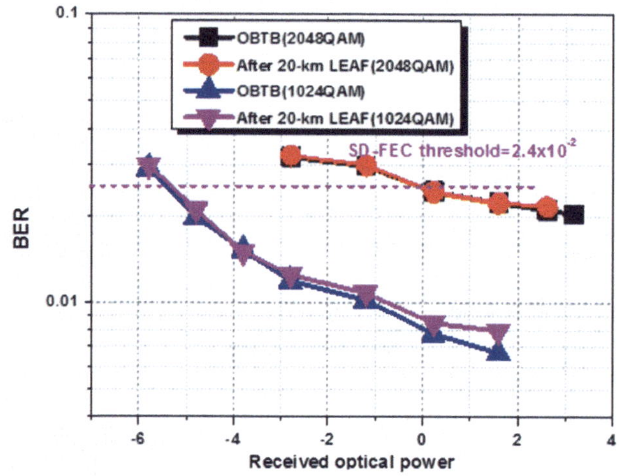

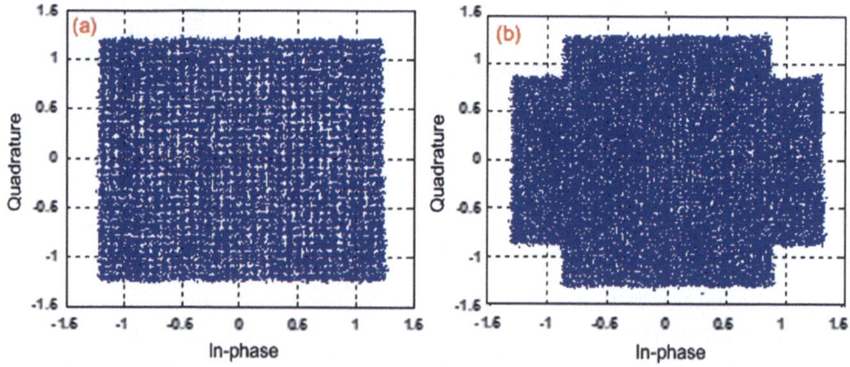

Fig. 13.13 Bit error rate and received power curve and constellation diagram of 1024QAM/2048QAM-OFDM signal. **a** Constellation diagram of 1024QAM-OFDM signal after receiving power of 1.6 dBm after 20 km large effective fiber transmission. **b** Constellation diagram of 2048QAM-OFDM signal after receiving power of 2.6 dBm after 20 km large effective fiber transmission

of 20% forward error correction coding, the actual transmitted 2048QAM-OFDM information rate is 26.4 Gbit/s (actual information rate = original information rate × 1/(1 + 0.2) ≈ 26.4 Gbit/s). The constellation diagram of the 1024QAM-OFDM signal with a received power of 1.6 dBm after 20 km of effective area fiber transmission is given in Fig. 13.13a, and the 2048QAM-OFDM signal is transmitted after 20 km of effective area fiber transmission. The constellation diagram at a received power of 2.6 dBm is given in Fig. 13.13b.

13.3.3 Summary

In this section, we implement the highest order QAM-OFDM signal transmission in the DDO-OFDM system. In this intensity modulated direct detection system, electro-optical modulation is achieved by direct modulation of the laser. The error of the 31.7 Gb/s 2048QAM-OFDM is still less than 20% of the error correction performance threshold of the forward error correction coding after 20 km of large effective area fiber transmission. The key factors for improving the performance of high order QAM-OFDM signal transmission in DDO-OFDM systems in this experiment are: using ISFA to improve channel estimation accuracy, increasing the size of FFT to improve the ability of the signal to resist ISI, to improve the frequency domain resolution in channel estimation, and to improve the channel estimation accuracy, and reducing the peak-to-average power ratio of the OFDM signal by using the DFT-spread technique.

13.4 Short Distance Transmission of Large Capacity DDO-OFDM Based on DFT-Spread

13.4.1 Optimization of Training Sequences in Large Capacity DDO-OFDM Systems Based on DFT-Spread

The importance of channel estimation in frequency domain equalization based OFDM has been discussed in the previous section of this chapter. The DFT-spread technique can be used to reduce the PAPR of OFDM signals in a large capacity DDO-OFDM system. The schematic of the DDO-OFDM system of the conventional OFDM signal and the DFT-spread OFDM signal are, respectively, shown in Fig. 13.14a, b. In the OFDM signal modulation, the size of the FFT is N, and the L subcarriers at the positive frequency are used to carry the 32QAM signal, and the corresponding L subcarriers at the negative frequency are used to transmit the conjugate symmetric 32QAM signal corresponding to the positive frequency, so that the Hermitian symmetry is satisfied to generate a real OFDM signal. The remaining subcarriers in the frequency domain are zeroed to reserve DC offset and achieve

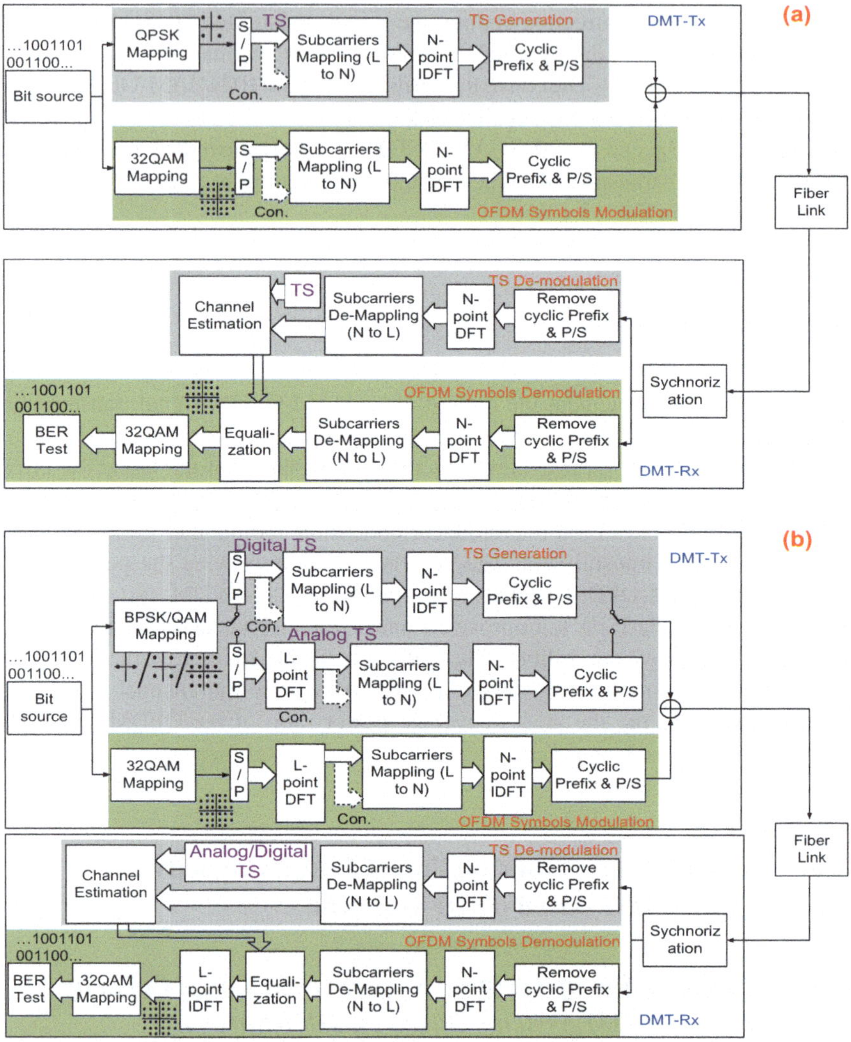

Fig. 13.14 Schematic of the DDO-OFDM system. **a** Traditional OFDM signal and **b** DFT-spread OFDM signal

oversampling. Compared with the traditional OFDM signal, the transmitting and receiving ends of the DFT-spread OFDM signal have an extra L-point DFT/IDFT.

In the frequency domain equalization scheme based on the training sequence (TS), it is theoretically required that the signal and the TS of DSP are preferably identical, so that the TS can accurately estimate the channel response of the DDO-OFDM system. At the transmitting and receiving ends of the conventional OFDM signal, the DSP and OFDM symbols of the TS are completely identical and the modulation format is selected as QPSK in the TS. In the transmission scheme of the DFT-spread OFDM signal, since a set of additional L-point DFT/IDFT is added at transmitting and receiving ends of the OFDM symbol, the DSP for the TS needs to be discussed. The main concern is two aspects: whether an additional set of DFT/IDFT needs to be added to the TS at the transmitting end and the receiving end; which modulation format of the TS is optimal for the system. The TS can be divided into analog TS and a digital TS according to the type of the incoming OFDM modulated signal. In the DFT-spread OFDM scheme, if a L-point DFT/IDFT is added during TS generation and reception, an analog TS is generated, and if not added, a digital TS is generated. In the conventional OFDM scheme, all TS are digital TS. If a BPSK/QPSK/16QAM signal is entered into the OFDM modulation during the generation of the TS, such a TS is referred to as a digital TS. Once an additional DFT process is added at the transmitting end, the TS will become an analog TS. Through the extra DFT at the transmitter, the digital signal will be converted into an analog signal and then added to the OFDM modulation to generate a TS. The advantage of this TS is that its PAPR is guaranteed to be as low as the subsequent DFT-spread OFDM symbol. However, since the TS usually has a very small proportion in one frame of OFDM signals, this processing does not make much sense. Next, we will analyze and experimentally compare the performance of two different types of TS in the DFT-spread OFDM scheme in this section. In the digital TS schemes, BPSK/QPSK/16QAM-TSs will be discussed. In analog TS schemes, we only choose the analog TS generated by extra L-point DFT with QPSK for discussion as the distribution characteristics of all kinds of analog TSs are similar. Apart from this, conventional OFDM with QPSK TS is also discussed for comparison.

Figure 13.15 gives the probability distribution of amplitude of OFDM signal: DFT-spread OFDM with BPSK/QPSK/16QAM/analog TS symbol and conventional

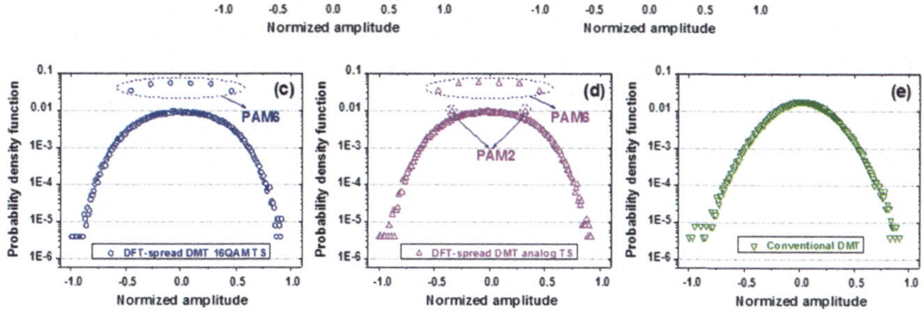

Fig. 13.15 Probability density function of **a** DFT-spread OFDM with BPSK TS, **b** DFT-spread OFDM with QPSK TS, **c** DFT-spread OFDM with 16QAM TS, **d** DFT-spread OFDM with analog TS, **e** Conventional OFDM

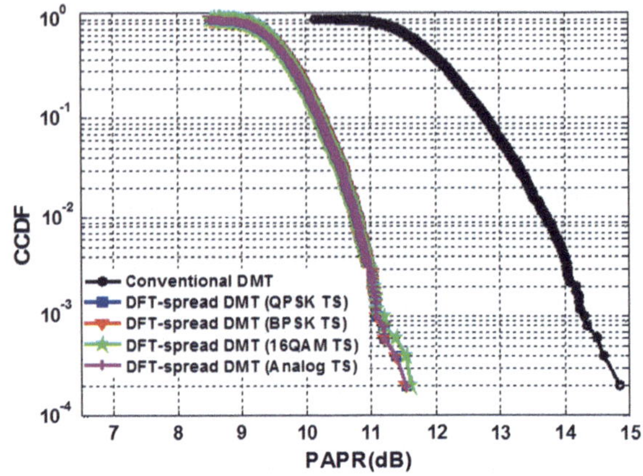

Fig. 13.16 CCDF curves of OFDM signals with different TS

OFDM. For the DFT-spread OFDM, 6 high discrete peaks appearing after that extra L-point DFT is applied for 32-QAM data. From this point of view, DFT-spread OFDM can be regarded as the combination of one high probability PAM-6 signal and one low probability analog signal, and it explains why the PAPR of DFT-spread OFDM is lower than conventional OFDM. The probability distributions of DFT-spread OFDM are quite similar except for slight differences in their TSs. In Fig. 13.15d, another two small discrete peaks like PAM-2 can be seen as extra L-point DFT and is applied for QPSK data in TS. The probability distribution of conventional OFDM in Fig. 13.15e is like Gaussian distribution. Cumulative distribution function (CCDF) denotes a probability distribution of the PAPR of the current OFDM symbol over a certain threshold. During the CCDF calculation, the N and L are 8192 and 2048, respectively. Figure 13.16 gives the calculated CCDF curves for conventional OFDM and DFT-spread OFDM with different types of TS. The PAPRs of DFT-spread OFDM with different types of TS are quite similar except for some slight differences in the TS and outperform that of conventional OFDM. This means the additional DFT/IDFT pair for TS symbol will not affect the PAPR of DFT-spread OFDM. A 3.4 dB PAPR improvement is attained at the probability of 2×10^{-4}.

The experimental setup for the DDO-OFDM signal transmission system with different TS is given in Fig. 13.17. The modulation of the DDO-OFDM signal is achieved by a DML. The OFDM signal is generated off-line in MATLAB and then uploaded into a Fujitsu DAC with 64-GSa/s sampling rate. The 3 dB bandwidth of the DAC is 12 GHz and the attenuation of the DAC at 16 GHz is 4.6 dB. Table 13.2 summarizes the parameters for the DFT-spread OFDM and conventional OFDM. The net bit rate is 79.86 Gbit/s. The OFDM signal from DAC is amplified to 2.4 V by an electrical amplifier (EA) with 20 dB gain. The optical carrier at 1295.43 nm from a commercial distributed feedback (DFB) DML is directly modulated by the OFDM signal. The DML with 10 GHz 3 dB bandwidth and 20 MHz linewidth is biased at 89 mA to produce 9.8 dBm average output power. The output signal is injected into 20 km SSMF. After fiber transmission, optical to electrical (O/E) conversion is implemented via an Agilent analog receiver. After a low-pass filter (LPF), the

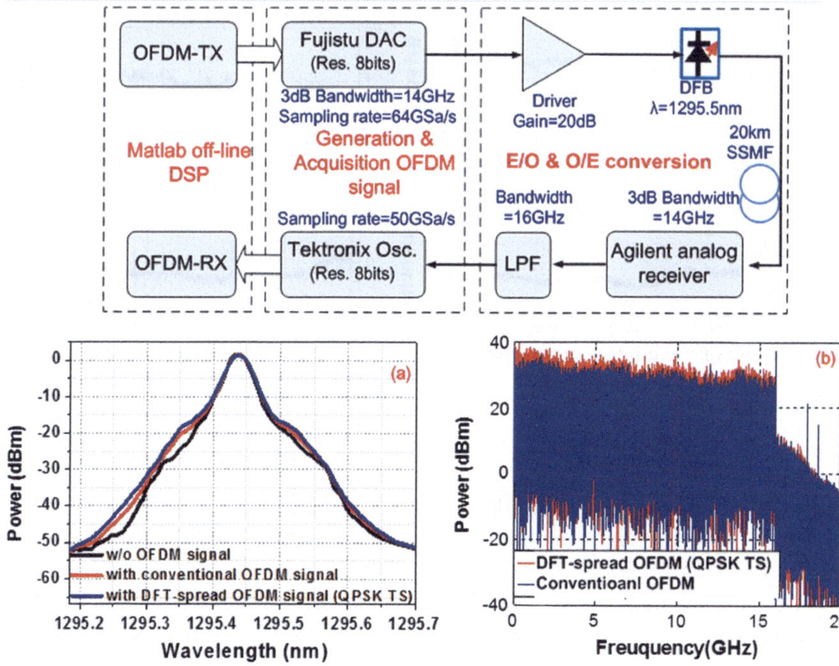

Fig. 13.17 Experimental setup. **a** Optical spectrum and **b** electrical spectrum

Table 13.2 Parameters for the DFT-spread OFDM and conventional OFDM modulation schemes

Parameter	Value
L-point IDFT/DFT	2048
N-point IDFT/DFT	8192
Symbol duration	128 ns
Total symbol duration	128.21875 ns
Electrical bandwidth (B) ($B = L \cdot Gs/N$)	16 GHz
Net bit rate (R) ($R = L \cdot SE/(N/Gs + t_{CP})$)	79.86 Gbit/s

signal is captured by a real-time Tektronix oscilloscope with 50 GSa/s sampling rate and subsequently processed by off-line DSP. The off-line DSP is listed in detail in Fig. 13.14. Figure 13.17a gives the optical spectra (0.02 nm resolution) of the optical carrier before and after modulation by conventional OFDM and DFT-spread OFDM signals. As the PAPR of DFT-spread OFDM signal is lower, the average electrical power of DFT-spread OFDM is larger than conventional OFDM. From Fig. 13.18a we can see that the optical signal-to-noise ratio (OSNR) of DFT-spread OFDM with QPSK TS is slightly larger than conventional OFDM. The corresponding electrical spectra of these two different OFDM signals with −2.73 dBm received optical power in optical back-to-back (OBTB) are shown in Fig. 13.4b, and the SNR of DFT-spread

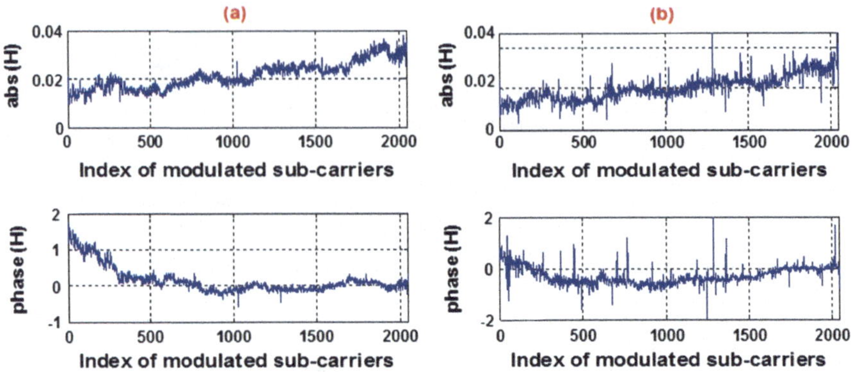

Fig. 13.18 Estimated channel response of DFT-spread OFDM with **a** digital QPSK TS and **b** analog TS

OFDM with digital QPSK TS is larger than that of conventional OFDM. In this section, BER was obtained by direct error counting with 95 OFDM symbols ($95 \times 2048 \times 5 = 972800$ bits).

The estimated channel response of DFT-spread OFDM signal with digital QPSK TS and analog TS is shown in Fig. 13.18a, b, respectively. Compared to the channel response obtained with digital QPSK TS, both the amplitude and phase of the channel response acquired with analog TS exhibit high frequency fluctuations as the analog TS is much more vulnerable than the digital TS to the noise in the optical fiber link. Thus, the performance of channel estimation with digital TS is much better than that with analog TS.

The error bits versus frequency for five types of OFDM signals in OBTB with −2.73 dBm received optical power are shown in Fig. 13.19. The corresponding constellations are given in Fig. 13.20. In the conventional OFDM shown in Fig. 13.19e there are some frequency bins with relatively high error ratio, which are caused by narrowband interference. And also the error ratio increases when the frequency increases, which is induced by power attenuation in high frequency bins. After DFT-spread is applied in the OFDM transceiver, the error bits become evenly distributed, as each QAM data symbol is carried onto all data subcarriers after an extra 2048-point DFT/IDFT. DFT-spread OFDM with BPSK/QPSK TS demonstrated the best performance among these DFT-spread OFDM signals, which means the digital TS outperforms analog TS and the best digital TSs are BPSK and QPSK modulated.

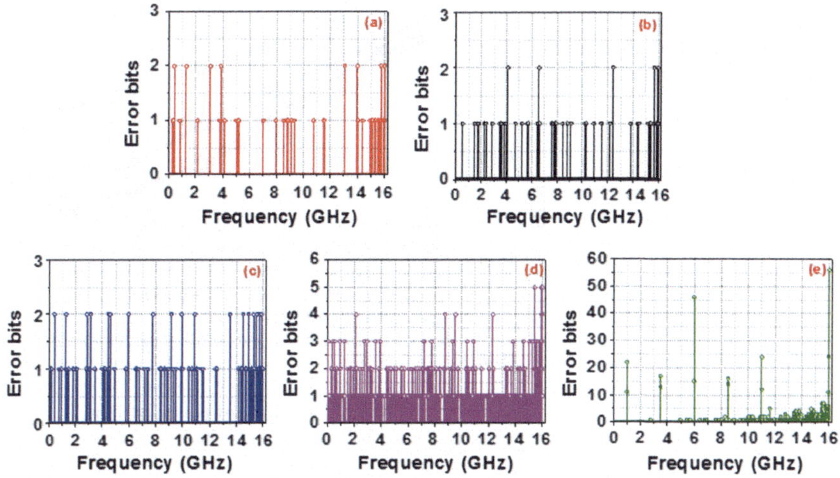

Fig. 13.19 Error distributions of **a** DFT-spread OFDM with BPSK TS, **b** DFT-spread OFDM with QPSK TS, **c** DFT-spread OFDM with 16QAM TS, **d** DFT-spread OFDM with analog TS, and **e** conventional OFDM

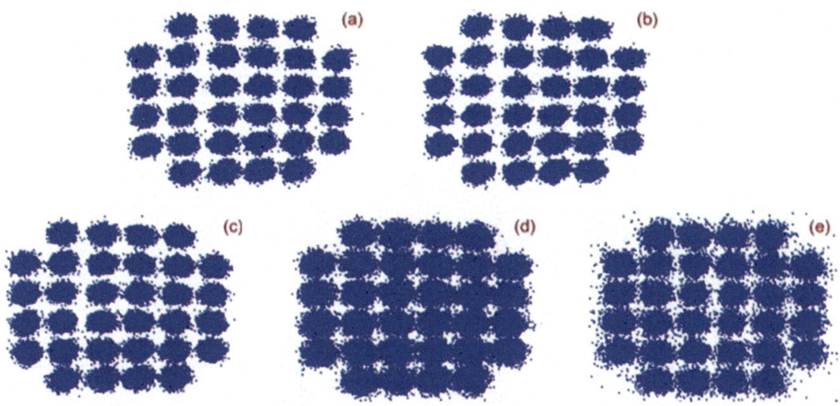

Fig. 13.20 Constellations of **a** DFT-spread OFDM with BPSK TS, **b** DFT-spread OFDM with QPSK TS, **c** DFT-spread OFDM with 16QAM TS, **d** DFT-spread OFDM with analog TS, and **e** conventional OFDM

The measured BER versus received optical power is shown in Fig. 13.21. The BER performance of DFT-spread OFDM with digital TSs is better than that of DFT-spread OFDM with analog TSs and conventional OFDM. DFT-spread OFDM with BPSK/QPSK TS symbol demonstrated the best performance. No power penalty is observed for 79.86 Gb/s DFT-spread 32QAM-OFDM signal after 20 km SSMF transmission and the BER is well below the requirement of the HD-FEC threshold (3.8×10^{-3}).

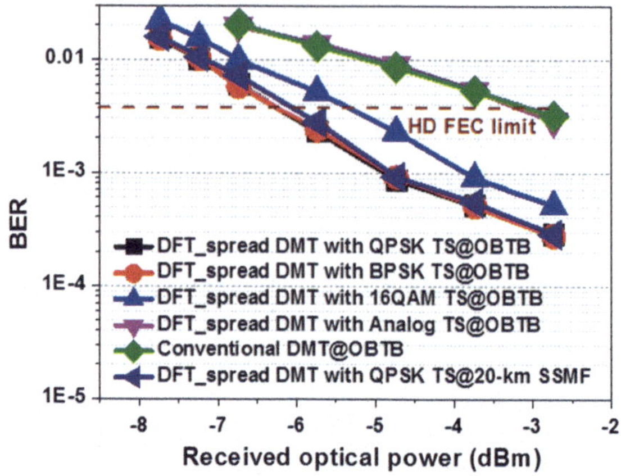

Fig. 13.21 BER versus received optical power for different OFDM signals with a rate of 79.86 Gb/s

13.4.2 Comparison of Pre-equalization and DFT-Spread Techniques in High-Capacity DDO-OFDM

High frequency attenuation is an inevitable problem in wideband DDO-OFDM systems. This section will discuss how to overcome the effects of high frequency attenuation on system performance and ultimately achieve a DDO-OFDM system with an information rate of 100 Gb/s. Figure 13.22 shows a DDO-OFDM experimental setup

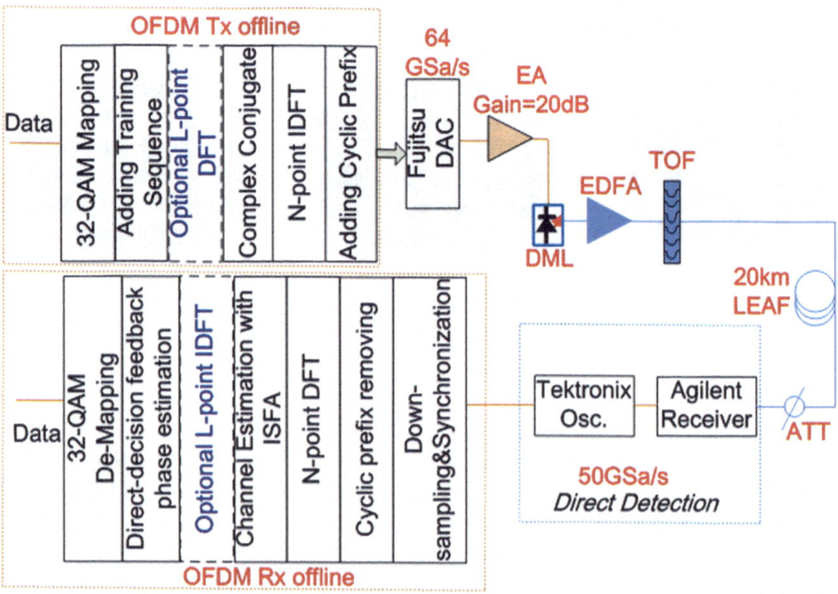

Fig. 13.22 Experimental setup for 32QAM DDO-OFDM signals with a rate of 100 Gb/s

of a 32QAM-OFDM signal at a rate of 100 Gb/s. The process of 100 Gb/s OFDM signal generation is the same as that of the OFDM signal generation in the previous section. Then the signal is uploaded into a Fujitsu DAC with 64 GSa/s sampling rate and 12 GHz 3 dB bandwidth. Here the DFT size for OFDM signal generation and demodulation is N, of which L low positive frequency subcarriers carry data and the 1st subcarrier of the IDFT array is set to zero for DC bias. Another L low negative frequency subcarriers are the complex conjugate of the aforementioned L low positive frequency subcarriers to enforce Hermitian symmetry in the input facet of the IDFT. Other high frequency subcarriers are set to zeros for oversampling. To obtain 100 Gbit/s raw bit rate with 32-QAM-OFDM, the bandwidth should be 20 GHz, the L should be $20/64 \times N$. One training sequence (TS) is inserted before OFDM symbols to realize time synchronization and obtain channel estimation. After cyclic prefix (CP) is added, the signal is uploaded into the DAC. The additional L-point DFT/IDFT is optional and only necessary for DFT-spread OFDM. The OFDM signal from DAC is amplified to 2.4 V by an EA with 20 dB gain. This signal is injected into a commercial DML to achieve E/O modulation. The wavelength and line width of signals generated by DML with a 3 dB bandwidth of 10 GHz are 1537.93 nm and 20 MHz, respectively. The laser was biased at 89 mA to produce an optical OFDM signal with a power of 8.6 dBm. An EDFA is cascaded behind the DML to amplify the signal and then generate a DSB/VSB optical OFDM signal through a tunable optical filter (TOF). Another EDFA cascaded behind the TOF ensures that the power of the signal injected into the 20 km large effective area fiber is 9 dBm. And then O/E conversion is implemented via an Agilent receiver with 14 GHz 3 dB bandwidth. And after a low-pass filter (LPF), the signal is captured by a Tektronix oscilloscope with 50 GSa/s sampling rate and 20 GHz maximum available bandwidth with DSP enhance technique, and then the signal is processed by off-line digital signal processing (DSP). The off-line DSP includes down-sampling, frame synchronization, CP removal, N-point DFT, channel estimation, optional L-point IDFT, direct decision feedback phase estimation, 32-QAM de-mapping, and error counting. The intra-symbol frequency-domain averaging (ISFA) is applied in the channel estimation, the optional L-point IDFT is only for the OFDM with additional DFT in the transmitter and direct decision feedback phase estimation is used to eliminate phase noise.

In order to compensate the high frequency attenuation with pre-equalization, we have to get the channel response with the aid of TS. In order to suppress the noise in the optical link during channel estimation, the time-domain averaging method is adopted. The length of TSs in the time-domain averaging method should set within a certain period when the channel response within can be regarded as static. In this section, 63 TSs are used to obtain channel response with time-domain averaging method. After obtaining the channel response, pre-equalization can be implemented in the frequency domain. Four types of OFDM signal including DFT-spread OFDM, DFT-spread OFDM with pre-equalization, conventional OFDM, and conventional OFDM with pre-equalization are tested in the system. Figure 13.23 gives the probability distribution of the amplitude of these four types of OFDM signals with 8192 total subcarriers and 2560 data subcarriers. The probability distribution of a conventional OFDM signal and conventional OFDM signal with pre-equalization follows Gaussian

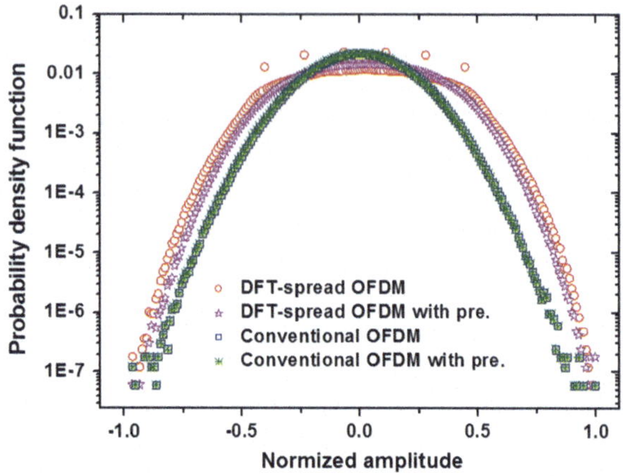

Fig. 13.23 Probability distribution of amplitude

distribution. However, in the two types of OFDM signals using DFT-spread technology, the probability distribution in small signals is significantly reduced, which means that the overall power of the signal is increasing. In the amplitude probability distribution of the DFT-spread OFDM signal, six relatively high peaks appear. From this point of view, the DFT-spread OFDM signal can be regarded as a high probability PAM-6 signal and a low probability analog signal combination. This also explains why the PAPR of the signal drops rapidly after DFT-spread. For the DFT-spread OFDM signal, 6 high probability discrete peaks disappear after pre-equalization, while the PAPR is still lower than conventional OFDM as the probability of signal with large amplitude is higher than that of conventional OFDM. CCDF denotes that a probability distribution of the PAPR of the current OFDM symbol is over a certain threshold. Figure 13.24 gives the calculated CCDF curves. The PAPRs of conventional OFDM and conventional OFDM with pre-equalization are nearly the same

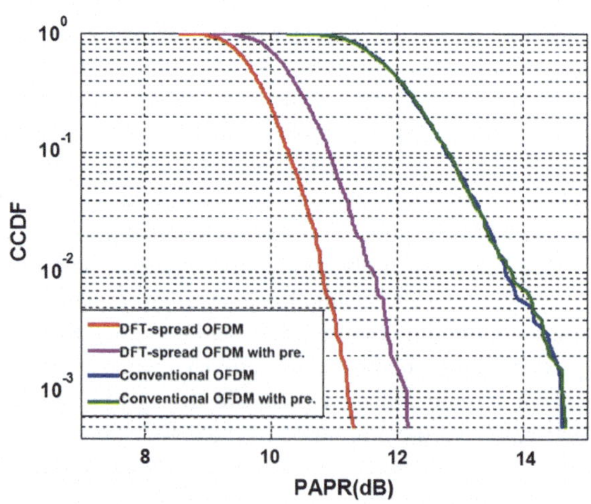

Fig. 13.24 CCDF curves for different OFDM signals with a bandwidth of 20 GHz

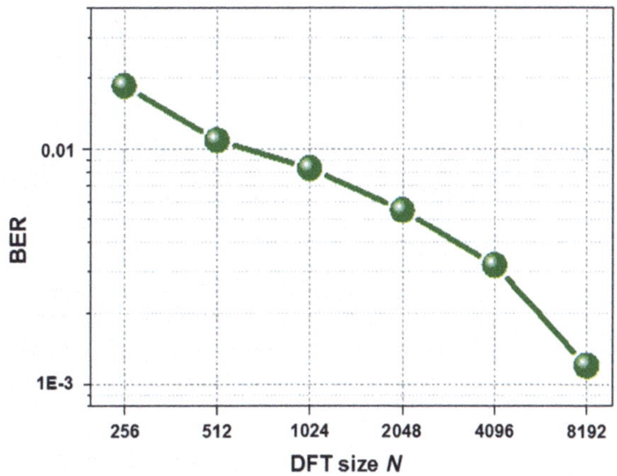

Fig. 13.25 BER versus DFT size N of DFT-spread OFDM in OBTB condition

and the PAPR of DFT-spread OFDM is increased after pre-equalization. The PAPR decrease 3.3 dB for DFT-spread OFDM compared with conventional OFDM at the probability of 1×10^{-3}.

Figure 13.25 gives the BER of DFT-spread OFDM versus DFT size N in optical back-to-back (OBTB) with 2.4 dBm received optical power and the raw bit rate of DFT-spread OFDM signal is maintained at 100 Gbit/s. The BER performance is improved when the DFT size N increases, which is shown in Fig. 13.25. In the rest of this section, N and L are set to 8192 and 2560, respectively, in order to avoid the problem of excessive computational complexity and limited data storage length in the DAC due to excessive DFT size. The length of CP is 4 and 1 TS is inserted every 62 DFT-spread OFDM symbols. The net bit rate is thus $62/(62 + 1) \times 2560/8192 + 4) \times 5 \times 64 = 98.36$ Gb/s. The performance is improved due to enhanced robustness to ISI and the improvement of frequency resolutions in the channel estimation.

Optical spectra (0.02 nm resolution) for four types of OFDM in OBTB are given in Fig. 13.26. Under the same received optical power, the OSNRs of DFT-spread OFDM and DFT-spread OFDM with pre-equalization are higher than conventional OFDM and conventional OFDM with pre-equalization, respectively. As the PAPR of DFT-spread OFDM is the lowest, the average electrical power and total OSNR of DFT-spread OFDM should be the highest. After pre-equalization, the PAPR of DFT-spread OFDM becomes higher, which leads to the OSNR degradation.

The error bits versus frequency for four types of OFDM signal in OBTB with 2.4 dBm received optical power are shown in Fig. 13.27. Some frequency bins with relatively high error ratio appear in conventional OFDM and conventional OFDM with pre-equalization in Fig. 13.27(c) and Fig. 13.27(d), this may be caused by narrowband interference from the clock tones of DAC and oscilloscope. While the error ratio of conventional OFDM increases when the frequency increases, which is induced by power attenuation at high frequency bins and can be compensated by

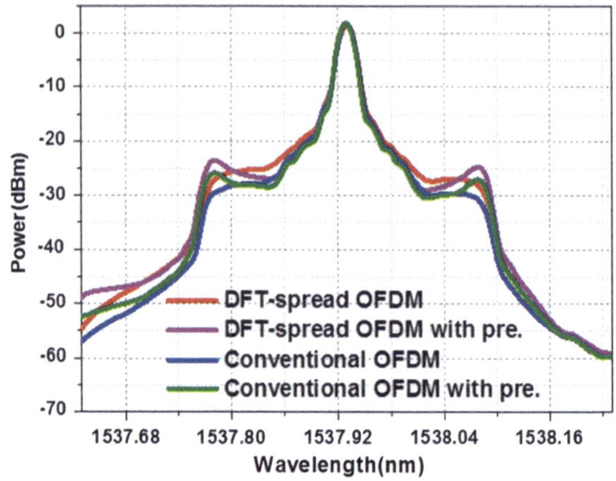

Fig. 13.26 Optical spectra of OFDM in OBTB. **a** DFT-spread OFDM, **b** DFT-spread OFDM with pre-equalization, **c** conventional OFDM, **d** conventional OFDM with pre-equalization

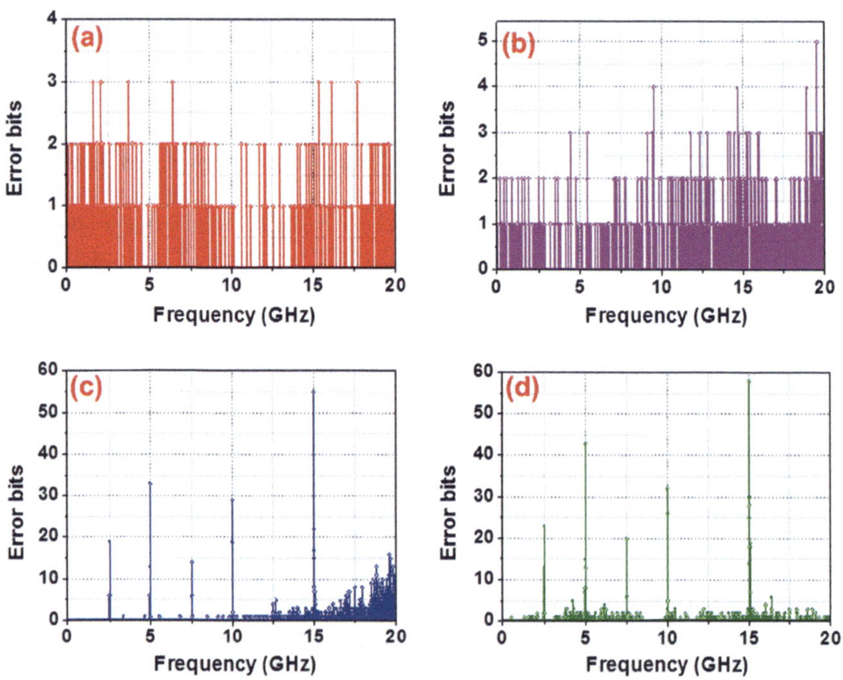

Fig. 13.27 Error bits distribution. **a** DFT-spread OFDM, **b** DFT-spread OFDM with pre-equalization, **c** conventional OFDM, and **d** conventional OFDM with pre-equalization

pre-equalization. However, the problem of excessive error distribution on some subcarriers caused by narrowband interference cannot be effectively solved in this way. After DFT-spread is applied in the OFDM transceiver, the error bits for DFT-spread OFDM in Fig. 13.27(a) and DFT-spread OFDM with pre-equalization in Fig. 13.27(b) become evenly distributed and the problem of excessive error distribution on some subcarriers caused by narrowband interference is also solved. The total BER of DFT-spread OFDM is the lowest. The corresponding constellations for four types of OFDM signals are shown in Fig. 13.28 and the constellation of DFT-spread OFDM is the most centralized. Figure 13.29 shows the BER versus received optical power for four types of OFDM signals in OBTB with a bandwidth of 20 GHz. The DFT-spread OFDM signal with the highest OSNR has a best BER performance. Moreover, such signal can resist the power attenuation at high frequency and eliminate narrowband effect.

In DDO-OFDM systems, frequency selective attenuation due to dispersion is also a problem that needs to be overcome. Since the bandwidth of the OFDM signal transmitted in this section is 20 GHz, even if the distance of the transmitted fiber is

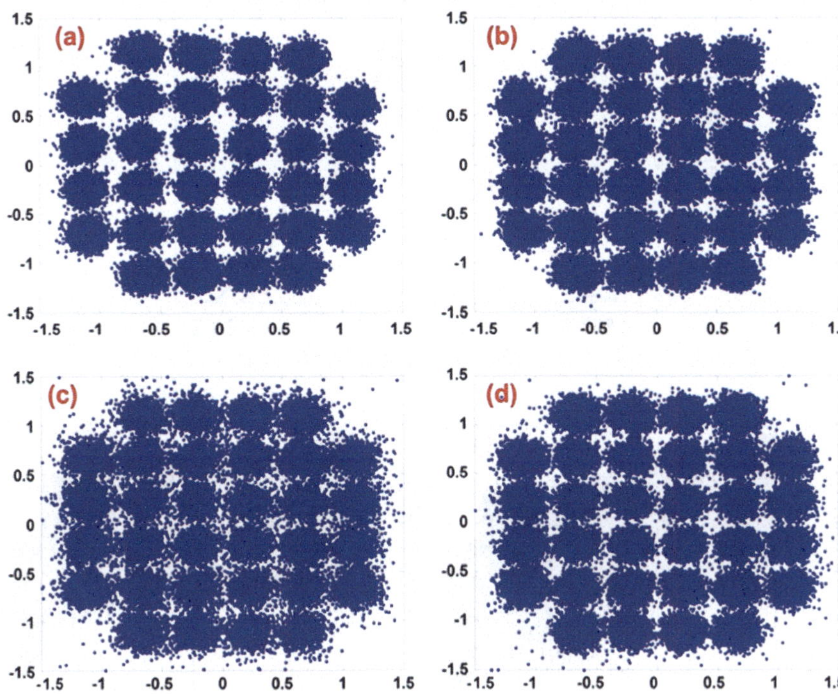

Fig. 13.28 Constellations. **a** DFT-spread OFDM, **b** DFT-spread OFDM with pre-equalization, **c** Conventional OFDM, **d** Conventional OFDM with pre-equalization

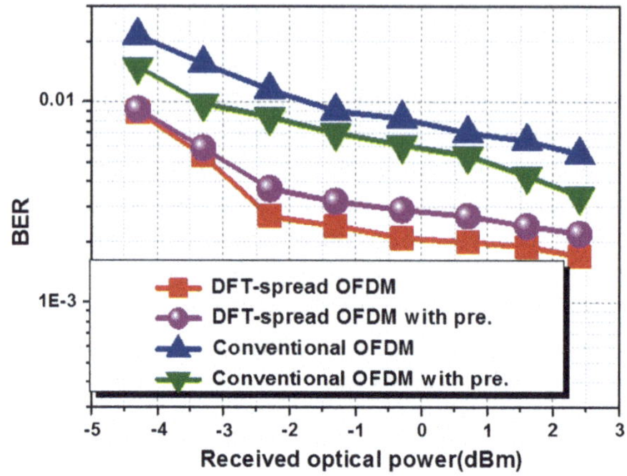

Fig. 13.29 BER versus received optical power in OBTB

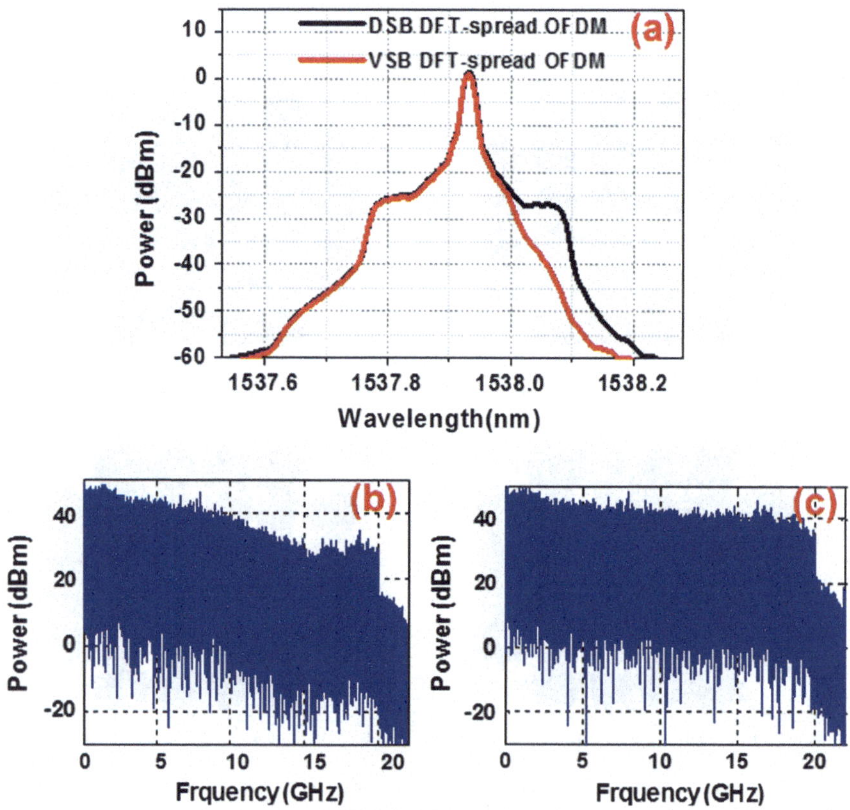

Fig. 13.30 **a** Optical spectra, Electrical spectra of **b** DSB DFT-spread OFDM and **c** VSB DFT-spread OFDM

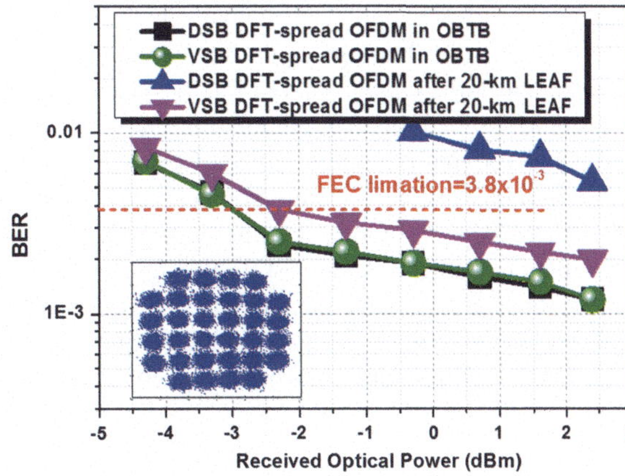

Fig. 13.31 BER versus received optical power for DSB/VSB DFT-spread OFDM

very short, the effect of frequency selective attenuation due to dispersion is already very obvious. The vestigial sideband technique is applied in this section to a 100G DDO-OFDM system to overcome the frequency selective attenuation caused by dispersion. The 20 GHz optical OFDM signal is first controlled by a TOF to produce a DSB or VSB OFDM signal before being injected into the fiber.

As DFT-spread OFDM demonstrates the best performance, we only test it in the fiber transmission. The optical spectra of optical DSB and VSB DFT-spread OFDM signal (0.02 nm resolution) are shown in Fig. 13.30a. The electrical spectra of received DSB and VSB DFT-spread OFDM signal after 20 km LEAF at 2.4 dBm received optical power are shown in Fig. 13.30b, c, respectively. We can see significant power fading in the optical DSB DFT-spread OFDM signal transmission case. The measured BER of DSB and VSB DFT-spread OFDM are shown in Fig. 13.31. The BER performance of DSB and VSB DFT-spread OFDM in OBTB are nearly the same. While after 20 km LEAF transmission, the BER performance of VSB DFT-spread OFDM is improved significantly compared to DSB DFT-spread OFDM and less than the HD-FEC limitation of 3.8×10^{-3}. The constellation of VSB DFT-spread OFDM after 20 km LEAF with 2.4 dBm received optical power is inserted in Fig. 13.31.

13.4.3 Summary

This section first discusses the optimization of the TS when transmitting DFT-spread OFDM signals in a large capacity DDO-OFDM system. The experimental results show that the DFT-spread OFDM signal has better BER performance when using BPSK/QPSK digital TS in channel estimation, compared with analog TS or high order modulation format TS. The transmission of the 79.86 Gb/s DFT-spread 32-QAM-OFDM signal is successfully achieved by this TS. Then, in this section, the DDO-OFDM system is used to transmit 100 Gb/s signals, and the performance of DFT-spread technology and pre-compensation technology in resisting high frequency attenuation and overcoming narrowband interference is compared. Experiments show that DFT-spread technology can not only reduce the PAPR of OFDM signal in short distance and large capacity DD-OFDM system, but also show better performance in resisting high frequency attenuation and overcoming narrowband interference. A scheme for implementing a VSB OFDM signal using a TOF is used to resist severe frequency selective attenuation caused by dispersion in fiber transmission in a DD-OFDM system. In the experiment, we successfully transmitted and received the 100 Gb/s DFT-spread 32QAM-OFDM signal. The BER of the OFDM signal after the 20 km large effective area fiber transmission is still lower than 3.8×10^{-3}. These experimental results show that DDO-OFDM using DFT-spread technology can be used to realize the future ultra-large capacity access network.

References

1. F. Li et al., Optimization of training sequence for DFT-spread DMT signal in optical access network with direct detection utilizing DML. Opt. Express **22**(19), 22962–22967 (2014)
2. F. Li et al., Transmission and reception of Quad-Carrier QPSK-OFDM signal with blind equalization and overhead-free operation. Opt. Express **21**(25), 30999–31005 (2013)
3. F. Li et al., High level QAM OFDM system using DML for Low-cost short reach optical communications. Photon. Technol. Lett. IEEE **26**(9), 941–944 (2014)
4. F. Li et al., Demonstration of DFT-spread 256QAM-OFDM signal transmission with cost-effective directly modulated laser. Opt. Express **22**(7), 8742–8748 (2014)
5. F. Li, et al. Blind equalization for dual-polarization 2-subcarrier coherent QPSK-OFDM signals. Opt. Lett. **39**(2), 201–204
6. F. Li et al., SSMI cancellation in direct-detection optical OFDM with novel half-cycled OFDM. Opt. Express **21**(23), 28543–28549 (2013)
7. F. Li et al., Fiber-wireless transmission system of PDM-MIMO-OFDM at 100 GHz frequency. J. Lightwave Technol. **31**(14), 2394–2399 (2013)
8. F. Li et al., Optimization of pilot interval design in direct-detected optical OFDM system. Opt. Commun. **285**(13), 3075–3081 (2012)
9. F. Li et al., Reducing the peak-to-average power ratio with companding transform coding in 60 GHz OFDM-ROF systems. IEEE/OSA J. Opt. Commun. Netw. **4**(3), 202–209 (2012)
10. F. Li et al., Reduction of frequency fading and imperfect frequency response with pre-emphasis technique in OFDM-ROF systems," Optics Communications, Vol. 284, Issue 19, pp. 4699–4705
11. F. Li et al., "Real-time Demonstration of DMT-based DDO-OFDM Transmission and Reception at 50 Gb/s," ECOC 2013, paper 6.2
12. F. Li et al., "SSMI Cancellation in Direct-detection optical OFDM with Novel Half-cycled OFDM," ECOC 2013, paper 3.3
13. F. Li et al., "Improving the Robustness to Timing Errors with Windowing Technique for 40 GHz 64-QAM OFDM-RoF System," OFC 2014, paper W2A.39
14. F. Li et al., "Transmission and Reception of Quad-Carrier QPSK-OFDM Signal with Blind Equalization," OFC 2014, paper Th2A.64
15. F. Li et al., "Realization of real-time 100G 16QAM OFDM signal detection," Accepted by SPIE Photonics West. (Invited Paper)
16. Xin Xiao, Fan Li et al., "100-Gb/s Single-band Real-time Coherent Optical DP-16QAM-OFDM Transmission and Reception," OFC 2014, paper PDP Th5C.6. (OFC PDP)
17. Z Cao, Fan Li, et al., "Synchronized signaling delivery for broadband 60 GHz in-building optical wireless network based on digital frequency division multiplexing and digital Nyquist shaping," Optics Express, 2013, Vol. 21, Issue 1, pp. 270–275

18. Zizheng Cao, Fan Li et al., "A synchronized signaling insertion and detection scheme for reconfigurable optical OFDM access networks," Lightwave Technology, Journal of, Vol. 30, Issue 24, pp. 3972–3979
19. Z. Cao, F. Li et al., Spatial filtering in a broadband in-home OFDM radio-over-fibre network. Photon. Technol. Lett. IEEE **26**(6), 575–578 (2014)
20. Z. Cao, Y. Jianjun, F. Li et al., Energy efficient and transparent platform for optical wireless networks based on reverse modulation. IEEE J. Selected Areas Commun. **31**(12), 804–814 (2013)
21. Y. Gao, J. Yu, J. Xiao, Fan Li et al., Direct-detection optical OFDM transmission system with pre-emphasis technique. J. Lightwave Technol. **29**(14), 2138–2145

Chapter 14
Intensity Modulation Direct Detection High-Speed Fiber Access System

14.1 Introduction

As mentioned in the introduction, with the rapid development of new services and new technologies such as high-definition video, cloud computing, Internet of Things, computing centers, etc., the short-distance communication, bandwidth requirements of access networks and metropolitan area networks are driven, and transmission capacity requirements reach 400 Gb/s even up to 1 Tb/s. Transmission distances also range from a few kilometers to tens of kilometers. In order to meet the increasing rate requirements, the signal modulation format of the access network has shifted from low spectral efficiency NRZ to high spectral efficiency high-order modulation formats [1–13], such as Orthogonal Frequency Division Multiplexing (OFDM) [1, 6–10], carrierless amplitude phase modulation (CAP) [5], single-carrier frequency domain equalization (SC-FDE) [1, 4, 12], and so on. At the same time, considering the simple structure, low price, low power consumption, and other factors, the techniques based on intensity modulation with direct detection (IM/DD) and high-order modulation are a more reasonable and practical solution for short-distance communication and high-speed optical access networks [1, 2, 4].

Figure 14.1 shows the basic structure of the IM/DD system [4]. The system structure is very simple, including external cavity laser (ECL), Mach-Zehnder modulator (MZM), power amplifier (EA), fiber optics, detector (PD), and the DAC that generates the data and the ADC module that collects the data. It is also possible to use a directly modulated laser instead of the external cavity laser and the MZM module. The data generated by the DAC is loaded by the MZM onto the continuous laser emitted by the ECL, and transmitted to the detector for photoelectric conversion after being transmitted through the optical fiber. The converted electrical signal is collected by the ADC and subjected to digital signal processing (DSP). The entire electro-optical-electric transmission and conversion process is subject to linear and nonlinear damage from the optoelectronic device and transmission link, which influ-

J. Yu and N. Chi, *Digital Signal Processing In High-Speed Optical Fiber Communication Principle and Application*,
https://doi.org/10.1007/978-981-15-3098-2_14

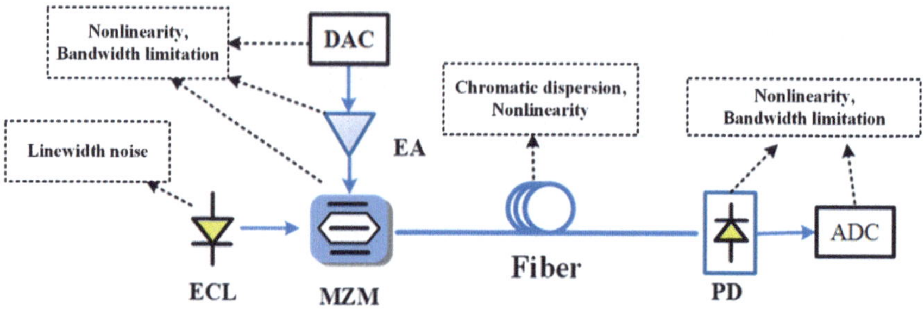

Fig. 14.1 Basic structure and damage mechanism of IM/DD system

ences a lot on signal quality. The generation of various damages therein has also been marked in Fig. 14.1.

In general, optical access systems based on intensity modulation and direct detection face three major challenges.

First, the bandwidth of the commercial transceivers in the system including direct modulated laser (DML), digital-to-analog converter (DAC), arbitrary waveform generator (AWG), detector (PD), and other devices is on the order of 10 GHz, especially the bandwidth of the device at the transmitter is more limited [7,13]. Second, digital-to-analog converters, electrical amplifiers, square-law detectors, and optical fiber introduce several nonlinear effects that severely limit the performance of higher order modulated signals. Third, for directly modulated double-sideband signals, the dispersion in the fiber (mainly the walk-off effect) causes the frequency fading of the directly detected electrical signal, which further limits the bandwidth of the signal [1].

These three problems will severely limit the transmission rate and distance of the IM/DD system. Especially for the transmission of optical fibers of 100 Gb/s or more and 80 km or even 20 km, it is necessary to consider a solution. This chapter will analyze and discuss these three issues one by one, and propose possible reasonable solutions. For the problem of insufficient bandwidth of system devices, this chapter proposes frequency domain pre-equalization techniques and modulation schemes with high spectral efficiency, such as Nyquist PAM, SC-FDE, and OFDM [7,14–16]. For the problem of nonlinearity, this chapter proposes the modulation technique DFT-S OFDM with low power peak-to-average ratio (PAPR) [7], the nonlinear compensation algorithm based on quasi-balanced detection and the nonlinear compensation algorithm based on volterra series. In view of the frequency fading caused by fiber dispersion, this chapter proposes a single sideband or vestigial sideband modulation technique based on optical filters. By adopting the above three technologies, this chapter successfully transmits a directly modulated 128 Gb/s 16QAM DFT-S OFDM signal with 320 km of single-mode fiber under direct detection. In addition, wavelength division multiplexing technology can improve the transmission capacity of the system. This chapter also combines wavelength division multiplexing technology to successfully implement 4 × 128 Gb/s 16QAM DFT-S OFDM signals transmitted in single-mode fiber for 320 km. To the best of our knowledge, this is the farthest distance for signal transmission above a single-wavelength of 100 Gb/s [4].

14.2 High Spectral Efficiency Modulation Technique

In view of the insufficient bandwidth of the device, the frequency domain pre-equalization technique has been proposed and elaborated in the foregoing sections. By pre-compensating the system response at the transmitting end, the function of improving the effective bandwidth of the system can be achieved. Please refer to Chap. 2 for details. This chapter will not repeat them. In the experimental verification of this chapter, this technology will also be used to broaden the bandwidth of the system. In addition to frequency domain pre-equalization, the transmission capacity of the system can be further improved on a limited bandwidth basis by adopting a modulation scheme with high spectral efficiency. There are two ways to improve the spectral efficiency of the system. One is the high-order modulation format, as shown in Fig. 14.2. The modulation format is enhanced from Quadrature Phase Shift Keying (QPSK) to 16QAM, 64QAM and even higher order modulation formats. One symbol of QPSK can carry a 2-bit signal, while one symbol of 16QAM can carry 4 bits, and 64QAM is up to 6 bits. Therefore, by adopting a higher order modulation format, the capacity of the system can be increased.

The second is to perform spectrum compression as much as possible, so that the signal bandwidth is close to the Nyquist limit, even beyond the Nyquist limit (Faster than Nyquist). Assuming that the signal symbol period is Ts, the signal symbol bandwidth is B = 1/Ts, and the spectrum of the signal is shown in Fig. 14.3. As can be seen from the figure, the signal main lobe occupies a bandwidth of 2B. After pulse shaping by different low-pass filters, the side lobes and some of the main lobe signals can be filtered out to obtain signals occupying different bandwidths.

The type of common filter is shown in Fig. 14.4. Figure 14.4a–c shows the frequency response of the four-order Bessel low-pass filter, Nyquist filter, and faster than Nyquist filter, respectively. The bandwidth of the four-order Bessel low-pass filter is larger than the symbol bandwidth B, and is related to the roll-off coefficient of the filter, which is generally about twice. The bandwidth of the Nyquist filter is

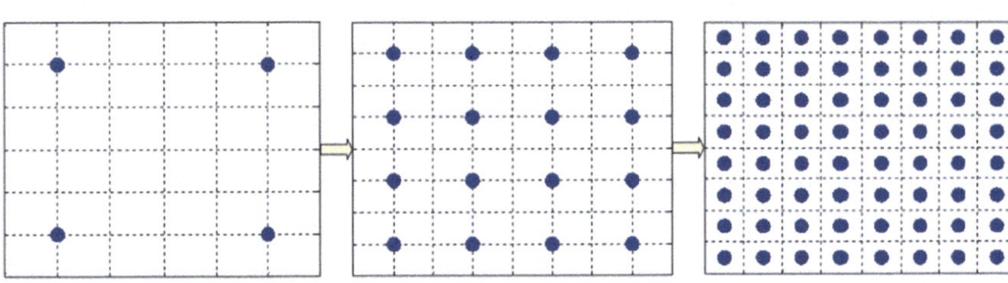

Fig. 14.2 The increasement of modulation order

Fig. 14.3 The spectrum of baseband signal with bandwidth B

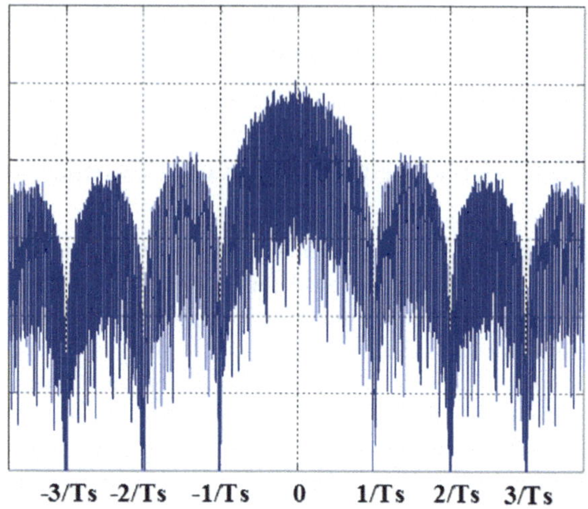

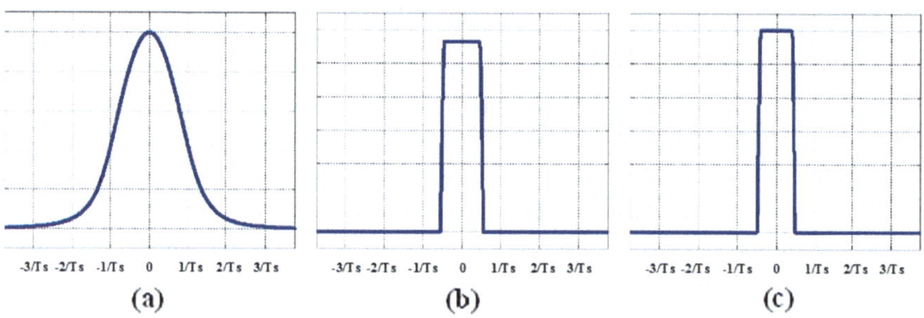

Fig. 14.4 The frequency response of different filters **a** Four-order Bessel filter **b** Nyquist filter and **c** Faster than Nyqujst filter

the same as the symbol bandwidth B. Here is the square root raised cosine filter. The roll-off factor is 0.01, and the effect of the roll-off factor on the system bandwidth is basically negligible. The bandwidth of the faster than Nyquist filter is smaller than the symbol bandwidth B. This chapter will focus on both Nyquist modulation techniques and Faster than Nyquist modulation techniques.

14.2.1 Nyquist Modulation Technique

Modulation techniques can be divided into single-carrier and multi-carrier modulation. Among them, Nyquist modulation technology based on single-carrier has carrierless amplitude and phase modulation (CAP) [1], Nyquist pulse amplitude modulation (PAM) [8], single-carrier modulation with Frequency domain equalization (SC-FDE) [15], half-cycle Nyquist subcarrier QAM modulation, etc. Nyquist modulation techniques based on multi-carrier include Orthogonal Frequency Division Multiplexing (OFDM) and its variants, such as Discrete Multitone Modulation (DMT) [7], adaptive bit and power loading based DMT, OFDM/OQAM, DFT-S OFDM, etc. Modulation techniques such as OFDM, SC-FDE, QBD-OFDM, which have been proposed and experimentally confirmed in the previous visible light communication sections [14–16], can also be applied to the system one by one. However, since the digital signal processing algorithms at the transceiver end are basically the same, the processing flow is similar, and the experimental systems are based on direct modulation and direct detection. The biggest difference is that one is transmitted in free space and the other is transmitted in the fiber link. In order to avoid repeated cumbersome, the above modulation format will not be discussed in this chapter.

This chapter introduces a Nyquist PAM8-FDE technique based on frequency domain equalization. With this technology, the 40-GBd PAM8 signal can be successfully compressed to 20 GBd, and the signal recovery is performed at the receiving end by using a simple zero-forcing frequency domain equalization technique [4]. The generation, transmission, and reception of this signal are verified by the experiment. The block diagram and overall test setup of Nyquist PAM8-SCFDE are illustrated in Fig. 14.5. The data generation process at the transmitting end is also shown in Fig. 14.5.

At the transmitter side, the original binary sequences are first mapped onto PAM8 signals. The training sequence is inserted for post FDE. The sequences are then upsampled to 2 samples per symbol in order to match the operating rate of the shaping filters, and turn to a raised cosine (RC) FIR filter for Nyquist shaping. The roll factor of the Nyquist filter is set to 0.1. These procedures are off-line accomplished by MATLAB, and then loaded into an 8 bit DAC with 80 GSa/s sampling rate and 3-dB

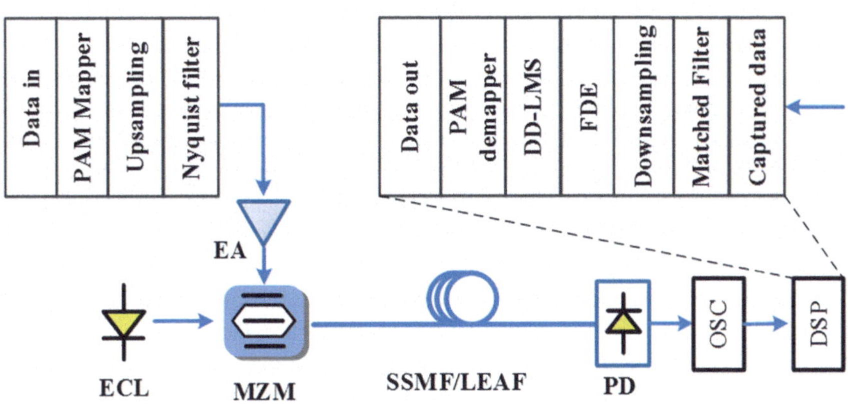

Fig. 14.5 Experimental principle and setup of direct modulation direct detection system based on Nyquist PAM8-FDE

bandwidth of 16 GHz. Thus, the signal baud rate is 40 GBd. After being compressed by the Nyquist filter, the actual occupied bandwidth is only 20 GBd, which can greatly save the bandwidth of the device. Each symbol can carry a 3-bit amount of information, so the total data rate is 120 Gb/s.

Subsequently, the Nyquist PAM8-SCFDE signal from DAC is amplified by a linear electrical amplifier (EA) with 19-dB gain and fed to an intensity modulator with 37 GHz bandwidth and 2.7 V half-wave voltage. The laser diode operates at 1541.4 nm with a linewidth of 100 kHz and an output power of 13 dBm. The output signal is injected into a span of 2-km SSMF or 20-km LEAF. A variable optical attenuator (VOA) is placed after fiber to adjust the received optical power. After fiber transmission, optical to electrical (O/E) conversion is implemented at the receiver side via an Agilent analog receiver. The 3-dB bandwidth of the commercial receiver is 14 GHz.

The detected signals are then captured by a real-time oscilloscope with 160-GSa/s sampling rate and 60-GHz bandwidth and subsequently processed by off-line DSP. The detailed off-line DSP flow chart is also listed in detail in Fig. 14.5. Synchronization and re-sampling to 2 sample per-symbol are firstly employed, and the matched Nyquist RC filter is then implemented to filter out the unwanted signals. After downsampling, training sequences are extracted and used for post frequency domain equalization. In this demonstration, one tap zero-forcing (ZF) algorithm is used to estimate the channel matrix for simplicity. In order to further improve the BER performance, decision-directed least mean square (DD-LMS) can be utilized to improve the accuracy of hard decisions in this experiment. The DD-LMS algorithm has also been explained in detail in the previous chapters.

The experimental setup of the 40-GBd traditional PAM8 signal generation and reception is also shown in Fig. 14.5. The difference between the transmitters is that the Nyquist filter is replaced by the 4th-order Bessel low-pass filter mentioned in the previous section. The difference between the receivers is that the time domain equalization (T/2 DD-LMS) is used instead of the frequency domain equalization, and the matched filter uses the same Bessel low-pass filter as the transmitter.

The electrical spectra of conventional PAM8 and Nyquist PAM8 are inserted in Fig. 14.6a, b, respectively. It can be easily seen that the signal power of the Nyquist PAM8 is concentrated at 0–20 GHz, and the signal of the conventional PAM8 is very scattered. In addition, the Nyquist PAM8 signal has a higher signal-to-noise ratio (SNR) than the traditional PAM8. At 20 GHz, the traditional PAM8 has an electrical SNR of only about 10 dB, while the Nyquist PAM8 has an SNR of more than 20 dB at 20 GHz.

The end-to-end channel frequency response of the overall system is measured and depicted in Fig. 14.7. As can be seen from the figure, the response of the high frequency part of the system drops dramatically. According to the measured results, the 10 dB bandwidth is about 10 GHz and the 20-dB bandwidth is around 20 GHz. By adopting Nyquist PAM8-SCFDE, the 40-GBd signal can be compressed to within 20 GHz, so the system can carry the transmission of 40-GBd Nyquist PAM8 signals.

A simulated comparison between conventional PAM8 and Nyquist PAM8-SCFDE is conducted firstly. In this simulation, the channel response is based on the measured

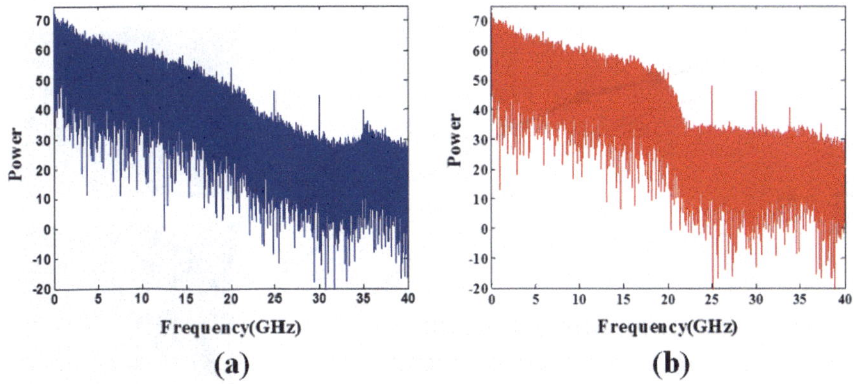

Fig. 14.6 The electrical spectra of 40-GBd PAM8 **a** conventional PAM8和 **b** Nyquist PAM8

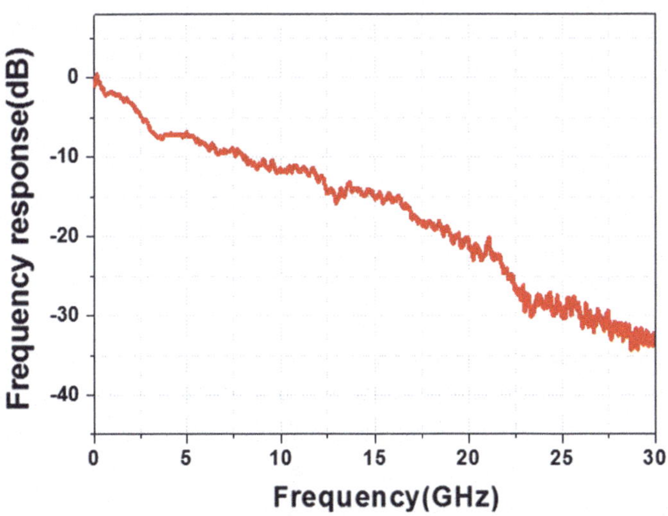

Fig. 14.7 The end-to-end channel frequency response of the overall system

frequency response illustrated in Fig. 14.7, as well as the additive white Gaussian noise (AWGN). In order to provide a fair comparison, these two modulation formats are tested in the same condition with a raw bit rate of 120 Gb/s and with the same digital processing algorithms. The simulated BER versus signal-to-noise ratio (SNR) is depicted in Fig. 14.8. The SNR is varied from 10 to 22 dB with a step of 2 dB and the BER performance also changes. The eye diagrams of these two modulation schemes at the SNR of 22 dB are also inserted in Fig. 14.8. From this figure, we can find that the receiver sensitivity can be enhanced by 4 dB at the BER of 1×10^{-2} by adopting Nyquist PAM8-SCFDE.

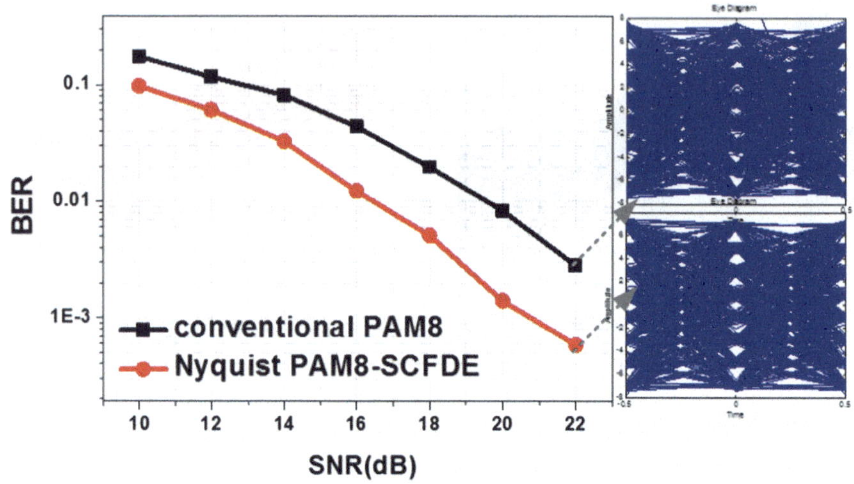

Fig. 14.8 BER versus SNR performance comparison between conventional PAM8 and Nyquist PAM8-SCFDE

The experimental setup for the 40-GBd Nyquist PAM8-SCFDE generation, transmission, and reception in IM/DD system is given in Fig. 14.5. First, a comparison of optical spectra at different conditions is carried out. Figure 14.9 shows the measured optical spectra with 0.02 nm resolution before and after modulated with 40-GBd Nyquist PAM8 and 40-GBd conventional PAM8 signals. The Nyquist PAM8 signals

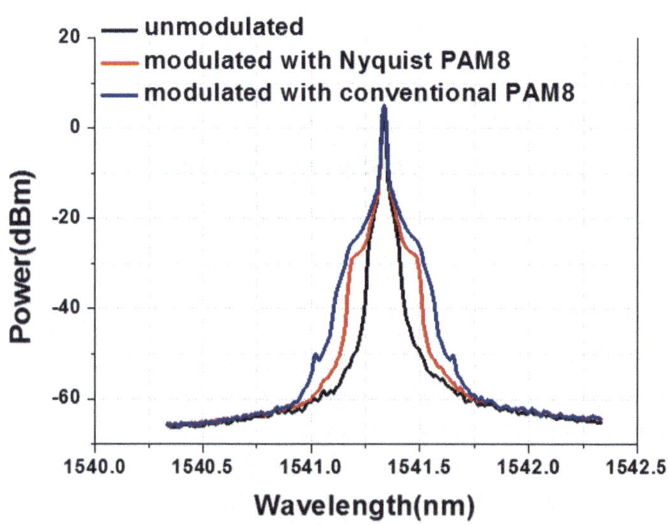

Fig. 14.9 Optical spectra in the case of before modulation and after modulated with Nyquist and conventional 40-GBd PAM8 (0.02 nm resolution)

are shaped by the Nyquist RC filter, and its occupied bandwidth is halved, which is much narrower than the conventional one. The difference can also be observed in Fig. 14.6.

Then, a comparison of BER performance between conventional PAM8 and Nyquist PAM8 is made under the same experimental setup and digital equalization algorithms in the case of optical back-to-back (OBTB). The BER versus received optical power is measured. The received optical powers are varied from −4 to 2 dBm, and the results are given in Fig. 14.10. According to the results, we can find the conventional PAM8 cannot reach the threshold value of 3.8×10^{-3}, but the BER of Nyquist PAM8-SCFDE can be under this threshold when the received optical power is larger than 0dBm. At the BER of 1×10^{-2}, the receiver sensitivity of Nyquist PAM8-SCFDE can be enhanced by 6 dB. The eye diagrams of these two modulation schemes at the power of 2 dBm are also shown in Fig. 14.10a, b. The

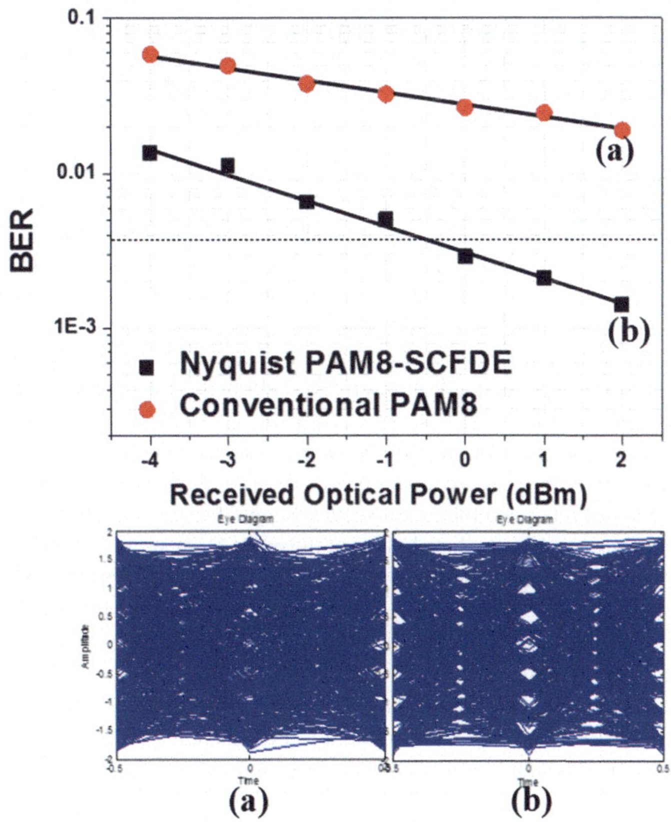

Fig. 14.10 Measured BER results versus received optical power in the case of Nyquist PAM8 and conventional PAM8. **a** Eye diagram of traditional PAM8 signal and **b** Eye diagram of Nyquist PAM8 signal at 2 dBm

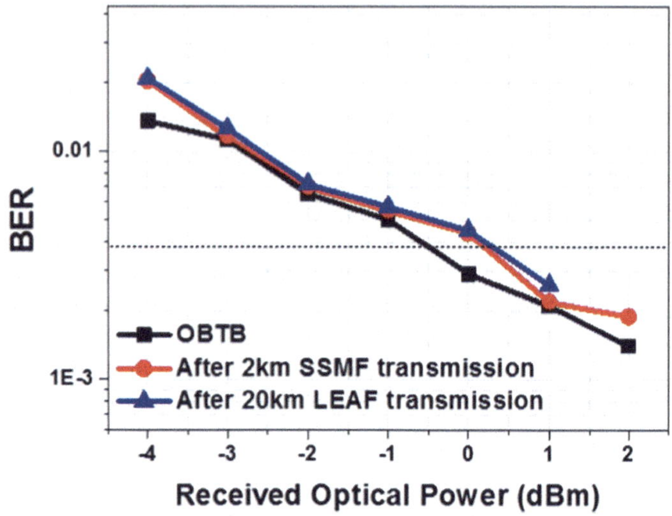

Fig. 14.11 BER performance of 40-GBd Nyquist PAM8 in case of optical back-to-back, 2 km SSMF, and 20 km LEAF transmission

eye diagrams are obtained after downsampling and equalization. The superiority of Nyquist PAM8-SCFDE can also be concluded from the eye diagrams.

Finally, we have successfully performed 2-km SSMF and 20-km LEAF transmission. The SSMF and LEAF are with the chromatic dispersion of 17 and 4 ps/km/nm and the attenuation coefficient of 0.2 and 0.21 dB/km, respectively. The BER performances versus received optical power in the case of OBTB, after 2-km SSMF and 20-km LEAF transmission are all measured and studied. The measured BER results are depicted in Fig. 14.4, 14.5, 14.6, 14.7, 14.8, 14.9, 14.10 and 14.11. The received power degradation induced by fiber transmission can be neglected. And the BER can be under 3.8×10^{-3} once the received power is larger than 0 dBm.

Finally, we also analyzed the Nyquist PAM8 signal recovered after 2 km of fiber transmission. The probability distribution of the eight-layer signal of PAM8 is shown in Fig. 14.12. The abscissa indicates that the amplitude of each layer of PAM8 is expressed as (−7, −5, −3, −1, 1, 3, 5, 7). It can be seen from the figure that the probability distribution of each layer of signals basically conforms to the normal distribution around the theoretical value of each layer, and the probability distribution of each layer of signals is the same, and the peak amplitude is basically the same. These results further prove that this modulation format can be well applied to short distances such as 2 km and high-speed access (>100 Gbit/s) applications.

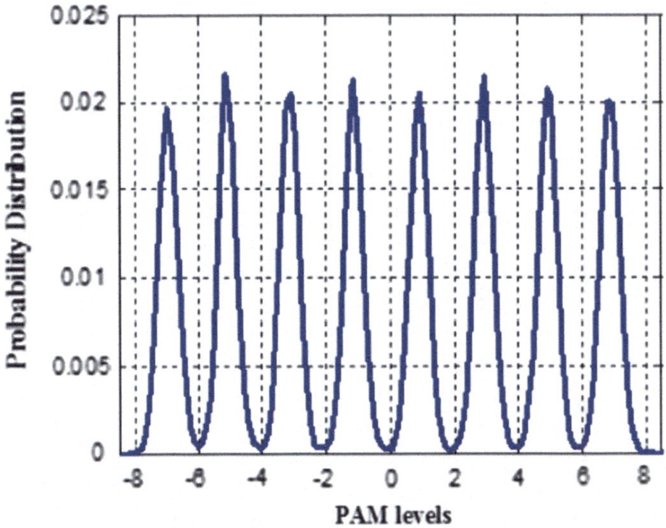

Fig. 14.12 Probability distribution of received Nyquist PAM8 signals

14.2.2 Faster Than Nyquist Modulation Technique

The previous section introduced the Nyquist modulation technique to improve spectral efficiency by compressing the signal bandwidth. If the signal bandwidth is further compressed on the basis of Nyquist, the signal of Faster than Nyquist can be obtained. However, this kind of super-Nyquist spectrum compression will have a serious impact on signal quality. This section will briefly introduce this technology.

The generation of the super Nyquist signal can be achieved by filtering two orthogonal electrical signals using a low-pass filter having a bandwidth less than the Nyquist bandwidth as described above. It is also possible to add an optical bandpass filter with a bandwidth smaller than the Nyquist bandwidth after the output of the optical domain modulator. After super Nyquist filtering, the binary signal becomes a duobinary signal. The change of the signal constellation is shown in Fig. 14.13. It can be seen from the figure that after the super Nyquist filtering, the four points of the QPSK signal will become 9 points, which is similar to the 9QAM signal and the 16 points of the 16QAM signal will become 49 constellation points which are similar to the 49QAM signal. Figure 14.13a–d shows the constellation diagrams for the four modulation formats such as QPSK, 9QAM, 16QAM, and 49QAM, respectively. In addition, from the radius of the generated signal, the four constellation points of QPSK are located on a circle of the same radius, and the double binary QPSK (9QAM) is located on three circles of different radii. Similarly, 16QAM is also located on three circles of different radii, while the 49QAM constellation points are distributed over 10 circles of different radii. The more signal layers, the higher

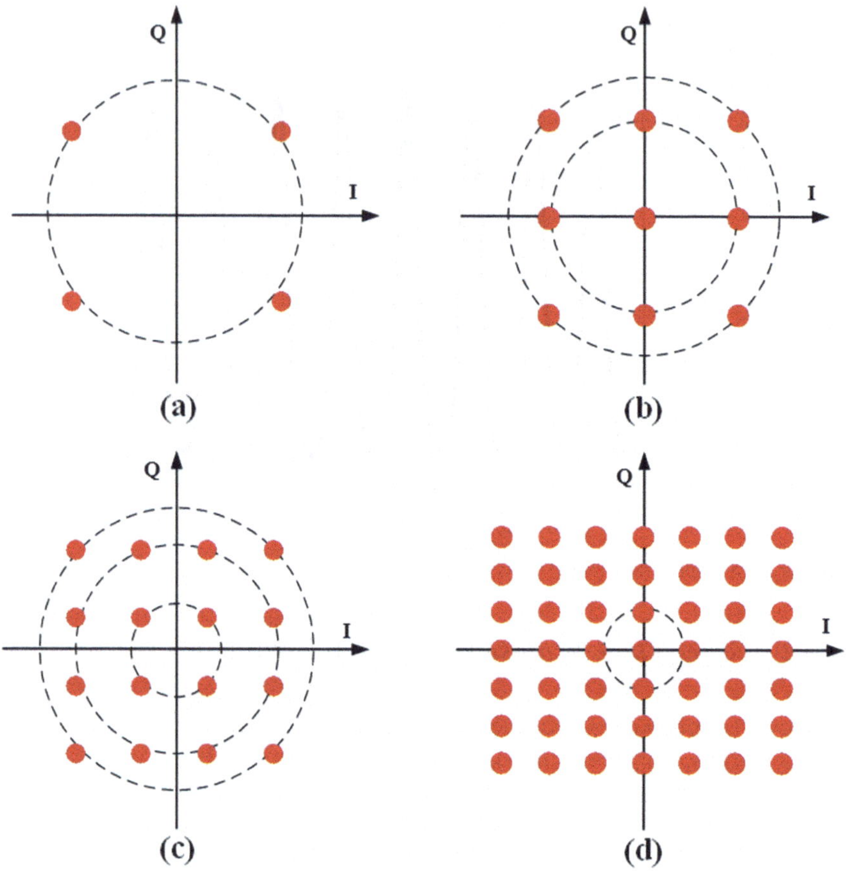

Fig. 14.13 The constellation diagrams of signals **a** QPSK, **b** 9QAM, **c** 16QAM, **d** 49QAM

the resolution required by the signal, the more stringent the system's required SNR is.

The generation of the super Nyquist signal proposed in this subsection is based on a double binary delay addition filter. The z-transform response of the filter can be expressed as

$$H(z) = 1 + z^{-1}. \tag{14.1}$$

The operation process is the addition of the symbols before and after the signal. For vector signals, the same operation is required. However, at the receiving end, in order to recover the original binary signal from the 9QAM or 49QAM signal, the signal needs to be differentially encoded at the transmitting end. The detailed complete Nyquist signal generation process is shown in Fig. 14.14. It includes mapping the binary signal into QPSK or 16QAM signal, separately differential encoding the I or

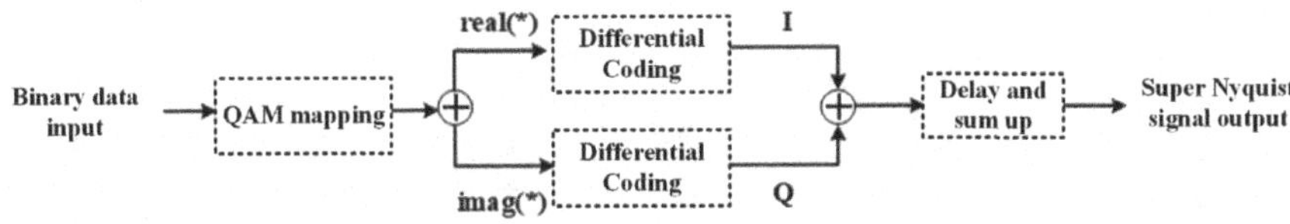

Fig. 14.14 The signal generation process of super Nyquist

Q signal, and then recombining into a vector signal after differential encoding, and then obtaining the super Nyquist signal through the delay adding module.

It is assumed that the I and Q paths of the kth symbol after the 2^M-QAM mapping of the signal can be expressed as i_k and q_k, and the I and Q of the kth symbol d_k after differential encoding can be expressed as I_k and Q_k, the differential codec process can be expressed by the following formula

$$I_k = i_k - I_{k-1} \text{ mod } M. \tag{14.2}$$

$$Q_k = q_k - Q_{k-1} \text{ mod } M. \tag{14.3}$$

$$d_k = I_k + jQ_k\,. \tag{14.4}$$

After differential encoding, the super Nyquist signal D_k generated by the delay addition can be expressed as

$$D_k = d_k + d_{k-1}\,. \tag{14.5}$$

Similarly, the corresponding differential decoding needs to be performed at the receiving end.

After all the recovery processes have processed the hard decision, the decoding process only needs to perform the modulo M processing on the I and Q paths of the recovered signal. In this way, the generation and reception of super Nyquist signals can be successfully achieved.

This section compares the performance between Nyquist 16QAM SC-FDE and Faster than Nyquist 49QAM SC-FDE. The block diagram and experimental setup are shown in Fig. 14.15. The flow of the two modulation formats is shown in the block diagram. The Faster than Nyquist signal generation includes differential encoding, delay additive filtering, and differential decoding at the receiving end as shown in Fig. 14.14. The square root raised cosine filter is used in Nyquist 16QAM SC-FDE modulation. After being generated in the off-line software and imported into the 64 GSa/s DAC, the 15-GBd Nyquist 16QAM and Faster than Nyquist 49QAM signals can be, respectively, obtained and amplified by a linear power amplifier to 20 dBm to drive the direct modulation laser (DML). The DML bandwidth used in this experiment is on the order of 10 GHz with a center wavelength of 1548.5 nm and an output power of 8 dBm. The modulated optical signal is transmitted to the

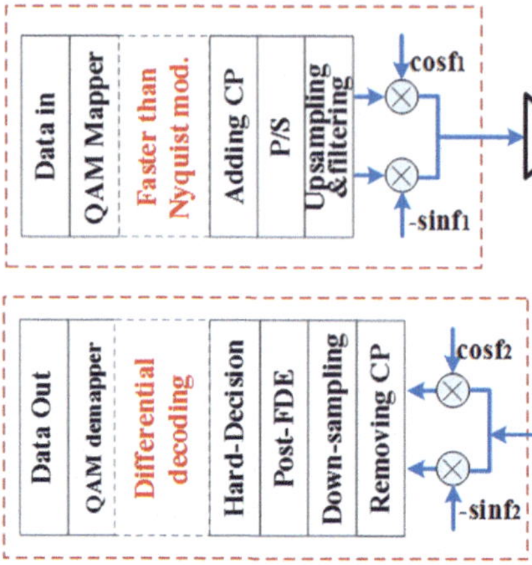

Fig. 14.15 The experimental setup of Nyquist 16QAM SC-FDE and Faster than Nyqusit 49QAM SC-FDE

receiving end through a 2 km single-mode optical fiber, photo-electrically converted via an intensity detector with a 3 dB bandwidth of 13 GHz, and then acquired by a real-time oscilloscope of 80 GSa/s, and then subjected to digital signal processing at the receiving end.

The spectrum of the 15-GBd Nyquist 16QAM and 15-GBd Faster than Nyquist 16QAM signals is shown in Fig. 14.16. As can be seen from the figure, the Nyquist signal occupies a bandwidth of 15 GHz, while the Faster than Nyquist signal occupies only about half of the Nyquist bandwidth, much less than 15 GHz. From a bandwidth perspective, Faster than Nyquist is more spectrally efficient.

At the receiving end, we tested the BER of 15-GBd Nyquist 16QAM and Faster than Nyquist (FTN) FTN 49QAM. The BER of Nyquist 16QAM is 9.41×10^{-4}, and the error rate of FTN 49QAM is 7.13×10^{-3}. The constellation diagram is shown

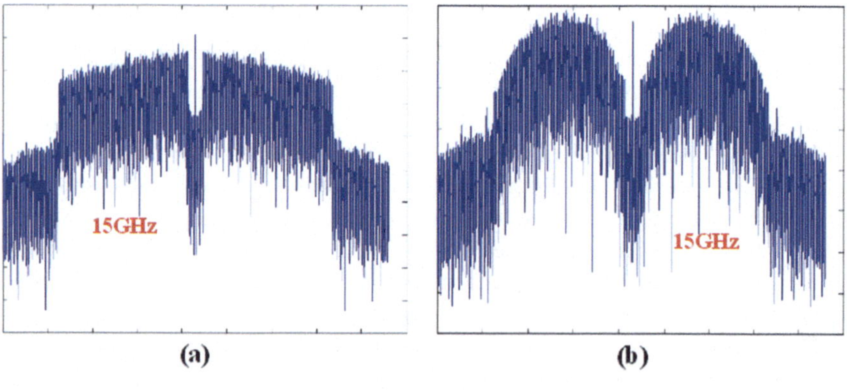

Fig. 14.16 The electrical spectra of **a** 15-GBd Nyquist 16QAM and **b** Faster than Nyquist 49QAM

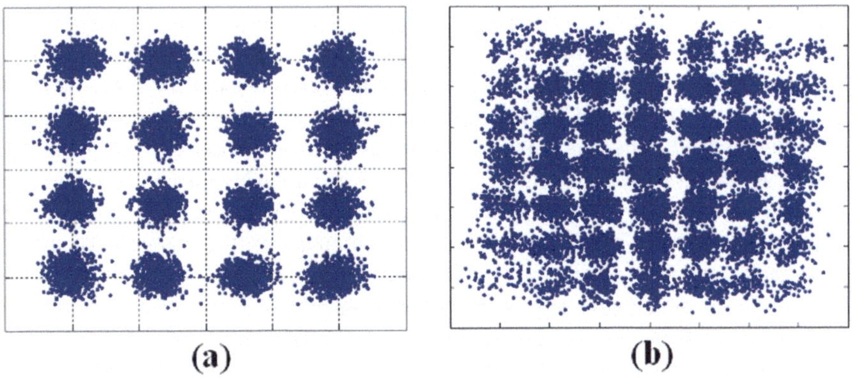

Fig. 14.17 The constellation diagram of **a** Nyquist 16QAM signals and **b** Faster than Nyquist 49QAM signals

in Fig. 14.17. The clear 16QAM constellation point can be seen from Fig. 14.17a, while the 49QAM constellation diagram in Fig. 14.17b is indeed obscure. Moreover, the constellation points with small intermediate radius are relatively large, and the constellation points with large radius are relatively small, which is related to the probability of occurrence of each constellation point of 49QAM after delay superposition. It can be seen from the BER and the constellation diagram that since the Euclidean distance of the 49QAM constellation point is small, the number of signal layers is as many as 10 layers, and although there is high spectral efficiency, it is at the expense of the performance of the system. Figure 14.18a, b shows the constellation diagrams of the 12-GBd and 10-GBd FTN 49QAM signals, respectively. The corresponding BER are 6.89×10^{-4} and 4.57×10^{-4}, respectively.

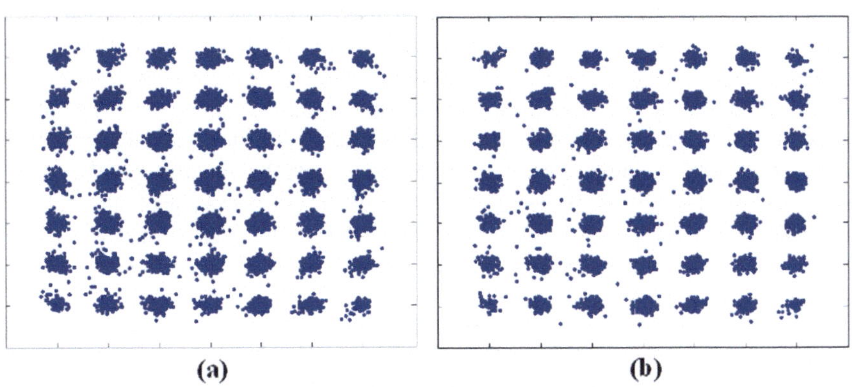

Fig. 14.18 The constellation diagram of **a** 12-GBd 和, **b** 10-GBd Faster than Nyquist 49QAM signal

The previous section compares the performance of the Nyquist PAM8 with the traditional PAM8. Simulations and experimental results confirm the superior performance of the Nyquist PAM8. In this section, the performance of Nyquist 16QAM and Faster than Nyquist 49QAM modulation formats are compared under the same experimental conditions. It is found that the performance of Nyquist 16QAM is better from the aspect of BER performance. Although Faster than Nyquist has been proven to be more resistant to WDM signal crosstalk, higher spectral efficiency, and more resistant to dispersion in coherent optical communications, there is no particularly strict channel spacing concept in direct modulation communication systems. Further compression of the Nyquist signal into a Faster than Nyquist signal does not make much sense and can cause system performance to deteriorate.

Through the above two experiments, Nyquist or quasi-Nyquist (Nyquist filter used in the system allows a certain roll-off coefficient, which will cause the signal bandwidth to be larger than the Nyquist bandwidth) is better than the traditional modulation and faster than Nyquist modulation in the short-distance direct modulation direct detection communication system. It is a more feasible modulation method for short-distance access to balance spectrum efficiency and system performance.

14.3 Nonlinear Compensation Technique

In addition to the lack of bandwidth mentioned in the previous section, the nonlinear effect is another factor in direct modulation direct detection systems that seriously affects system performance, especially factors that are based on the performance of higher order modulation systems. As explained above, the nonlinear effects mainly come from the nonlinearity of the optoelectronic device, the nonlinearity of the fiber transmission and the nonlinearity of the square-law detection. All three of these will result in the square term of the original signal appearing in the received signal, even higher order or higher order harmonics. In addition to the intrinsic nonlinear noise of devices such as digital-to-analog converters (DACs), the system can be operated at its optimum state by adjusting the electrical power, optical power, or operating range of the device. For example, the output power of the electric power amplifier and the bias of the modulator can be adjusted to operate the modulator in the quasi-linear region. It is also possible to reduce the effects of nonlinear effects such as stimulated Brillouin scattering in the fiber by adjusting the fiber input power of the fiber. Similarly, adjusting the optical power injected into the detector at the receiving end can optimize the performance of the detector.

By adjusting some key parameters in the system, the nonlinear effects can be minimized. It is worth noting that reducing the nonlinear effect will lead to a decrease in the signal-to-noise ratio of light or electricity, which will lead to the deterioration of the system. Therefore, when adjusting the above parameters, it is not adjusted to the condition that the nonlinear effect is the smallest, but by taking into account two factors of nonlinearity and signal-to-noise ratio, the parameters are selected to make the system work in the optimal state.

14.3.1 Nonlinear Compensation Technology Based on Volterra Series

The direct modulation laser (DML) is selected below to analyze the nonlinear effects of the device. Figure 14.19 shows the relationship between the DML input voltage and the output power. It can be seen from the figure that the relationship between the two is nonlinear, and the linear working range is very small, which causes the modulation factor to be small, which will also worsen the signal-to-noise ratio of the system, resulting in deterioration of system performance. For common nonlinear systems, the Volterra series can be used for expansion. If the system is treated as a black box and the input is *x(t)*, then the output signal *r(t)* can be viewed as the number of *p*th order of the input signal, which can be expressed as [16]

$$\begin{aligned} r(t) =& h_0 + \sum_{k=0}^{M-1} h_1(k)x(t-k) + \sum_{k_1=0}^{M-1}\sum_{k_2=k_1}^{M-1} h_2(k_1,k_2)x(t-k_1)x(t-k_2) \\ &+ \cdots + \sum_{k_1=0}^{M-1}\cdots\sum_{k_p=k_{p-1}}^{M-1} h_p(k_1,\cdots,k_p)x(t-k)_1\cdots x(t-k_p). \end{aligned} \tag{14.6}$$

where M is the memory length of the system, h_0 is a constant, and $h_p(\bullet)$ represents the pth Volterra nonlinear term coefficient. As the memory length and nonlinear term increase, the nonlinear parameters increase exponentially. Therefore, it is not realistic to consider and compensate nonlinear problems with the complete Volterra

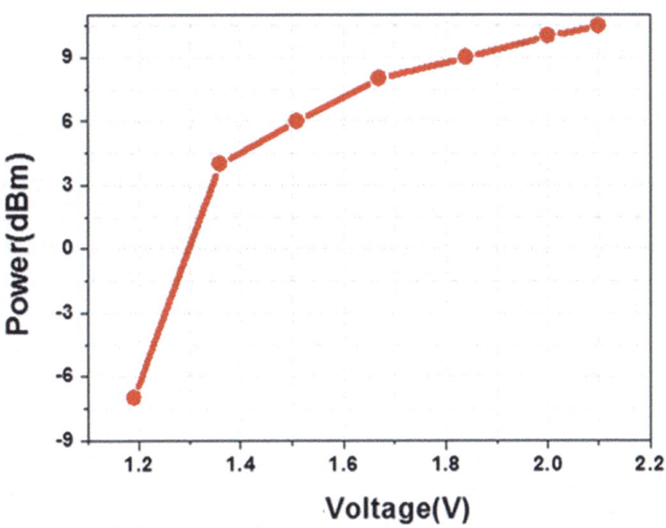

Fig. 14.19 The curve of output optical power versus DML input voltage

series. Generally, simplified Volterra series is used. The choice of memory length and nonlinear order can be adjusted according to the actual situation of the system.

Depending on whether the memory length M is 0, the nonlinear system can be divided into two categories: non-memory and memory nonlinear systems. For non-memory systems, the above formula can be simplified to

$$r(t) = h_0 + h_1 x(t) + h_2 x^2(t) + \cdots + h_p x^P(t). \tag{14.7}$$

The meaning of the parameters is the same as the above formula. For such systems, an adaptive equalizer can be used at the receiving end to eliminate squared or higher order terms. The output of the adaptive filter can be expressed as

$$\mathrm{y}(t) = w_0 + w_1 x(t) + w_2 x^2(t) + \cdots + w_L x^L(t). \tag{14.8}$$

where $x(t)$ is the signal collected by the receiver, that is, the input of the adaptive equalizer. $(w_0, w_1, w_2, \cdots w_L)$ is the coefficient of the adaptive filter. L is the order of the filter. By using the adaptive filter, the influence of the nonlinear effect can be eliminated to some extent.

For memory nonlinear systems, the processing is more complicated. Its expression is as shown in Eq. (14.6), and it can also be compensated by the adaptive equalizer. The expression after the adaptive equalizer is as follows:

$$\begin{aligned} y(t) = w_0 &+ \sum_{k=0}^{N-1} w_1(k) x(t-k) + \sum_{k_1=0}^{N-1} \sum_{k_2=k_1}^{N-1} w_2(k_1, k_2) x(t-k_1) x(t-k_2) \\ &+ \cdots + \sum_{k_1=0}^{N-1} \cdots \sum_{k_p=k_{p-1}}^{N-1} w_L(k_1, \cdots, k_p) x(t-k_1) \cdots x(t-k_p). \end{aligned} \tag{14.9}$$

The parameters are similar in meaning to the above equation, and the only ones that can be simplified are the memory length and the order of the nonlinear term. The general memory length is selected to be 10 orders of magnitude, and the nonlinear term is selected as two.

A nonlinear compensation algorithm based on the squared difference of adjacent sampling signals is proposed in the literature. The specific expression is as follows:

$$\mathrm{y}_i = \sum_{k=-M}^{M} c_k x_{i-k} + \sum_{p=-N}^{N} \sum_{k=-N}^{p-1} c_{kp} (x_{l-p} - x_{i-k})^2. \tag{14.10}$$

where x is the input to the equalizer, M is the order of the linear term filter, N is the order of the nonlinear term filter, c_k and c_{kp} are the tap coefficients of linear and nonlinear filters, respectively. The update of the above coefficients can be based on the training sequence and then updated using the LMS principles. Experiments based on this algorithm will be given in Sect. 14.4 of this chapter, and later on, this algorithm will greatly improve the performance of the system.

14.3.2 Nonlinear Compensation Technique Based on Quasi-balanced Coding and Detection

This section will present a nonlinear compensation technique based on quasi-balanced coding and detection. For direct modulation direct detection system, the $x(t)$ in Eq. (14.7) can be expressed as

$$x(t) = \mathrm{Re}\left(1 + \alpha e^{j2\pi \Delta f t} E_s(t)\right). \tag{14.11}$$

where Δf is intermediate frequency, $E_s(t)$ is the baseband complex signal, α is a coefficient that relates the relationship between optical carrier and signal power. Substituting the above formula into Eq. (14.7), you can get

$$\begin{aligned} r(t) =& h_0 + h_1 \mathrm{Re}\left(1 + \alpha e^{j2\pi \Delta f t} E_s(t)\right) + h_2\left(\mathrm{Re}\left(1 + \alpha e^{j2\pi \Delta f t} E_s(t)\right)\right)^2 \\ &+ h_3\left(\mathrm{Re}\left(1 + \alpha e^{j2\pi \Delta f t} E_s(t)\right)\right)^3 \\ =& \beta_0 + \beta_1 \mathrm{Re}\left(\alpha e^{j2\pi \Delta f t} E_s(t)\right) + \beta_2\left(\mathrm{Re}\left(\alpha e^{j2\pi \Delta f t} E_s(t)\right)\right)^2 \\ &+ \beta_3\left(\mathrm{Re}\left(\alpha e^{j2\pi \Delta f t} E_s(t)\right)\right)^3 . \end{aligned}$$

β_0, β_1, β_2 and β_3 are the coefficient associated with the nonlinear term. The first item in the above formula is a DC term, which can be ignored. The second item is the item proportional to the signal, which is also the signal we need to obtain. The third term is a second-order nonlinear term, and the fourth term is a third-order nonlinear term, which is generally small and can be ignored.

The block diagram of the class balance code proposed in this section is shown in Fig. 14.20. At the transmitting end, the single-carrier signal is first divided into blocks, and in each signal, there are two symbols. The second symbol is the inverse of the first symbol, i.e., the values are exactly the same, but the signs are opposite. At the receiving end, a photocurrent signal is generated after being detected by the square law. The two symbols $I_{2k-1}(t)$ and $I_{2k}(t)$ in the kth block can be expressed as

Fig. 14.20 The block diagram of quasi-balanced coding

$$I_{2k-1}(t) = \beta_0 + \beta_1 \mathrm{Re}\left(\alpha e^{j2\pi \Delta f t} E_{2k-1}(t)\right) + \beta_2 \left(\mathrm{Re}\left(\alpha e^{j2\pi \Delta f t} E_{2k-1}(t)\right)\right)^2. \tag{14.13}$$

$$I_{2k}(t) = \beta_0 - \beta_1 \mathrm{Re}\left(\alpha e^{j2\pi \Delta f t} E_{2k-1}(t)\right) + \beta_2 \left(\mathrm{Re}\left(\alpha e^{j2\pi \Delta f t} E_{2k-1}(t)\right)\right)^2. \tag{14.14}$$

Subtracting the above two equations, the photocurrent signal of the kth block can be obtained, as follows:

$$I_k = I_{2k-1}(t) - I_{2k}(t) = 2\beta_1 \mathrm{Re}\left\{\alpha e^{j2\pi \Delta f t} E_{2k-1}(t)\right\}. \tag{14.15}$$

As can be seen from the above equation, the second-order nonlinear term can be eliminated, and the sensitivity of the receiver can also be improved by 3 dB.

Then we verified the feasibility of the above scheme through experiments. The experimental setup for signal generation, transmission, and detection of a 256QAM Nyquist SC-FDE based on quasi-balanced coding is shown in Fig. 14.21. At the transmitting end, the binary is first mapped to an M-QAM signal, followed by a serial-to-parallel conversion. After the serial-to-parallel conversion, the two groups are processing quasi-balanced coding. The cyclic prefix is then added, up-sampled, and then filtered by the Nyquist filter. To achieve greater spectral efficiency, the filter has a roll-off factor of zero. Then, the baseband complex signal is uploaded to the intermediate frequency signal and introduced into the DAC with a sampling rate of 30 GSa/s. The resulting signal bandwidth is 10 GHz. It is amplified by an amplifier to drive a direct modulation laser. The DML has a working wavelength of 1295.43 nm, a bandwidth of 10 GHz, a linewidth of 20 MHz, a bias current of 60 mA, and an output power of 10dBm. The modulated optical signal is injected into a 20-kilometer single-mode fiber, and transmitted through an optical fiber, and then sent to a 3 GHz bandwidth detector for photoelectric conversion. The electrical

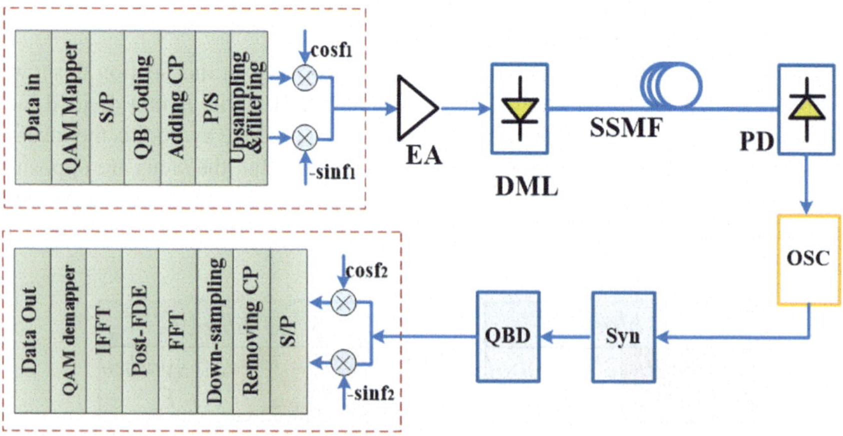

Fig. 14.21 The experiment setup of M-QAM SC-FDE DML-based quasi-balanced coding and detection

signal is then acquired by a real-time oscilloscope with a sampling rate of 40 GSa/s and a bandwidth of 13 GHz. The digital signal processing module at the receiving end is also shown in Fig. 14.21.

First, the transmitter produces a 3 GHz sine wave signal after quasi-balanced coding. The original signal, the signal collected by the optical back-to-back transmission, and the signal after the quasi-balanced detection are shown in Fig. 14.22. It can be seen from Fig. 14.22b that after the optical back-to-back transmission, the 6 GHz of second harmonic, the 9 GHz of third harmonic, and 12 GHz of fourth harmonic frequency components are generated. The power difference between the first harmonic and the second harmonic is 15 dB, which indicates that the nonlinear effect of the system is strong. The power of the third harmonic is small and can be neglected. After quasi-balanced detection, it can be found from Fig. 14.22c that the power difference between the first harmonic and the second harmonic can be reduced to 45 dB. This means that with this type of quasi-balanced coding and detection scheme, the nonlinear effect can be reduced by 30 dB. At the same time, it can be found that the powers of the second harmonic and the third harmonic are basically the same and can be ignored.

Then, the BER performance driven by different electrical signal powers is tested. The test results are shown in Fig. 14.23. The electrical signal power starts from 3 to 15 dBm, and one point is tested for every 3 dB increase. From the figure, we can

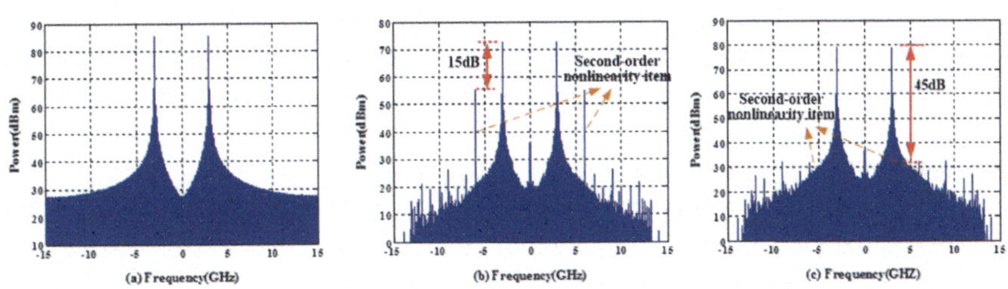

Fig. 14.22 The electrical spectra of **a** quasi-balanced encoded 3 GHz original signal, **b** Optical BTB signal and **c** signal after quasi-balanced detection

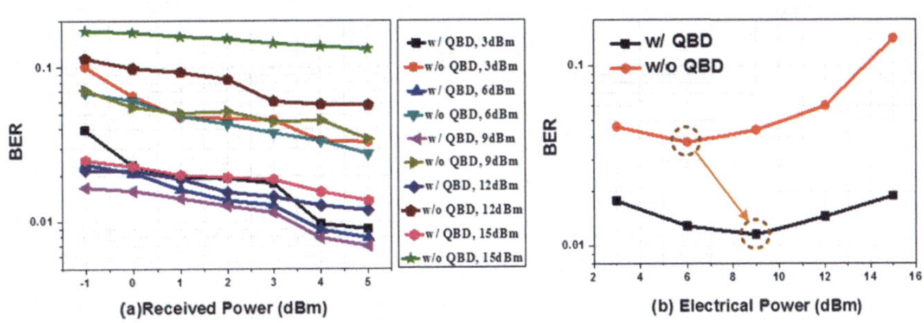

Fig. 14.23 The BER performance **a** versus received power with or without quasi-balanced detection and **b** versus electrical power with or without quasi-balanced detection

find that after using the quasi-balanced technique, the BER can be greatly reduced. Figure 14.23a shows the BER versus received optical power with and without quasi-balanced detection. Figure 14.23b shows a curve of BER performance versus electrical signal power at a received optical power is 4 dBm. It can be seen from the figure that the optimal electrical signal input power is 6 dBm when the quasi-balanced detection technique is not used, and the best electrical signal power can be increased to 9 dBm after the quasi-balanced detection technique. This shows that the tolerance of nonlinearity is stronger after adopting this technology. Moreover, optimal BER can be reduced from 0.04 to 0.01.

The BER performance of SC-FDE with and without quasi-balanced coding was compared experimentally. The experimental results are shown in Fig. 14.24a, b. Both have effective transmission rates of 40 Gb/s. The electrical signal input power was tested at 6 and 15 dBm. The values of these two electrical signal powers represent weak nonlinear effects and strong nonlinear effects, respectively. In the case of weak nonlinear effects, the scheme using the quasi-balanced coding detection can improve the sensitivity of the receiver by 4 dB. In the case of strong nonlinear effects, the sensitivity of the receiver can be increased by at least 5 dB. This not only shows that the scheme using quasi-balanced coding and detection is better than unused one, but also shows that this scheme is more effective for strong nonlinear effects.

Figure 14.25 shows the BER performance curves for the SC-FDE signal with and without the quasi-balanced coding for back-to-back and 20-kilometer single-mode fiber transmission. Since the operating wavelength of DML is at 1295.43 nm, the performance degradation caused by dispersion is basically negligible. Figure 14.25a, b shows the constellation diagrams for 256QAM without and with quasi-balanced detection, respectively. It can be found easily that the constellation diagram utilizing quasi-balanced detection is clearer.

Finally, the BER performance of different modulation orders is also compared experimentally. The experimental results are shown in Fig. 14.26. Experiments tested

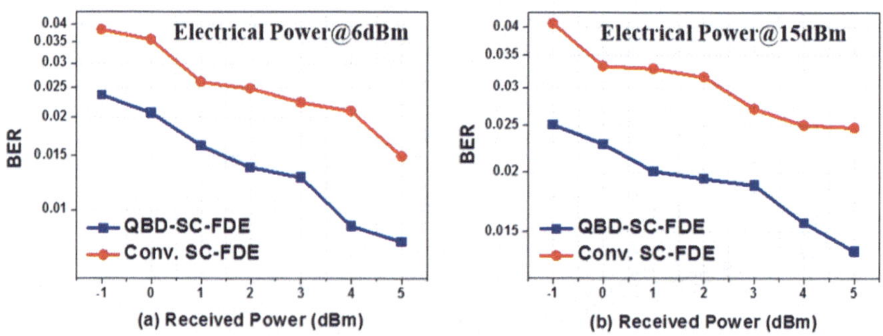

Fig. 14.24 Comparison of BER performance of SC-FDE with and without quasi-balanced coding detection **a** when electric signal power is 6 dBm and **b** when electric signal power is 15 dBm

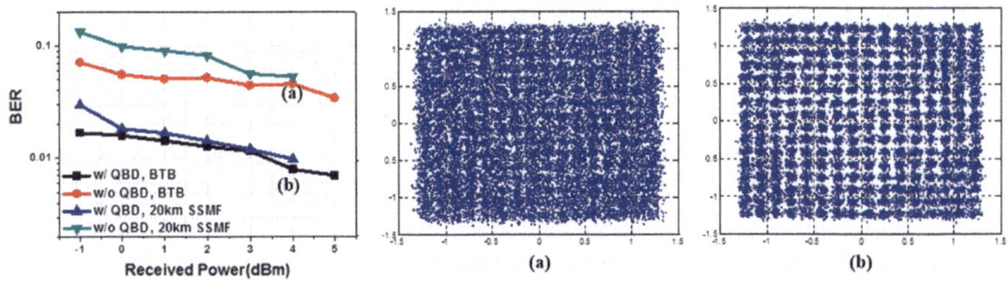

Fig. 14.25 BER performance with and without quasi-balanced coding signals in back-to-back and fiber-optic transmission

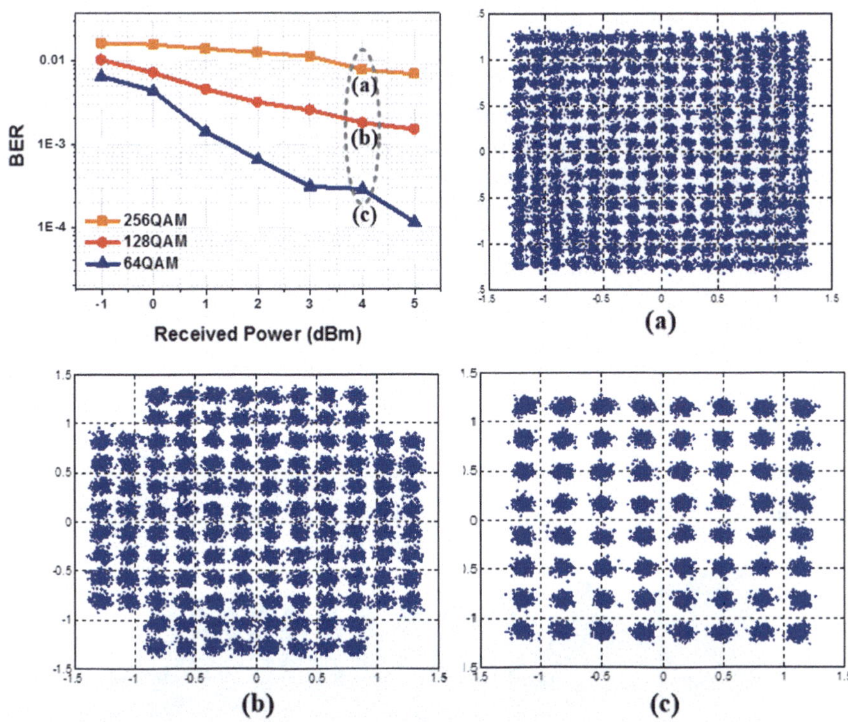

Fig. 14.26 The BER performance of different modulation order

the BER curves of 64QAM, 128QAM, and 256QAM modulation formats. The effective signal bandwidth of the three modulation orders is 5GBd, so the effective transmission rates of the three are 30 Gb/s, 35 Gb/s, and 40 Gb/s. As can be seen from Fig. 14.26, the BER performance of the three modulation orders can be lower than 2.4×10^{-2}. In addition, Fig. 14.26a–c, respectively, show constellation diagrams of 5-GBd 256QAM, 128QAM, and 64QAM signals.

14.4 Single Sideband Modulation System

Intensity modulation based on a one-arm Mach-Zehnder modulator produces a double-sided band signal (DSB). After the signal of the double-sideband is transmitted through the optical fiber, due to the influence of the dispersion of the optical fiber, the frequency fading effect is introduced after the direct detection, which seriously limits the bandwidth and transmission performance of the system. Figure 14.27 shows the spectrum of a 32-GBd DFT-S OFDM signal transmitted over 80-km single-mode fiber. It can be seen from the figure that the received electrical signal will cause severe frequency fading and cannot be recovered by the existing digital signal processing algorithms. One solution is to add a dispersion compensating fiber to the transmission system, but this only applies to links with a fixed transmission distance. For access networks or short-range communication systems, the transmission distance is generally variable due to user differences. This network architecture is bound to have limited effect on dispersion compensating fibers.

Alternatively, it seems that the practical method is to use single sideband modulation (SSB) or vestigial sideband modulation (VSB). By removing one of the sideband signals, the effect of eliminating frequency fading caused by dispersion is achieved. The single sideband signal can be generated by an I/Q modulator or a dual-arm driven Mach-Zehnder modulator, which is described in detail in the next chapter. This chapter mainly introduces the scheme of generating single sideband or vestigial sideband signals based on optical filters. Based on, the high spectral efficiency modulation and nonlinear compensation algorithm mentioned in the previous section, a 32 Gb/s 32-GBd 16QAM DFT-S OFDM signal is successfully transmitted

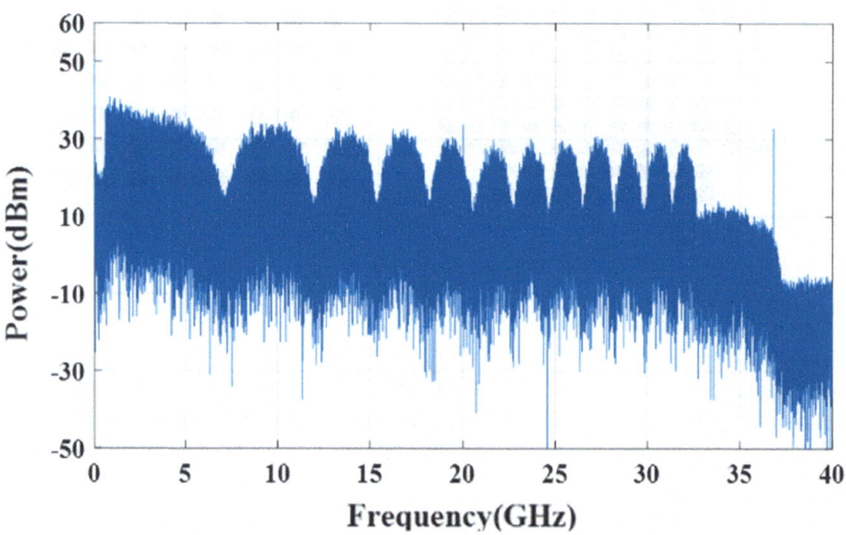

Fig. 14.27 Electrical spectra of 32-GBd DFT-S OFDM signal acquired after transmission over 80 km of single-mode fiber

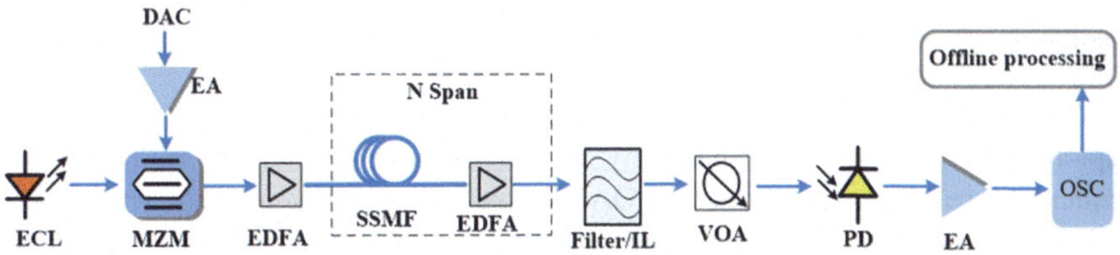

Fig. 14.28 Experimental setup of 16 QAM DFT-spread OFDM transmission system based on intensity modulation and direct detection

in single-mode fiber for 320 km. The BER after transmission may be lower than the soft decision forward error correction code threshold 2×10^{-2}. To the best of our knowledge, this is the farthest transmission distance in the international intensity modulation direct detection system at speeds above 100 Gb/s.

The experimental setup used in this experiment is very similar to the experimental system of direct modulation direct detection described above, except that the optical filter is used for filtering before or after transmission through the optical fiber [7]. A tunable wavelength laser was used in this experiment. Therefore, the optical filter here can use a filter with adjustable center wavelength and adjustable filter bandwidth, or an interleave with fixed center wavelength and bandwidth. The experimental setup for the 16 QAM DFT-S OFDM transmission and reception is illustrated in Fig. 14.28. At the transmitter side, an external cavity laser (ECL) with a linewidth less than 100 kHz is used as an optical source. The operating wavelength is located at 1548.5 nm and the output power is 13 dBm. The continuous-wavelength light wave output from ECL is modulated by 32-GBd 16 QAM DFT-S OFDM signals, which is generated off-line in MATLAB and then loaded into a digital-to-analog converter (DAC) with 80-GSa/s sampling rate and 16-GHz 3-dB bandwidth. Before modulated onto the optical carrier, the output signals from DAC are boosted by an electrical amplifier (EA) to 20 dBm to drive the modulator. The used single-drive Mach–Zehnder modulator (MZM) in this demonstration has a 3-dB bandwidth of 36 GHz, 2.8-V half-wave voltage at 1 GHz and 5-dB insertion loss. Before being transmitted over several spans of 80-km SSMF with Erbium-doped fiber amplifier (EDFA)-only amplification in the absence of optical dispersion compensation, the output optical signals from MZM are amplified to 9 dBm using an EDFA to compensate for modulation and insertion loss. It is then injected into the fiber. The length of the fiber is divided into many segments. For every 80 km, the transmission loss of the 80-km fiber is about 17.5 dB. Each segment uses only erbium-doped fiber amplifiers for loss compensation without the use of dispersion compensating fibers. After fiber transmission, a narrow bandwidth tunable optical filter or an interleave with 50-GHz grid is implemented to realize single sideband or vestigial sideband optical filtering. This part can be implemented before or after fiber transmission which depends on the actual application scenario and needs. Before optical to electrical (O/E) conversion (one 40 GHz photodiode and with one power electrical amplifier with 32-dB gain), a variable optical attenuator (VOA) is placed to adjust the received optical power. At last, the received electrical signals are captured by a real-time digital storage oscilloscope (OSC) with 80-GSa/s sampling rate and 36-GHz electrical bandwidth and subsequently processed by off-line digital signal processing.

The detailed signal processing flow charts at the transmitter and receiver side are depicted in Fig. 14.29a, b, respectively. For the transmitter side, the original binary sequences are first mapped onto 16 QAM signals, and then sent to DFT-S OFDM modulation model, including training sequence adding, DFT, zero padding, inverse DFT (IDFT), cyclic prefix (CP) insertion, and up-conversion to intermediate frequency. The DFT size is 2048. The CP of 32 samples is added to mitigate the CD induced distortions. Due to the insufficient bandwidth of DAC, pre-equalization at the frequency domain is needed. The digital signal processing flow at the receiver side is shown in Fig. 14.29b. Synchronization is firstly employed by convolution. Subsequently, the nonlinearity equalization algorithm is implemented to compensate the nonlinear distortions induced by VSB, optical modulator, electrical amplifiers fiber transmission, and square-law detection. The nonlinear compensation algorithm employed here is mentioned in the literature and has been introduced in the previous section. The signals then turn to the DFT-S demodulation model, which are the inverse steps of the DFT-S modulation at the transmitter, including down-conversion to baseband, CP removing, DFT, zero removing, training sequence extraction, post frequency domain equalization, and additional IDFT. During the post-equalization, one tap zero-forcing (ZF) algorithm is used to estimate the channel matrix for simplicity. After the obtained channel curve, the curve smoothing in the frequency domain is needed to reduce the noise introduced by the signal jitter, so that the signal estimation is more accurate. In order to further improve the BER performance, decision-directed least mean square (DD-LMS) can be utilized at frequency domain. It can be used for inter carrier interference mitigation, and precise decision. After equalization, the output 16QAM signals are decoded to binary stream for BER estimation. The raw data rate for 32-GBd 16 QAM DFT-S OFDM signal is 128-Gb/s. Counting the training sequence 1/30, CP 1/65, and 20% FEC overhead, the net data rate is $32 \times 4 \times (29/30) \times (64/65)/1.2 = 101.52$ Gb/s.

In terms of PAPR, a comparison between conventional OFDM and DFT-S OFDM with and without pre-equalizations is made. The calculated complementary cumulative distribution function (CCDF) curves have been added in Fig. 14.30. CCDF indicates the probability that PAPR exceeds a certain value. It can be seen from the figure that the threshold of PAPR of OFDM, pre-equalized OFDM, DFT-S OFDM, and pre-equalized DFT-S OFDM is 12 dB, 12.5 dB, 10 dB, and 11 dB, respectively,

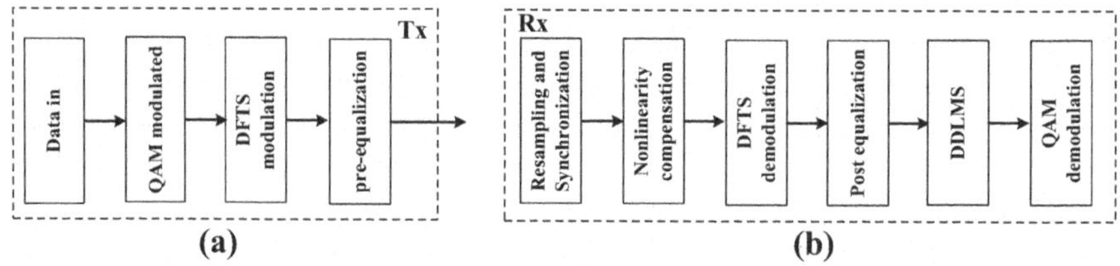

Fig. 14.29 Signal processing algorithms structures at **a** transmitter side, **b** receiver side

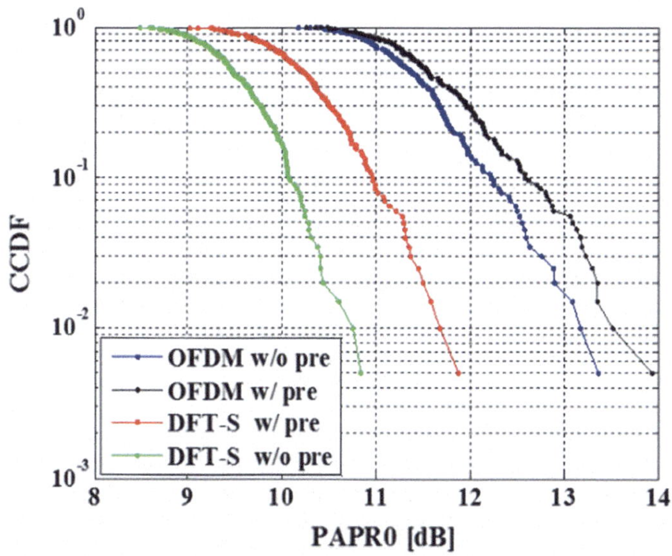

Fig. 14.30 The CCDF curves of conventional OFDM and DFT-S OFDM with or without pre-equalization

when the probability is 1×10^{-2}. Although the PAPR of DFT-S OFDM has a 1 dB increase after pre-equalization, it can be improved by 1 dB and 1.5 dB compared with OFDM or pre-equalized OFDM. It shows that the PAPR performance of DFT-S OFDM is indeed superior, and the introduced nonlinear effects are also relatively small.

In this section, we will mainly introduce the transmission of single-wavelength. The wavelength division multiplexing system will be introduced in the next section. Before conducting an experimental data test, some key parameters including fiber launch power, optical signal-to-noise ratio, transmission distance, and transmission capacity are measured and discussed to render this system work at the optimal condition. For the VSB modulation, both TOF and cost-effective interleave with 50-GHz grid are tested. First, all results are measured at the narrow bandwidth TOF case. The optical spectra of 32-GBd DFT-S OFDM signals before and after TOF are measured. The results are depicted in Fig. 14.31. Figure 14.31a shows the optical spectra without pre-equalization, and Fig. 14.31b shows the optical spectra with pre-equalization. The pre-equalized high frequency part of the spectrum is higher, mainly used to compensate for the lack of bandwidth of the receiver. From this figure, we can find the right band has been suppressed with a suppression ratio of around 20 dB compared to DSB. It should be noted that higher suppression ratio at right band can cause lower carrier-to-signal-power ratio (CSPR) at left band, which will induce relatively large signal-to-signal beating noise (SSBN). Thus the CSPR should be maintain at least 15–20 dB. In our demonstration, the CSPR is selected during the above range according to the minimal BER results.

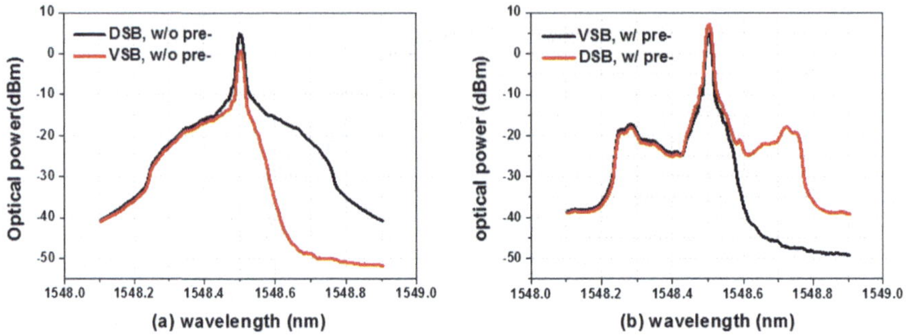

Fig. 14.31 Optical spectra of DFT-S OFDM signals before and after tunable optical filter at the cases of **a** without pre-equalization and **b** with pre-equalization

After setting the suppression ratio and CSPR, the optimal fiber launch power is also measured to keep the trade-off between SNR and nonlinearity. The BER performance versus fiber launch power ranging from 5 to 10 dBm with the varying step of 1 dB is depicted in Fig. 14.32a. The optimal injected power is 9 dBm. For the rest of the measurements, the fiber launch power for each span of 80-km SSMF is set to 9 dBm. The 80 km fiber loss is 17.5 dB. The required OSNR at BTB and 80-km SSMF transmission cases are also measured and the results are shown in Fig. 14.32b. At the SD-FEC limit, the required OSNRs for BTB and 80-km SSMF transmission are 30 dB and 34 dB, respectively, which can conclude that the 80-km fiber transmission can induce 4-dB OSNR penalty. The distortions mainly come from EDFA and fiber. The noise figure of EDFA is about 5 dB, which will introduce noise during amplification. As for fiber, the chromatic dispersion and nonlinearity will also cause OSNR penalty.

Next, the effects of different equalization algorithms on the system were tested. The digital signal processing algorithms used in this experiment include frequency domain pre-equalization, nonlinear compensation, DD-LMS, frequency domain post-equalization, and frequency domain smoothing. Due to the bandwidth limit

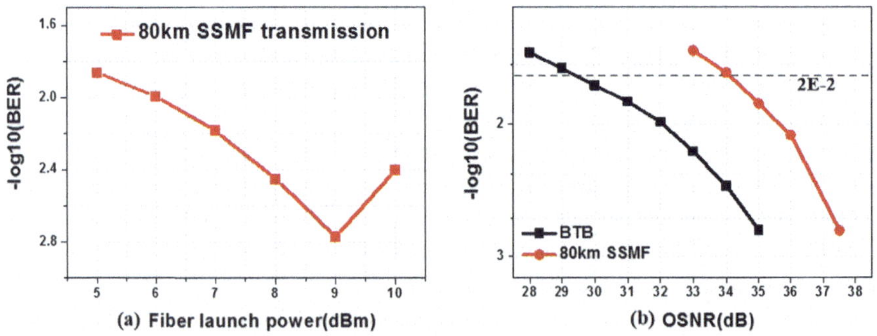

Fig. 14.32 BER versus **a** fiber launch power at 80 km SSMF transmission case and **b** OSNR at BTB and 80 km SSMF transmission cases with TOF

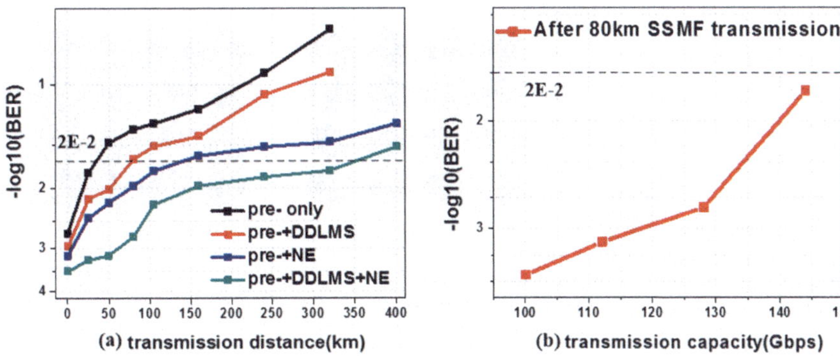

Fig. 14.33 Measured BER performances versus **a** transmission distance with different algorithms and **b** different transmission capacity at 80 km SSMF transmission case

of the DAC, the signals without pre-equalization cannot be recovered at all, so the pre-equalization is requisite. Therefore, on the basis of frequency domain pre-equalization, post-equalization, and smoothing, the influence of the combination of one or more algorithms of the other three algorithms on system performance are compared. The contributions of different equalization algorithms are compared and summarized in Fig. 14.33a. Adopting pre-equalization, the 32-GBd signals can be transmitted over 30-km. With the aid of pre-equalization and DD-LMS, it can be transmitted over 80 km. 160-km SSMF transmission can be achieved with the assistance of pre-equalization and nonlinearity equalization. After employing these three algorithms, 320-km SSMF transmission can be realized. For this system, nonlinear compensation plays an important role, followed by DD-LMS. Figure 14.33b shows the BER versus different transmission capacity at the 80-km SSMF transmission case. In this case, the modulation format is 16 QAM, and the signal bandwidth is 25 GBd, 28 GBd, 30 GBd, and 36 GBd, respectively. So the corresponding transmission data rate is 100 Gb/s, 112 Gb/s, 128 Gb/s, and 144 Gb/s, respectively. It can be seen from the figure that the BER of the 144 Gb/s signal transmitted through the 80 km single-mode fiber can also be lower than the BER threshold of the hard decision forward error correction code of 2×10^{-2}. To the best of our knowledge, this is also the maximum rate of single-wavelength single-polarization after 80 km transmission in a direct modulation direct detection system.

The mismatch between the laser frequency and optical filter central frequency will impact the system performance, which is experimentally investigated. In this measurement, the bandwidth of the optical filter is fixed at 36 GHz, the center frequency of the laser is also fixed, and the center frequency of the tunable filter is adjusted for testing. The measured results are shown in Fig. 14.34. As the left sideband is selected in this demonstration, carrier-to-signal-power-ratio will be reduced with the increasing of the optical filter central frequency. To a certain value, the optical carrier will be filtered out and the optic-to-electric conversion at the PD is dominated by the signal-to signal beating, which will severely impact the signal. As shown in

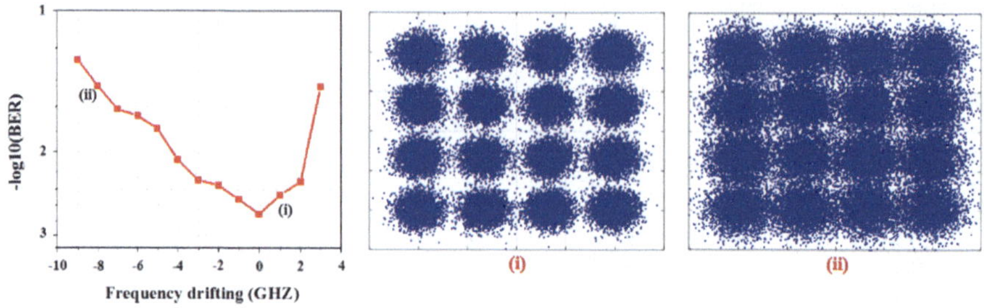

Fig. 14.34 BER vs. optical central frequency drifting

Fig. 14.34, the center frequency of the TOF increases by more than 2 GHz, the BER increases sharply, and the performance deteriorates severely. On the other hand, as the frequency of the TOF center decreases, the high frequency portion of the left band is gradually filtered out. Moreover, the low frequency part of the signal on the right side will gradually increase, which will cause the frequency fading effect similar to the double-sideband signal beat frequency. For a single sideband system with the left sideband selected, the reduction in the center frequency of the filter, i.e., the degradation of the dispersion introduction, is much less than the performance degradation caused by the filtering of the optical center carrier (the drop in CSPR). From Fig. 14.34, we can find the tolerance of the central frequency drifting is from −7 to 2 GHz, which is acceptable for recently commercially used laser and optical filter. Inset (i) and (ii) in Fig. 14.34 shows the constellations of 1 and −8 GHz frequency shifting of 32-GBd DFT-S OFDM signals. As can be seen from the figure, the constellation points at 1 GHz are concentrated, while the constellation at −8 GHz is blurred.

Due to the high cost of the tunable filter for the system, interleave with a fixed channel spacing of 50 GHz can also be used instead of the tunable optical filter. As the wavelength response of the interleave is fixed, the wavelength of ECL should be finely adjusted to match the response before the following measurement. Figure 14.35

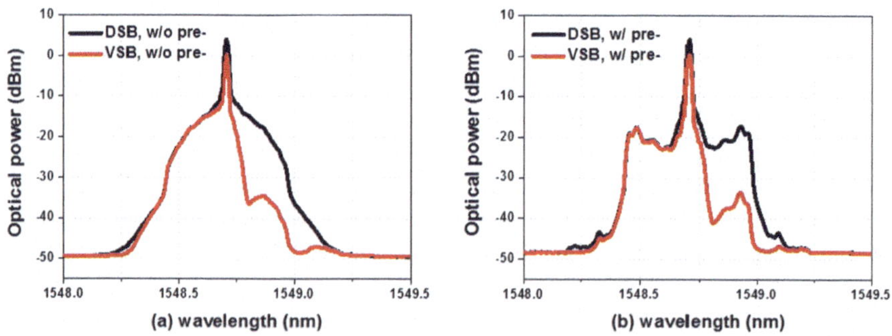

Fig. 14.35 Optical spectra of DFT-S OFDM signals before and after interleave at the cases of **a** without pre-equalization and **b** with pre-equalization

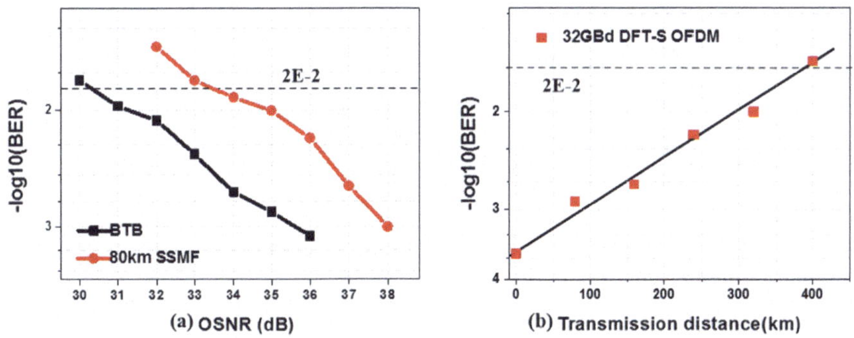

Fig. 14.36 BER versus **a** OSNR at BTB and 80 km SSMF transmission with interleave filter and **b** transmission distance for 32 GBd DFT-S OFDM signals

shows the spectrum before and after the Interleave filter. The measured optical spectra without and with pre-equalization are shown in Fig. 14.35a, b, respectively. The resolution for these optical spectra is set to 0.02 nm. From Fig. 14.35a, b, we can find the suppression ratio of the right band is about 18 dB, and the right part cannot be totally suppressed, which is determined by the extinct ratio of the interleave.

Figure 14.36a shows the measured BER versus OSNR. The test results were measured in optical back-to-back and 80-km fiber transmission. To meet the SD-FEC threshold of 2×10^{-2}, the required OSNR at BTB and 80-km SSMF transmission cases are 30 and 33.5 dB, which maintains the same performance with the OSNR requirement at TOF case. The 80-km SSMF transmission will induce 3.5 dB OSNR penalty. The transmission distance is 0, 80, 160, 240, 320, and 400 km, respectively. As the distance increases, the BER also increases. After 320-km SSMF transmission, the BER can be under the SD-FEC limit of 2×10^{-2}. These results fully confirm that a single sideband modulation system can achieve good transmission distance and transmission capacity after adopting a series of digital signal processing algorithms.

14.5 High-Speed Wavelength Division Multiplexing System

The previous section introduced a single-wavelength 128 Gb/s 320 km fiber transmission system based on advanced digital signal processing algorithms. This section will be expanded into a wavelength division multiplexing system based on the previous section. Figure 14.37 shows the experimental setup for 4 × 128-Gb/s WDM 16 QAM DFT-S OFDM signals with direct detection over 320-km SSMF transmission. At the transmitter, four ECLs with the linewidth less than 100 kHz and the output power of 13 dBm is used for 4 WDM channels. The operating wavelengths of ECL1 to ECL4 are ranging from 1541.0 to 1544.6 with carrier spacing of 150 GHz (The

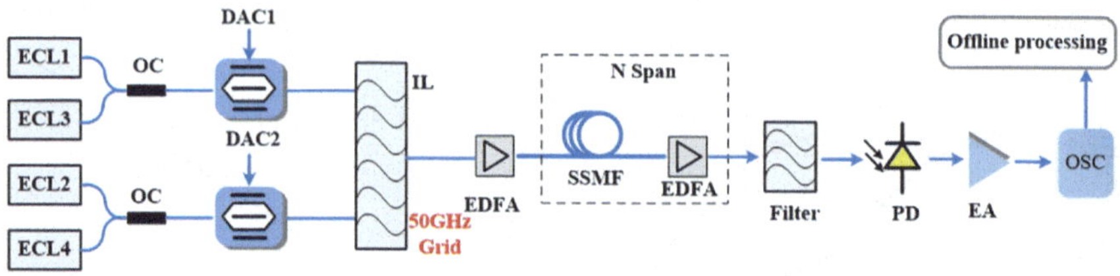

Fig. 14.37 Experimental setup for 4 × 128-Gb/s WDM 16 QAM DFT-S OFDM signals with direct detection over 320-km SMF-28 transmission

channel spacing can be narrower, but considering the frequency response characteristics of Interleave, this value is chosen). The odd (ECL 1 and 3) and even (ECL 2 and 4) channels are combined with two sets of polarization-maintaining optical couplers (PM-OCs) before intensity modulation. The modulator has the same bandwidth as the modulator used in the previous single-wavelength experiment, with a bandwidth of 36 GHz and an insertion loss of 5 dB.

After intensity modulation, four 32-GBd optical 16 QAM DFT-S OFDM signals are spectrally filtered and combined using an IL with 50-GHz grid.

Before being transmitted over 4 spans of 80-km SSMF with EDFA-only amplification in the absence of optical dispersion compensation, the output optical signals from IL are amplified to 9 dBm using an EDFA to compensate for modulation and insertion loss. After fiber transmission, a narrow bandwidth TOF is implemented to select the desired channel and adjust the suppression ratio to render the system work at the optimal condition. The detailed signal processing of WDM system are the same as the single-channel transmission system depicted in Fig. 14.29.

The optical spectra at different points of the overall system are shown in Fig. 14.38. Figure 14.38a, b shows the optical spectra of four wavelength channels before and after 32-GBd DFT-S OFDM signals modulation, respectively. Figure 14.38c shows the optical spectrum for channel 4 of 128-Gb/s DFT-S OFDM signals after 320-km SSMF transmission. The resolutions for the measured optical spectra in Fig. 14.38 are 0.02 nm.

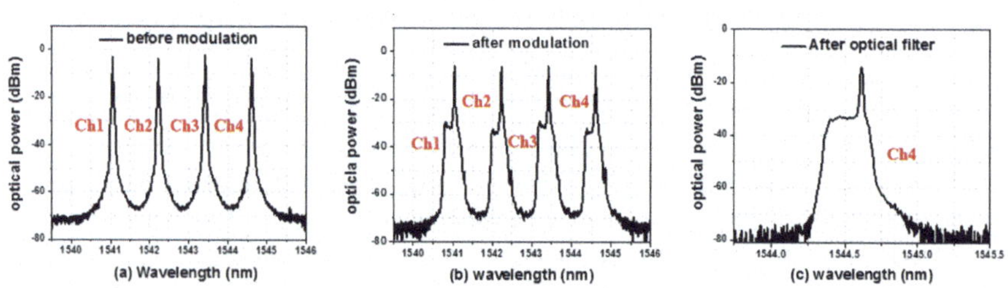

Fig. 14.38 Optical spectra for **a** four channels before modulation, **b** four channels after modulation, and **c** channel 4 after optical filter at the receiver side

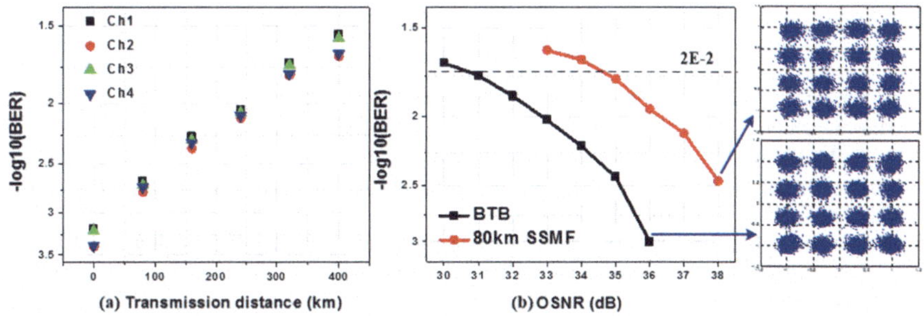

Fig. 14.39 BER performance versus **a** transmission distance for four channels and **b** OSNR for channel 4 at BTB and 80 km SSMF transmission cases

Finally, we have measured the BER performances versus transmission distances for each wavelength channel. The four curves are depicted in Fig. 14.39a. By adjusting the bias voltages of modulators and the TOF, the performances of the four channels can be almost the same. Each channel can realize 128 Gb/s 16QAM DFT-S OFDM signal over 320-km SSMF transmission, and after 400-km SSMF transmission the BERs are around SD-FEC threshold of 2×10^{-2}. The required OSNR for channel 4 at BTB and 80-km SSMF transmission cases are also measured and the results are shown in Fig. 14.39b. From this figure, we can find 3.5-dB OSNR penalty will be induced by 80-km SSMF transmission compared to the BTB case. The constellations after equalizations are also inserted in Fig. 14.39b. These results clearly show the feasibility of the proposed 16 QAM DFT-S OFDM modulation scheme with algorithms for beyond 100 Gbit/s with 320-km reach.

14.6 Conclusions

This chapter systematically introduces the effects of insufficient bandwidth, strong nonlinear effects and fiber dispersion in the intensity modulation direct detection system. And the corresponding solutions are proposed from the aspects of system structure, modulation and coding technology, and digital signal processing algorithms. By utilizing the low-peak-to-average power ratio, the high frequency-efficiency modulation technique DFT-S OFDM introduced in this chapter combines a series of advanced digital signal processing algorithms, including pre-equalization, nonlinear compensation, and DD-LMS to achieve 100 Gb/s single sideband signal transmitted 320 km in a single-mode fiber. To the best of our knowledge, this is the farthest transmission distance of 100 Gb/s in the international IM/DD system. The 144 Gb/s single wavelength single-polarization transmission is also the highest rate IM/DD system over 80 km single-mode fiber in the international. At the same time, it is verified by experiments that the 4×128-Gb/s WDM signal is successfully transmitted in a single-mode fiber for 320 km.

References

1. K. Zhong, X. Zhou, T. Gui et al., Experimental study of PAM-4, CAP-16, and DMT for 100 Gb/s short reach optical transmission systems. Opt. Express **23**(2), 1176 (2015)
2. J. Yu, M. Huang, D. Qian, L. Chen, G.K. Chang, Centralized lightwave WDM-PON employing 16-QAM intensity modulated OFDM downstream and OOK modulated upstream signals. IEEE Photon. Technol. Lett. **20**(18) (2008)
3. M. Chagnon, M. Osman, M. Poulin et al., Experimental study of 112 Gb/s short reach transmission employing PAM formats and SiP intensity modulator at 1.3 μm. Opt. Express **22**(17), 21018–21036 (2014)
4. Y. Wang, J. Yu, N. Chi, G.K. Chang, Experimental demonstration of 120-Gb/s Nyquist PAM8-SCFDE for short-reach optical communication. Photon. J. **7**(4) (2015)
5. M.I. Olmedo, T. Zuo, J.B. Jensen et al., Multiband carrierless amplitude phase modulation for high capacity optical data links. IEEE Lightwave Technol. J. **32**(4), 798–804 (2014)
6. T. Tanaka, M. Nishihara, T. Takahara, et al, Experimental demonstration of 448-Gbps + DMT transmission over 30-km SMF, in *Proceedings of Optical Fiber Conference (OFC) 2014*, paper M2I.5
7. F. Li, X. Li, J. Zhang, J. Yu, Transmission of 100-Gb/s VSB DFT-spread DMT signal in short-reach optical communication systems. Photon. J. **7**(5) (2015)
8. Q. Hu, D. Che, Y. Wang, W. Shieh, Advanced modulation formats for high performance short-reach optical interconnects. Opt. Express **23**(3), 3245 (2015)
9. S. Randel, D. Pilori, S. Chandrasekhar, G. Raybon, P. Winzer, 100-Gb/s discrete-multitone transmission over 80-km SSMF using single-sideband modulation with novel interference-cancellation scheme, in *Proceedings of ECOC 2015, Mo4.5.2*
10. R. Okabe, B. Liu, M. Nishihara, T. Tanaka, T. Takahara, L. Li, Z. Tao, J.C. Rasmussen, Unrepeated 100 km SMF transmission of 110.3 Gbps/lambda DMT signal, in *Proceedings of ECOC 2015, P5.18*
11. L. Zhang, T. Zuo, Q. Zhang, E. Zhou, G. N. Liu, X. Xu, Transmission of 112-Gb/s + DMT over 80-km SMF enabled by twin-SSB technique at 1550 nm, in Proceedings of ECOC 2015, We4.6.4
12. Y. Wang, Demonstration of 575-Mb/s downlink and 225-Mb/s uplink bi-directional SCM-WDM visible light communication using RGB LED and phosphor-based LED, Optics Express **21**(1), 1203–1208
13. W. Yan, et al., 100 Gb/s optical IM-DD transmission with 10G-class devices enabled by 65 G Samples/s CMOS DAC core, in *Presented at OFC 2013, Paper OM3H*
14. K. Ying, Z. Yu, R. J. Baxley, G.T. Zhou, Nonlinear distortion mitigation in visible light communications. Wireless Commun. **22**(2) (2015)
15. Y. Wang, X. Huang, J. Zhang, Y. Wang, N. Chi, Enhanced performance of visible light communication employing 512-QAM N-SC-FDE and DD-LMS. Opt. Express **22**(13), 15328 (2014)
16. W. Yan, B. Liu, L. Li, Z. Tao, T. Takahara, J.C. Rasmussen, Nonlinear distortion and DSP-based compensation in metro and access networks using discrete multi-tone, in *Presented at ECOC 2012*, Paper No. 1.B.2

Chapter 15
High-Speed Fiber Access System Based on Direct Detection of I/Q Modulation

15.1 Introduction

In the previous chapter, the high-speed fiber access system based on direct detection of intensity modulation is introduced in detail. This system is simple in structure, low in cost, and convenient in implementation. It is indeed a promising technology in short-distance fiber access systems. However, the left and right sidebands produced by this technique carry the same signal and do not fully utilize the full effective bandwidth around an optical carrier. Single-sideband signals can be generated using an I/Q modulator or a dual-arm Mach-Zehnder modulator (DD-MZM). The spectral efficiency of this method can be improved compared to the previous one, but only one sideband can be generated around one optical carrier. This chapter introduces a scheme based on an I/Q modulator to generate two sidebands carrying different information around the same optical carrier. At the receiving end of the scheme, a tunable optical filter is used to separate the two independent sidebands. The cooperative digital signal processing is performed after direct detection and data acquisition, respectively. With this scheme, the rate can be doubled compared with the above two.

The I/Q imbalance caused by the mismatch of the signal strength, phase, and delay of the I and Q signals in the I/Q modulator will cause the Image effect, that is, the left sideband signal will be mapped to the right. Similarly, the right sideband signal will be mapped to the left, and the aliasing of the signal will seriously affect the performance of the system. This chapter theoretically deduces the impact of Image phenomenon on system performance, and proposes two image elimination algorithms based on training sequence and adaptive blind equalization, which can be used for frequency-domain equalization systems, such as OFDM, DFT-S OFDM, SC-FDE, etc., or systems for time-domain equalization, such as Subcarrier QAM, CAP, etc.

J. Yu and N. Chi, *Digital Signal Processing In High-Speed Optical Fiber Communication Principle and Application*,
https://doi.org/10.1007/978-981-15-3098-2_15

It is verified by experiments that both equalization algorithms can effectively eliminate the influence of Image effect, and can successfully generate and receive 64 Gb/s CAP4 of two sidebands. In addition, combined with the pre-equalization, DD-LMS and nonlinear compensation algorithms mentioned in the previous chapter, 16QAM DFT-S OFDM signals with two sidebands with a total rate of 240 Gb/s are successfully realized for 160 km.

15.2 Independent Sideband Modulation Direct Detection System Based on I/Q Modulator

To generate two optical independent sidebands (O-ISBs) around the same optical carrier, a dual-arm Mach-Zehnder modulator (DD-MZM) or an I/Q modulator can be used. The generation of independent sidebands based on DD-MZM requires a Hilbert transform on the signal, which has been reported in [5]. This section describes an independent sideband generation scheme based on I/Q modulators. The block diagram is shown in Fig. 15.1. Figure 15.1a shows the principle of independent sideband generation based on I/Q modulator. The transmitter of the system includes external cavity laser, electric amplifier, and I/Q modulator. As can be seen from the figure, the I/Q modulator includes three integrated sub-modulators, MZ-a, MZ-b, and

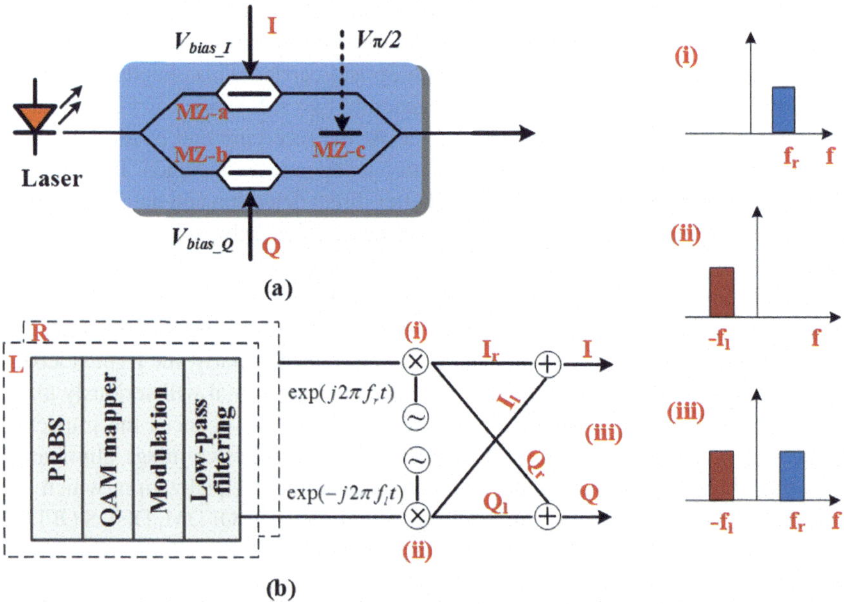

Fig. 15.1 **a** Schematic diagram of independent sideband generation based on I/Q modulator; **b** Flow chart of generation of left and right side charged signals

MZ-c, where MZ-c is biased at the orthogonal point to ensure the phase difference between the two arms is 90°. the bias of MZ-a and MZ-b is such that the power ratio of the center subcarrier and the two sidebands is around 20 dB. Figure 15.1b shows the flow generation process for the left and right bandside. First, each is modulated by QAM mapping and modulation, such as CAP, OFDM, etc.

Two different baseband signals can be obtained after modulation, which is represented as $A(t)\exp(j\varphi_l(t))$ and $B(t)\exp(j\varphi_r(t))$, respectively, where $A(t)$ and $B(t)$ are the amplitudes of the baseband signals, $\varphi_l(t)$ and $\varphi_r(t)$ are the phase of the baseband signal. Then upload the complex RF signal to the left and right sidebands. The left and right sidebands $E_l(t)$ and $E_r(t)$ can be as follows:

$$E_l(t) = A(t)\exp(j\varphi_l(t))\exp(-j2\pi f_l t). \tag{15.1}$$

$$E_r(t) = B(t)\exp(j\varphi_r(t))\exp(j2\pi f_r t). \tag{15.2}$$

where f_l and f_r are the intermediate frequency signals of the left and right sidebands and both are positive values. $-f_l$ indicates that the signal is uploaded to the negative frequency of the center carrier. The electrical signal generated after the simple addition of the two signals is a signal containing two sidebands, which can be written as

$$E(t) = E_l(t)+E_r(t). \tag{15.3}$$

The diagrams of right and left sideband signals are shown in the illustrations (i) and (ii) in Fig. 15.1, respectively. The schematic diagram containing the two sidebands after the addition is shown in Fig. (iii). Then, the signals after addition are taken as the real part and the imaginary part, respectively, and used as the driving signals of the I and Q channels of the I/Q modulator. They are loaded onto the stable and continuous laser generated by the external cavity laser, and in this way, an optical signal is generated that contains two separate sidebands. Figure 15.2 shows the spectrum of single right sideband, single left sideband, and the two independent sidebands, where the left and right sidebands are named in frequency scale, so the left and right are opposite in terms of wavelength.

At the receiving end, the beam is first split into two paths via a 3 dB optical coupler (OC), and two single optical filters are required to separate the single-wavelength double-sideband signal, as shown in Fig. 15.3. Filter 1 and Filter 2 are used to separate the right sideband signal and the left sideband signal, respectively, where the filter can be a tunable wavelength filter or Interleaver. Since the response of the filter is unlikely to be the same as shown in the figure, whose roll-off factor is 0, a certain frequency interval needs to be left in the middle of the double-sideband signal. The appropriate interval width is related to the signal bandwidth and the filter response. The power of the center carrier can be adjusted by adjusting the bias voltages of the two modulators of MZ-a and MZ-b, and the carrier-to-signal ratio (CSPR) is maintained at about 20 dB. The left and right sideband signals separated by the filter

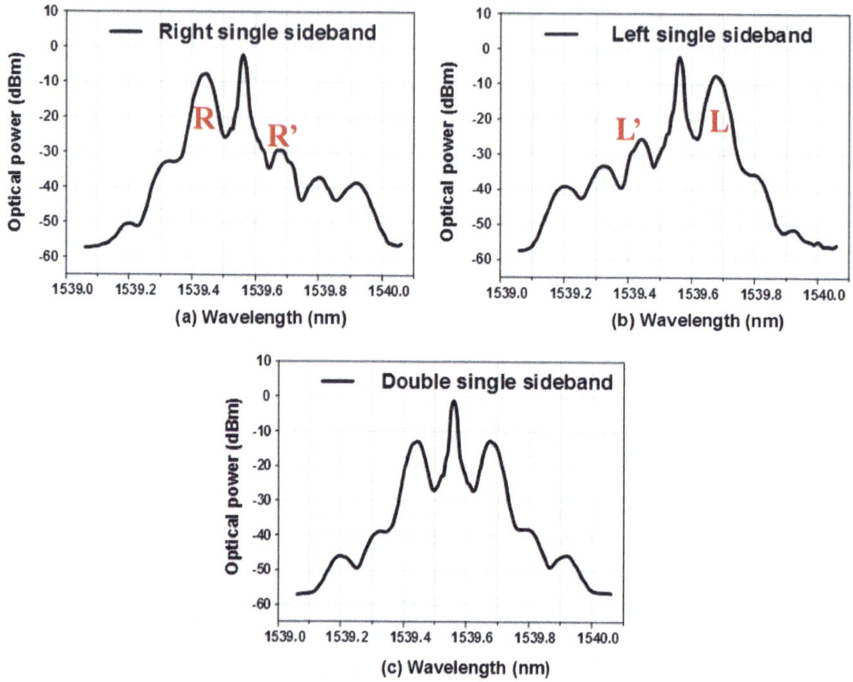

Fig. 15.2 **a** Spectrum of right sideband, **b** spectrum of left sideband, **c** spectrum on the left and right sidebands

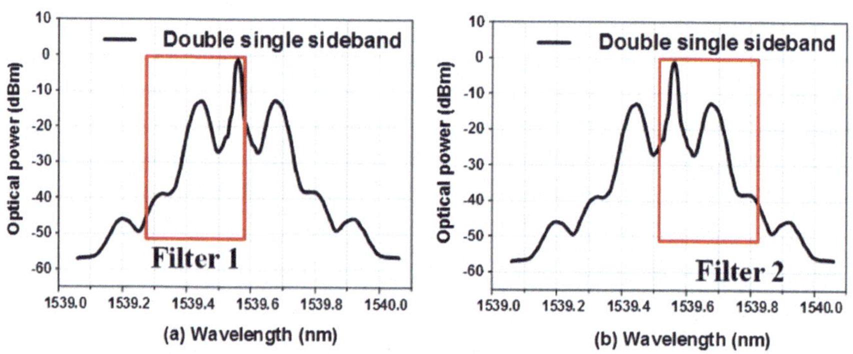

Fig. 15.3 **a** Right sideband signal separation, **b** Left sideband signal separation

and the center carrier can generate electrical signals of the corresponding sidebands after frequency beaten by the directly detected detectors, and respectively, processed to restore the original signals. In this way (O-ISB), using a modulator and a set of detectors, the amount of information that can be transmitted is twice that of a single-sideband (SSB) scheme or a double-sideband scheme (DSB) in theory.

However, there are some limitations in this scheme. Besides the fact that a certain bandwidth interval must be reserved between the two sideband signals, another serious influencing factor is the mismatch of amplitude, phase and delay parameters between the I and Q modulators, which causes the Image effect. That is, the left sideband signal will be mapped to the right sideband. Similarly, the right sideband signal is mapped to the left band. The Image effect can be observed in Fig. 15.2a, b, where L and R represent the left and right sideband signals, L' and R' represent Image signals generated on the right and left sides of the left and right sidebands, respectively. In this actual detection, the left sideband link collects the added signal of L and R', and the right sideband link collects the added signal of R and L'. The aliasing of the signal can seriously affect the performance of the system.

Assume that the frequency domain of the baseband signal can be expressed as $X(f)$, the time domain can be expressed as $x(t)$. The signal after uploading to the intermediate frequency of left sideband can be expressed as $X(f+f_l)$. Due to the Image effect, a mirrored signal is generated on the right sideband, which can be expressed as $(-X(f-f_l))^*$. That is, a signal that is symmetric and conjugate with the left sideband signal that is generated. According to the principle of Fourier transform, the signal generated in the right sideband is the conjugate signal of the original signal in the time domain, i.e., $(x(t))^*$. According to the characteristics of the Image signal in the time domain and the frequency domain, a corresponding Image elimination algorithm can be developed. This chapter proposes two algorithms based on training sequence and adaptive blind equalization to eliminate the effect of Image effect and to improve system performance.

15.3 Image Elimination Algorithm Based on Training Sequence

15.3.1 Principle of Image Elimination Algorithm Based on Training Sequence

As discussed above, from the time domain, the resulting Image signal is the conjugate signal of the original left/right sideband signal. The electrical signal of the double-sideband generated by the intensity detector and the real part of the original sideband complex signal, i.e., $\mathrm{Re}(A(t)\exp(j\varphi_l(t))\exp(-j2\pi f_l t))$ or $\mathrm{Re}(B(t)\exp(j\varphi_r(t))\exp(j2\pi f_r t))$, are completely consistent. With this property, Image noise can be removed with a training sequence.

In addition, in this system, the same nonlinear effects as those discussed in the previous chapter are also generated. The nonlinear effects also come from devices such as electrical amplifiers, modulators, optical fibers, optical filters, and square-law detectors. The effects of nonlinear effects on signals are severe, especially high-order modulated signals. This chapter proposes a joint Image and nonlinear effect elimination algorithm. Using this algorithm, Image effect and nonlinear effect can

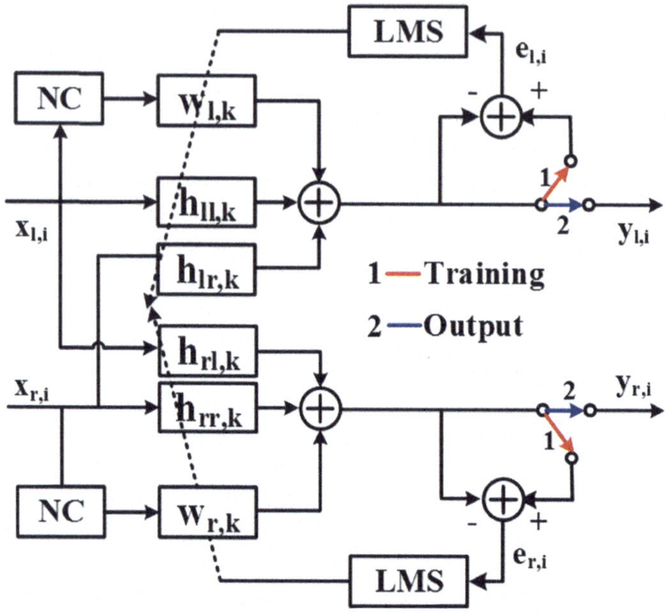

Fig. 15.4 Joint image and nonlinear effect elimination algorithm framework

be eliminated at the same time, which can greatly improve the performance of the system. The structure of the algorithm is shown in Fig. 15.4. It can be seen from the figure that the algorithm is a butterfly structure, and there are six real-adaptive equalization FIR filters in the algorithm, the coefficients are h_{ll}, h_{lr}, h_{rl}, h_{rr}, w_l, w_r, respectively, where h_{ll} and h_{rr} are linear damage terms of their own sideband signals, h_{lr} and h_{rl} are the Image damage items from the right and left sidebands, w_l and w_r are the noise terms of their own sidebands, and NC represents the construction of nonlinear terms. The specific equalization process can be divided into two levels. The first uses the training sequence to update the coefficients of the FIR filter, obtains the stable coefficients, and then performs the second-level equalization, and uses the updated coefficients to perform unified operations on the subsequent sequences. The equalization based on the training sequence will make the whole process converge fast.

x_l and x_r are the left and right sideband signals collected by the oscilloscope. They are also the input of the proposed Image and nonlinear equalizer. After the first level of the equalizer, the output is as follows:

$$y_{l,i} = \sum_{k=-M}^{M} h_{ll,k} x_{l,i-k} + \sum_{k=-M}^{M} h_{lr,k} x_{r,i-k} + \sum_{p=-N}^{N} \sum_{k=-N}^{p-1} w_{l,k} (x_{l,i-p} - x_{l,i-k})^2 \tag{15.4}$$

$$y_{r,i} = \sum_{k=-M}^{M} h_{rl,k}x_{l,i-k} + \sum_{k=-M}^{M} h_{rr,k}x_{r,i-k} + \sum_{p=-N}^{N}\sum_{k=-N}^{p-1} h_{r,k}(x_{r,i-p} - x_{r,i-k})^2 \tag{15.5}$$

$y_{l,i}$ and $y_{r,i}$ representing the ith left and right sidebands signal after equalization, M and N represent the order of the linear and nonlinear equalization filters, respectively. After comparing with the training sequence, the error values obtained are as follows:

$$e_{l,i} = y'_{l,i} - y_{l,i} \tag{15.6}$$

$$e_{r,i} = y'_{r,i} - y_{r,i} \tag{15.7}$$

$y'_{l,i}$ and $y'_{r,i}$ represent the left and right training sequences, respectively. The Least mean square (LMS) algorithm is used to update the coefficients of the filter, where the linear terms are updated as follows:

$$h_{ll,i+1} = h_{ll,i} + \mu e_{l,i}x_{l,i-k} \tag{15.8}$$

$$h_{lr,i+1} = h_{lr,i} + \mu e_{l,i}x_{r,i-k} \tag{15.9}$$

$$h_{rl,i+1} = h_{rl,i} + \mu e_{r,i}x_{l,i-k} \tag{15.10}$$

$$h_{rr,i+1} = h_{rr,i} + \mu e_{r,i}x_{r,i-k} \tag{15.11}$$

The value of μ can be adjusted according to the order and the actual effect, and the coefficient of the nonlinear term is also updated according to the LMS. In this way, six different FIR coefficients can be obtained, and the six FIR filters are used to equalize the subsequent data.

15.3.2 Experimental Setup

Figure 15.5 shows the system schematic and experimental setup of the direct detection of O-ISB modulation based on I/Q modulator. In this experiment, the left and right sidebands each carry a 30GBaud 16QAM DFT-S OFDM signal, resulting in a total transmission rate of 240 Gb/s. The generation process of the electrical signal is shown in Fig. 15.5a. The generation process is similar to the one in the previous section. The only difference is the DFT-S OFDM signal generation process, which will not be described here. The resulting two independent sidebands 30GBaud 16QAM DFT-S OFDM signals are introduced into the 80GS/s DAC and converted into analog signal outputs, which are amplified by the amplifier and used as I/Q modulator I and Q drive

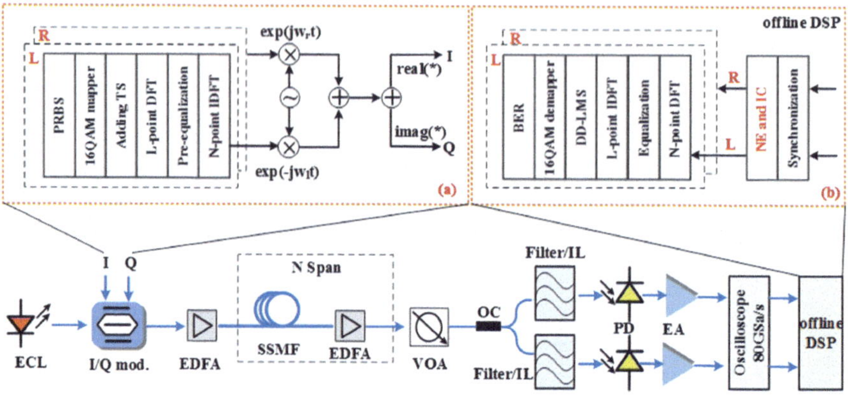

Fig. 15.5 Experimental setup of I-Q modulator based O-ISB DFT-S OFDM modulation direct detection system

signals. It is loaded onto a continuous wavelength (CW) lightwave generated by an external cavity laser (ECL) with an operating wavelength of 1550.1 nm and output power of 13 dBm. The lightwave modulated by the above signal is amplified by an erbium-doped fiber amplifier (EDFA) and then transmitted in a single-mode fiber. The signal is amplified by an EDFA for every 80 km transmission. At the receiving end, it is first divided into two paths of light through a 3 dB optical coupler (OC), each of which needs to pass through the tunable optical filter to select the required left/right sideband signals, and then sent to the intensity detector for photoelectric conversion. The photoelectrically converted electrical signal is processed by an oscilloscope with a sampling rate of 80 GS/s and a maximum bandwidth of 36 GHz, and then performs back-end data processing.

As shown in Fig. 15.5b, the back-end data processing flow firstly performs frame synchronization on the collected data, and then uses the algorithm proposed in this chapter to perform image elimination and nonlinear compensation. After compensation, the left sideband and the right sideband are separated. The data is taken and processed separately, and the process is completely opposite to the DFT-S OFDM modulation process. It is worth noting that in this experimental system, no dispersion compensation algorithm, dispersion compensation fiber, or even cyclic prefix is used at all to eliminate the influence of dispersion.

The spectrum of the test at different operating points in the experimental system is shown in Fig. 15.6. Figure 15.6a–d, respectively, identifies the spectrum of the output of the I/Q modulator, passing the left sideband filter, passing the right sideband filter, and the double-sideband after 80 km of fiber transmission. By adjusting the bias of the I/Q modulator and the center wavelength of the filter, the CSPR of the single-sideband can be guaranteed to be at an optimum value, which is guaranteed to be around 20 dB in this experiment. It can be seen from the figure that after 80 km of fiber transmission, the optical signal-to-noise ratio (OSNR) of the signal will decrease to a considerable extent, which can be seen from the performance of the bit error rate.

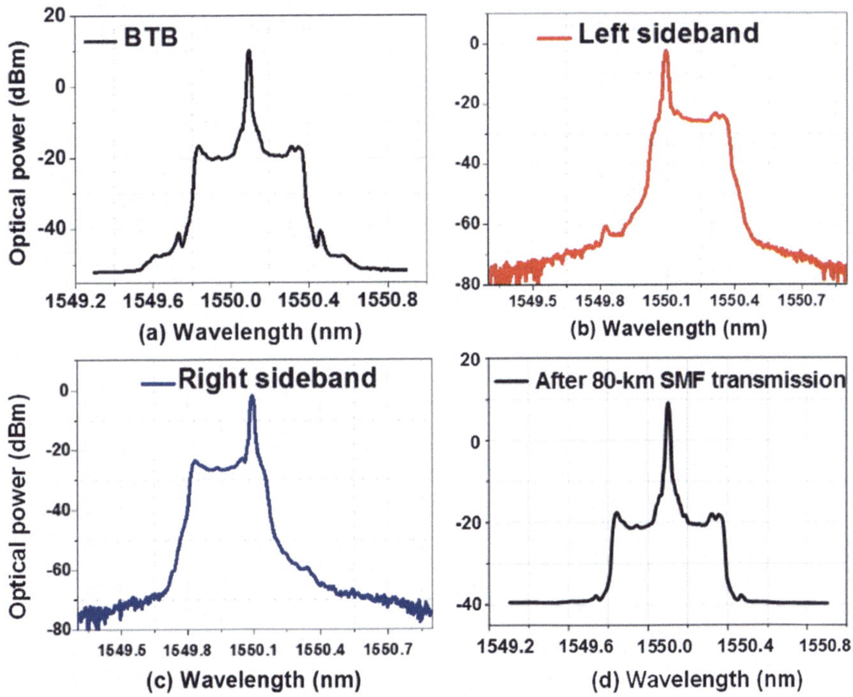

Fig. 15.6 **a** Spectrum of the double-sideband at the back-to-back transmission end, **b** Spectrum of the left sideband at the back-to-back receiving end, **c** Spectrum of the right sideband at the back-to-back receiving end, **d** Spectrum of the double-sided band after 80 km of fiber transmission

15.3.3 Experimental Result

First, load the single-sideband signal with only the left sideband or only the right sideband to adjust the system parameters. By adjusting the bias and delay of the I/Q modulator, the Image effect on the left and right sides is minimized, and the spectrums of the left and right sidebands are shown in Fig. 15.7a, b. It can be seen from the figure that the component and noise power ratio of the Image signal is about 10 dB, and the power ratio of the signal to the Image is about 20 dB. As can be seen from the values, Image can have a serious impact on the system.

Then we compare the improvement of system performance by different algorithms, including not using the nonlinear compensation and Image elimination algorithms, only using the nonlinear compensation algorithm and the joint nonlinear compensation and using Image elimination algorithms mentioned in this chapter. The comparison results are shown in Fig. 15.8a, b. Figure 15.8a shows the bit error rate performance as a function of the information carried by the left sideband, and Fig. 15.8b shows the bit error rate performance of the right sideband. The single-sideband information capacity is increased from 60 to 120 Gb/s. It can be seen from

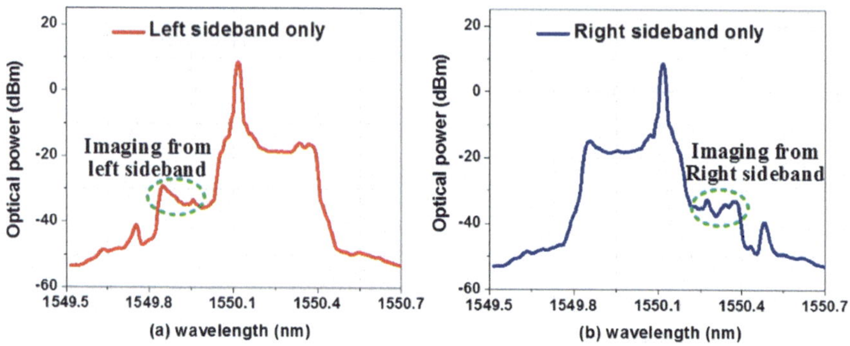

Fig. 15.7 **a** the spectrum of the left sideband modulated, **b** the spectrum of the right sideband modulated

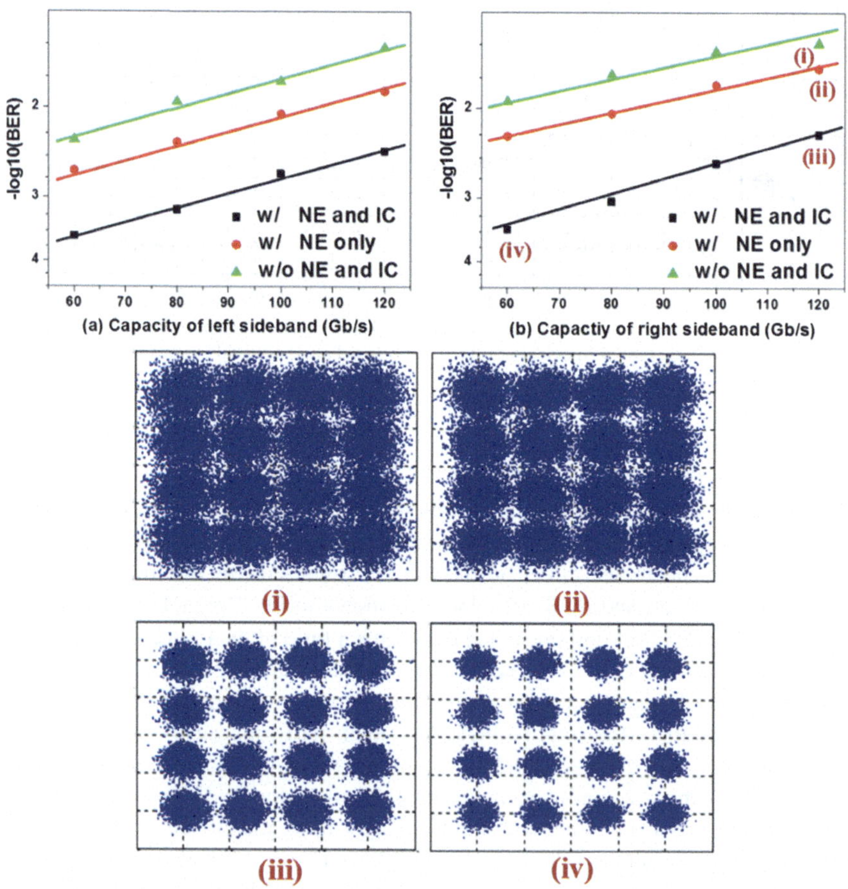

Fig. 15.8 BER performance as a function of transmission capacity with different algorithms **a** left sideband signal and **b** right sideband signal

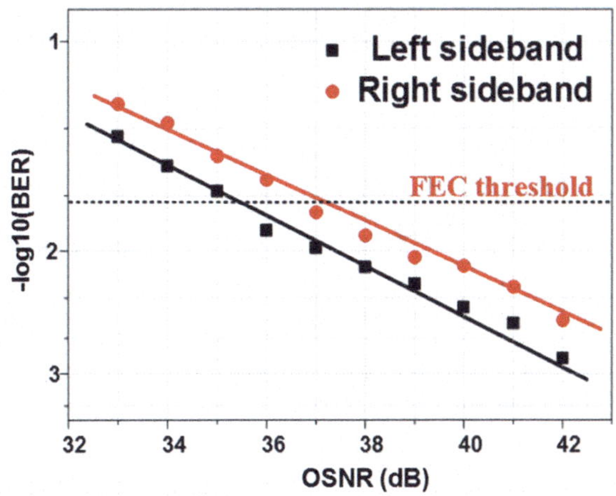

Fig. 15.9 The relationship graph of left and right sideband signal error rate performance and optical signal to noise ratio

the figure that if the nonlinear compensation and Image elimination algorithm is not used, the left/right sideband rate is 60 Gb/s. When the nonlinear compensation algorithm is used alone, the rate can be increased to 80 Gb/s, and the joint nonlinear compensation and Image elimination algorithm can be used to increase the speed to 120 Gb/s. The different transmission rates and constellations recovered with different equalization algorithms are also given in Fig. 15.8. The illustrations (i)–(iii) show the constellations recovered from the 120 Gb/s right sideband signal without the equalization algorithm, using only the nonlinear algorithm and the joint nonlinearity and Image elimination algorithm. Illustration (iv) shows the constellation recovered from 60 Gb/s right sideband signal by joint nonlinear compensation and Image elimination. As can be seen from the figure, the clarity of the constellation of illustration (iv) >illustration (iii) >illustration (ii) >illustration (i). This is also consistent with the value of the bit error rate. These experimental results can verify the effectiveness of the training sequence-based algorithms proposed in this chapter.

Figure 15.9 shows the relationship between the bit error rate performance of the left and right sideband signals and the optical signal to noise ratio. To achieve a 2×10^{-2} bit error rate, the minimum OSNR values required for the left and right sidebands are 36 dB and 37 dB, respectively. The slight difference between the left and right sidebands is mainly due to the inconsistency of the detectors used, and the performance of the detectors on the left sideband is slightly superior.

After the back-to-back (BTB) performance test is completed and the system parameters are adjusted to the optimal state, the fiber transmission capability of this scheme is also evaluated. It is necessary to test the optimum fiber input power. By changing the fiber input fiber power (5–11 dBm), different bit error rate values are tested to obtain the minimum fiber error power corresponding to the bit error rate value. The test results are shown in Fig. 15.10a. The optimal fiber-input power of the left and right sideband signals is 8 dBm. If lower than 8 dBm, the signal-to-noise ratio (SNR) is low, while if above 8 dBm, the system would introduce strong fiber nonlinear effects.

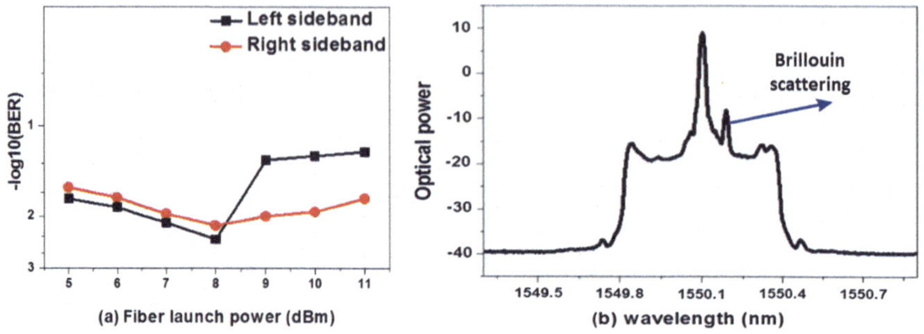

Fig. 15.10 **a** BER performance as a function of incident fiber power and **b** signal spectrum above 8 dBm incident power

The left sideband signal degradation is very serious. This is mainly due to the fact that the Brillouin scattering effect caused by the fiber insertion power is too high, which seriously affects the performance of the system. Figure 15.10b shows the spectrum of the signal after 80 km of fiber transmission at a fiber insertion power of 9 dBm. A Brillouin frequency shift occurs at around 10 GHz from the center subcarrier. Since Brillouin scattering mainly occurs in backscattering, the Brillouin frequency shift occurs below the center subcarrier frequency, so the left sideband signal is very severely affected. In the next tests, the fiber input power was fixed at 8 dBm to make the system work at its best.

Figure 15.11 shows the BER performance before and after transmission over 80 km of single-mode fiber as a function of receiver optical power. As can be seen from the figure, the 80 km optical fiber transmission introduces a 3 dB deterioration in receiver sensitivity regardless of whether the signal is left or right sideband. At the receiver optical power of −3.5 dB or more, the bit error rate of the left and right sideband signals can reach the soft decision FEC threshold. Subsequently, we

Fig. 15.11 Graph of BER performance and receiver optical power before and after transmission of 80 km fiber

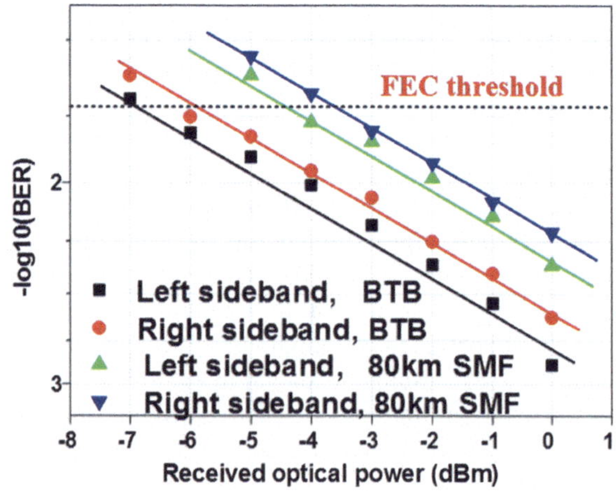

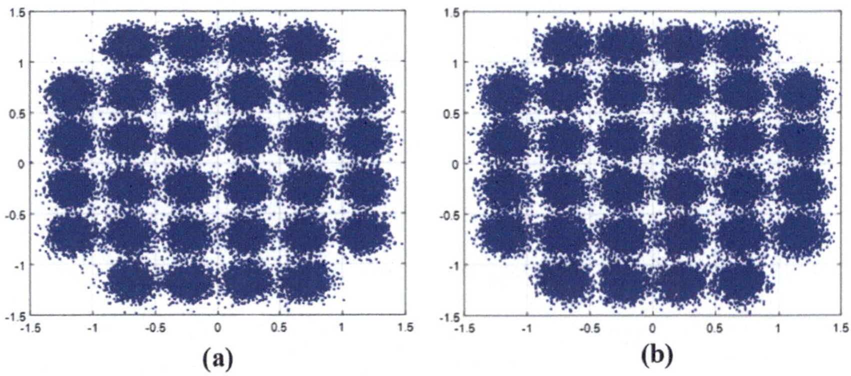

Fig. 15.12 **a** 30 Gbaud left sideband 32QAM constellation and **b** 30Gbaud right sideband 32QAM constellation

extended the length of the fiber to 160 km. After 160 km of fiber transmission, the error rates of the left and right sidebands were 1.34×10^{-2} and 1.57×10^{-2}, respectively, which were also below the soft decision FEC threshold.

We also tested the performance of the higher-order modulation format 32QAM in the system. The experimental block diagram is also shown in Fig. 15.5. The left and right sidebands have a bandwidth of 30Gbaud, so the total rate is 300 Gbps. In the case of back-to-back light, after pre-equalization, image elimination, nonlinear compensation, phase compensation, DD-LMS, and other algorithms, the constellations of the recovered left and right sideband 32QAM DFT-S OFDM are shown in Fig. 15.12a, b, respectively. The bit error rates of the left and right sidebands are 9.4×10^{-3} and 1.05×10^{-2}, respectively, which are also below the soft decision FEC threshold.

Wavelength division multiplexing technology is a key technology for improving the transmission capacity of the system in the access network. The independent sideband modulation based on I/Q modulator in this scheme can realize two sidebands around a single wavelength, and only one set of transmitters is used, thereby saving the overhead of the transmitter. Combined with wavelength division multiplexing technology, the transmission capacity of the system can be naturally expanded. We tested a 4×240 Gb/s wavelength division multiplexing system based on this scheme, and the system schematic is shown in Fig. 15.13.

For the ch1 and ch3 channels, the two continuous wavelength beams generated from the external cavity lasers ECL1 and ECL3 first coupled via a 3 dB optocoupler. It is then modulated by an in-phase/quadrature phase (I/Q) modulator for signals containing sidebands of two 30 GBaud 16QAM DFT-S OFDM. The operating wavelength of ECL1 is 1547.71 nm, the operating wavelength of ECL3 is 1549.31 nm, and the frequency interval between ECL1 and ECL3 is 200 GHz. The same process is also used for the ch2 and ch4 channels. The operating wavelength of ECL2 is 1548.51 nm, the operating wavelength of ECL4 is 1550.11 nm, and the frequency interval between ECL2 and ECL4 is 200 GHz. The spectrum of the four channels

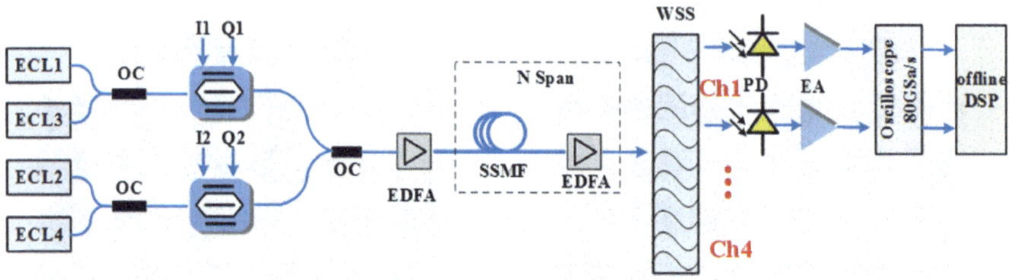

Fig. 15.13 Wavelength division multiplexing system based on independent sideband modulation direct detection

after coupled by a 3 dB optocoupler is shown in Fig. 15.14. The frequency interval between adjacent channels of the four optical waves is 100 GHz. The sum of the left and right sideband bandwidths is about 70 GHz, which is less than the adjacent channel spacing of 100 GHz, so the crosstalk between adjacent channels is substantially negligible. However, the common transmission of multiple channels leads to a decrease in optical OSNR, which affects the performance of the system. In the experimental test, the maximum OSNR of the single channel is 38 dB when four channels simultaneously transmit, and the left and right sideband error rates of Ch1 tested in back-to-back conditions are 0.0070 and 0.0113.

Fig. 15.14 Spectrum after four-channel channel multiplexing

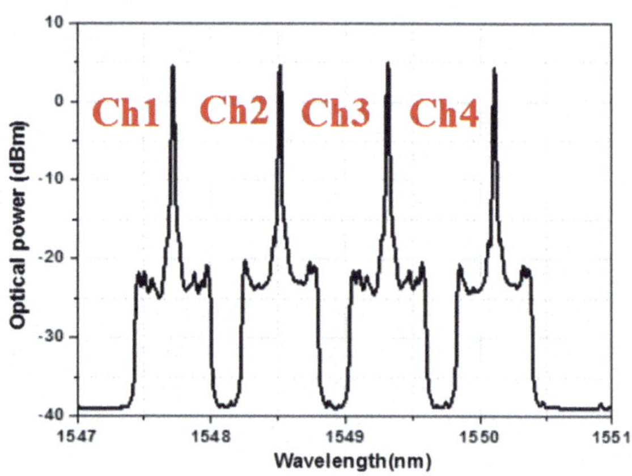

15.4 Image Elimination Algorithm Based on Adaptive Blind Equalization

15.4.1 Principle of Image Elimination Algorithm Based on Adaptive Blind Equalization

The previous section elaborates the image elimination algorithm based on the training sequence. The algorithm acts after synchronization before the signal is down-converted to baseband. It is mainly for communication systems based on training sequence equalization like SC-FDE and variants of this modulation format such as PAM SC-FDE, Nyquist PAM SC-FDE, and OFDM and its modified modulation formats such as DFT-S OFDM, DMT, OFDM/OQAM, etc. For single-carrier such as CAP, QAM, PAM, and adaptive blind equalization based communication systems such as CMA, CMMA, although training sequence equalization can also be adopted, it will increase the overhead of the system, which is not feasible. This chapter proposes an image elimination algorithm that is seamlessly integrated with the current adaptive blind equalization. This algorithm does not require a training sequence after the signal is down-converted to the baseband signal.

The system block diagram is shown in Fig. 15.5. The two independent optical sideband signals modulated by the I/Q modulator are transmitted through the optical fiber, separated by the adjustable center carrier and bandwidth filter at the receiving end, and then passed through two intensity detectors for photoelectric conversion. Assume that the electrical left and right sideband signals collected by the oscilloscope are r_l and r_r, these two signals are resampled separately. After downconverting to baseband, they can be expressed as C_l and C_r. The signal C_l with the left side down-converted to the baseband is composed of two parts, one part is the baseband signal c_l of the initial left sideband, and the other part is the Image signal E of the initial right sideband signal c_r on the left side, $c_r' = (c_r)^*$,where $(\cdot)^*$ represents conjugates. Similarly, the right sideband signal C_r that is down-converted to the baseband is also composed of two parts, which are the baseband signal c_r of the initial right sideband, and the other part is the Image signal c_l' of the initial left sideband signal c_l on the left side, where $c_l' = (c_l)^*$.

This section proposes a joint phase compensation and Image effect elimination algorithm. Using this algorithm, the Image effect and the compensation phase can be eliminated at the same time, which can greatly improve the performance of the system. The structure of the algorithm is shown in Fig. 15.15. It can be seen from the figure that the algorithm is a butterfly structure, and there are four complex adaptive equalization FIR filters in the algorithm, the coefficients are h_{ll}, h_{lr}, h_{rl} and h_{rr} respectively. Where h_{ll} and h_{rr} are the linear phase damage terms of the self-sideband signals, h_{lr} and h_{rl} are the Image damage terms from the right and left sidebands, respectively, and $(\cdot)^*$ represents the construction of the conjugate term. The specific equalization process is to update the coefficients of the FIR filter using CMA or CMMA.

Fig. 15.15 Joint phase compensation and Image effect elimination algorithm framework

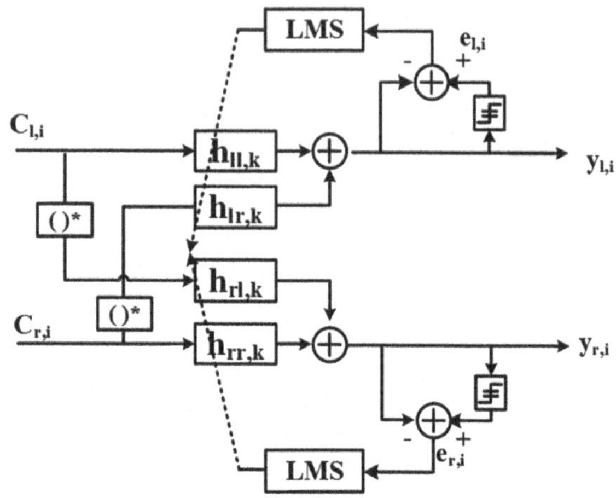

The down-converted left and right sideband signals collected by the oscilloscope are also inputs to the equalizer of the proposed Image and phase compensation. The output after the equalizer is as follows:

$$\mathrm{y}_{l,i} = \sum_{k=-M}^{M} h_{ll,k} C_{l,i-k} + \sum_{k=-M}^{M} h_{lr,k} (C_{r,i-k})^* \tag{15.12}$$

$$\mathrm{y}_{r,i} = \sum_{k=-M}^{M} h_{rl,k} (C_{l,i-k})^* + \sum_{k=-M}^{M} h_{rr,k} C_{r,i-k} \tag{15.13}$$

$\mathrm{y}_{l,i}$ and $\mathrm{y}_{r,i}$, respectively, indicate the ith signal after equalization of the left sideband and the right sideband, and M represents the order of the equalization filter. After using the CMA blind decision, the error values obtained are as follows:

$$\mathrm{e}_{l,i} = R - \mathrm{y}_{l,i} \tag{15.14}$$

$$\mathrm{e}_{r,i} = R - \mathrm{y}_{r,i} \tag{15.15}$$

R is the radius of QPSK. The Least mean square (LMS) algorithm is used to update the coefficients of the filter, and the coefficients of the four complex terms are updated as follows:

$$\mathrm{h}_{ll,i+1} = \mathrm{h}_{ll,i} + \mu e_{l,i} C'_{l,i-k} \tag{15.16}$$

$$\mathrm{h}_{lr,i+1} = \mathrm{h}_{lr,i} + \mu e_{l,i} C_{r,i-k} \tag{15.17}$$

$$\mathrm{h}_{rl,i+1} = \mathrm{h}_{rl,i} + \mu e_{r,i} C_{l,i-k} \tag{15.18}$$

$$h_{rr,i+1} = h_{rr,i} + \mu e_{r,i} C'_{r,i-k} \tag{15.19}$$

The value of μ can be adjusted according to the order and the actual effect. In this way, four different FIR coefficients can be obtained, thereby achieving the elimination of linear phase noise and Image effect.

15.4.2 Experimental Setup and Results

The previous section elaborates the principle of Image elimination algorithm based on adaptive blind equalization, and can be used to compensate phase noise damage. This section mainly verifies the feasibility of the above algorithm through experiments. In this experiment, two independent 16 Gbaud 4-ary carrierless amplitude phase modulation signals at 10 GHz and −10 GHz were generated by an integrated I/Q modulator, so the total rate was 64 Gb/s. The interval between the two sidebands is 4 GHz.

Figure 15.16 shows the experimental setup for direct detection of O-ISB CAP modulation based on I/Q modulators. Similar to Fig. 15.5, the experimental setup is identical, including an external cavity laser, an integrated I/Q modulator, fiber optics, erbium-doped fiber amplifier, two tunable optical filters, and two detectors. The detected electrical signals are sent to an oscilloscope with a sampling rate of 80 GSa/s, followed by back-end digital signal processing. The bias of the modulator is the same as in the previous section, the only difference being that the digital signal processing algorithms and processes at the transmitter and receiver are different. The generation process of the electrical independent sideband modulation signal at the transmitting end is as shown in Fig. 15.16a, including first mapping the random binary data into a QPSK signal, and then 4 times upsampling to achieve a data bandwidth

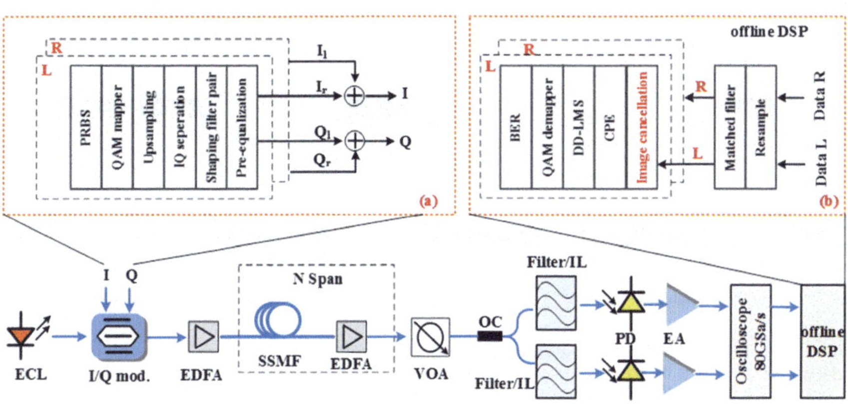

Fig. 15.16 Experimental setup for direct detection of O-ISB CAP modulation based on I/Q modulator

of 16 Gbaud. The baseband signal I and Q are then separated and filtered by a filter that forms a Hilbert transform pair. The center frequency of the filter is 10 GHz and the roll-off factor is 0.1.

In this way, the CAP4 signal can be generated. The I path of the two sidebands is added as the I signal, and the Q path is added as the Q signal, and then the two channels are, respectively, introduced into the digital-to-analog converter (DAC) with a sampling rate of 64GSa/s. The two electrical signal outputs are, respectively, amplified by an independent linear amplifier to a power of about 20 dBm, and then the I and Q paths of the I/Q modulator are driven. Figure 15.17a–c, respectively, shows only the 16 Gbaud CAP4 left sideband signal at −10 GHz, the 16 Gbaud CAP4 right sideband signal at 10 GHz, and the 16 Gbaud CAP4 signal with two separate sidebands.

The optical signal modulated by the I/Q modulator contains two independent sidebands, which are then transmitted to the receiving end through a single-mode fiber. At the receiving end, the two optical sidebands are separated using an optical filter with adjustable bandwidth and two center carriers. After separation, photoelectric conversion is performed using an intensity detector with a bandwidth of 30 GHz. The electrical signal is sent directly into a real-time digital oscilloscope (OSC) with a sampling rate of 80 GSa/s and an electrical bandwidth of 36 GHz. Subsequent digital signal processing (DSP) processes are shown in Fig. 15.16b, including resampling, IF downconversion, Image elimination, phase noise compensation, DD-LMS, QPSK decoding, and BER calculation. The DD-LMS can be used to further compensate for the residual phase of the image effect elimination and phase noise compensation combined with the previous step. Figure 15.18 shows the constellation diagram after processing by different algorithms. Figure 15.18a, b shows the left and right sidebands without the image elimination of the constellation, Fig. 15.18c, d is the constellation of left and right sideband signals after the Image effect is removed. It can be seen from the figure that after using the algorithm proposed in this section, the performance of the system is greatly improved, and the two sideband signals can be received without error.

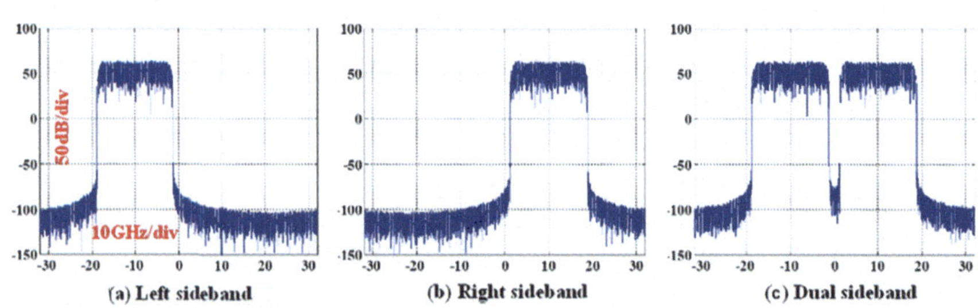

Fig. 15.17 Electrogram of 16 Gbaud CAP4 **a** left sideband, **b** right sideband, **c** left and right sidebands

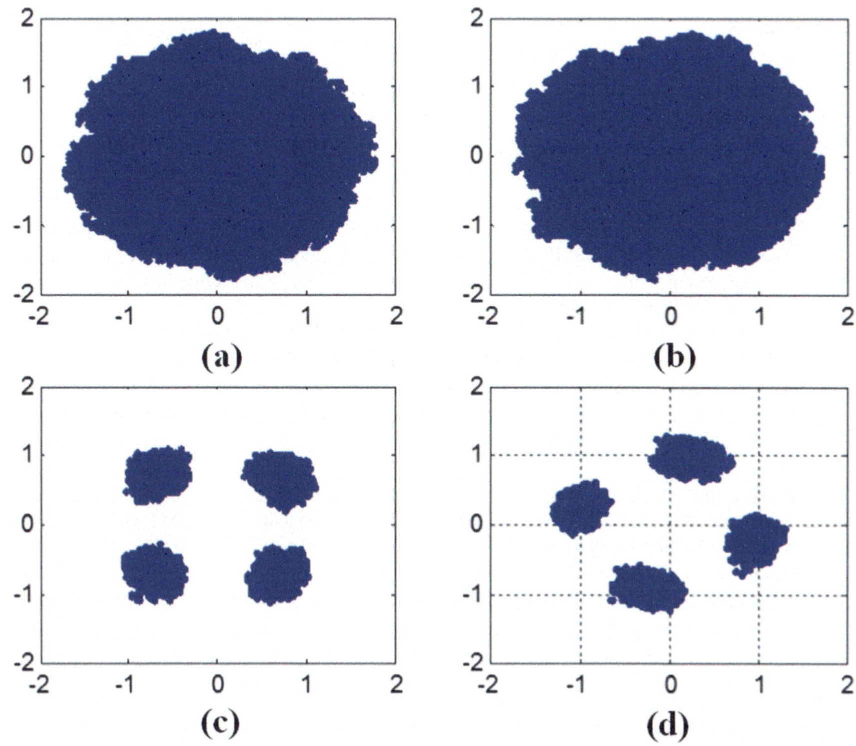

Fig. 15.18 Constellation diagram of 16Gbaud CAP4 **a** The left sideband signal not processed by the Image elimination algorithm, **b** The right sideband signal side not processed by the Image elimination algorithm, **c** The left sideband signal processed by the Image elimination algorithm and **d** The right sideband signal processed by the Image elimination algorithm

15.5 Summary

This chapter introduces an optical access system architecture based on I/Q modulator to generate independent double-sideband signals and direct detection at the receiving end. By using the new Image elimination algorithm, single-carrier 64 Gb/s CAP4 signals and multi-carrier 300 Gb/s DFT-S OFDM signal generation and detection based on two independent sidebands can be successfully implemented. The 240 Gb/s 16QAM DFT-S OFDM signal based on two independent sidebands can transmit 160 km in single-mode fiber without dispersion compensation. As far as we know, this is the record with the highest transmission rate in the single polarization direct detection optical communication system, and it is also the record that directly detects the highest distance capacity product in the optical communication system.

References

1. J. C. Rasmussen et al., Digital signal processing for short reach optical links, in *Proceedings of IEEE Eur. Conf. Opt. Communications (ECOC)*, Sep. 2014, pp. 1–3
2. W. Yan et al., 80 km IM-DD transmission for 100 Gb/s per lane enabled by DMT and nonlinearity management, in *Proceedings of Optical Fiber Communication Conference Exhibition (OFC)*, Mar. 2014, pp. 1–3
3. S. Randel, D. Pilori, S. Chandrasekhar, G. Raybon, P. Winzer, 100-Gb/s discrete-multitone transmission over 80-km SSMF using single-sideband modulation with novel interference-cancellation scheme, in *Proceedings of IEEE European Conference Optical Communication (ECOC)*, Oct. 2015, pp. 1–3 (2015)
4. Y. Wang, J. Yu, N. Chi, Demonstration of 4 times 128-Gb/s DFTS OFDM signal transmission over 320-km SMF with IM/DD. IEEE Photon. J. **8**(2), 1–9 (2016)
5. Y. Wang, J. Yu, H.C. Chien, X. Li, N. Chi, Transmission and direct detection of 300-Gbps DFT-S OFDM signals based on O-ISB modulation with joint image-cancellation and nonlinearity-mitigation, in *Proceedings of VDE, ECOC 42nd European Conference Optical Communication* (2016)
6. L. Zhang, T. Zuo, Q. Zhang, E. Zhou, G. N. Liu, X. Xu, C-band single wavelength 100-Gb/s IM-DD transmission over 80-km SMF without CD compensation using SSB-DMT, in *Proceedings of European Conference Optical Communication (ECOC)*, March 2015, pp. 1–3
7. L. Zhang, E. Zhou, Q. Zhang, G. N. Liu, and T. Zuo, "C-band single wavelength 100-Gb/s IM-DD transmission over 80-km SMF without CD compensation using SSB-DMT," in *Proc. Opt. Fiber Commun. Conf. Opt. Soc. Amer.*, 2015
8. L. Zhang, T. Zuo, Q. Zhang, J. Zhou, E. Zhou, and G. N. Liu, Single wavelength 248-Gb/s transmission over 80-km SMF based on twin- SSB-DMT and direct detection, in *Proc. VDE 42nd Eur. Conf. Opt. Commun. (ECOC)*, Sep. 2016, pp. 1–3
9. P.R. King, S. Stavrou, Low elevation wideband land mobile satellite MIMO channel characteristics. IEEE Trans. Wireless Commun. **6**(7), 2712–2720 (2007)
10. Y. Wang, L. Tao, X. Huang, J. Shi, N. Chi, Enhanced performance of a high-speed WDM CAP64 VLC system employing Volterra seriesbased nonlinear equalizer. IEEE Photon. J. **7**(3), 1–7 (2015)
11. M. Li et al., Low-overhead low-power-consumption LDPC-based FEC solution for next-generation high-speed optical systems, in *Proceedings of Optical Fiber Communication Conference*. Optical Society of America, 2015, p. Th3E-2

Chapter 16
Forward Error Correction

16.1 Introduction

Radio-over-fiber (RoF) technology can combine the high-speed, high-capacity, high-bandwidth, and low-cost performance of fiber with the mobility of wireless communications. In the RoF system, the transmitted millimeter-wave signal is not only damaged by chromatic dispersion, polarization mode dispersion, and fiber nonlinear effects in the fiber link, but also by multipath fading in the wireless link. OFDM technology has advantages in both wireless communication systems and optical fiber communication systems. OFDM technology has the advantages of high spectral efficiency, simple implementation, good anti-multipath fading effect, and interference capability in wireless systems. On the other hand, OFDM can effectively combat inter-symbol interference (ISI) and has the advantages of high spectral efficiency, resistance to fiber dispersion (CD) and polarization mode dispersion (PMD), low system complexity, and low-cost in fiber transmission systems. However, as optical network transmission rates and transmission distances continue to increase, the requirements for optical signal-to-noise ratio (OSNR), chromatic dispersion, polarization mode dispersion, and fiber nonlinearity are becoming more and more stringent and it can be difficult to solve the impact of the above reasons for OOFDM.

In order to alleviate the above-mentioned effects on high-speed optical fiber communication, a series of measures are required. Among them, the introduction of forward error correction (FEC) technology in optical fiber communication signals is one of the most effective methods. FEC technology finds and corrects errors caused by dispersion and nonlinearity in optical transmission by adding a small amount of redundant information to the signal, and reduces factors such as dispersion and nonlinearity in the optical link. In this way, FEC can reduce the OSNR tolerance at the receiving end by sacrificing the transmission rate of the signal, thereby obtaining coding gain, reducing the bit error rate, and improving the reliability of the communication system. The FEC technology combined with OFDM technology can be

J. Yu and N. Chi, *Digital Signal Processing In High-Speed Optical Fiber Communication Principle and Application*,
https://doi.org/10.1007/978-981-15-3098-2_16

applied in the optical fiber communication system to improve the transmission performance of the optical fiber communication system [1–7]. The FEC technology can make the symbols, which are originally independent of each other, to generate a certain correlation by adding relevant special identifiers to the information sequence, so that the system can obtain certain ability to automatically detect errors or correct errors by decoding the specific correlation at the receiving end. In channel coding, the selection of code words is generally for a specific channel situation. The general code words are difficult to meet the system requirements in the case where the channel conditions are relatively complicated.

From the construction method, the error correction code can be divided into block code and convolutional code [8]. In channel coding, the transmitted sequence of n bits is combined by an information sequence of k bits and a check sequence with a length of $n - k$ bits, which is added according to a certain rule. If the check symbols is only related to the k information bits of this group, and independent on the information bits of the other groups, such codes are called block codes, and (n, k) is used to represent the block code set. The generation of the check symbols of this group is not only related to the information bits of this group, but also related to the information bits of the encoder input before this time, which means that the information of the previous code group is also used for decoding, and such codes are called convolutional codes. Besides, (n, k, m) is commonly used to represent a set of convolutional codes, and m is the constraint length.

The performance of the Turbo code and LDPC code is more excellent in the forward error correction code, and has been widely studied in recent years. Compared with the Turbo code, the LDPC code is a linear block code, its check matrix has sparsity, and its decoding complexity is lower, which can realize the advantages of parallel processing and short decoding delay. Based on the basic principles of several common channel coding, the application of the Turbo code and LDPC code in the RoF system is mainly introduced.

16.2 Block Code

If an error occurs during transmission, the receiver can determine the error. The receiving end generally decodes by using the maximum likelihood algorithm to select the nearest symbol as the judgment result. In general, the larger the minimum distance, the better the performance of the code. For a normal block code, as long as the mapping between information bits and check bits is defined, this linear coding method is determined.

For linear block codes, the mapping between information bits and check bits can be represented by a matrix. The linear block code is to divide the information into information code blocks of length k, each of which is coded as a code word of length n. For a block of length k, there are totally 2^k kinds of combinations, and the code word corresponding to these blocks is called an allowable code group.

This section will mainly introduce the basic concepts and principles of linear block codes.

16.2.1 *Linear Block Code*

A block code of length n, whose information bit number is k, can be called (n, k) block code, and its supervised symbol number is $r = n - k$. A block code is also called a code word, and a code word of length n has 2^n combinations, from which 2^k combinations are selected as the allowable code group. For the (n, k) block code, the code rate is $R = k/n$.

A. **Generating matrix and parity-check matrix**

When the code word of the n-bit block code is linearly mapped to the k-bit information bit, it can be called a linear block code. Here we suppose the k-bit information bits are $d_1, d_2,\ldots, d_k$, which can be expressed as k-dimensional vectors $\mathbf{d}$

$$\mathbf{d} = (d_1, d_2, \ldots, d_k) \tag{16.1}$$

The corresponding code word is represented by an n-dimensional vector $\mathbf{c} = (c_1, c_2, \ldots, c_k, c_{k+1}, \ldots, c_n)$.

In the block code, the first k bits of the n-dimensional code word vector are just the information bits, i.e., $c_1 = d_1$, $c_2 = d_2$, …,$c_k = d_k$. While the supervised bits of length $r = n - k$ are generated linearly by information bits

$$\begin{aligned} c_{k+1} &= d_1 h_{11} + d_2 h_{21} + \ldots + d_k h_{k1} \\ c_{k+2} &= d_1 h_{12} + d_2 h_{22} + \ldots + d_k h_{k2} \\ &\ldots \\ c_n &= d_1 h_{1n} + d_{2n} h_{2n} + \ldots + d_k h_{kn} \end{aligned} \tag{16.2}$$

Equation (16.2) can be expressed in matrix form as

$$\mathbf{c} = \mathbf{dG} \tag{16.3}$$

where $\mathbf{G}$ is the generator matrix of k rows and n columns

$$\mathbf{G} = \begin{bmatrix} g_{11} & \cdots & g_{1n} \\ \vdots & \ddots & \vdots \\ g_{k1} & \cdots & g_{kn} \end{bmatrix} \tag{16.4}$$

Since the linear block code is a system code, the first k bits of the generated code words c_k are information bits, so the first k columns of the generation matrix constitute the identity matrix $\boldsymbol{I}_k$, and the latter $n - k$ columns are the weight matrix formed by the coefficient g_{ik}, which is recorded as $\mathbf{P}$, thus the generator matrix can be expressed as

$$\mathbf{G} = [\boldsymbol{I}_k \mathbf{P}] \tag{16.5}$$

$$\mathbf{P} = \begin{bmatrix} h_{11} & \cdots & h_{r1} \\ \vdots & \ddots & \vdots \\ h_{1k} & \cdots & h_{rk} \end{bmatrix} \tag{16.6}$$

The receiving end uses a parity-check algorithm to perform error detection and error correction on the received code word. The generation of the transmitted code word is based on the generator matrix, i.e.,

$$\mathbf{c} = \mathbf{dG} = \mathbf{d}[\boldsymbol{I}_k\mathbf{P}] = [\mathbf{d}\boldsymbol{c}_P] \tag{16.7}$$

where $\boldsymbol{c}_p$ are r-bit supervised bits. If the code word received by the receiver is correct, then

$$\boldsymbol{c}_P \oplus \boldsymbol{dP} = 0 \tag{16.8}$$

$$[\mathbf{d}\boldsymbol{c}_P]\begin{bmatrix} \boldsymbol{P} \\ \boldsymbol{I}_{n-k} \end{bmatrix} = 0 \tag{16.9}$$

where $\oplus$ denotes modulo 2 plus, and $\boldsymbol{I}_{n-k}$ denotes a n-k order identity matrix. Thus, Eq. (16.9) can be written as

$$\mathbf{c}\boldsymbol{H}^T = 0 \tag{16.10}$$

$$\boldsymbol{H}^T = \begin{bmatrix} \boldsymbol{P} \\ \boldsymbol{I}_{n-k} \end{bmatrix} \tag{16.11}$$

where $\boldsymbol{H} = \left[\boldsymbol{P}^T\boldsymbol{I}_{n-k}\right]$ is defined as a supervisory matrix. For the correct code word $\mathbf{c}$

$$\mathbf{c}\boldsymbol{H}^T = 0 \tag{16.12}$$

When an error occurs in the transmission of the code word, the code word received by the receiver is

$$\mathbf{r} = \mathbf{c} \oplus \boldsymbol{e} \tag{16.13}$$

where $\mathbf{e}$ denotes an error vector, also referred to as an error pattern, with the bit where the error occurs being 1 and the remaining bits being 0. Using the supervision matrix to verify, we have

$$\mathbf{r}\boldsymbol{H}^T = (\mathbf{c} \oplus \boldsymbol{e})\boldsymbol{H}^T = \mathbf{e}\boldsymbol{H}^T = \mathbf{S} \tag{16.14}$$

where **S** is an r-dimensional vector called a syndrome. When **S** is a non-zero vector, it indicates that one or more bits have an error, but all-zero **S** does not indicate that there is no error because the syndrome cannot detect all the error patterns.

Assuming that an error occurred in the i-th bit of the transmitted code word, there is

$$\mathbf{S} = \mathbf{e}\boldsymbol{H}^T = [h_{i1}h_{i2}\ldots h_{ik}]^T \tag{16.15}$$

where the i-th column of the **P** matrix is just the syndrome.

B. **Minimum code distance of linear block code**

In a group of code words, the number of different bits in the corresponding digits of the two code words is called the code distance, also known as the Hamming distance, which can measure the degree of difference between the code words. While the minimum value of all the distances for a group of code words is called the minimum code distance, also known as the minimum Hamming distance.

16.2.2 Cyclic Code

Prange discovered the cyclic code in 1957. Cyclic code is the most widely used class of code in linear codes. Its code structure can be constructed and analyzed by the algebraic method and has closed linearity. The linear sum of any allowed code group is still the allowed code. At the same time, due to its cyclic nature, it enables coding and companion calculations through a simple feedback shift register. And the minimum code weight is just the minimum code distance.

A. **Definition of cyclic code**

For any code vector of a (n, k) linear code group $C = (C_{n-1}, C_{n-2} \ldots\ldots C_1, C_0)$, whether it is shifted left or right, and no matter how many bits are shifted, the resulting vector $C_{n-2}, C_{n-3} \ldots C_0, C_{n-1}, C_{n-3}, C_{n-4} \ldots C_{n-1}, C_{n-2}, \ldots$ is still a code vector, then we call the linear code group a cyclic code.

B. **Polynomial of cyclic code**

For the convenience of calculation and representation, the code vector is usually represented as a polynomial, and the components of the code vector are used as the coefficients of the polynomial. The code vector C_1 is shifted to the left by one bit to obtain the code vector C_2. The expression is as follows:

$$\begin{aligned} C_1(x) &= C_{n-1}x^{n-1} + C_{n-2}x^{n-2} + \ldots + C_1x + C_0 \\ C_2(x) &= C_{n-2}x^{n-1} + C_{n-3}x^{n-2} + \ldots + C_1x^2 + C_0x + C_{n-1} \end{aligned} \tag{16.16}$$

Multiplied by x and divided by $(x^n + 1)$ on both sides, Eq. (16.16) can be converted to the following equation

$$xC_1(x) = C_{n-2}x^{n-1} + C_{n-3}x^{n-2} + \ldots + C_1x^2 + C_0x + C_{n-1} + C_{n-1}\left(x^n - 1\right) \quad (16.17)$$

$$xC_1(x) = C_2(x) + C_{n-1}\left(x^n + 1\right) \quad (16.18)$$

which can be written as

$$C_2(x) = xC_1(x)mod\left(x^n + 1\right) \quad (16.19)$$

Here we can conclude that the code polynomial, which is once cycled, is the remainder of the original code polynomial multiplied by x and divided by $(x^n + 1)$. Similarly, the i-cycle shift of the cyclic code vector is equivalent to the remainder of the code polynomial multiplied by x^i and divided by $(x^n + 1)$.

C. **Generating matrix and generating polynomial**

In the 2^k code polynomials of the (n, k) cyclic code, the code polynomial with the first $k - 1$ bits being 0 is taken, and then $k - 1$ times cyclic shift is performed to obtain k code words as follows: $g(x), xg(x), \ldots, x^{k-1}g(x)$.

Therefore, the independent k code words can form the generating matrix of the cyclic code

$$G(x) = \begin{bmatrix} x^{k-1}g(x) \\ x^{k-2}g(x) \\ \vdots \\ xg(x) \\ g(x) \end{bmatrix} \quad (16.20)$$

The (n, k) cyclic code is determined by the polynomial $g(x)$, which is, therefore, called the generating polynomial of the cyclic code.

The code word corresponding to the generating polynomial $g(x)$ is a code word in which the first k bits are all 0 in the 2^k code words of the (n, k) cyclic code. The degree of the polynomial is the lowest and is $r = n - k$. The code polynomial corresponding to other code words is a multiple of $g(x)$. It can be seen that the polynomial $g(x)$ has the following characteristics: $g(x) = x^r + g_{r-1}x^{r-1} + \ldots + g_2x^2 + g_1x + g_0$, where $g_0 \neq 0, r = n - k$.

The generating polynomial $g(x)$ is the lowest degree polynomial of the (n, k) cyclic code, i.e., the k code words $xg(x), x^2g(x), \ldots, x^{k-1}g(x)$ are all independent, so it can be used as a set of code generation bases, so that each code polynomial is a linear combination of the set of bases. Therefore, to find the proper $g(x)$ is the key to constructing a cyclic code.

D. **Check matrix and check polynomial**

The generating polynomial $g(x)$ must be the factor of $x^n + 1$, so that

$$x^n + 1 = g(x)h(x) \quad (16.21)$$

where $h(x)$ is called check polynomial, which is a k-th degree polynomial. We suppose the check polynomial is

$$h(x) = h_k x^k + h_{k-1} x^{k-1} + \ldots + h_1 x + h_0 \tag{16.22}$$

Then the first line of the check matrix of the cyclic code consists of the coefficient of the inverse polynomial $h^*(x)$ of the check polynomial and n-k-1 zeroes, and the second line is the first line shifted to the right by 1 bit, and so on. The resulting check matrix is of $r \times n$ order.

For any information block $M = (m_{k-1}, \ldots, m_1, m_0)$ of length k, the corresponding information polynomial is

$$M(x) = m_{k-1} x^{k-1} + \ldots + m_1 x + m_0$$

which can be multiplied by $g(x)$ to form an $(n - 1)$-th degree code polynomial

$$C(x) = M(x)g(x) = \left(m_{k-1} x^{k-1} + \ldots + m_1 x + m_0\right) g(x) \tag{16.23}$$

If we want to obtain an n-bit system code with the first k bits being information bits, and the last $r = n - k$ bits being the supervised bits, we can first multiply the information polynomial $M(x)$ with x^{n-k} and then divide it with $g(x)$, i.e.,

$$\frac{x^{n-k} M(x)}{g(x)} = q(x) + \frac{r(x)}{g(x)} \tag{16.24}$$

When checking errors:
When the received code block has no error code, the received code block $R(x)$ can be divisible by $g(x)$, i.e.,

$$R(x)/g(x) = Q(x) + r(x)/g(x) \tag{16.25}$$

where the remainder $r(x)$ of the formula should be zero, otherwise, there are bit errors in this code block.

When the number of error codes in the received code group is too large, beyond the error detection capability of the code, the received code group with the error code may also be divisible by $g(x)$. At this time, the wrong codes cannot be detected.

When correcting errors:
The received code block $R(x)$ is divided by the generating polynomial $g(x)$ to obtain the remainder $r(x)$;

According to the remainder formula $r(x)$, the error pattern $E(x)$ is obtained by using the method of look-up-table or calculation;

Subtracting $E(x)$ from $R(x)$, we can obtain the original transmitted code block, in which the error codes have been corrected.

16.2.3 BCH Code

The BCH code is a kind of cyclic code developed from 1959 that can correct multi-bit errors. It is also one of the best linear cyclic error correction codes discovered so far, and its error correction capability is strong, especially in the case of short and medium codes. Its performance is close to the theoretical value of Shannon. It was discovered by Bose, Chaudhuri, and Hochquenghem in around 1960.

The BCH code has strict algebraic structure, and can be described by the roots of generating polynomial $g(x)$. The generating polynomial of the code is related to the minimum Hamming distance. It is easy to directly determine the structure of the code according to the error correction capability requirements, so the structure is convenient, the coding is simple, and it is a type of widely used error control codes.

It divides the information sequence into information blocks with a length of K bits, and then independently converts each information blocks into a binary block of n bits ($n > K$), which is called a code word. If the number of information blocks is M (obviously $M \geq 2$), the code words obtained are called block codes with a code length of n and a number of information of M. The process of transforming an information block into a code word is called encoding, and the inverse process is called decoding.

For any positive integer m and t ($m \geq 3, t < 2m - 1$), there is a BCH code that can correct the error within t. The parameters are as follows:

Code length: $n = 2^m - 1$.
Number of supervisory bits: $n - k = mt$.
Minimum code distance: $d \geq 2t + 1$.

Its generating polynomial $g(x)$ is the least common multiple of the smallest polynomial $m_1(x), m_2(x), \ldots, m_{2t}(x)$, that is

$$g(x) = LCM[m_1(x), m_2(x), \ldots, m_{2t}(x)] \quad (16.26)$$

Since $d = 2t + 1$, Eq. (16.26) can also be expressed as

$$g(x) = LCM[m_1(x), m_2(x), \ldots, m_d(x)] \quad (16.27)$$

where d is the number of errors to be corrected, $m_d(x)$ is the smallest polynomial, and LCM stands for the least common multiple.

The BCM code is a type of cyclic code that can correct multiple random errors. It can be divided into two categories. If the $d - 1$ consecutive roots of $g(x)$ contain the primitive root, the BCH code generated by $g(x)$ is called primitive BCH code; if the $d - 1$ consecutive roots of $g(x)$ are all non-primitive roots, then the BCH code of $g(x)$ is called the non-primitive BCH code.

The code length n of the non-primitive BCH code is a factor of $2m - 1$. Its generating polynomial $g(x)$ doesn't contain the primitive polynomial with the highest m-th degree.

Construction step of the primitive BCH code:

(1) Determine m according to the code length $n = 2m - 1$, then find m-th degree primitive polynomial $p(x)$ by using look-up-table method, and construct the extended domain $GF\ (2^m)$.
(2) Take the primitive root ∂, determine the root of $g(x)$, $\partial^l, \partial^{2l}, \ldots, \partial^{2tl}$, according to the desired number of error corrections t, and find the minimum polynomial of the root, $\left[M_l(x), M_{3l}(x), \ldots, M_{(2t-1)l}(x)\right]$, by look-up-table.
(3) Calculate the least common multiple of the above minimum polynomial to obtain the generating polynomial $g(x)$.

Construction step of the non-primitive BCH code:

(1) Determine the minimum value of m satisfying $n|(2m - 1)$, find m-th degree primitive polynomial $p(x)$ by using look-up-table method, and construct the extended domain $GF\ (2^m)$.
(2) Find a n-order root $\beta = \alpha^l$ among the extended domain $GF\ (2^m)$, where l can take a value of $(2^m - 1)/n$. Determine the root of $g(x)$, $\partial^l, \partial^{2l}, \ldots, \partial^{2tl}$, according to the desired number of error corrections t, and find the minimum polynomial of the root, $\left[M_l(x), M_{3l}(x), \ldots, M_{(2t-1)l}(x)\right]$, by look-up-table.
(3) Calculate the least common multiple of the above minimum polynomial to obtain the generating polynomial $g(x)$.

We suppose that $g(x)$ is the generating polynomial of (n, k, d) BCH code, and uses $\beta^1, \beta^2, \ldots \beta^{2t}$ as the continuous roots. $C(x) = c_{n-1}x^{n-1} + c_{n-2}x^{n-2} + \ldots + c_1x + c_0$ is the code polynomial of this code, then we have $C(x) = m(x)g(x)$. Therefore, $\beta^1, \beta^2, \ldots \beta^{2t}$ are also the roots of $C(x)$.

The following formula can be obtained $HC^T = 0$.

$$H = \begin{bmatrix} \beta^{n-1} & \beta^{n-2} & \ldots & \beta & 1 \\ \beta^{2(n-1)} & \beta^{2(n-2)} & \ldots & \beta^2 & 1 \\ \ldots & \ldots & \ldots & \ldots & \ldots \\ \beta^{2t(n-1)} & \beta^{2t(n-2)} & \ldots & \beta^{2t} & 1 \end{bmatrix} \tag{16.29}$$

where H is the check matrix represented by the roots of generating polynomial $g(x)$.

Since β^i and β^{2i} belong to a conjugate root system, the check matrix H can be simplified to

$$H = \begin{bmatrix} \beta^{n-1} & \beta^{n-2} & \ldots & \beta & 1 \\ \beta^{3(n-1)} & \beta^{3(n-2)} & \ldots & \beta^3 & 1 \\ \ldots & \ldots & \ldots & \ldots & \ldots \\ \beta^{(2t-1)(n-1)} & \beta^{(2t-1)(n-2)} & \ldots & \beta^{2t-1} & 1 \end{bmatrix} \tag{16.30}$$

The elements of the check matrix H are just the elements in the $GF\ (2^m)$. Each element among them can be expressed as a column vector of weight m. Only $n - k$ rows in the matrix are linearly independent, and the $n - k$ linearly independent row vectors can form the binary check matrix of the BCH code.

Since the BCH code is a subclass of the cyclic codes, the encoding of the BCH code can be performed in the same manner as the cyclic code. In practical applications, the system code format is usually adopted. Its coding equation is

$$C(x) = x^{n-k}m(x) + \left[x^{n-k}m(x)\right]_{Modg(x)} \tag{16.31}$$

The actual application of the BCH code is usually a high code-rate code, and it can be implemented by a level coding circuit

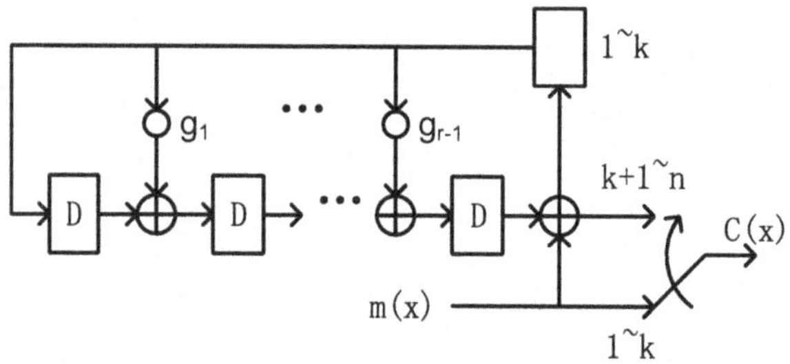

The decoding methods of the BCH code can be divided into two types: time-domain decoding and frequency-domain decoding. Frequency-domain decoding is to treat each code group as a digital signal, perform discrete Fourier transform (DFT) on the received signal, and then use the digital signal processing technology to decode in the frequency domain, and finally perform the Fourier inverse Transform to obtain the decoded code group. While time-domain decoding is performed by directly using the algebraic structure of the code in the time domain. There are many time domain decoding methods for BCH, but the BCH code decoding algorithm for correcting multiple errors is very complicated. Common time-domain BCH decoding methods include Peterson decoding, iteration decoding, and so on.

In 1960, Peterson proposed binary BCH code decoding. Soon Gorenstien and Zierler extended it to the multi-ary case. In 1968, Berlekamp proposed an iteration decoding algorithm for the first time. In 1975, Euclidean decoding was proposed.

The Peterson decoding step can be summarized as follows:

(1) Calculate the adjoint formula $S(s1, s2, \ldots s2t)$ according to the received code block $R(x)$.
(2) Calculate the root xi, $i = 1, 2, \ldots, e$ of the error location polynomial $\sigma(x) = \left(1 - x^1x\right)\left(1 - x^2x\right)\ldots(1 - x^ex) = 1 + \sigma^1x + \sigma^2x^2 + \ldots + \sigma^ex^e$. Then get the error location according to x^{i-1}.
(3) $\tilde{C}(x) = R(x) + E(x)$.

We can transfer the problem of solving the error position into the solution of the linear equations, but when the designed error correction number is large, we must continuously reduce the order of the coefficient matrix until we find a square matrix with full rank, so it is very complicated and is less practical in practice.

Berlekamp proposed an iterative algorithm for solving the error location polynomial in 1968, which fundamentally solved the complexity of the BCH code decoding algorithm and has been widely applied.

16.2.4 Rs Code

The Rs code is an abbreviation for Reed-Solomon code, which is a type of non-binary code. It originated from a paper from S. Reed and G. Solomon from the MIT Lincoln Laboratory in 1960 [9]. It is one of the best linear block codes. At the same time, the Rs code is simple to be encoded and has a strict algebraic structure. It is suitable for medium code length and short code, and its performance is close to the theoretical value.

In the (n, k) Rs code, each block consists of k symbols and each symbol consists of m bits. Rs code is a type of primitive BCH code of $n = q - 1$ length in the finite field $GF(q)$. An Rs code that can correct t symbol errors has the following parameters

Code length: $n = 2^m - 1$ symbols or $m(2^m - 1)$ bits.
Information block size: k symbols or mk bits.
Supervisory block size: $n - k = 2t$ symbols or $m(n - k)$ bits.
Minimum code distance: $d = 2t + 1$ symbols or $m(2t + 1)$ bits.

The Rs code is ideal for correcting sudden errors. For an Rs code of length $2^m - 1$ each symbol can be viewed as an element in the finite field GF. The generating polynomial of the Rs code whose minimum code distance is d symbol has the following form

$$g(D) = (D + \alpha)\left(D + \alpha^2\right)\ldots\left(D + \alpha^{d-1}\right) \tag{16.32}$$

At present, for the Rs encoder, there is mainly an encoder based on a multiplication form, an encoder based on a division form, or an encoder based on a check polynomial form.

For an encoder based on a multiplication form, the code word can be expressed as

$$c(x) = \mathrm{m}(x)g(x) \tag{16.33}$$

For an encoder based on a division form, the code word can be expressed as

$$\begin{aligned} \frac{x^{n-k}}{g(x)} &= b(x) + \frac{r(x)}{g(x)}; \\ c(x) &= x^{n-k}a(x) + r(x) \end{aligned} \tag{16.34}$$

For an encoder based on the check polynomial $h(x) = (2^n+1)/g(x)$, the code word can be expressed as

$$c_{n-k-1} = -\sum_{j=0}^{k-1} c_{n-i-j}h_j$$

where $i = 1, 2, 3, \ldots, n - k$.

For the decoding algorithm of the Rs code, since Rs code is a subclass of cyclic codes, any standard decoding algorithm for cyclic codes is applicable to Rs code. At present, the decoding algorithms for Rs code mainly include PGZ algorithm, BM algorithm, and Forney algorithm. The PGZ algorithm is a general-purpose algorithm that solves the BCH decoding problem, which is easy to understand and simple to implement. The error correction process of PGZ algorithm can be divided into three steps

(1) Calculate the syndrome.
(2) Calculate the error position.
(3) Calculate the error value.

Since the Rs code is a non-binary BCH code, only the specific value of the polynomial coefficient can be found to complete the error polynomial construction and achieve error correction.

However, when we use PGZ decoding algorithm to find $\varepsilon(x)$ and f_p, the calculation amount is really large, and the calculation amount is proportional to the cube of the order of the coefficient matrix. Therefore, two simplified algorithms, i.e., BM algorithm and Forney algorithm, are proposed. In view of this, the PGZ algorithm is mainly applicable to short codes.

Process of BM decoding algorithm

We firstly calculate $\varepsilon^{(1)}$ based on the first equation $S_1 + \varepsilon_1 = 0$, and take $\varepsilon^{(1)}$ into the second equation for verification. If the equation is satisfied, take $\varepsilon^{(2)} = \varepsilon^{(1)}$. If not, we need to correct $\varepsilon^{(2)}$. Then we take $\varepsilon^{(2)}$ into the third equation for verification until the resulting $\varepsilon^{(2t)}$, which is just the polynomial of the wrong position.

In 1969, J. L. Massey simplified the relationship between the iteration decoding algorithm and the shortest linear shift register of the sequence. Since then, this algorithm has become a BM iteration decoding algorithm. Compared with the general decoding algorithm, the iteration decoding algorithm has the advantages of simple calculation and no increase in calculation amount.

16.2.5 Parity-Check Code

Parity-check is a simple and effective way to check code words. This method makes the code distance become 2 by adding a parity bit to the code to make the number of '1's in the code being an odd or even. After using the odd parity (or even parity), the code of the odd bit error in the code can be detected, but the case of the even bit error cannot be found, that is, when the odd bit in the legal code is wrong (when the '1' in the code becomes '0' or '0' becomes '1'), the parity of the number of '1' in the code will change, so that an error can be found. Which calibration to be used is prescribed in advance. Parity can detect a part of the errors during the transmission of information (one bit error can be detected, but two or more errors cannot be detected). At the same time, it can't correct errors. After an error is found, only retransmissions can be requested. However, due to its simple implementation, it is still widely used.

Here we have a binary sequence $C_7C_6C_5C_4C_3C_2C_1$. If the modulo 2 plus is performed on each bit and the sum is 1, the binary sequence is an odd sequence. While if the sum is 0, the binary sequence is an even sequence. Here the parity indicates the inherent nature of this binary bit string. Oddness indicates that the binary digit string has an odd number of '1's, for example, 1101101 has 5 '1's, which is odd. While evenness indicates that the binary digit has an even number of '1's or none. 1 For example, 1101100 has 4 '1's, and 0000000 has no '1'.

How can we specifically implement parity checking? The most common method is to use error control coding. Before the data information bit is transmitted to the channel, a certain redundancy bit is added according to a certain relationship to form a code word. This process is called an error control coding process. After receiving the code word, the receiving end checks the relationship between the information bits and the additional redundant bits to check whether an error has occurred during the transmission. This process is called a checking process. According to this principle, the sender takes a redundancy bit C_0 for the binary bit string $C_7C_6C_5C_4C_3C_2C_1$ for verification.

Here we have two methods to generate C_0. The first method is that $C_0 = C_7 \oplus C_6 \oplus C_5 \oplus C_4 \oplus C_3 \oplus C_2 \oplus C_1$ and the second one is that $C_0 = C_7 \oplus C_6 \oplus C_5 \oplus C_4 \oplus C_3 \oplus C_2 \oplus C_1 + 1$, where $\oplus$ is modulo 2 plus symbol. The C_0 generated by the first method is called even check code, and the C_0 generated by the second method is called odd check code. We can find the relationship between C_0 and binary sequence $C_7C_6C_5C_4C_3C_2C_1$ according to the generation process of C_0.

Under the first method, when the binary sequence $C_7C_6C_5C_4C_3C_2C_1$ is odd, C_0 is also odd, that is C_0 takes a value of '1'. At this time, the new binary digit string $C_7C_6C_5C_4C_3C_2C_1C_0$ after C_0 is programmed into the binary digit string satisfies $C_7 \oplus C_6 \oplus C_5 \oplus C_4 \oplus C_3 \oplus C_2 \oplus C_1 \oplus C_0 = 0$. Similarly, when the binary digit string is even, C_0 is also even, that is C_0 takes a value of zero. At this time, the new binary digit string $C_7C_6C_5C_4C_3C_2C_1C_0$ after C_0 is programmed into the binary digit string also satisfies $C_7 \oplus C_6 \oplus C_5 \oplus C_4 \oplus C_3 \oplus C_2 \oplus C_1 \oplus C_0 = 0$.

For the second method, when the binary sequence $C_7C_6C_5C_4C_3C_2C_1$ is odd, C_0 is even, that is C_0 takes a value of '0'. At this time, the new binary sequence $C_7C_6C_5C_4C_3C_2C_1C_0$ after C_0 is programmed into the binary digit string which satisfies $C_7 \oplus C_6 \oplus C_5 \oplus C_4 \oplus C_3 \oplus C_2 \oplus C_1 \oplus C_0 = 1$. Similarly, when the binary sequence is even, C_0 is odd, that is C_0 takes a value of '1'. At this time, the new binary sequence $C_7C_6C_5C_4C_3C_2C_1C_0$ after C_0 is programmed into the binary digit string also satisfies $C_7 \oplus C_6 \oplus C_5 \oplus C_4 \oplus C_3 \oplus C_2 \oplus C_1 \oplus C_0 = 1$. After receiving the binary sequence $C_7C_6C_5C_4C_3C_2C_1C_0$, the receiving end checks the relationship between the information bits and the redundant bits to check if an error has occurred during the transmission.

We use the first method to check whether $C_7 \oplus C_6 \oplus C_5 \oplus C_4 \oplus C_3 \oplus C_2 \oplus C_1 \oplus C_0 = 0$ can be satisfied, and this check method is called even check. While the method to check whether $C_7 \oplus C_6 \oplus C_5 \oplus C_4 \oplus C_3 \oplus C_2 \oplus C_1 \oplus C_0 = 1$ is called odd check.

In addition to the one-dimensional parity-check codes mentioned above, some researchers have proposed the idea of multi-dimensional parity-check codes. On the basis of one-dimensional error detection, the multi-dimensional parity-check codes also have the function of error correction. In Ref. [10], a multi-dimensional parity-check code with strong error-correcting ability is proposed. This code is a type of error-correcting code with unique rules and complex rotation theory, it has advantages of clear like block code, correlation like convolutional, and simplicity like parity code. This code also uses large number logic decoding and maximum likelihood decoding. Reference [11] constructs a multi- dimensional parity-check product code, which has advantages in performance and bit rate, and its performance is related to the code length, bit rate, and dimension of component code. When the code length of each component is the same, the increase of dimension will improve the performance, but it will also reduce the bit rate. When the code rate and dimension are the same, the performance of different code lengths will be similar. When the code rate is the same and the dimension is different, the product code with higher dimension can be selected to achieve better performance at the same information transmission efficiency. Of course, the increase of dimension will increase the decoding complexity in practical application. But for parity-check codes, the decoding complexity is low, and with the gradual improvement of hardware level and the acceleration of operation speed, the decoding complexity is no longer a constraint. The performance and analysis of other kinds of multi-dimensional parity-check codes will not be discussed here.

16.3 Turbo Code

In 1993, Berrou et al. proposed the Turbo code [12–14], which brought channel coding theory into a new era. Berrou et al. combined convolutional codes with random interleavers and adopted soft-output iteration decoding algorithm to approximate maximum posteriori probability decoding, which achieved extraordinary performance.

16.3.1 Decoding of Turbo Code

The principle of the Turbo encoder is shown in Fig. 16.1. The Turbo encoder consists of two convolutional codes, RSC1 and RSC2, an interleaver, a deletion unit, and a multiplexing unit [15, 16].

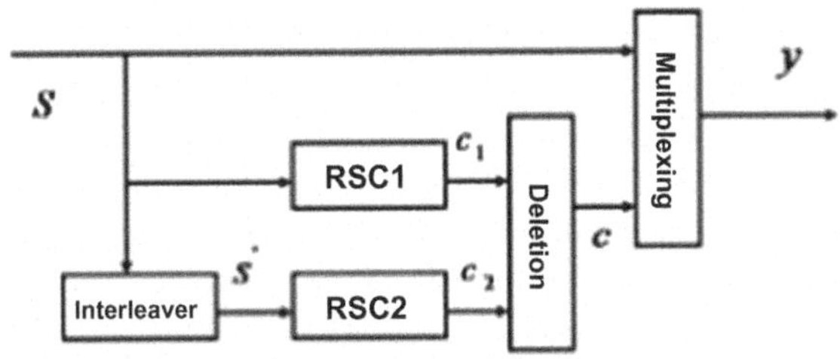

Fig. 16.1 Principle of Turbo encoder

The input information sequence is $S = \{s_1, s_N, \ldots s_N\}$. One path is directly input to the convolutional encoder RSC1, which generates the check sequence c_1 after encoding, and another path is input to the convolutional encoder RSC2 after interleaving to generate the check sequence c_2. In order to improve the bit rate, the Deletion unit deletes some check bits from the two check sequences according to a certain format, so as to get a higher bit rate. The deleted sequence is multiplexed with the original information sequence to form the final transmission sequence. This method uses drilling technology to obtain different coding rates for the same encoder. Parallel concatenation is adopted in Turbo codes, which not only provides better distance and weight spectrum characteristics, but also facilitates global decoding. The sub-coder adopts the form of system code, and only needs to send a system sequence of the encoder, thus improving the code rate. In Turbo coding, the interleaver outputs the input information sequence from front to back after random permutation.

In the Turbo coding, the interleaver mainly rearranges the signal sequence in some way to get a new sequence, so as to reduce the correlation of input information between the two RSC coders and make the coding process independent of each other. Interleaved coding can randomize and homogenize the code, and shape the weight distribution of the code. The performance of interleaved coding has a great impact on the performance of the Turbo code. In the decoding process, if one of the sub-decoders cannot correct the errors, the other decoder will be dispersed after interleaving to make it possible to correct the error. From the point of view of the weight distribution of codes, the function of interleaving is to rearrange the sequence of small weights to increase the Hamming distance of code words. Besides, the reverse process of interleaving is de-interleaving. There are three main interleaving modes, including regular interleaving, irregular interleaving, and random interleaving. Rule interleaving uses row writing and column reading, which is relatively ineffective. Random interleaving distributes the position of the signal randomly, which has the best effect on the system. However, since the position information of random interleaving needs to transmit to the decoder of the receiver, the coding efficiency is reduced. In engineering applications, irregular interleaving is usually used. First, the signal is divided into blocks, and then fixed interleaving is used for the signal block. However, the interleaver structure between blocks is different, which improves the decoding efficiency. If we want to obtain a high coding gain, the structure of interleaver should be optimized and the interleaver length should be increased appropriately.

16.3.2 Iteration Decoding of Turbo Code

The iteration decoding principle of Turbo codes is shown in Fig. 16.2. Turbo iterative decoder consists of decoder 1 (DEC1) and decoder 2 (DEC2), an interleaver, and a de-interleaver. Decoders DEC1 and DEC2 correspond to RSC1 and RSC2 of the Turbo encoder. The iterative Turbo Decoder firstly separates the received signals from each other, which is contrary to the functions of the multiplexer and redundant at the transmitter side. The received signal r_k is divided into signal sequences x_k, y_{1k}, and y_{2k}. This function interpolates the received check sequence according to the deletion rule, and adds '0' to the deleted bits to ensure the integrity of sequence. DEC1 and DEC2 adopt soft-input and soft-output decoding, each iteration has three channels of information, including one path of information code x_k, two-path check code y_{1k} and y_{2k}, and external information. The soft-output includes not only the hard estimation of the received code word by this decoding, but also the credibility of these estimates [16–18].

In the process of Turbo decoding, information sequence x_k and the check sequence y_{1k} generated by convolutional encoder RSC1 are fed into soft-output decoder DEC1. The information sequence $L_1(d_k)$ is interleaved into a sequence $L_1(d_n)$, which is the input of soft-output decoder DEC2. Information sequence x_k is also to DEC2 after interleaver, and the check sequence y_{2k} generated by convolutional encoder RSC2 is the third input of DEC2. Then the output sequence $L_2(d_n)$ of DEC2 is de-interleaved into sequence $L_2(d_k)$.

$L_2(d_k)$ is input to DEC1 as feedback. Repeat the above decoding cycle until the performance improvement of the decoding output is not obvious. The final result is output by DEC2, then the output signal is de-interleaved and determined. In this iteration decoding, the front and back decoders make use of not only their own information bits and check bits, but also the information provided by other decoders. The two decoders process together to improve the credibility of the whole decoding results. This kind of iteration decoding has complex decoding process and long decoding time, which is limited in the fast processing real-time system.

Turbo decoding algorithms are based on maximum a posteriori (MAP) algorithm or soft-output Viterbi algorithm (SOVA). In the low SNR transmission system, the

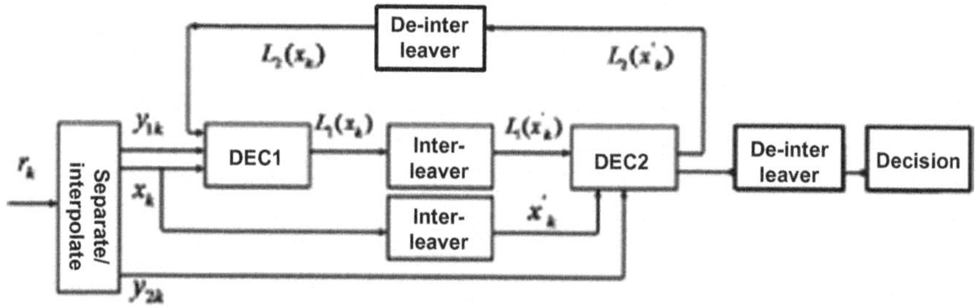

Fig. 16.2 Principle of Turbo iterative decoder

performance of the system using the MAP decoding algorithm is better than that using the SOVA algorithm, but the MAP algorithm needs to consider all the paths in the decoding process. Moreover, the operation of the MAP algorithm is multiplication and exponential operation, and the complexity of this algorithm is higher, but the accuracy is better.

16.3.3 MAP Decoding

The process of MAP decoding is as follows:

First, calculate the conditional probability $\lambda_k^i = \Pr\{x_k = i|r_1^N\}$, where $\forall k = 1, \ldots N$, $\forall i = \{0, 1\}$ and supposing that the received sequence is $r_1^N = \{r_1, r_2, \ldots, r_N\}$. Logarithmic likelihood ratios (LLR) of the k-th bit in an information block with length N is defined as follows [16–18]:

$$\begin{aligned}\Lambda(x_k) &= \log\left(\frac{P(x_k = 1|r_1^N)}{P(x_k = 0|r_1^N)}\right) = \log\left(\frac{P(x_k = 1, r_1^N)/P(r_1^N)}{P(x_k = 0, r_1^N)/P(r_1^N)}\right)\\ &= \log\left[\frac{\sum_{s'} P(s_{k-1} = s', x_k = 1, r_1^N)/P(r_1^N)}{\sum_{s'} P(s_{k-1} = s', x_k = 0, r_1^N)/P(r_1^N)}\right]\end{aligned} \tag{16.35}$$

where $S_{k-1} = S'$ means that the state of k-1 moment is S'. Equation (16.35) can be simplified as

$$\begin{aligned}\Lambda(x_k) &= \log\left[\frac{\sum_{s'} P(s_{k-1} = s', r_1^N)P(x_k = 1, r_k, r_{k+1}^N/s_{k-1} = s', r_1^{k-1}/P(r_1^N))}{\sum_{s'} P(s_{k-1} = s', r_1^N)P(x_k = 0, r_k, r_{k+1}^N/s_{k-1} = s', r_1^{k-1}/P(r_1^N))}\right]\\ &= \log\left[\frac{\sum_{s'} P(s_{k-1} = s', r_1^N)P(x_k = 1, r_k, r_{k+1}^N/s_{k-1} = s'/P(r_1^N))}{\sum_{s'} P(s_{k-1} = s', r_1^N)P(x_k = 0, r_k, r_{k+1}^N/s_{k-1} = s'/P(r_1^N))}\right]\end{aligned} \tag{16.36}$$

$$\begin{aligned}P(x_k = x, r_k, r_{k+1}^N = s') &= P(r_{k+1}^N/s_{k-1} = s', x_k = x, r_k)P(x_k = x, y_k/s_{k-1} = s')\\ &= P(r_{k+1}^N/s_{k-1} = s)P(x_k = x, r_k/s_{k-1} = s')\end{aligned} \tag{16.37}$$

BCJR algorithm can be applied to get

$$\begin{aligned}\alpha_k(s) &= P(s_k = s, r_1^k) = \sum_{s'} \alpha_{k-1}(s')\gamma_k(s', s)\\ \beta_{k-1}(s') &= P(r_k^N/s_k = s) = \sum_{s'} \beta_k(s')\gamma_k(s', s)\\ \gamma_k(s', s) &= P(x_k = x, r_k/s_{k-1} = s') = P(r_k/s_{k-1} = s', x_k = x)P(x_k = x/s_{k-1} = s')\end{aligned} \tag{16.38}$$

where $\alpha_k(s)$ is a forward recursive vector, s is the current state, $\beta_{k-1}(s')$ is a backward recursive vector and s' is the previous state. $\gamma_k(s', s)$ is the transition probability from state s' to state s. By substituting Eq. (16.37) and (16.38) into Eq. (16.36), we can get

$$\Lambda(x_k) = \log\left[\frac{\sum_{x+} \alpha_{k-1}(s')\beta_k(s)\gamma_k(s', s)}{\sum_{x-} \alpha_{k-1}(s')\beta_k(s)\gamma_k(s', s)}\right] \tag{16.39}$$

The Turbo code can be divided into information code and check code parts, which can be expressed as $r_k = (r_k^s, r_k^p)$, where r_k^s is the information bit and r_k^p is the check bit. Under the condition that $x_k = i$, r_k^s and r_k^p are statistically independent and not related to information bits and states, i.e.,

$$P(r_k/s_{k-1} = s', x_k = x) = P(r_k^s/x_k = i)P(r_k^p/x_k = i, x_k = s, s_{k-1} = s') \tag{16.40}$$

According to Eq. (16.38), (16.39), and (16.40), we can get

$$\begin{aligned}\Lambda(x_k) =& L(x_k) + \log\frac{P(r_k^s|x_k = 1)}{P(r_k^s|x_k = 0)} \\ &+ \log\frac{\sum_s \sum_{s'} P(r_k^p|x_k = 1, s_{k-1} = s)\alpha_{k-1}(s')\beta_k(s)}{\sum_s \sum_{s'} P(r_k^p|x_k = 0, s_{k-1} = s)\alpha_{k-1}(s')\beta_k(s)}\end{aligned} \tag{16.41}$$

The decoding result is just the decision result of x_k, which can be obtained by the signs of $\Lambda(x_k)$, i.e.,

$$\hat{x}_k = sign[\Lambda(x_k)] = \begin{cases} 1 & \Lambda(x_k) \geq 0 \\ 0 & \Lambda(x_k) < 0 \end{cases} \tag{16.42}$$

While the Log-MAP algorithm is to replace the likelihood of the MAP algorithm with logarithmic likelihood, so the multiplication operation in the MAP algorithm becomes the addition operation. At the same time, the input and output of the decoder are changed to the logarithmic likelihood ratio accordingly. There is also the maximum Log-MAP algorithm, which ignores logarithm in Log-MAP additions and makes likelihood additions become maximum values for operation. Although it is simple, it loses some accuracy of information.

16.3.4 MAP Equalization Technology

Turbo equalization technology was proposed by C. Donillard et al. in 1995, and then improved by G. Bauch and V. Franz et al. Turbo equalization technology is similar to Turbo iteration decoding, but the external information in Turbo equalization technology is acquired serially rather than in parallel as Turbo iteration decoding. In Turbo iteration equalization technology, the equalizer and decoder cannot perform hard decision in the iteration process, but only perform hard decision after the termination of iteration. It combines equalization technology with Turbo iteration decoding technology to make the equalizer in Turbo have the function of processing input prior information and output posterior information. The equalizer in Turbo exchanges information with Turbo decoder to reduce the algorithm complexity of Turbo encoding technology. Turbo equalization technology retains the interleaving, decoding, and iteration functions of Turbo encoding. In a high-speed optical transmission system, Turbo equalization technology improves the transmission performance significantly compared with the traditional transmission system, which separated equalizer from the decoder. Traditional equalizers usually use hard decision, which makes the latter decoder only use hard-input decoding, while Turbo equalization technology uses soft decoding, and the soft-output value of the decoding system is fed back to the former equalization module as the prior information of the next iteration, so that the cyclic decoding and equalization can be achieved. However, the gain obtained by each iteration is decreasing, and the performance is basically stable after certain number of iterations. Therefore, the soft-input soft-output performance of Turbo iteration decoding technology is better than the hard-input decoding of the traditional equalizer. This technique can reduce the inter-symbol interference caused by dispersion and nonlinearity in optical fiber links, and also reduce the influence of channel estimation error [19–24].

16.3.5 Turbo Iterative Equalization of OFDM Signal

In order to improve the performance of the transmission system, we can use the combination of coding and equalization techniques, the advantages of coding technology and equalization technology can be used for iterative coding equalization. The ISI interference of the transmission channel can be reduced by using equalization technology. The coding technology adds some redundant information to the desired information, and improves the transmission performance of the system by sacrificing some transmission rate of the system. Traditional coding technology and equalization technology are processed independently, which can reduce the complexity of the system. The structure of the traditional Turbo coded for OOFDM system is shown in Fig. 16.3. However, this method of separating coding technology from equalization technology will damage the system performance. Turbo iteration equalization technology combines Turbo encoding and decoding technology with channel equalization technology to improve the performance of the transmission system, and reduces the complexity of Turbo iteration equalization technology by adopting appropriate equalization technology.

Turbo equalization is a method that combines Turbo principle with equalization technology. Through multiple iterations, the performance of the system can be

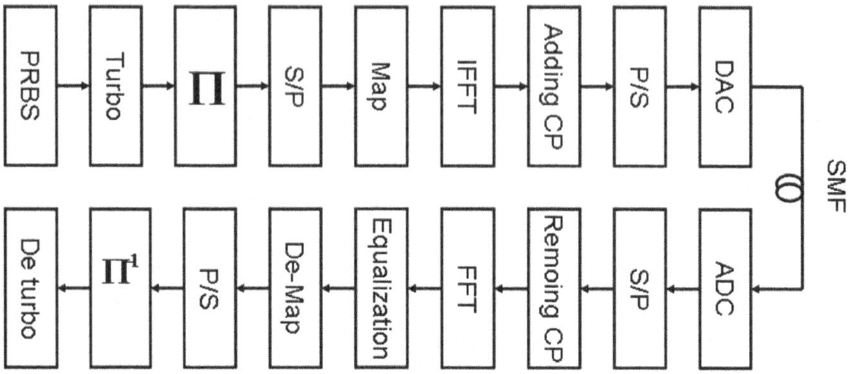

Fig. 16.3 Principle diagram of traditional Turbo coded OOFDM system

improved by fully exchanging external information between equalizer and decoder. The principle diagram of Turbo iteration equalization system is shown in Fig. 16.4.

In Turbo equalized OOFDM transmission system, the pseudo-random bit sequence (PRBS) is firstly encoded by Turbo encoder. The principle of Turbo coding has been shown in the previous Sect. 16.2.1. The coded sequence after interleaver is serial-to-parallel converted, and then the signal is mapped according to the need

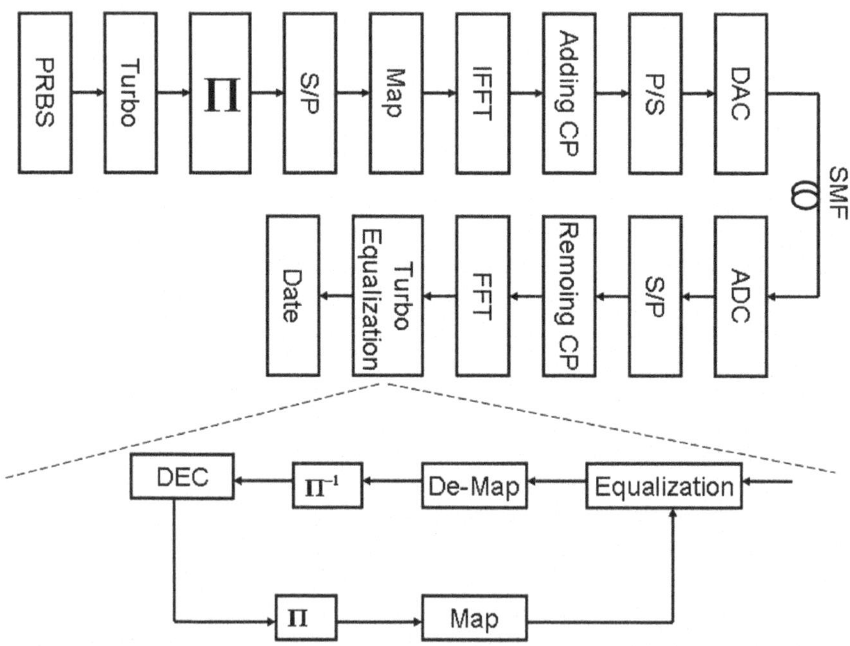

Fig. 16.4 Principle diagram of Turbo iteration equalization in OOFDM system

of the transmission system. Then IFFT transform is performed to obtain the OFDM signal, and a cyclic prefix is added before each OFDM signal, which can reduce the influence of dispersion and nonlinearity on the delay of the OFDM signal. Then the parallel-to-serial conversion is performed. The resulting sequence is transmitted to the optical modulator by an arbitrary waveform generator (WAG) to form an optical OFDM signal. The optical signal is transformed into an electrical signal by photo-electric detection at the receiving end after transmission over fiber. The real-time oscilloscope is used for sampling. After serial-to-parallel conversion, FFT transformation and Turbo equalization are carried out for the OFDM signal. The principle of the whole system process can be reflected in Fig. 16.4.

Turbo iteration equalization technology is composed of equalizer and decoder which work based on iteration. The equalizer and decoder use soft-input and soft-output method. Firstly, the signal processed by FFT is processed by an equalizer and then the soft-output of information is obtained. After de-mapping and corresponding processing, the external information of the signal is obtained. After the information is de-interleaved, the prior information needed by the decoder in equalization technology is obtained. After de-interleaving, the decoder uses the prior information needed by the decoder to calculate the external information of soft-output. After interleaver, the prior information of equalizer can be obtained. Moreover, the equalizer can use prior information and received signals again to process the corresponding information and obtain external information, so as to carry out new iteration processing. After several iterations, the performance of the system is basically stable, and the output signal is determined from the decoder.

16.3.6 MIMO-CMA Equalization Algorithm

The performance of an optical fiber transmission system is not only affected by chromaticity dispersion, but also by polarization mode dispersion (PMD), which widens the transmitted signal and distorts the signal. Polarization mode dispersion causes different group delay (DGD) and phase shift between two polarization states, while different group delay causes optical pulse signals on two components to reach the receiving end at different times. The transmission rate of two signals with different polarizations is different, which affects the transmission performance of optical fiber links. In the polarization multiplexing transmission system, the signal will suffer a series of damage from polarization mode dispersion and polarization correlation loss in the transmission process of the optical fiber link, resulting in crosstalk between the two signals at the receiving end, which becomes the main factor affecting the performance of the polarization multiplexing system.

16.3.6.1 MIMO-CMA Turbo Equalization Technology in PM-OOFDM System

In the polarization multiplexing OOFDM system, the OOFDM signal is divided into two different OOFDM signals in arbitrary polarization state by polarization beam splitter (PBS). The MIMO-CMA iteration equalization technology is similar to linear least mean square error (LMMSE) equalization technology. In the LMMSE equalization system, the equalizer uses the LMMSE filter. The LMMSE filter only equalizes one path of signals, which can reduce the influence of dispersion on the signal. In the polarization multiplexing OOFDM system, signals are transmitted in both of two polarization states in fiber, so that the system is subjected to polarization mode dispersion and polarization crosstalk, which is larger than that of the single-polarization transmission. According to the adaptive equalizer technology of polarization multiplexing signal mentioned above, MIMO-CMA iteration equalization technology is used in polarization multiplexing OOFDM system. The MIMO-CMA equalization technique can compensate for the damage caused by polarization mode dispersion and polarization crosstalk in the transmission system. Moreover, the complexity of the algorithm is lower than that of MAP-based equalization technology and higher than that of LMMSE equalization technology, but the performance of the system is better than that of LMMSE equalization technology.

Compared with LMMSE equalization technology, the equalizer using MIMO-CMA iterative equalization technology uses a CMA structure filter, which is a two-way joint equalizer and consists of four filters. While LMMSE equalization technology uses only one filter. Therefore, in Fig. 16.6, CMA equalizer is used to replace LMMSE equalizer in the principle diagram of OOFDM system and the other algorithms are all the same. The main difference between LMMSE equalization technology and MIMO-CMA equalization technology is that LMMSE equalization technology only equalizes one signal, while MIMO-CMA equalization technology equalizes two polarized signals jointly, considering the influence of chromaticity dispersion and polarization mode dispersion in the transmission system. However, LMMSE equalization technology does not consider the effects of polarization mode dispersion and polarization crosstalk.

Figure 16.5 gives the principle diagram of polarization multiplexing DDO-OFDM transmission system based on MIMO-CMA Turbo iteration equalization technology. The generated OFDM signals at the transmitter side are the same as that of the former two equalization techniques. At the transmitter side, the polarization beam splitter is used to divide the two signals into two orthogonal polarization states. Then the two signals are coupled through the polarization combiner to form polarization multiplexed DDO-OFDM signals. After transmitting over fiber, the received signal is processed at the receiving end. Unlike the traditional signal processing system without polarization multiplexing structure, in which signal processing system only processes one signal, while the polarization multiplexing structure transmission system processes two polarization signals, namely, polarization demultiplexing. Because of chromaticity dispersion and polarization mode dispersion in polarization multiplexing structure, MIMO-CMA equalization technology is used at the receiver.

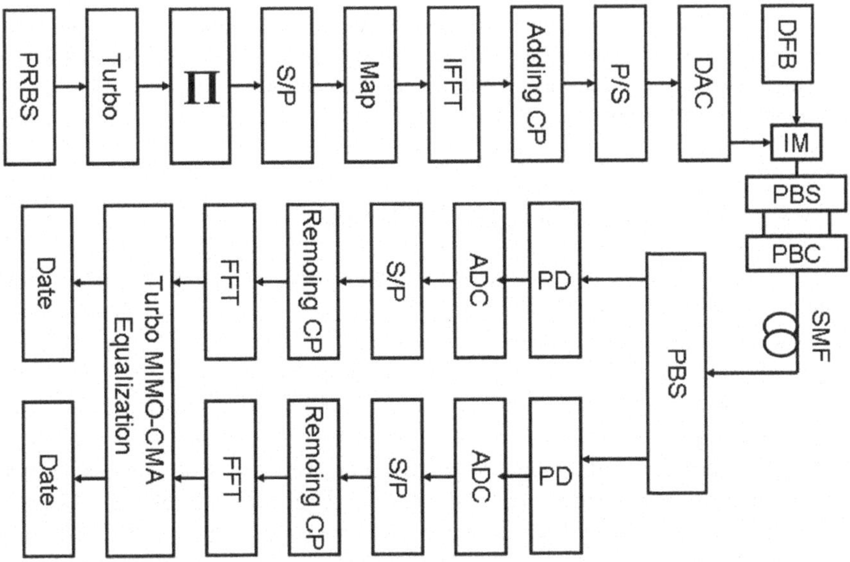

Fig. 16.5 Principle diagram of DDO-OFDM system based on MIMO-CMA equalization algorithm

Figure 16.6 shows the principle of MIMO-CMA equalizer. MIMO-CMA equalization technology receives and processes two signals in the polarization multiplexing system separately. After serial-to-parallel conversion, two signals are processed by

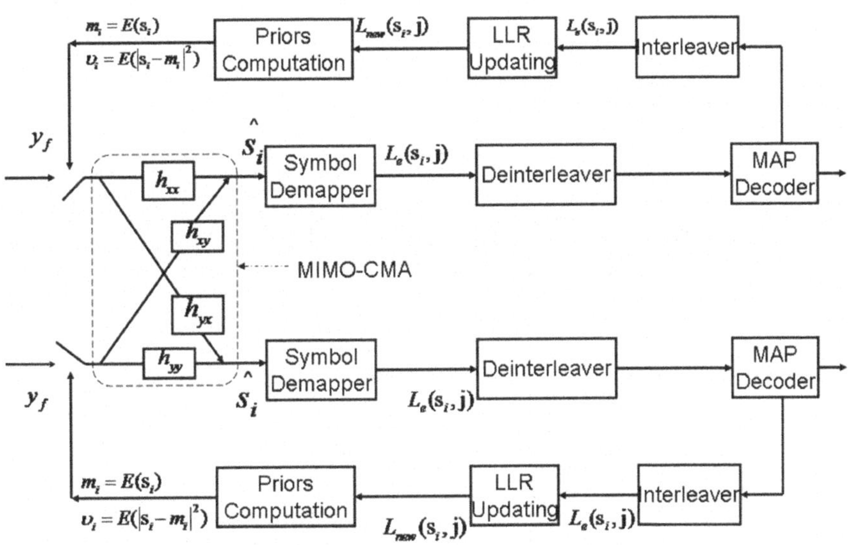

Fig. 16.6 Principle diagram of MIMO-CMA equalization algorithm

FFT and channel estimation, and two signal sequences are obtained. Then MIMO-CMA iteration equalization is applied to these two signals. By MIMO-CMA iteration equalization, the influence of PMD and polarization crosstalk on polarization multiplexed (PM) signal can be reduced.

16.3.6.2 Experiment on PM CO-OFDM System

The experimental setup of the PM CO-OFDM system based on Turbo iteration equalization is shown in Fig. 16.7.

At the transmitter, I and Q paths of Turbo coded OFDM signals from an arbitrary waveform generator (AWG) are transmitted to the I/Q modulator to generate optical CO-OFDM signals, which consist of in-phase and quadrature components. The bandwidth of ECL laser is less than 100 kHz and the optical power is 15 dBm. The CO-OFDM signal is divided into two polarizations by a polarization beam splitter and one of them is delayed to make the two polarizations uncorrelated. Then the dual-polarization CO-OFDM signals are coupled through a polarization beam combiner to form a polarization multiplexed CO-OFDM signal. After transmission over 200-km SSMF, the signal out-of-band noise is filtered by the tunable optical filter (TOF) with 1-nm bandwidth, which is amplified by Erbium-doped Fiber Amplifier (EDFA), and then input to polarization diversity 90° hybrid. The received optical signal is mixed with the local oscillator in the hybrid. The polarization diversity hybrid separates input optical signals into two polarization states. Then the output signals are fed into four photodetectors (PD) with the bandwidth of 10 GHz. Four electrical signals output from PD are sampled by a real-time oscilloscope with a sampling rate

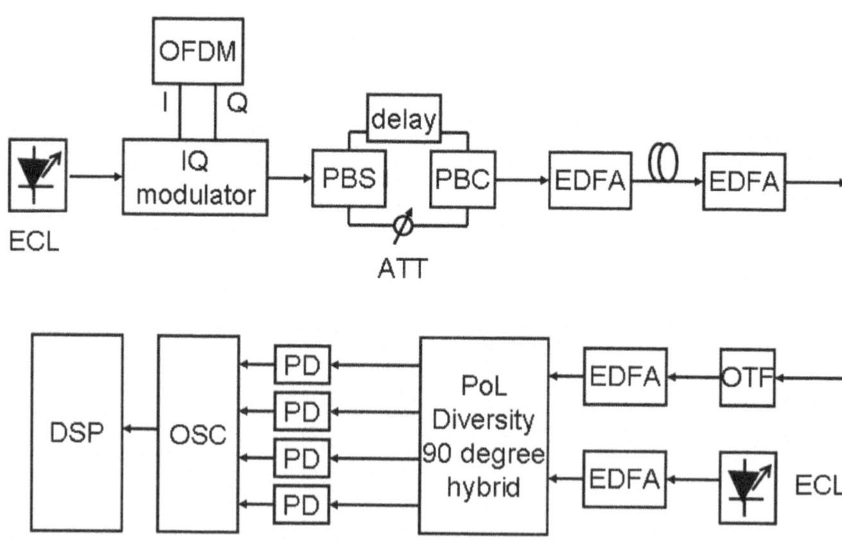

Fig. 16.7 Experimental setup of PM CO-OFDM system

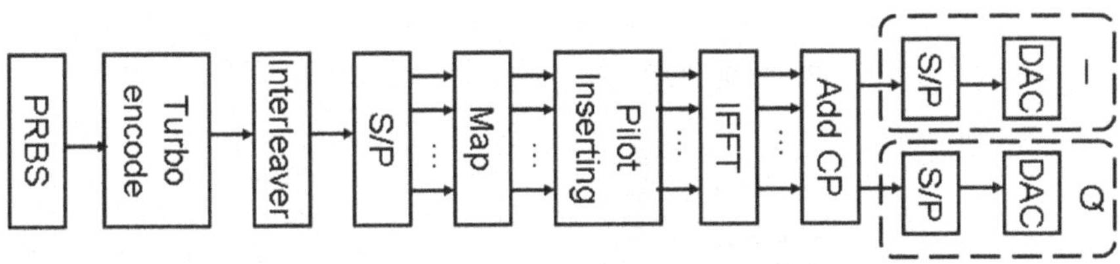

Fig. 16.8 Generation of Turbo iteration equalization CO-OFDM signal

of 20-GS/s and a bandwidth of 8 GHz. The sampled signals are processed by off-line DSP.

Figure 16.8 shows the generation principle of the Turbo iteration equalization CO-OFDM signal. A PRBS sequence is encoded by a Turbo encoder and interleaved by an interleaver, which can disperse continuous burst signals and reduce errors during decoding. After 16QAM mapping, pilots are inserted and IFFT processing is performed. Moreover, by adding cyclic prefix, the influence of dispersion and fiber nonlinearity on the delay of OFDM signal can be reduced. Then the I and Q-path CO-OFDM signals from an arbitrary waveform generator (AWG) are transmitted to the I/Q modulator to generate optical CO-OFDM signals, which consist of in-phase and quadrature components. In this experimental system, the number of subcarriers of CO-OFDM signal is 256, including 196 effective subcarriers, 8 pilots, 56 numeric bits, and 32 cyclic prefixes. Besides, the bandwidth of I/Q modulator is around 8 GHz.

Figure 16.9 shows the off-line DSP of received Turbo iteration equalization polarization multiplexing CO-OFDM signal based on MAP or LMMSE equalization technology. Two fixed digital filters to compensate for the chromatic dispersion (CD) of the normalized CO-OFDM signals in two polarizations. The frequency offset between the two laser sources is estimated and eliminated by the corresponding carrier recovery algorithm, and then the Turbo iteration equalization process is carried out and Turbo iteration equalization process has been described in detail in the previous sections. In addition, the dual-polarization signals are equalized separately in this CO-OFDM system.

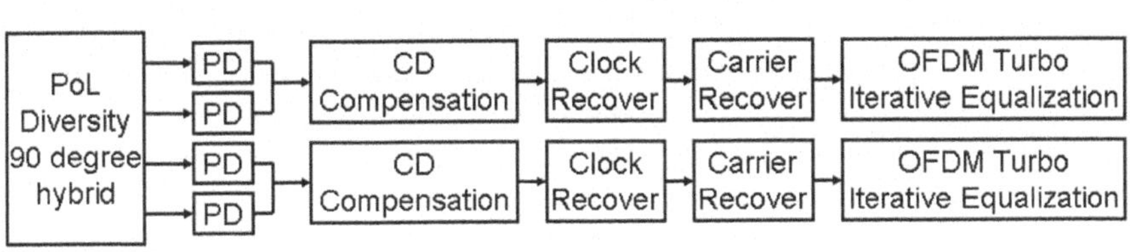

Fig. 16.9 Principle of DSP for CO-OFDM signal based on MAP or LMMSE iterative equalization

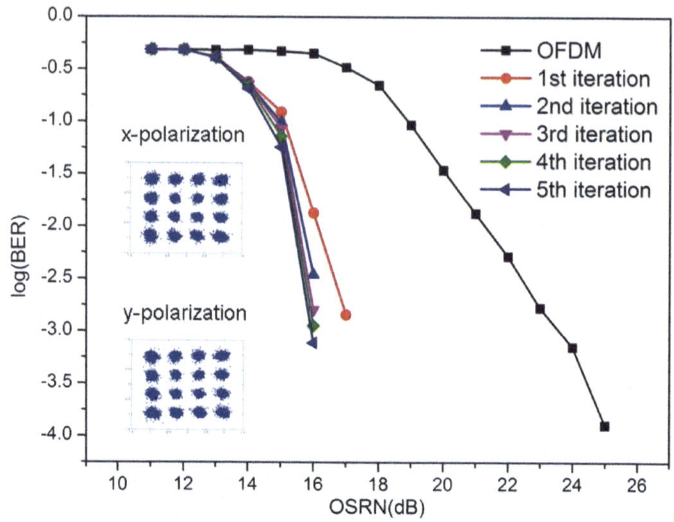

Fig. 16.10 BER performance of MAP equalization technology

16.3.6.3 Experimental Results of PM CO-OFDM System

Figure 16.10 gives the BER performance curve of CO-OFDM signal based on MAP equalization technology after transmission over 200-km SSMF. It can be seen that the BER performance of CO-OFDM system using MPA equalization is much better than that without MPA equalization. With the increase of iteration times, the performance of CO-OFDM system becomes better, but the improvement of system performance is gradually decreasing after each iteration, mainly because that the gain obtained by iteration is also decreasing, and the system performance is basically stable after three times of iteration. The insets in Fig. 16.10 also show the 16QAM signal constellations at 25 dB optical signal-to-noise ratio (OSNR) at two polarization states, respectively.

The performance of CO-OFDM system based on LMMSE equalization and MIMO-CMA equalization is similar to that of CO-OFDM system based on MAP equalization. As the number of iterations increases, the system performance improves, but the improvement of the system performance decreases after each iteration. The gains of Turbo iteration obtained by LMMSE equalization and MIMO-CMA equalization both decreases gradually, so the system performance based on these two equalization algorithms is similar.

Figure 16.11 shows the BER performance curve of CO-OFDM system after five iterations based on three different equalization techniques. It can be seen that under the same times of iteration, the system performance based on MAP equalization technology is the best, followed by MIMO-CMA equalization technology, and the performance of LMMSE equalization technology is slightly worse. In the iteration equalization system based on the MAP equalization, the MAP equalization algorithm uses the optimal criterion, so the performance is the best. The algorithm complexity

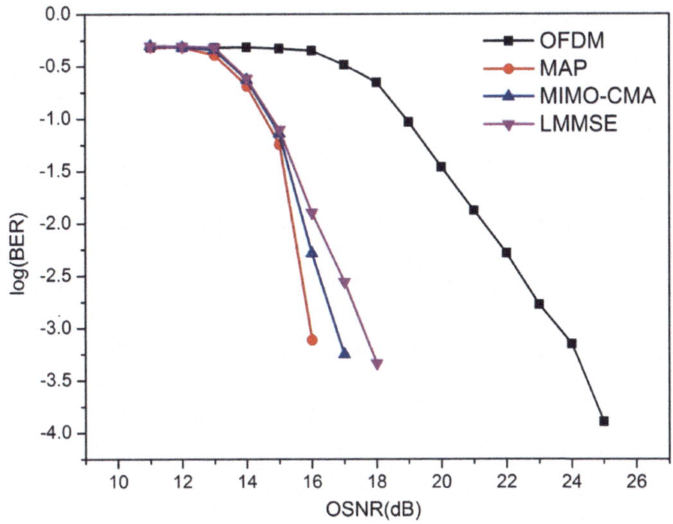

Fig. 16.11 BER performance of different equalization technologies

of MIMO-CMA equalization and LMMSE equalization is much lower than that of MAP equalization, but the system performance is worse. The performance of MIMO-CMA equalization technology is better than that of LMMSE equalization technology, mainly because that MIMO-CMA equalization technology uses the joint equalization of two signals in different polarization states to eliminate the interference of PMD and polarization crosstalk, while LMMSE equalization technology can only process two polarized signals independently. Therefore, the performance of CO-OFDM system based on LMMSE equalization technology is poor, and the reception sensitivity is not as good as that of MAP and MIMO-CMA equalization technology. However, as the number of iterations increases, the performance of these algorithms approaches more and more.

In polarization multiplexing transmission system, the Turbo iteration equalization method based on MIMO-CMA equalization is used to eliminate the influence of PMD. The experimental results show that the CO-OFDM system based on Turbo iteration MIMO-CMA equalization can effectively reduce the influence of chromaticity dispersion and polarization mode dispersion and improve the transmission performance of CO-OFDM system. Moreover, MIMO-CMA equalization technology can simultaneously process dual-polarization signals, which can improve the system performance and reduce the computational complexity of signal processing, and is suitable for the high-speed optical real-time transmission system.

16.4 LDPC Code

Turbo code and low-density parity-check (LDPC) code have excellent performance in forward error correction codes, and more and more research has been focused on them in recent years. Compared with the Turbo code, LDPC code is a linear block code with the sparse check matrix and low decoding complexity, which can realize parallel processing and short decoding delay. In 2003, Bane Vasic et al. applied the LDPC code and iteration decoding method to the long-distance optical transmission system, which improved the performance of the system [25]. In 2005, Ivan B et al. proposed Generalized Low-Density Parity-Check (GLDPC) code, which was applied to an optical communication system. The performance of GLDPC code was also compared with that of the LDPC code [26, 27]. In 2007, Djordjevic et al. proposed a coded modulation algorithm, which combines bit interleaved coded modulation with LDPC code to realize ultra-high-speed transmission in coherent detection system.

16.4.1 Principle of LDPC Code

The LDPC code is a kind of linear block error correction code based on the sparse check matrix. Most elements of the check matrix are '0', only a few elements are '1', and the number of '1' in each row of check matrix is far less than the number of columns of the check matrix. Depending on whether the number of rows or columns in the sparse check matrix of LDPC codes is the same, LDPC codes are divided into LDPC regular code and LDPC irregular code. If the number of rows or columns in the sparse check matrix is the same, it is LDPC regular code, and if the number is different, it is LDPC irregular code. The degree of all message nodes or bit nodes in the binary graph corresponding to LDPC regular code is equal. The encoding and decoding processes of LDPC are operated by the sparse check matrix $\boldsymbol{H}$. The sorting and number of '1' in the sparse check matrix $\mathbf{H}$ are both important factors that determine the complexity of LDPC coding and decoding and the performance of LDPC code. The degree of nodes in LDPC irregular codes varies widely. Choosing a better check matrix of LDPC irregular codes can make irregular codes achieve better performance than regular codes.

We suppose that the information bit length is K, the code length after LDPC encoding is N, and the generated check bit length is $M = N\text{-}K$, that is to say, a (N, K) linear block code $\mathbf{C}$ is obtained and it is uniquely determined by the $M \times N$ order check matrix $\boldsymbol{H}$. The check matrix $\boldsymbol{H}$ can be transformed to the generation matrix $\boldsymbol{G}$, and then the transmitted code word is obtained by multiplying information code $\boldsymbol{m}$ by the generation matrix $\boldsymbol{G}$.

The LDPC code generally uses the Log-likelihood Ratio Belief Propagation (LLR-BP) algorithm to perform decoding. The decoding principle is that: we suppose that the output sequence $\{x_k\}$ of the LDPC encoder is transmitted through the channel after binary modulation $\{u_k = 2x_k - 1\}$, and at the receiving end, suppose that the input signal sequence of the decoder is $\{y_k = u_k + n_k\}$, where n_k is noise [28].

The binary LLR value of the random variable U can be defined as

$$L(U) = \log \frac{p(U = -1)}{p(U = 1)}$$

This LLR value can be interpreted as: we judge whether a random variable is 1 or -1 according to the sign of $L(U)$, its absolute value represents the belief level of -1 and 1, that is, the greater the absolute value of $L(U)$, the greater the possibility of being -1 or 1.

We define the following formula

$$\begin{aligned} \lambda_{n\to m}(u_n) &= \log \frac{q_{\mathrm{nm}}(-1)}{q_{\mathrm{nm}}(1)} \\ \Lambda_{m\to n}(u_n) &= \log \frac{r_{\mathrm{mn}}(-1)}{r_{\mathrm{mn}}(1)} \end{aligned} \tag{16.43}$$

where $\lambda_{n\to m}(u_n)$ is the belief level of the check node m obtained by the information node n, and $q_{\mathrm{nm}}(-1)$ is the belief level of the information node n in state '-1' based on the information provided by other checking nodes related to the information node n besides m. Similarly, $q_{nm}(1)$ is the belief level of the information node n in state '1' based on the information provided by other checking nodes related to the information node n besides m. While $\Lambda_{m\to n}(u_n)$ is the belief level of the check node n obtained by the information node m, and $r_{nm}(-1)$ is the probability of check node m under the condition that the information node is in '-1' state and the state of other check nodes besides m is known. Similarly, $r_{nm}(1)$ is the probability of check node m under the condition that the information node is in '1' state and the state of other check nodes besides m is known.

The decoding process of LDPC code is as follows:

(1) Nodes initialization: Firstly, each information node n is assigned a value, and the posterior LLR value of each information node is obtained

$$L(u_n) = \log\{p(u_n = \pm 1 | y_n)/p(u_n = -1 | y_n)\}.$$

In the equal probability input channel, $L(u_n) = 2y_n/\sigma^2$, where σ^2 is the noise variance. Initialize each pair (m, n) of $H_{m,n} = 1$

$$\lambda_{n\to m}(u_n) = L(u_n),\ \Lambda_{m\to n}(u_n) = 0$$

(2) Iteration process:
Check node m update: For each m and n, calculate the belief level of information node n obtained from check node m

$$\Lambda_{m\to n}(u_n) = 2artanh\{\prod_{n'\in N(m)\backslash n} \tan h[\lambda_{n'\to m}(u_n)/2]\}$$

Information node n update: For each m and n, calculate the belief level of information node m obtained from check node n

$$\lambda_{n\to m}(u_n) = L(u_n) + \overset{\Sigma}{\sum_{m'\in N(n)\backslash m}} \Lambda_{m'\to n}(u_n)$$

For each information node n

$$\lambda_n(u_n) = L(u_n) + \sum_{m\in M(n)} \Lambda_{m\to n}(u_n) \tag{16.44}$$

Then the output result is decided. If $\lambda_n(u_n) \geq 0$, then $x_n = 1$, if $\lambda_n(u_n) < 0$, then $x_n = 0$. The condition that decided the end of LDPC iteration is whether the value of XH^T is zero. If $XH^T = 0$, the iteration will stop and X will be output. While if $H^T \neq 0$, the iteration decoding process will continue, but the iterative decoding is not endless. If the satisfied code word is not found when the limited number of iterations is reached, the decoding fails.

16.4.2 Principle of LDPC-TCM OFDM Optical Millimeter-Wave Signal Transmission System at 60 GHz

This system is a 60-GHz ROF system based on the LDPC-TCM OFDM signal using optical double-sideband modulation format, which is shown in Fig. 16.12. At the center station, the 60-GHz optical millimeter wave is generated by using the carrier-suppression technology. Then the LDPC encoded and 16QAM TCM modulated

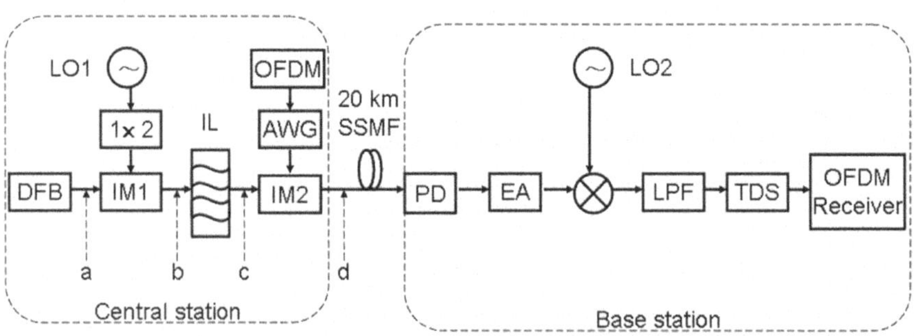

Fig. 16.12 Experimental setup of LDPC-TCM OFDM transmission at 60 GHz

LDPC-TCM OFDM optical signal is transmitted over 20-km SSMF. At the receiving end, i.e., the base station, the received optical signal is transferred into electrical millimeter-wave signal by using PDs with the bandwidth of 60 GHz. At last, the baseband LDPC-TCM OFDM signal is obtained by mixing the 60-GHz electrical signal with the local oscillator signal and low-pass filtering. In this experiment, the LDPC code length is 1024, and the code rate is 1/2. For the TCM code modulation, we use 3/4 code rate and 16QAM mapping.

The structure of the transmitted LDPC-TCM OFDM signal is shown in Fig. 16.13. Firstly, the pseudo-random bit sequence (PRBS) signal is LDPC encoded. After the serial-to-parallel conversion, the 16QAM TCM modulation is performed. After inserting the pilot as the channel estimation, we perform IFFT processing on the data to obtain orthogonal subcarrier sequences, and use the rear sequence of the OFDM symbol as the cycle prefix (CP) to overcome the delay caused by the dispersion and nonlinearity. In this experiment, the number of OFDM subcarriers is 256, including 192 subcarriers of transmitted signal data, 8 pilots, 56 intervals, and 32 cyclic prefixes. The electrical LDPC-TCM OFDM signal generated from Tektronix arbitrary waveform generator (AWG) is fed into the optical modulator to generate an optical LDPC-TCM OFDM signal and loaded onto the optical fiber link. Besides, the peak-to-peak voltage ($V_{p\text{-}p}$) value of the electrical OFDM signal is 2 V.

The user unit performs corresponding decoding and demodulation processing on the received baseband LDPC-TCM OFDM signal, and the processing process is shown in Fig. 16.14. The first step is synchronization and the start position of the OFDM signal is determined by performing high-correlation of the training sequence. Then the guard prefix is removed from the OFDM signal and the channel is estimated by the training sequence and the pilot sequence. Then, the OFDM signal is subjected to FFT processing, de-TCM decoding processing, and LDPC decoding processing to recover the transmitted signal sequence. At last, the received signal data is compared with the signal data at the transmitting end to obtain a bit error rate of the transmission system. The decoding and demodulation process of the LDPC-TCM OFDM signal is just a reverse processing process of encoding and modulation.

In this experiment, the 60 GHz LDPC-TCM OFDM optical millimeter-wave signal transmission system includes the central station and the base station. The function of the central station is the generation of optical millimeter wave. The distributed feedback laser (DBF-LD) generates the continuous light wave with 7-dBm optical

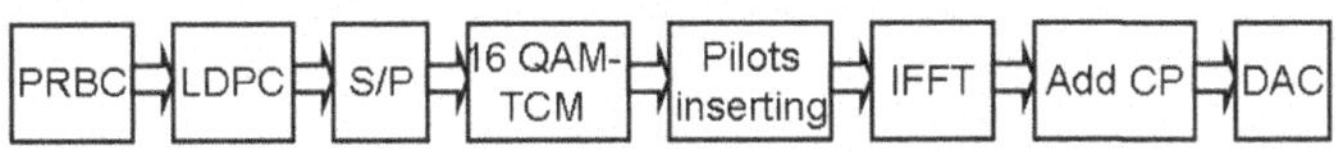

Fig. 16.13 Encoding and modulation principle of transmitted LDPC-TCM OFDM signal

Fig. 16.14 Decoding and demodulation principle of received LDPC-TCM OFDM signal

power and 1542.8-nm wavelength as the optical carrier. Its spectrum is shown in Fig. 16.15a. While the radio-frequency (RF) signal is generated by an analog radio-frequency signal generator with 16-dBm electrical power and 14.5-GHz frequency. After an electric amplifier, the 29-GHz RF signal is generated by a frequency doubler. Then the 29-GHz RF signal is loaded on the Mach-Zehnder single-arm modulator with a 3-dB bandwidth of 20 GHz, an extinction ratio of more than 25 dB, and the half-wave voltage of 7.8 V. When the DC bias voltage of the optical modulator is set to 2.9 V, the optical modulator outputs a double-sideband modulation signal with a first-order sideband interval of 58 GHz, whose spectrum is shown in Fig. 16.15b. The output optical signal of the first Mach-Zehnder single-arm modulator contains the central carrier and the first-order optical sideband signal. Then the 50/100 GHz cross-multiplexer is used to filter out the central carrier and the optical millimeter wave with a frequency interval of 58 GHz, which is shown in Fig. 16.15c. At the same time, the influence of other high-order sideband power on the system can be neglected because of the small power of other high-order sidebands. LDPC-TCM OFDM signals are generated off-line via Tektronix Arbitrary Waveform Generator (AWG) and loaded on the optical link to generate OFDM optical millimeter-wave signals through the second optical intensity modulator. The transmission rate of the AWG is 2.5 GSa/s, $V_{P\text{-}P}$ is 2 V, and the DC bias voltage of the second intensity modulator is 6.8 V. The spectrum of the optical signal generated by the second optical

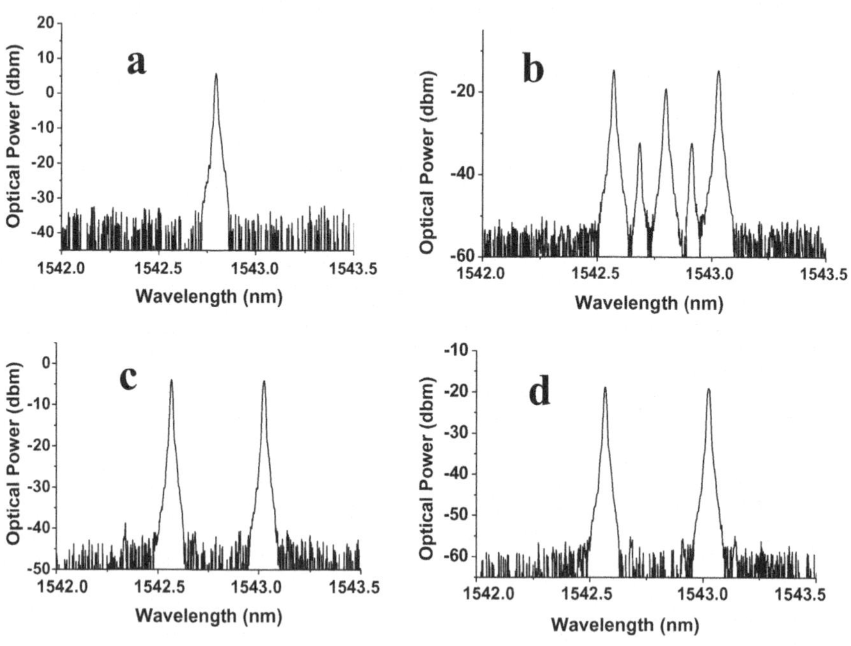

Fig. 16.15 Spectra measured at different points

intensity modulator is shown in Fig. 16.15d. The optical OFDM signal is amplified by EDFA, and the input power into fiber is 8 dBm. Then it is transmitted by 20-km standard single-mode optical fiber to the base station. The loss of standard single-mode optical fiber is 0.19 dB/km, and the dispersion coefficient is 17 ps/(nm km).

In the base station, the optical millimeter-wave signal is first converted into the electrical millimeter-wave signal by a high-speed photoelectric detector with a bandwidth of more than 60 GHz, then amplified by an electric amplifier (EA) with a 3-Db bandwidth of 60 GHz. Then the generated RF signal is mixed with a local oscillator (LO) at 58 GHz, which is generated from the 14.5 GHz sinusoidal wave signal after frequency doubling by quadruple, and then the baseband OFDM signal is obtained by a low-pass filter (LPF) with a bandwidth of 7.5 GHz. The OFDM baseband signal is sampled by Tektronix TDS6804B real-time digital oscilloscope with a sampling rate of 10GSa/s. Then the decoding and demodulation of the sampled OFDM signal are carried out, and the BER performance of the transmission system is obtained by comparing the processed signal with the transmitted signal. The decoding and demodulation process of the LDPC-TCM OFDM signal is performed by using off-line MATLAB according to Fig. 16.14.

16.4.3 Experimental Results and Discussion

Based on this transmission system, three kinds of signals are compared, including un-encoded OFDM signal, LDPC coded OFDM signal, and concatenated LDPC-TCM coded OFDM signal. Here, LDPC code is an irregular code with a code rate of 0.5, and the number of iteration decoding is 8. The check matrix is constructed by PEG and Logarithmic Likelihood Belief Propagation algorithm is used for decoding. TCM consists of 16 states with a bit rate of 3/4. Under the condition of transmitting over 20-km single-mode optical fiber and other parameters of the experimental system is unchanged. The BER performance of the above three signals under different receiving power is analyzed and we can further analyze the anti-dispersion performance of the LDPC coding and LDPC-TCM coding combined with OFDM technology in the optical fiber link. The BER performance of three kinds of signals is shown in Fig. 16.16.

From Fig. 16.16, it can be seen that after transmission over 20-km SSMF, the received power of LDPC-TCM coded 16QAM OFDM signal, LDPC coded 16QAM OFDM signal, and un-coded 16QAM OFDM signal are -33 dBm, -32 dBm, and -17 dBm, respectively, when BER is 1×10^{-3}. It can be concluded that the LDPC-TCM coded OFDM signal and LDPC coded OFDM signal can get 16 dB and 15 dB receiving sensitivity gain compared with the un-coded OFDM signal. In other words, the performance advantages of combining LDPC-TCM coding with OFDM technology is the most obvious for the optical fiber transmission system, that is, a high-quality information transmission can be obtained at the cost of losing a certain bandwidth and increasing a certain amount of algorithm complexity.

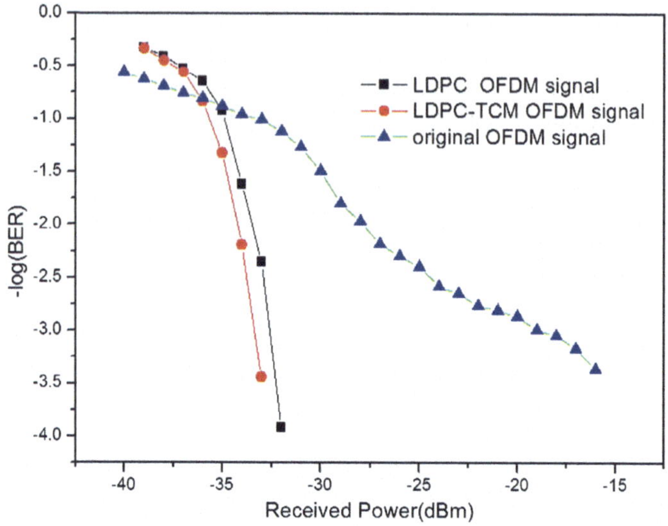

Fig. 16.16 BER performance of LDPC-TCM OFDM system

Since LDPC-TCM is to add the modulation by TCM coding based on the LDPC code, it does not cause much damage to LDPC coded signals, and only increases the mapping distance of adjacent LDPC signals without additional overhead. But the BER performance of LDPC-TCM coding is better than that of LDPC-only coding. It shows that the LPDC-TCM OFDM signal has stronger error correction ability than the LDPC-only OFDM signal, and can better resist the dispersion and nonlinear effect in optical fiber transmission.

16.4.4 Experimental Conclusion

In this experiment, we have demonstrated an optical millimeter-wave transmission at 60 GHz. The carrier-suppression method is used to generate 60 GHz optical millimeter-wave signal. We also compared the BER performance of un-encoded OFDM signal, LDPC coded OFDM signal, and concatenated LDPC-TCM coded OFDM signal after transmission over the RoF system. After transmission over 20-km SSMF, LDPC-TCM coded 16QAM OFDM signal, and LDPC-only coded OFDM signal can get 16 dB and 15 dB receiving sensitivity gain compared with the un-coded OFDM signal, respectively. The experimental results show that using LDPC coding technology and LDPC-TCM coding and modulation technology in the ROF system can effectively suppress the influence of inter-beat interference between OFDM subcarriers and reduce the effects of optical fiber dispersion. Moreover, TCM can obtain the coding gain of the system without reducing the signal frequency band and power, which can solve the impact of insufficient bandwidth utilization. However, if the system performance needs to be further improved, large code length LDPC codes can be used, but for large code length LDPC codes, there are too many factors to consider in the selection and performance improvement of check matrix and it is difficult to construct high performance check matrix.

16.5 Concatenated Coding

For systems with multiple encoding, the concatenated code can be regarded as a whole encoding with several levels, which is called concatenated coding. The initial idea of concatenated coding is to further improve the asymptotic performance. While in practical systems, it can also improve the performance at low OSNR. When a concatenated code is formed by two kinds of codes, the encoding in the generalized channel is called internal code, and the channel encoding in the generalized channel is called external code. Burst errors inevitably occur in decoding results of internal codes. Therefore, there is usually a layer of interleaver between the internal and external codes. The structure of concatenated encoder is shown as follows (Fig. 16.17).

The common concatenation method is to use the convolutional codes as internal codes, and Rs codes as external codes. This is mainly to make full use of Viterbi decoding of convolutional codes and soft decision decoding. While Rs code has a better ability to correct burst errors.

The encoding relationship of serial concatenated code is as follows:

$$C_1 = f(x), \text{ which is external code,}$$

$$C_2 = g(C_1), \text{ which is internal code.}$$

However, the information about symbol x output from external decoding does not directly provide soft information about input C_2 of internal decoding when the convolutional codes are used as both internal and external codes. While when internal codes can output soft information, Rs codes can meet the requirements of interleaver when they are external codes. For convolutional codes, if non-recursive convolutional codes are used as component codes, the increase of interleaver length will not improve the performance of codes.

The structure of the decoder is shown in Fig. 16.18.

Fig. 16.17 Structure of concatenated encoder

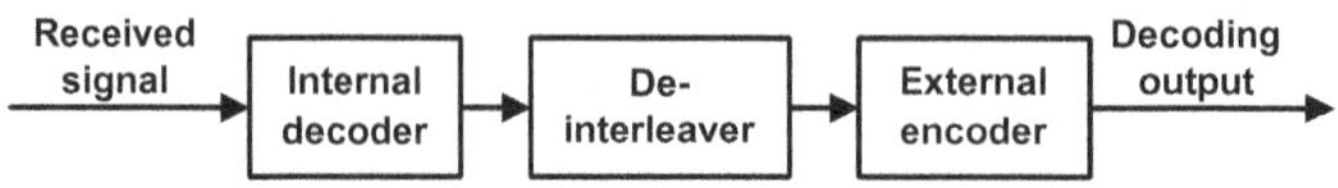

Fig. 16.18 Structure of concatenated decoder

As mentioned above, one of the reasons for using convolutional codes as internal codes is that it can perform soft decision decoding, thus providing soft decision gains of 2–3 dB. Furthermore, if the internal decoding output is also a soft decision output, the external decoding can also be soft decision decoding, thus improving the overall performance.

From another point of view, if the outer code is decoded by soft decision, convolutional codes are usually used, so they can only be designed to correct random errors. For this reason, when choosing an internal decoding algorithm, the criterion should be the lowest output symbol error rate, but not the lowest output sequence error rate. Therefore, Viterbi decoding is no longer the optimal algorithm, while symbol-by-symbol decoding algorithm should be adopted.

Although concatenated codes greatly improve the error correction ability, most of the improvement comes from the reduction of coding efficiency. From the E_b/N_0 point of view, the benefits of the cascade are not too great, but one obvious advantage is that, when the channel quality is slightly better, the error code can be very low.

In addition, the BCH code can be concatenated with interleaving technology to improve the anti-interference ability of the HF channel. The principle of the BCH-interleaver concatenated system is shown in Fig. 16.19.

The core idea of the interleaver cascade system is: firstly, source code words are encoded to get code words, and then code words are interleaved to improve the error correction ability. Then the digital signals pass through channels, and then the received signal is de-interleaved and decoded.

In 2008, Ivan B et al. published a paper that combines LDPC coding with Turbo equalization technology to alleviate nonlinearity in optical transmission systems [29]. In 2009, Mizuochi T et al. proved that the NCG of a two-bit soft decision-based LDPC (9216,7936) and Rs (992,956) FEC coding can reach 9.0 dB when the output BER is 10^{-13} and the transmission rate is 31.3 Gb/s [30]. Based on the idea of concatenated coding and iteration decoding, some scholars have proposed some

Fig. 16.19 Structure of BCH-interleaver concatenated system

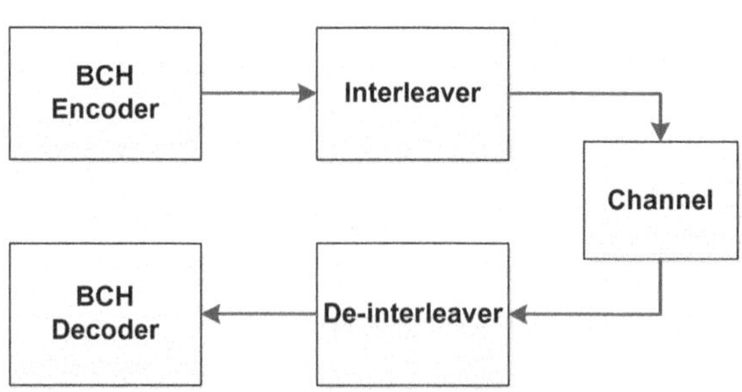

concatenated codes with other structures, such as Mc Eliece, H. Jin and Divsalar's Repeat-Accumulate Codes [31–35]; Concatenated Tree (CT) Codes [36], Concatenated Zigzag Codes [37], and Concatenated Zigzag-Hadamard Codes [38] proposed by Li Ping; and Richardson's Multilateral LDPC Codes [39].

16.6 Summary

This chapter mainly introduces several common encoding techniques about FEC algorithms. The application of LDPC coding and cascade TCM coding technology in the OFDM-ROF system is mainly introduced. We also use the Turbo equalization technology to reduce the dispersion effect of the optical transmission system. The Forward Error Correction technology can discover and correct the code errors caused by dispersion and nonlinearity in optical transmission by adding a small amount of redundant check bits to the information signal, which can reduce the influence of dispersion and nonlinearity in optical fiber link and reduce the OSNR tolerance of the receiver by sacrificing the transmission rate of the signal. In this way, the coding gain can be obtained, the bit error rate can be reduced and the reliability of the communication system can be improved.

References

1. J. Yuan, Research on Ultra-Strong FEC Code Type in High Speed and Ultra-Long Distance Optical Communication System, Doctoral Dissertation of Chongqing University, pp. 11–17 (2007)
2. Z. Cao, Y. Jianjun, M. Xia et al., Reduction of inter-subcarrier interference and frequency-selective fading in OFDM-ROF systems. J. Light Wave Technol. **28**(16), 2423–2429 (2010)
3. T. Mizuochi, Recent progress in forward error correction and its interplay with transmission impairments. IEEE J. Selected Topics Quant. Electron. **12**(4), 544–554 (2006)
4. A. Tychopoulos, O. Koufopavlou, I. Tomkos, FEC in optical communications-A tutorial overview on the evolution of architectures and the future prospects of out band and inland FEC for optical communications. IEEE Circuits Devices Mag. **22**(6), 79–86 (2006)
5. P. Moro, D. Candiani, 565-Mb/s optical transmission system for repeater less sections up to 200 km, in *IEEE International Conference on Conference Record*, pp. 1217–1221 (1991)
6. S. Yamamoto, H. Takahira, M. Tanaka et al., 5 Gbit/s optical transmission terminal equipment using forward error correcting code and optical amplifier. Electron. Lett. **30**(3), 254–255 (1994)
7. G. Ungerboeck, Channel coding with multilevel/phase. IEEE Trans. Informat. Theor. J **28**(3), 55–67 (1982)
8. R.E. Blahut, *Principles and Practice of Information Theory* (Addison-wesley, MA, 1987)
9. I.S. Reed, G. Solomon, Polynomial codes over certain finite fields. J. Soc. Ind. Appl. Math. **8**(2), 300–304 (1960)
10. F. Jin, W. Luo, Discussion on new multiwich parity check code. Southwest Jiaotong University, pp. 1–26 (1985)
11. Ying Huang, Jing Lei, Performance analysis of multiwich parity check product codes. J. Univ. Electron. Sci. Technol. **39**(2), 214–218 (2010)

12. C. Berrou, A. Glavieux, Near optimum error correcting coding and decoding: turbo-codes, IEEE Trans. Comm. **44**(10), 1261–1271 (1996)
13. C. Berrou, A. Glavieux, P. Thitimajshima, Near Shannon limit error-correcting coding and decoding: turbo-codes, in *Proceedings of ICC'93*, May, pp. 1064–1070 (1993)
14. J. Hagenauer, The turbo principle: Tutorial introduction and state of the art, in Proc. Int. Symp. Turbo Codes and Related Topics, Brest, France, 1997, Sept., pp. 1–11
15. Xiaoqun Zhao. Modern coding theory. Huazhong University of Science and Technology Press, 2007, 193–195
16. Shuangshuang Han. Iterative Receiving Technology in Wireless Communication System [Master's Degree Thesis of Shandong University], 2009, 9–19
17. Tianfang Luo. Turbo Code and Turbo Equalization in Communication System [Ph.D. Dissertation of Harbin Engineering University], 2003, 30–40
18. Yidan Jin. Research on Turbo Iterative Receiving and Key Technologies in Mobile Communication System [PhD Dissertation of Beijing University of Posts and Telecommunications], 2006, 34–35
19. C. Douillard, M. Jezequel, C. Berrou et al., Iterative correction of intersymbol interference: turbo equalization. Eur. Trans. Telecommun **6**(5), 507–511 (1995)
20. Bauch G, Khorram H, Hagenauer J. Iterative Equalization and Decoding in Mobile Communications Systems. IRG TATHBERICHT, 1997, 307–312
21. Proakis J G. Digital Communications. 5 ed. McGraw Hill, 2007
22. Marcin Sikora, Daniel J, Costello, Jr. A new SISO algorithm with application to Turbo equalization. IEEE Proceedings International Symposium on Information Theory, 2005, 2031–2035
23. Fabian Bogelbruch, Sven Haar.Receded Complexity Turbo Equalization by Means of Hard Output Channel Decoding. Signals. In: Conference on Systems and Computers, 2001, 290–294
24. Thomas Woerz, Joachim Hagenauer. Multistage Coding and Decoding for an M-psk System.In: Multistage Coding and Decoding for an M-psk System, 1990, 698–703
25. B. Vasic, I.B. Djordjevic, R. Kostuk, Low-density parity check codes and iterative decoding for long-haul optical communication systems. IEEE/OSA Lightwave Technol. **21**, 438–446 (2003)
26. I.B. Djordjevic, O. Milenkovic, B. Vasic, Generalized low-density parity-check codes for optical communication systems. IEEE/OSA Light wave Technology **23**(16), 1939–1946 (2005)
27. I.B. Djordjevic, M. Cvijetic, L. Xu et al., Using LDPC-coded modulation and coherent detection for ultra-high-speed optical transmission. IEEE/OSA Light wave Technology **25**, 3619–3625 (2007)
28. Zhongpei Zhang, Zhiping Shi and Chuandan Wang. Modern coding theory and Application. Electronic Industry Press, 2007, 105–107
29. Djordjevic I B, Minkov L L,Batshon H G. Mitigation of linear and nonlinear impairments in high-speed optical networks by using LDPC-coded turbo equalization[J].IEEE Optical Communication Newt, 2008, 26(6):73–83
30. T. Mizuochi, Y. Konnishi, Y. Miyata et al., Experimental demonstration of concatenated LDPC and RS codes by FPGAs emulation. IEEE Photonics Technol. Lett. **21**(18), 1302–1304 (2009)
31. D. Divsalar, H. Jin, and R. J. Mc Eliece, "Coding theorems for 'Turbo-like' codes," in Proc. 36th Allerton Conf. on Communication, Control, and Computing, 1998, Sep., pp. 201–210
32. H. Jin, A. Khandekar, and R. Mc Eliece, "Irregular repeat-accumulate codes," in Proc. 2nd International Symposium on Turbo Codes, 2000, pp. 1–8
33. S. Ten Brink, G. Kramer, Design of repeat-accumulate codes for iterative detection and decoding, IEEE Trans. Signal Process **51**(11), 2764–2772 (2003)
34. G. Yue, X. Wang, Optimization of irregular repeat accumulates codes for MIMO systems with iterative receivers. IEEE Trans. Wireless Commun. **4**(6), 2843–2855 (2005)
35. A. Abbasfar, D. Divsalar, K. Yao, Accumulate repeat accumulate codes, in *IEEE Proceedings of Globecom 2004*, Dallas, TX, USA, 2004, pp. 509–513
36. L. Ping, K.Y. Wu, Concatenated tree codes: a low complexity, high performance approach. IEEE Trans. Inform. Theory special issue on codes on graphs 47(2): 791–799

37. L. Ping, X. Huang, and N. Phamdo, "Zigzag codes and concatenated zigzag codes," IEEE Trans. Inform. Theory special issue on codes on graphs, 2001, Feb., 47(2): 800–807
38. W.K. Leung, G. Yue, L. Ping, X. Wang, Concatenated Zigzag Hadamard codes, IEEE Trans. Inform. Theor. **52**(4), 1711–1723
39. T.J. Richardson, *Multi-edge type LDPC codes, presented at the Workshop honoring Prof. Bob Mc Eliece on his 60th birthday,* California Institute of Technology, 2002, May 24–25

Chapter 17
High Spectral Efficiency Optical Four-Dimensional Modulation

17.1 Introduction

Since the twenty-first century, information science and technology have been in the ascendant. They are the leading force for sustained economic growth. More than 97% of China's information is transmitted through optical communication systems with broadband transmission capabilities. Therefore, optical transmission systems have become an irreplaceable bearer platform for information transmission and exchange in national information infrastructure. The rapid development of communication technology and the emergence of new services have led to a global increase in the amount of information, which puts higher demands on the transmission bandwidth of the backbone network. According to statistics, from 1990 to 2010, the global optical communication capacity increased by 100,000 times. The Spectral Efficiency (SE) increased from 0.25 in 1990 to 2 bit/s/Hz in 2010 which is expected to reach 20 bit/s/Hz by 2020. Ultra-wideband and large-capacity optical transmission technology have become an urgent demand in the global information field. From an international perspective, major developed countries such as Europe, the United States, and Japan attach great importance to the development of ultra-wideband large-capacity backbone transmission technology. The US government is actively pursuing a broadband incentive program to achieve the construction of a large-capacity backbone network across the United States, which is listed as the focus of the economic stimulus plan. The EU continues to focus on research in this area through its 7th Framework Program (FP7) and takes it as an opportunity to promote the exploration and development of large-capacity transmission technology in Europe. Japan has further refined and established the infrastructure of high-capacity broadband backbones by developing national plans such as "Next Generation Broadband Plan 2010" and "i-Japan Strategy 2015". Under the background of globalization, China has firmly grasped the historical opportunity of the upgrading of UWB large-capacity backbone transmission technology, and proposed the national strategy of "Broadband China" as the

J. Yu and N. Chi, *Digital Signal Processing In High-Speed Optical Fiber Communication Principle and Application*,
https://doi.org/10.1007/978-981-15-3098-2_17

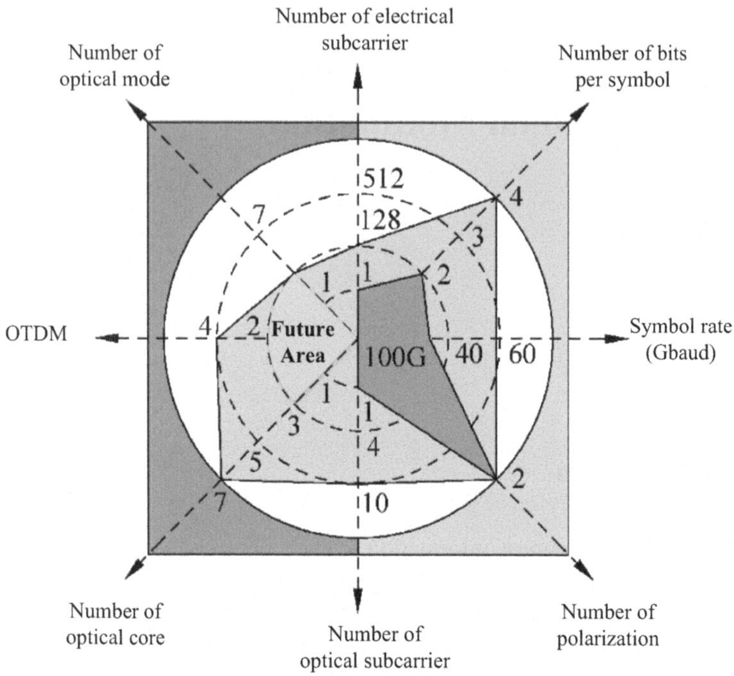

Fig. 17.1 Technical methods for improving the capacity of optical communication system

focus of future development of the national information field. Ultra-wideband large-capacity optical transmission technology has become the top priority of the national information development strategy (Fig. 17.1).

In order to increase the overall capacity of the network, there are three natural or direct choices: increasing the symbol transmission rate, increasing the number of carriers or using a modulation scheme with higher spectral efficiency. Simply increasing the symbol rate is not a good choice because the spectral efficiency has not changed. Moreover, faster symbol rates increase the transmission rate requirements for optical and electrical related devices, especially for Digital to Analog and Analog to Digital Converters (DAC, ADCs) and corresponding electrical signal processing components. Another option is to use multiple carriers instead of single carriers. Under the given target transmission rate and transmission distance, the number of carriers can be selected according to the modulation format and the adjustable parameters to optimize the overall network capacity. In combination with Nyquist filtering or Orthogonal Frequency Division Multiplexing (OFDM) techniques, multi-carrier transmission can be very close to or equal to the symbol rate, forming a so-called "super channel." This channel has high spectral efficiency and low wavelength division multiplexing crosstalk. The last option is to use a more advanced modulation format, such as 16QAM because this approach increases spectral efficiency. For example, the single-carrier 50 Gbaud PM-16-QAM modulation mode can achieve a transmission rate of 400 Gbit/s. However, increasing the modulation order is at the expense of a higher Optical Signal-to-Noise Ratio (OSNR) because the minimum distance between each constellation point is reduced. In addition to using QAM modulation, the use of optimized Four-Dimensional (4D) modulation is another option. In these formats, all four dimensions (amplitude and phase of the two polarization states) of the light field are used for modulation. This is slightly different from the polarization multiplexing format, where the amplitude and phase are jointly modulated in only one polarization and the polarization is used to multiplex two separate polarization branches. Optimized 4D modulation guarantees a better noise margin than traditional PM formats [1].

Therefore, regardless of the increase in optical communication capacity or scientific research, the research on high-spectrum efficiency optical four-dimensional modulation is of great scientific value and practical significance. This is also the focus and commanding height of international and domestic academic competition.

17.2 Constellation Point Distribution and Performance Analysis of Two-Dimensional and Three-Dimensional Constant Modulus Modulation

A common concern in data transmission is the more efficient use of the transport channel. Under the constraints of existing signal power, bandwidth, distortion, and noise, signals with high data rates and low bit error rates are required with reasonable complexity and price. In order to increase the capacity, the modulation order can be increased. It is necessary to keep $d_{\min}$ unchanged to ensure the Bit Error Rate (BER) because the probability of error essentially depends on the distance between the two constellation points in the constellation. To keep $d_{\min}$ unchanged, the signal transmission power needs to be increased, but the high power and high Peak-Average Power Ratio (PAPR) will cause the system to have less tolerance to fiber nonlinearity and the transmission distance will drop sharply.

A coherent optical communication system using an Optical Amplifier (OA) can be well approximated as an Additive White Gaussian Noise channel (AWGN channel). If there is no Inter-Symbol Interference (ISI) during transmission, the transmission vector can be optimally estimated from the received vector. The Maximum Likelihood receiver (ML receiver) performs a symbol decision based on which constellation point is the smallest Euclidean distance from the received signal point. Sometimes it is difficult to accurately calculate the SER due to the complexity of the decision area. We can approximate the SER [2] with the union bound.

$$\mathrm{SER} \leq \frac{1}{M}\sum_{k=1}^{M}\sum_{\substack{j=1 \\ j\neq k}}^{M}\frac{1}{2}erfc\left(\frac{d_{kj}}{2\sqrt{N_0}}\right) \tag{17.1}$$

where d_{kj} is the distance between the constellation point k and the constellation point j, and $erfc(\cdot)$ is the complementary error function? In (17.1), the term containing $erfc\left(\frac{d_{\min}}{2\sqrt{N_0}}\right)$ plays a decisive role, where $d_{\min} = \min_{j\neq k}\{d_{kj}\}$ is the minimum distance of the constellation.

For simplicity, we consider the relationship between constant mode modulation, modulation order, and $d_{\min}$ (Figs. 17.2 and 17.3).

In Figs. 17.2 and 17.3, the distance from each constellation point to the origin is 1 (the radius of the circumscribed circle and the circumscribed sphere is 1), so the average symbol energy of the signal is $E_S = 1$.

In two- and three-dimensional modulation, many systems use one of amplitude, phase, or frequency modulation. But combined amplitude and phase modulation are contemplated in view of the need for information transmission. Considering that a signal in a two-dimensional space has a peak or average transmission power limitation, the literature [3] uses a vector representation signal of a four-dimensional European

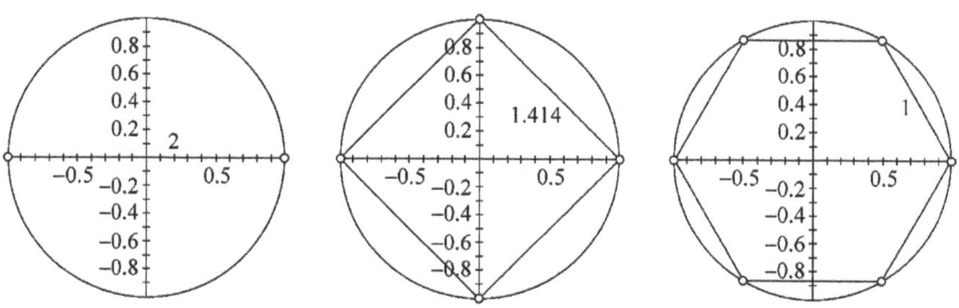

Fig. 17.2 Two-dimensional constant modulus modulation constellation

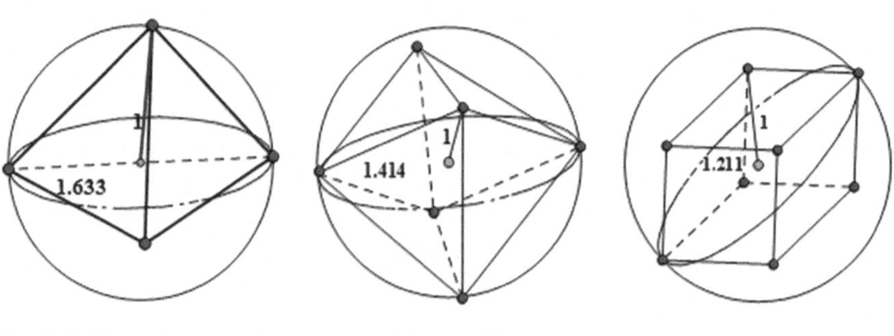

Fig. 17.3 Three-dimensional constant modulus modulation constellation

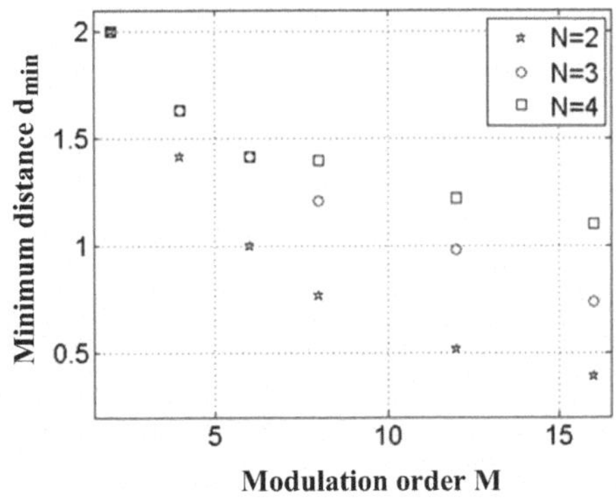

Fig. 17.4 Minimum distance $d_{\min}$ of the constant modulus constellation with different dimensions and orders

space. It considers both variables of two amplitudes and two phases simultaneously with a fixed total power of the signal. It introduces quaternions to represent signal vectors, using quaternion algebras to define the distance between signal vectors and the equivalent representation between signal symbols.

The signal symbol structure designed in [3] is based on a conventional four-dimensional polyhedron, using two types of phase modulated signals as reference symbols. The symbol is evaluated by calculating the distance distribution and the error probability hypothesis. The evaluation found that the group code is optimized by 1.3–3.4 dB over the average power than the optimal phase modulation code, thus demonstrating the superiority of the four-dimensional modulation.

The calculations show that as the modulation order M increases, the $d_{\min}$ between the constellation points decreases more slowly as the number of dimensions N increases, which is shown in the following Fig. 17.4.

The results show that in order to achieve a high transmission rate and appropriate signal transmission power, we can increase the modulation dimension N of the signal.

There are two key issues from the above analysis:

① How to create high-dimensional (four-dimensional) space for high-dimensional modulation?
② How to design a constellation diagram to minimize the average symbol energy E_S without reducing $d_{\min}$?

At the same time, to compare the modulation formats of different orders at the same bit rate R_B, (17.1) is rewritten as

$$erfc\left(\sqrt{\gamma\frac{P}{R_B N_0}}\right) = erfc\left(\gamma\frac{E_b}{R_B N_0}\right) \tag{17.2}$$

where

$$\gamma = \frac{d_{\min}^2}{4E_b} \tag{17.3}$$

and

$$E_b = \frac{\mathrm{P}}{R_B} = \frac{E_S}{\log_2 M} \tag{17.4}$$

E_b is the average bit energy.

The parameter γ is often expressed in dB, called asymptotic power efficiency. When approaching high SNR, the required power is proportional to $1/\gamma$ to achieve the specified SER. We can also understand γ as the sensitivity gain of a certain modulation format compared to BPSK when transmitting data at the same bit rate. Because of BPSK, γ of QPSK and DP-QPSK is 0 dB. In fact, many common modulation formats have a power penalty compared to BPSK. Therefore, it is necessary to consider which γ is the highest for a given N, M, and progressive SNR.

So far, we point out the contradiction between the requirements of high-rate optical fiber communication systems and the low power limitations of fiber nonlinearity, which can be solved by increasing the dimension of the signal modulation. We will answer the two questions ① and ② below.

17.3 Principle and Implementation of Four-Dimensional Multi-level Modulation

As noted above, it is of advantage to increase the modulation dimension N. Next, let's answer the first question, "How to create high-dimensional (four-dimensional) space for high-dimensional modulation?". This section firstly describes the basic properties of an electromagnetic field and how we interpret it as a four-dimensional signal. Next, the implementation of four-dimensional multi-order modulation is introduced. Finally, a four-dimensional modulation codeword is given.

17.3.1 Basic Principles of Four-Dimensional Multi-order Modulation

The electromagnetic field has two orthogonal states in the two polarization components. Therefore, it shares four degrees of freedom across a four-dimensional signal space. The electric field amplitude of the light wave can be written as

$$\vec{E} = \begin{pmatrix} E_{x,r} + iE_{x,i} \\ E_{y,r} + iE_{y,i} \end{pmatrix} = \begin{pmatrix} \|E_x\| e^{i\varphi_x} \\ \|E_y\| e^{i\varphi_y} \end{pmatrix} \tag{17.5}$$

wherein the indices x and y represent the polarization components, and r and i represent the real and imaginary parts of the field, respectively. The values of the phases φ_x and φ_y are $[\pi, -\pi]$. An electric field can equivalently describe its phase, amplitude, and polarization state (the latter being the relative phase and amplitude between the x and y field components) as

$$\begin{aligned}\vec{E} &= \left\|\vec{E}\right\| \exp(i\varphi_a)\vec{J} \\ &= \left\|\vec{E}\right\| \exp(i\varphi_a)\begin{pmatrix} \cos\theta \exp(i\varphi_r) \\ \sin\theta \exp(-i\varphi_r) \end{pmatrix}\end{aligned} \quad (17.6)$$

where $\left\|\vec{E}\right\|^2 = |E_x|^2 + \left|E_y\right|^2$, $\theta = \sin^{-1}\left(\left|E_y\right|/\left\|\vec{E}\right\|\right)$, J represents a Jones vector, which is usually normalized. Note the difference in the relative phase $\varphi_r = (\varphi_x - \varphi_y)/2$ between the absolute phase $\varphi_a = (\varphi_x + \varphi_y)/2$ of the magnetic field and the field vector component. The relative phase $\varphi_r \in (-\pi, \pi]$ describes the ellipticity of the polarization state. In special cases, $\varphi_r = 0, \pm\pi/2, \pi$ represents linear polarization and $\varphi_r = \pm\pi/4, \pm 3\pi/4$ represents circular polarization. Other cases are called elliptical polarization. The angle $\theta \in [0, \pi/2]$ is often referred to as the azimuth because it describes the orientation of the XY plane of the linear polarization state, or more generally the long axis of the elliptical polarization.

Finally, the signal can be represented as a four-dimensional form with a real number:

$$\vec{S} = \begin{pmatrix} E_{x,r} \\ E_{x,i} \\ E_{y,r} \\ E_{y,i} \end{pmatrix} = \begin{pmatrix} \|E\| \cos\varphi_x \sin\theta \\ \|E\| \sin\varphi_x \sin\theta \\ \|E\| \cos\varphi_y \cos\theta \\ \|E\| \sin\varphi_y \cos\theta \end{pmatrix} \quad (17.7)$$

The transmitted optical power is $P = \|S\|^2 = \left\|\vec{E}\right\|^2 = E_{x,r}^2 + E_{x,i}^2 + E_{y,r}^2 + E_{y,i}^2$. Note that the four-dimensional vector should not be confused with the description of the Stokes vector of the polarization state, which is a completely different way and intensity ratio, rather than linear. In the 1990s, three-dimensional (Stokes) space was used as the signal space for polarization shift keying modulation [4]. However, due to the lack of absolute phase description, the constellation points have different absolute phases. But the same polarization coincidence point exists in the Stokes space, so the signal space is not very useful in additive noise coherent communication systems. However, the Stokes space description of the light field is useful when discussing the polarization properties of different modulation formats [2].

17.3.2 Implementation of Four-Dimensional Multi-order Modulation

17.3.2.1 Implementation of Four-Dimensional Modulation of the Electrical Domain

People have already studied "how to create a 4-D constellation space for 4-D modulation". They usually use 2 different carriers or 2 different time slots to achieve 4-D space. Here is a brief description of the case of implementing [3, 5] with dual carriers. Reference [5] designed a 4-D signal based on permutation codes and compared the signal with two-dimensional amplitude and phase modulation (2-D APK) to illustrate 4-D for a given information rate. Modulation can save 1 dB of average power over 2-D modulation. [3] uses the results of regular four-dimensional polytopes to design codes for joint phase and amplitude modulated signals, and uses quaternions to represent the signals. The algorithm of the numerator is used to calculate the distance between the signal points. The proposed in [5] signal has many energy levels. [3] proposes only one energy level, which is superior to [5].

r_1 and r_2 are two amplitude variables, and ϕ_1 and ϕ_2 are two phase variables. During a time period T, the transmitted signal can be expressed as

$$s(t) = r_1\sqrt{\frac{2}{T}}\sin(\omega_1 t + \phi_1) + r_2\sqrt{\frac{2}{T}}\sin(\omega_2 t + \phi_2) \tag{17.8}$$

$\frac{\omega_1 T}{2\pi}$ and $\frac{\omega_2 T}{2\pi}$ must be two different integers to make the two terms in equation (17.8) orthogonal. Equation (17.8) can be expanded as

$$s(t) = a_1\sqrt{\frac{2}{T}}\sin\omega_1 t + b_1\sqrt{\frac{2}{T}}\cos\omega_1 t + a_2\sqrt{\frac{2}{T}}\sin\omega_2 t + b_2\sqrt{\frac{2}{T}}\cos\omega_2 t \tag{17.9}$$

This is a linear combination of vectors in an orthogonal basis with combination coefficients a_1, b_1, a_2, b_2. The transmitted signal is written in the form of a vector, which is $\vec{s} = (a_1, b_1, a_2, b_2)^T$, and written in the form of a quaternion, which is $\sigma = a_1 e + b_1 j + a_2 k + b_2 l$.

17.3.2.2 Four-Dimensional Modulation Coding of the Optical Domain

Although the implementation of 4-D modulation in the electrical domain is complicated, fortunately, it has unique advantages in optical communication systems. Since the coherent optical communication system allows data to be transmitted using two mutually orthogonal polarization directions, in-phase and quadrature components in each direction, it has formed four degrees of freedom (DOF). The method of four-dimensional coding of the optical domain is given below.

Considering two amplitudes and two phase variables simultaneously when the total power of the signal is fixed. The signal is represented as a vector $\vec{s}$ of four-dimensional European space V_4. A signal symbol M is a finite set $\vec{s}_i$ of vectors of unit length in V_4 or equivalent to a series of points in the sphere space S^3. To find a good signal coding, a regular geometry called polytope is used in V_4. It is summarized as follows based on [6, 7].

We firstly define the operation H for the quaternion. In operation H, the addition of quaternions is the addition of components, and the multiplication is determined by the following rules

$$e \cdot \sigma = \sigma \cdot e = \sigma \quad \forall \sigma \in H;$$

$$\begin{aligned} & j^2 = k^2 = l^2 = -e \\ & jk = -kj = l, \ kl = -lk = l, \ lj = -jl = k \end{aligned} \tag{17.10}$$

The element e is the unit element of H, which is equivalent to 1. The element j is equivalent to $\sqrt{-1}$. A quaternion can be written as

$$\sigma = z_1 k + z_2 k = (z_{1,} z_2) \tag{17.11}$$

where $z_1 = a_1 + jb_1$, $z_2 = a_2 + jb_2$, or $z_1 = r_1 \exp\{j\varphi_1\}$, $z_2 = r_2 \exp\{j\varphi_2\}$. Rule (17.11) shows that quaternion algebra operations are not commutative. When the quaternion is $\tau = u_1 + u_2 k$, there is

$$\sigma \cdot \tau = (z_1 u_1 - z_2 \bar{u}_2) + (z_1 u_2 + z_2 \bar{u}_1)k \tag{17.12}$$

When the complex conjugate of σ is defined as $\bar{\sigma} = a_1 e - b_1 j - a_2 k - b_2 l = \bar{z}_1 - \bar{z}_2 k$, then there is

$$\sigma \cdot \bar{\sigma} = \bar{\sigma} \cdot \sigma = z_1 \bar{z}_1 + z_2 \bar{z}_2 = r_1^2 + r_2^2 \tag{17.13}$$

We introduce symbols $N(\sigma) = \sigma \cdot \bar{\sigma}$. Due to $\sigma\sigma^{-1} = \sigma^{-1}\sigma = 1$, each element σ in H has an independent inverse $\sigma^{-1} = \bar{\sigma}/N(\sigma)$. σ satisfied $N(\sigma) = 1$ corresponds to the point on a unit circle in the four-dimensional vector space V_4.

We define the Euclidean distance between two quaternions σ and τ as $d(\sigma, \tau)$.

$$d^2(\sigma, \tau) = N(\sigma - \tau) \tag{17.14}$$

From the above calculations, there is $N(\sigma^{-1}) = N(\sigma)^{-1}, N(\sigma \cdot \tau) = N(\sigma)N(\tau)$. We define two mappings $h_1(\sigma)$ and $h_2(\sigma, \mu)$:

$$h_1(\sigma) : \tau \rightarrow \sigma\tau\sigma^{-1} \tag{17.15}$$

$$h_2(\sigma, \mu) : \tau \rightarrow \sigma\tau\mu^{-1} \tag{17.16}$$

where $\tau \in H, \sigma, \mu \in S^3$. These conversions provide the distance characteristics from the original set $\{\tau_1, \ldots, \tau_m\}$ to the converted set.

For a finite vector set $\{\vec{s}_i\}$, $\vec{s}_i \in V_4$, when all vectors are the same length, that is $\|\vec{s}_i\| = 1$, the vector set is called a signal symbol. References [8] and [9] studied the regular four-dimensional polytopes and used symmetry and rotation in four-dimensional space to classify them. If the distance distribution of two symbols C and C' is the same, these two symbols are the same. The two mappings $h_1(\sigma)$ and $h_2(\sigma, \mu)$ mentioned above retain the distance, so they can be used to build classes of equivalent symbols $h_1(\sigma)C$ and $h_2(\sigma, \mu)C_1$.

Next, we take a double-loop symbol as an example of how to generate a symbol. When $\mu = \cos\alpha \cdot j + \sin\alpha \in S^3$ is converted by $h_1(\mu)$, the result is

$$\sigma = \Big(\cos\tfrac{\pi v}{p} + \cos 2\alpha \sin\tfrac{\pi v}{p} \cdot j, \sin 2\alpha \sin\tfrac{\pi v}{p} \cdot j, v = 0, 1, \ldots, 2p-1$$
$$\Big(-\cos\tfrac{\pi v}{p} + \sin 2\alpha \sin\tfrac{\pi v}{p} \cdot j, -\cos 2\alpha \sin\tfrac{\pi v}{p} \cdot j\Big), v = 2p, 2p+1, \ldots 4p-1$$

When $\alpha = 0, v = 0, 1, \ldots, 2p-1$, the signal is a phase modulated signal. When $v = 2p, \ldots, p-1$, the signal is an amplitude modulated signal. When $\alpha = \pi/4, \sigma_1, \ldots, \sigma_{2p-1}$, it is amplitude modulation. When $\alpha = \sigma_{2p}, \ldots, \sigma_{4p-1}$, it is phase modulation in z_1 plane and is zero in z_2 plane. Signals with amplitude and phase modulation can be generated in two time intervals by selecting other μ values.

In terms of the classification of symbols, we record the symbol elements as m, and directly convert the classification of polytopes into the classification of signal symbols according to the symbol elements. These classes are partially overlapping, so one symbol may belong to more than one class. For any symbol, all signal vectors have the same distance distribution for the other vectors, so all symbols are regular.

1. The symbol of regular crystal graphics, such as $m = 5, 8, 16, 24$.

For single layer structure encoding with $m = 5$, the following symbols are selected $\sigma_1 = (1, 0), \sigma_2 = (u, v), \sigma_3 = (u, -v), \sigma_4 = (\bar{u}, \bar{v}), \sigma_5 = (\bar{u}, -\bar{v})$, and $u = -\frac{1}{4} + j\frac{\sqrt{5}}{4}, u = \frac{5}{4} + j\frac{\sqrt{5}}{4}$. The distance from each vector to another vector is equal to $d = \sqrt{5/2}$ (Fig. 17.5).

For the 16 polytope $G(8)$, when $m = 8$, the simplest way to represent this polytope is to use $\pm e, \pm j, \pm k, \gamma l$. These elements make up a group called a quaternary array

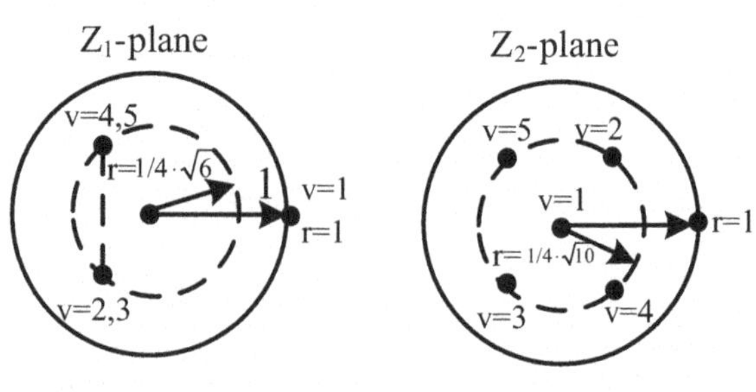

Fig. 17.5 Single layer code when $m = 5$

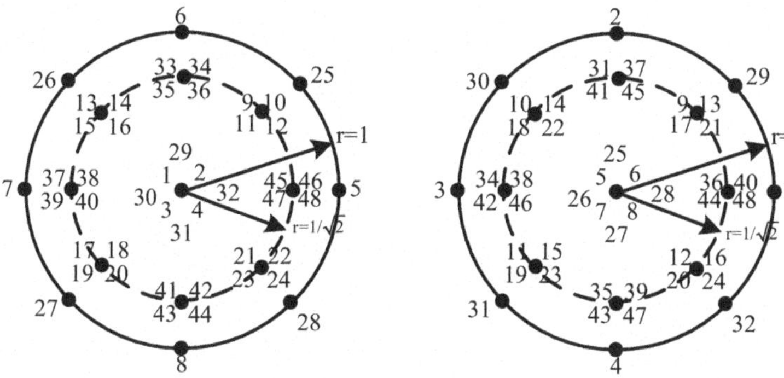

Fig. 17.6 Group code G (8), elements from 1 to 8; group code G (24), elements from 1 to 24; group code G (48), elements from 1 to 48

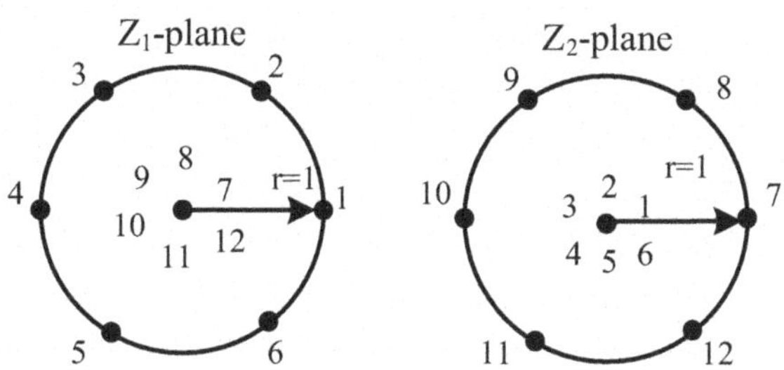

Fig. 17.7 Double ring group code when m = 12

$G(8)$. The vector signal is shown in Fig. 17.6. The symbol generates a set of bi-orthogonal signals. The distance distribution is as follows (Fig. 17.7):

$d:$	$\sqrt{2}$	2
number :	6	1

For the encoding of the cube structure, when $m = 16$, the coordinates of the 16 vertices can be obtained by adding the base vectors of all the positive and negative symbols

$$\frac{1}{2}[\pm e \pm j \pm k \pm l] = \frac{1}{2}[\pm 1 \pm j, \pm 1 \pm j],$$

The distance distribution is as follows (Fig. 17.8):

$d:$	1	$\sqrt{2}$	$\sqrt{3}$	2
number :	4	6	4	1

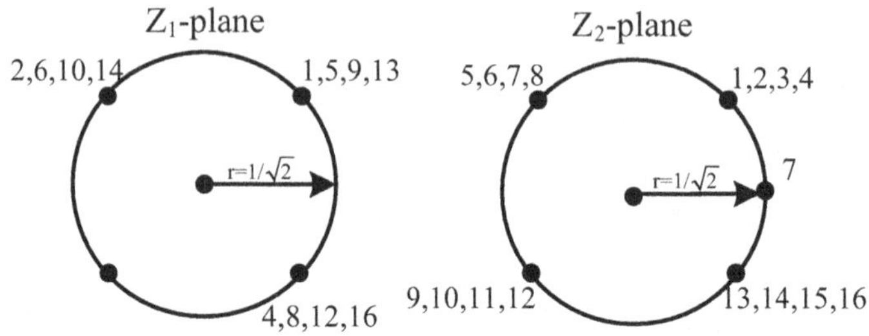

Fig. 17.8 Square code when $m = 16$

2. Double prism symbol, $m = p^2, p = 3, 4, \ldots$

For double prism (phase) encoding, when $m = p^2, p = 3, 4, \ldots$, an element in one symbol $m = p^2$ can be represented as $\sigma(\mu, v) = \left(\frac{1}{\sqrt{2}}e^{\mu}, \frac{1}{\sqrt{2}}e^{v}\right); \mu, v = 0, 1, \ldots, p-1$, where $e = \{j2\pi/p\}$. This is a phase modulation for both carriers with independent phase values. When $p = 4$, the result is a symbol equivalent to the cube, whose distance distribution is a simple modification of the double ring group code.

3. Group code, $m = p, 2p, 4p, 24, 48, 120, p = 2, 3, \ldots$

Due to the performance and algebraic characteristics, group code symbols are the most important classification. Some symbols with regular crystal graphics are also group codes. Group codes can be subdivided into five subclasses:

(1) Cyclic group code: $m = 2, 3, 4, \ldots$

For the cyclic group: $m = 2, 3, 4, \ldots$, the generator of the group code is $\sigma = (e, 0), e = \exp\{j2\pi/m\}$, and the elements in the group can be recorded as $\sigma_v = (e^v, 0), v = 0, 1, \ldots, m-1$. This is a degradation of the signal vector to the two-dimensional subspace, where the signal is pure phase modulation.

(2) Double ring group code: $m = 4p, p = 2, 3, \ldots$

This group code forms an extension of a cyclic group code:

$$\sigma_v = \begin{cases} (e^v, 0);\ v = 0, \ldots, 2p-1 \\ (0, e^v);\ v = 2p, 2p+1, \ldots 4p-1 \end{cases}$$

where $e = \exp\{2\pi j/2p\}$. The generated elements are $\sigma = (\varepsilon, 0)$ and $\tau = (0, \varepsilon)$. It is a special case when $p = 2$. The four elements from σ_1 to σ_4 are $\pm e, \pm j$, and the other four elements are $\pm h, \pm l$. So the double ring group code of $m = 8$ is equivalent to the quaternion code group of $G(8)$. The distance distribution is as follows:

$$\begin{array}{ll} k: & 1, 2, \ldots, p-1 \\ d: & 2\sin\frac{k\pi}{2p}\ 2\ \sqrt{2} \\ \text{number}: & 2\ 1\ \ 2p \end{array}$$

The transmitted signal can be described as a phase modulated signal.

(3) Binary tetrahedron group code $m = 24$

For the binary tetrahedron group, when $m = 24, \omega = \frac{1}{2}(e+j+k+l) = \frac{1}{2}(1+j, 1+j)$ is added into the quaternary array $G(8)$. The result is the binary tetrahedron group $G(24)$, which consists of $G(8)$ and the cosets of $\omega G(8)$ and $\omega^2 G(8)$,where $\omega^3 = (-1, 0)$. When $e = 1/2(1 + j)$, the elements of $G(24)$ are as follows:

$$\begin{aligned} &\pm e, \pm j, \pm k, \pm l \equiv G(8) \\ &(\pm e, \pm e), (\pm \bar{e}, \pm \bar{e}) \equiv \omega G(8) \\ &(\pm \bar{e}, \pm e), (\pm e, \pm \bar{e}) \equiv \omega^2 G(8) \end{aligned}$$

It can be seen that for each carrier combined with amplitude and phase modulation, the amplitude values used are and the phase values used are eight equidistant. So there is a symbol with regular crystal graphics equivalent to $G(24)$.

The distance distribution is as follows:

$$\begin{array}{ll} d: & 1 \ \sqrt{2} \ \sqrt{3} \ 2 \\ \text{number}: & 8 \ 6 \ \ 8 \ \ 1 \end{array}$$

(4) Binary octahedron group code $m = 48$

For the binary octahedron group, when $m = 48, \omega_1 = \frac{1}{\sqrt{2}}(e+j) = \frac{1}{\sqrt{2}}(1, 1)$ is added into $G(24)$. The new group $G(48)$ consists of $G(24)$ and the coset $\omega_1 G(24)$.

The amplitude values of $G(48)$ and $G(24)$ with amplitude modulation are $0, 1/\sqrt{2}$ and 1 with 8 equidistant phase values. The distance distribution is as follows:

$$\begin{array}{lcccccc} d: & \sqrt{2-\sqrt{2}} & 1 & \sqrt{2} & \sqrt{3} & \sqrt{2+\sqrt{2}} & 2 \\ \text{number}: & 6 & 8 & 18 & 8 & 6 & 1 \end{array}$$

(5) Binary icosahedron group code $m = 120$

For the binary octahedron group, when $m = 120$, The group $G(120)$ consists of

$$\begin{aligned} \sigma_1 &= \frac{1}{2}(\gamma e + \gamma^{-1} j + l) = \frac{1}{2}(\gamma + \gamma^{-1} j, j) \\ \sigma_2 &= \frac{1}{2}(e + \gamma^{-1} k + \gamma l) = \frac{1}{2}(1, \gamma^{-1} + \gamma j) \\ \sigma_3 &= 1 = (0.j) \end{aligned}$$

where $\gamma = (\sqrt{5} + 1)/2$. The distance distribution is as follows:

$$\begin{array}{lcccccccc} d: & \sqrt{2-\gamma} & 1 & \sqrt{3-\gamma} & \sqrt{2} & \sqrt{1+\gamma} & \sqrt{3} & \sqrt{2+\gamma} & 2 \\ \text{number}: & 12 & 20 & 12 & 30 & 12 & 20 & 12 & 1 \end{array}$$

When there are three amplitudes on a phase plane, the symbol vector is as follows [6]:

$$\begin{array}{lll} (e^{2\mu+!}, 0) & \mu = 0, 1, \ldots, 9 & \text{Number of vertices} : 10 \\ (e^{2\mu}\cos\lambda, e^{2v}\sin\lambda) & u + v \; is \text{ even} & \text{Number of vertices} : 50 \\ (e^{2\mu}\sin\lambda, e^{2v}\cos\lambda) & u + v \; is \text{ even} & \text{Number of vertices} : 50 \\ (0.e^{2v+!}) & \mu = 0, 1, \ldots, 9 & \text{Number of vertices} : 10 \end{array}$$

where $e = \exp\{j\pi/10\}, 2\lambda = ar\tan 2$

4. Symbols generated by finite groups
5. Symbols generated by finite rotation groups
6. Symbols generated from a limited single group $m = ns$. The elements of the symbol are $\frac{1}{\sqrt{2}}(e_1^{qv+\mu} \cdot e_2^{-(qv'+d\mu)}, e_1^{qv+\mu} \cdot e_2^{qv'+d\mu})$,

where $e_1 = \exp(j\pi/n), e_2 = \exp(j\pi/r), v = 0, 1, \ldots, p-1, v' = 0, 1, \ldots, s-1, \mu = 0, 1, \ldots, q-1$ and $2n = q \cdot p$, $2r = qs$, d must be compatible with q. These symbols essentially have the same structure as the bi-phase code, but also have more degrees of freedom in phase values. How this type of symbol is converted to a class with phase and amplitude modulation is shown in [7].

17.3.2.3 Four-Dimensional Coding Example

A simple 4-D modulation method uses two polarization states, X-pol and Y-pol, and transmits 2-D constellation points in each polarization state, such as PM-16QAM with 8bit/symbol.

Coelho and Hanik [10] first introduced Ungerboeck's Set-Partitioning (SP) method [11] in 4-D modulation to transmit PM-16QAM signals, which they call 32-level Set-Partitioning QAM (32-SP-QAM) and 128-level Set-Partitioning QAM (128-SP [12] QAM). The Ungerboeck's set-partitioning (SP) method is simply a set-partitioning principle. The QAM constellation points are treated as collection elements. Each time a set is divided into two subsets and the elements in the subset are reduced by half. The Euclidean distance of the adjacent constellation points is $\sqrt{2}$ times the original. Compared with the traditional PM-QAM, SP-QAM has a significant improvement in energy efficiency with minor spectral efficiency cost.

Another advantage of SP-QAM is that the hardware structure of the traditional PM-QAM signal can be utilized with minor adjustments without additional cost. Figure 17.9 illustrates the transmission schematic of the PM-16QAM and 128-SP-QAM signals.

Here, PM-16QAM and SP-128QAM can be regarded as four-dimensional modulation signals due to the use of two polarization states of X-pol and Y-pol. They have a similar transmission structure. The last 1 bit of SP-128QAM is the XOR of the other 7 bits, which is regarded as the check bit. The modulation efficiency

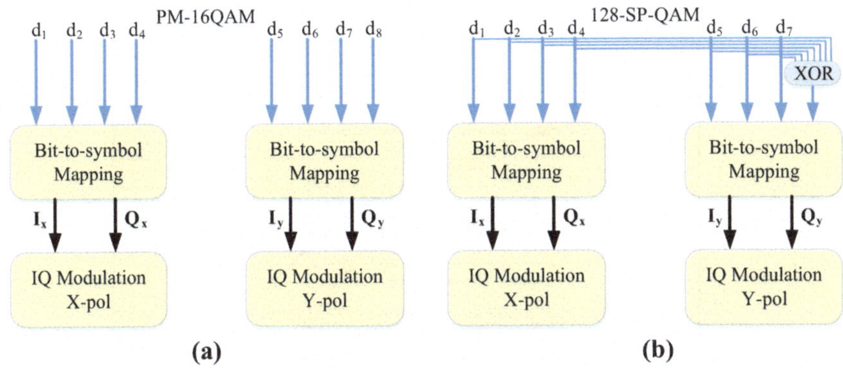

Fig. 17.9 **a** PM-16QAM **b** 128-SP-QAM transmission diagram

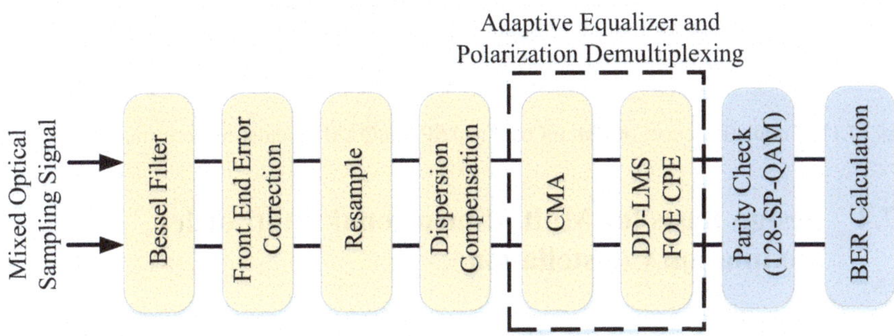

Fig. 17.10 Signal processing block diagram of 128-SP-QAM and PM-16QAM

of PM-16QAM is 8bit/symbol, and the modulation efficiency of SP-128QAM is 7bit/symbol (Fig. 17.10).

Bit check should be considered when decoding SP-128QAM. When the bit check bit finds an error, it needs to modulate the decision constellation point and move the constellation point into the adjacent position. Due to the 4-dimensional modulation, the quaternary constellation points $(X1, X2, X3, X4)$ are considered, where $Xi \in \{\pm 1, \pm 3\}$. When error bits are found by the decoding check, the constellation points have to be re-modulated at the position with the largest error, and then re-decoded. The specific algorithm refers to the literature [11].

The MATLAB equivalent baseband model Monte Carlo simulation is used. The channel is the AWGN channel. Figure 17.11 shows the simulation curves of PM-16QAM and SP-128QAM. It can be seen that the bit error rate is reduced because the Euclidean distance of the SP-128QAM is $\sqrt{2}$ times the distance of the conventional 256QAM, and the bit check and the error correction of constellation points are considered.

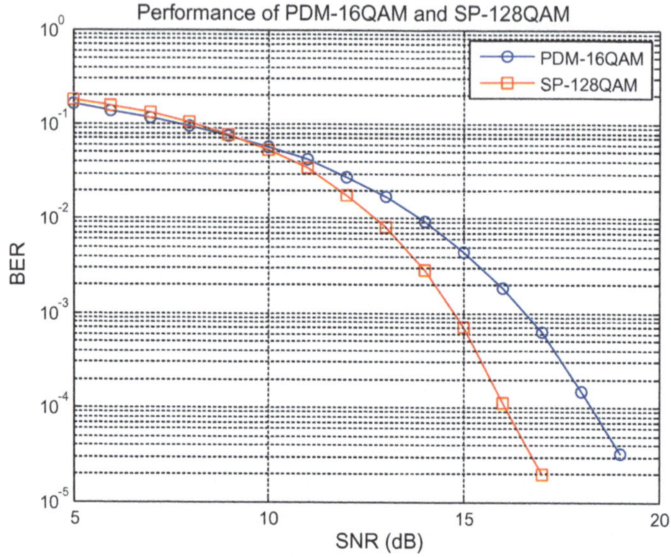

Fig. 17.11 Simulation curve of PM-16QAM and SP-128QAM modulation performance

17.4 Design Basis for Multi-dimensional Multi-order Modulation Constellation

After successfully implementing 4D modulation coding, the second problem raised above is solved. Then we need to consider which γ is the highest for a given dimension N, modulation order M and progressive SNR. This is the core of the second question. This problem is equivalent to a physical problem, which is the Sphere-Packing Problem of M N-D spheres, also known as Kepler Conjecture, or equivalent to Newtonian Kissing-number Problem.

"Most dense" can be understood as minimizing the maximum distance from the center of the sphere to the origin or minimizing the mean distance from the center of the sphere to the origin. In the sphere-packing problem, the former is called the "cluster" problem, and the latter is called a "ball" problem. It is difficult to find the M-order optimal constellation in a given N-D space. A known optimal constellation usually has no formal mathematical proof, and comes from experience which means the better constellation has not been found yet. Since the analytical method is quite difficult, the numerical method is often used to design the optimal constellation. In fact, people usually create thousands of constellations and choose the best one as the optimal constellation.

17.4.1 "Cluster" Problem

"Cluster" can be understood as to how to place M balls to occupy the smallest space. According to the definition of [13], density is

$$\begin{aligned}\Delta \overset{\text{def}}{=} \text{the proportion of the ball in space} &= \frac{\text{volume of the ball}}{\text{volume of the basic area}} \\ &= \frac{\text{volume of the ball}}{\sqrt{\det(\Lambda)}}\end{aligned} \tag{17.17}$$

where Λ represents lattice. A detailed description can be found in [14].

For the 1-D case, as shown in Fig. 17.12, obviously, a line segment can obtain

$$\Delta(1) = 100\% \tag{17.18}$$

These spheres form a 1-D lattice $\mathbb{Z}$.

For the 2-D case, this is the Kepler conjecture on the plane. "The hexagonal arrangement is the highest density on the plane. "This conclusion is often referred to as the Axel Thue's theorem, as shown in Fig. 17.13. The cluster problem on the plane has been well studied, and the optimal constellation of the design is often arranged in a regular hexagon around the origin.

According to the arrangement of Fig. 17.14, the 2-D cluster sphere-packing can be obtained.

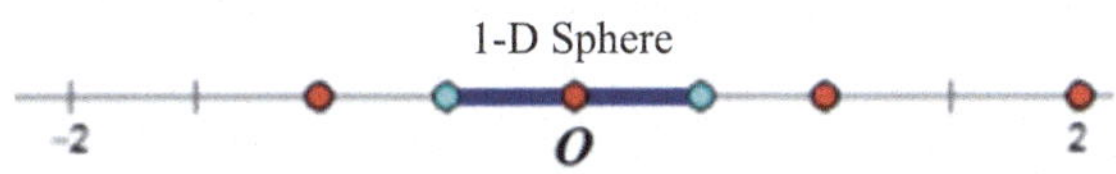

Fig. 17.12 1-D cluster sphere-packing

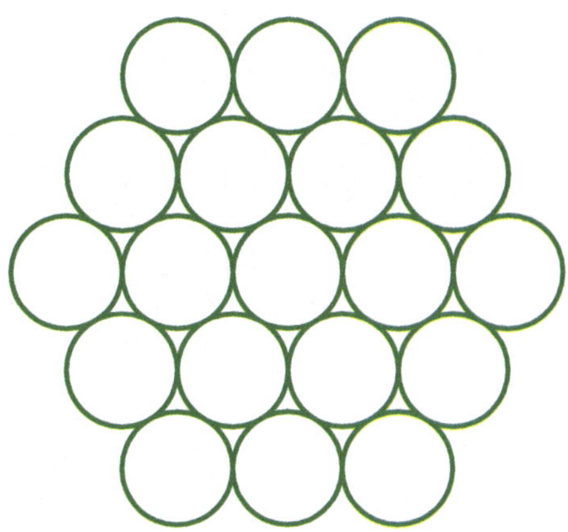

Fig. 17.13 Diagram of the Thue's theorem

Fig. 17.14 2-D cluster sphere-packing

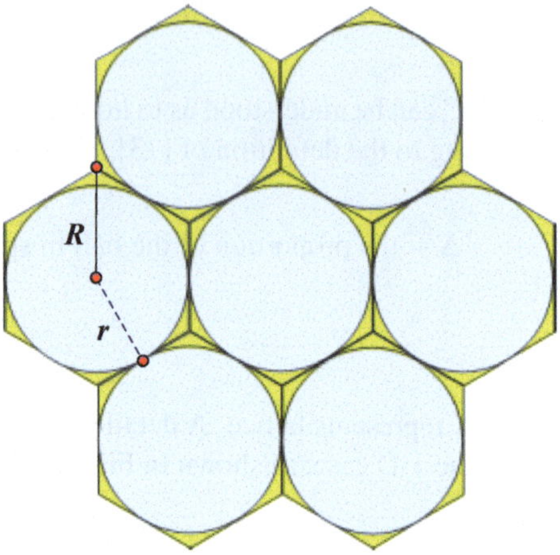

$$\Delta(2) = \frac{\pi}{2\sqrt{3}} = 0.9069 \tag{17.19}$$

For the 3-D case, it is the Kepler conjecture shown in Fig. 17.15. If there are a bunch of small balls in front of you, how to place the most space-saving? Based on experience and intuition, the upper layer should be staggered to the adjacent recesses of the lower layer. It is more reasonable than to place it one by one.

Stack the same size balls in the box and use the Face-Centered Cubic (FCC) stacking method (the upper spheres are placed in the recesses of the lower spheres) to maximize space utilization [14]. The density is

$$\Delta(3) = \frac{\pi}{3\sqrt{2}} = 0.7405 \tag{17.20}$$

Fig. 17.15 Diagram of the Kepler conjecture

For 3-D case, the checkerboard lattice $\mathcal{D}_4$ is the densest arrangement in 4-D space, the density of which is

$$\Delta(4) = \frac{\pi^2}{16} = 0.6169 \tag{17.21}$$

17.4.2 "Ball" Problem

"Ball" problem can be understood as filling M balls of unit size into a larger ball with the smallest radius. Such a description may cause a misunderstanding of the spherical problem. The "ball" structure is like the previous "cluster" structure, but with differences. The "cluster" must satisfy the condition of "occupying the smallest space", which requires the E_s of the signal constellation to be the smallest. The "ball" does not necessarily require the smallest space occupied. Instead, the distance from the center of the sphere which is the farthest from the origin to the origin is required to be the smallest among all the constellations specifying N and M. It is the request for $E_{s,\max}$ of the signal constellation. Readers can gradually understand the difference from the following.

For the "ball" structure, many optimal results in the 2-D case can be found in [15]. We are more concerned about the situation of high-dimension. For N $\geq$ 3, the results of spherical codes can be used to construct a "possible" optimal constellation [16]. In spherical coding, all points are equal in distance from the origin, and this distance is the smallest. This can be understood as the Newtonian kissing-number Problem. The Newton number of 2-D is $\mathcal{K}_2 = 6$. The Newton number of 3-D is $\mathcal{K}_2 = 12$. The Newton number of 4-D is $\mathcal{K}_4 = 24$. Similarly, the Newtonian kissing-number problem proves to be quite difficult. The conjecture of 2-D, 3-D, and 4-D Newton numbers has a long history. Its proof has just been completed recently [14] (Figs. 17.16 and 17.17).

17.5 Typical Multi-dimensional Multi-order Constellation Performance Analysis

Based on the conclusions of "Sphere-Packing Problem" and "Newtonian Kissing-number Problem", we analyze the performance of some typical constellations here. We first analyze the 2-D situation and then analyze the 4-D situation. In the N-D space, the constellation of M points forms a cluster, represented by $\mathcal{C}_{N,\mathrm{M}}$ while the ball is represented by $\mathcal{B}_{N,M}$. The minimum distances of all constellations are set to $d_{\min} = 2$. In this way, all the spheres to be packaged are unit spheres.

Fig. 17.16 Diagram of 2-D Newton number

Fig. 17.17 Diagram of 3-D Newton number

Therefore, (17.3) can be written as

$$\gamma = \frac{d_{\min}^2}{4E_b} = \frac{2^2}{4\frac{E_S}{\log_2 M}} = \frac{\log_2 M}{E_S} = 10\log_{10}\left(\frac{\log_2 M}{E_S}\right)\text{dB} \tag{17.22}$$

17.5.1 N = 2

17.5.1.1 M = 2, 3, 4

These optimal constellations are known as BPSK, 3-PSK, QPSK (Table 17.1).

Table 17.1 Constellation point distribution and basic performance of BPSK, 3-PSK, QPSK

	M = 2	M = 3	M = 4
Modulation format name	BPSK	3-PSK	QPSK
Constellation	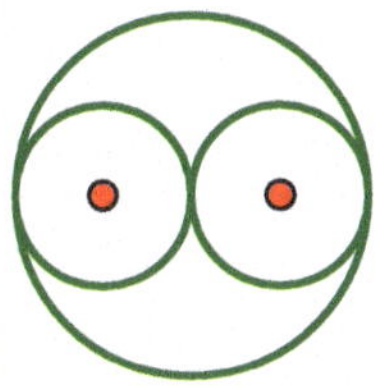		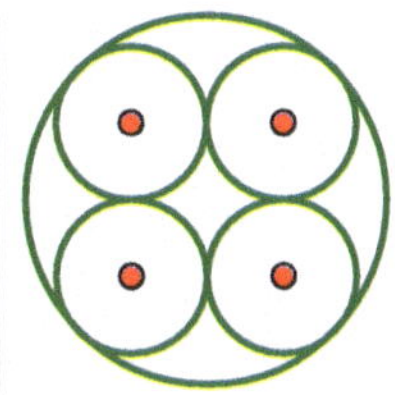
$E_{S,\max}$	1	$\frac{4}{3} = 1.333$	2
E_S	1	$\frac{4}{3} = 1.333$	2
SE (bits/symbol)	1	$\log_2 3 = 1.5850$	2
γ (dB)	0	$\frac{3}{4}\log_2 3 \to 0.75$	0

17.5.1.2 M = 6, 7

M = 7 is the Newtonian kissing-number problem in the 2-D case, $\mathcal{B}_{2,7} = \mathcal{C}_{2,7} = \left\{(0,0), \left(\pm\sqrt{3}, \pm 1\right), (0, \pm 2)\right\}$.

When M = 6, the constellation of the ball structure removes the constellation point at the center of the constellation of M = 7, while the constellation of the cluster structure removes any constellation point around and re-adjusts the center of the constellation.

The point $(0, -2)$ can be removed. To find the new center of the constellation $\mathcal{O}'$ to minimize E_S, let $\mathcal{O}'$ coordinates be (x, y), as shown below (Fig. 17.18).

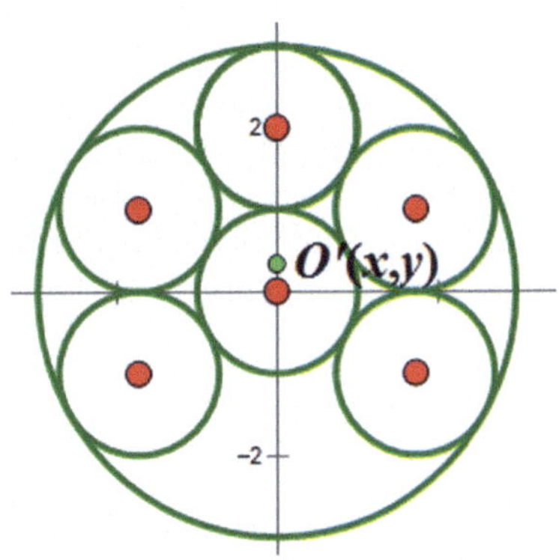

Fig. 17.18 Finding the center of the cluster structure $\mathcal{O}'$ when $M = 6$

$$
\begin{aligned}
\mathrm{E}_S = \mathrm{E}_S(x, y) = \frac{1}{6} \\
\times \Big\{ \left[(x-0)^2 + (y-0)^2\right] + \left[\left(x+\sqrt{3}\right)^2 + (y-1)^2\right] \\
+ \left[\left(x-\sqrt{3}\right)^2 + (y-1)^2\right] + \left[\left(x+\sqrt{3}\right)^2 + (y+1)^2\right] \\
+ \left[\left(x-\sqrt{3}\right)^2 + (y+1)^2\right] + \left[(x-0)^2 + (y-2)^2\right] \Big\}
\end{aligned} \tag{17.23}
$$

When

$$\frac{\partial \mathrm{E}_S(x, y)}{\partial x} = \frac{12x}{6} = 0 \tag{17.24}$$

$$\frac{\partial \mathrm{E}_S(x, y)}{\partial y} = \frac{12\mathrm{y} - 4}{6} = 0 \tag{17.25}$$

$(0, 1/3)$ is found as the only stagnation point of the function. Further, the Hessian Matrix can further verify that the stagnation point is the minimum point of the function and also the minimum point. Therefore, the coordinates of $\mathcal{O}'$ are $(0, 1/3)$, and $\mathrm{E}_S = \mathrm{E}_S(0, 1/3) = 29/9 = 3.22$ (Table 17.2).

17.5.1.3 M = 19

This figure is a good illustration of the subtle differences between the cluster structure and the ball structure. As shown above, when M $= 19$, both structures have the symmetry of a regular hexagon, and the middle is a $\mathcal{B}_{2,7}$ structure.

The outermost circle of the cluster structure forms a regular hexagon to satisfy the condition of "occupying the smallest space" which minimizes E_S. The outermost circle of the ball structure forms a circle. It satisfy that the distance from the center of the outermost sphere to the origin is the smallest among all the constellations specifying N and M, which minimizes $E_{s,\max}$.

Figures 17.19 and 17.20 are as large as the dotted circle, which indicates that the $E_{S,\max}$ of the cluster structure constellation is larger.

17.5.2 N = 4

Due to the limitations of our vision, the constellation in the 4-D case is difficult to visualize.

However, some algebraic methods can be used to represent the 4-D case. For example, a 4-D vector (a_1, a_2, a_3, a_4) or a quaternion $\sigma = a_1 e + a_2 j + a_3 k + a_4 l$ is used to represent a constellation point, where the quaternion representation is like

Table 17.2 Constellation point distribution and basic performance when N = 2, M = 6, 7

	M = 6		M = 7
	Cluster	Ball	
Constellation	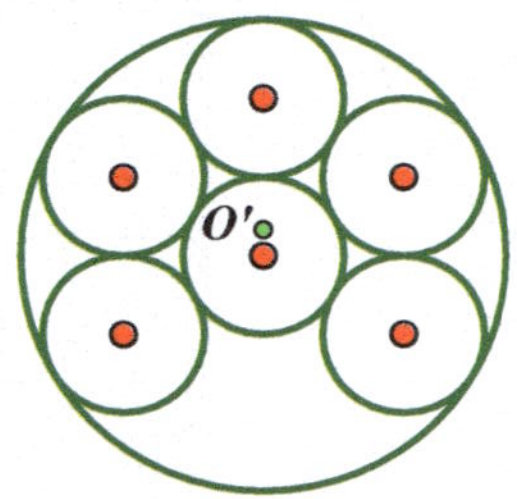	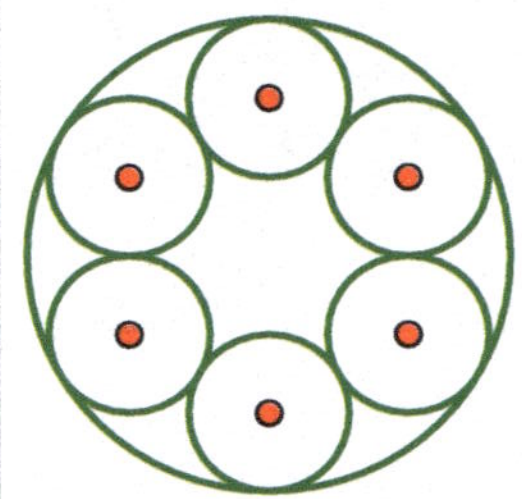	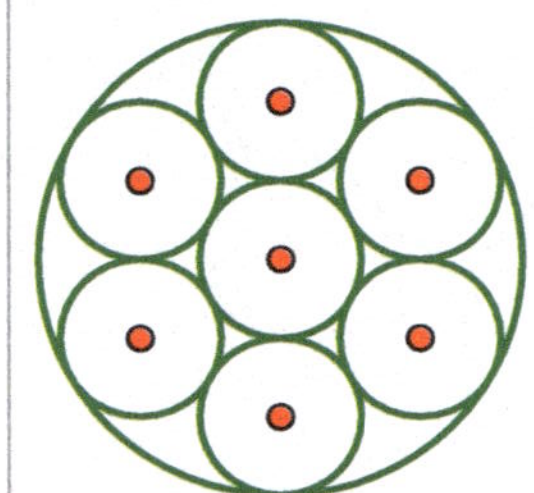
$E_{S,\max}$	4		
E_S	$\frac{29}{9} = 3.22$	$\frac{24}{6} = 4$	$\frac{24}{7} = 3.43$
SE (bits/symbol)	$\log_2 6 = 2.5850$		$\log_2 7 = 2.8074$
γ (dB)	$\log_2 6/E_s \to -0.96$	$\log_2 6/E_s \to -1.90$	$\log_2 7/E_s \to -0.87$

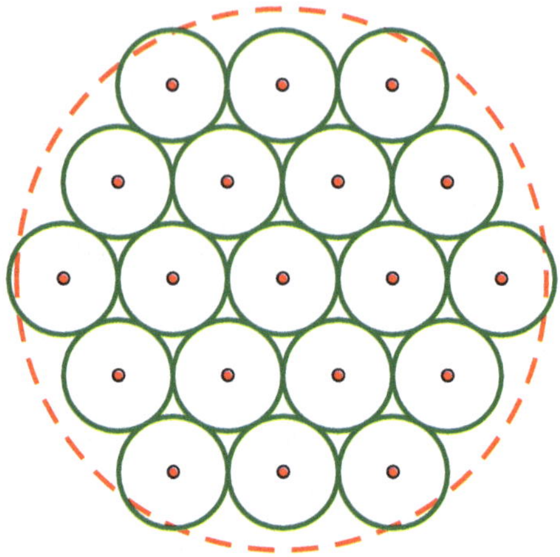

Fig. 17.19 M = 19, cluster structure

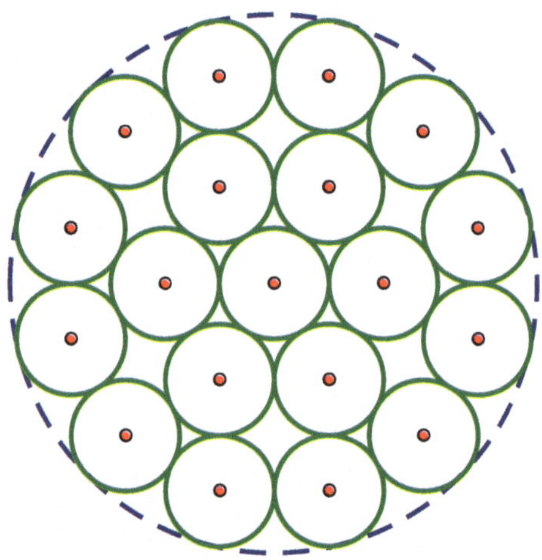

Fig. 17.20 M = 19, ball structure

using a complex number $z = x + yi$ to represent a constellation point on a plane. Similarly, some geometric methods can be used to describe the 4-D case, such as using two orthogonal planes to represent two dimensions (Figs. 17.14, 17.15), or using projections to express the object in 4-D space (like drawing the projection of a cube on the flat paper) (Figs. 17.21 and 17.22).

17.5.2.1 M = 8 (PS-QPSK)

$\mathcal{C}_{4,8}$ is the optimal 4-D constellation that minimizes the average bit energy [17]. $\mathcal{B}_{4,8} = \mathcal{C}_{4,8}$ because all its constellation points are on the 4-D sphere. Its eight constellation points follow the bi-orthogonal form $\mathcal{B}_{4,8} = \mathcal{C}_{4,8} =$

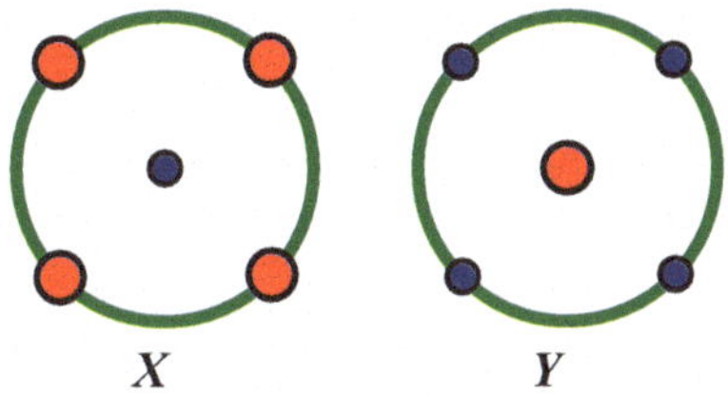

Fig. 17.21 Constellations of PS-QPSK for two orthogonal polarization states. The large red dot indicates that QPSK transmits on the X- polarization and is zero on the Y- polarization. The small blue dot indicates that QPSK transmits on the Y- polarization and is zero on the X- polarization

Fig. 17.22 Constellations of DP-QPSK in two orthogonal polarization states

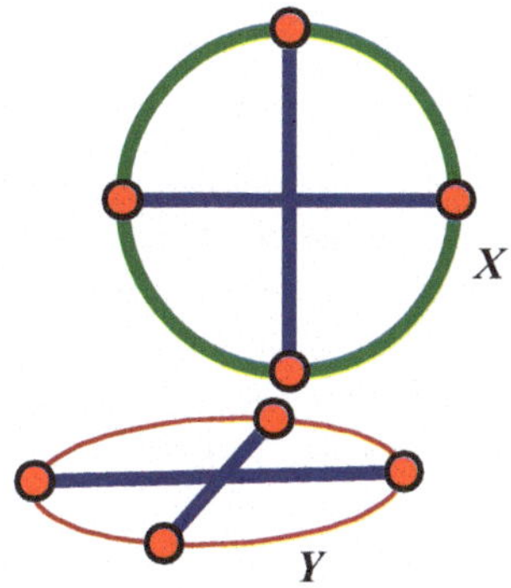

$\left\{\left(\pm\sqrt{2}, 0, 0, 0\right), \ldots, \left(0, 0, 0, \pm\sqrt{2}\right)\right\}$. This structure is called a cross-polytope. After rotating the absolute phase of the PS-QPSK by 45°, another form of coordinates can be obtained. $\mathcal{C}'_{4,8} = \{(\pm 1, \pm 1, 0, 0), (0, 0, \pm 1, \pm 1)\}$.$\mathcal{C}'_{4,8}$ shows that PS-QPSK can only form QPSK in one direction of x or y, unlike DP-QPSK, which can have components in both polarization directions. Therefore, this modulation format is called PS-QPSK (polarization-switched QPSK).

PS-QPSK uses only half points of DP-QPSK, so the SE is reduced to 3 bits/symbol. However, γ is increased by 1.76 dB since its $d_{\min}$ is $\sqrt{2}$ times that of DP-QPSK.

Reference [17] discussed in detail the overall performance and coding of PS-QPSK. References [2, 17] introduced the transmitter, receiver and bit-to-symbol mapping of PS-QPSK.

17.5.2.2 M = 16 (DP-QPSK)

The constellation point coordinates of DP-QPSK are $\mathcal{D}_{4\text{cube}} = \{(\pm 1, \pm 1, \pm 1, \pm 1)\}$, and the constellation points form a cubic. Although it is not the optimal constellation which minimizes $E_{s,\max}$ and E_s. However, due to its symmetry, it is easy for generation and detection. Therefore, it is a popular modulation format in current coherent optical systems.

17.5.2.3 M = 24 (6P-QPSK, 25

M = 25 is the Newtonian kissing-number problem in the 4-D case, where $\mathcal{B}_{4,25} = \mathcal{C}_{4,25} = \mathcal{B}_{4,24} \cup \{(0, 0, 0, 0)\}$. The center of the sphere forms a subset of the lattice $\mathcal{D}_4$.

For M = 24, in order to ensure the symmetry of the constellation, (0, 0, 0, 0) can be removed. The modulation format is called 6P-QPSK, the constellation of which is $\mathcal{B}_{4,24}$. $\mathcal{B}_{4,24}$ is defined as 24 vertices of a 4-D regular polyhedron. The 4-D regular polyhedron is called the 24-cell (icositetrachoron) and is the only regular polyhedron in the 4-D space.

$\mathcal{B}_{4,24}$ has two representations. The first one is $\mathcal{B}_{4,24} = \mathcal{D}_{4\text{cube}} \cup \sqrt{2}\mathcal{B}_{4,8} = \{(\pm1, \pm1, \pm1, \pm1), (\pm2, 0, 0, 0), \ldots, (0, 0, 0, \pm2)\}$, which indicates that DP-QPSK can be extended to a new modulation format 6Pol-QPSK with 24 constellation points, and constant E_s and $d_{\min}$. The second is $\mathcal{B}'_{4,24} = \left\{\sqrt{2}(\pm1, \pm1, 0, 0), \ldots, \sqrt{2}(0, 0, \pm1, \pm1)\right\}$. References [2, 13] introduced the implementation of $\mathcal{B}_{4,24}$ by 9-bit mapping to two adjacent symbols, the γ of which is increased by 0.59 dB than DP-QPSK.

Typically, the light field is elliptically polarized, but there are three important combinations of amplitude and phase, called degenerate polarization states. The first one is linearly horizontal/vertical polarized light (LHP/LVP). The second one is linear $\pm45^{\circ}$ polarized light (L + 45P/L-45P). The third one is right/left circularly polarized light (RCP/LCP). Their mathematical formulas and the corresponding diagrams for each case are as follows (Table 17.3):

For the modulation format with constellation $\mathcal{B}_{4,24}$, its polarization is just the above three degenerate polarization states, so it is called 6P-QPSK. A detailed introduction of the field, polarization, etc., can be found in Ref. [18]. $\mathcal{C}_{4,24}$ can be obtained by removing a non-zero point of $\mathcal{C}_{4,25}$ and translating the constellation. Its γ is 0.79 dB higher than DP-QPSK. We list some typical representatives of optical 4-D modulation formats: DP-QPSK, PS-QPSK, and 6P-QPSK. A summary of the above typical 4-D modulation constellations is shown in Table 17.4.

Nowadays, with the widespread deployment of EDFAs, fiber capacity is less affected by low optical power. However, nonlinearities like self-phase modulation (SPM) and cross-phase modulation (XPM) are becoming limiting factors. When the dispersion is neglected and the damage to signals caused by the SPM and XPM is studied alone, it indicates that the SPM usually has little to do with the equal-amplitude format, while the phase shift introduced by XPM is proportional to the instantaneous power in all WDM channels. The format with large γ allows the transmission power to be reduced, thus introducing a smaller nonlinear error. When the bit rate is the same, the power required by PS-QPSK is 1.76 dB smaller than DP-QPSK. Some studies also show that PS-QPSK has better tolerance to XPM nonlinearity than DP-QPSK.

It is worth noting that since SPM and XPM are dependent on instantaneous power, when the nonlinear phenomenon is obvious, it is better to minimize $E_{S,\max}$ than the average symbol power E_S. It means that the spherical constellation performs better.

Table 17.3 Degenerate polarization states (Ref. [18])

LHP: $E_{0y} = 0$ 	LVP: $E_{0x} = 0$ 
L+45P: $E_{0x} = E_{0y} = E_0, \delta = 0$ 	L-45P: $E_{0x} = E_{0y} = E_0, \delta = \pi$ 
RCP: $E_{0x} = E_{0y} = E_0, \delta = \pi/2$ 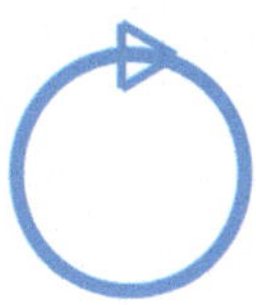	LCP: $E_{0x} = E_{0y} = E_0, \delta = -\pi/2$ 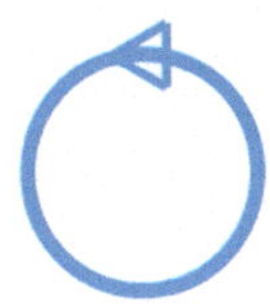

Table 17.4 Constellation point distribution and basic performance of DP-QPSK, PS-QPSK, and 6P-QPSK

	M = 8	M = 16	M = 24
Modulation format	PS-QPSK	DP-QPSK	6Pol-QPSK
Projection	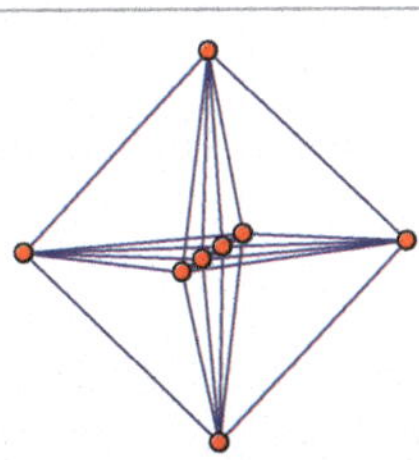 Cross polyhedron	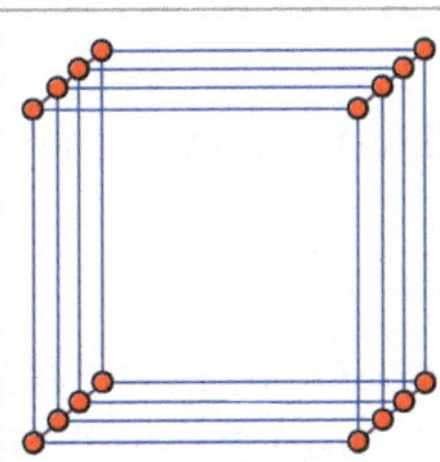 Cubic	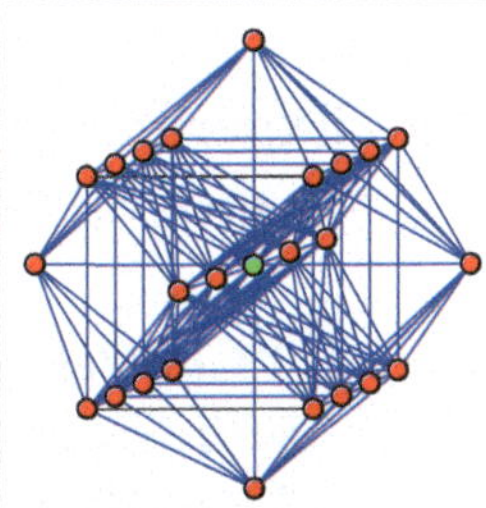 Icositetrachoron
$E_{S,\max}$	2	4	4
E_S	2	4	4
SE (bits/symbol)	3	4	$\log_2 24 = 4.5850$
γ (dB)	$3/2 \rightarrow 1.76$	0	$\log_2 24/E_s \rightarrow 0.59$

(continued)

Table 17.4 (continued)

	M = 8	M = 16	M = 24
Coordinates	$\mathcal{B}_{4,8} = \mathcal{C}_{4,8} = \left\{ \begin{array}{c} \left(\pm\sqrt{2}, 0, 0, 0\right), \\ \cdots \\ , \left(0, 0, 0, \pm\sqrt{2}\right) \end{array} \right\}$ $\mathcal{C}'_{4,8} = \left\{ \begin{array}{c} (\pm 1, \pm 1, 0, 0), \\ (0, 0, \pm 1, \pm 1) \end{array} \right\}$	$\mathcal{D}_{4\text{cube}} = \{(\pm 1, \pm 1, \pm 1, \pm 1)\}$	$\mathcal{B}_{4,24} = \mathcal{D}_{4\text{cube}} \cup \sqrt{2}\mathcal{B}_{4,8} = \left\{ \begin{array}{c} (\pm 1, \pm 1, \pm 1, \pm 1), \\ (\pm 2, 0, 0, 0), \\ \ldots, \\ (0, 0, 0, \pm 2) \end{array} \right\}$ $\mathcal{B}'_{4,24} = \left\{ \begin{array}{c} \sqrt{2}(\pm 1, \pm 1, 0, 0), \\ \ldots, \\ \sqrt{2}(0, 0, \pm 1, \pm 1) \end{array} \right\}$

Considering $E_{S,\max}$, many cluster constellations perform much worse, while spherical constellations perform well in $E_{S,\max}$ and E_S. In contrast, the cost of using a spherical constellation when the E_S is limited is less than using a cluster constellation when the $E_{S,\max}$ is limited. Therefore, when $E_{S,\max}$ and E_S are both limited, the robustness of spherical constellations is better than cluster constellations for most nonlinear damage.

17.6 Experimental Demonstration Based on Geometric Shaped Signal

17.6.1 A Constellation-Designed Method for 32QAM with Geometric and Probabilistic Shaping (G/PS 32QAM)

We proposed a constellation-designed method for 32QAM with geometric and probabilistic shaping (G/PS 32QAM), which is not limited by symmetry [19]. The G/PS-32QAM design method is introduced using pairwise optimization (PO) algorithm and constant composition distribution matcher (CCDM) implemented by arithmetic coding. It could improve the generalized mutual information (GMI) of PS-32QAM by ~0.06 bits/symbol at SNR = 13 dB. The constellation diagram of the G/PS-32QAM and the probability of each point are depicted in Fig. 17.23.

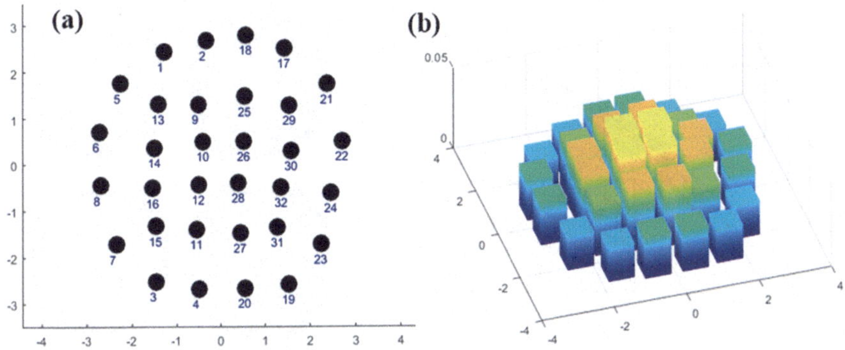

Fig. 17.23 Constellation diagram **a** and graphical illustration **b** of G/PS-32QAM

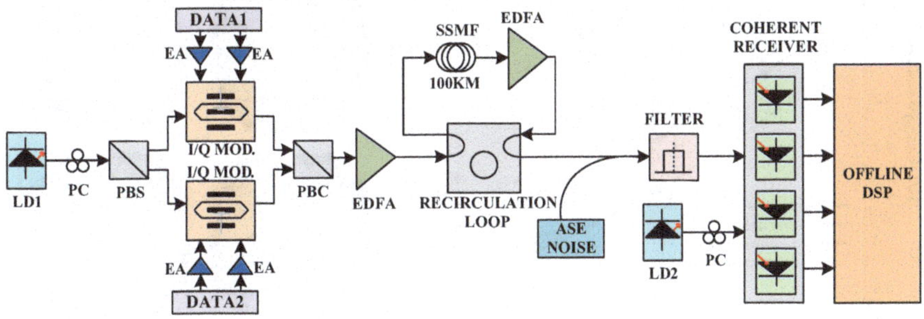

Fig. 17.24 Simulation setup for PDM multi-span single-carrier system

The proposed G/PS 32QAM is adapted to a 50 Gbaud polarization division multiplexed (PDM) multi-span optical fiber system, which is established on the simulation platform shown in Fig. 17.24. The performance of the designed G/PS 32QAM is analyzed at the measured OSNR region from 17.5 dB to 20.5 dB, with transmission distance from 1500 to 2300 km, and with transmission rate from 20 GBaud to 60 GBaud. As shown in Fig. 17.25, compared with PS-32QAM, GS-32QAM, and R-32QAM, the designed G/PS 32QAM can always outperform the three formats under the measured circumstances. The simulation results show that the proposed 32QAM with GS and PS has a gain of 0.04bits/symbol/pol and 0.16bits/symbol/pol over PS-32QAM and R-32QAM at 19 dB OSNR, respectively, and when the net data rate reach 211Gbit/s, G/PS-32QAM is capable of achieving about 0.3 dB and 0.1 dB OSNR gain over PS-32QAM and R-32QAM, respectively. When the net data rate reaches 213Gbit/s, G/PS-32QAM can be transmitted ~ 60 km and ~ 210 km longer than PS-32QAM and R-32QAM, respectively. Moreover, when OSNR is set to 19 dB and signal bandwidth is 40Gaud, the net rate of G/PS-32QAM is 2Gbit/s and 7.5Gbit/s larger than PS-32QAM and R-32QAM, respectively.

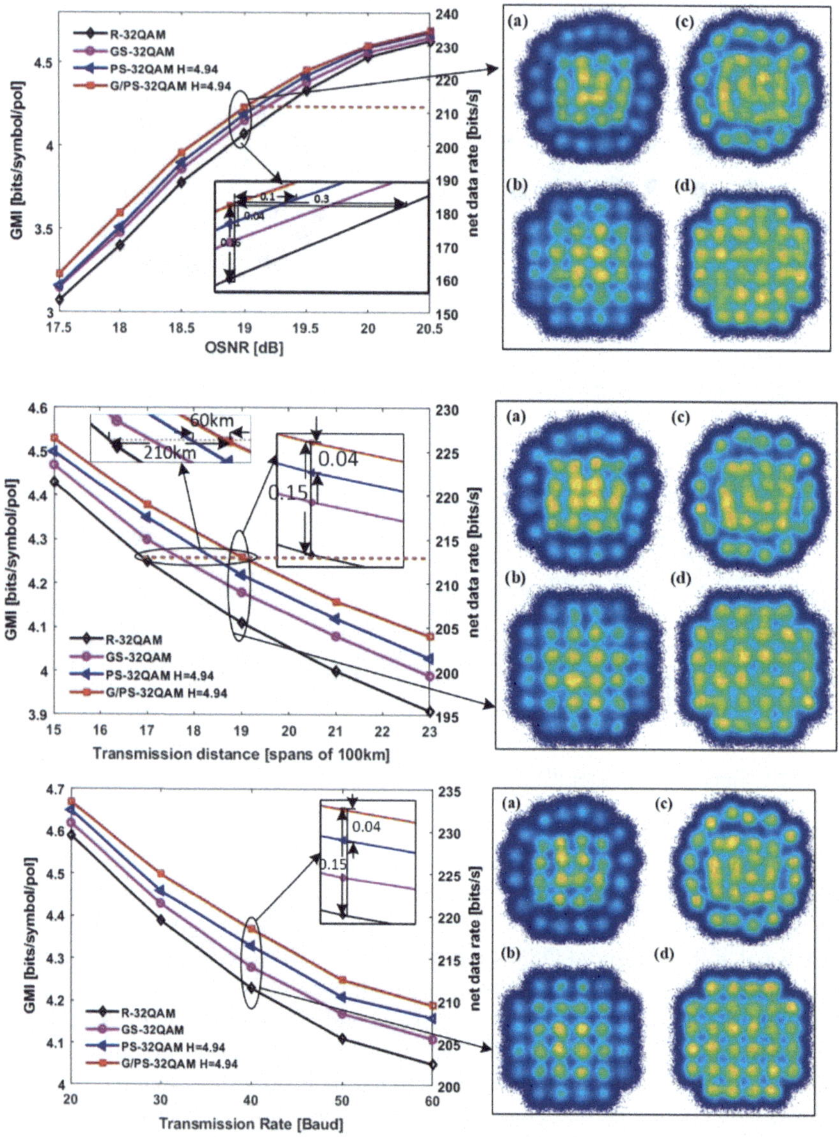

Fig. 17.25 GMI versus OSNR/transmission distance/transmission rate with recovered constellation for **a** G/PS-32QAM, **b** PS-32QAM, **c** GS-32QAM, **d** R-32QAM

17.6.2 *Experimental Demonstration Based on PS-Star-16QAM*

We investigate and compare the performance of star-shaped 16QAM (Star-16QAM) and square-shaped 16QAM (Square-16QAM) in the probabilistic shaping (PS) and uniform schemes with coherent detection [20]. Figure 17.26 shows the probabilities of signal points in constellations for Uniform/PS-Square-16QAM and Uniform/PS-Star-16QAM, respectively. PS technology tactfully transmits lower-energy signal points near the origin more than higher-energy signal points far from the origin to reduce the average transmitted power. In general, the input signal points are usually distributed with uniform probabilities such as Uniform-Square-16QAM shown in Fig. 17.26a and Uniform-Star-16QAM shown in Fig. 17.26c. However, for a PS scheme, the input signal points are transmitted with nonuniform probabilities as illustrated in Fig. 17.26b, d.

The BER performance of 160 Gb/s PDM-16QAM signal at different situations is measured in Fig. 17.27. The curves of BER versus the OSNR input into optical hybrid are measured in eight different cases. As shown in Fig. 17.27, in the case

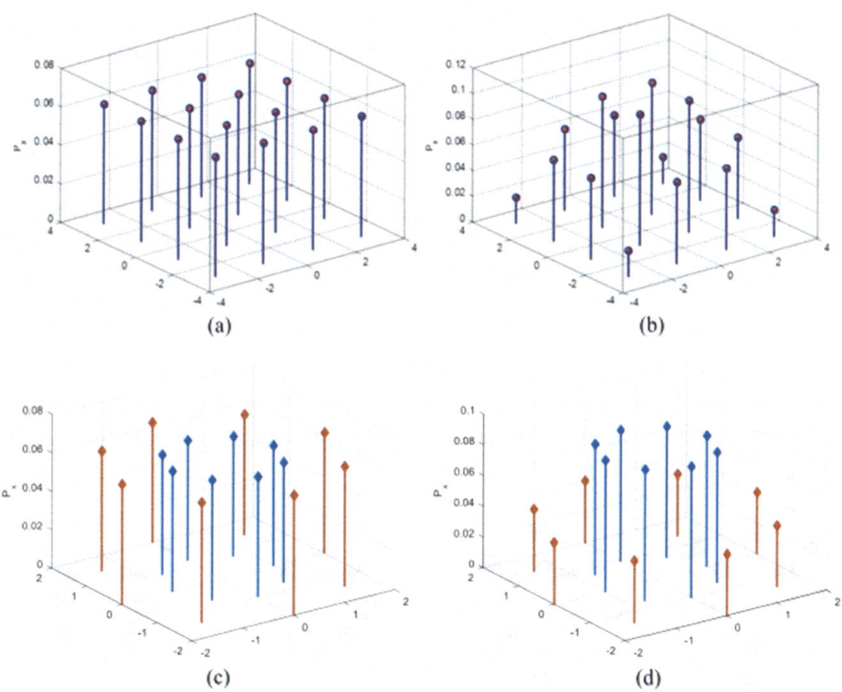

Fig. 17.26 Probabilities of signal points in constellations for Uniform-Square-16QAM and PS-Square-16QAM: **a** Uniform-Square-16QAM constellation, **b** PS-Square-16QAM constellation, **c** Uniform-Star-16QAM constellation, and **d** PS-Star-16QAM constellation

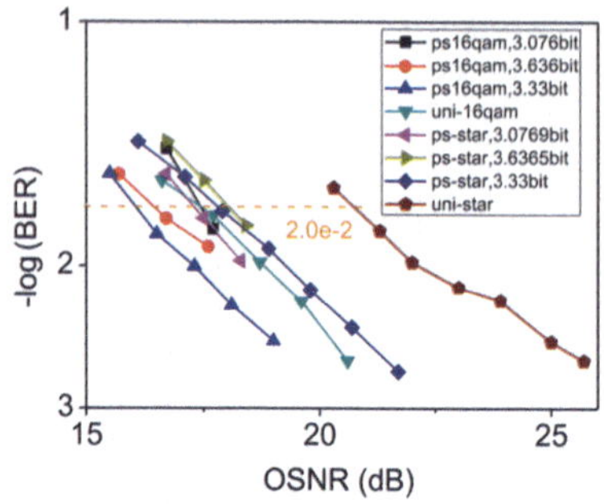

Fig. 17.27 BER versus OSNR input into optical hybrid

of the same bit rate, PS-16QAM shows superiority over Uniform-16QAM in BER performance for both Square-16QAM modulated schemes and Star-16QAM modulated schemes. In the Star-16QAM schemes, the BER performance will be better when the bits per symbol are fewer. What is more, with the use of PS technology, the BER of Star-16QAM modulated schemes can be improved by about 3 dB. While the BER improvement of Square-16QAM modulated schemes is only 1.5 dB by employing PS. Therefore, the performance improvement introduced by PS technology is more remarkable in Star-16QAM modulated schemes. This fully proves the great significance of applying PS in Star-16QAM modulated schemes. As can be seen from Fig. 17.27, PS-Square-16QAM can achieve a better BER performance than PS-Star-16QAM. The reason is that, when the average power of the signal is fixed, the minimum Euclidean distance of the square-16QAM is larger than that of the star-16QAM which leads to a better white Gaussian noise resisting ability. Therefore, the PS square-16QAM can achieve a better BER performance than PS-star-16QAM. The recovered constellations of the 160 Gb/s Uniform/PS-Square-16QAM signal and the 160 Gb/s Uniform/PS-Star-16QAM signal are shown in Figs. 17.28 and 17.29, respectively.

With the help of PS technology, the BER improvement achieved in the PS-Star-16QAM scheme is greater than that of the PS-Square-16QAM when compared with the uniform schemes in our numerical simulation and experiment. Therefore, the

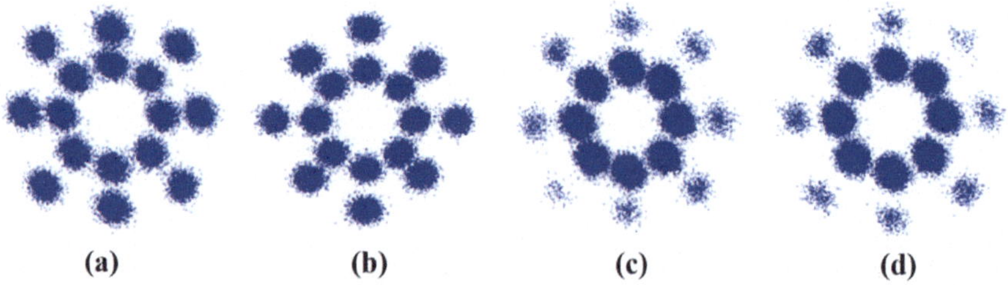

Fig. 17.28 The recovered constellations of the 160 Gb/s Uniform/PS-Square-16QAM signal

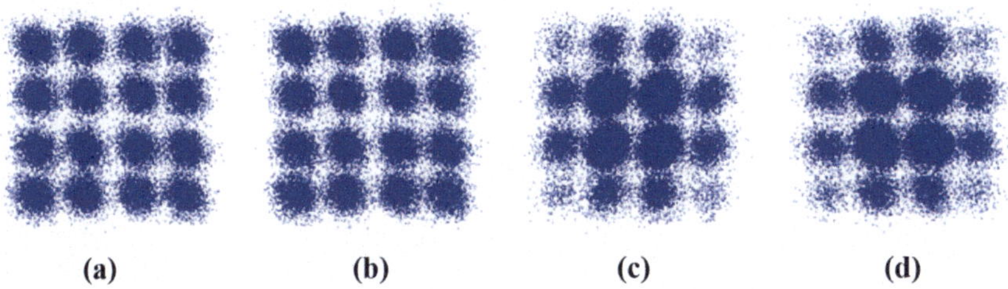

Fig. 17.29 The recovered constellations of the 160 Gb/s Uniform/PS-Star-16QAM signal

PS-Star-16QAM shows superiority over the PS-Square-16QAM in terms of BER improvement.

17.6.3 CAP-PS-9QAM Gray Mapping Scheme in IM/DD System

Based on the comprehensive consideration of shaping gain and coding impairment, we proposed a probabilistically shaped 9QAM scheme based on Gray mapping coding to reduce coding impairment, and adopted Carrier-less Amplitude Phase (CAP) in a single-carrier IM/DD system. The modulation method generates a probabilistic shaped 9QAM signal and verifies its performance. In this scheme, PS-9QAM is combined with Huffman coding. The process of constellation mapping is implemented by the Huffman tree as shown in Fig. 17.30a. As seen in Fig. 17.30b, the code of each constellation point is generated by the Huffman tree and the corresponding probability histogram is shown in Fig. 17.30c. Let L be the length of the symbol to be transmitted. The random binary bit sequence is first transmitted into the $4 \times L$ interleaver. Unlike the traditional 9QAM shaping, the Gray mapping scheme does not need to write redundant bits in the empty space of the interleaver. Instead, the data of each column is mapped into a bit sequence of 4 bits in length, and the proposed

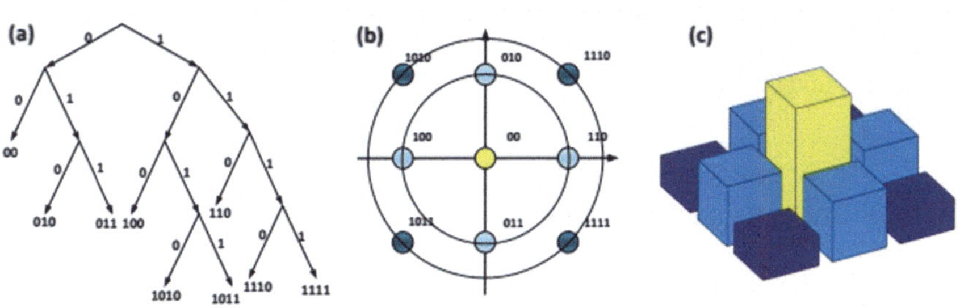

Fig. 17.30 **a** 9QAM Huffman tree **b** 9QAM constellation diagram **c** 9QAM constellation point probability histogram

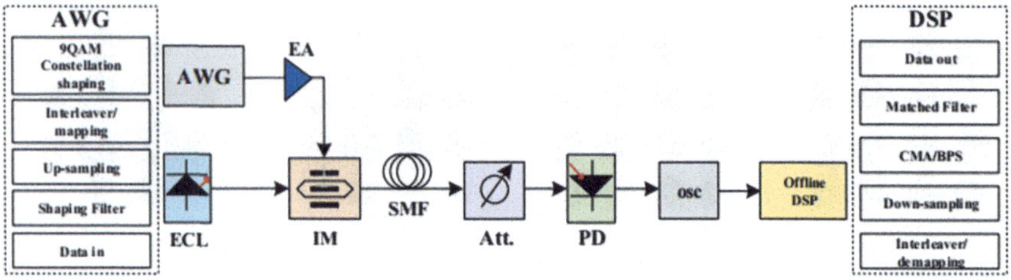

Fig. 17.31 IM/DD experimental setup

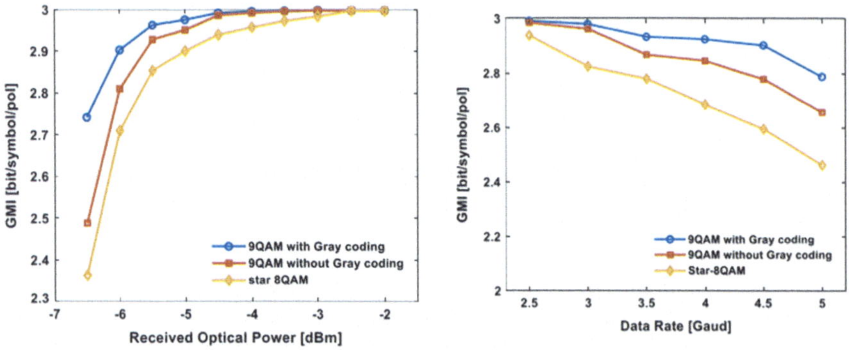

Fig. 17.32 Comparison of GMI performance under different schemes

constellation mapping scheme has the same spectral efficiency and complexity as the conventional scheme. The coding impairments of 9QAM with Gray coding is smaller than 8QAM, so it has better noise tolerance.

In order to verify the performance of the proposed scheme, we built up a 2-5 GBaud single-carrier IM/DD system as shown in Fig. 17.31. The GMI performance of different schemes under different received optical power is shown in Fig. 17.32. Compared with 9QAM and Star 8QAM which do not use Gray coding, the GMI of 9QAM with Gray coding increases by 0.034 bits/symbol and 0.111 bits/symbol, respectively. The GMI performance of different schemes under different baud rates is also shown in Fig. 17.32. The GMI of the proposed scheme is increased by 0.131 bits/symbol and 0.324 bits/symbol.

17.7 Summary and Outlook

In order to meet the demand for high-speed optical communication systems, we can use higher SE and higher-order modulation to increase system capacity. But the signal transmission power can't be too high due to the fiber nonlinearity. To ensure a larger minimum constellation distance $d_{\min}$, the dimension of the modulated signal can be increased. This paper briefly reviews the method of implementing 4-D modulation in the electric domain with dual carrier and optical domain with polarization states. It shows that the optimal constellation with given dimension N and modulation order M which satisfies the minimum E_S and $E_{S,\max}$ is equivalent to the sphere-packing problem and Kepler conjecture. This paper summarizes the optimal constellations of $M = 8, 16, 24, 25$ when $N = 4$ and $M = 2, 3, 4, 6, 7, 19$ when $N = 2$ and compares their progressive power efficiency. It also discusses 4-D modulation tolerance to fiber nonlinearity and robustness.

The popularity of multi-dimensional and multi-order modulation has made this topic a hot topic for researchers. Regardless of the improvement of optical communication capacity or scientific research, the research on optical four-dimensional modulation with high-spectrum efficiency is of great scientific value and practical significance. It is also the focus of international and domestic academic competition. As research progresses, we must change our mindset, from the previous design of the optimal constellation to the comprehensive consideration of bit rate, baud rate, spectral efficiency, power efficiency, power cost, tolerance to nonlinearity, and robustness. They are used to design a modulation format that is suitable for a communication system. It is crucial to choose an appropriate modulation format to build a flexible, low-cost, and high-capacity communication system. The ideas adopted in this paper can provide some reference value for future research on the overall performance of multi-dimensional multi-order modulation formats and comprehensive analysis of a specific modulation method.

References

1. C. Schubert, New trends and challenges in optical digital transmission systems. Optical Communications (ECOC), Netherlands We.1.C.1 (2012)
2. E. Agrell, M. Karlsson, Power-efficient modulation formats in coherent transmission systems. Lightwave Technol. **27**(22), 5115–5126 (2009)
3. L. Zetterberg, H. Brändström, Codes for combined phase and amplitude modulated signals in a four-dimensional space. IEEE Trans. Commun. COM-25(9), 943–950 (1977)
4. S. Benedetto, P. Poggiolini, Theory of polarization shift keying modulation[J]. IEEE Trans. Commun. **40**(4), 708–721 (1992)
5. G. Welti and J. Lee. Digital transmission with coherent four-dimensional modulation[J]. IEEE Trans. Inf. Theory, 1974, IT-20(4):497–502
6. H. Briindstrom and L. H. Zetterberg: Four-dimensional amplitude-phase modulation based on Lie groups, Technical report no 103,August 1975, Telecommunication Theory, Royal Inst. of Techn.,Stockholm, Sweden
7. H. Brandstrom. Classification of codes for phase and amplitude modulated signals in four dimensional base space, Technical report no 105, January 1976, Telecommunication Theory, Royal Inst. of Techn., Stockholm, Sweden
8. H. S. M. Coxeter: Regular complex polytopes. Cambridge University Press 1974
9. P. duVal: Homographies, quaternions and rotations. Oxford University Press, 1964
10. G. Ungerboeck, Channel coding with multilevel/ phase signals. IEEE Trans. Inf. Theory **28**, 55–67 (1982)
11. J. Conway, N. Sloane, Fast quantizing and decoding and algorithms for lattice quantizers and codes. IEEE Trans. Inf. Theory **28**, 227–232 (1982)

12. L. Coelho, N. Hanik. Global otimization of fiber-optic communication systems using four-dimensional modulation formats. European Conference and Exhibition on Optical Communication, (2011), paper Mo.2.B.4
13. D. Ding, J. Zhang, Investigation of a Novel Modulation Scheme of 6PolSK-QPSK Signal in High-speed Optical Transmission System, in *8th International ICST*, Guilin, pp. 745–747
14. H. Conway, N.J.A. Sloane. *Sphere Packings*, Lattices and Groups, pp. 1–62 (1999)
15. E. Specht, *The Best Known Packings of Equal Circles in the Unit Circle* (2009). http://hydra.nat.uni-magdeburg.de/packing/cci/cci.html
16. M. Karlsson, E. Agrell, Power-efficient modulation schemes. OFCR **7**, 219–252 (2011)
17. M. Karlsson, E. Agrell, Which is the most power-efficient modulation format in optical links? Opt. Exp. **17**(13), 10814–10819 (2009)
18. Edward Collett, *Field Guide to Polarization* (Washington, USA, Bellingham, 2005), pp. 7–8
19. Kai Lv, Qi Zhang et al., Constellation-designed 32QAM with geometric and probabilistic shaping for optical fiber system. Opt. Eng. **58**(01), 71 (2019)
20. B. Liu, Y. Zhang, K. Wang, et al. Performance comparison of PS Star-16QAM and PS square-shaped 16QAM (Square-16QAM). IEEE Photon. J. PP(99), 1–1 (2017)

Chapter 18
Machine Learning Algorithm in the Optical Communication System

18.1 Introduction

High spectral-efficient modulation format and digital signal processing (DSP) are the fundamental components of the modern optical communication network. Digital signal processing algorithms flexibly compensate for various impairments in linear systems such as Chromatic Dispersion (CD), Polarization Mode Dispersion (PMD), and the phase noise of the laser [1]. Communication service providers, content providers, and researchers are also working to increase the capacity of existing fiber systems (up to 1 Tbps), to support data center connectivity, cloud computing, and the Internet of Things. Unfortunately, despite the rapid development of digital signal processing technology in recent years, the transmission performance of optical networks has been limited by the Kerr nonlinear effect. Though the digital backpropagation (DBP) technology is the most advanced, it can only partially solve the nonlinear effects caused by noise interference, CD, and PMD [2]. Therefore, the nonlinear effect of fiber is still the fundamental factor limiting the system capacity.

In addition to advanced modulation formats and high-speed signal processing transceivers, the network architecture of fiber-optic communication systems is still undergoing major changes, developing complexly, transparently, and dynamically. Real-time estimation of various channel impairments in dynamic networks, i.e., optical performance monitor (OPM), is indispensable for reliable and complex networks. OPM is also a key technology in elastic optical networks (EON). The EON relies on OPM to know the network status in time and adaptively adjust various network parameters, such as modulation format, data rate, spectrum planning, Forward Error Correction, etc. [3]. Nonetheless, detecting various channel impairment parameters independently, in a timely and cost-effective manner is a difficult task because the various impairments in the network are often aliased and physically indistinguishable [4]. Another requirement for an EON is the ability to automatically identify the modulation format of the transmitted signal at the receiver since the receiver's demodulation algorithm can be correlated with the signal modulation format.

J. Yu and N. Chi, *Digital Signal Processing In High-Speed Optical Fiber Communication Principle and Application*,
https://doi.org/10.1007/978-981-15-3098-2_18

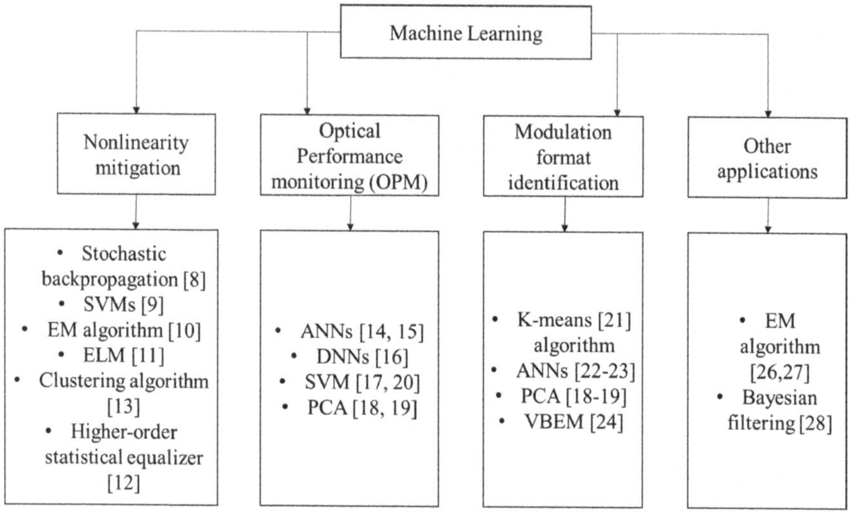

Fig. 18.1 Some key applications of machine learning in optical communication

These require a new signal processing technique to solve the nonlinear damage and accurate estimation of various channel parameters in the optical network. Based on this, machine learning technology is viewed as a new direction to address the above challenges in optical communication systems. In the past decade, machine learning technology has been successfully applied to solve problems such as prediction, classification, pattern recognition, and data mining, and has shown great potential in computer vision, speech recognition, and medical fields [5]. At the same time, machine learning algorithms sometimes show excellent performance for problems such as physics and mathematics that are difficult to accurately describe or analyze, as shown in Fig. 18.1.

Machine learning has provided several powerful techniques, such as estimating parameters from noise data, determining complex mappings between input and output data, inferring probability distributions, predicting output, and performing classifications from historical input data [5, 6]. Choosing the right machine learning algorithm mainly depends on the problem you need to solve.

Nonlinear Suppression: Recently, some studies have applied machine learning algorithms to solve nonlinear damage. These methods learn various characteristics of nonlinear damage from the received signal and establish a probability model to compensate or quantify various damages. Literature [7] proposes that random backward propagation is superior to the traditional backward propagation algorithm (DBP) in suppressing nonlinear phase noise (NLPN). Similarly, in [8], it is proposed to use the Support Vector Machine (SVM) algorithm to learn the optimal decision region, thereby suppressing nonlinear phase noise (NLPN). Other machine learning algorithms for reducing nonlinear effects include expectation maximization algorithm (EM) [9], extreme learning algorithm (ELM) [10], and high-order statistical equalizers [11], clustering algorithms [12]. Although these algorithms show performance improvements in a single channel, the problem of multi-channel fiber nonlinear compensation in long-haul WDM systems is still challenging.

Optical Network Performance Monitoring (OPM): Machine learning algorithms have also been successfully applied to damage monitoring in optical networks. These algorithms include artificial neural networks (ANNs) [13, 14], deep neural networks [15], support vector machines (SVM) [16], principal component analysis (PCA) [17, 18], and various nuclear methods [19]. However, these distributions are currently concentrated in the dispersion management system. Performance monitoring for a dispersion-free system remains to be studied.

Modulation Format Identification (MFI): The modulation format of the signal is identified in the coherent receiver, which is necessary for selecting the appropriate carrier recovery module for the blind equalization. Also, the modulation format information in the OPM device can help select the appropriate OSNR/CD/PMD detection technique [20]. Traditional machine learning algorithms have been used to address modulation format recognition problems such as K-means [21], artificial neural networks (ANNs) [22, 23], principal component analysis (PCA) [18, 19], and variable bayesian expectation maximization algorithms (VBEM) [24]. In [25], a feature learning method based on signal amplitude distribution is proposed to solve the modulation pattern recognition problem in digital coherent receivers.

Other applications: Machine learning algorithms are also used for carrier synchronization [26], reducing laser linewidth effects based on EM algorithms [27] and laser phase noise cancelation based on Bayesian filtering [28]. This chapter will introduce some initial attempts to apply machine learning algorithms to the field of optical communications, and hope to explore more potential of machine learning technology in the field of optical communications.

18.2 Support Vector Machines

The SVM method is a new machine learning method proposed by Vapnik et al. based on statistical learning theory in the early 1990s. It minimizes the actual risk of the learning machine based on the principle of structural risk minimization, by appropriately selecting the function subset and the discriminant function in the subset. This algorithm ensures that the small error classifier obtained through the limited training samples still has the small measurement error of the test set.

The basic idea of SVM is to find the optimal classification hyperplane of two types of samples in the original space in the case of linear separability. In the case of linear inseparability, add the slack variable for analysis, and input the low-dimensional input by using nonlinear mapping. The spatial sample is mapped to the high-dimensional attribute space to make it linear, which makes it possible to analyze the nonlinearity of the sample in a high-dimensional attribute space by using a linear algorithm, and find the optimal classification hyperplane in the feature space, such as Fig. 18.2.

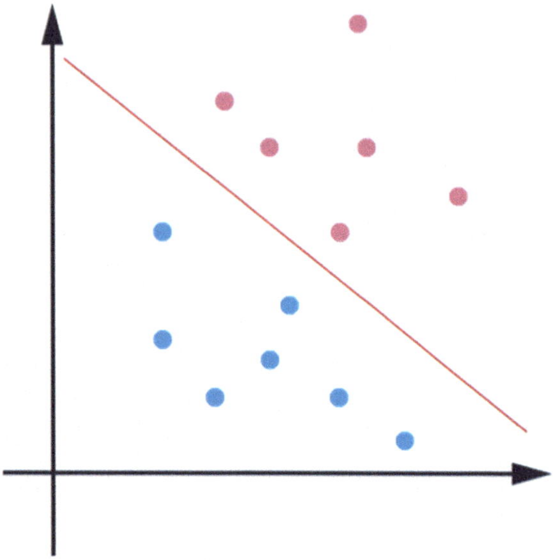

Fig. 18.2 Linear separable sample

18.2.1 Margin and Support Vector

For a given training sample set $D = \{(x_1, y_1), (x_2, y_2), \ldots, (x_m, y_m)\}$, $y_i \in \{-1, +1\}$, The basic idea of the classifier is to find a separate hyperplane in the feature space based on the training set, which can distinguish different samples as much as possible. The division hyperplane corresponds to the following linear equation,

$$w^T x + b = 0 \tag{18.1}$$

where w is the normal vector, which determines the direction of the hyperplane, and b is the intercept, which determines the distance between the hyperplane and the origin. The hyperplane can be determined by the normal vector w and the intercept b. Suppose the hyperplane is able to correctly classify the training samples, for $(x_i, y_i) \in$D, if $y_i = 1$, then $w^T x_i + b > 0$. If $y_i = -1$, then $w^T x_i + b < 0$. Assuming:

$$\begin{cases} w^T x_i + b > +1, \ y_i = +1 \\ w^T x_i + b \leq -1, \ y_i = -1 \end{cases} \tag{18.2}$$

As shown in Fig. 18.3, the training sample points closest to the hyperplane make the Eq. (18.2) equal. They are called the "support vector". The sum of the two heterogeneous support vectors from the hyperplane is

$$\gamma = \frac{2}{\|w\|} \tag{18.3}$$

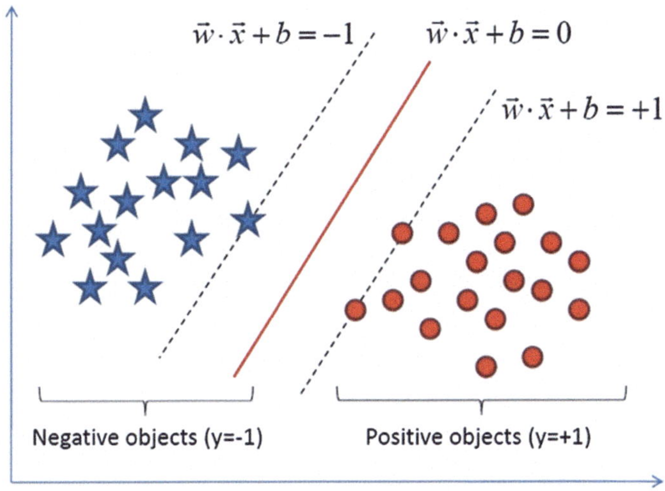

Fig. 18.3 Support vector and margin

γ is called the margin.

To find the dividing hyperplane with the largest interval, that is, to satisfy:

$$\max_{w,b} \frac{2}{\|w\|} \tag{18.4}$$

$$\text{s.t.}\, y_i\left(w^T x_i + b\right) \geq 1, \text{i} = 1, 2, \ldots, \text{m}$$

Notice that the maximized $\frac{2}{\|w\|}$ and minimized $\frac{1}{2}\|w\|^2$ is equivalent. So, the optimization problem of the linear separable support vector machine is

$$\min_{w,b} \frac{1}{2}\|w\|^2 \tag{18.5}$$

$$\text{s.t. } y_i\left(w^T x_i + b\right) \geq 1,\ \text{i} = 1, 2, \ldots, \text{m}$$

This is the most basic model of the support vector machine [29]. In addition, this is a convex quadratic programming problem.

18.2.2 Dual Problem

We can get its "dual problem" by the Lagrange multiplier method, in which the Lagrange multiplier $\alpha_i \geq 0$,

$$\max_{a} \sum_{i=1}^{m} \alpha_i - \frac{1}{2} \sum_{i=1}^{m} \sum_{j=1}^{m} \alpha_i \alpha_j y_i y_j x_i^T x_j$$

$$\text{s.t.} \sum_{i=1}^{m} y_i \alpha_i = 0 \tag{18.6}$$

$$\alpha_i \geq 0, i = 1, 2, \ldots, m$$

where $\alpha = (\alpha_1, \alpha_2, \ldots, \alpha_m)$ solves a, this model is got as,

$$f(x) = w^{\mathrm{T}} x + b = \sum_{i=1}^{m} \alpha_i y_i x_i^{\mathrm{T}} x + b \tag{18.7}$$

18.2.3 Kernel Function

The nonlinear problem can be transformed into a linear problem in a high-dimensional space by nonlinear exchange, and the optimal classification hyperplane is obtained in the transform space. This transformation can be complicated, so this idea is not easy to achieve under normal circumstances. However, in the above dual problem, whether it is the optimization objective function (18.6) or the classification function (18.7) only involves the inner product operation between the training samples $(x_i \cdot x_j)$. There is a nonlinear mapping $\phi : R^d \rightarrow H$ maps the samples of the input space into the high-dimensional (possibly infinite-dimensional) feature space H. When constructing the optimal hyperplane in the feature space H, the training algorithm uses only the dot product in space. That is, $\varphi(x_i) \cdot \varphi(x_j)$, without A separate $\varphi(x_i)$ appears. If we can find a function of K that meets,

$$\mathrm{K}(x_i \cdot x_j) = \varphi(x_i) \cdot \varphi(x_j) \tag{18.8}$$

$\mathrm{K}(x_i \cdot x_j)$ is the Kernel Function. In this way, in the high-dimensional space, only the inner product operation is only needed, and the inner product operation can be realized by the function in the original space, and we do not even need to know the form in the transformation. According to the relevant theory of the functional, if a kernel function $\mathrm{K}(x_i \cdot x_j)$ satisfies the Mercer condition, it corresponds to the inner product in certain transform space. Therefore, the linear classification of a nonlinear transformation can be achieved by using the appropriate inner product function $\mathrm{K}(x_i \cdot x_j)$ in the optimal hyperplane, but the computational complexity is not increased. At this point, the objective function (18.6) becomes

$$\sum_{i=1}^{n} \alpha_i - \frac{1}{2} \sum_{i,j=11}^{n} \alpha_i \alpha_j y_i y_j K(x_i \cdot x_j) \tag{18.9}$$

And the corresponding classification function also becomes

$$f(x) = \sum_{i=1}^{m} \alpha_i y_i K\left(x_i \cdot x_j\right) + b \tag{18.10}$$

The other conditions of the algorithm are unchanged, which is the SVM in the form of a kernel function. In conclusion, SVM maps an input vector to a high-dimensional feature space through some pre-selected nonlinear mapping, and constructs an optimal classification hyperplane in this feature space.

18.2.4 SVM-Based Modulation Format Recognition

SVM can use kernel functions to manipulate high-dimensional data features. This idea can be applied to modulation format recognition. In [30], eight characteristics of the modulated signal are extracted, and the modulation format is identified by SVM. The signal is unsampled by 10 times and divided into 50 frames of 80,000 symbols per frame. Through the eye diagram observed at the receiving end, the mean and variance are calculated in the eye diagram at point A and the point B to obtain four sets of features. The other four sets of features are the overall mean of points A and B, the difference of variance, the ratio of variance, and the mean value of the last frame. Selected features are related to the modulation format. The SVM classifier is trained by different modulation signals, and finally, the modulator learns the mapping function of the data features to the modulation format.

During the training process, we add the 8D vector of the feature and the modulation format tag to the SVM classifier. It is hoped that it has a good generalization ability for new data features while reducing the training error.

Figure 18.4 shows the modulation format recognition results in the feature space. The reason for choosing feature 5 and feature 8 is that these two features have the best distinguishing ability for the modulation format. It can be seen from the figure that the 16QAM and 64QAM signals cannot be distinguished by only using these two features, and the above various modulation formats can be well distinguished only by fully utilizing the eight-dimensional feature vectors.

18.3 BP Neural Network

The BP neural network was proposed by a group of scientists led by Rinehart and McClelland in 1986. It is a multi-layer feedforward network trained by an error inverse propagation algorithm and is one of the most widely used neural network models. The BP network can learn and store many input-output modes mapping relationships without the need to reveal mathematical equations describing such mapping relationships in advance. Its learning rule is to use the steepest descent

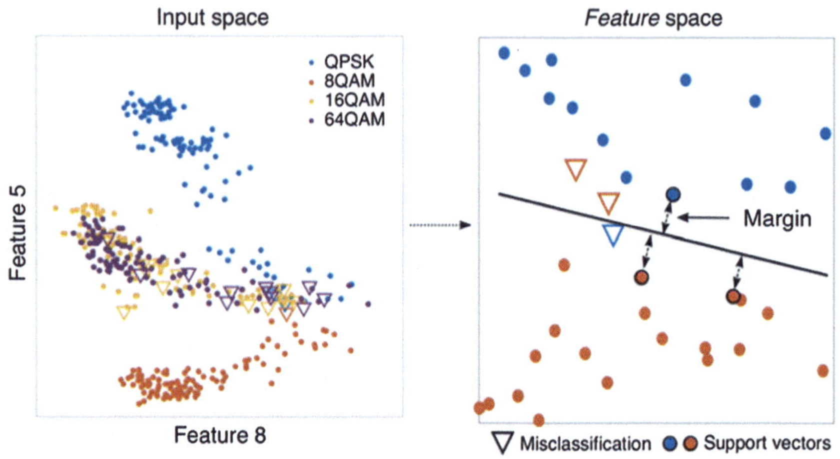

Fig. 18.4 Results of modulation format identification based on feature 5 and feature 8

Fig. 18.5 Schematic of BP neural network

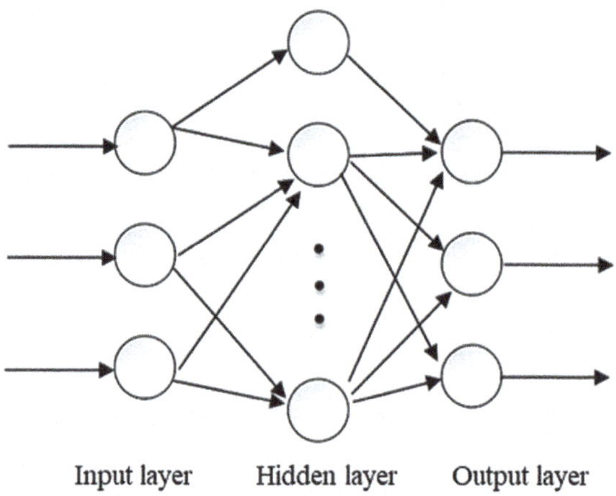

method to continuously adjust the weights and thresholds of the network through backpropagation to minimize the sum of squared errors of the network. As shown in Fig. 18.5, the BP neural network model topology includes an input layer, a hidden layer, and an output layer.

18.3.1 BP Neuron

Figure 18.6 shows the jth basic EP neuron (node), which mimics only the three most basic and important functions of biological neurons: weighting, summation, and transfer. Where $x_1, x_2, x_3, \ldots, x_i, \ldots, x_n$ represent the inputs from neurons 1,

Fig. 18.6 BP neural

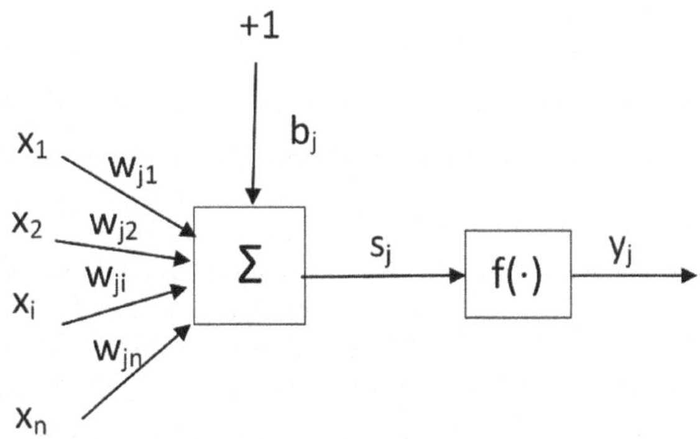

2, …, I, …, n, respectively. and $w_{j1}, w_{j2}, \ldots, w_{ji}, \ldots w_{jn}$ represent the connection strengths of neurons 1, 2, …, i, …, n and the jth neuron, respectively. The connection strength is namely weights; b_j is the threshold; f(·) is the transfer function; y_i is the output of the jth neuron. The net input value S_j of the jth neuron is

$$S_j = \sum_{i=1}^{n} w_{ji} \cdot x_i + b_j = W_j \mathbf{X} + b_j \tag{18.11}$$

where $\mathbf{X} = [x_1, x_2, \cdots, x_n]^{\mathrm{T}}, \mathbf{W}_j = \left[\mathbf{w}_{j1}, \mathbf{w}_{j2}, \cdots, w_{jn}\right]^{\mathrm{T}}$. If assuming $x_0 = 1$, $w_{j0} = b_j$, i.e., X and $\mathbf{W}_j$ include x_0 and w_{j0} , $\mathbf{X} = [x_0, x_1, x_2, \cdots, x_n]^{\mathrm{T}}$, $\mathbf{W}_j = \left[w_{j0}, w_{j1}, w_{j2}, \cdots, w_{jn}\right]^{\mathrm{T}}$. The net input value S_j of the jth neuron is

$$S_j = \sum_{i=0}^{n} w_{ji} \cdot x_i = W_j X \tag{18.12}$$

Net input through transfer function f (•). After that, the output of the jth neuron is obtained.

$$y_j = f(s_j) = f\left(\sum_{i=0}^{n} w_{ij} \cdot x_i\right) = f(\mathbf{W}_j \mathbf{X}) \tag{18.13}$$

In the formula, f (•) is a monotonically rising function, and must be a bounded function, because the signal transmitted by the cell cannot be increased indefinitely, and there must be a maximum value.

18.3.2 BP Network

The BP algorithm consists of two processes: the forward calculation of the data stream (forward propagation) and the backpropagation of the error signal. In the case of forward propagation, the direction of propagation is the input layer → hidden layer → the output layer, and the state of each layer of neurons affects only the next layer of neurons. If the desired output is not obtained at the output layer, the reverse propagation process of the error signal is turned. Through the alternation of these two processes, the error function gradient descent strategy is executed in the weight vector space, and the dynamic iterative search for a set of weight vectors makes the network error function reach a minimum, thus completing the information extraction and memory process.

Let the input layer of the BP network have n nodes, the hidden layer has q nodes, the output layer has m nodes, the weight between the input layer and the hidden layer, and the weight between the hidden layer and the output layer are v, such as Fig. 18.7 shows. The transfer function of the hidden layer is $f_1(\cdot)$, and the transfer function of the output layer is $f_2(\cdot)$, then the output of the hidden layer node is (the threshold is written into the summation term)

$$z_k = f_1\left(\sum_{i=0}^{n} v_{ki}x_i\right) \quad k = 1, 2, \ldots, q \tag{18.14}$$

The output of the output layer node is

$$y_j = f_2\left(\sum_{k=0}^{q} w_{jk}z_k\right) \quad j = 1, 2, \ldots, m \tag{18.15}$$

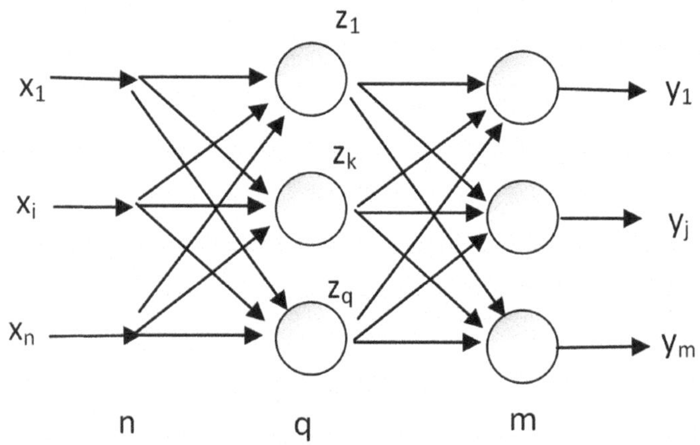

Fig. 18.7 Three-layers BP neural network

So far, the BP network has completed the approximate mapping of the n-dimensional space vector to the towel dimension space.

18.3.3 OSNR Estimator Based on BP Neural Network

A nonlinear regression model based on feedforward neural network is proposed in [30] for OSNR estimation. The specific structure is shown in Fig. 18.8. Inside the neural network, a series of weighted inputs pass through the nonlinear activation function h () and then be weighted and summed to the output. Here the nonlinear activation function is tanh (). The output expression of the single hidden layer structure neural network in Fig. 18.8 is

$$y(x, w) \equiv \sum_{j=1}^{3} w_{1j}^{(2)} h\left(w_{j1}^{(1)} x\right) \tag{18.16}$$

where x is the input variable, as obtained from the signal eye diagram. $w = \left[w_{11}^{(1)}, w_{21}^{(1)}, w_{31}^{(1)}, w_{11}^{(2)}, w_{12}^{(2)}, w_{13}^{(2)}\right]$ is the weight vector. The superscript represents the number of layers of the network. These weights are adaptive and the coefficients are constantly adjusted by the optimization algorithm.

As shown in Fig. 18.8, in order to get the best weight vector w, the neural network needs to be trained in the supervised learning mode. Consider the data set of N observations $S = \left\{\left(\sigma_n^2, OSNR_n\right) | n = 1, \cdots, N\right\}$, each test set consists of the variance σ_n^2 and the signal-to-noise ratio $OSNR_n$ of the eye diagram. For regression analysis, if the target value $t_n = OSNR_n$ conforms to the Gaussian distribution, and its output expression is

$$t_n = y\left(\sigma_n^2, w\right) + \varepsilon \tag{18.17}$$

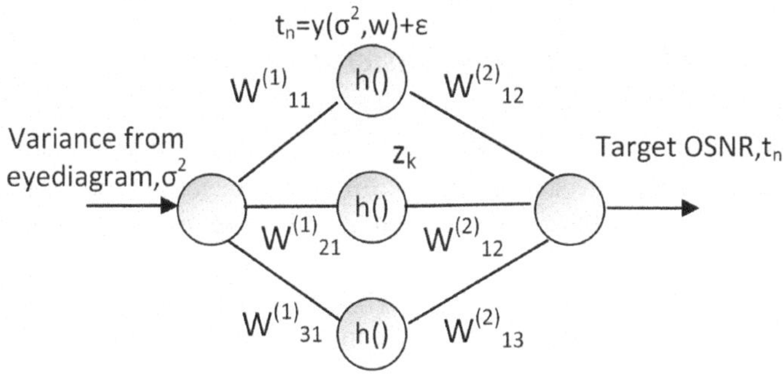

Fig. 18.8 OSNR estimator base on neural network

where ε is a zero mean variance and a Gaussian random variable of β^{-1}, the maximum likelihood estimation parameter w is used. Considering the data set $\mathbf{X} = \left[\sigma_1^2, \cdots, \sigma_N^2\right]$ and the target $\mathbf{t} = [t_1, \cdots, t_N]$, the maximum likelihood function of the neural network with respect to the weight vector w is

$$p(t|\mathbf{X}, w, \beta) = \prod_{n=1}^{N} \frac{1}{2\pi\beta^{-1}} e^{\frac{\beta}{2}\left(y(6_n^2, w) - t_n\right)^2} \tag{18.18}$$

To obtain the optimal parameter vector w, it is necessary to maximize the likelihood function and to minimize the squared error function by collating

$$E(w) = \frac{1}{2}\sum_{n=1}^{N}\left\{y\left(\sigma_n^2, w\right) - t_n\right\}^2 \tag{18.19}$$

Because an accurate error function expression cannot be obtained, an alternative method is to continuously correct the weight vector by the gradient descent method, and finally reduce the error function below the expected threshold.

18.4 Clustering Algorithm

Objects are clustered and people are grouped. Everything that finds something like itself is clustering. The clustering algorithm is a statistical analysis method for studying the classification problem of samples (samples or indicators), and it is also a classic algorithm in the field of artificial intelligence. Common clustering algorithms include K-means, K-medoids, GMM, DBSCAN, and others. Before being applied to the field of communication, clustering algorithms have been widely used in many scenarios such as news classification, user grouping, and commodity classification.

18.4.1 Principle of K-Means Clustering Algorithm

The K-Means algorithm is an unsupervised learning called K-Means because it spontaneously aggregates many samples into k different categories. Each category is a cluster, and the center of the cluster is calculated from the mean of all points in the cluster.

Given a sample set $D = \{x_1, x_2, \cdots, x_n\}$, $\mathbf{x}_i$ is a vector of m dimensions representing each sample in the sample set, where m is the number of attributes of sample x.

The purpose of clustering is to group similar samples in sample set D into the same set. We refer to the divided set as a cluster, denoted by G, where the number of

G is represented by humans. Each cluster has a center point, the center of all points in the cluster is called the centroid, represented by u_k.

Therefore, the K-Means algorithm can be expressed as a process of dividing $D = \{x_1, x_2, \cdots, x_n\}$ into $G = \{G_1, G_2, \cdots, G_k\}$. The sum of the squares of the distances between every point and the centroid of each divided cluster is called the sum of squared errors, i.e., SSEC (sum of squared error)

$$\mathrm{SSE} = \sum_{i=1}^{k} \sum_{x \in G} \|x - \mu_k\|^2$$

Therefore, the K-Means algorithm should achieve a large sample similarity within $G_1, G_2, \ldots, G_k$, and between clusters, that is, the value of SSE is reduced as much as possible.

The input is: sample set D, and the number of clusters k

The output is: $G = \{G_1, G_2, \cdots, G_k\}$, which is k divided clusters.

18.4.2 Algorithm Flow

(1) First, select the value of k.

(2) In sample set D, k points are randomly selected as the initial centroid, i.e., $\{\mu_1, \mu_2, \cdots, \mu_k\}$.

Calculate the distance from each sample x_i to the centroid of each centroid μ_j in D. The formula for calculating the distance is as follows:

$$l_i = \left(x_i - \mu_j\right)^2 \tag{18.20}$$

(3) If the distance of l_i is the smallest, mark x_i as the sample in the cluster G_j, i.e., $G_j = \{x_i\}$.

After all the sample points are assigned to different clusters, the new centroid is calculated, that is, the average of all the points in G_j, and calculate the sum of squared errors. Among them, the average value is calculated as

$$\mu'_j = \frac{1}{|G_j|} \sum_{x \in G_j} x \tag{18.21}$$

(4) Compare the difference between the squared sum of the two errors before and after and the set threshold. If it is greater than the threshold, repeat the steps (3)–(5).

(5) If the change in the square of the error is less than the set threshold, the clustering is completed.

18.4.3 Algorithm Display and Analysis

For ease of presentation, a commonly used two-dimensional data set: 4k2_far (a public machine learning data set) will be used as a test sample. See, for example, Table 18.1

where $x1, x2$ represents the attributes of the sample in the dataset. The approximate distribution of the data set is shown in Fig. 18.9.

In addition, some classic data sets, such as Iris, Wine, Glass are also applicable to test data as a clustering algorithm. Figure 18.10 shows the simulation results, and red is the cluster center point.

In the figure, we can observe that with constant iteration, the centroid is constantly approaching the center of each cluster.

The advantages of K-Means are mainly reflected in the simple and easy implementation of algorithms. In the actual situation, k-Means has some obvious shortcomings to be aware of.

Table 18.1 A part of samples of train set

$X1$	$X2$
7.1741	5.2429
6.914	5.0772
7.5856	5.3146
6.7756	5.1347
…	…

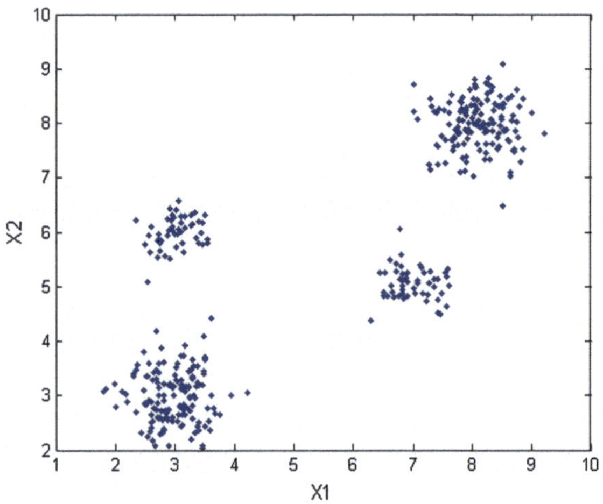

Fig. 18.9 Data set

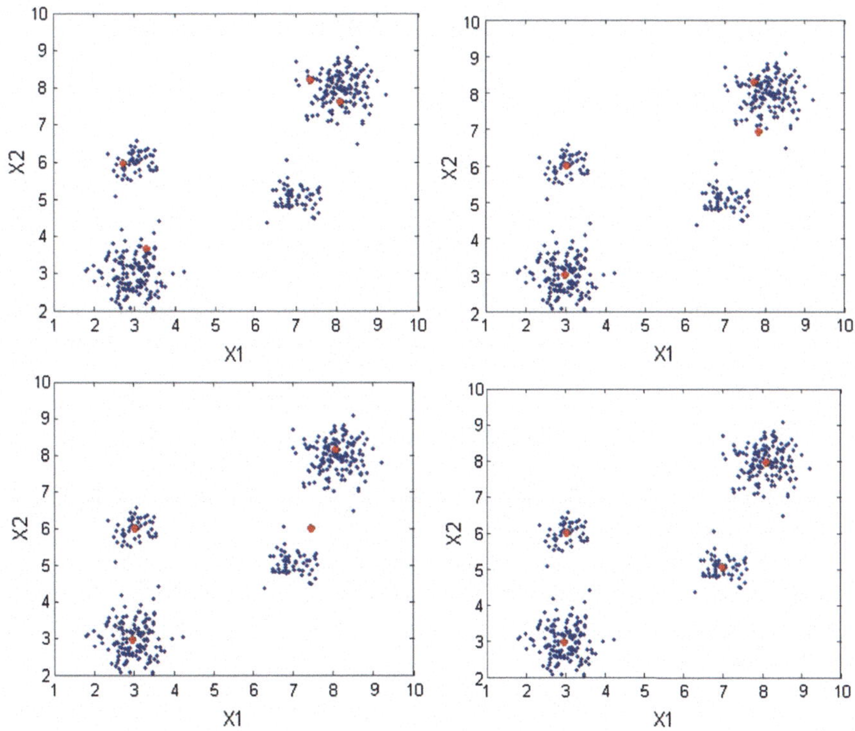

Fig. 18.10 Iteration processing of K-means algorithm

(1) Selection of *k* value.

Since the human value needs to be set by the user, it is difficult to determine that the data set should be clustered into several categories in the data set of the high-dimensional attribute, and the selection of the k value has a great influence on the clustering effect.

Under normal circumstances, it is common to use multiple changes in k values to observe the clustering effect. However, this method does not apply to large data sets.

(2) Choice of centroid.

In the selection of the initial centroid, a random method is generally used, which also affects the effect of clustering. If the randomly selected centroid is too deviated, there will even be an empty cluster phenomenon.

One common technique for dealing with the initial centroid problem is to run multiple times, each time using a different set of random initial centroids, and then selecting the cluster with the smallest SSE. This strategy, while simple, is time-consuming and may not work well depending on the number of data sets and clusters being sought.

18.5 Application of Clustering Algorithm in Nonlinearity

18.5.1 Application Principle

In general, a communication system is a combination of some interacting subsystems. Nonlinear effect in an optical communication system includes Kerr nonlinearity in Fiber Channel, nonlinear effects from photodetectors, transmitter drive circuits, and amplifiers, so on. Therefore, some studies use the clustering algorithm (k-means) in machine learning to compensate for the comprehensive nonlinear phenomena existing in the optical communication system that is suitable for direct inspection (IM-DD).

Clustering algorithm-based perception decision (CAPD) applies to the compensation of the distorted signal using induction rather than deduction. Machine learning models include regression models, classification models, and cluster models. Taking into account the characteristics and practicality of the system, clustering models in machine learning are more appropriate.

The CAPD is a clustering algorithm that only considers the statistical laws of the received data itself, and does not care which part of the system leads to the generation of nonlinear effects. The algorithm flow chart is shown in Fig. 18.11 and includes the following three steps.

(1) From the received data r_x, select the subsequence of the specified length as the training sequence selection $\{x_1, x_2, \cdots x_i, \cdots, x_n\}$ if the system has a slow change characteristic, the subsequence can be greatly affected without affecting the performance of the system and reduce complexity.
(2) The IQ signal is clustered by the K-Means clustering algorithm. The specific steps are

Initializing the center point $\{c_1, c_2, \cdots, c_j, \cdots, c_k\}$ (each point is a complex number containing IQ components), where k is equal to the modulation order, such as k = 16 in the CAP16 system; using Eq. (18.22). Calculate the shortest distance from each x_i to c_j,

Among them, each x^i is an IQ two-dimensional vector $[x_i(n), x_q(n)]$, c_j is the current cluster center point, $f(x, c)$ is a distance function, which can be defined as Euclidean distance, Manhattan distance or Mahala Nobis distance, etc. In the example of Euclidean distance, as shown in Eq. (18.22)

$$f\left(x^i, c_j\right) = \left|x^i - c_j\right|^2 \tag{18.22}$$

When the value of the cost function e does not exceed the preset threshold E, the iteration formula (18.22) is repeated to the formula (18.23). Among them, the iteration (18.24) of the update center point is as follows:

$$c_j \rightarrow \frac{\sum_{i=1}^{m} 1\{d^i = j\}x^i}{\sum_{i=1}^{m} 1\{d^i = j\}} \tag{18.23}$$

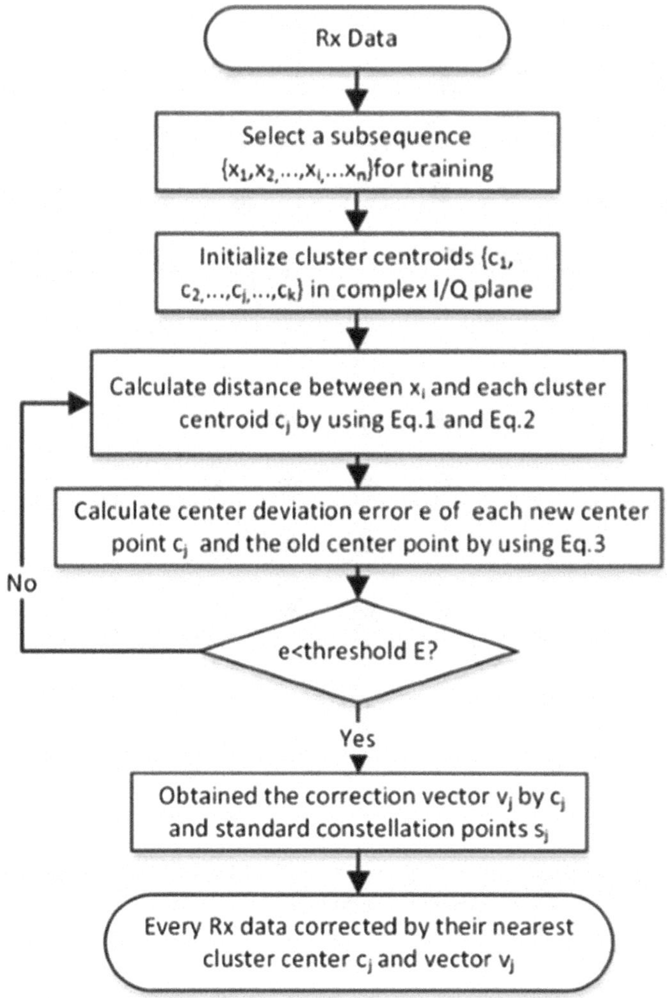

Fig. 18.11 CAPD based on clustering algorithm

Computation correction vector v_j

$$v_j = s_j - c_j \tag{18.24}$$

where c_j is the clustering center point, S_j is the standard constellation point; For each received data R_x, the compensation data is obtained through multiplying the IQ signal and the real part of the correction vector v_j, and multiplying the IQ signal and

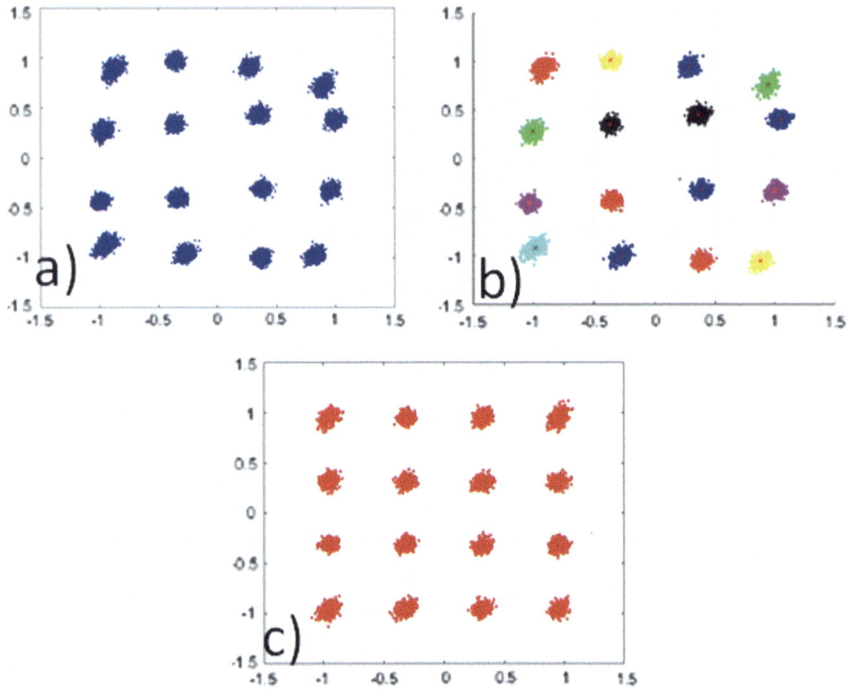

Fig. 18.12 **a** original constellation at Rx. **b** Results of K-means clustering algorithm (Different clusters are represented by different colors. The Red Cross in the cluster is the center of the cluster). **c** Final results corrected by CAPD algorithm

the imaginary part of the correction vector v_j belonging to the nearest center point c_j. Finally, the compensation data is outputted.

18.5.2 Result Analysis

As shown in Fig. 18.12, the original constellation points are corrected by the CAPD, and the performance is significantly improved. To further test the anti-linearity of CAPD, the literature [13] compared CAPD and Volterra equalizers. Strictly speaking, CAPD is a decision-making method, not an equalizer. Due to the high complexity, the Volterra equalizer only achieves 2 or 3 orders [7] in the experiment, and the fitting

Table 18.2 Complexity comparison of Volterra Equalizer and CAPD

Compensation method (CAP16)	Volterra equalizer (2-Order)	Volterra equalizer (3-Order)	CAPD
Multiplier	$2N^2$	$3N^3 + 2N^2$	$(L^2/2)$
Adder	N^2-1	$N^3 + N^2-2$	L
Comparator	0	0	32
Iterations	All Rx data	All Rx data	I*L (L = 2000–3000)
Data aided	No	No	No

performance is limited in more complex nonlinearities. For nonlinear compensation, CAPD outperforms Voltaire equalizers.

Table 18.2 compares the complexity of Volterra equalizers and CAPD. Where N is the length of the nonlinear equalizer, L is the length of the training sequence, and I is the number of iterations. In the experiment, I is usually between 3 and 7. Similarly, CAPD does not require data assistance as same as a blind equalizer. In general, the complexity of the CAPD is lower than that of Volterra equalizers.

References

1. E. Ip et al., Coherent detection in optical fiber systems. Opt. Exp. **16**(2), 753–791 (2008)
2. E. Ip, J.M. Kahn, Compensation o£ dispersion and nonlinear effects using digital backpropagation. J. Lightwave Technol. **26**(20), 3416–3425 (2008)
3. Z. Dong, F.N. Khan, Q. Sui, et al. Optical performance monitoring: a review of current and future technologies. J. Lightwave Technol. **34**(2), 525–543 (2016)
4. C. Lu, et al. Optical performance monitoring techniques for high capacity optical networks, in *International Symposium on Communication Systems Networks and Digital Signal Processing*, Newcastle upon Tyne, pp. 678–681 (2010)
5. C.M. Bishop, *Pattern recognition and machine learning* (Springer, New York, 2006)
6. Z. Ghahramani, Probabilistic machine learning and artificial intelligence. Nature **521**(7553), 452–459 (2015)
7. N. Jiang et al. Stochastic backpropagation for coherent optical communications, in *Proceedings of European Conference on Optical Communications* Geneva, 2011, Paper We. 10. Pl. 81
8. M. Li, S. Yu, J. Yang, et al. Nonparameter nonlinear phase noise mitigation by using M-ary support vector machine for coherent optical systems. IEEE Photon J. **5**(6) (2013)
9. D. Zibar et al., Nonlinear impairment compensation using expectation maximization for dispersion managed and unmanaged PDM 16 QAM transmission. Opt. Exp. **20**(26), B181–B196 (2012)
10. T.S.R. Shen, A.P.T Lu, Fiber nonlinearity compensation using extreme learning machine for DSP-based coherent communication systems, in *Proc Opto Electronics and Communications Conference, Kaohsiung* 2011: 816–817
11. AKINO T K, et al. High-order statistical equalizer for nonlinearity compensation in dispersion-managed coherent optical communications. Opt. Exp., 2012, 20 (1 4): 15769–15780
12. LU X, WANG K, QIAO L, et al. Nonlinear compensation of multi-CAP VLC system employing clustering algorithm based perception decision. IEEE Photonics Journal, 2017, 9(5)

13. F.N. Khan, T.S.R. Shen, Y. Zhou et al., Optical performance monitoring using artificial neural networks trained with empirical moments of asynchronously sampled signal amplitudes. IEEE Photon. Technol. Lett. **24**(12), 1041–1135 (2012)
14. KHAN F N, YU Y, TAN M C, et al. Simultaneous OSNR monitoring and modulation format identification using asynchronous single channel sampling [C]. Proc. Asia Communications and Photonics Conf., Hong Kong, 2015, Paper AS4F. 6
15. TANIMURA T, HOSHIDA T, RASMUSSEN J C, et al. OSNR monitoring by deep neural networks trained with asynchronously sampled dataQ [C]. Proc. Opto Electronics and Communications Conference > Niigata, 2016, Paper TuB3–5
16. R.A. Skoog et al., Automatic identification o£ impairments using support vector machine pattern classification on eye diagramsE [J]. IEEE Photon. Technol. Lett. **18**(22), 2398–2400 (2006)
17. M.C. Tan et al., Simultaneous optical performance monitoring and modulation format/ bit-rate identification using principal component analysis [J]. Optical Communications and Networking **6**(5), 441–448 (2014)
18. F.N. Khan, Y. Yu, M.C. Tan et al., Experimental demonstration of joint OSNR monitoring and modulation format identification using asynchronous single channel sampling. Opt. Exp. **23**(23), 30337–30346 (2015)
19. ANDERSON T B et al. Multi impairment monitoring for optical networks [J]. J. Lightwave Technol. » 2009, 27(16) : 3729–3736
20. KHAN F N, DONG Z, LU C, et al. Optical performance monitoring for fiber-optic communication networks. / / Zhou X, Xie C. Enabling Technologies for High Spectral- Efficiency Coherent Optical Communication Networks. New York: Wiley, 2016, Chap. 14
21. GONZALEZ N G, ZIBAR D, MONROY I T. Cognitive digital receiver for burst mode phase modulated radio over fiber links E[C]. Proc. European Conf, on Optical Commun., Torino, 2010, Paper P6. 11
22. F.N. Khan, Y. Zhou, A.P.T. Lau, C. Lu, Modulation format identification in heterogeneous fiber-optic networks using artificial neural networks. Opt. Exp. **20**(11), 12422–12431 (2012)
23. F.N. Khan, Y. Zhou, Q. Sui, A.P.T. Lau, Non-data-aided joint bit-rate and modulation format identification for next-generation heterogeneous optical networks. Optical Fiber Technol. **20**(2), 68–74 (2014)
24. BORKOWSKI R, ZIBAR D, CABALLERO A, et al · Stokes space-based optical modulation format recognition in digital coherent receivers. IEEE Photon. Technol. Lett. 2013, **25**(21), 2129–2132 (2013)
25. KHAN F N, ZHONG K, AL-AR ASH I W H, et al. Modulation format identification in coherent receivers using deep machine learning. IEEE Photon. Technol. Lett., 2016, 28(17), 1886–1889
26. D. Zibar et al., Joint iterative carrier synchronization and signal detection employing expectation maximization [J]. J. Lightwave Technol. **32**(8), 1608–1615 (2014)
27. ZIBAR D, et al. Nonlinear impairment compensation using expectation maximization for dispersion managed and unmanaged PDM 16 QAM transmission [J]. Opt. Exp., 2012, 20(26): B181-B196
28. ZIBAR D, et al. Bayesian filtering for phase noise characterization and carrier synchronization of up to 192 Gb/ s PDM 64 QAM EC]. Proc. European Conf, on Optical Commun., Cannes, 2014, Paper Tu. 1. 3.1
29. ZH Zhou. Machine Learning, Beijing:Tsinghua University Press,2016
30. ZIBAR D, THRANE J, WASS J, et al. Machine learning techniques for optical performance monitoring from directly detected PDM-QAM signals [J] . Journal of Lightwave Technology? 2017, 35(4): 868–875

Chapter 19
Kramers-Kronig Receiver in Direct Detection Systems

19.1 Introduction

In this chapter, we introduce and analyze the principle of the Kramers-Kronig receiver and its application in direct detection systems. There is an increasing demand for ultra-high data rates due to the exponential traffic growth in optical interconnect networks such as metro networks, mobile front-haul, access networks, and data-center interconnects (DCIs) [1–5]. Nowadays, with the rapid development of new applications such as Cloud Computing, Internet of Things, Artificial Intelligence, Augmented and Virtual Reality, researchers have focused more on short-reach links in DCIs with the distance range from hundreds of meters to around 80 or 100 km [6, 7]. Since 400 Gigabit Ethernet (400 GbE) has been standardized as the task force by IEEE 802.3 group [6–8], 4-λ100 Gb/s WDM was proposed and considered as a promising solution for 400 GbE standard [9–12]. In recent years, 4-level pulse amplitude modulation (PAM4) is applied to beyond 400 Gb/s transmission. However, the schemes utilizing PAM4 are based on either large bandwidth devices or complex digital signal processing (DSP). In such a bandwidth-hungry era, the main limitation using PAM4 is that the achievable largest spectrum efficiency (SE) is only 4 bit/s/Hz even with Nyquist shaping. For next generation specifications of 800 GbE or even 1.6 TbE, the continuous usage of PAM4 has to adopt much larger bandwidth devices and more complicated DSP leading to a huge increase in cost and latency which is inconsistent with the cost-sensitive requirement for short-reach data-center applications [6, 17]. Thus, applying advanced modulation formats such as higher-order QAM with higher SE is a hopeful direction for the future beyond 800 GbE standardization. Compared with coherent detection, DD is preferred for typical data-center applications due to lower cost, lower power consumption, and smaller footprint [13–17]. To mitigate spectrally selective fading distortion introduced by chromatic dispersion (CD) in conventional double-sideband (DSB) DD systems, SSB signaling is widely applied in DD optical systems and it also doubles SE [18–20]. Either a Mach-Zehnder modulator (MZM) based IQ modulator or a

J. Yu and N. Chi, *Digital Signal Processing In High-Speed Optical Fiber Communication Principle and Application*,
https://doi.org/10.1007/978-981-15-3098-2_19

dual-drive MZM (DD-MZM) can be used to generate an SSB signal. Applying a DD-MZM is attractive because of lower cost, however, better SSB signals are generated with an MZM based IQ modulator [18]. The SSBI caused by square-law detection severely deteriorates system performance. To overcome this problem, many effective techniques are proposed and demonstrated, such as iterative SSBI estimation and cancelation scheme [21], single-stage linearization filter [22], two-stage linearization filter [23], iterative linearization filter [24], Volterra filter [25], and Hilbert superposition cancelation and modified single-stage linearization filter [26]. Recently, KK receiver [14] is proposed as a promising method to inherently remove the SSBI by reconstructing the complex optical field from the photo-detected amplitude with the minimum-phase condition. Compared with the above SSBI mitigation schemes, applying KK receiver also enables receiver-side digital CD compensation and lower DSP complexity [27]. The reduction of DSP complexity indicates KK receiver's potential in future DCI applications. Thus, combining KK receiver and higher-order QAM modulation can make future data-center applications more achievable with beyond 800 GbE standardization.

In recent years, the outstanding performance of applying KK receiver has been verified through simulations and experimental demonstrations. Four-channel 28 Gbaud 16QAM (112 Gb/s) over 240 km of SSMF and 28 Gbaud 64QAM (168 Gb/s) with 35 GHz channel spacing over 80 km of SSMF SSB KK transmissions using IQ modulators were demonstrated in [28] and [20], respectively which confirmed KK receiver's superior capability to remove SSBI. But the theoretical upper bound of net SE in these experimental demonstrations was only 4.61 b/s/Hz. Furthermore, a 30 Gbaud 64QAM (180 Gb/s) SSB KK transmission over 125 km of SSMF utilizing a DFB laser was demonstrated in [29]. A net data rate of 279.4 Gb/s transmission over 100 km of SSMF with KK detection and probabilistically shaped (PS) discrete multitone modulation (DMT) 256QAM was presented in [30]. Applying KK receiver in higher-order QAM systems requires more on optical signal noise ratio (OSNR), effective number of bit (ENOB) of both digital-to-analog converter (DAC) and analog-to-digital converter (ADC), and it is more sensitive to nonlinearity. The main limitation of KK receiver in 256QAM and higher order QAM application has not been investigated until now. A simplified AC-coupled PD based KK receiver was proposed in [16] and verified in a DD-MZM PAM system. Compared with a DC-coupled PD, an AC-coupled PD has the advantages of avoiding DC-drift, reducing quantization noise, and increasing voltage gain [16, 31, 32].

Single channel 20 Gbaud 128QAM (140 Gb/s) and 256QAM (168 Gb/s) over 20 km of SSMF in a 25 GHz grid employing an AC-coupled PD based KK receiver was experimentally demonstrated in [33]. Compared with single channel transmission, WDM transmission is more promising to achieve much higher data rates, but it has a higher total power which may lead to higher nonlinearity, and it can be affected by inter-channel interference (ICI). Thus, the performance of applying KK receiver with higher-order QAM modulation in a single channel system and a WDM system may be different. Thus, exploration of the system performance with higher-order QAM format and KK receiver through both simulations and experimental demonstrations is carried out in the following part, and performance comparison between a single channel system and a WDM system is also presented. A comprehensive system performance assessment of 128QAM, 256QAM and 512QAM formats are studied with the same simulation and experiment setups. 25 GHz-spaced 4 × 140 Gb/s WDM 128QAM and 4 × 160 Gb/s WDM 256QAM SSB KK transmissions over 20 km of SSMF are successfully achieved using an AC-coupled PD with sufficient OSNR, ENOB of DAC and ADC, and system linearity. And the main limitation of achieving 512QAM with KK receiver is also analyzed in the following part.

19.2 Principle of KK Receiver

A baseband QAM signal is up-converted to an intermediate frequency to generate an SSB signal. To fulfill the minimum-phase condition of KK receiver, an optical tone with an amplitude larger than that of the SSB signal is required. The most common way to generate the optical tone for SSB transmission is offsetting the bias of an IQ modulator while driving the modulator with an SSB signal as shown in Fig. 19.2. Thus, the complex optical field envelope of the transmitted signal before impinging upon the PD can be expressed as

$$E(t) = E_0 + s(t)e^{j\pi f_i t} \tag{19.1}$$

$$f_i = \alpha \times f_s \tag{19.2}$$

where E_0 is the amplitude of the optical carrier, $s(t)$ is the baseband QAM signal, f_{i} is the intermediate frequency, f_{s} is the symbol rate, α is the gap factor which determines the frequency gap between the signal and the optical carrier. The signal $I(t)$ detected by a PD without DC blocking can be written as

$$I(t) = E_0^2 + E_0 s(t)e^{j\pi f_i t} + E_0 s^*(t)e^{-j\pi f_i t} + |s(t)|^2 \tag{19.3}$$

where $s^*(t)$ is the conjugate term, $E_0 s(t)e^{j\pi fit}$ is the desired useful signal which can be obtained by filtering, $|s(t)|^2$ is the SSBI component which is an essential part to apply KK receiver and reconstruct complex optical field. Over the years, SSBI mitigation algorithms have become a hot research area. KK scheme is one of the most effective ways to solve this problem.

The DC component of the detected signal includes E_0^2 and the DC component of the $|s(t)|^2$ term, and it will be almost zero if an AC-coupled PD is utilized. Both a DC-coupled PD and an AC-coupled PD can be utilized to achieve KK detection. According to [31], it's clear that there are some low-speed fluctuations due to the imperfections of the devices with high launch power. This problem is called as DC-drift by the authors. And a DC-filtering step is added to stabilize the signal sequence. This step can be skipped if they use an AC-coupled PD with a better suppression of the low frequency component around DC. Thus, an AC-coupled PD is applied here to carry out KK detection. The signal detected by an AC-coupled PD is defined as

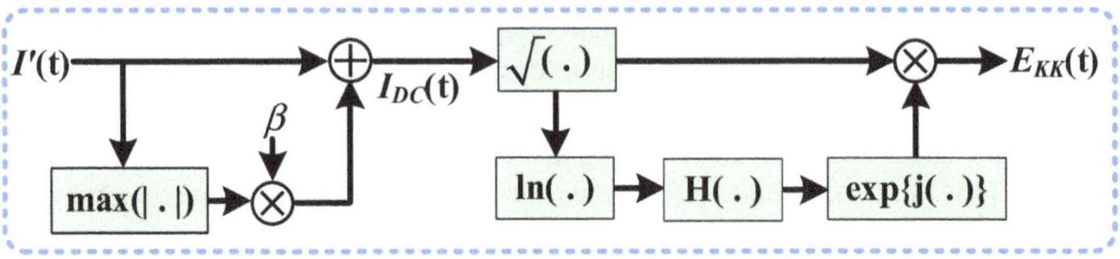

Fig. 19.1 Block diagram of KK receiver

$I'(t)$. Thus, a positive value acting as the virtual DC component should be added to $I'(t)$ which can be expressed as

$$I_{DC}(t) = \beta \times |I'(t)|_{\max} + I'(t) \tag{19.4}$$

where β is the scaling factor which is larger than 0 to adjust the value of the virtual DC component. The phase of λ the signal can be retrieved as

$$\phi(t) = H\left[\ln(\sqrt{I_{DC}(t)}\right] \tag{19.5}$$

where H[.] is the Hilbert transform operator and ln(.) is the logarithm operator. Therefore, the reconstructed SSB signal $E_{KK}(t)$ can be written as

$$E_{KK}(t) = \sqrt{I_{DC}(t)}e^{j\phi(t)} \tag{19.6}$$

The block diagram of KK receiver is shown in Fig. 19.1. Three necessary conditions are required to achieve KK receiver. The first one is an optical tone with sufficient power to fulfill the minimum-phase condition. In other words, a sufficiently large CSPR is required, which is defined as

$$\text{CSPR} = 10\log(|E_0|^2 / |s(t)|^2) \tag{19.7}$$

In simulations and experimental demonstrations, CSPR is measured by turning on/off the radio frequency (RF) signal to obtain the signal power PS/carrier power PC and thus CSPR can be calculated as [34]

$$\text{CSPR} = 10\log(P_C / P_S) \tag{19.8}$$

Secondly, a sufficiently high digital up-sampling rate is important to cope with the spectral broadening due to square-root and logarithm operations in the KK scheme. Up-sampling number values from 4 to 6 is found to be the optimum according to recent experimental results [35]. Down-sampling is then carried out after KK detection for the subsequent DSP procedures. Thirdly, the transmitted signal should be an SSB signal.

19.3 Simulation Setup and Results

19.3.1 Simulation Setup

Single channel simulations are firstly carried out to test the system performance with the KK scheme. Figure 19.2 shows the simulation setup and Table 19.1 shows the key parameters in the simulation setup. Simulations are carried out utilizing VPI Design Suite 9.8 and Matlab R2018b. At the transmitter part, the bit sequence is generated by the random function of Matlab for both simulations and experimental demonstrations. The sequence is then mapped into 216-length 128/256/512QAM symbols with 20 Gbaud symbol rate and up-sampled by a factor of 4. The signal then passes through a root-raised-cosine (RRC) pulse shaping filter with a roll-off factor of 0.01 and up-converted to the intermediate frequency fi with α equal to 0.51. The real and imaginary parts of the signal generated by the Matlab program are subsequently loaded into a DAC with 80 GSa/s sampling rate and 20 GHz bandwidth. Its ENOB is set to 5. The parameter values of the DAC are set according to the DAC used in the experimental demonstration. Then, the output signals drive an IQ modulator biased on the null point. The half-wave DC and RF voltages of the IQ modulator are set to 5 V with an operating temperature of 25 °C. CSPR values are further controlled by changing the bias point of the IQ modulator. The continuous wave light is generated by an external cavity laser (ECL) with 100 kHz linewidth at 1552.52 nm. An Erbium-doped fiber amplifier (EDFA) with 4 dB noise figure is applied to increase the optical signal's launch power. 20 km of SSMF with 16 ps/nm/km dispersion coefficient at

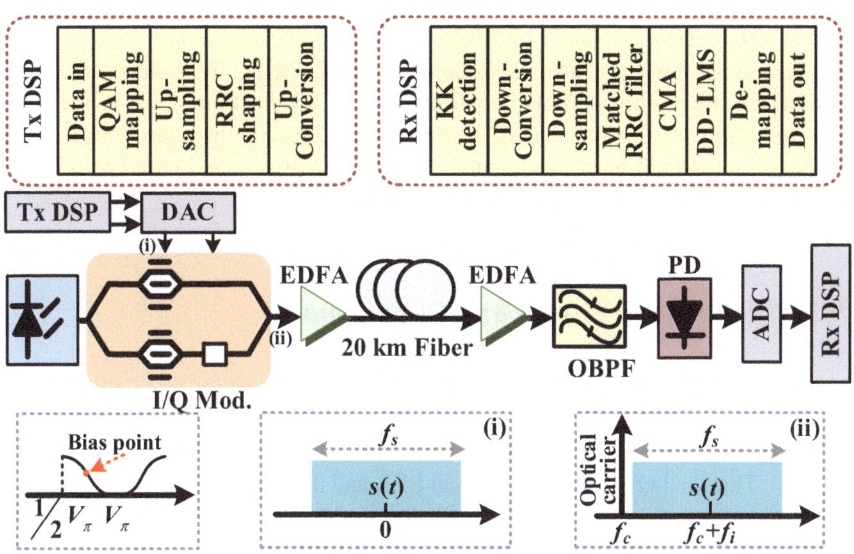

Fig. 19.2 Simulation setup. Insets: (i) baseband QAM signal, (ii) transmitted optical signal

Table 19.1 Key parameters in simulation setup

Parameters	Values	Parameters	Values
DAC sampling rate	80 Gsa/s	DAC ENOB	5
ECL wavelength	1552.52 nm	ECL linewidth	100 kHz
ECL average power	1 mW	VpiRF	5 V
VpiDC	5 V	Extinction ratio	90 dB
Operating temperature	25 °C	EDFA noise figure	4 dB
EDFA noise bandwidth	4 THz	Dispersion	16 ps/nm/km
Dispersion slope	0.08 ps/nm^2/km	Core area	80 μm^2
Attenuation coefficient	0.2 dB/km	Nonlinear index	2.6×10^{-20}
Max step width	50 km	Mean step width	50 km
OBPF transfer function	Gaussian	OBPF bandwidth	25 GHz
OBPF noise dynamic	3 dB	OBPF Gaussian order	1
OBPF noise resolution	5 GHz	PD responsivity	1 A/W
PD thermal noise	10 pA/Hz$^{0.5}$	PD bandwidth	50 GHz
ADC sampling rate	80 Gsa/s	ADC ENOB	5

1552.52 nm, 0.2 dB/km attenuation coefficient, 0.08 ps/nm^2/km dispersion slope, and 80 μm^2 core area is used for transmission.

At the receiver side, the received optical power is adjusted to 0 dBm using the EDFA. A 25 GHz bandwidth Gaussian-shaped optical bandpass filter (OBPF) is used to simulate a 25 GHz channel spacing. The filtered signal is then detected by a 50 GHz PD with 1.0 A/W responsivity and followed by an ADC with 80 GSa/s sampling rate, 33 GHz bandwidth, and 5-bit ENOB. These parameters are also set according to experimental devices. After all these procedures, the received electrical signal is processed by an offline Matlab program. After KK detection, down-conversion, down-sampling, and matched RRC filtering, constant modulus algorithm (CMA) and decision-directed least mean square (DD-LMS) are carried out. The original data stream is then recovered after QAM de-mapping.

19.3.2 Simulation Results and Discussions

The simulated bit error ratio (BER) and OSNR performance versus CSPR applying KK detection for 128QAM are shown in Fig. 19.3. The theoretical BER results assuming additive white Gaussian noise (AWGN) is marked as BER Theory shown in Fig. 19.3. The gap between the theoretical BER and the simulated BER is caused by limiting factors existing in a real simulation system such as ENOB and sampling rate of DAC and ADC, nonlinearity, and dispersion of SSMF, thermal noise, and responsivity of PD. And the BER results versus CSPR for 256QAM and 512QAM are depicted in Fig. 19.4a, b, respectively. According to the results, it is clear that apply-

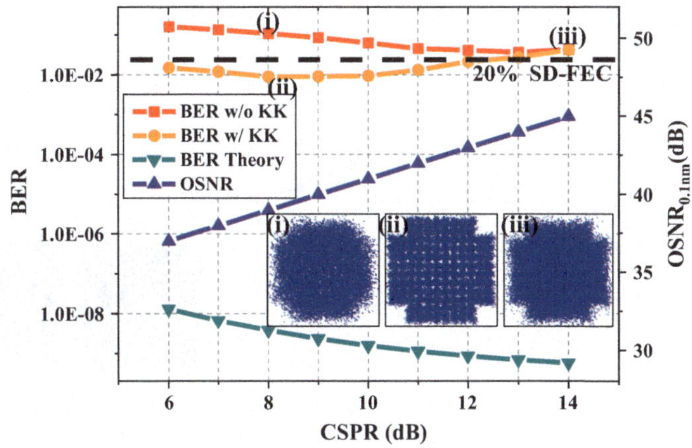

Fig. 19.3 BER and OSNR versus CSPR for 128QAM. Insets: received 128QAM constellations (i) w/o KK receiver w/8 dB CSPR, w/KK receiver w/(ii) 8 dB, and (iii) 14 dB CSPR

ing the KK scheme can provide significant system performance improvement. The optimum system performance can be achieved by sweeping CSPR values. By adjusting the bias point of the IQ modulator from the null point to obtain a higher CSPR value, the transmitted optical signal power also increases. Thus, OSNR increases as CSPR increases. It can be seen that the optimum CSPR applying KK receiver is around 8 dB for all the three QAM formats. The optimum CSPR values utilizing KK receiver are about 4–5 dB smaller than those without KK receiver. For example, the optimum CSPR is 13 dB for 128QAM system without KK receiver. But the optimum CSPR reduces to 8 dB because of applying KK receiver. That is to say, a relatively small CSPR can be used to achieve good system performance with KK receiver which can reduce system power consumption. There is a trade-off between SSBI and excessive optical carrier power. With smaller CSPR, the system performance is restricted by SSBI, and therefore, applying KK receiver can achieve higher system performance improvement. However, when CSPR is further increased, the system performance is restricted by excessive optical carrier power. Therefore, even utilizing KK receiver can't improve system performance anymore. This conclusion can also be verified through Fig. 19.5. With the optimum CSPR, BER decreases from 0.1108 to 0.0091, from 0.1562 to 0.0326 and 0.1837 to 0.0583 for 128QAM, 256QAM, and 512QAM, respectively. It is noted that the lowest BER is still higher than the 25% SD-FEC threshold of 0.04 for 512QAM even with the optimum CSPR.

Therefore, BER versus ENOB is further investigated as shown in Fig. 19.4c to explore the effect of DAC and ADC ENOB on 512QAM system performance. CSPR is set to 8 dB to get the optimum system performance applying KK receiver and the corresponding measured OSNR is 39 dB. The red line represents the BER performance variation by changing the ENOB of DAC and ADC simultaneously. And the other two lines stand for the BER variation versus the ENOB of DAC/ADC and fixing the ENOB of ADC/DAC as 5-bit. BER is lower than the 25% SD-FEC threshold

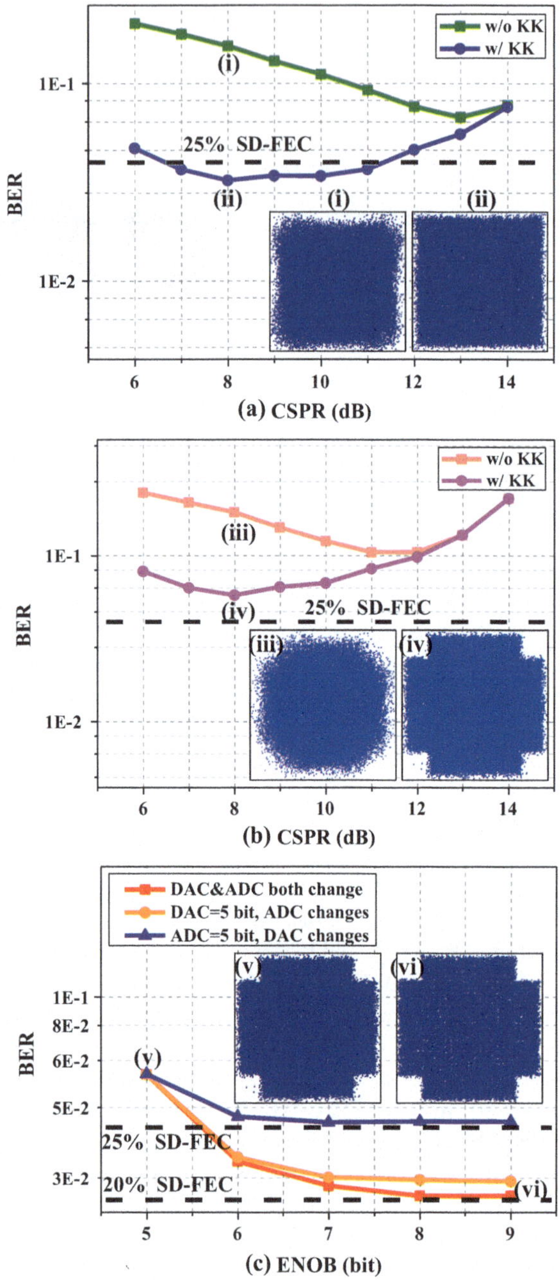

Fig. 19.4 BER versus CSPR for **a** 256QAM and **b** 512QAM, **c** BER versus ENOB for 512QAM. Insets: received constellations for 256QAM (i) w/o and (ii) w/KK receiver; 512QAM (iii) w/o and (iv) w/KK receiver; 512QAM (v) w/KK receiver w/ENOB of 5, (vi) w/KK receiver w/ENOB of 9

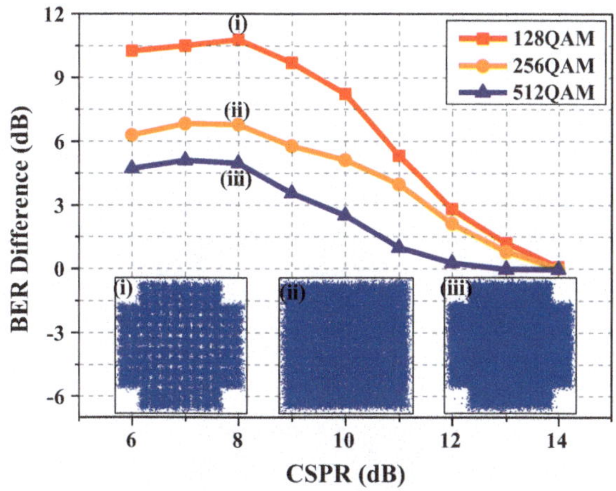

Fig. 19.5 BER difference value variation w/and w/o KK receiver versus CSPR. Insets: received constellations for (i) 128QAM, (ii) 256QAM and (iii) 512QAM with the optimum CSPR of 8 dB using KK receiver

of 0.04 if ENOB of ADC is higher than 6, and it is approaching the 20% SD-FEC threshold of 0.024 if the ENOB of DAC and ADC are both higher than 8. That is to say, the ENOB of DAC and ADC in a real experimental demonstration may be the main obstacle of achieving a 512QAM transmission system based on KK receiver because the ENOB of the DAC and ADC used in the following experimental part is only 5. And Compared with DAC, the ENOB of ADC has a greater impact on system performance. Because the ENOB of ADC should not be smaller than that of DAC, otherwise, the system performance cannot be further improved. The BER difference value variation with and without KK receiver versus CSPR for high-order QAM systems is shown in Fig. 19.5. The overall improvement after applying KK receiver increases at first because the starting CSPR is too low to fully meet the minimum-phase condition. Once the minimum-phase condition is fulfilled, the improvement brought by KK receiver becomes stable in a certain range. After this range, the improvement then decreases rapidly because the excessive optical carrier power is turning to be the main limitation for the system.

19.4 Experimental Setup and Results

19.4.1 Experimental Setup

Figure 19.6 shows the experimental setup for WDM high-order QAM transmission utilizing an AC-coupled PD based KK receiver. The odd and even WDM channels are generated by two 30 GHz bandwidth IQ modulators seeded by four ECLs with less than 100 kHz linewidth at 1551.88, 1552.08, 1552.28, and 1552.48 nm utilizing two

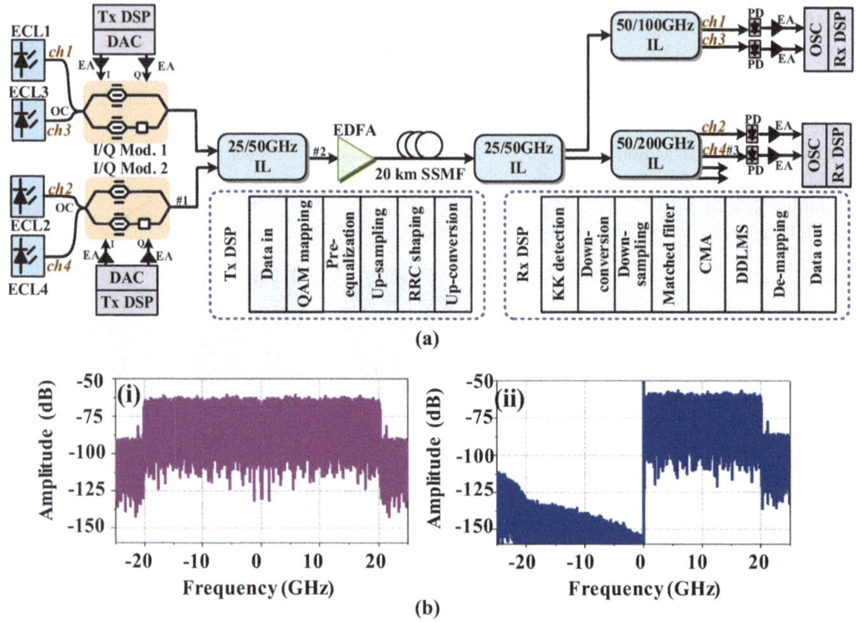

Fig. 19.6 **a** Experimental setup of WDM high-order QAM transmission with KK receiver; **b** spectra of (i) detected DSB signal and (ii) reconstructed SSB signal after KK receiver

optical couplers (OCs). In the Tx DSP, four different 20 Gbaud 128/256/512QAM baseband signals are generated at first. Then the QAM signals are pre-equalized based on a linear CMA equalizer. The finite impulse-response (FIR) tap coefficients of CMA equalizer without pre-equalization is adopted as a feedback to the transmitter for pre-equalization. The details of this pre-equalization scheme are presented in [36]. After that, the signals are up-sampled, RRC filtered and up-converted with the same parameters used in the simulation. The four signals are loaded into two DACs with 80 GSa/s sampling rate, 20 GHz bandwidth, and 5-bit ENOB. Each output signal from the DACs is boosted by a cascaded Electrical Amplifier (EA) with 5 dB noise figure and then used to drive the IQ modulators. The IQ modulators are biased above the null points to generate the optical carrier required by KK receiver and the bias point is further manually adjusted to get different CSPR values. The four channels are multiplexed by a 25/50 GHz interleaver (IL) with 25 GHz WDM channel spacing. The transmission is performed by two EDFAs and 20 km of SSMF.

After 20 km of SSMF transmission, the received signal is fed into a 25/50 GHz IL to separate the WDM signal into two branches. Subsequently, a 50/100 GHz and a 50/200 GHz ILs are applied to filter out the two groups of two-branch signals into four individual signals. There are four output ports for the 50/200 GHz IL but only two of them are used to get the ch2 and ch4 signals. Four photodetectors (PD) with 50 GHz bandwidth are used to detect the signals. Note that the detected signals are real-valued and double-sideband due to square-law detection as shown in Fig. 19.6

inset (i). The signals are sampled by an 80 GSa/s digital real-time oscilloscope (OSC) with 33 GHz bandwidth after amplified by EAs. In the Rx DSP, the optical information is recovered by KK receiver thus the SSB signal is reconstructed as shown in Fig. 19.6 inset (ii). After that, the signal is further processed by down-converting, down-sampling, filtering, CMA, DD-LMS, and de-mapping to get the BER performance.

19.4.2 Experimental Results and Discussions

1. **Single Channel Transmission**

Single channel transmission is carried out at first to test the system performance. The spectra of the detected DSB and reconstructed SSB signals are shown in Fig. 19.6b which clearly validates that the complex-valued SSB signal can be reconstructed with KK detection. Only one 25/50 GHz IL is applied to simulate 25 GHz WDM channel spacing. The BER performance with different received optical power using KK receiver for single channel 128QAM system in back-to-back (btb) case is investigated at first as shown in Fig. 19.7. The overall system performance improves when the received optical increases from −2 to 1 dBm. But the system performance is impaired due to the saturation effect of PD when the received optical power is further increased from 1 to 2 dBm. Thus, the received optical power is set as 1 dBm for further experimental demonstrations.

The influence of the 25/50 GHz interleaver is then tested with and without KK detection using 1 dBm received optical power and 128QAM format as shown in Fig. 19.8. The spectra of the optical SSB signal with and without the interleaver is shown in Fig. 19.9. CSPR values from 11 to 18 dB are controlled by changing the bias point of the IQ modulator and fixing the signal power before loading into the IQ

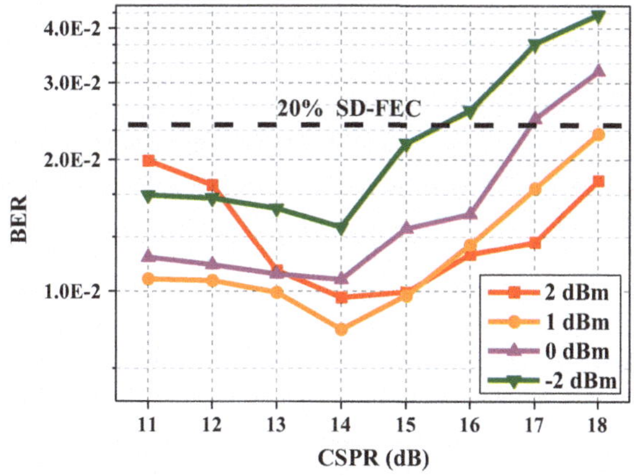

Fig. 19.7 BER versus CSPR with different received optical power using KK

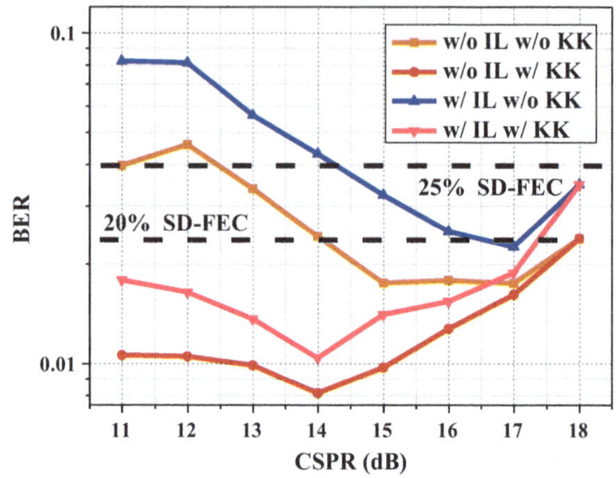

Fig. 19.8 BER versus CSPR w/o, w/KK detection, w/o and w/the interleaver (the interleaver is represented as IL in the figures)

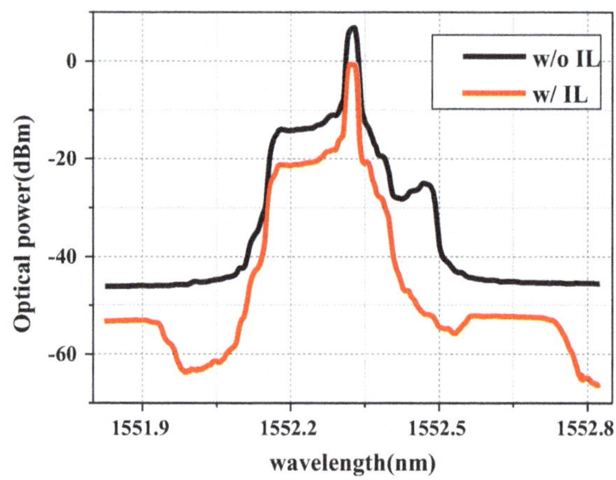

Fig. 19.9 Spectra of the optical SSB Signal w/o and w/25/50 GHz interleaver

modulator. The system performance becomes worse due to the limited bandwidth caused by the interleaver.

Figures 19.10 and 19.11 show the BER performance versus CSPR for single channel KK systems with 128QAM and 256QAM, respectively. Note that the results for 512QAM transmission are not provided here because the achieved lowest BER is still higher than the 25% SD-FEC threshold of 0.04. The optimum CSPR with KK receiver is 14 dB which is 3 dB lower than the optimum CSPR without KK receiver for 128QAM. BER reduces from 0.057 to 0.011 under the 20% SD-FEC threshold after 20 km of SSMF transmission with the optimum CSPR. Thus, a single channel 140 Gbit/s (net data rate 117 Gbit/s) 128QAM transmission over 20 km of SSMF with 25 GHz channel spacing is successfully achieved. The system performance applying 256QAM is then evaluated. BER reduces from 0.090 to 0.037 under the 25% SD-FEC threshold after 20 km of SSMF transmission with 14 dB CSPR. Therefore,

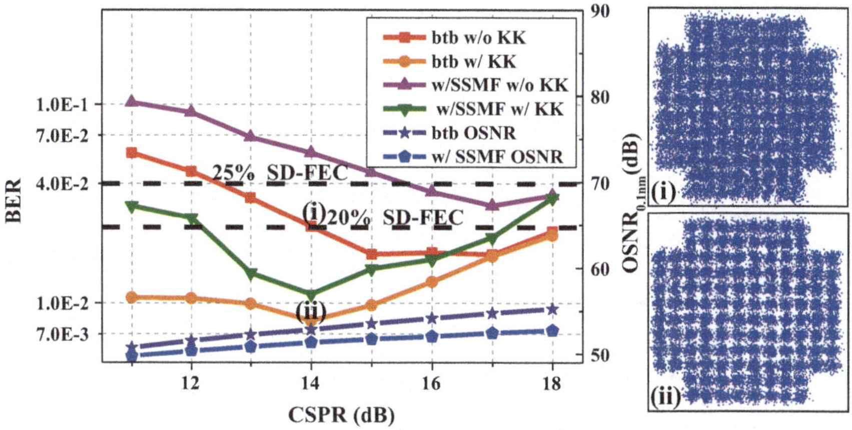

Fig. 19.10 BER and OSNR versus CSPR for the single channel 128QAM system. Insets: received constellations (i) w/o and (ii) w/KK receiver

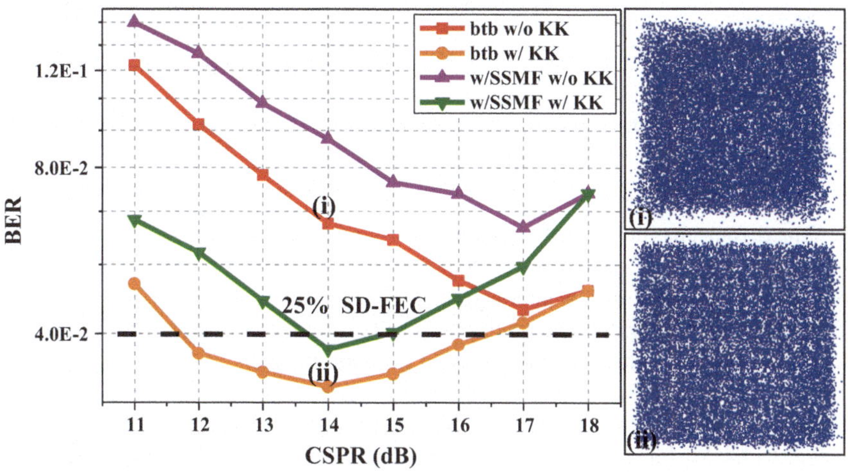

Fig. 19.11 BER versus CSPR for the single channel 256QAM system. Insets: received constellations (i) w/o and (ii) w/KK receiver

a single channel 160 Gbit/s (net data rate 128 Gbit/s) 256QAM transmission over 20 km of SSMF with 25 GHz channel spacing is also achieved.

2. **WDM Transmission**

Figure 19.12a shows the combined optical spectrum of ch2 and ch4 at marked position #1 in Figs. 19.6 and 19.12b show the optical spectrum of ch2 and ch4 at position #2 after the 25/50 GHz IL when IQ modulator 1 is turned off. ch2 and ch4 are under 25 GHz bandwidth limitation because of applying the IL. Figure 19.12c shows the

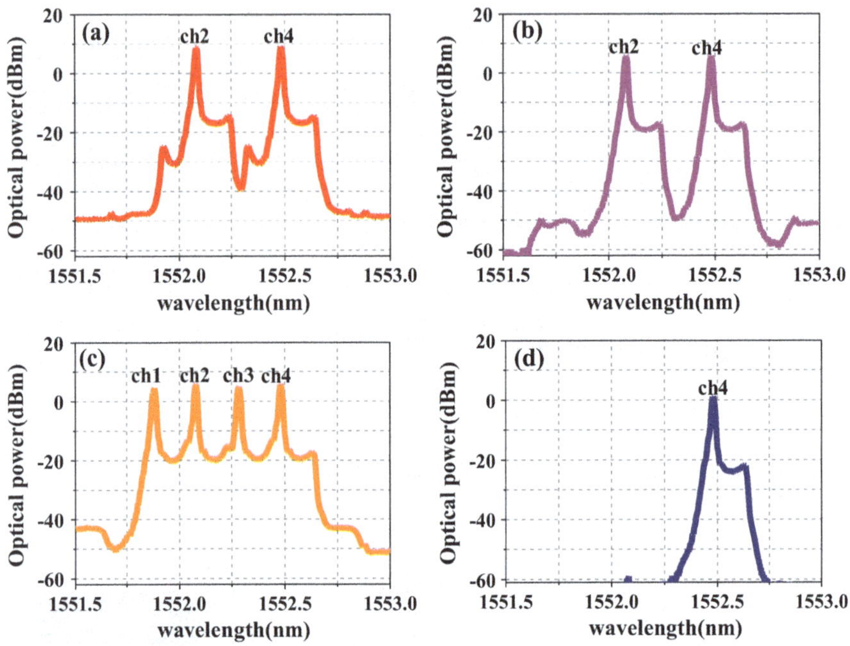

Fig. 19.12 Optical spectra (0.01 nm resolution) of **a** ch2 and ch4 at #1 in Fig. 19.6, **b** ch2 and ch4 at #2 in Fig. 19.6 with IQ modulator 1 off; **c** combined four-channel at #2 and **d** ch4 at #3

combined four-channel WDM signal at position #2 when both two IQ modulators are on. There's no high frequency attenuation due to the use of pre-equalization. The four branches of the four-channel signal can be separated after the 25/50 GHz and 50/200 GHz ILs. Only the separated optical spectrum of ch4 is shown in Fig. 19.12d as an example due to the high similarity of the four channels.

Figure 19.13 shows the BER performance versus CSPR with and without KK receiver in both btb and 20 km of SSMF transmission cases for WDM 128QAM system at ch4. We have verified that ch1, ch2, ch3, and ch4 had the same performance, and therefore, the results for the other three channels are not provided in order to avoid duplication. BER decreases from 0.051 to 0.013 and from 0.069 to 0.018 which is under the 20% SD-FEC threshold applying KK receiver with the optimum CSPR of 11 dB in btb case and fiber transmission case, respectively. Therefore, a 25 GHz-spaced 4 × 140 Gb/s (net data rate 4 × 117 Gb/s) WDM 128QAM SSB KK transmission over 20 km of SSMF is achieved. Applying KK receiver can provide higher performance improvement with smaller CSPR. The improvement decreases when CSPR increases from the optimum value. It is consistent with the conclusions summarized in the simulation and single channel transmission results. The optimum CSPR value without KK receiver is about 2 dB higher than those with KK receiver.

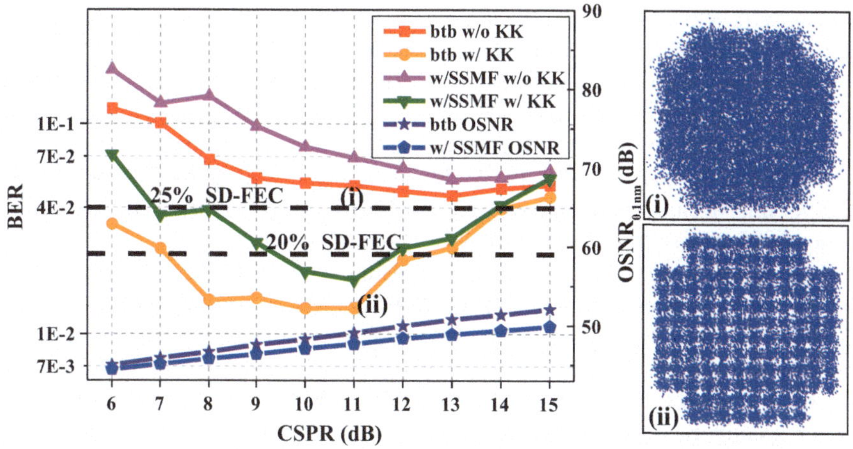

Fig. 19.13 BER and OSNR versus CSPR for the WDM 128QAM system. Insets: received constellations (i) w/o and (ii) w/KK receiver

The WDM system performance applying 256QAM is then investigated as shown in Fig. 19.14. For the btb case, BER is 0.101 which is above the 25% SD-FEC threshold without applying KK receiver. But the BER reduces to 0.034 under 25% SD-FEC threshold with the optimum CSPR of 9 dB using KK receiver. After 20 km of SSMF transmission, BER decreases from 0.099 to 0.039 with the optimum CSPR of 11 dB utilizing KK receiver. Thus, a 25 GHz-spaced 4 × 160 Gb/s (net data rate 4 × 128 Gb/s) WDM 128QAM SSB KK transmission over 20 km of SSMF

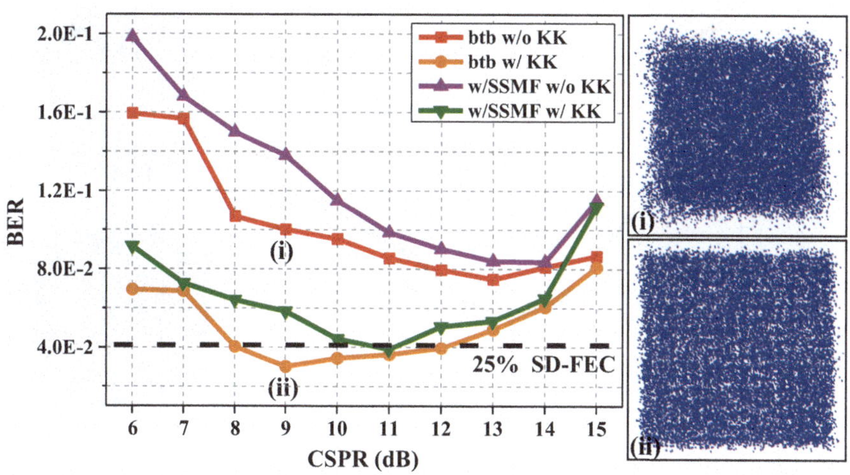

Fig. 19.14 BER versus CSPR for the WDM 256QAM system. Insets: received constellations (i) w/o and (ii) w/KK receiver

is achieved. When OSNR becomes lower, KK receiver can bring higher system performance improvement with lower CSPR. However, the optimum CSPR value with KK receiver for 20 km of SSMF transmission is 2 dB higher than that for btb case. Although applying KK receiver can provide higher performance improvement with lower CSPR for lower OSNR conditions, the system also requires higher optical carrier power to increase effective SNR. Thus, there is a trade-off between KK receiver improvement and effective SNR for CSPR values, and the optimum CSPR values for fiber transmission cases may be higher than btb case.

The performance is then investigated using 512QAM format as shown in Fig. 19.15. The achieved BER for 512QAM is far beyond the 25% SD-FEC threshold, but this part of the results is still presented to verify the conclusions drawn from the simulations. BER reduces from 0.167 to 0.080 and from 0.173 to 0.085 applying KK receiver with the optimum CSPR of 10 dB in btb case and fiber transmission case, respectively. It's clear that applying KK receiver with 512QAM can also provide performance improvement. Note that the demonstration of 512QAM may be restricted by the limited 5-bit ENOB of the DAC and ADC used in the experimental demonstrations according to the simulation results. It is worth mentioning that the optimum CSPR values become smaller for WDM transmission compared with single channel transmission. The transmitted optical power becomes lower due to the utilization of OCs and ILs, and thus the overall OSNR becomes lower which means that ASE noise has a greater effect on system performance. Therefore, lower CSPR would be better to increase the effective SNR with lower OSNR. OSNR is not kept as a constant value in this paper to pursue the best system performance and achieve WDM 256QAM transmission based on KK receiver. Further investigation should be

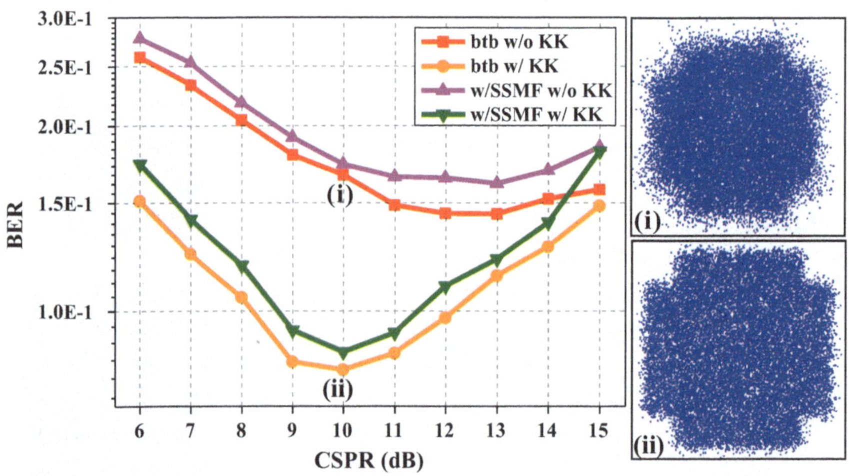

Fig. 19.15 BER versus CSPR for the WDM 512QAM system. Insets: received constellations (i) w/o and (ii) w/KK receiver

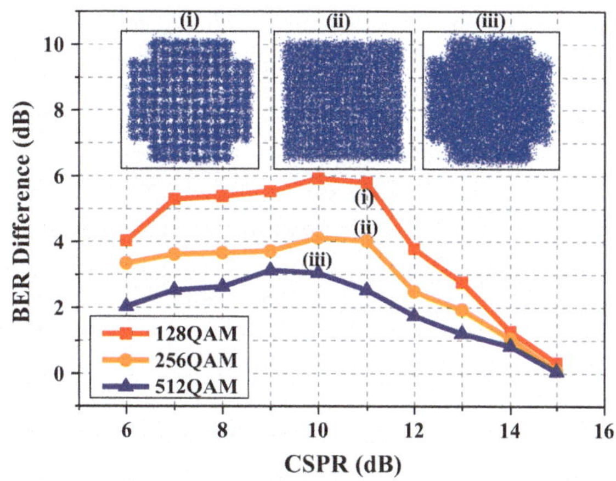

Fig. 19.16 BER difference value variation w/and w/o KK receiver versus CSPR. Insets: received constellations for (i) 128QAM, (ii) 256QAM and (iii) 512QAM with the corresponding optimum CSPR using KK receiver

carried out in the future based on a constant OSNR by using experimental equipment with better performance.

The BER performance improvement with and without KK receiver versus CSPR for experimental demonstrations is shown in Fig. 19.16. The basic variation trend is consistent with the simulation results. For 128 and 256 QAM, the lowest BER is achieved with CSPR of 11 dB which is referred to as the optimum CSPR with KK receiver and the CSPR with the highest performance improvement is 10 dB. And for 512QAM, these two CSPR values are 10 dB and 9 dB, respectively. Sometimes, these two CSPR values may not be the same since SSBI is not the only reason which can affect system performance. The system performance is also limited by other factors such as ASE noise, nonlinearities, and ENOB. Once the minimum-phase condition is fulfilled, the improvement brought by KK receiver is very close in a certain CSPR range. The transmitted optical power and the OSNR remain the same for 128/256/512QAM which indicates that the optimum CSPR for all the three modulation formats should be very close. The optimum CSPR value for 512QAM is slightly lower than that for 128 and 256QAM, but all the three modulation formats have the same approximate optimum CSPR range of 10–11 dB. For the simulation results, the optimum CSPR is almost equal to the CSPR which can achieve the highest performance improvement because SSBI is the main influence in a simulation system considering the relatively ideal simulation equipment and environment compared with the experimental demonstration.

19.5 Conclusions

We present the research of KK receiver in DD systems. 4 × 140 Gb/s (net data rate 4 × 117 Gb/s) WDM 128QAM and 4 × 160 Gb/s (net data rate 4 × 128 Gb/s) WDM 256QAM SSB KK transmissions with 25 GHz channel spacing over 20 km of SSMF are experimentally demonstrated. A constellation size as high as 256QAM with a net SE of 5.12 b/s/Hz is experimentally demonstrated in a WDM system with KK receiver.

Besides, a detailed system performance assessment of applying KK receiver in 128, 256 and 512QAM systems is also presented with both simulation and experimental results. The effect of adjusting CSPR on the performance of 128, 256, and 512QAM systems is investigated. Once the minimum-phase condition is met, the improvement brought by KK receiver is very close in a certain CSPR range. After this range, the improvement decreases rapidly because the excessive optical carrier power is turning to be the main limitation for the system. Sometimes, the optimum CSPR achieving the system's lowest BER may not be the same as the CSPR achieving the highest performance improvement with KK receiver since SSBI is not the only reason which can affect system performance. The WDM 512QAM system based on KK receiver is not successfully demonstrated probably because of the limited ENOB of DAC and ADC according to the simulation results. In the future, it is possible to demonstrate 512QAM transmission based on KK receiver if the ENOB of DAC and ADC is sufficiently high with suitable CSPR and OSNR values.

References

1. L. Shu, J. Li, Z. Wan, Z. Yu, X. Li, M. Luo, S. Fu, K. Xu, Single-photodiode 112-Gbit/s 16-QAM transmission over 960-km SSMF enabled by Kramers-Kronig detection and sparse I/Q Volterra filter. Opt. Express **26**(19), 24564–24576 (2018)
2. C. Antonelli, A. Mecozzi, M. Shtaif, X. Chen, S. Chandrasekhar, P.J. Winzer, Polarization Multiplexing With the Kramers-Kronig Receiver. J. Light. Technol. **35**(24), 5418–5424 (2017)
3. S.T. Le, K. Schuh, F. Buchali, M. Chagnon, H.Bulow, 1.6 Tbps WDM direct detection transmission with virtual-Carrier over 1200 km, in *Proc. Opt. Fiber Commun. Conf.*, Mar. 2018, paper Tu2D.5
4. J. Shi, J. Zhang, Y. Zhou, Y. Wang, N. Chi, J. Yu, Transmission performance comparison for 100-Gb/s PAM-4, CAP-16, and DFT-S OFDM with direct detection. J. Light. Technol. **35**(23), 5127–5133 (2017)
5. J. Shi, Y. Zhou, J. Zhang, Y. Wang, N. Chi, J. Yu, Enhanced performance utilizing joint processing algorithm for CAP signals. J. Light. Technol. **36**(16), 3169–3175 (2018)
6. D. Li, L. Deng, Y. Ye, Y. Zhang, H. Song, M. Cheng, S. Fu, M. Tang, D. Liu, Amplifier-free 4 × 96 Gb/s PAM8 transmission enabled by modified Volterra equalizer for short-reach applications using directly modulated lasers. Opt. Express **27**(13), 17927–17939 (2019)
7. K. Zhong, X. Zhou, J. Huo, C. Yu, C. Lu, A. Lau, Digital signal processing for short-reach optical communications: a review of current technologies and future trends. J. Light. Technol. **36**(2), 377–400 (2018)
8. IEEE P802.3bs 400 Gb/s Ethernet Task Force, accessed on Oct. 1, 2018. http://www.ieee802.org/3/bs/
9. Lau F. Suhr, J.J. Vegas Olmos, B. Mao, X. Xu, G. N. Liu, I. Tafur Monroy, 112-Gbit/s × 4-lane duobinary-4-PAM for 400GBase, in *Proc. 2014 Eur. Conf. Opt. Commnu.*, Sep. 2014, paper Tu.4.3.2
10. T. Chan, I. Lu, J. Chen, W.I. Way, 400-Gb/s transmission over 10- km SSMF using discrete multitone and 1.3-μm EMLs. IEEE Photon. Technol. Lett. **26**(16), 1657–1660 (2013)

11. I.O. Miguel, T. Zuo, B.J. Jesper, Q. Zhong, X. Xu, I.T. Monroy, Towards 400GBASE 4-lane solution using direct detection of multiCAP signal in 14 GHz bandwidth per lane, in *Proc. Opt. Fiber Commun. Conf./Nat. Fiber Opt. Eng. Conf.*, Mar. 2013, Paper PDP5C.1
12. P. Dong, J. Lee, Y. Chen, L. L. Buhl, S. Chandrasekhar, J.H. Sinsky, and K. Kim, Four-channel 100-Gb/s per channel discrete multi-tone modulation using silicon photonic integrated circuits, in *Proc. Opt. Fiber Commun. Conf.*, Mar. 2015, Paper Th5B.4
13. T. Tanaka, M. Nishihara, T. Takahara, W. Yan, L. Li, Z. Tao, M. Matsuda, K. Takabayashi, and J. C. Rasmussen, Experimental demonstration of 448-Gbps + DMT transmission over 30-km SMF, in *Proc. Opt. Fiber Commun. Conf.*, Mar. 2014, Paper M2I.5
14. A. Mecozzi, C. Antonelli, M. Shtaif, Kramers–Kronig coherent receiver. Optica **3**(11), 1220–1227 (2016)
15. J. C. Rasmussen, T. Takahara, T. Tanaka, Y. Kai, M. Nishihara, T. Drenski, L. Li, W. Yan and Z. Tao, Digital signal processing for short reach optical links, in *Proc. 2014 Eur. Conf. Opt. Commnu.*, 2014, paper Tu.1.1.3
16. M. Zhu, J. Zhang, H. Ying, X. Li, M. Luo, X. Huang, Y.Song, F. Li, X. Yi, and K. Qiu, 56-Gb/s Optical SSB PAM-4 Transmission over 800-km SSMF Using DDMZM Transmitter and Simplified Direct Detection Kramers-Kronig Receiver, in *Proc. Opt. Fiber Commun. Conf.*, Mar. 2018, Paper M2C. 5
17. T. Gui, L. Yi, C. Guo, C. Lu, A. Lau, Single-photodiode 112-Gbit/s 16-QAM transmission over 960-km SSMF enabled by Kramers-Kronig detection and sparse I/Q Volterra filter. Opt. Express **26**(20), 25934–25943 (2018)
18. X. Ruan, K. Li, D.J. Thomson, C. Lacava, F. Meng, I. Demirtzioglou, P. Petropoulos, Y. Zhu, G.T. Reed, F. Zhang, Experimental comparison of direct detection Nyquist SSB transmission based on silicon dual-drive and IQ Mach-Zehnder modulators with electrical packaging. Opt. Express **25**(16), 19332–19342 (2017)
19. H. Chen, N. Kaneda, J. Lee, J. Chen, Y. Chen, Experimental comparison of direct detection Nyquist SSB transmission based on silicon dual-drive and IQ Mach-Zehnder modulators with electrical packaging. Opt. Express **25**(6), 5852–5860 (2017)
20. Z. Li, M.S. Erkılınç, K. Shi, E. Sillekens, L. Galdino, T. Xu, B.C. Thomsen, P. Bayvel, R.I. Killey, Spectrally Efficient 168 Gb/s/λ WDM 64-QAM Single-Sideband Nyquist-Subcarrier Modulation With Kramers-Kronig Direct-Detection Receivers. J. Light. Technol. **36**(6), 1340–1346 (2018)
21. W.R. Peng, X. Wu, K.-M. Feng, V.R. Arbab, B. Shamee, J.-Y. Yang, L.C. Christen, A.E. Willner, S. Chi, Spectrally efficient direct-detected OFDM transmission employing an iterative estimation and cancellation technique. Opt. Express **17**(11), 9099–9111 (2009)
22. S. Randel, D. Pilori, S. Chandrasekhar, G. Raybon, and P. J. Winzer, 100-Gb/s discrete-multitone transmission over 80-km SSMF using single-sideband modulation with novel interference-cancellation scheme, in *Proc. 2015 Eur. Conf. Opt. Commnu.*, Sep. 2018, paper 0697
23. Z. Li, M. Sezer, R. Maher, L. Galdino, K. Shi, B.C. Thomsen, P. Bayvel, R.I. Killey, Two-stage linearization filter for direct-detection subcarrier modulation. IEEE Photon. Technol. Lett. **28**(24), 2838–2841 (2016)
24. K. Zou, Y. Zhu, F. Zhang, Z. Chen, Spectrally efficient terabit optical transmission with Nyquist 64-QAM half-cycle subcarrier modulation and direct-detection. Opt. Lett. **41**(12), 2767–2770 (2016)
25. X. Li, S. Zhou, H. Ji, M. Luo, Q. Yang, L. Yi, R. Hu, C. Li, S. Fu, A. Alphones, W. Zhong, and C. Yu, Transmission of 4 × 28-Gb/s PAM-4 over 160-km single mode fiber using 10G-Class DML and Photodiode, in *Proc. Opt. Fiber Commun. Conf.*, Mar. 2016, Paper W1A.5
26. M. Zhu, J. Zhang, X. Yi, Y. Song, B. Xu, X. Li, X. Du, K. Qiu, Hilbert superposition and modified signal-to-signal beating interference cancellation for single side-band optical NPAM-4 direct-detection system. Opt. Express **25**(11), 12622–12631 (2017)
27. L. Shu, J. Li, Z. Wan, F. Gao, S. Fu, X. Li, Q. Yang, K. Xu, Single-lane 112-Gbit/s SSB-PAM4 transmission with dual-drive MZM and Kramers-Kronig detection over 80-km SSMF. IEEE Photon. J. **9**(6), 1–9 (2017)

28. Z. Li, M.S. Erkılınç, K. Shi, E. Sillekens, L. Galdino, B.C. Thomsen, P. Bayvel, R.I. Killey, SSBI mitigation and the Kramers-Kronig scheme in single-sideband direct-detection transmission with receiver-based electronic dispersion compensation. J. Light. Technol. **35**(10), 1887–1893 (2017)
29. K. Schuh and S. T. Le, 180 Gb/s 64QAM Transmission Over 480 km Using a DFB Laser and a Kramers-Kronig Receiver, in *Proc. 2018 Eur. Conf. Opt. Commnu.*, Sep. 2018, paper Tu3G.4
30. X. Chen, J. Cho, S. Chandrasekhar, P. J. Winzer, C. Antonelli, A. Mecozzi, and M. Shtaif, Single-wavelength, single-polarization, single-photodiode Kramers-Kronig detection of 440-Gb/s entropy-loaded discrete multitone modulation transmitter over 100-km SSMF, in *Proc. IEEE Photon. Conf.*, Oct. 2017, Paper Th5B.6
31. M. Zhu, J. Zhang, X. Yi, H. Ying, X. Li, M. Luo, Y. Song, X. Huang, K. Qiu, Optical single side-band Nyquist PAM-4 transmission using dual-drive MZM modulation and direct detection. Opt. Express **26**(6), 6629–6638 (2018)
32. T. M. Hoang, Q. Zhuge, Z. Xing, M. Sowailem, M. Morsy-Osman, and D. V. Plant, Single wavelength 480 Gb/s direct detection transmission over 80 km SSMF enabled by Stokes vector receiver and reduced-complexity SSBI cancellation, in Proc. Opt. Fiber Commun. Conf., Mar. 2018, paper W4E.7
33. Y. Zhou, J. Yu, Y. Wei Y, R. Deng, L. Zhao, N. Chi, G.K. Chang and Y. Zhang, 160 Gb/s 256QAM Transmission in a 25 GHz Grid Using Kramers-Kronig Detection, in *Proc. Opt. Fiber Commun. Conf.*, Mar. 2019, Paper Th2A. 46
34. Y. Zhu, K. Zou, F. Zhang, C-Band 112 Gb/s Nyquist Single Sideband Direct Detection Transmission Over 960 km SSMF. IEEE Photon. Technol. Lett. **29**(8), 651–654 (2017)
35. T. Bo, H. Kim, Kramers-Kronig receiver operable without digital upsampling. Opt. Express **26**(11), 13810–13818 (May 2018)
36. J. Zhang, J. Yu, N. Chi, and H. Chien, Time-domain digital pre-equalization for band-limited signals based on receiver-side adaptive equalizers, *Opt. Express* **22**(1)

The manufacturer's authorised representative in the EU is Springer Nature Customer Service Centre GmbH, Europaplatz 3, 69115 Heidelberg, Germany. If you have any concerns regarding our products, please contact ProductSafety@springernature.com

Printed and bound by CPI Group (UK) Ltd, Croydon, CR0 4YY
30/09/2026
02225557-0015